Topics in Applied Physics Volume 60

Topics in Applied Physics Founded by Helmut K. V. Lotsch

Volume 57 **Strong and Ultrastrong Magnetic Fields** and Their Applications
Editor: F. Herlach

Volume 58 **Hot-Electron Transport in Semiconductors**
Editor: L. Reggiani

Volume 59 **Tunable Lasers** 2nd Edn.
Editors: L. F. Mollenauer, J. C. White, and C. R. Pollock

Volume 60 **Ultrashort Laser Pulses** Generation and Applications 2nd Edn.
Editor: W. Kaiser

Volume 61 **Photorefractive Materials and Their Applications I**
Fundamental Phenomena
Editors: P. Günter and J.-P. Huignard

Volume 62 **Photorefractive Materials and Their Applications II**
Survey of Applications
Editors: P. Günter and J.-P. Huignard

Volume 63 **Hydrogen in Intermetallic Compounds I**
Electronic, Thermodynamic, and Crystallographic Properties, Preparation
Editor: L. Schlapbach

Volume 64 **Sputtering by Particle Bombardment III**
Characteristics of Sputtered Particles, Technical Applications
Editors: R. Behrisch and K. Wittmaack

Volume 65 **Laser Spectroscopy of Solids II**
Editor: W. M. Yen

Volume 66 **Light Scattering in Solids V**
Superlattices and Other Microstructures
Editors: M. Cardona and G. Güntherodt

Volume 67 **Hydrogen in Intermetallic Compounds II**
Surface and Dynamic Properties, Applications
Editor: L. Schlapbach

Volume 68 **Light Scattering in Solids VI**
Recent Results, Including High-TC Superconductivity
Editors: M. Cardona and G. Güntherodt

Volume 69 **Unoccupied Electronic States**
Editors: J. C. Fuggle and J. E. Inglesfield

Volume 70 **Dye Lasers: 25 Years**
Editor: M. Stuke

Volume 71 **The Monte Carlo Method in Condensed Matter Physics**
Editor: K. Binder

Volumes 1-56 are listed on the back inside cover

Ultrashort Laser Pulses

Generation and Applications

Edited by W. Kaiser

Second Edition

With Contributions by

D. H. Auston K. B. Eisenthal R. M. Hochstrasser
C. K. Johnson W. Kaiser A. Laubereau
D. von der Linde A. Seilmeier C. V. Shank
W. Zinth

With 197 Figures

Springer-Verlag
Berlin Heidelberg New York
London Paris Tokyo
Hong Kong Barcelona
Budapest

Professor Dr. Dr. h. c. Wolfgang Kaiser

Fakultät für Physik, Technische Universität München, Physik-Department E11, James-Franck-Str., W-8046 Garching/München, Fed. Rep. of Germany

Managing Editor:

Dr. Helmut K. V. Lotsch

Springer-Verlag, Tiergartenstraße 17
W-6900 Heidelberg, Fed. Rep. of Germany

ISBN 3-540-55877-2 Springer-Verlag Berlin Heidelberg New York
ISBN 0-387-55877-2 Springer-Verlag New York Berlin Heidelberg

Library of Congress Cataloging-in-Publication Data. Ultrashort laser pulses: generation and applications/edited by W. Kaiser; with contributions by D. H. Auston ... [et al.]. – 2nd ed. p. cm. – (Topics in applied physics; v. 60) Includes bibliographical references and index.
ISBN 0-387-55877-2 (New York). – ISBN 3-540-55877-2 (Berlin)
1. Laser pulses, Ultrashort. I. Kaiser, W. (Wolfgang), 1925- . II. Auston, D. H. III. Series.
QC689.5.L37U4693 1993 621.36'6–dc20

Printed in the United States of America

Typesetting: Asco Trade Typesetting Limited, Hong Kong
54/3140 - 5 4 3 2 1 0 - Printed on acid-free paper

Preface to the Second Edition

Since the publication of the first edition the field of ultrashort laser pulses and their applications has witnessed continuous rapid growth. Thousands of papers have appeared during the past four years covering numerous subjects in physics, engineering, chemistry, and biology.

A few fascinating discoveries of recent years should be mentioned very briefly: (i) With femtosecond laser pulses one is now in a position to study directly carrier-carrier interaction in semiconductors and metals. (ii) It is possible to deliver substantial energy to a crystalline solid in such a short time that a phase transition from the crystalline state to a disordered state occurs before true melting of the surface is observed. (iii) One should be aware that with the development of shorter and shorter pulses one is at the point where the pulse duration may be shorter than the oscillation time of inter- and intramolecular motions. Novel excitation methods and new analytical techniques have been demonstrated. (iv) Intermolecular proton transfer and electron transfer in molecules and proteins was followed directly on the time scale of molecular motion of approximately 100 fs.

It is impossible for a single scientist to discuss critically and comprehensively the many topics covered by this book. Once more, the contributors to the present book have joined forces to update the first edition. The eight addenda are directly related to the original chapters. In this way the reader should find quite readily the most recent publications on the subject he is most interested in. The eight addenda provide more than seven hundred references representing an enormous amount of recent information. It is hoped that the second edition in paper-back will help students and actively working scientists to participate in the rapid progress of the field.

The editor is grateful to the contributors who took their valuable time to update the first edition.

Munich, July 1992 *Wolfgang Kaiser*

Preface to the First Edition

Ten years ago, Stanley L. Shapiro edited the book entitled *Ultrashort Light Pulses* (Topics Appl. Phys., Vol. 18), which was written by eight experts in the field. Six years later, Charles V. Shank added a bibliography (1980–1983) in the second edition with approximately one thousand new references.

During the past decade the field has grown so rapidly that a completely new book had to be written. In particular, the reduction of the time scale of light pulses into the femtosecond range has opened up new experimental possibilities never even foreseen in the preceding literature.

The vast literature with countless ideas and applications makes it impossible for a single person to write a comprehensive review. Nine scientists, actively working in the field since its beginning, have decided to join forces to prepare a new book describing the present state of the art. Emphasis is placed on the generation and numerous applications of ultrashort laser pulses. This book covers a wide area of science: physics, engineering, chemistry, and biology. The various chapters and sections are prepared in each case such that the reader is given a brief introduction to the specific subject. Ample references for a more detailed study are given at the end of each chapter.

The book is intended for a variety of readers: for our colleagues in the field who wish to learn more about what can be done and what the "neighbors" are doing; for newcomers who want their problem solved with ultrashort light pulses and who need to know the present state of the art and the literature of previous work in their specific area of interest. Finally, we hope that numerous readers who simply browse through the book will be stimulated by the fascinating new possibilities offered by ultrashort pulses. One should be aware that dynamic processes can now be studied on the extremely short time scale of a few 10^{-15} seconds.

The editor wishes to thank the contributors for writing the various review articles, a difficult task in a rapidly expanding field.

Munich, January 1988 *Wolfgang Kaiser*

Contents

Contributors

Auston, David H.
Department of Electrical Engineering and Applied Physics, Columbia University, New York, NY 10027, USA

Eisenthal, Kenneth B.
Department of Chemistry, Columbia University, New York, NY 10027, USA

Hochstrasser, Robin M.
Department of Chemistry, University of Pennsylvania, Philadelphia, PA 19104, USA

Johnson, Carey K.
Department of Chemistry, University of Kansas, Lawrence, KS 66045, USA

Kaiser, Wolfgang
Physik-Department E 11, Technische Universität München James-Franck-Straße, D-8046 Garching, Fed.Rep. Germany

Laubereau, Alfred
Experimentalphysik III, Universität Bayreuth, Postfach 101251, D-8580 Bayreuth, Fed.Rep. Germany

Linde, Dietrich von der
Fachbereich Physik, Universität Essen, Universitätsstr.2, D-4300 Essen 1, Fed.Rep. Germany

Seilmeier, Alois
Experimentalphysik III, Universität Bayreuth, Postfach 101251, D-8580 Bayreuth, Fed.Rep. Germany

Shank, Charles V.
Lawrence Berkeley Laboratory, 1 Cyclotron Road, Berkeley, California, CA 94720, USA

Zinth, Wolfgang
Institut für Medizinische Optik, Universität München, Barbarastraße 16, D-8000 München 40,Fed.Rep. Germany

1. Introduction

Wolfgang Kaiser

At the beginning of this book a few brief comments on "where we stand" appear to be appropriate. The intensive effort of numerous researchers and laboratories has led to a certain maturing of the field with respect to the generation of *picosecond* laser pulses. The available wavelengths cover a wide range from the ultraviolet (~ 300 nm) to the mid-infrared ($\sim 12\,\mu$m). Pulse durations between one and one hundred picoseconds can be generated in various systems, and peak powers exceed the breakdown limit without difficulties. Equipment for the generation and analysis of picosecond laser pulses has left the laboratory stage and is now commercially available.

Quite different is the situation for pulses with a duration of 100 *femtoseconds* and less. Here the wavelength range is still limited and most systems operate at 620 nm, the wavelength of the dye rhodamine 6G. Reproducibility, stability, and sufficient peak power are topics of high priority in many laboratories. Work with femtosecond pulses requires special attention on the part of the experimentalist. For instance, it is not simple to maintain the pulse properties in a complex experimental system. On account of group velocity dispersion an optical pulse of duration 10^{-14} s doubles its width after passage through approximately 1 mm of glass or 4 m of air. In addition, very short pulses are limited in their energy content, e.g. a pulse of 10^{-6} Joule and 10 fs duration has the huge intensity of 1000 Gigawatts in a beam of 100 μm diameter. The corresponding field strength of 2×10^7 V/cm is extremely high and cannot be tolerated in most experiments. On the other hand, a rough estimate shows quite clearly that extremely short pulses are well suited for the investigation of fast optical nonlinearities. A pulse duration of 5 fs is close to the lower experimental limit in the visible part of the spectrum. The utilization of these short pulses in new experiments is a task for the immediate future.

What are the trends? Where will the major activities be? Part of the answers to these questions are found in the different chapters of this book. Numerous recent investigations have given new and valuable information. As is frequently the case, new data raise new questions. Extensive work is progressing in order to take full advantage of the new possibilities offered by ultrashort pulses. In other words, the field is wide open. Unforeseen and unexpected discoveries may be anticipated.

1.1 Organisation of the Book

The present state of the art of the generation of ultrashort pulses is discussed by *C.V. Shank* in Chap. 2. Active and passive mode-locking in different laser media is still the most common way to generate pulses in the picosecond time range. Continuous wave lasers and systems operating in the giant pulse mode with (isolated) single short pulses are the work horses in many laboratories. Fortunately, substantial progress in the quality of the ultrashort pulses can be reported. The introduction of ring lasers operating in the colliding-pulse mode (CPM) proved to be an important advance for the entire field. Laser pulses of approximately 50 fs are now available with fair stability and reproducibility. Further reduction of the pulse duration is achieved by pulse compression, where the nonlinear properties of optical fibers turned out to be of particular value. The shortest pulses obtained with optical pulse compressors have a duration of only 5 fs. This number is evidence of the tremendous progress made during the past few years.

The high peak intensities of ultrashort pulses often lead to optical nonlinearities. In Chap. 3 *A. Laubereau* summarizes our present knowledge of different short pulse nonlinearities. Experimentalists working with ultrashort pulses must be aware of a variety of nonlinear effects which may occur and which frequently limit the peak intensities permitted in the investigation of interest. On the other hand, nonlinear optical processes are of great interest in their own right. New measuring techniques have become possible with nonlinear optical methods. For instance, new observations on surfaces and interfaces have been made with the help of nonlinear optical techniques, and the up-conversion of weak infrared signals is now possible with subpicosecond time resolution. Of major importance are nonlinear processes for the generation of ultrashort pulses at new frequencies. Different nonlinearities, e.g. harmonic generation, frequency mixing, parametric amplification, stimulated Raman scattering, and continuum generation, provide us with ultrashort pulses covering the frequency range from the vacuum ultraviolet to the far infrared.

In Chap. 4, *D. von der Linde* reports on numerous new and exciting investigations in solid-state physics. In particular, the need to know more about fast relaxation processes of hot carriers in microelectronic devices has prompted intensive studies of semiconductors of practical importance. The chapter is divided in four sections. (i) Exciton dynamics – at low and high density – has provided new information on the interaction between excitons and phonons, excitons and excitons, and excitons with a hole-electron plasma. (ii) For a long time the semiconductor physicists have speculated about various ultrafast relaxation processes of hot charge carriers. Optical techniques are an ideal way to study these processes directly and they allow one to follow the return of hot carriers to thermal equilibrium. (iii) The very short lifetimes of optical phonons in crystals require ultrashort pulses for time-resolved investigations. It is now possible to directly observe the very short relaxation times and to make a distinction between coherent phonons (generated coherently in a few modes with known wave vectors close to the center of the Brillouin zone) and incoherent phonons (distributed over many modes with random phases). (iv) Under intense irradiation with ultrashort pulses the melting, evaporation, and resolidification of solid surfaces show unusual properties. For instance, an unexpected superheating of the solid of several hundred degrees has been observed.

Recent progress in ultrafast optoelectronics is reviewed by *D.H. Auston* in Chap. 5. In optoelectronics the high speed of short optical pulses is applied to the design of new and faster electronic devices and measurement systems. Of special importance in this respect are photoconductors and electro-optical materials, both of which are essential for ultrafast optoelectronics. A wide range of novel ultrafast optoelectronics allows the generation and measurement of electronic transients between microwaves and the far infrared. Non-invasive probing techniques are now available which permit the analysis of high-speed photodetectors, high-speed transistors, and integrated circuits with picosecond pulses. A beautiful example of the potential of optoelectronics is a recent investigation of subpicosecond mobility transients in GaAs single crystals.

Investigations where advantage is taken of the coherent properties of ultrashort light pulses are discussed in Chap. 6 by *W. Zinth* and *W. Kaiser*. In recent years a series of coherent techniques have been developed. Information on dynamic processes are obtained in the time and frequency domain. Electronic dipole transitions and Raman type transitions have been studied extensively. Single and multiple excitation processes with appropriate time sequences have been successfully applied to gain valuable information about material parameters, mainly in the condensed phases. One is now in the position to determine the dephasing time of homogeneously and inhomogeneously broadened transitions. It is possible (with pulses of 60 fs) to determine very short dephasing times of less than 1 ps with good accuracy and to distinguish between pure dephasing and energy relaxation times. In a number of cases the mechanism and the pathway of energy relaxation could be elucidated. Of special interest are experiments where spectral line narrowing allows the observation of transition frequencies in congested spectral regions. Very recently, the beating of coherently excited and widely separated modes of $\sim 300\,\mathrm{cm}^{-1}$ or 10 THz has been observed. The frequency differences of these modes may be determined with high precision.

Ultrashort intramolecular and intermolecular vibrational energy transfer of polyatomic molecules in liquids is discussed by *A. Seilmeier* and *W. Kaiser* in Chap. 7. Several excitation and probing techniques had to be developed to study these very fast, hitherto unknown molecular processes. The vibrational population lifetime T_1 and the vibrational redistribution has been elucidated in a series of molecules. It turns out that in larger molecules the intramolecular vibrational redistribution is very fast (below 1 ps). It is possible to assign a transient internal temperature to the excited molecules. Evidence is obtained for internal temperatures as high as 1000 K. These hot molecules are short-lived since the intermolecular transfer of energy to the surroundings proceeds very quickly, within times of the order of 10 ps. The introduction of large molecules as a local molecular thermometer allowed the observation of the fast microscopic energy equalization in liquids. Most experiments were made with molecules in the electronic ground state S_0. Recent investigations of the first excited electronic state S_1 indicated that the intramolecular vibrational relaxation is very short indeed, approximately 450 fs.

Ultrashort chemical reactions in the liquid state are reviewed by *K.B. Eisenthal* in Chap. 8. Numerous experimental observations convincingly show the very rapid chemical events, which take place in solutions. Ultrashort pulses are an ideal tool for monitoring a sequence of chemical processes. Substantial progress has been made in the understanding of photoisomerisation. Of special interest to the chemist are chemical intermediates which – frequently short-lived – have remained elusive in previous investigations. New information is obtained on the role of charge transfer in chemical reactions, on carbenes and nitrenes, and on the photofragmentation of singlet oxygen. Intra- and intermolecular proton transfer continues to be an area of intense endeavor. Transient absorption and fluorescence spectroscopy in the visible has given valuable new data on the very rapid proton transfer taking place in the condensed phases. Very recently, spectroscopic data with picosecond infrared pulses showed very directly how fast the proton moves and where it resides.

Our present knowledge of ultrafast biological processes is discussed by *R.M. Hochstrasser* and *C.K. Johnson* in Chap. 9. This chapter concentrates on three biological topics: Hemeproteins, reaction centers (photosynthesis), and Rhodopsin. These subjects are of great significance for our very existence. We recall that human life requires oxygen transport (mediated by hemoglobin), needs hydrocarbons and oxygen (generated in photosynthesis) and heavily relies on vision (made possible by rhodopsin). Biological processes frequently occur in a series of individual sheps that cover a wide time range from picoseconds to many seconds. It was certainly a great surprise to biologists to find that the primary processes are exceedingly fast, in some cases below 1 ps. For instance, the first event in oxygen transport by hemoglobin, the dissociation of the ligand O_2 or CO, occurs in approximately 350 fs. Major progress in the understanding of photosynthesis stems from the recent determination of the protein structure of one bacterial reaction center. It is now possible to identify the different absorption bands with specific pigments in the reaction center and to study the electron transfer between the different pigments on a subpicosecond time scale. Rhodopsin and bacteriorhodopsin both employ retinal for their biological function. There is strong experimental evidence that the isomerisation of retinal is the primary photochemical event. Detailed transient absorption studies with laser pulses of 100 fs duration exist for bateriorhodopsin. The data suggest that the isomerisation of retinal in the protein pocket occurs with a time constant of 430 fs, very quickly initiating the subsequent biological processes.

2. Generation of Ultrashort Optical Pulses

Charles V. Shank

With 20 Figures

2.1 Introduction

It has been two decades since the era of ultrashort optical pulse generation was ushered in with the report of passive model-locking of the ruby laser by *Mocker* and *Collins* [2.1]. Shortly, followed the generation of the first optical pulses in the picosecond range with Nd:glass laser by *DeMaria* [2.2]. Since that time we have witnessed dramatic advances in pulse generation. As is seen in Fig. 2.1 the shortest optical pulse width has fallen at an exponential rate, a reduction of more than three orders of magnitude since 1966. Optical pulse widths as short as 6 fs have been generated, approaching the fundamental limits of what is possible in the visible region of the spectrum. In addition, ultrashort optical pulse lasers have become more useful as advances have taken place in pulse energy, tunability, repetition rate, reliability and ease of operation. A bewildering array of techniques and lasers systems are currently available to meet a diverse set of needs.

The goal of this chapter will be to summarize the current state of the art in ultrashort optical pulse generation. We will limit our attention to laser systems and techniques that produce optical pulses of less than 10 ps. As a consequence, we will not discuss the active mode-locking of gas lasers, Nd:YAG lasers and other lasers which do not meet this criteria. Lasers of this type have been extensively reviewed elsewhere [2.4].

A major advance in the generation of ultrashort optical pulses has been the process of mode-locking. The basic concepts of mode-locking will be described in the first section of this chapter. This will be followed by a description of the operation of some of the most important laser systems for producing ultrashort optical pulses.

A second, conceptually different, approach to generating short pulses, the process of pulse compression, will then be discussed along with the relevant experimental work. Finally we will direct our attention to novel pulse generation schemes that do not fit into the above categories.

2.2 Mode-locking

2.2.1 Basic Concepts

A laser is typically constructed with a pair of mirrors which enclose a gain medium. The range of frequencies over which a laser will oscillate is determined

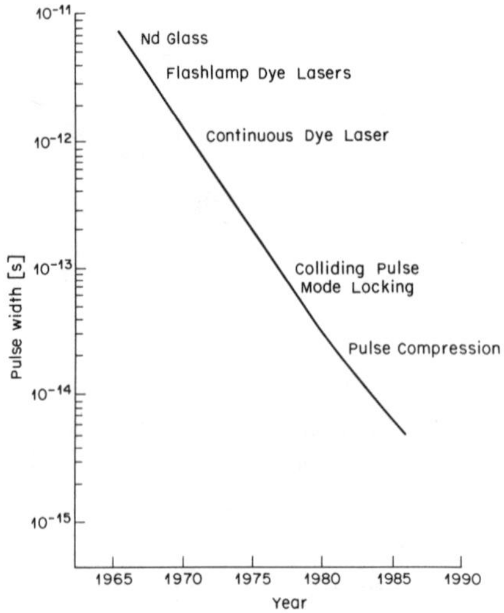

Fig. 2.1. Plot of the logrithm of the shortest reported optical pulse width versus year. Note that each reduction in pulse width has been accompanied by an advance in technology

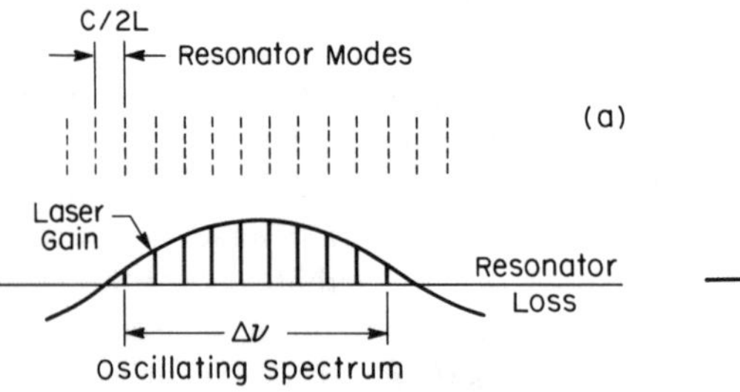

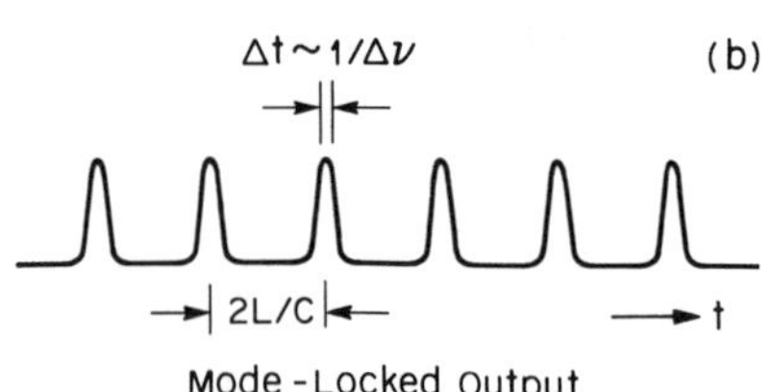

Fig. 2.2. (**a**) The resonator modes which oscillate are determined by the gain profile and the resonator loss. (**b**) The temporal output of the laser with all modes locked in the proper phase

by the gain bandwidth and the loss in the cavity. The precise frequencies which oscillate within this range are determined by the resonator. The modes of oscillation of a laser are the self reproducing field distributions which are established in the optical cavity. A variety of such laser modes exist, some of which have different field distributions normal to the resonator axis; these are termed transverse modes. Each of these transverse modes has an infinite set of eigenfrequencies or longitudinal modes separated in frequency by $c/2L$ where L is the optical length of the resonator and c is the speed of light. In most useful mode-locked laser systems only the lowest order fundamental mode having a Gaussian profile is permitted to oscillate. The corresponding set of longitudinal modes consists of a picket fence of regularly spaced modes separated by $c/2L$. The number of modes which oscillate is limited by the bandwidth $\Delta\nu$ over which the laser gain exceeds the loss of the resonator. This is illustrated in Fig. 2.2a.

Unless some mode selecting element is placed in the laser resonator, the output consists of a sum of frequency components corresponding to the oscillating modes. The electric field may then be written

$$E(t) = \sum_{n} \alpha_n \exp i[(\omega_0 + n\delta\omega)t + \phi_n] \tag{2.1}$$

where α_n is the amplitude of the nth mode and $\delta\omega$ is the mode spacing. In general, the relative phases between the modes are randomly fluctuating. If nothing fixes the relative phases, the laser output will vary in time although the average power will remain relatively constant. On the other hand, if the modes are forced to maintain a fixed phase and amplitude relationship, the output of the laser will be a well defined periodic function of time and the laser will be considered to be "mode-locked". Mode-locking can produce an output consisting of a train of regularly spaced pulses. The pulses have width, $\Delta\tau$, which is approximately equal to the reciprocal of the total mode-locked bandwidth, $\Delta\nu$, and the temporal periodicity $T_p = 2L/c$. The ratio of the pulse width to the period is approximately equal to the number of locked modes. In Fig. 2.2b a mode-locked train is illustrated. In this case the physical picture corresponds to a single pulse traveling back and forth between the laser resonator mirrors. It is also possible to produce mode-locking with N pulses in the cavity spaced by a multiple of $c/2L$.

2.2.2 Active Mode-locking

An intuitively obvious way to mode-lock a laser is to insert a shutter into the optical cavity which opens for the desired pulse duration and remains closed for a cavity round trip time. Such a shutter has yet to be constructed. A rough approximation has been to insert an intracavity phase or loss modulator into the optical cavity driven at the frequency corresponding to the mode spacing [2.5]. The principle of active mode-locking with a sinusoidal modulation is shown in Fig. 2.3. In this situation an optical pulse is likely to form in such a way as to minimize the loss from the modulator. The peak of the pulse adjusts in phase to be at the point of minimum loss from the modulator. The slow variation of the sinusoidal modulation provides only a weak mode-locking effect making this technique unsuitable for generating ultrashort optical pulses. In a similar way, phase modulation can also produce a mode-locking effect [2.6].

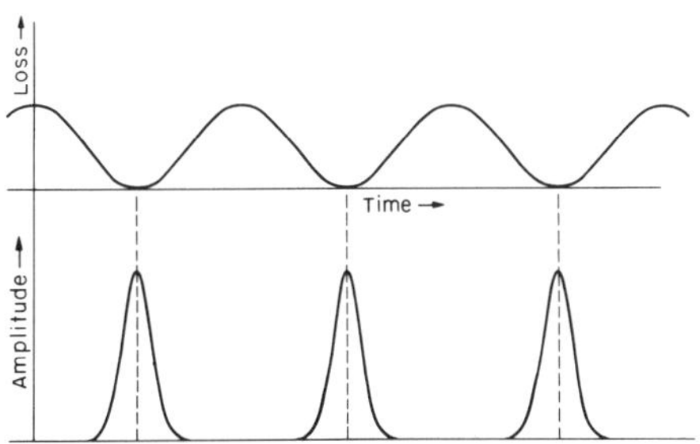

Fig. 2.3. Diagram illustrating the process of active mode-locking by loss modulation. The laser pulse positions itself in the cavity to minimize the loss

The active mode-locking approach has been found particularly useful for mode-locking the Nd:YAG laser and gas lasers such as the Argon laser. A detailed discussion of these laser systems is outside the scope of this chapter. A number of excellent review papers have been written on this subject [2.3,7].

2.2.3 Synchronous Pumping

Another approach to active mode-locking is to use a mode-locked laser as an optical pumping source. The pumping source acts as a master laser while the optically pumped laser is the slave. In this scheme it is very important to carefully match the cavity lengths of the master and slave lasers. The gain of the slave laser is then modulated at a cavity round trip time. Rapid modulation of the gain medium by the short optical pumping pulse provides the coupling mechanism to phase lock the cavity modes. The rise time of the gain modulation is approximately determined by the time integral of the optical pumping pulse and is thus much more effective in mode-locking the laser than the sinusoidal loss modulation described in the previous section. As a consequence, the optical pulse of the slave laser is typically much shorter than the pumping pulse.

A diagram of a synchronously mode-locked dye laser is shown in Fig. 2.4. The output of the pump laser of length L is a train of optical pulses spaced in time by $2L/c$. The pumping pulses turn on the gain rapidly in a time determined by the pump pulse width. When the dye laser and the pump laser have very closely matched cavities, the gain modulation is synchronized with a pulse that bounces back and forth in the dye laser cavity. When the pulse in the cavity passes through the gain medium, stimulated emission depletes the upper state dye population and reduces the gain to a value equal to the cavity loss. The temporal evolution of the gain and of the pulse are shown in Fig. 2.5. The rapid rise and fall of the gain provides a shuttering action which is the pulse shortening mechanism. The operation of this laser has been described in a number of theoretical treatments. *Yasa* and *Teschke* [2.8] have developed a theory based on a rate equation approach. The self reproducing profile technique developed by *Kuizenga* and *Seigman* [2.9] was applied by several authors to model the

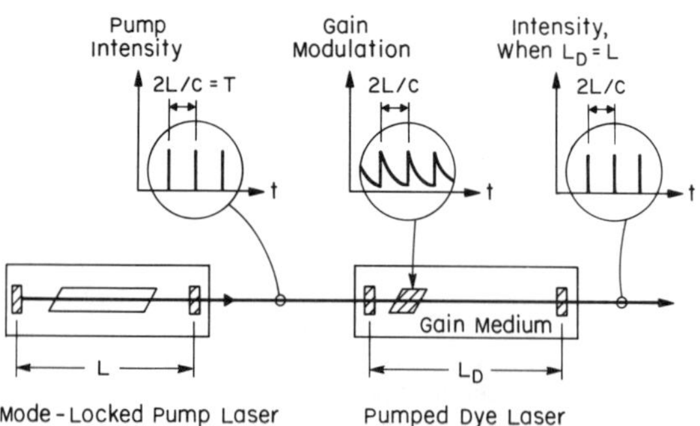

Fig. 2.4. Experimental arrangement of mode-locking by synchronous pumping

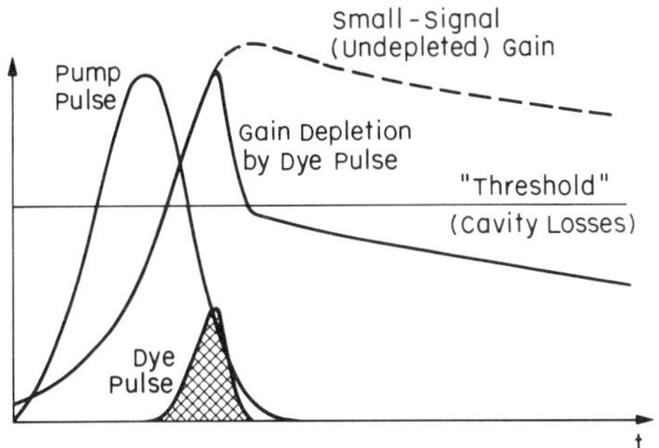

Fig. 2.5. Temporal evolution of the laser gain and optical pulse in a synchronously pumped dye laser

behavior of synchronously pumped dye lasers [2.10–12]. *Catherall* et al. [2.13a] have described a stepping model which allows the generation of steady state pulse profiles without including a bandwidth limiting element in the optical cavity.

The case of synchronously pumping infrared dyes [2.13b] differs from the situation encountered for visible dyes described above because the life time of the excited state can be less than the pump pulse width. In this case the gain follows the pump pulse time profile resulting in a lower gain because excited dye molecules recover while pumping is taking place unlike the situation described in Fig. 2.5. This also places a more stringent demand the syncrhonization of the pump pulse and the mode-locked dye laser pulse.

2.2.4 Passive Mode-locking

Passive mode-locking utilizes the insertion of a saturable absorbing element inside the optical cavity of a laser. The saturable absorber is typically an organic dye but gases and solids have been used as well. This approach to generating ultrashort optical pulses has been extremely successful. The very first optical pulses in the picosecond time domain were obtained using this method [2.2] as were the first optical pulses in the femtosecond time domain [2.14].

Passive mode-locking has been applied to several different laser systems. Extensive experimental investigation over the last two decades has created a large body of knowledge on this subject. Passively mode-locked lasers can be divided into two distinct classes of operation: type 1, which I refer to as burst mode or giant-pulse lasers and type 2, which correspond to continuous or quasi-continuous operation. Historically, passive mode-locking was first observed in the burst mode in ruby lasers [2.1] and in Nd:glass lasers [2.2]. Continuous passive mode-locking is observed primarily in dye laser systems and was first described theoretically by *New* [2.15] in 1972. As we shall see, entirely different requirements are placed on the selection of saturable absorbers depending upon which type of mode-locking occurs.

a) Type 1: Giant-Pulse Lasers

For giant pulse lasers such as ruby or Nd:glass the upper laser level is long lived, typically hundreds of microseconds. Pulse generation occurs in a highly transient manner in a time much shorter than the upper level population response. A model which describes the operation of these lasers is the "fluctuation model" proposed by *Letokhov* [2.16,17] and independently by *Fleck* [2.18,19]. The model begins with a description of the very first stages of laser action. At the start of the flashlamp pumping pulse, spontaneous emission excites a broad spectrum of laser modes within the optical cavity. Since the modes are randomly phased, a fluctuation pattern is established in the cavity with a periodic structure corresponding to a cavity round trip time $T = 2L/c$. When the gain is sufficient to overcome the linear and nonlinear losses in the cavity, laser threshold is reached and the fields in the cavity initially undergo linear amplification. At some point the field becomes intense enough to enter a phase where the random pulse structure is transformed by the nonlinear saturation of the absorber and by the laser gain saturation. As a result, one of the fluctuation spikes grows in intensity until it dominates and begins to shorten in time. As the short pulse gains intensity it reaches a point where it begins to nonlinearly interact with the glass host and the pulse begins to deteriorate. At the beginning of the pulse train as recorded on an oscilloscope, the pulses are a few picoseconds in duration and nearly bandwidth limited [2.20,21]. Later, pulses in the train undergo self-modulation of phase and self-focusing which leads to temporal fragmentation of the optical pulse.

The statistical nature of the pulse generation process in giant-pulse mode-locked lasers leads to a nondeterministic mode-locking. The stochastic features in the performance of passively mode-locked giant-pulse lasers has been investigated both experimentally and analytically [2.22,23]. Procedures have been developed to optimize the laser reproducibility but shot to shot variations have not been eliminated completely.

The optical configuration for a mode-locked giant pulse laser is shown in Fig. 2.6. In the design of the cavity it is important to eliminate sub-cavity resonances and spurious reflections which may cause the formation of subsidiary pulse trains. The problem of satellite pulses is reduced by placing the absorber in optical contact with one of the cavity mirrors [2.24–26] as shown in the figure.

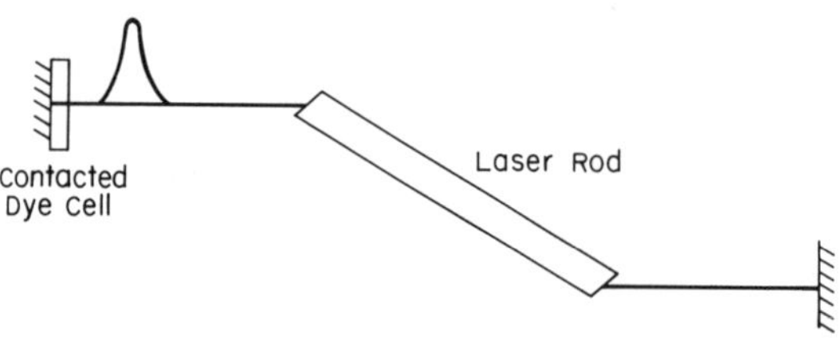

Fig. 2.6. Configuration of a passively mode-locked solid state laser. The dye cell is optically contacted with the laser mirror on one end of the cavity

The role of the saturable absorber in the flucutation model is to select a noise burst that is amplified and ultimately becomes the mode-locked laser pulse. As a consequence, the relaxation time of the absorber, T_a, sets an approximate limit to the duration of mode-locked pulses [2.27,25]. The saturable absorber dyes often used are the Kodak dyes A9740 and A9860 which have relatively short lifetimes, 7 picoseconds and 11 picoseconds respectively [2.28]. Dyes with even shorter lifetimes have been investigated [2.29,30].

b) Type 2: Continuous Lasers

Passive mode-locking of continuous lasers such as dye lasers, which we have previously referred to as type 2 mode-locking, involves a very different physics of pulse formation from that just described. The mode-locked pulse duration is typically much shorter than either the lifetime of the amplifying or gain medium, T_{1a}, or the saturable absorber recovery time, T_{1b}. The discussion which follows here will also apply to flashlamp pumped passively mode-locked dye lasers which operate in a quasi-continuous fashion since the energy storage time is less than a cavity round trip time.

Using a rate equation analysis, *New* [2.15] first described the conditions for pulse formation in a type 2 system. He showed that under an appropriate set of circumstances a stable short pulse could develop through the nonlinear interaction of the amplifying medium and the saturable absorber. Analytical and numerical techniques were then applied to describe the transient formation of an ultrashort optical pulse [2.31–33]. *Haus* [2.34] was able to obtain a closed formed solution by assuming a cavity bandwidth and a hyperbolic secant pulse shape.

To obtain some insight into the analysis of mode-locking with a "slow" absorber [2.15] it is instructive to discuss the situation depicted in Fig. 2.7. The

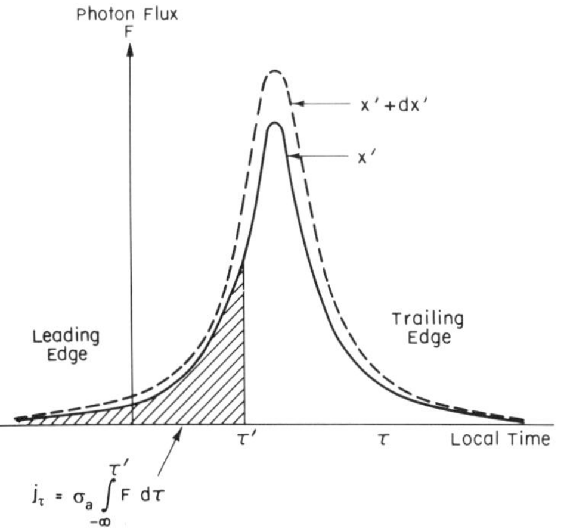

Fig. 2.7. Profile of an optical pulse in a continuously mode-locked laser

pulse profile shown in the figure has developed from a situation where the energy was roughly uniformly distributed in the cavity and the pulse duration which is assumed to be less than the recovery time of both the nonlinear absorbing and gain media. If the mode-locking process is to continue and the pulse peak to be further enhanced relative to the background, there must be a net round trip loss at L and T on the leading and trailing edges of the pulse and yet have a net round trip gain near the peak P.

The above conditions can be satisfied provided that the relaxation time of the amplifying medium T_{1a} is of the same order as the cavity round trip time T_{RT} and that the ratio of the cross section of the absorption to the cross section for amplification, $s = \sigma_a/\sigma_b$ be greater than two [2.35].

The requirement that $T_{\mathrm{RT}}/T_{1a} = \xi$ be on the order of unity ensures that the amplifier does not fully recover between transits resulting in a net loss at L. If ξ is small nonlinear amplification does not play a role in pulse shortening.

The requirement that $s \geq 2$ ensures that the absorber saturates before the amplifier. The unsaturated absorption clips the leading edge of the optical pulse while the saturated amplification clips the trailing edge of the optical pulse. The peak of the pulse continues to grow if the absorber is saturated while the amplifier is unsaturated.

The general equation describing a pulse traveling through an amplifying medium is given by

$$\partial F/\partial x = n_a \sigma_a F \ , \tag{2.2}$$

where F is the photon flux, n_a is the population inversion of the amplifying medium, σ_a is the emission cross section and time is in local coordinates $\tau = t - (x/c)$.

The rate equation for n is

$$\dot{n}_a = \frac{(n_{a0} - n_a)}{T_{1a}} - n_a \sigma_a F \tag{2.3}$$

where n_{ao} is the fully pumped value of n_a at $F = 0$.

For a "slow" amplifying medium ($T_{1a} \gg$ pulse duration) the above equation can be solved to give

$$n_a(\tau) = n_{ai} \mathrm{e}^{-j_\tau} \qquad \text{where} \tag{2.4}$$

$$j_\tau = \sigma_a \int_{-\infty}^{\tau} F(\tau') d\tau' \tag{2.5}$$

and n_{ai} is the value of n_a before the arrival of the pulse.

Now consider the passage of the optical pulse through a slab of amplifying material of thickness d. Far out on the leading edge, j_τ is nearly zero and the gain factor g_{L} is thus

$$g_{\mathrm{L}} = \mathrm{e}^{n_{ai} \sigma_a d} \ . \tag{2.6}$$

The gain for the trailing edge can be determined from Eq. (2.3) with $\tau = \infty$. The final value of j at $x = d$ is given by

$$j_f = \ln[1 + g_L e^{j_i - 1}] \quad (2.7)$$

which can be used to calculate an expression for the trailing edge gain in terms of the leading edge gain and the initial and final value of j.

$$g_T = g_L e^{j_i - j_f} . \quad (2.8)$$

The above relations can be calculated for the absorber replacing n_a and σ_a by the absorber parameters n_b and σ_b and replacing j with sj where s is the cross section ratio defined previously.

For a cavity containing saturable gain and absorption as well as linear loss, the above equations can be used to calculate the energy of a pulse passing back and forth in the cavity as it passes through each element. After several round trips g_L and g_T approach a steady value, while the gain at the center of the pulse approaches unity. In a regime where g_L and g_T are less than one, pulse compression occurs. In this condition both the leading and trailing edges of the pulse experience net loss while the constancy of the pulse energy ensures gain at the pulse peak. For the case when g_L and g_T are greater than unity the pulse spills out either on the leading edge or the trailing edge. *New* [2.36] has shown that for values of the cross section ratio's, less than 2, no stable regime exists.

2.3 Short Pulse Laser Systems

2.3.1 Dye Lasers

Organic dyes have proven to have nearly ideal properties for the generation of ultrashort optical pulses. Possibly the most significant feature of dyes from the perspective of generating short optical pulses is the set of broadened electronic energy levels that permit optical gain over a frequency range as large as a few thousand cm^{-1}. This broad gain spectrum is a consequence of the rapid thermalization of the vibrational and rotational manifolds of the ground and excited electronic states of dye molecules [2.37]. The thermalization of the upper and lower laser levels is sufficiently rapid to permit the generation of optical pulses in the femtosecond time regime. In addition, organic dyes span the frequency spectrum from the ultraviolet to the infrared and are readily optically pumped. Numerous approaches have been taken to mode-lock the dye laser including both active and passive techniques.

a) Synchronous Mode-locking

Synchronous mode-locking has been a very successful approach to mode-locking dye lasers. In fact mode-locking of the Rhodamine 6G dye laser was first achieved by *Soffer* and *Lin* in 1968 using a mode-locked Nd:glass laser as an optical pump [2.38]. The output of the Nd:glass laser was frequency doubled and used to pump rhodamine 6G and rhodamine B lasers. The optical resonator of the dye laser was adjusted in length to provide synchronism with the pumping mode-locked pulses. When the length of the dye laser resonator was made equal to (or a submultiple of) the length of the pumping resonator, trains of pulses were observed with a period equal to (or a submultiple of) the pump pulse period providing the dye cell is placed at one end of the cavity. With the pumping and dye laser resonators equal in length and the dye cell placed at intermediate harmonic positions, i.e., at a fractional cavity length, multiple pulses were observed. Numerous other workers used mode-locked ruby lasers [2.39–41] and Nd:YAG lasers [2.42,43] to produce efficient ($\simeq 35\%$) tunable picosecond pulse dye lasers using this approach.

Mode-locking of the continuous dye laser by synchronous pumping was first attempted by *Dienes* et al. in 1971 [2.44] in a cavity arrangement in which the Rhodamine 6G cell was placed within the cavity of a passively mode-locked argon laser. A somewhat simpler arrangement using mode-locked pulses from a mode-locked argon laser to pump a dye laser was described by *Shank* and *Ippen* in 1974 [2.45]. Taking advantage of advances in acousto-optic mode-locking of the argon laser, *Chan* and *Sari* [2.46] succeeded in obtaining pulses as short as 2.5 picoseconds using a configuration similar to that described above. This laser has formed the basis for virtually all of the mode-locked dye lasers that are now available commercially.

Synchronous pumping techniques were extended into the subpicosecond regime by *Heritage* and *Jain* [2.47] by pumping a Rhodamine B dye laser with 5 picosecond pulses from a synchronously pumped Rhodamine 6G dye laser. By improving the dispersive properties of the cavity, optical pulses of less than a picosecond have been obtained with a single stage of optical pumping of a dye laser with a mode-locked argon laser [2.47, 48].

The sensitivity of synchronously pumped lasers to cavity length mismatch and stability of the master pumping laser have been investigated extensively [2.48–51]. It was found that a change in cavity length by only a micron would seriously affect the pulse quality. The sensitivity to cavity mismatch was found to increase as the optical pulse width becomes shorter. [2.52]

Streak camera pulse width measurements of optical pulses have shown possible pitfalls in using the autocorrelation technique as the sole determination of optical pulse quality. *Shapiro* et al. [2.53] found that under some circumstances of cavity mismatch, a sharp spike in the autocorrelation function on top of a pedestal was observed. This has often been interpreted incorrectly as a short pulse with wings. In fact according to the measurements of *Shapiro* et al. the streak camera pictures showed more than one pulse in the cavity with a relative spacing that varies in time. The autocorrelation measurement just gives an average of the possible pulse configurations in the cavity.

Ring laser cavities have also been used with synchronous pumping with the initial emphasis on obtaining unidirectional operation [2.54–56]. Subsequent experiments demonstrated that higher quality optical pulses could be obtained by simplifying the design and permitting bidirectional ring oscillation [2.57,58].

Most of the results described in this section thus far have used the mode-locked argon laser as the pumping source. Advances in mode-locked Nd:YAG laser technology have made these lasers very useful for synchronously pumping dye lasers [2.59] and color center lasers [2.60,61] in the infrared. In the visible region of the spectrum the frequency doubled Nd:YAG laser has been found to be a particularly attractive optical pumping source [2.62,63]. New nonlinear optical materials, such as $KTiOPO_4$ [2.64], have permitted high efficiency doubling (in excess of 20%) of the Nd:YAG laser output. In addition, the short pulse width of the doubled Nd:YAG laser (less than 40 picoseconds) allows convenient generation of tunable subpicosecond optical pulses [2.65].

Synchronous pumping is naturally suited to the task of generating synchronized pulse trains with different wavelengths. A single pumping laser is used to excite two or more optical cavities with matched cavity lengths each having independent tuning capability. This has been demonstrated experimentally [2.47,66] with excellent results. Pulse synchronism to within a fraction of a pulse width has been routinely observed in the picosecond range.

b) Passive Mode-locking

Passive mode-locking of the dye laser was first demonstrated with a flashlamp pumped configuration. The first report was by *Schmidt* and *Schafer* in 1968 [2.67] who observed mode-locking of a flashlamp pumped rhodamine 6G dye laser using an organic dye DODCI (3,3'-diethyloxacarbocyanine iodide). *Bradley* and *O'Neill* confirmed these results using both rhodamine 6G and rhodamine B as active medium [2.68].

The experimental arrangement of the passively mode-locked dye laser is very similar to that shown in Fig. 2.6 for the passively mode-locked solid state laser. This is the configuration reported by *Bradley* et al. [2.69,70]. The gain dye cell is pumped with a flashlamp. All surfaces are either at the Brewster angle or wedged to prevent etalon resonances. The mode-locking dye solution is placed next to the mirror at the end of the resonator [2.71].

In the early 1970s, techniques were developed for passively mode-locking continuously pumped dye lasers. The first report of passively mode-locking the cw dye laser was in 1972 in a cw rhodamine 6G dye laser with DODCI as the saturable absorber [2.72]. Optical pulses as short as 1.5 ps with peak powers of 100W were observed. Somewhat later *O'Neill* [2.73] measured optical pulses as short as 4 picoseconds using a streak camera. In 1974 *Shank* and *Ippen* [2.74] reported the first generation of optical pulses less than a picosecond and kilowatt peak power using a cavity dumped and passively mode-locked rhodamine 6G dye laser. A further reduction in optical pulsewidth to 0.3 ps was achieved by analyzing the phase structure on the optical pulse and using a grating compressor [2.75]. *Diels* et al. [2.76] removed the bandwidth limiting prism from the cavity and replaced it with a dielectric coating and in the process were able to generate and measure a 0.2 ps optical pulse.

The advent of the colliding pulse mode-locked dye laser [2.14] pushed pulse generation methods into the femtosecond time domain with the report of the first pulses less than 0.1 ps. The colliding pulse passively mode-locked laser scheme provides an improvement in passively mode-locked laser performance over previous configurations. The central idea is to utilize the interaction, or "collision," of two pulses in an optical cavity to enhance the effectiveness of the saturable absorber. Figure 2.8 illustrates this general process for two, three, and four optical pulses in the laser cavity. With two optical pulses in a simple optical cavity, the saturable absorber must be placed precisely in the center of the cavity so that the two oppositely directed pulses will be able to interact in the saturable absorber at the same time. Since both pulses are coherent, they interfere with each other, creating a standing wave. At the antinodes of the wave the intensity is greatest, more completely saturating the absorber and minimizing loss. At the nodes of the field the absorber is unsaturated, but then, of course, the field is a minimum, again minimizing the loss. The net effect of using the standing wave field to saturate the absorber rather than the fields of the two pulses separately is a reduction in the energy required to saturate the absorber by a factor of approximately 1.5 [2.27]. Since the gain medium is being pumped continuously, each pulse reaches the gain medium at a point in time when the gain medium is fully recovered. The result is that when the two pulses meet in the saturable absorber there is twice as much energy to saturate the absorber than when there is only one pulse in the optical cavity. Thus, the effective saturation parameter is increased by approximately a factor of three over that for a conventional passively mode-locked dye laser.

The linear configuration shown in Fig. 2.8 has been demonstrated experimentally and performs well for pulse generation, but is difficult to align. The saturable absorber stream must be placed at precisely an integer submultiple of the cavity length for a pulse collision to take place. The precision required is on the order of 10 μm. Figure 2.9 depicts a ring cavity configuration that allows a pulse collision to take place without this alignment problem. In the ring configuration the pulse collision occurs between oppositely directed pulses traveling around the cavity. To minimize the energy loss to the absorber, the pulses meet in the saturable absorber. In effect, the ring cavity allows the pulses to automatically synchronize and removes the requirement for precise positioning of the saturable absorber stream. This scheme greatly alleviates the alignment difficulty found in accurately positioning the saturable absorber in the linear configuration.

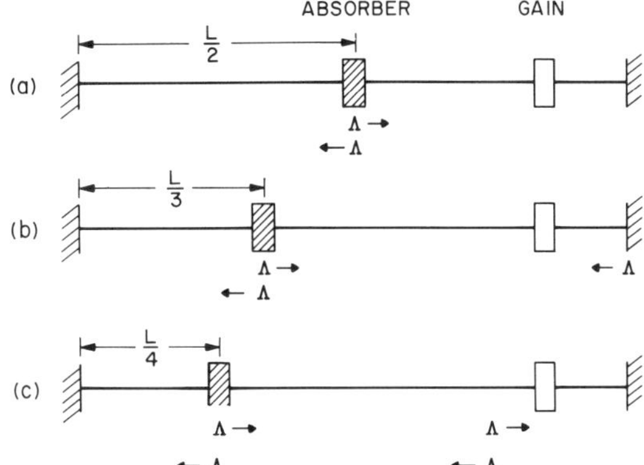

Fig. 2.8. Diagram illustrating the colliding pulse mode-locking configuration for (**a**) two, (**b**) three, and (**c**) four optical pulses in the cavity. In a conventional passively mode-locked laser, only a single pulse is present in the cavity since the saturable absorber is not placed at a submultiple of the laser cavity

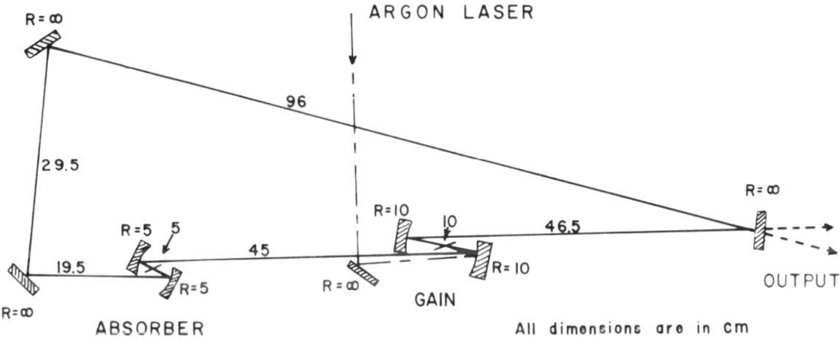

Fig. 2.9. Cavity configuration for the colliding pulse mode-locked laser pumped by an argon laser

Proper positioning of the absorber and gain jets in the ring cavity can enhance the stability of the laser operation. Figure 2.10 illustrates the virtues of placing the absorber and gain medium at approximately one-quarter of the round trip around the ring. Since the two oppositely directed pulses meet in the absorber, they will draw power from the gain medium with the same time delay corresponding to a one-half cavity round-trip time. In this manner both pulses see the same gain, since the continuously pumped gain dye has the same time to recover following each pulse. This has the effect of reducing the formation of extra pulses in the cavity.

Optimum performance of a femtosecond dye laser has been found to be critically dependent on the group velocity dispersion in the optical cavity. Empirical selection of mirrors has been used to minimize the effects of dispersive mirror coatings on pulse width. A more elegant approach has been to compensate the group velocity dispersion in the optical cavity by inserting an element with adjustable negative group velocity dispersion. *Fork* et al. [2.77] have devised a

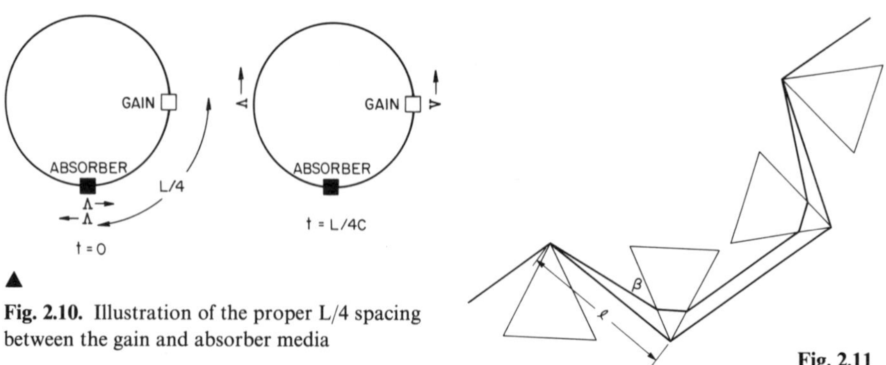

▲ **Fig. 2.10.** Illustration of the proper L/4 spacing between the gain and absorber media

Fig. 2.11

Fig. 2.11. Four-prism sequence having negative dispersion. The prisms are set at the minimum deviation angle and oriented so that the rays enter and leave at the Brewster angle. Movement of any of the four prisms normal the prism base adjusts the magnitude of the negative dispersion

novel sequence of prisms which can provide an adjustable amount of group velocity dispersion. The configuration is shown in Fig. 2.11. The dispersion can be adjusted simply by moving a prism along a normal to the prism base. *Valdmanis* et al. [2.78] inserted this device into the optical cavity of a colliding pulse ring dye laser and adjusted the prisms to compensate the cavity group velocity dispersion. The result was the generation of optical pulses with a duration of 27 femtoseconds.

Femtosecond pulses have been generated over a large portion of the visible spectrum. The Krypton ion laser has been used to pump mode-locked cw dye lasers in the region 700–778nm [2.79]. More recently, new active and passive dye combinations have been used to cover the spectral range 550–700nm [2.80].

c) Hybrid Mode-locking Techniques

The goal of combined mode-locking is to receive the benefit of synchronization supplied by a master laser and yet take advantage of the pulse shortening due to the action of a saturable absorber. Synchronous pumping also allows the use of a Nd:YAG pumping source which must be mode-locked to be efficiently frequency doubled.

A straightforward approach is to add saturable absorber directly to the gain dye solution. Minor improvements in stability and pulse shape were reported using this approach [2.50]. *Sizer* et al. added a separate saturable absorber cell internal to the sychronously pumped optical cavity and reported the generation of optical pulses of less than 100 fs [2.81]. *Johnson* [2.65] using a synchronously pumped ring configuration containing both saturable absorber and gain cells reported the generation of 150 fs optical pulses.

A more elegant approach has been to design a laser with some of the attributes of a colliding pulse mode-locked dye laser. One such approach has

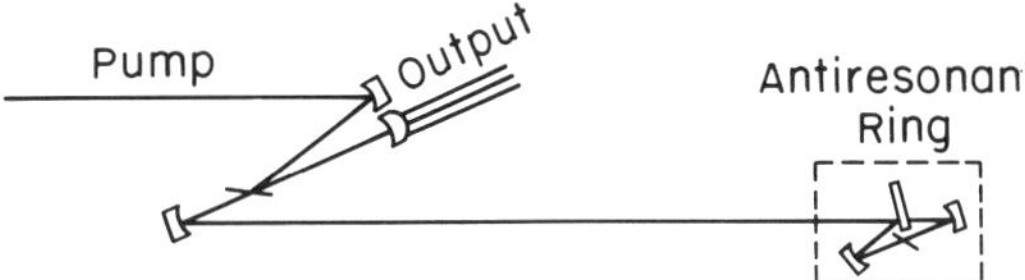

Fig. 2.12. Diagram of an antiresonant ring mode-locked dye laser

been to use an antiresonant ring interferometer which includes a saturable absorber cell at the end of the optical cavity. Experiments of this sort were reported for flashlamp pumped Nd:YAG lasers by *Siegman* [2.82] and *Vanherzeele* et al. [2.83,84]. This approach has been extended to a synchronously pumped dye laser with the cavity configuration shown in Fig. 2.12 [2.84]. The antiresonant ring is made up of a 50% beam splitter and two high reflecting focusing mirrors and a dye jet. The nozzle is adjusted to be exactly at the point in the ring where the two oppositely directed pulses collide in the dye stream. Optical pulses as short as 130 fs were reported for this configuration.

Another approach demonstrated by *Nuss* et al. [2.85] was to directly pump the standard CPM ring dye laser cavity [2.14] with an appropriate pump-pulse sequence. Previous attempts [2.65] encountered an instability as two pairs of counter propagating pulses built up in the laser cavity. This difficulty was overcome by using a pump-pulse sequence which favors a single pair of pulses. Optical pulse widths of less than 100 fs were reported.

d) Amplification

Optical pulses generated by continuous mode-locking techniques have energies in the nanojoule range. This is an insufficient energy for many applications. To obtain higher energies optical amplifiers have been devised. Optically pumped organic dyes have proven quite useful for amplifying ultrashort optical pulses. The large emission cross section of organic dyes allows gains in excess of 10–100 cm^{-1} to be achieved with modest excitation energies. The very high gain provides a number of constraints on dye laser amplifier design. Amplified noise or amplified spontaneous emission can rapidly deplete the gain of a dye amplifier. This effect combined with the very short spontaneous emission times of organic dyes (10^{-9} s) limits the energy storage in this type of amplifier.

The problem of amplifying ultrashort optical pulses is greatly complicated by the presence of amplified spontaneous emission. A population inversion in a high gain medium must be maintained for a sufficient time to integrate the energy of the pumping pulse without dumping the population inversion with amplified spontaneous emission [2.86].

In an elementary way we can gain some insight into the amplification process by considering the rate equations describing a pulse, $I(x,t)$, traveling through an amplifying medium with a population inversion, $n(x,t)$.

$$\frac{dn(x,t)}{dt} = \frac{n_0 - n(x,t)}{T_1 \sigma(n(x,t)I(x,t))} \quad \text{and} \tag{2.9}$$

$$\frac{dI}{dx} = \sigma n(x,t)I(x,t) \ . \tag{2.10}$$

For an ultrashort pulse the pulsewidth $t_p \ll T_1$ the population inversion is given by

$$n(x,t) = n_1 \exp[-\sigma \int I(x,t')dt'] \ . \tag{2.11}$$

From (2.11) we can define a saturation energy density as

$$E_s = 1/\sigma \ .$$

If the input pulse energy is greater than E_s, the stored energy will be swept out by the pulse and the amplifier will become nonlinear and will saturate. Nonlinear amplification can give rise to distortion of the pulse by preferentially amplifying the leading edge of the pulse. This effect can be controlled by passing the pulse through a saturable absorber.

To efficiently pump a short pulse dye amplifier, it is desirable for the pump duration to be comparable or less than the energy storage time, T_1. For most dyes T_1 is in the range of a few nanoseconds. For a multistage amplifier it is often advantageous to isolate the stages of the amplifier with saturable absorbers to minimize loss to amplified spontaneous emission.

A frequency doubled and Q-switched Nd:YAG laser has proven to be a useful excitation source for a dye laser amplifier. A diagram of such an amplifier is shown in Fig. 2.13 [2.87]. The amplifier consists of four stages which are pumped by a frequency doubled Nd:YAG laser and operates at a 10 Hz repetition rate. Each stage is isolated with a saturable absorber. A grating pair is used to correct for the group velocity dispersion that arises when the optical pulse passes through the amplifier components. This amplifier produces gigawatt optical pulses while maintaining a pulse duration of 70 fs.

Rolland and *Corkum* [2.88] have reported the operation of an amplifier similar to that above but pumped with a XeCP laser [2.89,90]. One advantage of an eximer laser pumped system is that repetition rates of several hundred Hertz appear possible. They achieved powers of greater than 7 GW with a duration of 70 fs at 100 Hz. repetition rate.

Even higher repetition rates are possible using a copper vapor laser as an amplifier pumping source [2.91,92]. Currently copper vapor lasers are available commercially with repetition rates in the tens of kilohertz. A multipass configuration shown in Fig. 2.14 has been designed to achieve amplification from the nanojoule level from a CPM dye laser to the microjoule level. A grating pair compressor [2.75] or a prism pair [2.77] is used to compensate for the dispersion caused by the dye amplifier components. Currently, in the laboratory, optical

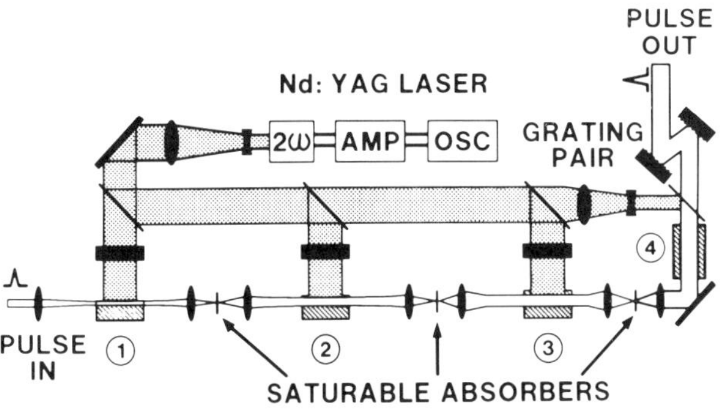

Fig. 2.13. Diagram of a four stage optically pumped femtosecond pulse amplifier

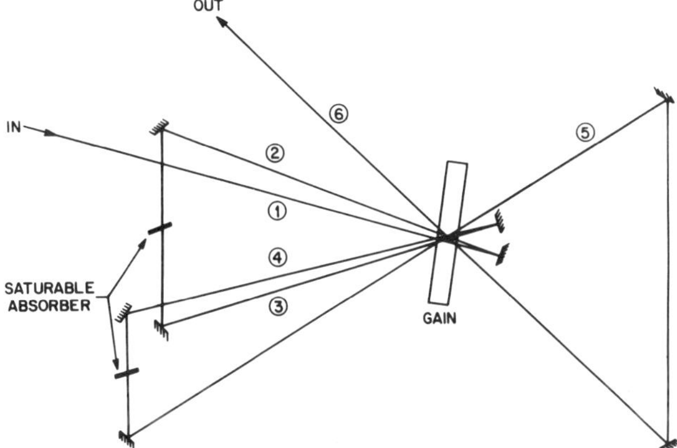

Fig. 2.14. Experimental arrangement for a multi-pass femtosecond pulse amplifier pumped by a copper vapor laser

pulses of less than 50 fs have been amplified to several microjoules at a repetition rate near 10 KHz.

Another approach to amplifying short optical pulses at a moderate repetition rate is to use a regenerative Nd:YAG amplifier [2.93]. This approach requires a complete system design that consists of a synchronously pumped CPM dye laser, a regenerative Nd:YAG amplifier and a dye amplifier chain. A schematic of this system is shown in Fig. 2.15. This system has the advantage of synchronously pumping the amplifier chain with short 90 ps pump pulses. This approach can potentially increase efficiency because the pumping pulses are much shorter than the energy storage time of the amplifier dye. The dye laser pulses must be closely sychronized with the pump pulses. This creates the requirement that the dye laser must be sychronously pumped with pulses derived from the same Nd:YAG laser pumping source. Pulses from the master oscillator are split into two parts. One part is frequency doubled and used to pump the synchronous dye laser and the other part is injected into the regenerative Nd:YAG amplifier. The output of the amplifier is then also frequency doubled and injected into the dye amplifier chain with the appropriate synchronization. Optical pulses as short

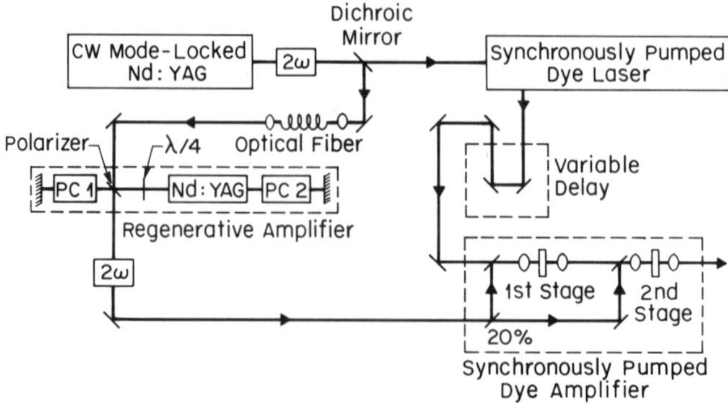

Fig. 2.15. Synchronous dye-amplifier system. The Nd:YAG regenerative amplifier configuration is detailed inside the dotted line

as 90 fs have been amplified to the microjoule level at a repetition rate of 1 KHz using this system.

2.3.2 Color Center Lasers

Color centers are electron or hole trapping defects that produce optical absorption and emission bands in insulating crystals. In some ways color center lasers [2.94] are closely analogous to lasers based on organic dyes. The absorption and emission bands of color centers are broad and diffuse. The Stokes shifted emission can have high quantum efficiency. Efficient tunable lasers have been constructed using an optical cavity configuration very similar to cw dye lasers with a thin slab of crystal containing the laser-active centers substituted for the dye jet stream. Although laser action has been demonstrated in the uv, the majority of color centers operate in the infrared covering the tuning range $0.8\,\mu\text{m} \leqslant \lambda \leqslant 4\,\mu\text{m}$. Most color centers must be operated at cryogenic temperatures. An optical cavity configuration illustrating the focusing optics and dewar is shown in Fig. 2.16.

Mode-locking of the color center laser [2.95] provides some unique challenges. Unlike the case of a synchronously pumped dye laser, the upper state lifetime of a color center does not recover in a cavity round trip time. This means that the gain seen between successive pulses does not change significantly. This leads to some rather unusual behavior. Two thresholds are observed for a synchronously mode-locked color center laser [2.61]. At low pumping powers and low intracavity power the first threshold for continuous lasing is observed. A second threshold at higher pump power is observed for mode-locking. The intracavity power must be sufficiently intense to deplete the gain by stimulated emission to allow stable mode-locking to occur. Optical pulse widths of a few picoseconds are readily attainable throughout the near infrared using synchronously pumped color center lasers.

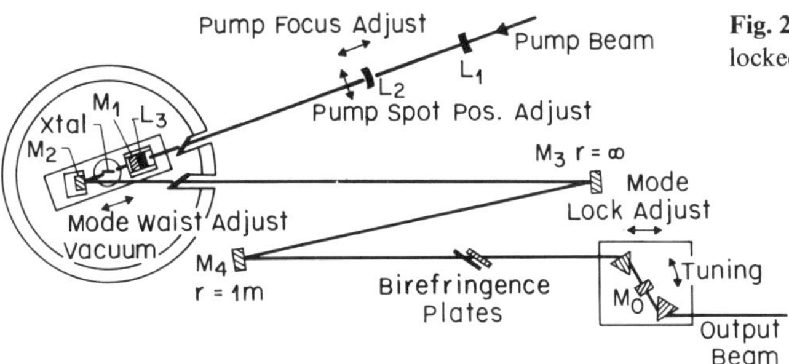

Fig. 2.16. Schematic of a mode-locked color center laser

2.3.3 Novel Pulse Generation Techniques

Although mode-locking is the dominant method of ultrashort pulse generation some intriguing and clever approaches are worth mentioning. *Bor* [2.96] has demonstrated a technique for generating picosecond pulses using the self Q-switching of the distributed feedback dye laser [2.44,96]. The major distinction between a distributed feedback laser and a conventional laser is that the distributed feedback laser uses feedback from spatially periodic perturbations of the medium susceptibility. The wavelength of the laser is determined by the Bragg condition

$$\lambda = 2\Lambda\eta \ ,$$

where λ is the wavelength in the vacuum, η is the index of refraction and Λ is period of modulation.

Distributed feedback oscillation is then achieved by pumping a dye medium with two interfering beams which excite the dye and produce a periodic modulation of gain with a period that satisfies the Bragg condition within the gain profile. For the case of pure gain modulation the cavity decay time τ_c is proportional to $\alpha_1^2(t)$ where $\alpha_1(t)$ is the amplitude of the gain modulation.

The pulse formation mechanism is similar to pulse formation by relaxation oscillation with the important difference that the cavity decay time is not a constant but $\tau_c \propto n^2$ where n is the population inversion. Thus τ_c has a large value during the rising half a of self-Q-switched pulse and a small value during the decaying half.

Bor has shown that the self-Q-switched distributed feedback dye laser can be a source of tunable picosecond pulses. An improved version of this technique has been reported that is capable of generating an optical pulse of less than one picosecond [2.98].

Another very clever method of generating picosecond optical pulses is to use traveling wave excitation to pump a dye laser [2.59,98–100,103]. The moving excitation creates a pulse that builds up from spontaneous emission and moves along in synchrony with the pump energy.

A traveling wave excitation beam for transverse pumping a dye cell can be created by imaging the pumping beam onto a diffraction grating. For each pitch of the grating, an optical delay corresponding to the pump wavelength is introduced. In this way, a continuous spatial delay is created across the diffracted beam [2.101,102]. With proper adjustment of the grating angle, the transverse excitation pumping beam can be synchronized with the amplified spontaneous emission in the dye cell. With picosecond pulse excitation the duration of the amplified spontaneous emission pulse was found to be 3–10 times shorter in duration than the pumping pulse [2.99].

A tunable, highly monochromatic, picosecond light source has been demonstrated using a slight modification of this approach [2.100]. In this case, two dye cells were used. The first cell was pumped using the traveling excitation just described. The broadband amplified spontaneous emission short pulse was then passed through a tunable grating filter to select a narrow frequency band. The filtered pulse was then amplified with a second traveling wave pumped cell. Picosecond optical pulses were obtained which were tunable across the entire gain bandwidth of the dye as the grating was rotated.

The traveling wave pumping scheme has been used to advantage for pumping dyes with low quantum efficiencies [2.103]. Dyes in the infrared have quantum efficiencies on the order of 10^{-4} and lifetimes of 5–12 picoseconds. Using traveling wave excitation, single picosecond optical pulses were obtained covering the range from 1.4 to 1.8 μm.

2.3.4 Diode Lasers

Mode-locked semiconductor diode lasers hold promise for applications to high-speed communications and optical signal processing. Such lasers are already compatible with the wavelength and power requirements for distortion-free transmission over optical fibers.

A semiconductor diode laser cavity is typically a few hundred microns in length. Straightforward application of mode-locking techniques would result in a pulse repetition rate of several hundred GHz. Not only is this an inconvenient pulse repetition rate but the gain recovery time is much to slow for optimum pulse shaping. To overcome these problems *Ho* et al. [2.104] formed an extended resonator by placing a GaAlAs diode laser at the focus of an external convex mirror to form a cavity of about 5 cm in length. Mode-locking was achieved by modulating the laser current at a frequency of 3 GHz, corresponding to the inverse of the cavity round trip time. Optical pulses in these initial experiments were on the order of twenty picoseconds. Similar results were obtained with InGaAsP operating at 1.21 μm [2.105,106].

A more highly evolved actively mode-locked laser configuration is shown in Fig. 2.17 where an optical fiber is incorporated into the resonator structure. This single mode fiber resonator combines compact construction with the flexibility of a standard biconic fiber connector at the output port. Optical pulses of less

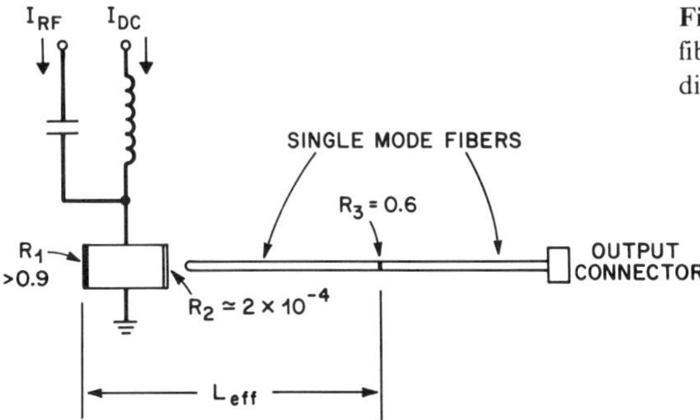

Fig. 2.17. Diagram of a single mode fiber composite cavity mode-locked diode laser

than 5 ps in duration were generated at a repetition rate of 2 GHz. The peak pulse power was 56 mW.

As with dye lasers, the shortest diode laser pulses have been achieved by passive mode-locking *Ippen* et al. [2.107]. Initially the saturable absorber was formed by damage induced during the aging process. Later saturable absorption was introduced into the exit face of the laser diode using proton bombardment [2.108]. In another approach, *Silberberg* et al. [2.109] used a GaAs/GaAlAs multiple quantum well as a saturable absorber in an external resonator. Optical pulses of less than 1 picosecond have been achieved using these satruable absorber techniques.

2.4 Pulse Compression

More than two decades ago, *Gires* and *Tournois* [2.110] and *Giordmaine* et al. [2.111] proposed that optical pulses be shortened by adapting microwave pulse compression techniques to the visible spectrum. Optical pulse compression is accomplished in two steps. In the first step, a "chirp", or frequency sweep is impressed on the pulse. The pulse is then compressed by using a dispersive delay line. A chirp can be impressed on an intense optical pulse simply by passing the pulse through an optical Kerr medium [2.112]. In general when an intense optical pulse is passed through a nonlinear medium, the refractive index, n, is modified by the electric field, E.

$$n = n_0 + n_2\langle E^2\rangle + \cdots \,. \tag{2.13}$$

A phase change, $\delta\phi$, is impressed on the pulse

$$\delta\phi \approx n_2\langle E^2\rangle \frac{\omega z}{c} \,, \tag{2.14}$$

where ω is the frequency, z is the distance traveled in the Kerr medium, c is the velocity of light. As the intensity of the leading edge of the optical pulse rises rapidly, a time-varying phase or frequency sweep is impressed on the pulse carrier. Similarly, a frequency sweep in the opposite direction occurs as the intensity of the pulse falls on the trailing edge. The amount of frequency sweep is given approximately by

$$\delta\omega \approx \frac{\omega z n_2}{c}\frac{d}{dt}\langle E^2(t)\rangle \ . \tag{2.15}$$

A more rigorous approach to this problem requires a solution of the wave equation with the addition of a nonlinear term to account for the Kerr nonlinearity. This will be discussed in more detail in the next section.

Optical fiber waveguides are a nearly ideal optical Kerr medium for pulse compression [2.113,114]. Problems of self-focusing, nonuniform intensity profile and diffraction are overcome with the use of an optical fiber. Also the Kerr nonlinearities are large enough that only modest optical powers are required.

Once the chirp has been applied to a pulse, the pulse is passed through a dispersive delay to reassemble all its frequency components in order to achieve compression. An experimentally convenient compression device is a pair of parallel gratings [2.75]. Each wavelength passing through the grating pair is diffracted at a different angle and follows a different path giving rise to a wavelength dependent optical path delay. By properly adjusting the grating spacing it is possible to provide the right amount of group delay to form the compressed pulse.

2.4.1 Theory

As discussed in the previous section, the propagation of an optical pulse in a Kerr medium can be described by the wave equation with the inclusion of the Kerr nonlinearity. Such an equation can be reduced to a nonlinear Schrodinger equation [2.113,115].

$$\mathrm{i}\frac{\partial V}{\partial(z/z_0)} = \frac{\pi}{4}\left(\frac{\partial^2 V}{\partial(t/t_0)^2} + 2|V|^2 V\right) , \tag{2.16}$$

where the complex variable V is the pulse amplitude. The time variable t is a retarded time and is defined such that for any distance z along the fiber, the center of the pulse is at $t = 0$, and the input pulse envelope is assumed to be of the form

$$V(z=0,t) = A\,\mathrm{sech}\left(\frac{t}{t_0}\right) . \tag{2.17}$$

The normalized length z_0 and peak amplitude A are defined by

$$z_0 = 0.322\frac{\pi^2 c^2 \tau^2}{|D(\lambda)|\lambda} \qquad \text{and} \tag{2.18}$$

$$A = \sqrt{\frac{P}{P_1}} \quad \text{where} \tag{2.19}$$

$$P_1 = \frac{nc\lambda A_{\text{eff}}}{16\pi z_0 n_2} \times 10^{-7}\,\text{W} \tag{2.20}$$

and τ_0 is the intensity FWHM of the input pulse ($\tau_0 = 1{,}76 t_0$), $D(\lambda)$ is the group velocity dispersion in dimensionless units, n is the refractive index and n_2 is the nonlinear coefficient in electrostatic units, c is the velocity of light in centimeters per second and λ is the vacuum wavelength. The peak input pulse is given by P, and the quantity A_{eff} is the effective area. In the case of an optical fiber waveguide A_{eff} is the area of the optical mode.

The above equation can be solved for both positive and [2.91] negative group velocity dispersion [2.113]. For the case of a negative group velocity dispersion the nonlinear Schrodinger equation can be solved in closed form yielding soliton solutions. Solitons in optical fibers have been observed and have been the subject of extensive study [2.94]. In the region of positive group velocity dispersion, solitons are not possible and numerical techniques must be used.

Solutions to the problem of an intense optical pulse with an amplitude $V(z, t)$ propagating along an optical fiber are shown in Fig. 2.18. In this figure a perspective plot of the pulse temporal shape ($|V(t)/A|^2$) is shown as a function of length with $A = 5$ for the case of normal group velocity dispersion. Note that as the pulse propagates along the fiber, it spreads and develops a rectangular profile with steep leading and trailing edges [2.116]. Because of the positive group velocity dispersion, the red-shifted light generated at the leading edge of the pulse travels faster than blue-shifted light at the trailing edge. This leads to a pulse spreading and the rectangular pulse shapes observed in the figure.

As described in the introduction, the next step of the pulse compression process is to reassemble the frequency chirped pulse with a compressor. The action of the compressor is most easily described in the frequency domain since it is simply a frequency-dependent time delay. The Fourier transform of the pulse can be expressed in the form

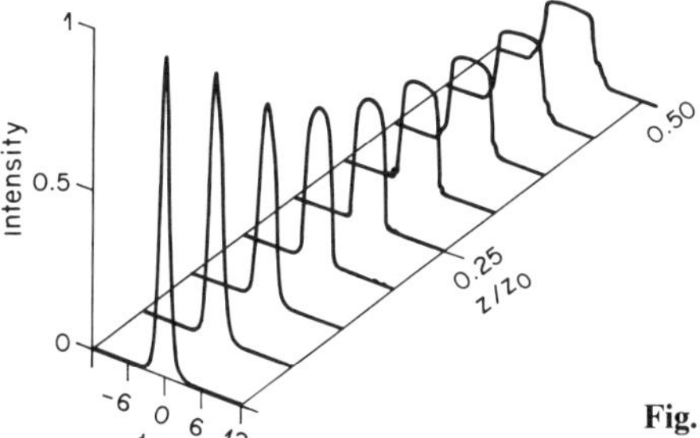

Fig. 2.18. Perspective plot of the temporal shape of a pulse as a function of normalized distance along a fiber

$$V(z,\omega) = A(\omega)e^{i\Phi(\omega)} \tag{2.21}$$

where $A(\omega)$ and $\Phi(\omega)$ are the amplitude and phase. The effect of the compressor can be described by a phase function $\Phi_c(\omega)$ so that the Fourier transform of the compressed pulse is given by

$$V_c(z,\omega) = A(\omega)\exp\{i[\Phi(\omega) + \Phi_c(\omega)]\} \ . \tag{2.22}$$

If $\Phi_c(\omega) = -\Phi(\omega)$, then all the frequency components of the pulse will be in phase and will thus create the pulse with the maximum peak amplitude.

One of the most useful types of compressors is a grating pair compressor. A grating pair compressor has a delay function that is approximately of the form [2.75]

$$\Phi_c(\omega) = \Phi_0 - a\omega^2 \ . \tag{2.23}$$

The grating compressor constant a can be adjusted by varying the grating separation and thus is a directly accessible experimental parameter.

The expression for the grating pair constant is given by

$$a = \frac{b\lambda^3}{4\pi c^2 d^2 \cos^2\gamma} \ , \tag{2.24}$$

where b is the center to center distance between the two gratings, d is the groove spacing, λ is the center wavelength of the pulse and γ is the angle normal to the input grating and the diffracted beam at λ.

Some useful expressions for estimating optimum achievable compressions, optimum fiber lengths and grating parameters are given below.

$$\frac{\tau_0}{\tau} \approx 0.63A \qquad \frac{z_{opt}}{z_0} \approx \frac{1.6}{A} \qquad a \approx 0.52\tau_0^2 A \ . \tag{2.25}$$

It is interesting to discuss the role group velocity dispersion plays in the pulse compression process. *Grischkowsky* and *Balant* [2.116] have shown that group velocity dispersion acts to linearize the chirp over most of the length of the pulse so that almost all the input power appears in the compressed pulse. The penalty to be paid for higher pulse quality is that the compressed pulses are not as short as could be obtained in the absence of group velocity dispersion and that optimum compression becomes a critical function of the fiber length.

For the case of long input pulses the optimum fiber length and the required compressor factor may become inconveniently large, and it may be necessary to use shorter fibers. Under these conditions operation can be in a regime where group velocity has very little effect. In this case, numerical results yield the following approximate expressions

$$\frac{\tau_0}{\tau} \simeq 1 + 0.9\frac{A^2 z}{z_0} \qquad a \approx 0.081\tau_0^2\frac{z_0}{A^2 z} \ . \tag{2.26}$$

The above expressions are useful for choosing a fiber length and input power to achieve a desired pulse compression. These expressions have been confirmed in experiments covering the visible and the near infrared.

2.4.2 Compression Experiments

Pulse compression using optical fibers as a Kerr medium was first investigated experimentally by *Mollenauer* et al. [2.113] in their work on soliton compression of optical pulses from a color center laser. In this work, the wavelength of the optical pulses ($\lambda = 1.3\,\mu m$) was in the anomalous or negative dispersion region. A separate compressor is not needed in these experiments because the dispersive properties of the fiber material self compress the pulse. When the group velocity dispersion is negative ($\partial v_g/\partial\lambda < 0$) the group velocity increases for increasing frequency. The trailing edge of a chirped pulse containing the higher frequencies is advanced while the leading half of the pulse containing the lower frequencies is retarded. The pulse then tends to collapse upon itself. Using this compression technique [2.117], a 7 ps optical pulse was compressed to 0.26 ps with a 100 m length of single mode fiber. This dramatic 27-fold reduction in pulse width was achieved with just a few hundred Watts peak power at the input of the fiber.

Mollenauer and *Stolen* [2.118] extended the ideas of fiber soliton pulse compression to form a new type of mode-locked color center laser called a soliton laser. In the soliton laser an optical fiber of the appropriate length is added to the feedback path of a synchronously pumped color center laser. The soliton properties of the fiber feedback force the laser itself to produce pulses of a definite shape and width. Using this approach optical pulses of less than 50 fs in duration have been produced.

Optical fiber compression in the positive group velocity dispersion regime ($\lambda \leqslant 1.3\,\mu m$) was initiated by Nakatsuka and Grischkowsky who used an optical fiber for the chirping process and anomalous dispersion from an atomic vapor as the compressor. *Shank* et al. [2.119] replaced the atomic vapor compressor with a grating pair compressor [2.75] and achieved pulse compression to 30 fs optical pulse width. Considerable activity at several laboratories [2.120–122] has led to remarkable progress in this area leading to the generation of an optical pulse less than 10 fs. The experimental apparatus for pulse compression in the femtosecond time domain is shown in Fig. 2.19. Single optical pulses of 40 fs in

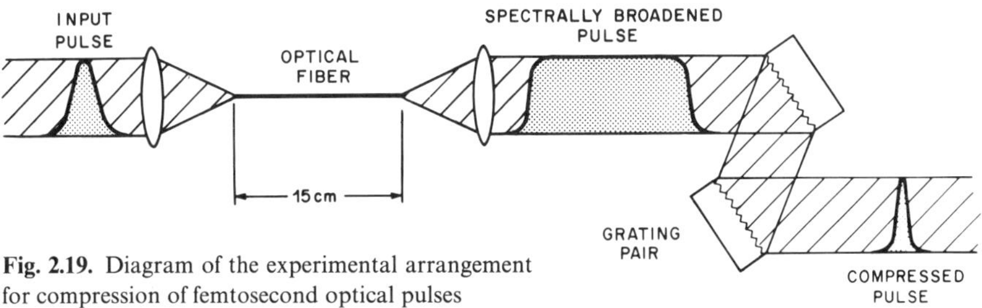

Fig. 2.19. Diagram of the experimental arrangement for compression of femtosecond optical pulses

duration were obtained from a CPM dye laser amplifier combination [2.91] and focused with a 10× microscope objective into a short 7 mm length of optical fiber. The light exiting the fiber was collimated with a microscope objective and passed through a grating pair compressor. An optical pulse of 8 fs was generated.

It is interesting to explore the limits of pulse compression. When an optical pulse propagates through any dielectric medium, group velocity dispersion broadens the pulse. For example, an 8 fs optical pulse will have its width doubled by passage through ~1 mm of glass or ~3 m of air. These linear propagation effects are not fundamental and can in principle be corrected with a compressor. However, the grating pair compressor has an inherent bandwidth limitation that was pointed out by *Treacy* [2.75] in his original paper. In the large bandwidth case, higher order terms in the phase response of the grating pair compressor become important. A "real compressor" has a phase response of the form

$$\Phi(\omega) = -a(\omega - \omega_0)^2 - b(\omega - \omega_0)^3 \ . \tag{2.27}$$

The ratio of the cubic to the quadratic terms becomes larger as the pulse bandwidth is increased for shorter optical pulses. Recently *Tomlinson* and *Knox* [2.123] have calculated the distorting effects of the cubic terms for the 8 fs compression experiment described above. Their calculations show that in this experiment the limiting factor preventing the generation of an even shorter pulse is the cubic phase distortion caused by the grating pair compressor. Most recently further progress in short pulse generation has been achieved by noting that the cubic phase distortion for a grating pair is of the opposite sign of that for a prism pair. Using a grating pair followed by a prism sequence, a 6 fs pulse has been generated and measured [2.124]. To accurately measure a pulse of such short duration it is necessary to match wave fronts by using an interferometric autocorrelation technique [2.125] to determine the pulse width. A plot of the experimentally measured interferometric autocorrelation function for a 6 fs optical pulse [2.124] is shown in Fig. 2.20.

Pulse compression techniques have effectively been applied to the compression of 1.06 μm pulses from the Nd:YAG laser. The 80 ps optical pulses from a

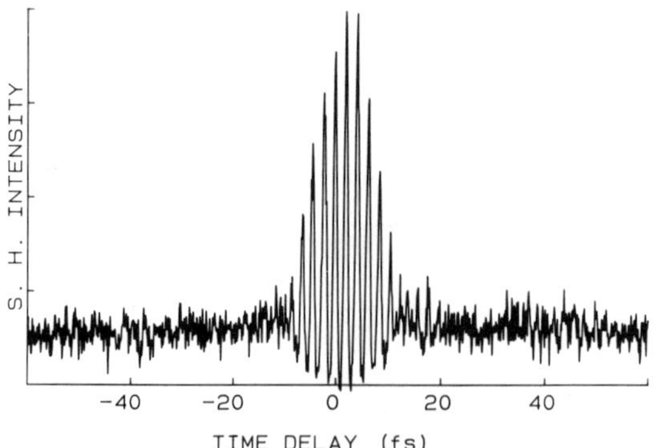

Fig. 2.20. Plot of the interferometric autocorrelation function measured for a 6 femtosecond optical pulse

mode-locked Nd:YAG laser were compressed to a few picoseconds using 300 m of optical fiber and a grating pair. Improvcments in quality of the optical pulse have been obtained using spectral masking techniques in the grating pair compressor [2.126]. Most recently very high quality optical pulses have been obtained by taking advantage of the nonlinear birefringence of the fiber and the polarizing properties of the grating [2.127]. Pulse wings of a 1.2 ps compressed pulse were measured to fall exponentially over four orders of magnitude.

Recently, pulse compression techniques have been applied to the amplification of high energy 1.06 μm pulses. The onset of self-focusing limits the amplification of ultrashort optical pulses. To overcome this limitation *Mourou* et al. [2.128] used an optical fiber to stretch a short optical pulse, amplify, and then recompress using a grating pair. Since the stretched pulse is amplified, the energy density can be increased, more efficiently extracting the stored energy in the amplifier. The authors assert that it should be possible to amplify 1.06 μm pulse to the joule level while maintaining a pulsewidth of a fraction of a picosecond.

References

2.1 H. Mocker, R. Collins: Appl. Phys. Lett. **7**, 270 (1965)
2.2 A.J. DeMaria, D.A. Stetser, H. Heyman: Appl. Phys. Lett. **8**, 22 (1966)
2.3 P.W. Smith. Proc. IEEE **58**, 1342 (1970)
2.4 G.H.C. New: Rep. Prog. Phys. **46**, 877 (1983)
2.5 L.E. Hargrove, R.L. Fork, M.A. Pollack: Appl. Phys. Lett. **5**, 4 (1964)
2.6 S.E. Harris, R. Targ: Appl. Phys. Lett. **5**, 205 (1964)
2.7 S.E. Harris: Proc. IEEE **54**, 1401 (1966)
2.8 A. Yasa, O. Teschke: Opt. Commun. **15**, 169 (1975)
2.9 D. Kuizenga, A. Seigman: IEEE J. Quantum Electron. QE-**6**, 673 (1970)
2.10 N. Frigo, T. Daley, H. Mahr: IEEE J. Quantum Electron, QE-**13**, 101 (1977)
2.11 D. Kim, M. Kuhl, R. Lambrich, D. von der Linde: Opt. Commun. **27**, 123 (1978)
2.12 C. Auschnitt, R. Jain, J. Heritage: IEEE J. Quantum Electron. QE-**15**, 912 (1979)
2.13 J. Catherall, G.H.C. New, P. Rodmore: Opt. Lett. **7**, 319 (1982)
2.14 R.L. Fork, B.I. Greene, C.V. Shank: Appl. Phys. Lett. **38**, 671 (1981)
2.15 G.H.C. New: Opt. Commun. **6**, 188 (1972)
2.16 V. Letokhov: Soviet Phys. JETP **28**, 562 (1969)
2.17 V. Letokhov: Soviet Phys. JETP **28**, 1026 (1969)
2.18 J. Fleck: Appl. Phys. Lett. **12**, 178 (1968)
2.19 J. Fleck: J. Appl. Phys. **1969**, 3318 (1968)
2.20 D. von der Linde, O Bernecker, W. Kaiser: Opt. Commun. **2**, 149 (1970)
2.21 W. Zinth, A. Lauberau, W. Kaiser: Opt. Commun. **22**, 161 (1977)
2.22 G.H.C. New, T. O'Hare: Phys. Lett. **68A**, 27 (1978)
2.23 J. Hermann, F. Weidner, B. Wilhelmi: Appl. Phys. **20**, 237 (1979)
2.24 H. Weber: J. Appl. Phys. **39**, 6041 (1968)
2.25 D. Bradley, G.H.C. New, S. Caughey: Phys. Lett. **30A**, 78 (1970)
2.26 D. von der Linde: IEEE J. Quantum Electron. QE-**8**, 328 (1972)
2.27 E. Garmire, A. Yariv: IEEE J. Quantum Electron. QE-**3**, 222 (1967)
2.28 D. von der Linde, K. Rodgers: Opt. Commun. **8**, 91 (1973)

2.29 B. Kopinsky, W. Kaiser, K. Drexhage: Opt. Commun. **32**, 451 (1980)
2.30 R. Alfano, N. Schiller, G. Reynolds: IEEE J. Quant. Electron QE-**17**, 290 (1981)
2.31 G.H.C. New, D.H. Rea: J. Appl. Phys. **47**, 3107 (1976)
2.32 B. Garside, T. Lim: J. Appl. Phys. **44**, 2335 (1973)
2.33 B. Garside, T. Lim: Opt. Commun. **12**, 8 (1974)
2.34 H. Haus: IEEE J. Quantum. Electron. QE-**11**, 736 (1975)
2.35 P. Kryukov, V. Letokhov: Soviet Phys. Usp. **12**, 641 (1970)
2.36 G.H.C. New: IEEE J. Quantum Electronics QE-**10**, 115 (1974)
2.37 C.V. Shank: Rev. Modern Phys. **41**, 649 (1975)
2.38 B.H. Soffer, L.W. Linn: J. Appl. Phys. **39**, (5859) 1968
2.39 D.J. Bradley, A. Durrant, G. Gale, M. Moore, P.D. Smith: IEEE J. Quantum Electron. QE-**4**, 707 (1968)
2.40 D.J. Bradley, A. Durrant: Phys. Lett. **27A**, 73 (1968)
2.41 M.R. Topp, P.M. Rentzepis: Phys. Rev. A **3**, 358 (1971)
2.42 L. Goldberg, C. Moore: Appl. Phys. Lett. **27**, 217 (1975)
2.43 C. Moore, L. Goldberg: Opt. Commun. **16**, 21 (1976)
2.44 H. Kogelnik, C.V. Shank: Appl. Phys. Lett. **20**, 152 (1971)
2.45 C.V. Shank, E.P. Ippen: In *Dye Lasers*, ed. by F.P. Schaefer, Topics Appl. Phys. Vol. 1 (Springer, Berlin, Heidelberg 1972) p. 121.
2.46 C.K. Chan, S.O. Sari: Appl. Phys. Lett. **25**, 403 (1974)
2.47 J.P. Heritage, R. Jain: Appl. Phys. Lett. **32**, 101 (1978)
2.48 A. Ferguson, J. Eckstein, T. Hansch: J. Appl. Phys. **49**, 5389 (1978)
2.49 R. Jain, J. Heritage: Appl. Phys. Lett. **32**, 41 (1978)
2.50 J. Ryan, L. Goldberg, D. Bradley: Opt. Commun. **24**, 127 (1978)
2.51 M. Adams, D. Bradley, W. Sibbett, J. Taylor: Philos. Trans. R. Soc. London **298**, 217 (1980)
2.52 R. Johnson: IEEE J. Quant. Electron. QE-**15**, 84 (1979)
2.53 S. Shapiro, R. Cavanagh, J. Stephenson: Opt. Lett. **6**, 470 (1981)
2.54 S. Blit, C. Tang: Appl. Phys. Lett. **36**, 16 (1980)
2.55 G. Mitchell, W. Proffitt: IEEE J. Quantum Electron. QE-**17**, 52 (1981)
2.56 J. Halbout, A. Olsson, C. Tang. Appl. Phys. Lett. **39**, 463 (1981)
2.57 J. Heritage, E. Issacs: IEEE J. Quant. Electron. QE-**17**, 52 (1981)
2.58 P. May, W. Sibbett, J. Taylor: Appl. Phys. **26**, 179 (1981)
2.59 A. Seilmeier, W. Kaiser, B. Sens, K. Drexhage: Opt. Lett. **8**, 205 (1983)
2.60 L.F. Mollenauer, D. Bloom: Opt. Lett. **4**, 427 (1979)
2.61 L.F. Mollenauer, N.D. Vieria, L. Szeto: Opt. Lett. **7**, 414 (1982)
2.62 G. Mourou, T. Sizer: Opt. Commun. **41**, 47 (1982)
2.63 T. Sizer, G. Mourou, R. Rice: Opt. Commun. **37**, 207 (1982)
2.64 F. Zumsteg, J. Bierlein, T. Grier: J. Appl. Phys. **47**, 498 (1976)
2.65 A. Johnson, W. Simpson: Optics Lett. **8**, 554 (1983)
2.66 T. Royt, C. Lee: Appl. Phys. Lett. **30**, 332 (1977)
2.67 W. Schmidt, F.P. Schafer: Phys. Lett. **26A**, 558 (1968)
2.68 D.J. Bradley, F. O'Neill: Opto-Electron. **1**, 69 (1969)
2.69 D.J. Bradley, B. Liddy, A.G. Roddie, W. Sibbett, W. Sleat: Opt. Commun. **3**, 426 (1971)
2.70 E.G. Arthurs, D.J. Bradley, A.G. Roddie: Appl. Phys. Lett. **20**, 125 (1972)
2.71 D.J. Bradley, A.J. Durrant, F. O'Neill, B. Sutherland: Phys. Lett. **30A**, 535 (1969)
2.72 E.P. Ippen, C.V. Shank, A. Dienes: Appl. Phys. Lett. **21**, 348 (1972)
2.73 F. O'Neill: Opt. Commun. **6**, 360 (1972)
2.74 C.V. Shank, E.P. Ippen: Appl. Phys. Lett. **24**, 373 (1974)
2.75 E.B. Treacy: J. Quantum Electron. QE-**5**, 454 (1969)
2.76 J. Diels, E. Van Stryland, G. Benedict: Opt. Commun. **25**, 93 (1978)
2.77 R.L. Fork, O.E. Martinez, J.P. Gordon: Appl. Phys. Lett. **9**, 150 (1984)
2.78 J.A. Valdmanis, R.L. Fork, J.P. Gordon: Opt. Lett. **10**, 131 (1985)
2.79 K. Smith, N. Langford, W. Sibbett, J.R. Taylor: Opt. Lett. **10**, 559 (1985)

2.80 P.W. French, J.R. Taylor: In *Ultrafast Phenomena V*, ed. by G. Fleming and A.E. Siegman, Springer. Ser. Chem. Phys., Vol. 46 (Springer, Berlin, Heidelberg 1987)
2.81 T. Sizer, J. Kafka, A. Krisiloff, G. Mourou: Opt. Commun. **39**, 259 (1981)
2.82 A.E. Siegman: Optics Lett. **6**, 334 (1981)
2.83 H. Vanherzeele, J. Van Eck, A.E. Siegman: Appl. Opt. **20**, 3483 (1981)
2.84 H. Vahnerzeele, R. Torti, J.-C. Diels: Applied Optics, **23**, 4182 (1984)
2.85 M. Nuss, R. Leonhardt, W. Zinth: Optics Lett. **10**, 16 (1984)
2.86 A. Migus, C.V. Shank, E.P. Ippen, R.L. Fork: IEEE Journ. of Quantum Electron. QE-**18**, 101 (1982)
2.87 R.L. Fork, C.V. Shank, R.T. Yen: Appl. Phys. Lett. **41**, 223 (1982)
2.88 C. Rolland, P.B. Corkum: Optics Commun. **59**, 64 (1986)
2.89 H. Egger, T.S. Luk, K. Boyer, D.F. Muller H. Plummer, T. Srinivasan, C.K. Rhodes: Appl. Phys. Lett. **41**, 1032 (1982)
2.90 P.B. Corkum, R.S. Taylor: IEEE J. Quantum Electron. QE-**18**, 1962 (1982)
2.91 W.J. Tomlinson, R.H. Stolen, C.V. Shank: J.O.S.A. **B1**, 139 (1984)
2.92 A.H. Waldeck, W.T. Lotshaw, D.B. Mcdonald, G. Fleming: Chem. Phys. Lett. **88**, 257 (1982)
2.93 I.N. Duling III, T. Norris, T. Sizer II, P. Bado, G.A. Mourou: J. Opt. Soc. Am. **B2**, 616 (1985)
2.94 L.F. Mollenauer: In *Laser Handbook Vol. 4*, ed. by M.L. Stitch, M. Bass (Elsevier, Amsterdam 1985)
2.95 L.F. Mollenauer, D.M. Bloom, A.M. Del Gaudio: Opt. Lett. **3**, 48 (1978)
2.96 Z. Bor: IEEE Journ. Quant. Electron. **16**, 517 (1980)
2.97 A. Dienes, E.P. Ippen, C.V. Shank: Appl. Phys. Lett. **19**, 258 (1971)
2.98 G. Szabo, B. Racz, Z. Bor, B. Nikoloaus, A. Muller: In *Ultrafast Phenomena IV*, ed. by D.H. Auston and K.B. Eisenthal, Springer Ser. Chem. Phys. H, Vol. 38 (Springer, Berlin, Heidelberg 1984)
2.99 Z. Bor, S. Szatmari, A. Muller: Applied Physics, **B32**, 101 (1983)
2.100 S. Szatmari, F.P. Schafer: Optics Commun. **49**, 101 (1983)
2.101 G. Scabo, Z. Bor, A. Muller: Appl. Phys. **B31**, 53 (1983)
2.102 R. Wyatt, E. Marinero: Appl. Phys. **25**, 297 (1981)
2.103 P.O. Scherer, A. Seilmeir, W. Kaiser: J. Chem. Phys. **83**, 3948 (1985)
2.104 P.T. Ho, L.A. Glasser, E.P. Ippen, H.A. Haus: Appl. Phys. **33**, 241 (1978)
2.105 L.A. Glasser: Electron. Lett. **14**, 725 (1978)
2.106 J.P. Van der Ziel, R.M. Mikulyak: J. Appl. Phys. **51**, 3033 (1980)
2.107 E.P. Ippen, D.J. Eilenberger, R.W. Dixon: Appl. Phys. Lett. **37**, 267 (1980)
2.108 J.P. Van der Ziel, W.T., Tsang, R. Logan, R.M. Mikulyak, W.M. Augustyniak: Appl. Phys. Lett. **39**, 525 (1981)
2.109 Y. Silberberg, P.W. Smith, D.J. Eilenberger, D.A.B. Miller, A.C. Gossard, W. Wiegmann: Optics Lett. **9**, 507 (1984)
2.110 F. Gires, P. Tournois: C.R. Acad. Sci. Paris **258**, 6112 (1964)
2.111 J.A. Giordmaine, M.A. Duguay, J.W. Hansen: IEEE J. Quantum Electron. QE-**4**, 252 (1968)
2.112 R.A. Fischer, P.L. Kelly, T.K. Gustafson: Appl. Phys. Lett. **14**, 140 (1969)
2.113 L.F. Mollenauer, R.H. Stolen, J.P. Gordon: Phys. Rev. Lett. **45**, 1095 (1980)
2.114 H. Nakatsuka, D. Grischkowsky, A.C. Balant: Phys. Rev. Lett. **47**, 1910 (1981).
2.115 A. Hasegawa, F. Tappert: Appl. Phys. Lett. **23**, 142 (1973)
2.116 D. Grischkowsky, A.C. Balant: Appl. Phys. Lett. **41**, 1 (1982)
2.117 L.F. Mollenauer, R.H. Stolen, J.P. Gordon, W.J. Tomlinson: Opt. Lett. **8**, 289 (1983)
2.118 L.F. Mollenauer, R. Stolen: Opt. Lett. **9**, 13, Jan. 1984. See also, F.M. Mitschke, L.F. Mollenauer: IEEE JQE, **22**, 2242, Dec. 1986.
2.119 C.V. Shank, R.L. Fork, R. Yen, R.H. Stolen, W.J. Tomlinson: Appl. Phys. Lett. **40**, 761 (1982)
2.120 J.G. Fujimoto, A.M. Weiner, E.P. Ippen: Appl. Phys. Lett. **44**, 832 (1984)
2.121 J.M. Halbout, D. Grischkowsky: Appl. Phys. Lett. **45**, 1281 (1984)
2.122 W.H. Knox, R.L. Fork, M.C. Downer, R.H. Stolen, C.V. Shank, J. Valdmanis: Appl. Phys. Lett. **46**, 1120 (1985)

2.123 W.J. Tomlinson, W.H. Knox: Private Communication
2.124 R.L. Fork, C.H. Brito Cruz, P.C. Becker, C.V. Shank: Opt. Lett. **12**, 483 (1986)
2.125 J. Diels, J. Fontaine, I. McMichael, F. Simoni: Appl. Opt. **24**, 1270 (1985)
2.126 J.P. Heritage, R.N. Thruston, W.J. Tomlinson, A.M. Weiner, R.H. Stolen. Appl. Phys. Lett. **47**, 87 (1985)
2.127 H. Roskos, A. Seilmeier, W. Kaiser, J.D. Harvey: Opt. Commun. (To be published)
2.128 G. Mourou, D. Strickland, S. Williamson: Laser Focus **22**, 104 (1986)

3. Optical Nonlinearities with Ultrashort Pulses

Alfred Laubereau

With 27 Figures

Picosecond and femtosecond pulses provide a unique tool for the study of ultrafast processes accompanying light-matter interaction. Novel measuring techniques, on the one hand, expanded the experimental capabilities in this field while, on the other hand, a great demand for suitable pulses in various spectral regions was created that cannot be directly fulfilled by available laser systems. A basic solution of this problem is offered by quantum optics which can convert pulse frequencies with the help of nonlinear optical effects. A second motivation for the study of short pulse nonlinearities is the high power level necessarily connected with shorter and shorter pulses of finite energy content. Consequently, nonlinear effects cannot be ignored in spectroscopic applications for an adequate understanding of the coupling between the short pulse and the dynamical processes under investigation. Without exaggeration one can say that nonlinear optics is an inherent and fundamental part of the generation, analysis and application of ultrashort laser pulses.

It is beyond the scope of the present chapter to review the abundant literature on nonlinear optics. The following discussion will be restricted to investigations with pulses shorter than 1 ns. Special emphasis will be placed on applications e.g. conversion of the pulses frequencies. Related work with longer time scales will only be briefly mentioned. For background information the reader is referred to text books on quantum optics, to the second volume of the *Laser Handbook* (Arecchi and Schulz-Dubois 1972), to the monograph on parametric processes by Reintjes (1984) and to other volumes of the series *Quantum Electronics – Principles and Applications*. Applications of ultrafast nonlinear optics to mode-locking techniques and to spectroscopic investigations are outlined in other chapters of this volume.

3.1 Nonlinear Polarization

The concept of nonlinear optics is considerably older than the laser. Interactions involving two or more quanta, e.g. two-photon absorption and stimulated Raman scattering were described theoretically as early as 1931 by Goeppert-Mayer. Other nonlinear optical interactions such as the saturation of an electronic transition of atoms were also observed before the invention of the laser.

At the high light intensities of $\lesssim 10^{12}$ W/cm^2 which, with ultrashort pulses, can be applied to condensed matter without irreversible damage, nonlinear phenomena turn out to be amazingly efficient.

The intense photon flux favours the use of the rather simple classical theory of electromagnetism. The response of the medium to the incident optical radiation is formulated in terms of the induced macroscopic polarisation. Familiar processes like (linear) absorption, refraction and scattering arise from polarizations proportional to the first power of the optical field. To include nonlinear phenomena higher order terms in the Taylor expansion of the polarization have to be considered. In the electric dipole approximation one may write

$$\boldsymbol{P}(\boldsymbol{E}) = \chi^{(1)}\boldsymbol{E} + \chi^{(2)}\colon\boldsymbol{EE} + \chi^{(3)}\colon\boldsymbol{EEE} + \cdots \,. \tag{3.1}$$

The constant term has been omitted in (3.1) since it cannot contribute to nonlinear processes. $\chi^{(2)}$ and $\chi^{(3)}$ denote the second- and third-order susceptibilities describing three- and four-wave interactions, respectively. Similar to the well-known linear susceptibility $\chi^{(1)}$, the higher order coefficients $\chi^{(n)}$ display a distinct frequency dependence, corresponding, in general, to time-dependent nonlinear susceptibilities in the time domain. Simple use of (3.1) for very short pulses is therefore possible only for special cases with negligible frequency dependency, i.e. far-off resonances.

For an understanding of the nonlinear interaction, the macroscopic susceptibilities have to be related to a microscopic picture of the physical mechanisms. Quantum mechanical derivations were discussed by Armstrong et al. (1962) and more recently by Flytzanis (1975). For short pulse nonlinearities, e.g. close to material resonance it is often convenient to avoid the χ-formalism and to treat the transient response of the medium directly by microscopic equations of motion. Typical time constants characterizing the dynamics are 10^{-13}–10^{-10} s in condensed matter (and somewhat longer in gases) and are directly measurable with femto- and picosecond pulses.

Equation (3.1) allows a classification of nonlinear processes into three- or four-wave interactions etc. This ordering will be used in Sects. 3.2, 3. In Sect. 3.4 particular physical systems that have attracted great interest in current research work and which promise to be important in the future will be described.

3.2 Three-Wave Interactions

3.2.1 Second Harmonic Generation

After its first observation by Franken et al. (1961), second harmonic (SH) generation quickly became an important process for the extension of the wavelength range accessible with available laser sources because of its simple, reliable application and good conversion efficiency. The process is of second order in the electromagnetic field amplitude representing the optical radiation, i.e. is connected with the nonlinear polarization (see for example Yariv 1976a).

$$P_i(2\omega) = \sum_{j,k} d_{ijk} E_j(\omega) E_k(\omega) \ , \tag{3.2}$$

where $i, j, k = 1, 2, 3$ represent Cartesian coordinates. The nonlinear coupling coefficients d_{ijk} used conventionally instead of χ_{ijk} represent a similar third rank tensor. Symmetry arguments show that systems with (macroscopic) inversion symmetry, e.g. liquids and gases under normal conditions, possess only vanishing second-order coefficients, $d_{ijk} = 0$. Second harmonic generation, like other three photon processes based on $\chi^{(2)}$, is therefore restricted to

(i) single crystals of certain crystal classes,
(ii) surfaces and,
(iii) systems where the inversion symmetry is (partially) broken by an electrostatic field or a density gradient.

For a discussion of the d-tensor and further details, the reader is referred to the literature.

For plane monochromatic waves, i.e. for ideal cw fields and neglecting pump depletion, the second harmonic intensity I_{SH} is evaluated from the nonlinear wave equation to be

$$I_{SH} = \mathrm{const} \cdot d_{\mathrm{eff}}{}^2 l^2 I_L{}^2 \left(\frac{\sin(\Delta kl/2)}{\Delta kl/2} \right)^2 . \tag{3.3}$$

Here I_L is the intensity at the fundamental frequency and l denotes the crystal length. Of particular importance in (3.3) is the phase mismatch Δkl originating from the mismatch between the SH and fundamental wave vectors, $\Delta k = k_{SH} - 2k_L$. The equation readily shows that perfect phase matching $\Delta k = 0$ is required for optimum frequency doubling, allowing large values for the interaction length l. In practical cases, approximate phase matching

$$|\Delta k_{\mathrm{tol}}| l \leqslant 2.78 \tag{3.4}$$

may be sufficient. The numerical factor on the rhs. of this expression describes the half width of the factor $(\sin x/x)^2$ in (3.3). Since $k(\omega) = n(\omega)\omega/c$ (refractive index $n(\omega)$) we have $\Delta k \neq 0$ in general. In fact, for condensed matter in the visible spectrum, $\Delta k \simeq 10^3\,\mathrm{cm}^{-1}$ is a typical figure limiting the useful interaction length to tens of microns [see (3.4)]. Efficient harmonic generation is not possible under these conditions.

There are a few possibilities to overcome the color dispersion of matter and to achieve a long interaction length. The common trick is to make use of the polarization and orientation dependence of the refractive index in birefringent crystals (Giordmaine 1962; Maker et al. 1962). For details the reader is referred to textbooks of quantum optics.

For practical applications, large conversion efficiencies are of interest where pump depletion has to be included. Assuming phase matching, $\Delta k = 0$, and monochromatic fields, one finds for the intensity conversion

$$\eta = \frac{I_{SH}(l)}{I_L(0)} = \tanh^2[\sqrt{\text{const}\, d_{\text{eff}}^2 I_L(0) l^2}\,] \ . \tag{3.5}$$

According to the $\tanh^2$ dependence, the second harmonic intensity increases more slowly for large conversion factors >0.1. One hundred percent conversion is approached asymptotically for $l \to \infty$. The effective interaction length is limited for real systems by the walk-off angle between the second harmonic and the fundamental field components because of the finite beam diameter.

The plane wave approximation of (3.3) does not take into account the beam divergence of focused beams; in birefringent crystals, Δk is angle dependent, $\Delta k = \Delta k(\theta)$. As a consequence, phase matching cannot be exactly adjusted for the whole divergent beam. For approximate phase matching of the total beam with divergence $\delta\theta$, (3.4) requires the crystal to be sufficiently short, $l \propto 1/\delta\theta$. Only a small divergence of the order of 10^{-3} rad is tolerable for crystals of $l = 1$ cm for common nonlinear crystals (Hagen and Magnante 1969). The harmonic generation of focussed Gaussian beams was discussed by Boyd and Kleinman (1968). For a given nonlinear crystal (length, walk-off angle) an optimum focusing condition for the fundamental beam is obtained when the confocal parameter of the beam waist is comparable to the crystal length. More recently SHG of arbitrary beam profiles has been worked out (Agrawal 1981).

Comparing second harmonic generation of ultrashort pulses with more conventional situations, e.g. giant laser pulses, the following points have to be considered:

(i) the higher damage threshold of shorter pulses allows higher peak intensities favoring the nonlinear process;

(ii) the short pulse length limits the interaction length in the nonlinear crystal because of group velocity dispersion. In contrast to the phase velocity matching for $\Delta k = 0$, the fundamental and harmonic pulses propagate with different group velocity. The group delay t_g (difference of transit times) at the exit window of the crystal is given by (Glenn 1969)

$$t_g = \frac{l}{c}\left[\lambda_L\left(\frac{dn}{d\lambda}\right)_L - \lambda_{SH}\left(\frac{dn}{d\lambda}\right)_{SH}\right] , \tag{3.6}$$

where c denotes the speed of light. The interaction length is limited by the condition $t_g \leqslant t_p$ (pulse duration t_p).

For the Nd laser wavelength $\lambda_L = 1.06\,\mu\text{m}$ and the so-called type I phase-matching, the group delay is 0.08 ps/cm for KDP (potassium dihydrogen phosphate) and 6 ps/cm for $LiNbO_3$. In the limit $t_g \ll t_p$ and for small conversion efficiency the SH pulse has minimum duration; laser pulses of Gaussian shape

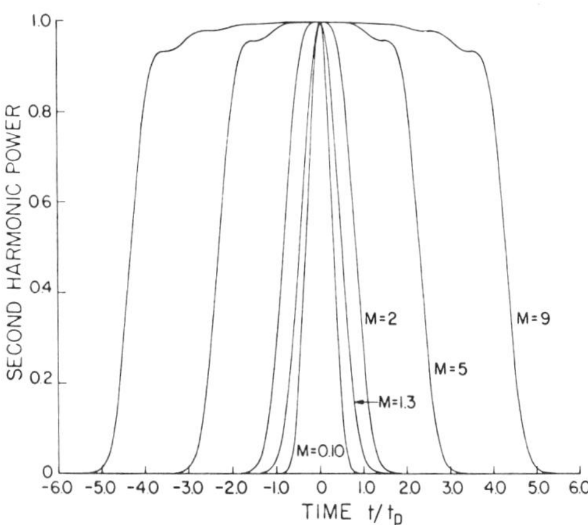

Fig. 3.1. Calculated second harmonic pulse shape for various crystal lengths. The SH power is plotted vs time, normalized to the same peak power; M denotes the crystal length in units of the distance over which the group delay between the SH and fundamental pulses equals the pulse duration, $t_g = t_p$ [see (3.6); after Comly and Garmire, 1968]

accomplish a minor shortening by a factor $\sqrt{2}$ in the nonlinear process. For $t_g > t_p$, the second harmonic intensity no longer increases with crystal length but a pulse stretching occurs.

Finally for $t_g \gg t_p$, the duration of the frequency doubled pulse approximately equals the group delay. The SH pulse shape is depicted in Fig. 3.1 for various crystal lengths.

(iii) The finite frequency width of a short pulse requires phase matching not only at the frequency center ν_L but over the whole spectral intensity distribution for efficient harmonic generation. In the frequency picture, we see that color dispersion leads to $\Delta k = \Delta k(\nu)$, and a finite phase matching bandwidth originates from the condition of approximate phase matching $|\Delta k(\omega - \omega_L)| \leqslant 2.78/l$ (4.4). Converting to frequencies in units of cm^{-1}, one can readily derive (Miller 1968)

$$\Delta\nu = |\nu - \nu_L| \leqslant \frac{0.221}{l}\left[\lambda_L\left(\frac{dn}{d\lambda}\right)_L - \lambda_{SH}\left(\frac{dn}{d\lambda}\right)_{SH}\right]^{-1} . \tag{3.7}$$

The finite phase matching bandwidth $\Delta\nu$ has two consequences: For a short SH pulse, the SH spectrum has to be correspondingly broad and a large value of $\Delta\nu$ (i.e. short crystal) is required as compared to the frequency width of the fundamental pulse; actually this fact has already been discussed in the context of (3.6) using time-domain arguments. The second conclusion, which is readily visualized in the frequency picture, is that a high conversion efficiency also necessitates a large bandwidth $\Delta\nu$, since only Fourier components within this frequency interval participate in the nonlinear process. For a KDP crystal of 1 cm length, the phase matching bandwidth is approximately $20\,cm^{-1}$ for $\lambda_L = 1.06\,\mu m$. For a $LiNbO_3$ crystal of the same length, $\Delta\nu$ is smaller by a factor of almost 40. The latter material is therefore less suitable for short pulses, in spite of its tenfold larger nonlinearity.

The relationship between crystal length and pulse duration of the second harmonic radiation was studied in an early investigation by Shapiro (1968) using short Nd:glass laser pulses. Varying the length of a $LiNbO_3$ crystal between 1.2 mm and 9.2 mm, corresponding changes in the SH duration from 1.5 ps to 9 ps were reported.

Mode-locked laser pulses often deviate notably from the Fourier transform limitation. The broadened spectrum (compared to a bandwidth-limited pulse of the same duration) makes efficient frequency doubling more difficult. SHG of chirped pulses was studied theoretically by Glenn (1969), Akhmanov (1969), and Orlov et al. (1970). In agreement with the arguments presented above, a decrease of the conversion efficiency is found when the spectral width of the frequency modulated pulse exceeds the phase-matching bandwidth. Furthermore, in the case of a mismatch between the group velocities, the phase modulation destroys the optimum phase relation between the fundamental and harmonic fields and the SH energy flows back to the fundamental frequency radiation (Karamzin and Sukhorukov 1975). On the other hand, an unexpectedly high conversion of $\simeq 10\%$ was reported for broadband dye laser pulses of 500 ps and $\simeq 40\,\mathrm{cm}^{-1}$ compared to a phase matching width of $\simeq 13\,\mathrm{cm}^{-1}$ of a 2 cm KDP crystal (Kwok and Chiu 1985). The observations are explained by sum-frequency generation of the side bands of the pulse in addition to the SH process improving the efficiency by more than a factor of 2.

The high intensity level attainable in solid material without damage, allows us to overcome, to a large extent, the limitations of ultrashort SHG; in other words, equally favorable conversion efficiencies have been reported for ultrashort pulses as for nanosecond radiation. In an early investigation, Kung et al. (1972) demonstrated SHG with an impressive energy conversion of up to 80% in KDP for 50 ps pulses of a Nd:YAG laser. More recently, Reintjes and Eckardt (1977) reported a conversion of 85% from 532 nm to 266 nm in ADP (ammonium dihydrogen phosphate) and of 75% in KD*P (deuterated potassium dihydrogen phosphate) for pulses of 30 ps and $3.2\,\mathrm{cm}^{-1}$ width. Crystals of 4 mm and a peak intensity of 8 $\mathrm{GW/cm}^2$ were used. The green pulse at 532 nm was derived in turn from a Nd:YAG laser by SHG in KDP with 70% conversion. Even better numbers were reported for the Nd:glass laser (Gulamov et al. 1983).

Second harmonic generation of tunable red and infrared pulses derived from a stimulated parametric device (see Sect. 3.2.4) was investigated by Seilmeier and Kaiser (1980) in $LiIO_3$ and KDP (lengths 1.4 and 8 mm, respectively). Broadly tunable uv pulses of 5 ps were obtained in the wavelength range 300–400 nm with 1% energy conversion.

Recently, efficient frequency doubling of the pulse train of the cw mode-locked YAG laser gained practical importance for the generation of picosecond and femtosecond pulses. The peak power of these pulses of 1.2 kW is considerably smaller than for pulsed laser systems. Due to the favorable properties of a 5 mm long $KTiOPO_4$ crystal (KTP) allowing phase matching of a tightly focused beam with focal length 7.6 cm, the 50 ps pulses were converted with 25% power efficiency to 532 nm (Johnson and Simpson 1983, 1986). The green cw pulse train subsequently served as a pump source for a synchronously coupled dye laser. Conversion of considerably shorter pulses to the uv was reported by Ippen and Shank (1976) and Shank et al. (1977). In spite of the low power of a cw Rhodamine 6G laser, tight focusing of the subpicosecond pulses into a 0.25 mm $LiIO_3$ crystal yielded 15% energy efficiency of the SH process.

A versatile tool to meet the intensity demand of the nonlinear effect is the intracavity arrangement. For high quality cavities in particular, the intensity inside the laser is considerably larger. In addition, the multiple passes of the second harmonic radiation tend to increase the effective interaction length provided that the phase matching requirements can be fulfilled. The latter point was studied theoretically and experimentally by Apanasevich et al. (1982, 1984). An early intracavity experiment was performed by Rabson et al. (1972) with a 1.3 cm CDA crystal (CsH_2AsO_4) installed in the folded resonator of a mode-locked Nd:glass laser. The material which belongs to the KDP crystal family allows the so-called uncritical phase matching (90° phase matching angle) at 45°C, particularly favorable because of the larger effective *d*-coefficient and the smaller angular dispersion compared with other crystal orientations. SH peak intensities of 1 GW/cm^2 were reported, the effective pump power density being larger only by a factor of 4.

More recently, intracavity SHG was demonstrated for a synchronously mode-locked cw dye laser by Welford et al. (1980). Pulse trains at the fundamental wavelength of 590 nm and at the SH position, 295 nm, were simultaneously generated. For a 3 mm ADP crystal, average output powers of 5.1 mW and 0.13 mW were measured for the visible and uv components, respectively; the laser pulse duration was 3.9 ps. Similar results were reported for a $LiIO_3$ crystal where the group delay of 0.7 ps/mm required a shorter crystal of length 0.3 mm.

At the high intensity level of ultrashort SH pulses, additional nonlinearities, which may interfere with the generation process, have to be taken into account. A possible competing effect limiting the conversion efficiency of SHG is two-photon absorption occurring when the fourth harmonic frequency coincides with a uv absorption band of the nonlinear crystal. A second mechanism hindering the SH process is degenerate parametric amplification of the fundamental frequency component. Volosov et al. (1976) made theoretical and experimental studies of KDP and similar materials to seek ways to suppress this perturbation. A tunable spectral component, dependent on crystal orientation, was observed in the emission from $LiNbO_3$ and $LiIO_3$ crystals (Atanesyan et al. 1976) and explained by higher order frequency mixing. In a similar study for KDP, a frequency shift of $\simeq 20\,\mathrm{cm}^{-1}$ was reported for the SH radiation and associated with the cubic nonlinearity of the medium (Telegin and Chirkin 1982).

In summary, we note that second harmonic generation is routinely used in the field of ultrashort pulses. Without special efforts, conversion efficiencies above 10% are obtained with high power picosecond lasers. For subpicosecond pulses similar results have been reported; thin crystals are used that make tight focusing possible for (partial) compensation of the more stringent requirements of the nonlinear interaction in this time domain.

3.2.2 Sum-Frequency Generation and Its Application for UV Pulses

Sum-frequency generation (SFG) is a generalization of the second harmonic process where two identical photons combine to produce a new photon of twice the energy. In SFG two photons of different frequencies ω_1 and ω_2 are converted in the nonlinear interaction to produce radiation at the new frequency $\omega_3 = \omega_1 + \omega_2$.

The process has become of considerable interest for non-centrosymmetric crystals and, more recently, in the context of the nonlinear optical spectroscopy of surfaces (see Sect. 3.3). For monochromatic waves and neglecting depletion of the pump components with intensities I_1 and I_2 respectively new light at frequency ω_3 is produced with an intensity

$$I_3 = \text{const} \cdot {d_{\text{eff}}}^2 l^2 I_1 I_2 \left(\frac{\sin \Delta k\, l/2}{\Delta k\, l/2} \right)^2 . \tag{3.8}$$

The same k-matching argument applies as for frequency doubling with the mismatch now being defined by $\Delta k = |\boldsymbol{k}_3 - \boldsymbol{k}_2 - \boldsymbol{k}_1|$. It is important to note that in the general case, a noncollinear k-vector geometry is possible. Apart from these changes, the discussion of Sect. 3.2.1 of the influence of beam divergence, group delay and phase-matching bandwidth also applies qualitatively for sum-frequency generation. A comparison of the phase-matching situation of the collinear case with an off-axis geometry in a 2 mm $LiNbO_3$ crystal is presented in Fig. 3.2; sum-frequency generation of infrared (ν_1) and Nd:laser radiation ($\nu_2 \simeq 9500\,\text{cm}^{-1}$) is considered. For the non-collinear situation with $\alpha = 5°$, the solid angle of acceptance with approximate phase matching is reduced by a factor of two (Fig. 3.2c) while the phase-matching bandwidth is practically unchanged (Fig. 3.2b).

The quantitative discussion of SFG is of course more elaborate than the SH case since three fields are involved which differ in frequency and direction. The interaction of two focused beams with unequal confocal parameters was treated by Guha and Falk (1980). Group delay effects were studied in detail by Tomov et al. (1982). Gandelman et al. (1984) gave a theoretical discussion of pulse lengthening and shaping of the upconverted component with respect to the primary pulses.

For spectroscopic purposes, short wavelength pulses in the uv part of the spectrum are urgently required, since available laser sources, e.g. for femtosecond pulses, are restricted to the longer wavelength range. Filling this gap is an important application of SFG. The combination of frequency doubling and the original laser pulse may lead to the production of higher harmonic frequency components in multiple step processes. Since the individual stages are performed with high efficiency, the cascade process is quite often superior to a single-step

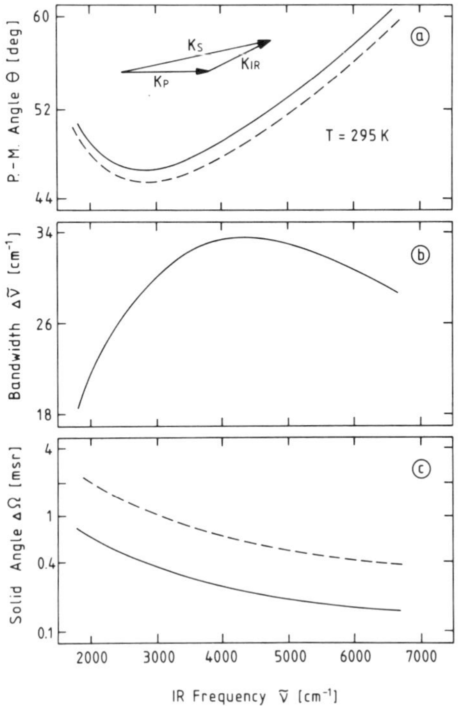

Fig. 3.2a–c. Sum-frequency generation in $LiNbO_3$ at 295 K using Nd:laser radiation as a pump source; (**a**) calculated phase-matching angle between optical axis of the crystal and the up-converted beam vs infrared frequency; (——): off-axis geometry with $\alpha = 5°$; (---): collinear geometry ($\alpha = 0$); inset: k-vector diagram; (**b**) frequency width of ir radiation accepted by the up-conversion process vs frequency; the two values $\alpha = 5°$ and $0°$ yield approximately the same curve; (**c**) calculated solid angle of acceptance of the up-conversion process vs ir frequency; (——): off-axis geometry with $\alpha = 5°$; (---): collinear case, $\alpha = 0$; after Hartmann and Laubereau (1981)

nonlinear interaction of higher order. Besides the two-fold frequency doubling to the fourth harmonic mentioned above, the simplest process of this sort is third harmonic generation (THG) by frequency mixing of the laser and SH wavelength. The single-step THG will be discussed in Sect. 3.3.1.

One and a half decades ago, Kung et al. (1972) demonstrated efficient SFG of the second harmonic (532 nm) and the fundamental 50 ps pulse (1.06 μm) of a Nd:YAG laser for the subsequent generation of the sixth, seventh and ninth harmonic (118 nm). Similar results with 60% energy conversion for the frequency mixing were reported by Attwood et al. (1975) for somewhat longer pulses of 150 ps. Special polarization schemes were developed to optimize the overall efficiency of the two-step interaction (Craxton 1980). Figure 3.3 shows the details of the so-called polarization mismatch scheme. By adjusting the orientation of the polarization plane of the fundamental input beam, the conversion of the first step (SHG) can be controlled and the overall efficiency optimized. An energy yield of 80% was obtained by Seka et al. (1980) for 140 ps (and also 700 ps) pulses of a Nd:glass laser system working with 12 mm KDP crystals at the high intensity level of 2–3 GW/cm^2. Up to 30 J per pulse were generated in the uv at 351 nm. Similar results were reported by Gulamov et al. (1983). The efficient conversion of high power sub-nanosecond pulses to the uv is important for laser fusion applications.

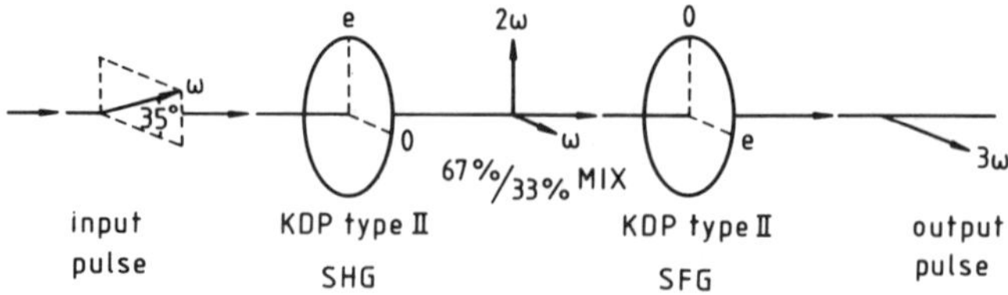

Fig. 3.3. Frequency tripling in an efficient two-step process. Input polarization at 35° with respect to the ordinary polarization plane of the first crystal provides optimum overall efficiency. The second crystal serves for SFG (after Seka et al. 1980)

For the modest peak power of cw mode-locked, high repetition rate laser systems it is difficult to meet the intensity requirements of the nonlinear mixing process. THG of the 25 ps emission of a cavity-dumped synchronously mode-locked dye laser, tunable around 630 nm was investigated by McDonald (1984). The pulses were amplified with a copper vapor laser at 5 kHz repetition rate. Using KD*P and the uv transparent material KB5 for the SHG and frequency mixing steps, respectively, tunable pulses around 210 nm were generated. Overall efficiencies were 0.5–1% for 100 μJ input pulses.

Broadband or tunable ultrashort pulses of moderate peak power on the one hand, and available high power laser pulses at fixed wavelength on the other hand, may be combined for the generation of tunable uv pulses by the help of efficient sum-frequency generation. An important requirement is the proper pulse synchronization, since the intensities I_1 and I_2 in (3.8) have to be simultaneously present. Sarkisyan (1978) studied the up-conversion of the picosecond continuum generated in D_2O (see Sect. 3.3) with the help of intense 10 ps pulses of a Nd:glass laser. A 3 cm long $LiIO_3$ specimen provided the SFG while the phase-matching bandwidth selected the up-converted fraction of the broad input spectrum. Changing the crystal orientation with a corresponding shift of the phase-matching wavelength yielded megawatt pulses in the tuning range 330–700 nm. In a similar investigation, the much weaker frequency side bands of frequency-modulated Nd:glass laser pulses were up-converted in ADP into the blue part of the spectrum at kilowatt power level (Van Laak et al. 1980).

Up-conversion of the intense, widely tunable pulses of stimulated parametric devices (see Sect. 3.2.4) leads to correspondingly large tuning ranges in the blue and uv. This scheme is of great interest for spectroscopic applications. Tanaka et al. (1978) studied sum-frequency generation of red and infrared parametric pulses ($\simeq$10 MW peak power) derived from a Nd:YAG laser with the 18 ps output of the same source at 1.06 μm. Picosecond pulses with a peak power of $\simeq$300 KW were obtained in the 390–490 nm spectral range. Frequency mixing of the parametric light with the second harmonic of the YAG laser yielded tunable uv radiation at 255–310 nm with 30 kW peak power. Pulse durations were estimated to be 30–40 ps.

The feasibility of sum-frequency generation on the femtosecond time scale was demonstrated very recently by Mokhtari et al. (1987). The 60 fs pulses of a cw mode-locked Rhodamine 6G laser at 620 nm, synchronously pumped by a Nd:YAG laser source, were upconverted to 390 nm by frequency mixing in a 100 μm thick $LiIO_3$ specimen. The residual 85 ps pulse train at 1064 nm of the laser pump served for the SFG. The long pulse duration of the second component does not affect the duration of the up-converted radiation [compare with (3.8)] but favorably softens the synchronization condition of the nonlinear interaction. For the short crystal mentioned, a minor pulse lengthening of only 10% was estimated. Tight focusing with a 1/e radius of the beam waist of 10 μm enabled 0.75% energy conversion (8% power conversion of the Nd:pulses) despite of the small peak powers of the non-amplified cw pulse trains (8 kW and 600 W respectively for 620 and 1064 nm).

Concluding this section, we stress the potential of sum-frequency generation for ultrafast time gates and ir detection schemes; these topics will be reviewed in Sect. 3.2.5 together with further applications.

3.2.3 Difference-Frequency Generation and Down-Conversion to the Far Infrared

A minor modification of the general three wave interaction leads to difference-frequency generation (DFG) which represents the inverse process of the up-conversion discussed in the preceding subsection. Again, two incident fields with frequencies ω_1 and ω_2 ($\omega_1 > \omega_2$) interact in the nonlinear medium to generate light at the new spectral position $\omega_3 = \omega_1 - \omega_2$. Equation (3.8) remains valid but with the constant factor being somewhat reduced because of the smaller frequency values involved (see text books on quantum optics, e.g. Yariv 1976a). The phase mismatch governing the efficiency of the process now reads $\Delta k = |\boldsymbol{k}_3 + \boldsymbol{k}_2 - \boldsymbol{k}_1|$, again necessitating birefringent crystals, in general, to adjust $\Delta k \simeq 0$ and to overcome normal color dispersion. Different crystal orientations are necessary to phase-match either the sum- or difference-frequency process. In comparison with the stimulated parametric emission discussed below, it is important to note that DFG refers to the case of small or moderate intensities of the high frequency component ω_1; if the intensity I_1 is increased, an induced amplification phenomenon occurs for the process $\omega_1 \rightarrow \omega_2 + \omega_3$ – in contrast to up-conversion where such amplification does not exist.

Since the difference frequency ω_3 can be smaller than both the primary frequencies ω_1 and ω_2, the major application of the process is the conversion of visible or near ir laser pulses to longer infrared wavelengths. An early investigation of this kind was performed by Moore and Goldberg (1976); the output of a pulsed synchronously mode-locked dye laser was difference-mixed with the pulses of a mode-locked Nd:YAG laser, yielding ir pulses of $\simeq$30 ps duration. A further development of this technique generating tunable pulses of a few ps duration at 1.2 to 1.6 μm was reported by Cotter and White (1984). Pulses from a picosecond dye laser (Rhodamine 6G) synchronously pumped by a cw mode-locked Ar^+ laser of 1.5–5 ps were mixed with the 100 ps pulses of a cw mode-locked Nd:YAG system in a $LiIO_3$ specimen. Pulse synchronization is achieved by phase-locking the acousto-optic mode-lockers of the Ar^+ and Nd:YAG lasers. For the low peak powers of 340 W from the dye laser and 20 W from the Nd:YAG system, ir signals with $\simeq 3$ mW peak power were generated. The investigated tuning range has practical importance for ultrashort pulse propagation in fibers with low attenuation and small group velocity dispersion (see Sect. 3.4.3).

Widely tunable pulses of 8 ps were obtained in the medium infrared at 3.9 to 9.4 μm by Elsaesser et al. (1985). The output of a mode-locked Nd:YAG laser and a travelling-wave dye laser tunable from 1.2 μm to 1.46 μm were mixed in a silver thiogallate crystal, $AgGaS_2$ (length 1.5 cm). This particular material is very transparent in the ir up to $\simeq 12$ μm and has excellent nonlinear and phase-matching properties. The Nd:YAG intensity was adjusted to be 2 GW/cm^2 in the crystal. Several percent of the Nd:YAG laser photons were converted into the ir. Some results on the wavelength dependence are depicted in Fig. 3.4. The observed pulse shortening of the generated ir radiation points to the contribution of stimulated parametric amplification in the investigation, i.e. the small intensity case assumed for (3.8) was abandoned for the sake of conversion efficiency. The frequency width of the ultrashort ir pulses was measured to be 6.5 cm^{-1}.

Picosecond far-infrared generation was performed by Campillo et al. (1979a). Difference-frequency generation of the signal and idler output pulses of a stimu-

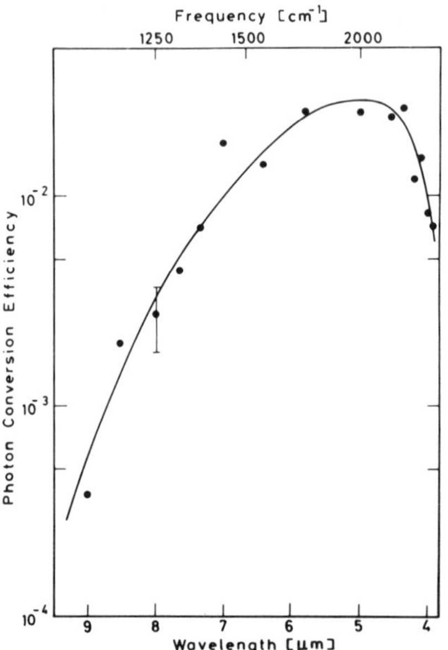

Fig. 3.4. Widely tunable picosecond pulses in the infrared by difference-frequency generation in silver thiogallate; measured photon conversion efficiency vs wavelength of the down-converted pulses. The pump pulse at 1.064 μm has 10^{16} photons (after Elsaesser et al. 1985)

lated parametric system, operated near the degenerate point, yielded tunable radiation over the region 14.8–18.5 μm with an estimated duration of 10–15 ps and 2 μJ pulse energy. Angle-tuning of the CaSe mixing crystal (length 1 cm) to readjust the phase-matching was not necessary for the spectral interval investigated. Even longer wavelengths in the far infrared were reported by Berg et al. (1985). A short dye laser pulse of 1–2 ps and 0.2 mJ energy was generated at a fixed wavelength of 589 nm and mixed with a long tunable pulse of 5 ns and 20 mJ energy content that was produced by a second dye laser in the wavelength range 590–596 nm. The far ir transparency window of $LiNbO_3$ allowed the generation of pulses with small difference frequencies of 20–200 cm^{-1} in this material. The quantum efficiency for down-conversion of photons of the visible ps pulse varied from 0.1% to 0.3%.

A demonstration of difference-frequency generation on the femtosecond time scale was given by Mokhtari et al. (1987). Using the same arrangement as discussed above (Sect. 3.2.2) for sum-frequency generation, the 60 fs dye laser pulses were down-converted to 1.5 μm in a thin $LiIO_3$ crystal. The low power level of the cw mode-locking system resulted in an energy conversion of 2×10^{-4}.

3.2.4 Stimulated Parametric Interaction

The basic three-photon interaction that causes stimulated parametric emission is the same as for difference-frequency generation, $\omega_1 \rightarrow \omega_2 + \omega_3$. The decay of one photon into two new quanta can be accelerated by induced emission since the second component ω_2 is enhanced simultaneously with the production of the new field ω_3. In this way a avalanche-type of growth is achieved for the two components termed "signal" (ω_2) and "idler" ($\omega_3, \omega_3 < \omega_2$). The stimulated amplification mechanism is readily derived from the coupled nonlinear wave equations for the field amplitude at frequencies ω_1 to ω_3 (see textbooks of quantum optics). Exponential growth of the signal and idler intensities is predicted for large pump intensities I_1 satisfying the differential equation

$$\frac{dI_j}{dz} = 2\gamma I_j (I_1)^{1/2} \; ; \qquad (j = 2, 3) \tag{3.9a}$$

where the gain factor γ for $\Delta k = 0$ has the maximum value

$$\gamma = \left(\frac{32\pi^3}{c^3} \frac{\omega_2 \omega_3}{n_1 n_2 n_3} d_{\mathrm{eff}}^2 \right)^{1/2} . \tag{3.9b}$$

An experimental verification of the strong intensity dependence of the stimulated parametric interaction is depicted in Fig. 3.5. The first demonstration of an optical parametric oscillator was that of Giordmaine and Miller (1965). Recent reviews have been given by Byer (1975) and Tang (1975).

The use of stimulated parametric emission for the generation of tunable ultrashort pulses was proposed by Glenn (1967) and Akhmanov et al. (1968a–c).

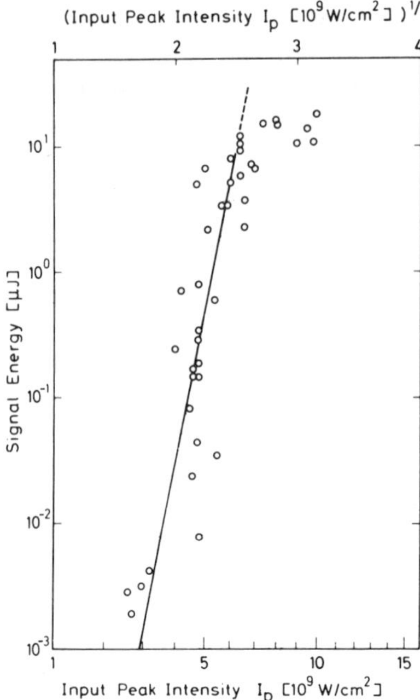

Fig. 3.5. Measured energy of signal pulses at 6500 cm^{-1} in a stimulated parametric generator ($LiNbO_3$) pumped by picosecond pulses at 9450 cm^{-1} from a mode-locked Nd:glass laser. The straight line is calculated using experimental parameters (after Laubereau et al. 1974)

These authors discussed the importance of group velocity matching and of pulse steepening on the picosecond time scale. The following experimental systems are currently used for ultrashort stimulated parametric emission:

(i) The optical parametric amplifier (OPA) with a well-defined input pulse at either the signal or the idler wavelength together with a synchronized pump pulse;

(ii) The stimulated parametric generator in a high gain traveling wave setup where the weak quantum noise serves as initial condition for the amplification process instead of an incident signal or idler pulse.

(iii) The optical parametric oscillator (OPO) where feedback of the parametric emission is performed by means of an optical resonator. For ultrashort pulses the cavity is much longer than the pulses. A train of pump pulses is required, as is matching of the cavity transit time of the parametric emission to the incident pulse train for repetitive pumping during the cavity transits.

We note here that no unique terminology has been developed in the original literature to distinguish the cases (i) to (iii); quite often parametric generators are termed oscillators and vice versa. The generation process is sometimes called parametric superluminescence.

Significant amplification of picosecond parametric emission from quantum noise was first observed by Rabson et al. (1972) and Burneika et al. (1972). In the former investigation the frequency-doubled pulse train of a mode-locked

Nd:glass laser was used to pump an OPO setup with a 5 mm nonlinear crystal of barium sodium niobate ($Ba_2NaNb_5O_{15}$). Low intensity emission from 0.96 to 1.16 μm was observed depending on the temperature tuning of the crystal. Optical damage limited the conversion efficiency to $\simeq 10^{-4}$. Estimates suggested that only two passes were sufficient to build up the parametric radiation.

a) Stimulated Parametric Generators

The advantage of single picosecond pulses allowing higher pump intensities up to 10 GW/cm^2 without crystal damage was demonstrated by Laubereau et al. (1974). In a traveling wave set-up with a single pass through a 2 cm $LiNbO_3$ crystal, intense parametric pulses were generated in the wavelength range 1.4–3.5 μm using amplified laser pulses at 1.06 μm and 6 ps duration as a pump source. An energy conversion of up to 3% was measured. The finite divergences of the pump ($\simeq$1 mrad) and of the parametric emission ($\simeq$20 mrad) lead to a spectral width of the tunable ir pulses of $\simeq 100\,\mathrm{cm}^{-1}$ as a consequence of non-collinear phase-matching. Dychyus et al. (1975) studied single-pass parametric amplification in a 4 cm α-HIO_3 crystal pumped by the frequency-doubled pulse train of a Nd:glass laser. Efficiencies of $\simeq$0.5% and angle tuning from 0.71–1.12 μm were reported.

The performance of the traveling-wave parametric generator was notably improved using two or more crystals (Kung 1974) or a double-pass through a single crystal (Liu 1979). The first stage of the experimental setup represents the generator producing parametric emission from quantum noise while the subsequent crystal passes achieve parametric amplification of the generated signal or idler pulses. Conversion efficiencies of up to 50% have been reported (Kabelka et al. 1979). It should be noted however that for high conversion $\geqslant$30% a complex temporal and spectral structure of the parametric pulses has to be expected because of self-phase-modulation and since the parametric process switches from amplification to stimulated loss in the extreme saturation region.

The multiple-stage parametric generator has the advantage of reduced divergence and correspondingly narrowed spectral width. For a crystal spacing of several tens of centimeters, a simple geometrical argument shows that off-axis phase-matching, which is a major source of spectral broadening, is largely suppressed (Seilmeier et al. 1978). In addition, the total interaction length in the nonlinear crystals is increased, thus reducing the bandwidth of the stimulated amplification of the collinear process via a smaller tolerable mismatch, similar to the approximate phase-matching of frequency mixing (3.4). Advanced systems apply separate pump pulses for the different generator-amplifier stages to minimize the perturbation of self-phase-modulation at the high pump intensity level of several GW/cm^2. By adjusting the synchronization of the second pump pulse, the duration of the signal and idler pulses can be reduced. In this way pulse shortening by more than a factor of ten to the subpicosecond time scale was obtained (Laubereau et al. 1978; Fendt et al. 1979).

A series of nonlinear crystals, crystal cuts and pump frequencies has been discussed in the literature (Kryukov et al. 1977; Danelyus et al. 1977; Ivanova et al. 1977; Kranitzky et al. 1980). The influence of the optical inhomogeneity of the nonlinear material, which reduces the maximum possible efficiency, was studied by Mironov and Filonenko (1982) and Magnitskii et al. (1982). Typical values for the parametric ir generation by the traveling wave system as used in spectroscopic experiments are: total quantum efficiency $\simeq 10\%$, frequency width 5–10 cm^{-1}, and pulse shortening by a factor of 2–3 compared to the primary Nd:glass or Nd:YAG laser pulses (Graener and Laubereau 1982). Because of the highly nonlinear generation mechanism, the parametric pulses display very steep slopes. As a result, starting with picosecond Nd:glass laser radiation, subpicosecond time resolution of a few 10^{-13} s is experimentally available (Wondrazek et al. 1983; Hartmann et al. 1985).

b) Optical Parametric Amplifiers

As demonstrated by the parametric generator system, large amplification factors of many orders of magnitude are provided by nonlinear crystals of several cm in length and a pump intensity of 10^9 W/cm^2. The reduced intensity requirements of OPA systems with smaller gain are readily met by available high power laser systems and/or by careful focusing of the pump beam. The only difficulty of parametric amplification, therefore, is the generation of a precisely synchronized second pulse at the proper frequency, which serves as the signal or idler input. Lack of tunability of the input pulses will result in the sacrifice of the tuning potential of the parametric interaction. In the light of this argument, it is not surprising that, apart from the generator-amplifier devices discussed above, stimulated parametric amplification has found only few applications.

Massey et al. (1976) have demonstrated the amplification of cw HeNe laser light of 25 mW at 633 nm in an ADP crystal pumped by the fourth harmonic of a mode-locked Nd:glass laser. Signal and idler pulses of $\simeq 10$ ps duration were generated with peak powers of 67 and 49 kW at the fixed wavelengths 459 and 633 nm, respectively. Tunable radiation of a semiconductor PbS laser emitting long, narrow-band pulses of 0.8 μs duration around 4 μm, was amplified in a 2.5 cm $LiNbO_3$ crystal that was pumped by picosecond pulses of a Nd:glass laser (Boichenko et al. 1984). The low power level of 2 mW of the input signal made competition against quantum noise difficult since this is simultaneously amplified, i.e. along with the generator process. Due to the high directionality of the semiconductor laser beam, the energy of the amplified pulse exceeded the amplified noise in the same solid angle by a factor of 100. In a similar investigation by Magnitskii et al. (1986) the tunable radiation of cw GaAs diode laser operating in single mode around 850 nm was amplified. Using a two-step parametric amplifier with $LiNbO_3$ crystals and Nd:YAG laser pulses, intense parametric emission at 1.4 μm with 18 ps duration and 1.2 cm^{-1} bandwidth was produced close to the Fourier limit. The maximum energy conversion was 25%.

Parametric amplification of a femtosecond continuum of $\simeq 150$ fs duration with full tunability within the 1.0–1.6 μm range was demonstrated by Ledoux et al. (1986). The input radiation was derived from an amplified colliding-pulse mode-locked dye laser (100 fs at 620 nm) which also pumped the parametric setup. A novel highly nonlinear organic crystal material, N-(4-nitrophenyl)-L-prolinol, of length 1.5 mm and in the so-called λ-noncritical phase-matching configuration was used to minimize group velocity dispersion. The amplified spectral components are selected by the phase-matching condition of the amplification process with gain factors of up to 10^4 corresponding to an ir intensity of 10^8 W/cm^2. Significant time broadening up to $\simeq 1$ ps was observed at this high output level.

c) Synchronously Coupled Parametric Oscillators

Repetitive multipass amplification at a lower gain level is possible if the nonlinear crystal is placed inside an optical resonator providing feedback of the signal and/or idler component. A sequence of pump pulses is required to carry out the parametric interaction for every cavity transit. This approach represents the direct extension of the nanosecond OPO's to the picosecond time domain. It has the potential of improved frequency width, pulse duration and beam divergence at the expense of reduced output power of the parametric pulses in comparison with the generator setup discussed above (Becker et al. 1974). The use of mode-locked pulse trains instead of single pulses imposes a serious disadvantage since the pulse properties vary within the train, with significant frequency broadening and pulse distortion towards the end of the pulse sequence, thus lowering the quality of the parametric pulses.

A singly resonant non-collinear $LiIO_3$ oscillator synchronously pumped by the second harmonic of a mode-locked, amplified Nd: glass laser was studied by Weisman and Rice (1976). Intense picosecond output, tunable from 1.4 to 3.8 μm, was observed with a frequency width of 20 to 25 cm^{-1}, but increasing to about 50 cm^{-1} near degeneracy. The pulse energy amounted to 150 μJ (a quantum efficiency of several percent). Similar results were reported for a $LiNbO_3$ OPO based on a mode-locked Nd:YAG laser (Tanaka et al. 1978). Intense tunable pulses of 30–40 ps duration were obtained with 10% energy conversion in the wavelength range between 630 nm and 3.6 μm by varying the crystal temperature over 400°C. Subsequent up-conversion of the parametric output yielded uv and visible pulses in the whole spectral range >240 nm. The performance of various crystal materials was studied in the OPO configuration (Babin et al. 1979; Bareika et al. 1980; Onishchukov et al. 1983).

The importance of properly adjusting the synchronization between the pump sequence and the counter-propagating parametric pulse for optimum shortening was emphasized by Bareika et al. (1983). For a KDP-OPO pumped by a Nd: glass laser (4 ps), pulses of duration $\geqslant 1$ ps were generated; most important, on account of the different group velocities of the signal and the idler wavelengths the minimum duration of the two components was observed for a different tuning of the cavity length.

d) Stimulated Parametric Emission in Extended Spectral Regions

Parametric generation of intense picosecond pulses has been successfully applied in recent years in several laboratories for investigations of molecular dynamics. In the following we focus our attention on the generation of frequencies above $16000\,cm^{-1}$ in the visible and uv and below $2500\,cm^{-1}$ in the mid- and far infrared.

In a pioneering investigation, Kung (1974) obtained efficient tunable pulses in the visible using two temperature-tuned ADP crystals (5 cm long) and a quadrupled mode-locked Nd:YAG laser amplifier system in a single-pass collinear geometry. The tuning range covered the whole visible spectrum, 420 to 720 nm with conversion efficiencies $>10\%$ ($>100\,\mu J$ parametric output). A related investigation was carried out by Wondrazek et al. (1983) starting from the third harmonic of a Nd:glass laser. The tuning range was shown to be superior to any other known light source, extending over $17\,000\,cm^{-1}$ with the help of angle tuning of two ADP crystals. The experimental tuning curve is shown in Fig. 3.6. For an estimated output duration of 4.5 ps, a frequency width of approximately $10\,cm^{-1}$ was reported outside the region of degeneracy, i.e. $|\nu - \nu_{deg}| > 2000\,cm^{-1}$, while the conversion efficiency was measured to be several 10^{-3}. At the degeneracy frequency, $\nu_{deg} = 14\,214\,cm^{-1}$, energy conversion of the parametric generator exceeded 1%.

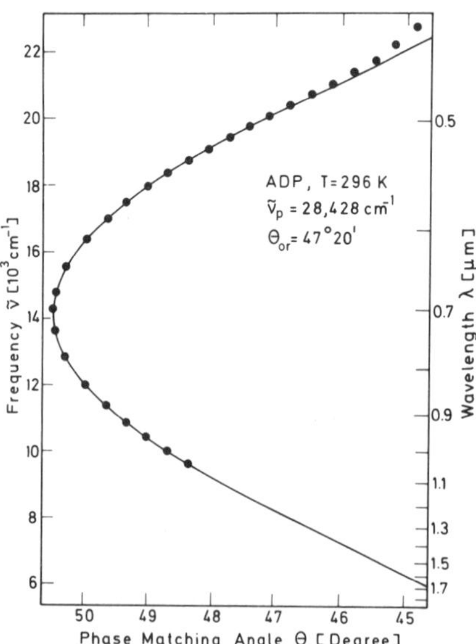

Fig. 3.6. Angular tuning curve of a stimulated parametric generator (experimental points and calculated curve). Orientation angle of the crystals is $\theta_{or} = 47°20'$. The large tuning range of $17\,000\,cm^{-1}$ should be noted extending from the NIR to the blue part of the spectrum (after Wondrazek et al. 1983)

Large tuning ranges of $\simeq 7000\,\mathrm{cm}^{-1}$ in the infrared were demonstrated with the help of new nonlinear materials with good transparency in the mid- and far infrared. Using a parametric generator-amplifier device with two proustite (Ag_3AsS_3) or silver thiogallate ($AgGaS_2$) crystals, Nd:YAG laser pulses were converted to the mid-infrared up to 8 and 10 μm, respectively (Elsaesser et al. 1983, 1984). Quantum efficiencies between 10^{-1} and 10^{-3} were found, depending on emission frequency, for a ir pulse duration of 8 ps.

In concluding this subsection on parametric emission, the generation of picosecond continua by stimulated three-photon interaction should be mentioned. For multichannel ultrafast absorption spectroscopy, such spectral sources are of practical importance. In the degeneracy region of the parametric interaction, i.e. at twice the wavelength of the pump radiation, the large angular dispersion of noncollinear phase-matching facilitates the generation of broadband radiation. For example, a uniform spectrum ranging from 1.92 to 2.38 μm was emitted with 4 mrad beam divergence and 200 μJ energy from a $LiNbO_3$ generator based on the Nd:YAG laser (Campillo et al. 1979) and subsequently used for parametric amplification or far ir difference frequency generation. In the visible, parametric continua extending over $5300\,\mathrm{cm}^{-1}$ (KDP) and $13\,000\,\mathrm{cm}^{-1}$ ($LiIO_3$) were derived with conversion efficiencies of up to 50% from the second to fourth harmonic of the Nd laser (Danelyus et al. 1978; Krylov and Papernyi. 1980; Bareika et al. 1983; Pokhsraryan 1985).

The high peak intensities involved in the parametric light generation require special attention. In particular, the oscillator systems are reported to be prone to crystal damage.

3.2.5 Further Applications

In the previous subsections three-photon interaction was discussed for the generation of ultrashort pulses at new frequencies and with reduced pulse duration. More general applications will be reviewed in the following.

a) Second Harmonic Probing

In physical systems with (macroscopic) inversion symmetry, second harmonic generation is not possible in the electric dipole approximation. Changes of the material properties may be detected by studying (weak) second harmonic generation if a (partial) breaking of the symmetry occurs. For example, the necessary nonlinear coefficient d_{eff} may be induced by a static external electric field leading to electric field induced SHG (Alexiewicz et al. 1978). Induced SHG of a mode-locked Nd:YAG laser in atomic sodium vapor was reported by Miyazaki et al. (1981) and explained by a spontaneous electric field which builds up in the interaction of an intense laser pulse with atomic media. Alternatively, the inversion symmetry of the nonlinear susceptibility can be removed by an inhomogeneous distribution of free carriers (Schwartz et al. 1975). A mechanism of this sort was proposed for SHG at 5.3 μm in tellurium (Johnson and Pratt 1978). The density gradient of electrons in the conduction band was produced by two-photon absorption of the incident laser pulse.

Second harmonic probing of dielectric breakdown in air, liquids and glasses has been demonstrated by Telle and Laubereau (1980). For the expected intensity level of $\simeq 10^{12}$ W/cm^2, a sharp increase in the SHG of an incident 10 ps pulse at 1.06 μm was observed and taken as direct evidence for the generation of quasi-free electrons by dielectric breakdown. The measured SH increase of three orders of magnitude was in agreement with theoretical estimates. Symmetry-allowed third harmonic generation under the same conditions did not show a discontinuity, so that a possible explanation in terms of nonlinear intensity changes (optical self-focusing) could excluded. On the basis of the SH studies, breakdown was found to occur in H_2O at notably higher power levels than inferred by visual inspection of broad-band luminescence used by other authors.

Second harmonic generation in laser-produced plasmas has also been investigated (Maaswinkel, 1980). Frequency doubling of Nd:YAG laser pulses of 30 ps and 10^{13} W/cm^2 was found to agree with theoretical predictions for linear-mode conversion in the plasma.

b) Ultrafast Parametric Light Gates

A fascinating application of sum- or difference-frequency generation with ultrashort laser pulses is the realization of light gates with exposure times $\geqslant 10^{-14}$ s in combination with conventional optoelectric detectors. The time resolution of the setup is solely determined by the duration of the up- or down-conversion pulse and by group delay arguments for the propagation of the interacting radiation in the nonlinear crystal. Compared to other light gates based on nonlinear absorption or birefringence, the parametric light gate – apart from its almost instantaneous electronic response – offers the specific advantage of converting the gated radiation to a spectral range where it is readily detectable.

The potential of parametric up-conversion for the sensitive detection of ir radiation was already recognized in the early days of nonlinear optics (Midwinter and Warner 1967). More recently, ir photon counting with picosecond and subpicosecond time resolution was demonstrated with parametric devices (Hartmann and Laubereau 1981). An example for the sensitive detection of photons at 3.4 μm with 10 ps pulse of a Nd:glass laser is depicted in Fig. 3.7. The linear intensity dependence of the up-conversion process on the ir signal is verified over eight orders of magnitude [cf. (3.8)]. A similar dynamic range is not available for any other known detection schemes with ultrashort time resolution. It should be noted that sum-frequency generation is free of quantum noise. This fact favors the up-conversion scheme for photon-counting applications compared to difference-frequency mixing.

Optical up-conversion with picosecond resolution was first demonstrated by Mahr and Hirsch (1975). Using a cw 20 ps dye laser, parametric light gating was

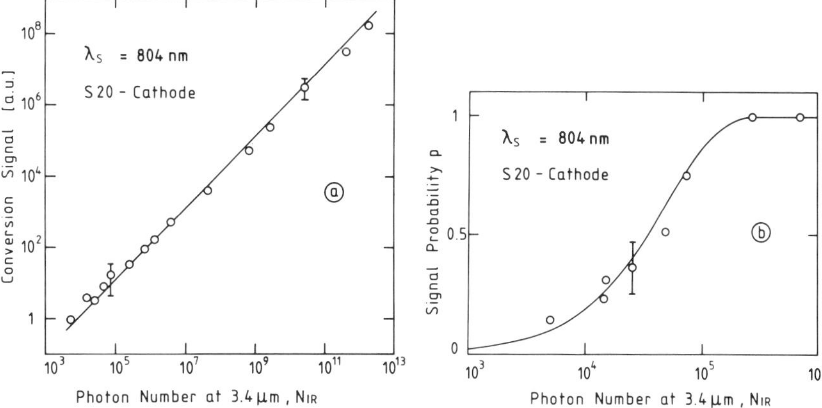

Fig. 3.7a, b. Sensitive detection of picosecond ir radiation by parametric up-conversion; (**a**) measured conversion signal vs input photon number at 3.4 µm; solid line, theoretical curve with slope 1; (**b**) measured probability for the observation of a photoelectron vs input photon number at 3.4 µm; theoretical curve (after Hartmann and Laubereau 1981)

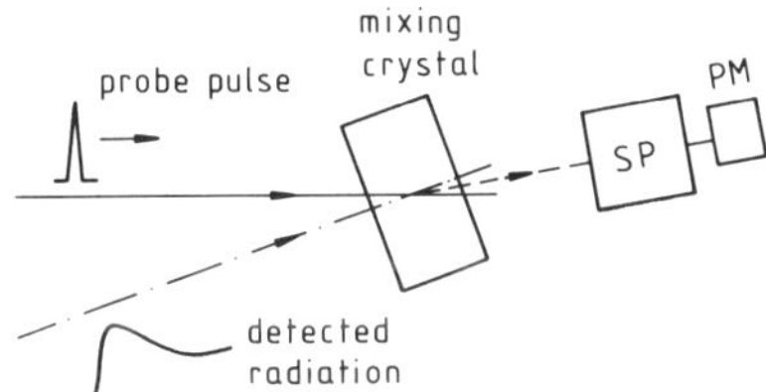

Fig. 3.8. Schematic of the experimental setup of the parametric light gate; synchronization between the detected radiation and the probing (up- or down-conversion) pulse is required

achieved with a small conversion efficiency but a high repetition rate. The time evolution of the incoherent luminescence of Rhodamine 6G excited by a mode-locked Ar^+ laser was measured. A schematic of the parametric light gate is shown in Fig. 3.8. Varying the time delay of the up-converting (or down-converting) pulse enables the time dependence of the incident signal to be mapped out. As with any sampling technique, repetitive measurements are performed, and this necessitates a time correlation between the signal event and the conversion pulse. Parametric light gates have been used in numerous measurements to study radiative and radiationless relaxation processes in organic molecules (Halliday and Topp 1977; Kopainsky and Kaiser 1978; Choi et al. 1980; Beddard et al. 1980, 1981); coherent pulse propagation on the picosecond and subpicosecond time scale has also been studied in the infrared (Hartmann and Laubereau 1983, 1984; Hartmann et al. 1984) and in the visible (Rothenberg et al. 1984) with parametric gate devices.

An apparatus for recording broadband infrared absorption spectra from 2.2 to 2.7 µm with subpicosecond time resolution was described by Glownia et al. (1987) combining an ir continuum generator (350 fs) with an ultrafast light gate.

Four-wave interaction in Rb vapor and a long up-conversion pulse of 15 ns were used for the frequency mixing. In a different experiment, up-conversion of the picosecond emission of a diode laser at 1.3 μm with a nanosecond pulse was used so that the pulse duration could be measured with a conventional streak camera sensitive in the visible part of the spectrum only (Onodera et al. 1983).

A special version of the parametric light gate was applied by Hulin et al. (1986). These authors used the down-conversion process in the high intensity regime, i.e. parametric amplification in a highly nonlinear organic crystal belonging to the family of paranitroaniline derivatives. Intense mode-locked dye laser pulses of 100 fs served as pump pulses. Amplification factors of up to 10^4 were reported in the spectral range 1–1.6 μm with a time resolution $\geqslant 0.2$ ps for small input signals where gain saturation did not occur.

c) Analysis of Pulse Shape

The invention of the passively mode-locked Nd:glass laser (DeMaria et al. 1966) produced a pressing need for techniques to measure the duration (and shape) of ultrashort optical pulses. Indirect methods were required since the combined use of photodetectors and oscilloscopes was no longer adequate for the temporal resolution of the pulses. Within a few months, several versions of a nonlinear technique related to second harmonic generation had been invented, by Maier et al. (1966), Weber (1967) and Armstrong (1967). Although designed many years earlier, they represent special cases of the parametric light gate depicted in Fig. 3.8. A part of the original pulse is split off (usually 50%) and this interrogates the remainder of the pulse by sum-frequency generation. The measurement is called the second-harmonic (or intensity) autocorrelation technique but should certainly not be confused with normal SHG since the two interacting beams I_1 and I_2 of the same frequency differ in optical phase and/or direction and/or polarization. Varying the time delay t_D the intensity autocorrelation function

$$G(t_D) = \int_{-\infty}^{\infty} I(t)I(t - t_D)dt \tag{3.10}$$

is directly obtained by a measurement of the time-integrated SH signal $\int I_3 dt$ [see (3.8) for $I_1 = I(t)$; $I_2 = I(t - t_D)$]. $G(t_D)$ does not contain the full information on the pulse shape; it is always symmetric, even for asymmetric pulses, and its halfwidth, τ, is correlated with the duration t_p of the pulse intensity (FWHH) only to within a numerical factor of the order of unity. In other words, additional information on the pulse shape is required to derive the pulse duration in a unique manner from intensity autocorrelation measurements. Table 3.1 gives examples for the conversion factor τ/t_p for several simple pulse shapes. Care is also necessary in the interpretation of $G(t_D)$ observed for frequency-modulated pulses since a coherence peak may affect the data. Comparison of the measured frequency width $\Delta\nu$ with the predicted width according to the bandwidth product $\Delta\nu t_p$ (Table 3.1) is an important indicator of the quality of the ultrashort pulse.

Table 3.1. Halfwidth τ of the intensity autocorrelation curve and the spectral bandwidth $\Delta\nu$ (FWHM) of several analytical pulse shapes

$I(t)$	τ/t_p	$\Delta\nu t_p$
1; for $0 \leqslant t \leqslant t_p$	1	0.886
$\exp[-(4\ln 2)t^2/t_p^2]$	1.414	0.441
$\mathrm{sech}^2(1.76t/t_p)$	1.55	0.315
$\exp[-(\ln 2)t/t_p]$; for $t \geqslant 0$	2	0.11

It is beyond the scope of the present chapter to present a complete review of the experimental methods of analyzing short pulses; in the following we confine ourselves to the most powerful techniques based on three-photon interaction; these have excellent time resolution, sensitivity and dynamic range and relatively low cost.

The second-harmonic autocorrelation technique mentioned above is generally applied for the measurement of picosecond and femtosecond pulses. For pulses shorter than 10^{-12} in particular, the original method of Maier et al. (1966) with noncollinear input beams of parallel polarization (phasematching type I) turns out to be convenient because group velocity dispersion is small for the two pulses. A geometric broadening effect is readily minimized for thin beams and small intersection angles ($\simeq 1°$). The measurement is routinely performed in practically all subpicosecond laborations around the world using automatic data acquisition. For high repetition rate lasers, e.g. cw mode-locked systems, measuring times < 1 s are achieved for the autocorrelation function. This allows one to monitor the ultrashort pulse "in real time" as an important aid to the experimentalist in aligning the laser system (Sala et al. 1980).

An interesting modification of the pulse analysis described so far, is the intensity-and-phase autocorrelator demonstrated by Kurobori et al. (1981) and Diels et al. (1985). For precisely collinear incident beams with transverse coherence, the phase of the electromagnetic field is constant over the beam cross section leading to an interference phenomenon in the nonlinear crystal analogous to a Michelson interferometer. The so-called interferometric autocorrelation function is obtained with a peak-to-background ratio of 8 to 1, which provides phase information otherwise not available. Frequency modulation and/or random frequency distribution is readily recognized from the shapes of the signal envelopes. An example is shown in Fig. 3.9. Needless to say, interferometric stability and precision are prerequisites for the autocorrelator and for the optical delay line if one is to observe the fs interference maxima and minima of the electromagnetic field.

Single-shot techniques to measure the pulse duration have gained in importance in recent years, particularly, for pulsed laser systems with modest repetition rates ($\leqslant 1$ Hz). Besides the two-photon fluorescence technique (Giordmaine et al.

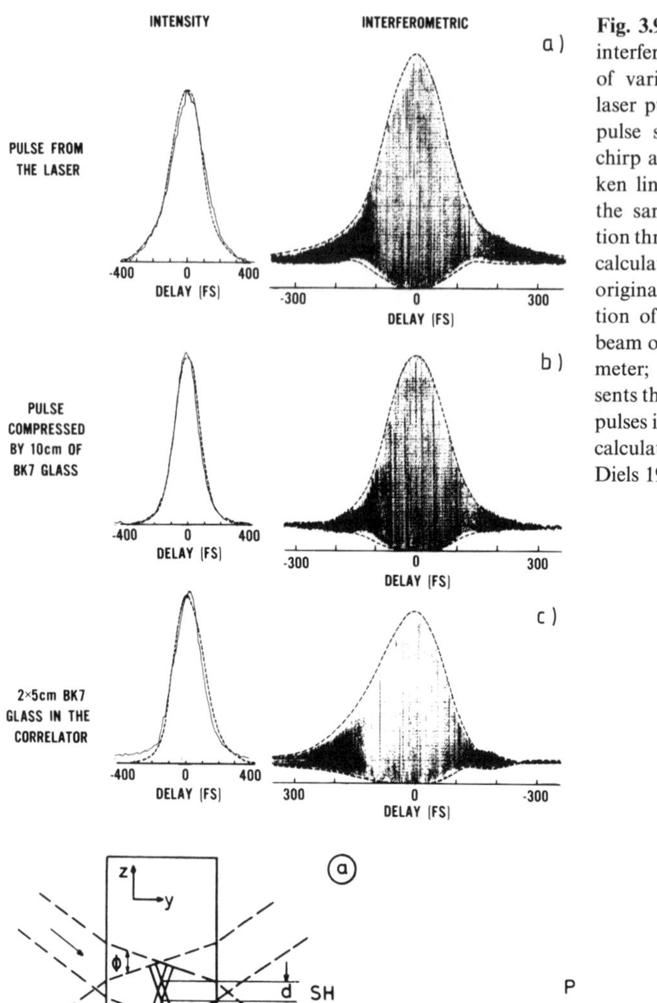

Fig. 3.9a–c. Intensity (*left*) and interferometric autocorrelations of various pulses. (**a**) original laser pulse; using the measured pulse spectrum, the frequency chirp and the pulse shape (broken lines) were determined; (**b**) the same pulse after propagation through 10 cm of BK7 glass; calculated broken lines; (**c**) the original laser pulse after insertion of 5 cm BK7 glass in one beam of the nonlinear interferometer; the measurement represents the cross-correlation of the pulses investigated in (**a**) and (**b**); calculated broken curves (after Diels 1985)

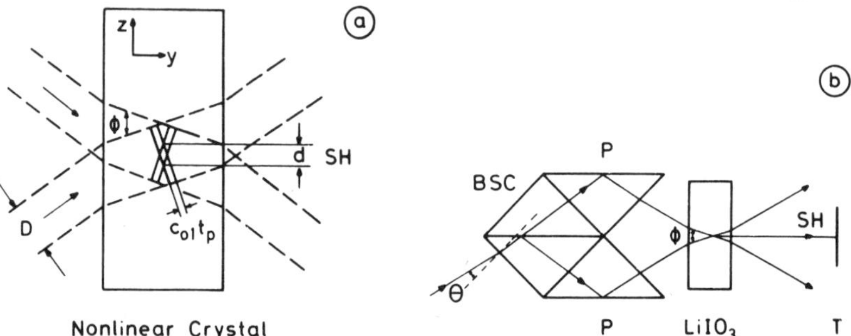

Fig. 3.10. (**a**) Schematic of the single-shot second harmonic autocorrelation technique. The two incident pulses with beam diameters D cross in the nonlinear crystal at angle ϕ. The spatial width d of the second harmonic light beam provides directly the duration of the pulses t_p. (**b**) Schematic of a practical imaging device consisting of a beam splitting cube BSC and two attached prisms P. The distribution of the second harmonic beam is measured at the position T by an optical multichannel analyser. The crossing angle ϕ can be changed by varying the entrance angle θ (after Kolmeder et al. 1979)

1967) and fast streak cameras (see Bradley 1977) a second harmonic method has been developed. For the noncollinear beam geometry (as discussed above) it was noticed that the time delay of the two overlapping beams varies over the beam cross section (Krivoshtshekov and Stroganov 1970). For large input beams, the measurement of the lateral intensity distribution of the second harmonic emission may yield the autocorrelation function in a single shot if multichannel detector arrays are used. The idea was theoretically worked out in detail by Janszky et al. (1977) and experimentally verified shortly afterwards (Gyuzalian et al. 1979). A schematic of the experimental set-up is shown in Fig. 3.10. A compact optical beam splitter and imaging system that avoids alignment problems was described by Kolmeder et al. (1979); these authors emphasized the superior sensitivity of the technique compared with two-photon fluorescence. Further details were discussed by Arakelyan et al. (1981, 1982). A related technique with tilted pulse fronts over the beam diameter has also been studied (Wyatt and Marinero 1981; Saltiel et al. 1981; Szabo and Bor 1983).

More detailed information on the pulse shape than that available from the intensity autocorrelation function may be supplied by a shorter gating pulse than the original laser pulse in the up-conversion experiment of Fig. 3.8; this is the basic idea of measurements of higher order cross-correlation functions. Different versions of such an experiment have been demonstrated; they vary in the particular method used to obtain a shorter probing pulse I_{pr}.

Instead of (3.10) we now measure

$$F(t_D) = \int_{-\infty}^{\infty} I(t) I_{pr}(t - t_D) \ . \tag{3.11}$$

If I_{pr} approaches a delta-function, $F(t_D)$ reproduces the investigated pulse shape $I(t)$.

Treacy (1969) and, very recently, Rothenberg and Grishkowsky (1987) have applied optical pulse compression to a part of the laser pulse to obtain a shorter probing pulse in the cross-correlation experiment. The fifth order correlation function $\int I(t) I^4(t - t_D) dt$ was generated in an experiment of Auston (1971) by four-photon frequency mixing of the laser pulse with the second harmonic in a methanol sample. A display of the approximate pulse shape and of the frequency sweep of the mode-locked Nd:glass laser pulses was obtained.

Von der Linde and Laubereau (1971) devised a different method which relies on the pulse shortening and steepening in stimulated Raman scattering in combination with coherent anti-Stokes Raman probing. A minor asymmetry and steep slopes were reported for the single picosecond pulses of a Nd:glass laser. More recently, a third order correlation technique was described by Albrecht et al. (1981, 1982) using the second harmonic pulse as a probe of the laser signal in the experimental setup of Fig. 3.8. This simple approach avoids some of the disadvantages of the second harmonic autocorrelation. The authors studied the temporal shapes of mode-locked YAG lasers over a large dynamic range, up to ten orders of magnitude. Further possibilities to incorporate higher order correlations in single-shot measurements have been discussed recently (Janszky and Corradi 1986).

3.3 Four-Wave Interactions

Four-wave interactions have become one of the most thoroughly investigated nonlinear optical effects. These processes are symmetry-allowed in all media and have found numerous important applications. For a physical understanding of the various phenomena, the frequency dependence of the third order susceptibility $\chi^{(3)}$ is often used for the theoretical description. For spectroscopic purposes, e.g. pump-and-probe experiments to study the dynamics of the medium, the frequency resonances of $\chi^{(3)}$ deserve particular interest, since they are related to various material excitations. For applications involving the generation of pulses at new frequencies, off-resonance situations are also important.

3.3.1 Sum-Frequency Generation

An extension of the sum-frequency process of Sect. 3.2.2 involves an additional (fourth) wave. The coupling constant is now the third order nonlinear susceptibility $\chi^{(3)}$ (3.1). In the photon picture, three photons of frequencies ω_1, ω_2, ω_3 combine to create a new quantum with frequency ω_4:

$$\omega_4 = \omega_1 + \omega_2 + \omega_3 \ . \tag{3.12}$$

As a higher order nonlinearity the interaction is generally weaker compared to the processes of the previous section. Experimentally, a higher intensity level may be used to compensate (in part) for the weaker coupling.

a) Third and Higher Order Harmonic Generation

Third harmonic generation (THG) is a straightforward generalization of the frequency doubling discussed above. With respect to (3.12) it represents the simple case $\omega_1 = \omega_2 = \omega_3$. The process can occur in systems with and without inversion symmetry including liquids and gases where it is the lowest order nonlinearity allowed by symmetry arguments. As a third order nonlinearity it is governed by a cubic power law with respect to the incident intensity I_L:

$$I_{TH} = \text{const}|\chi_3|^2 I_L^3 l^2 \left(\frac{\sin \Delta k\, l/2}{\Delta k\, l/2} \right)^2 \ . \tag{3.13}$$

Here χ_3 and l denote respectively the effective nonlinear coefficient and the interaction length of the medium. $\Delta k = k_{TH} - 3k_L$ is the wave vector mismatch. Similar to SHG, phase-matching ($\Delta k = 0$) can be accomplished in certain birefringent crystals. For many materials, however, this is not possible and other methods of phase-matching have been developed using:

(i) mixtures of normally and anomalously dispersive materials (Bey et al. 1968)
(ii) periodic structures (Freund 1968)
(iii) waveguides (Anderson and Boyd 1971).

Eckardt and Lee (1969) used THG to measure the duration of picosecond pulses (1.06 μm) via the third order autocorrelation function. Phase-matching was achieved in the liquid hexafluoroisopropanol (cell length 0.1 mm) by adding 37.5 g/l of the dye fuchsin red which displays anomalous dispersion (and absorption) at the TH position (353 nm). In addition to the phase-matching mechanism considered by Bey et al. (1968), Herman (1974) recognized the contribution of the added dye molecules to the nonlinear susceptibility χ_3 of the solution which was shown to be significant for various cyanine dyes.

Shelton and Shen (1970, 1971) studied THG of picosecond pulses in cholesteric liquid crystals. The layer structure of these materials with different molecular orientations produces a helical structure which can be used for the TH phase-matching of circularly polarized light. These authors also demonstrated phase-matching via the periodicity of the helical structure which supplies a lattice vector to compensate for the k-mismatch of waves travelling in opposite directions. In analogy to Bragg reflection at the Brillouin zone boundaries of a crystal, the process was called "umklapp third harmonic generation".

For investigating phase-matched THG in dye solutions, a detailed knowledge of the nonlinearity $\chi^{(3)}$ is desirable. Detailed work in this field was carried out by Thalhammer and Penzkofer (1983) and by Leupacher and Penzkofer (1985). From a quantitative study of THG with Nd:glass picosecond pulses in several dyes, the real and imaginary parts of the nonlinear susceptibility were determined in absolute units. A strong resonance enhancement due to the electronic transitions in the visible (leading to the well-known strong absorption bands) was observed as expected from the energy denominators that appear in the quantum mechanical derivation of $\chi^{(3)}$ (Armstrong et al. 1962). The measurements were also extended to rare gases and to N_2, using the same experimental approach (Lehmeier et al. 1985).

An extensive literature exists on THG in gases in the short wavelength range where suitable crystals are not available. Miles and Harris (1971, 1973) first discussed phase-matched THG in mixtures of metal vapors with noble gases, where the strong visible absorption of alkali metal atoms produces anomalous dispersion so that $\Delta k = 0$ can be achieved for the proper concentration ratio. The metal vapor also provides a strong enhancement by a factor of up to 10^6 of the nonlinearity compared to the inert gas. Conversion efficiencies of up to 10% were achieved experimentally for special heat pipes (Bloom et al. 1975; Puell et al. 1980).

Vacuum ultraviolet (vuv) generation of picosecond pulses by THG was first performed by Kung et al. (1972) in a mixture of cadmium and argon starting from the frequency-doubled (532 nm) and frequency-tripled (355 nm) pulses of a 50 ps Nd:YAG laser; radiation at 177 nm and 118 nm was produced representing the 6th and 9th harmonic of the laser wavelength, respectively. In a similar experiment with mixtures of Ar and Xe, strong phase-matched THG was achieved with 2.8% conversion to 118 nm, working with peak intensities of 6×10^{12} W/cm^2 of tightly focused beams (Kung et al. 1973). A variety of gaseous media have been investigated for the THG of ultrashort pulses (Taylor 1976; Ferguson and Arthurs 1976; Metchkov et al. 1977). Intense tunable pulses at around 200 nm were generated by frequency tripling of dye laser pulses with 12 ps duration in sodium vapor. A conversion efficiency of 8% was achieved (Drabovich et al. 1977).

The maximum intensity level in these experiments is determined by multiphoton or avalanche ionization, leading to dielectric breakdown in the sample. Additional limitations are connected with the phase-matching condition; e.g. density gradients reducing the interaction length or population of upper electronic states changing the refractive index and Δk.

Fifth harmonic generation of ultrashort pulses at 266 nm was investigated by Reintjes et al. (1978). The process is analogous to THG, involving the fifth order susceptibility $\chi^{(5)}$. Starting from the fourth harmonic of an Nd:laser at 266 nm, vuv radiation at 53.2 nm was produced in helium. Saturation of the harmonic conversion was observed and explained by a depletion of the fifth order susceptibility and a phase-matching offset via the nonlinear refractive index of the gas.

More recently Srinivasan et al. (1982) generated tunable picosecond pulses in the extreme ultraviolet (xuv). THG of mode-locked dye laser pulses in a strontium heat-pipe and subsequent amplification in an ArF* excimer laser-amplifier produced powerful 10 ps pulses of 30 mJ at around 193 nm, which served for further up-conversion steps. The lack of suitable window materials in the xuv was solved by using a small exit pin-hole for the chamber containing the nonlinear gas and differential pumping. Frequency tripling of the 193 nm pulses yielded in various gases ultrashort radiation at 64 nm with a conversion efficiency of up to $\simeq 10^{-4}$. Via harmonic generation and nonlinear mixing the 5th harmonic at 38 nm was also produced with conversions $\leqslant 10^{-6}$. Tunability of the xuv pulses was limited by the tuning range of the excimer laser amplifier of $\simeq 200\,\mathrm{cm}^{-1}$.

An extension of this work to even shorter wavelength values has been performed with intense pulse of $\simeq 1$ ps, 248.4 nm and 10–20 mJ energy content. The highest harmonic observed was the 13th in neon (19.1 nm) but with a very low intensity level of only a few photons per pulse (Luk et al. 1986).

b) Four-Wave Up-Conversion

Sum-frequency generation according to (3.12) can be used to convert radiation to the short wavelength range. For experimental reasons, the special case $\omega_1 = \omega_2$

has been used in practical applications, generating radiation beyond the second-harmonic of the fundamental component ω_1. Akhmanov et al. (1975) demonstrated four-wave up-conversion for fifth harmonic generation in calcite in a two-step process. The third harmonic ($\omega_3 = 3\omega_1$) of 1.06 µm picosecond pulses (ω_1) was combined with the fundamental radiation in the interaction $5\omega_1 = \omega_1 + \omega_1 + 3\omega_1$. Very weak signals of a few photons were detected in spite of the phase-matching possible with birefringence. A theory of the resonantly enhanced up-conversion process was developed by New (1976) and Elgin et al. (1980). A two-photon resonance of $\chi^{(3)}$ was considered, i.e. $2\omega_1$ corresponding to a (two-photon allowed) transition of the medium. Special attention was paid to the case of non-synchronized incident pulses at frequency ω_1 and ω_3. In earlier work Matsuoka et al. (1975) had already shown experimentally that in calcium vapour a coherent two-photon state is generated with finite lifetime so that interaction with a second pulse is possibly delayed by $\simeq 10^{-10}$ s in this system. The process is closely related to time-resolved coherent anti-Stokes Raman scattering (CARS).

Application of four-wave mixing to the generation of high power vuv pulses with 10 ps duration was reported by Hutchinson and Manning (1985). Starting with mode-locked ruby laser pulses (694 nm) the second harmonic ($\omega_2 = 2\omega_1$) was mixed in magnesium vapor with the fundamental (ω_1) in the process $4\omega_1 = \omega_1 + \omega_1 + 2\omega_1$, i.e. the fourth harmonic (174 nm) was generated. Output powers of 0.3 MW equivalent to conversion efficiencies of 0.2% were obtained by optimizing the mismatch Δk via the temperature control of the heat pipe.

An important property of up-conversion processes is, of course, the efficiency with which the incident radiation is transformed to the new frequency. A complete discussion of this point necessarily includes other processes (linear and nonlinear) that can affect the interaction. For the gases and liquids discussed in this subsection the competing effects are quite important. The following mechanisms should be considered here:

linear absorption at the frequencies ω_1 to ω_4

nonlinear absorption e.g. two-photon absorption of one (or more) frequency components involved

phase-matching offset via nonlinear refractive index changes proved to be important in many gases

pump depletion.

For short pulses allowing higher peak intensities without breakdown, the nonlinear perturbations tend to interfere more strongly than for other light sources. For a detailed discussion the reader is referred to the monograph by Reintjes (1984).

3.3.2 Four-Wave Difference-Frequency Mixing

Unlike its three-wave analogue, four-wave difference-frequency generation has gained only minor importance for the down-conversion of pulse frequencies.

The mechanism, however, has found numerous applications in spectroscopic investigations with short pulses, e.g. pump-and-probe experiments. Due to the generality of the $\chi^{(3)}$-formalism, very different physical mechanisms can be treated with this concept without necessarily specifying its physical nature. In this section our interest will be focused on the nonlinear properties of these interactions.

a) General Discussion

The four frequency components of the difference process combine [as in (3.12)] to generate new light with frequency

$$\omega_4 = \omega_1 + \omega_2 - \omega_3 \ . \tag{3.14}$$

The inherent four-photon interaction involves the annihilation of two quanta (ω_1, ω_2) and the creation of two new photons (ω_3, ω_4): $\omega_1 + \omega_2 \rightarrow \omega_3 + \omega_4$. The intensity of the new component (I_4) is determined by the incident intensities I_1 to I_3 according to

$$I_4 = \mathrm{const} \cdot |\chi_3|^2 I_1 I_2 I_3 l^2 \left(\frac{\sin \Delta k\, l/2}{\Delta k\, l/2} \right)^2 \tag{3.15}$$

which is a generalization of (3.13) for THG. The mismatch Δk now reads $\Delta k = |\boldsymbol{k}_1 + \boldsymbol{k}_2 - \boldsymbol{k}_3 - \boldsymbol{k}_4|$. The other quantities have their usual meaning. Equation (3.15) refers to the low intensity situation neglecting both depletion of the incident radiation and stimulated amplification (see Sect. 3.3.3). Usually two instead of three incident fields are coupled by the four-photon interaction; i.e. photons of one field act twice, $\omega_1 = \omega_2$ or $\omega_1 = \omega_3$ corresponding to $I_1 = I_2$ or $I_1 = I_3$ in (3.15). The specific role a frequency component plays in the interaction is determined by the mismatch Δk.

The general case $\omega_1 \neq \omega_3 \neq \omega_4$ involving three (or four) different frequency components is called non-degenerate frequency mixing. Equation (3.15) also applies for fully degenerate four-wave mixing (DFWM), $\omega_1 = \omega_2 = \omega_3 = \omega_4$, where the fields may still be distinguishable by their polarizations and/or beam directions. Obviously there exists a third possibility involving two different frequencies: $\omega_1 = \omega_3, \omega_2 = \omega_4$. This situation, which is sometimes termed "scattering from light-induced (phase) gratings", will be referred to in the following as "partially degenerate" four-wave mixing, because of its close relationship to the totally degenerate case. The wave-vector diagrams for the three different situations are depicted in Fig. 3.11. For partial or totally degenerate four-wave mixing, automatic phase-matching $\Delta k = 0$ can occur as a consequence of the frequency degeneracy. The broken arrows in Fig. 3.11 represent the k-vectors of the induced gratings which are often helpful for visualing the nonlinear process. A detailed theoretical discussion of four-wave mixing processes with double and triple resonances of the nonlinear susceptibility χ_3 has been presented by Oudar

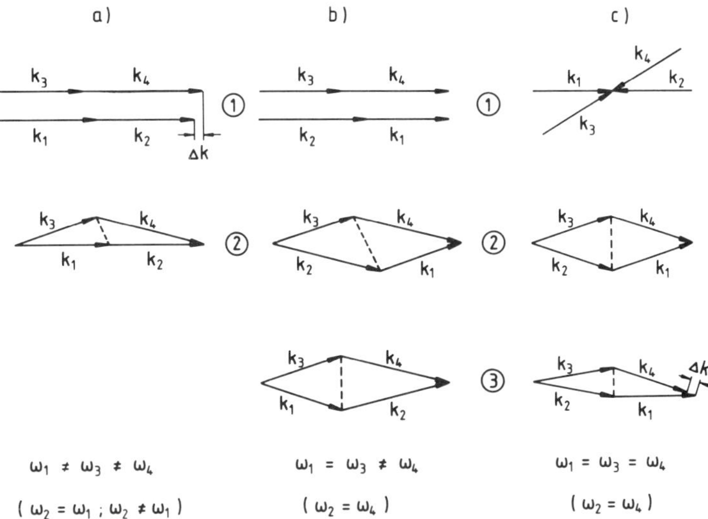

Fig. 3.11a–c. Wavevector diagrams of the mixing process $\omega_1 + \omega_2 \rightarrow \omega_3 + \omega_4$; (**a**) nondegenerate case with phase-matching $\Delta k = 0$ adjustable in the off-axis geometry (2); (**b**) partially degenerate four-wave mixing with automatic phase-matching in the one-beam (1) and two-beam (2) experiments, whereas in the four-beam geometry, $\Delta k = 0$ has to be adjusted (3); (**c**) fully degenerate case, DFWM, with automatic phase-matching with pairwise counterpropagating (1) or co-propagating beams (2) for arbitrary beam intersection angle. Broken lines indicate the k-vectors of several induced (non-propagating) phase and/or amplitude gratings

and Shen (1980). They also consider special polarization arrangements to suppress non-resonant contributions in spectroscopic applications. The transient case was treated by Ye and Shen (1982).

Auston (1971) demonstrated nondegenerate four-wave frequency mixing in various low-dispersion liquids. Using the second harmonic of a mode-locked Nd:glass laser ($\omega_1 = \omega_2 = 2\omega_L$) difference-frequency generation with the fundamental picosecond pulses ($\omega_3 = \omega_L$) yielded efficient radiation at $\omega_4 = 3\omega_L$. The process was used to measure the fifth order correlation function of the short pulses and to observe the frequency chirp by spatial resolution of the parametric off-axis emission.

The same process was investigated in water and aqueous solutions or crystals of alkali halides by Penzkofer et al. (1981, 1982). Using a non-collinear phase-matched geometry, picosecond pulses at the second harmonic ($\nu_1 = \nu_2 = 18\,960\,\text{cm}^{-1}$) and the fundamental frequencies ($\nu_3 = 9480\,\text{cm}^{-1}$) from a Nd:glass laser were mixed yielding intense pulses at $\nu_4 = 28\,440\,\text{cm}^{-1}$. Energy conversion in H_2O was up to 7% for peak intensities of $30\,\text{GW/cm}^2$ in a 2 cm sample cell. Somewhat smaller values were obtained for the aqueous alkali halide solutions at a lower intensity level; these were studied in a wide concentration range. Values of the tensor elements $\chi^{(3)}_{1111}$ and $\chi^{(3)}_{2112}$ of the nonlinear susceptibility were reported for the systems investigated; for the saturated solutions of LiCl, CsCl, KF and KI (6–15 mol/l) the nonlinear susceptibilities approached the values of the crystalline materials.

In the case of Raman (difference frequency $\omega_1 - \omega_3$) resonances of $\chi^{(3)}$, the four-wave mixing process $\omega_1 + \omega_1 \rightarrow \omega_3 + \omega_4$ is usually called coherent anti-Stokes Raman scattering (CARS); this situation is treated in other chapters of this volume devoted to spectroscopic studies.

b) Degenerate Four-Wave Mixing (DFWM)

In the partially or fully degenerate case ($\omega_1 = \omega_3$) far off single- or difference-frequency resonances of $\chi^{(3)}$, four-wave mixing is readily visualized in terms of the nonlinear change of refractive index connected with the real part of the third order nonlinear susceptibility and the two incident waves of equal frequency:

$$n = n_0 + n_2\langle E^2\rangle + \cdots , \qquad \text{where} \tag{3.16a}$$

$$n_2 \simeq \tfrac{3}{4} n_0 \mathrm{Re}\{\chi_3\} . \tag{3.16b}$$

In this way a nonlinear polarization term proportional to $(n_2 E_1 E_3^*)E_2$ is produced which oscillates with the second input frequency ω_3. For the off-axis geometry the wave vector difference $|\boldsymbol{k}_1 - \boldsymbol{k}_3|$ represents the k-vector of an induced phase grating from which the incident beam ($\boldsymbol{k}_2$) is scattered. Close to resonances of χ_3 the imaginary part also becomes important for degenerate four-wave mixing; i.e. an absorption grating is induced which is accompanied at the same time by population changes in the medium.

Important contributions to the susceptibility or, correspondingly, to the nonlinear refractive index n_2 arise from

(i) electronic transitions (Maker et al. 1964) in terms of the so-called electronic hyperpolarizability,
(ii) the orientational Kerr effect and intermolecular interactions (Hellwarth 1970).

Due to its slow response, the electrostrictive contribution to n_2 is negligible for ultrashort times. Further mechanisms were discussed in the literature (Gordon et al. 1965; Glass and Auston 1972). In liquids, the electronic contribution typically amounts to $n_2 \simeq 10^{-13}$ esu, which is smaller, in general, than the orientational Kerr effect with values up to 10^{-11} esu for strongly anisotropic molecules (e.g. CS_2, nitrobenzene, chlorobenzene etc.). For solids, values of n_2 in the range 10^{-12} to 10^{-13} esu were reported. The different relaxation times of the mechanisms offer the possibility of separating the various contributions. A detailed discussion has been given by Sala and Richardson (1975).

DFWM has been applied in a series of investigations to study the dynamical behavior of the nonlinear coupling coefficient (n_2 or χ_3) in pump-and-probe experiments. The most common way of performing these measurements is by monitoring the induced birefringence; i.e. non-collinear frequency-degenerate beams interact in the sample under special polarization conditions which are necessary to distinguish the incident probe pulse (k_2) from the scattered beam (k_4) (Duguay and Hansen 1969). These techniques detect the difference of the index changes $\delta n_{\parallel} - \delta n_{\perp}$, parallel and perpendicular to the pump pulse polarization. More recently, the transient refractive index changes were monitored with an interferometric technique where the index change $\delta n_{\parallel}$ is measured for parallel pump and probe polarizations (Halbout and Tang 1982). In these fully degenerate measurements, the scattered beam (k_4) travels parallel to the probe field (k_2, cf. Fig. 3.11c). More refined is the phase grating approach with non-collinear excitation (k_1, k_3) and probing (k_2, k_4) beams, where the pump and probe polarizations are freely adjustable yielding different relative contributions from the electronic and orientational parts of the nonlinear refractive index (Etchepare et al. 1986). In the degenerate experiments, a coherence peak artifact is to be expected (but not resolved experimentally) for short times with temporal overlap of pump and probe pulses (see Sect. 3.3.4b).[1]

Time-resolved studies of nonlinear refractive index changes have been performed for several organic liquids and glasses (Duguay and Hansen 1969; Reintjes et al. 1973; Etchepare et al. 1980). Detailed investigations were carried out for CS_2; an orientational relaxation time of 2.1 ps was measured from the orientational Kerr effect by Ippen and Shank (1975) in fair agreement with spectroscopic observations. Further measurements yielded values in the range of 1.4 to 2.2 ps (Etchepare et al. 1982; Green and Farrow 1982, 1983; Halbout and Tang 1982; Kenney-Wallace 1984). Most important measurements with femtosecond pulses $\leqslant 100$ fs revealed a fast component in the signal decay from which a second time constant of 330 fs (Halbout and Tang 1982), 240 fs (Greene and Farrow 1982, 1983) was inferred. Some results are shown in Fig. 3.12. The data have to be viewed with caution since the apparent absence of the electronic part of the nonlinear refractive index and of the coherence artifact are not fully understood. From a comparison of dc Kerr effect and Rayleigh scattering, the electronic contribution is estimated to be 11% of the total nonlinearity (Hellwarth 1977). A satisfactory explanation of the subpicosecond time constant is still lacking. Here, some basic criticism should be leveled against the exponential decay time assumed in the data analysis since a non-exponential time evolution of molecular dynamics is theoretically expected for sufficiently short times, i.e. on the time scale of 10^{-13} s for intermolecular interactions in liquids at room temperature (Madden 1984).

DFWM may be also carried out for an off-axis geometry with large angles between the interacting pulses. The special case of pairwise counter-propagating beams (see Fig. 311c) displays interesting properties with optical phase conjugation and aberration correction of the scattered signal ($\boldsymbol{k}_4$). For pump-and-probe experiments, this situation has been applied by several experimenters (see Sect. 3.3.3b). It should be noted that the nonlinear coupling mechanism is the same as for the general off-axis case, i.e. the same dynamical information is

[1] Note added in proof: The coherence peak was observed very recently by *Kalpouzos* et al. 1987 (J. Phys. Chem. **91**, 2028).

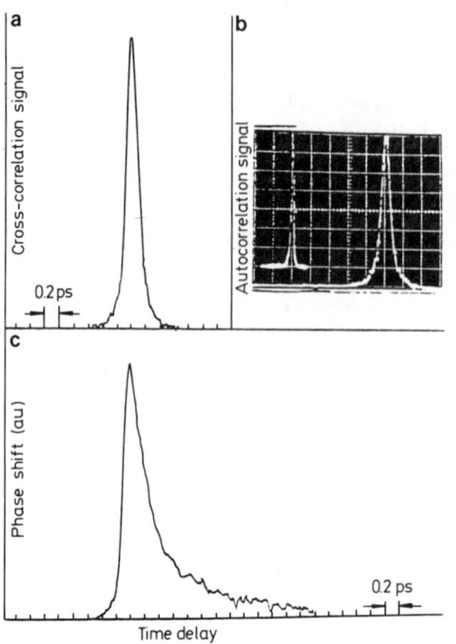

Fig. 3.12. (**a**) Intensity cross-correlation between the pump and probe pulses; FWHM 190 fs; (**b**) autocorrelation of the laser pulse; upper trace: 1 ps/div; lower trace: 0.2 ps/div.; FWHM 110 fs; (**c**) transient DFWM in CS_2; light-induced phase-shift is plotted vs delay time between pump and probe pulses. The signal first decays rapidly within a few 10^{-13} s, and then with a subsequent slower component attributed to orientational relaxation (after Halbout and Tang 1982)

available; however, the effect of the large geometrical group delay of counter-propagating pulses on experimental time resolution has to be taken into account, necessitating very short interaction lengths with correspondingly weak signals.

For the special case of a single photon resonance, where one incident frequency coincides with an absorption line of the medium, partially or fully degenerate four-wave mixing will be discussed below in Sect. 3.3.4, which is devoted to induced populating gratings.

Partially degenerate four-wave mixing ($\omega_1 + \omega_2 \rightarrow \omega_1 + \omega_2$) in the case of a Raman (difference frequency $\omega_1 - \omega_2$) resonance is usually termed coherent Stokes Raman scattering or the Raman-induced Kerr effect and will be discussed in other chapters of this volume in the context of spectroscopic applications.

3.3.3 Stimulated Four-Wave Interactions

The elementary four-photon mixing process of the preceding subsection, $\omega_1 + \omega_2 \rightarrow \omega_3 + \omega_4$, offers the possibility of stimulated amplification by high intensity components ω_1 and ω_2, since the (weak) third incident component ω_3 is enhanced by the coupling mechanism. In this way the wave at ω_3 may build up exponentially [compare (3.15)] until the depletion of the pump components ω_1 and ω_2 occurs. Experimentally, this process is generally achieved with a single intense pump beam, i.e. $\omega_1 = \omega_2$. Similar to the three-photon analogue (Sect. 3.2.4), the process may start from the zero-point fluctuations of the electromagnetic modes at ω_3 (or ω_4) leading to four-wave parametric fluorescence (spontaneous emission) at low pump intensity, and to stimulated four-wave emission for high intensities (Chiao et al. 1966). In the general case the interaction is termed stimulated four-wave parametric emission or amplification and may start from quantum noise or from an input signal $I_3(0)$ [or $I_4(0)$], respectively. The special conditions of full frequency degeneracy $\omega_1 = \omega_2 = \omega_3 = \omega_4$ and pairwise anti-parallel beams represent optical phase conjugation. Of particular interest are the difference-frequency (or Raman-type) resonances of the third order susceptibility $\chi^{(3)}$; this case is termed stimulated Raman scattering. The various situations will be treated in the following subsections.

a) Stimulated Four-Wave Parametric Emission

Solving the coupled nonlinear wave equations for the process $\omega_1 + \omega_1 \to \omega_3 + \omega_4$ the exponential growth of the waves at frequencies ω_3 and ω_4 is readily evaluated; neglect of pump depletion and linear absorption leads to the simple result (Penzkofer et al. 1973a, b)

$$I_4(z) = \text{const} \cdot I_3(0)e^{gz} \;, \qquad \text{where} \tag{3.17a}$$

$$g = \text{Re}\{(\gamma\chi_3 I_p^2 - 4\Delta k^2)^{1/2}\} \qquad \text{and} \tag{3.17b}$$

$$\gamma = 1024\pi^4 \frac{\omega_3\omega_4}{n_1 n_2 n_3 n_4 c^4} \;. \tag{3.17c}$$

A similar expression may be derived for $I_3(z)$. Equations (3.17) refer to the case $I_1 = I_2 = I_p$ and large gain, $gz \gg 1$. A non-vanishing real part of the susceptibility $\chi_3 \equiv \chi^{(3)}(-\omega_4; \omega_1, \omega_1, -\omega_3)$ is required for the process. Maximum gain occurs for $\Delta k = 0$ (3.17b).

Stimulated four-wave parametric emission in borosilicate glass BK-7 was first observed by Alfano and Shapiro (1970). Using second harmonic pulse trains derived from a mode-locked Nd:glass laser, intense up-converted (signal) and down-converted (idler) emission was photographically recorded extending over several thousand cm^{-1}. Off-axis phase-matching, $\Delta k = 0$, explained the observed angular frequency dependence of the parametric light in accordance with theoretical arguments taking into account self-focusing of the laser pump and nonlinear changes of refractive index.

Penzkofer et al. (1973a) quantitatively studied the stimulated parametric emission in water at different pump intensities with single picosecond pulses at 1.06 μm. Results are shown in Fig. 3.13. A significant spectral intensity distribution was measured for input pulses of $2-3 \times 10^{10}$ W/cm^2 [curves *c*) and *d*) in Fig. 3.13] which results from the frequency dependence of χ_3 because of one-photon resonances of the infrared idler component. At higher pump levels $\gtrsim 1 \times 10^{11}$ W/cm^2, broad spectra with little structure were recorded [curve *a*)] and explained by secondary four-photon mixing processes depleting the intense part of the primary signal and idler emission, and thus filling the spectral gaps. No evidence of self-focusing was observed so that self-phase modulation via the

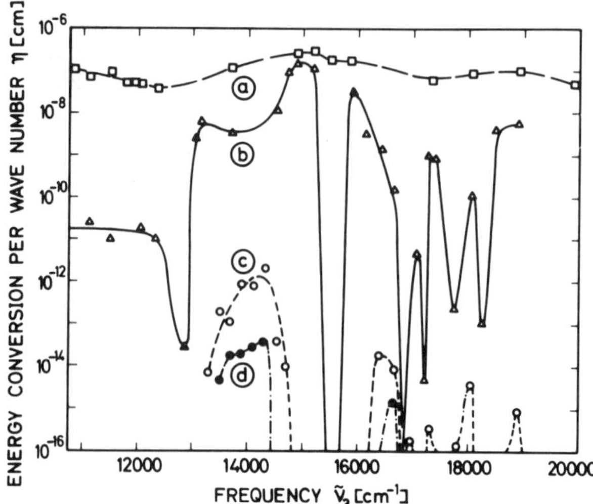

Fig. 3.13. Stimulated four-photon parametric generation in a 2 cm water cell pumped by an intense single picosecond pulse at 9450 cm^{-1}. The conversion efficiency η per frequency unit is plotted vs signal frequency for four pump intensities: (*a*) 1×10^{11}; (*b*) 5×10^{10}; (*c*) 3×10^{10}; (*d*) 2×10^{10} W/cm^2 (after Penzkofer et al. 1973)

nonlinear index changes could not account for the extreme spectral broadening extending from the ir to the uv. These conclusions were later supported by Sharma and Yip (1979). Penzkofer et al. (1973, 1977) obtained similar results in D_2O, fused silica and NaCl. These materials possess a small nonlinear refractive index n_2, low stimulated Raman gain and weak dispersion leading to small Δk's. The small n_2 is particularly important since it is possible to work with high quality beams at a high intensity level without optical self-focusing. Generation of broad continua in other materials with large nonlinear refractive index will be discussed below.

Stimulated four-wave parametric amplification was discussed theoretically by Dutta (1979) for the case of a two-photon resonance ($\omega_1 + \omega_1 = \omega_{21}$) of the nonlinear susceptibility under stationary and short pulse conditions. The time evolution of the process is modified by the transient response of the two-photon transition of the medium. A related investigation was carried out by Boyd et al. (1981) who considered the neighborhood of a one-photon resonance ($\omega_1 = \omega_{12} - \Delta$) of two-level atoms; additional frequency resonances were predicted when the frequency difference of signal and idler, $|\omega_3 - \omega_4|$, coincides with the generalized Rabi frequency of two-level systems.

b) Optical Phase Conjugation

Phase conjugate optics is a new and exciting area in nonlinear optics for real-time processing of optical fields. It is based on the phase reversal of the incoming electromagnetic wave. Pioneering work in this field was carried out by Zeldovich et al. (1972), Nosach et al. (1972), Yariv (1976b,c), Hellwarth (1977) and Yariv and Pepper (1977). The first observations of phase conjugation by four-wave mixing

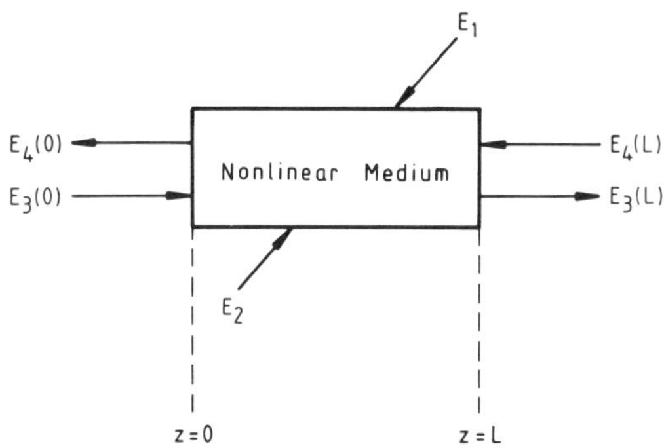

Fig. 3.14. Geometry for optical phase conjugation by degenerate four-wave mixing (schematic)

were reported by Jensen and Hellwarth (1978) and by Bloom and Bjorklund (1977) for CS_2. Other nonlinear effects have also been used for optical phase conjugation, e.g. stimulated Raman scattering and three-wave difference-frequency generation in the degenerate case $\omega_2 = \omega_3$. A comprehensive discussion of the effect is given in the volume *Optical Phase Conjugation*, edited by Fisher (1983).

In the following we concentrate on the case of degenerate four-wave mixing. For the generation of the phase conjugated wave, the geometry shown in Fig. 3.14 with two pairs of counterpropagating beams is generally used and where phase-matching is automatically ensured (see Fig. 3.11c-1). A small-angle non-collinear arrangement (Fig. 3.11a-2) is also possible but suffers from phase-matching constraints. Under stationary conditions and for counter-propagating pump fields E_1 and E_2 one finds (Yariv, 1978):

$$E_4(0) = \frac{-\mathrm{i}\kappa^*}{|\kappa|} E_3^*(0) \times \tan(|\kappa l|) \tag{3.18a}$$

$$E_3(l) = \frac{E_3(0)}{\cos(|\kappa l|)} \quad \text{with} \tag{3.18b}$$

$$\kappa = \frac{2\pi\omega}{nc} \chi^{(3)} E_1(0) E_2(0) \ . \tag{3.18c}$$

Pump depletion has been neglected in the derivation of (3.18). Inspection of the three equations reveals the major properties of optical phase conjugation:

phase conjugation of the output wave $E_4(0)$ for any incident field $E_3(0)$
amplification of the conjugate wave for $|\kappa l| > \pi/4$ and of the probe wave E_3
(stimulated) parametric oscillation for $|\kappa l| = \pi/2$ without mirror feedback.

The general treatment of phase conjugation, i.e. backward DFWM for $E_3(0)$, $E_4(l) \neq 0$, is quite complicated since simultaneous solutions of the two counter-propagating fields E_3 and E_4 have to be found. Time-dependent fields were first discussed by Marburger (1978) who emphasized the importance of optically thin nonlinear media since the interaction of the counter-propagating waves integrates over the transit time ln/c. An analytical solution of the short-pulse case with undepleted pump pulses was given by Shih (1986) who considered an instantaneously responding Kerr-like medium. Lavoine and Villaeys (1986) theoretically analysed the transient response of a two-level phase conjugator close to the one-photon resonance, $\omega = \omega_{12} (= \omega_1 = \omega_2 = \omega_3 = \omega_4)$, which is significantly effected by the dephasing time of the two-level transition.

The capability of the phase conjugation setup for wavefront and image reconstruction stems from automatic phase-matching, $\Delta k = 0$, for counter-propagating waves (backward scattering) in combination with a sufficiently large interaction length ($\geqslant 10^{-1}$ cm). As a result, every component of a divergent incident beam (E_3) is reflected back in the appropriate direction and with the correct phase. Significant stimulated amplification as considered in the previous subsection requires $|\kappa l| \rightarrow \pi/2$. The low intensity situation $|\kappa l| \ll 1$ is a special case of the four-wave mixing discussed in Sect. 3.3.2.

The potential of optical phase conjugation for the correction of aberration and for image reconstruction has found little application in the ultrashort time domain. The geometrical conditions for wavefront and image reconstruction in the backward scattering of picosecond pulses were discussed for stimulated Raman interaction (Ferrier et al. 1982, 1984) and other nonlinear mechanisms (Brekhovskikh et al. 1983). Picosecond pumping of passive phase-conjugate mirrors was studied by Cronin-Golomb et al. (1985) who emphasize the favorable insensitivity of the device to mechanical disturbances.

Various investigators in ultrafast spectroscopy have used degenerate four-wave mixing for the special geometry of Fig. 3.14. For detecting intensities (or pulse energies) of collimated beams, the phase conjugation aspect has only little relevance. With regard to the dynamical information, the close relationship to DFWM with arbitrary non-collinear geometry should be noted, e.g. in scattering from induced phase gratings of Kerr-like media or from induced population gratings for one-photon resonances of χ_3 (see Sect. 3.3.4). Conversion efficiencies of up to 50% have been reported for picosecond pulses of a mode-locked dye laser in DODCI and other saturable absorbers (Tocho et al. 1980, 1981). Similar work at lower intensity levels has been performed in ruby (Vorobiev et al. 1982). Different polarization geometries have been tested with green 35 ps pulses in CS_2 and Rhodamine 6G (dissolved in ethanol) by Wu et al. (1983). The nonlinear susceptibility of various materials was studied in a series of experiments, e.g. highly nonlinear soluble polydiacetylenes (Dennis et al. 1985, 1986), semiconductor-doped glasses (Roussignol et al. 1985), thin films of organic polymers (Rao et al. 1986; Prasad et al. 1986) and the electro-optic material BSO ($B_{12}SiO_{20}$) where partially degenerate FWM was also used (Ferrier et al. 1986).

Of special interest is an early demonstration of a subpicosecond optical gate with the phase conjugation setup by Bloom et al. (1978). A 50 μm thick CS_2 cell served as phase conjugator operated by amplified dye laser pulses (100 μJ, 0.5 ps), the second pump pulse (E_2) being generated by retroreflection of the first one

(E_1) with an immersed internal mirror. The effective interaction length is limited by the length of the counter-propagating pulses. Subpicosecond time resolution was demonstrated by SH cross correlation with the original pulse. High efficiency and a large acceptance angle (without a phase-matching limitation) make the device attractive. The light gate is frequency selective; the acceptance bandwidth was estimated to be approximately equal to the pump pulse frequency width. An extension of this gating application for single-shot pulse analysis was reported by Buchert et al. (1985). The investigated laser pulse is split into three beams overlapping both spatially and temporally in the nonlinear medium (CS_2). The probe pulse (E_3) with large beam diameter makes an angle of 90° with the two counter-propagating pump components. The pulse duration is determined from the spatial intensity distribution of the phase conjugate backward emission. Experimental results were obtained for single pulses ($\simeq$20 ps) of a mode-locked YAG laser.

To conclude this subsection, two remarks are called for:

(i) Phase conjugation is a general property of the interaction $\omega_4 = \omega_1 + \omega_2 - \omega_3$, possible in various kinds of partially and fully degenerate four-wave mixing experiments ($\omega_1 = \omega_3; \omega_2 = \omega_4$), as well as in the non-degenerate case sufficiently close to degeneracy ("nearly degenerate FWM"). Because the complex conjugate E_3^* of the "substracted" wave ω_3 enters the nonlinear wave equation for the conjugate wave E_4 (3.18a), two pairs of counter-propagating beams have to be used. This geometry also ensures phase matching for an arbitrary beam intersection angle.

(ii) Some care is necessary in interpreting the words "pump" and "probe" for short pulse spectroscopy. Above, these terms apply to the four-wave mixing process. With respect to the resonant material excitation, however, produced (and probed) in one-, two- or difference-frequency resonances of $\chi^{(3)}$, the assignment of components ω_3 and ω_4 as pump and probe waves, respectively, has to be interchanged. This peculiarity is just one aspect of the general finding that the $\chi^{(3)}$ formalism obscures the physical process for the sake of generality and completeness of the theoretical treatment.

c) Stimulated Raman Scattering and Related Raman Interactions

In the high intensity region, where stimulated amplification occurs and close to a difference frequency resonance of $\chi^{(3)}$, the partially degenerate four-wave interaction $\omega_1 + \omega_2 \rightarrow \omega_3 + \omega_4$ is termed stimulated Stokes Raman scattering (SRS). The degenerate components at frequencies $\omega_1 = \omega_3 = \omega_L$ and $\omega_2 = \omega_4 = \omega_S$ are usually termed "laser" and "Stokes", respectively, while the resonance condition requires $\omega_1 - \omega_2 \simeq \omega_{12}$. Like other resonant scattering processes, the formal four-photon interaction constitutes predominantly of a twofold two-photon process also involving a quantum transition (ω_{12}) of the medium, which may possess electronic, vibrational, or rotational character (Fig. 3.15b). The

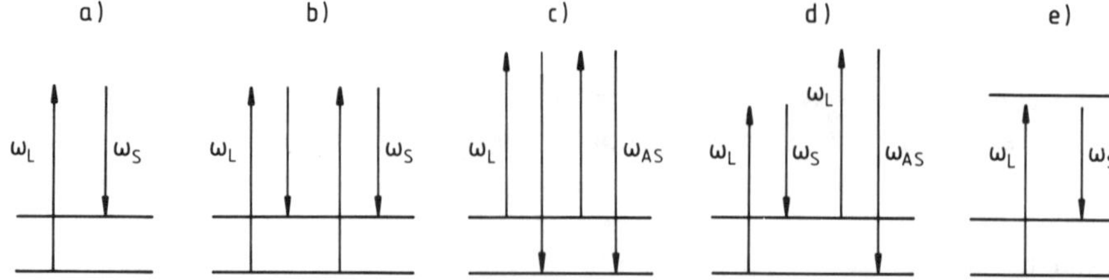

Fig. 3.15. Energy level diagrams for various non-resonant (**a–d**) and resonant (**e**) Raman scattering processes: (**a**) spontaneous Stokes scattering; (**b**) stimulated Stokes scattering; (**c**) stimulated anti-Stokes scattering; (**d**) stimulated Stokes-anti-Stokes generation; (**e**) (near-)resonant spontaneous Stokes scattering; vertical arrows denote photons, annihilated (up) or generated (down) in the interaction

elementary Raman process converting one laser photon ω_L to a Stokes photon ω_S and promoting the physical system to a higher energy state is shown in Fig. 3.15a. The inverse process of stimulated anti-Stokes Raman scattering (Fig. 3.15c) is also possible; this interaction necessitates a significant population of the excited energy level and has found only little interest in the past (Carman and Lowdermilk 1974). Attention should be paid to the competing non-degenerate four-wave process $\omega_L + \omega_L \rightarrow \omega_S + \omega_{AS}$ (Fig. 3.15d) representing parametric Stokes – anti-Stokes coupling, which inhibits SRS (Bloembergen 1967; Duncan et al. 1986).

To clarify the notation we should recognize that the situations shown in Figs. 3.15a–d with a Raman resonance of $\chi^{(3)}$ are called "non-resonant" in conventional Raman spectroscopy with respect to further energy levels of the physical system. The "resonant" case of conventional Raman scattering is shown for comparison in Fig. 3.15e where near-resonant coupling to a third energy level increases the scattering cross section. In stimulated scattering a corresponding enhancement is found which can be related to a double-resonance of the nonlinear susceptibility. In the following we focus our attention to the "non-resonant" case in the language of conventional Raman spectroscopy.

The nonlinear properties of stimulated (Stokes) Raman scattering are well established in the literature. Besides the amplitudes of the electromagnetic fields involved in the excitation, the coupled two-level systems are described by an amplitude Q and a relative population change N, equivalent to the off-diagonal and diagonal elements of the density matrix (Giordmaine and Kaiser 1966). Neglecting pump depletion and anti-Stokes or higher-order Stokes components ($\omega_L + \omega_{12}, \omega_L - 2\omega_{12}$ etc.) of the electromagnetic field, three coupled equations account for the build-up of the Stokes field and the material excitation (Akhmanov et al. 1968; Maier et al. 1969)

$$\left(\frac{\partial}{\partial z} + \frac{1}{v}\frac{\partial}{v\,\partial t}\right)E_S = \kappa_1 E_L Q^* \tag{3.19a}$$

$$\left(\frac{\partial}{\partial t} + \frac{1}{T_2}\right)Q = \kappa_2 E_L E_S^*(1 - 2N) \tag{3.19b}$$

$$\left(\frac{\partial}{\partial t} + \frac{1}{T_1}\right)N = \kappa_3 E_L E_S^* Q^* + \text{c.c.} \; . \tag{3.19c}$$

Equations (3.19) refer to a homogeneously broadened Raman line and to exact resonance, $\omega_L - \omega_S = \omega_{12}$. Parallel polarization and the slowly varying amplitude approximation are assumed for the laser and Stokes field. v is the group velocity of the Stokes pulse. The κ's include the coupling coefficient $\partial\alpha/\partial q$ and other parameters, where $\partial\alpha/\partial q$ denotes the Raman polarizability. The dephasing time T_2 and the population lifetime T_1 are of the order of 10^{-13} to 10^{-10} s in liquids and solids at room temperature, while larger values occur for gases depending on the number density. The short values of the time constants in condensed matter lead to small occupation changes, $N \ll 1$. As a consequence, (3.19a, b) may be solved independently of (3.19c).

Under stationary conditions, (3.19) yields exponential amplification of an initial Stokes intensity $I_S(0)$:

$$I_S(z) = I_S(0)e^{\gamma z}, \qquad \text{where} \tag{3.20a}$$

$$\gamma = \frac{16\pi}{cn}\kappa_1\kappa_2 T_2 I_L \; . \tag{3.20b}$$

A gain coefficient $\gamma \simeq 1\,\text{cm}^{-1}$ typically requires $I_L \simeq 10^9\,\text{W/cm}^2$ for a strong Raman line in condensed matter.

For transient SRS with short pulses of duration $t_p \lesssim 10\,T_2$ deviations from (3.20) occur; the interaction is less effective and requires higher pump intensities I_L to compensate. Analytic solutions of (3.19) were given by Carman et al. (1970) and Akhmanov et al. (1969, 1970) for arbitrarily shaped laser pulses neglecting pump depletion and group velocity dispersion. Some calculated results of the transient amplification $G_T = \ln(I_S(l)/I_S(0))$ versus steady-state gain γl are depicted in Fig. 3.16.

Due to the finite dephasing time T_2 in (3.19b), the interaction integrates over short pump pulses, $t_p \lesssim T_2$. As a result, the maxima of the material amplitude Q and of the generated Stokes pulse are delayed. A numerical example is shown in Fig. 3.17 for a high gain situation with an intense Stokes pulse built up from quantum noise. It is readily seen that the Stokes emission is notably delayed and shortened compared to the laser pulse (Carman et al. 1970).

For large gain and negligible dispersion, the transient stimulated amplification is independent of any frequency modulation of the laser pulse (Carman et al. 1970). This property originates from the fact that the Stokes field can adjust its phase to follow even rapid phase changes of the laser. In this way the product $E_L E_S^*$ in (3.19b) drives the material amplitude with (approximately) constant

Fig. 3.17 ▶

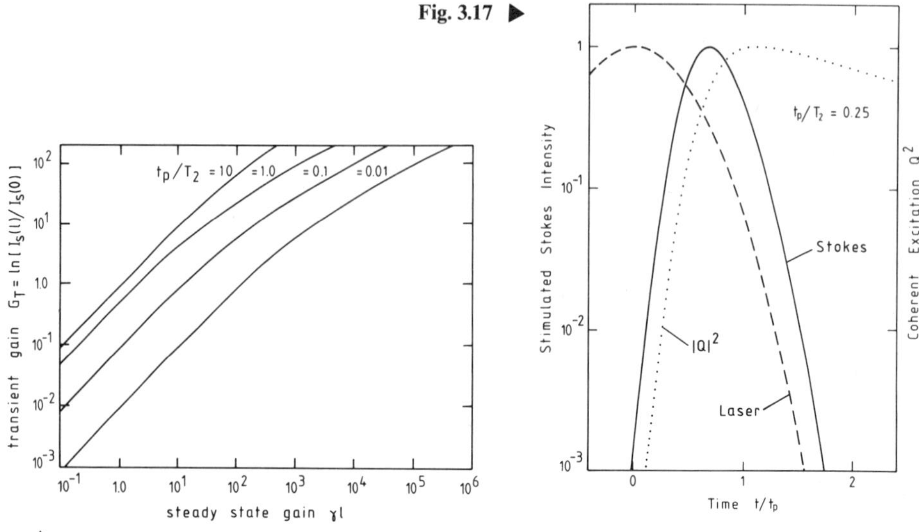

▲

Fig. 3.16. Calculated transient Raman gain for Gaussian laser input pulses and different values of t_p/T_2. The steady-state gain γl refers to the peak intensity of the laser pulse (after Carman et al. 1970)

Fig. 3.17. Semilog plot of the calculated time evolution of the Stokes pulse (——) and the coherent material excitation (···) generated by stimulated Raman scattering in the transient case $t_p/T_2 = 0.25$; Gaussian laser pump pulse (– – –); arbitrary units

phase. In a dispersive medium the situation is rather different. Since the laser and Stokes field do not maintain synchronism, the compensation of phase variations in the product $E_L E_S^*$ is destroyed and the build-up of the material excitation according to (3.19b) is reduced. The smaller gain of chirped pulses (in contrast to transform-limited pulses) is generally quite significant. The shortening of the interaction length because of group delay, i.e. the reduction of pulse overlap (Akhmanov et al. 1972) is frequently of less importance.

The extension of the theory to high conversion was treated by Darée and Kaiser (1974) who included depletion of the pump pulse (see also Darée, 1975; Neef 1975; Betin et al. 1975). Numerical calculations showed that gain saturation occurs in the transient case at higher pump intensities than under stationary conditions. In this region also anti-Stokes radiation and higher order Stokes components are simultaneously generated (Herrmann 1975; Valat et al. 1978).

For a detailed theoretical discussion of nonlinear Raman interactions and the relationship of the microscopic equations of motions, (3.19), to the macroscopic susceptibility $\chi^{(3)}$, the reader is referred to the review by Penzkofer et al. (1979). The transient material response including orientational effects was elucidated by Laubereau and Kaiser (1978).

The coherent material excitation of SRS represented by the amplitude Q was first observed by time-resolved CARS for molecular vibrations in liquids (von der Linde et al. 1971) and for optical phonons in solids (Alfano and Shapiro 1971; Laubereau et al. 1971) using picosecond pulses of Nd:glass lasers. The population of excited vibrational states via SRS was verified by spontaneous anti-Stokes scattering of probe pulses in hydrogen gas on a long time scale (DeMartini and Ducuing 1966) and in liquids on the picosecond time scale (Laubereau et al. 1972).

Carman and Mack (1972) experimentally demonstrated Stokes pulse delay and shortening of transient SRS in an investigation of SF_6 gas at 18 atmospheres with mode-locked ruby pulses of 15 ps. The pump intensity was carefully adjusted to avoid pump depletion. A delay of the Stokes emission of 6 ps and a duration of 9 ps were measured by the help of correlation techniques in agreement with theoretical expectations.

More recent streak-camera observations confirmed these results (Lowdermilk and Kachen 1975; Adrain et al. 1975). The ultrafast camera of Adrain et al. with sub-picosecond time resolution directly displayed the shortening of the Stokes pulses to 3 ps, produced by 12 ps dye laser pulses in ethanol (7.5 cm cell); to avoid longer Stokes pulses, the pump intensity had to be adjusted close to threshold. Even more drastic shortening of the Stokes pulses was found under special conditions in earlier investigations on the picosecond (Maier et al. 1966) and subpicosecond time scales (Colles 1971).

The automatic phase-matching of stimulated Raman scattering as a partially degenerate four-wave interaction excludes limitations on the beam geometry. In principle, any angle between pump and Stokes beams including backward scattering is possible. However, gain arguments [cf. (3.20a)] favor the longer interaction length of approximately parallel or anti-parallel beams. For ultrashort pulses, the sample is often longer than the pump pulse so that the forward scattering geometry is strongly favored compared to counterpropagating laser and Stokes beams with $l \leqslant t_p c/n$. In the collinear case, the interaction length in condensed matter is limited for picosecond pulses by group delay between laser and Stokes radiation to 1–10 cm in the visible (Colles 1969).

In close analogy with the stimulated parametric emission discussed in Sects. 3.2.4. and 3.3.3., three different experimental set-ups have been designed for SRS:

(i) the stimulated Raman amplifier with a well-defined input Stokes signal besides a laser pump pulse
(ii) the stimulated Raman generator where the Stokes pulse builds up from quantum noise in a high gain travelling-wave process
(iii) the stimulated Raman oscillator where an optical resonator provides feedback for the Stokes amplification starting from quantum noise.

The amplifier set-up (i) has frequently been applied (with small or moderately large gain) for the generation of the coherent material excitation (amplitude Q), the time evolution of which was studied with delayed probing pulses in time-resolved coherent Raman scattering, detecting the anti-Stokes (CARS) (Laubereau et al. 1973; Zinth et al. 1978; Kuhl and von der Linde 1982; Kosic et al. 1983). The time constant T_2 was directly obtained from these investigations.

In an alternative approach of Heritage (1979) the time evolution of the material excitation is followed by a measurement of the stimulated amplification of delayed pairs of laser and Stokes pulses.

A special version of SRS was developed by DeSilvestri et al. (1985). These authors used femtosecond laser and Stokes input pulses of the same carrier frequency for highly transient stimulated Raman amplification of low-frequency phonon-modes in a non-collinear beam geometry. The broad spectrum of the short input pulses provides Fourier components with proper frequency shifts for the Raman process. Since in the time domain the pulse duration was comparable to (or shorter than) the vibrational period $2\pi/\omega_{12}$, the interaction was termed impulsive stimulated Raman scattering.

A common application of stimulated Raman scattering is the generation of intense pulses at new frequencies. The amplification process is often repeated in several steps following the Raman generating stage to improve the peak intensity and beam divergence of the Stokes emission (Papernyl et al. 1982). The moderate intensity level of the Raman amplifier has the significant advantage that perturbing nonlinear effects can often be avoided.

For ultrashort pulses the stimulated Raman oscillator, case (iii), requires synchronous coupling to a train of pump pulses. For a very recent development related to fiber-soliton propagation, see Sect. 3.4.3.

In spite of its high intensity requirements the stimulated Raman generator (ii) was chosen by many experimentalists in various investigations of SRS. The obvious reasson is the simplicity of having only a single input pulse. Zero-point fluctuations provide the necessary Stokes input. The large amplification demands a gain coefficient of $G_T \simeq 25$ in order to obtain a conversion efficiency of 10^{-2}. The corresponding peak intensities of about 10^{10} W/cm^2 lead readily to accompanying nonlinear effects, e.g. self-focusing and self-phase-modulation. In fact, SRS of ultrashort light pulses in the generator setup is only rarely observed as an isolated nonlinear phenomenon. For short pulses, however, competing stimulated scattering processes that occur only on longer time scales are suppressed, e.g. stimulated Brillouin scattering (von der Linde et al. 1969).

Intense stimulated Stokes emission has been observed in various liquids, liquid mixtures, and crystals using pulse trains of mode-locked Nd:glass, Nd:YAG or ruby lasers (Shapiro et al. 1967; Bret and Weber 1968; Colles 1969; Carman et al. 1969; Rahn et al. 1969; Pochon and Bourene 1976; Maple and Knudtson 1978). Large conversion efficiencies exceeding 50% were measured at the peak of the Stokes pulse train. Similar results were also obtained for several high pressure gases both for vibrational and rotational transitions (Mack et al. 1970; Carman and Mack 1972; Chatelet and Oksengorn 1975; Loree et al. 1976). A list of selected materials and available frequency shifts is presented in Table 3.2.

Special attention has been paid in recent years to the generation of tunable, intense ir pulses by means of SRS from electronic transitions in metal vapors. In spite of the small number density in these systems of 10^{15} cm^{-3}, resonance enhancement of the spontaneous Raman process (see Fig. 3.15e) and, correspond-

Table 3.2. Substances and Raman shifts for frequency conversion of ultrashort laser pulses by stimulated Raman scattering

Substance	Shift[cm^{-1}]	Excitation	Reference
Liquids:			
$SnBr_4$	221	SH Nd:glass	1
$SnCl_4$	368	SH Nd:glass	1
$SiCl_4$	425	SH Nd:glass	1
CCl_4	459	SH Nd:glass	2, 3
Cl_2	552	ruby	4
CS_2	656	SH Nd:glass	3, 5, 6, 7
SF_6	775	ruby	4
d-benzene	944	ruby	8
benzene	992	SH Nd:glass	3, 5, 6, 7
bromobenzene	998	SH Nd:glass	6
chlorobenzene	1002	SH Nd:glass	5, 6
toluene	1004	SH Nd:glass, ruby	6, 8
N_2O	1289	ruby	4
nitrobenzene	1344	SH Nd:glass, ruby	5, 6, 8
CO_2	1392	ruby	4
CH_3OD	2200	SH Nd:glass	9
$N_2(T = 77\,K)$	2326	SH Nd:glass	10
HBr	2493	ruby	4
HCl	2814	ruby	4
methanol	2835	SH Nd:glass	3, 5
isopropylalcohol	2882	SH Nd:glass	5
acetone	2925	SH Nd:glass	5, 7
ethanol	2928	SH Nd:glass	2, 5
2,2-dichlorodiethylether	2938	SH Nd:glass	11
CH_3CCl_3	2939	SH Nd:glass	5, 12
CH_3F	2970	ruby	4
CH_2Cl_2	2989		13
H_2O	3450	SH Nd:glass	5, 7, 14
Solids:			
calcite	1086	SH Nd:glass	5
diamond	1332	SH Nd:glass	15
Gases and vapors:			
NH_3	334	ruby	16
Cl_2	556	ruby	16
SF_6 (18 atm)	775	ruby	16
N_2O (50 atm)	774	ruby	17
	1282		
C_2H_4 (55 atm)	1344	ruby	17
CO_2 (20–50 atm)	1385	ruby	17
O_2 (50–100 atm)	1550	ruby	17
SF_6 (15–20 atm)	1551	ruby	17
	2323		
NO	1877	ruby	16
CO	2145	ruby	16
N_2 (55–100 atm)	2330	ruby	17, 18
HBr (20 atm)	2558	ruby	17
HCl (35 atm)	2883	ruby	17

Table 3.2 (*continued*)

Substance	Shift[cm^{-1}]	Excitation	Reference
CH_4 (80–100 atm)	2916	ruby	17
C_3H_6 (90 atm)	2920	ruby	17
D_2	2991	ruby	8
H_2	4160	ruby	8, 18
Ba			
(Ar buffer gas			
1–300 Torr)	11395	552 nm	19
(1050°C, 10 Torr)	28300	308 nm	20
Cs			
(1200°C, 50 Torr)	14597	dye	21

References:
1: Laubereau et al. (1971); 2: von der Linde et al. (1971); 3: Alfano and Shapiro (1970); 4: Maple and Knudtson (1978); 5: Colles (1969); 6: Shapiro et al. (1967); 7: Bret and Weber (1968); 8: Bloembergen (1967); 9: Laubereau et al. (1973); 10: Laubereau (1974); 11: Telle and Laubereau (1983); 12: Laubereau et al. (1972); 13: Laubereau (unpublished data); 14: Rahn et al. (1969); 15: Laubereau et al. (1971); 16: Carman and Mack (1972); 17: Mack et al. (1970); 18: Chatelet and Oksengorn (1975); 19: Sapondzhyan and Sarkisyan (1983); 20: Glowina et al. (1987); 21: Harris et al. (1984).

ingly, of the Raman gain factor provide sufficient stimulated amplification. Pioneering work on stimulated electronic Raman scattering (SERS) was performed by Sorokin and Lankard (1973) and by Wyatt and Cotter (1980, 1981) who studied cesium vapor. Intense picosecond pulses at wavelengths as long as 20 μm were generated. More recently, stimulated electronic Raman scattering (SERS) of the $6s-5d$ transition of cesium vapor was investigated by Berg et al. (1984) with amplified pulses of 1–2 mJ and 1 ps derived from a commercial, synchronously pumped dye laser. Reducing the dimer concentration of the metal vapor by superheating the heat pipe to 1200°C and using a He buffer gas pressure of 50–60 Torr were essential factors in achieving a peak quantum efficiency of 4.6% at 2520 cm^{-1}. Variation of the laser wavelength from 570 to 610 nm with the help of different laser dyes resulted at a tuning range of 3040 to 1950 cm^{-1} (see Fig. 3.18). An ir pulse duration of 1–2 ps was estimated from the measured bandwidth of 6–10 cm^{-1}. Glownia et al. (1987) very recently reported an extension to the subpicosecond time domain for SERS in barium vapor. Intense broadband pulses of $\simeq$160 fs were obtained at 2.4 μm, starting from 350 fs laser pulses at 308 nm which were amplified to several mJ in XeCl excimer modules. In this investigation the SERS process in rubidium vapor also served to upconvert the ir radiation into the visible part of the spectrum.

Finally, some additional features of SRS should be mentioned. The relationship between spontaneous and stimulated Raman scattering was theoretically studied by Mostowski and Raymer (1981), Rzazewski et al. (1982) and Trippenbach et al. (1984). A full quantum-mechanical treatment of the build-up of

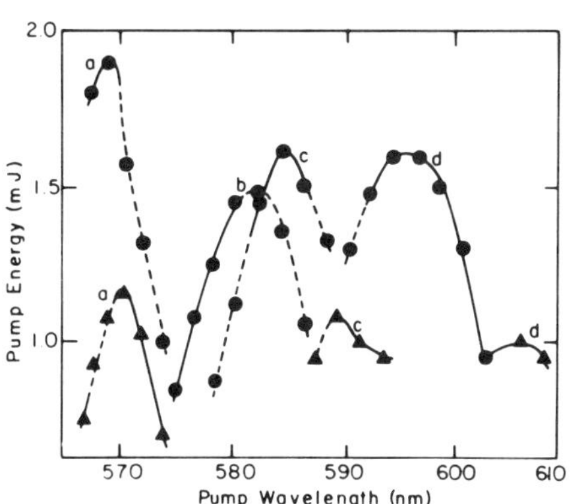

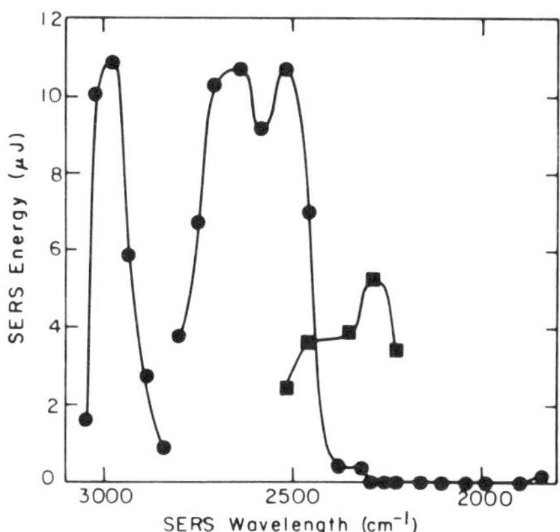

Fig. 3.18. Tunable picosecond pulses in the ir by stimulated electronic Raman scattering. Left: Energy of visible dye laser pump-pulse vs wavelength, using different laser dyes; a, R590; b, R610; c, KR620; d, R640; ●, methanol solutions; ▲, water solutions (no surfactants). Points used for best SERS generation are outlined with the solid portions of the curves. Right: ●, SERS energy generated with 1 ps pump pulses; ■, SERS energy generated with broad pulses (after Berg et al. 1984)

stimulated Raman scattering from quantum noise was presented that correctly describes the transition from spontaneous to stimulated Raman scattering. Besides minor deviations from earlier rate-equation results, a probability distribution was predicted in the transient case for the energy of the stimulated Stokes emission. The output fluctuations reflect the statistical situation of the quantum noise which serves as the initial condition. The predictions were verified experimentally by Fabricius et al. (1984) for transient SRS in hydrogen gas with 30 ps YAG laser pulses working at very low conversion efficiencies of $\simeq 10^{-10}$. These observations may be regarded as the two-photon counterpart to the phenomenon of superfluorescence (spontaneous emission of inverted two-level systems under transient conditions).

The frequency of the stimulated Stokes pulse often deviates notably (several cm^{-1}) from the value expected on the basis of spontaneous Raman data. This question was theoretically and experimentally tackled by Zinth and Kaiser (1980). These authors show that the inclusion of the nonresonant part of $\chi^{(3)}$ (i.e. nonlinear refractive index n_2) in (3.19) leads to a corresponding nonlinear phase modulation of the Stokes radiation following the time evolution of the laser pulse. For transient SRS the frequency modulation in combination with the Stokes pulse delay causes a shift to higher frequencies, accompanied by a spectral broadening of the laser pulse. Experimental data on CH_3CCl_3 with 7 ps pulses of a Nd: glass laser quantitatively agreed with theoretical estimates. An example is shown in Fig. 3.19. For stationary SRS the effect vanishes.

Further details of the molecular dynamics were incorporated into the theory of SRS in liquids. The rotational motion couples to the nonlinear interaction via the anisotropy of the Raman polarizability so that additional time constants have

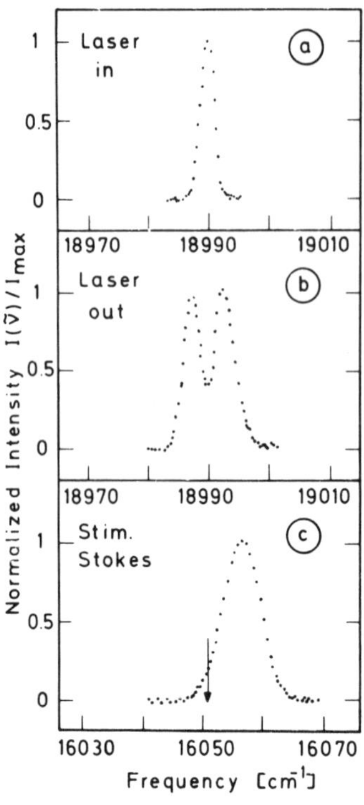

Fig. 3.19a–c. Effect of self-phase-modulation on transient stimulated Raman scattering in liquid CH_3CCl_3. (**a**) Spectrum of the incoming bandwidth-limited laser pulse centered at $\nu_L \simeq 18990\,cm^{-1}$. (**b**) Spectrum of the laser pulse transmitted through the Raman cell when Stokes light with an energy conversion of 1% was generated. The spectrum is centered around ν_L with a structure typical for self-phase-modulation. (**c**) Spectrum of the generated stimulated Stokes pulse. The maximum of the stimulated Stokes spectrum is shifted to higher frequencies. The position of spontaneous Raman scattering at $16\,051\,cm^{-1}$ is indicated by the arrow (after Zinth and Kaiser 1980)

to be introduced into the theoretical treatment (Laubereau and Kaiser 1978). For parallel polarization, these effects are generally quite small. Transient stimulated Raman amplification with perpendicular polarizations of the laser and Stokes pulse, however, differs distinctly (Kohles and Laubereau 1985, 1986, 1987). Besides the change of the coupling coefficient $\partial\alpha/\partial q$, the dynamical response of the liquid is considerably accelerated by the fast orientational relaxation of small molecules (shorter effective T_2). Very recent picosecond observations on CH_3I confirm the theoretical results.

General molecular dynamics arguments and also the results of spontaneous Raman spectroscopy suggest that the linear damping term for the material excitation in (3.19b) is no longer valid in liquids on the subpicosecond time scale. This point deserves special attention for transient SRS with femtosecond pulses. A generalization of the theory in this respect was presented by Telle and Laubereau (1983, 1984). These authors replace the equation of motion (3.19b) by an integral equation, where an autocorrelation function accounts for the correct dynamical response. The exponential dephasing time T_2 is contained in this formalism as a special case. The authors show that transient stimulated Raman amplification in the near-resonant case, $\omega_L - \omega_S \simeq \omega_{12}$, is particularly sensitive to the non-exponential dephasing even for picosecond pulses. A doublet structure was predicted and experimentally observed for the amplified (near-resonant) Stokes pulse from which a subpicosecond autocorrelation time was inferred, in agreement with other observations. This time constant characterizes the time scale of the elementary dephasing mechanism, e.g. the random translational motion of the molecules in their local environments.

Thus far, population changes (3.19c) have been assumed to be negligible in the stimulated Raman interaction, which may be described by the $\chi^{(3)}$ formalism only in this approximation. In the terminology of coherent effects, the Stokes pulse generation discussed above represents the two-photon analogue to small area pulse propagation in the one-photon case. A characteristic of such a phenomenon is the formation of a substructured pulse for a long propagation distance. Similarly, for transient SRS in the extreme saturation regime, an interplay of laser pulse depletion and material response result in a gain and loss modulation of the Stokes and laser pulses. Numerical calculations of Kachen and Lowdermilk (1976) were found to be in good agreement with subnanosecond experimental data on H_2 gas. At considerably higher intensities, population changes have to be taken into account, leading to an apparent intensity dependence of $\chi^{(3)}$, i.e. higher order susceptibilities have to be introduced. New propagation phenomena occur, e.g. the Raman analogue of self-induced transparency (Akhmanov et al. 1971; Shimoda 1970; Medvedev et al. 1971, 1974; Tan-no et al. 1975; Hoshimiya et al. 1977). In addition to the coupled equations (3.19), at least two more equations for the laser field and the anti-Stokes component now have to be solved, since population inversion in the pulse propagation may also introduce the inverse process of stimulated anti-Stokes scattering. Neglect of higher order Stokes as well as anti-Stokes components may be questionable under these conditions. Experimental observations have not yet been made. Extensions to sequences of input pulses are also possible, but are experimentally difficult, e.g. Raman-echos (Hu et al. 1976). Investigations of this kind represent a challenge to experimentalists in the future.

3.3.4 Further Four-Wave Interactions

a) Induced Population Gratings

Bragg scattering from induced gratings may be generally considered as a four-wave interaction. A special motivation to use this physical picture exists for resonances of the third order nonlinear susceptibility where the physical nature of the relevant material excitation is known and where the grating concept facilitates the theoretical discussion.

An example of this type is light scattering from induced amplitude (absorption) and phase (dispersion) gratings in absorbing media which has found various spectroscopic applications in the past. The physical origin of the phenomenon is the simple one-photon process with corresponding population changes. In the terminology of nonlinear optics, scattering off induced population gratings represents a four-wave mixing process in the fully or partially degenerate case with a

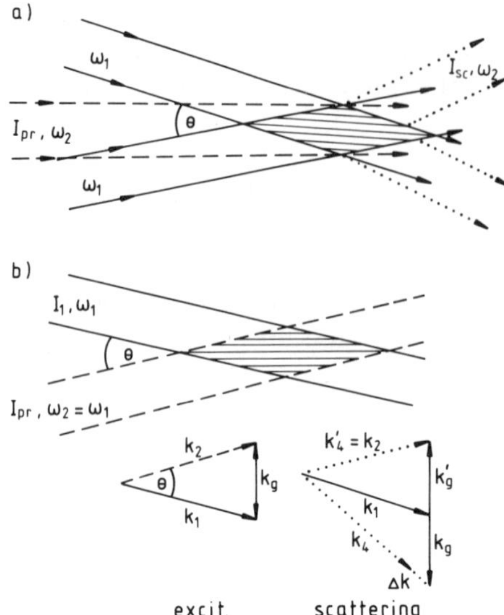

Fig. 3.20a, b. Schematic of the beam geometries for light scattering from induced population gratings; (**a**) three input beams for partially or fully degenerate pulse frequencies ($\omega_1 \neq \omega_2$ or $\omega_1 = \omega_2$); (**b**) two input beams in the fully degenerate case; k-vector diagrams for the excitation and probing processes are also shown; scattering components occur both in new directions (e.g. wave-vector k_4; mismatch Δk) and parallel to the input beams (e.g. k_4' with $\Delta k = 0$)

single-frequency resonant $\chi^{(3)}$ ($\omega_1 = \omega_3 \simeq \omega_{12}$; $\omega_2 = \omega_4$), provided that low pump intensities give rise to small population changes. The nonresonant contribution of $\chi^{(3)}$ and the refractive index changes via sample heating may also play a role in experimental investigations. Polarization effects have to be carefully considered since the material excitation generated by the linearly polarized fields is anisotropic as a consequence of the vector character of the electric dipole interaction. Figure 3.20 shows a schematic of the grating experiment. The general case with three incident beams is depicted in Fig. 3.20a. Two synchronized pump pulses of equal frequency ω_1 produce the material excitation via resonant (linear) absorption. The grating results from the interference of the two beams, which gives rise to a spatial modulation of the pump intensity I_L of the material excitation with k-vector $k_g = 2k_1 \sin(\theta/2)$. Linear Bragg scattering of a delayed probe pulse I_{pr} with the same or with different frequency ω_2 produces the detected scattering signal. Forward scattering with small angles θ is generally used for reasons of time resolution. In the fully degenerate case, $\omega_1 = \omega_2$, three-dimensional k-matching is advantageous (Tröger et al. 1985) to separate input and output beam directions similar to other cases of four-wave mixing (Laubereau et al. 1973). The time evolution of the induced grating is monitored by varying the delay time of the probe pulse.

Some physical insight can be gained from the following theoretical example. The simple case of an ensemble of two-level molecules with identical transition frequency ω_{12} is treated. The molecules are oriented parallel to the electric field vectors of the excitation (E_1, E_2) and probing pulses (E_{pr}) so that the vector character of the field amplitudes and of the induced polarizations may be omitted. A phase-matched, nearly collinear geometry with pulse propagation close to the

z-axis is considered. The small spatial dependence perpendicular to the z-axis is included in the amplitude factors. For a population grating with amplitude N_{12} in transverse direction, the Bragg scattered field E_{sc} of the probe pulse E_{pr} is described by the nonlinear wave equation

$$\left(\frac{\partial}{\partial z}+\frac{1}{v}\frac{\partial}{\partial t}\right)E_{sc}=\mathrm{i}\lambda_1 P_{sc}\ , \tag{3.21a}$$

where the induced polarization P_{sc} producing the dipole emission is derived from a quantum mechanical treatment of the two level systems (see, for example, Yariv 1976a).

$$\left(\frac{\partial}{\partial t}+\frac{1}{T_2}+\frac{\mathrm{i}(\omega_{12}^2-\omega_1^2)}{2\omega_L}\right)P_{sc}=-2\mathrm{i}\lambda_2 E_{pr}N_{12}\ . \tag{3.21b}$$

The population grating is generated by the mutual interaction of the pump fields with the two-level transition:

$$\left(\frac{\partial}{\partial t}+\frac{1}{T_2}+\frac{\mathrm{i}(\omega_{12}^2-\omega_1^2)}{2\omega_1}\right)P_j=\mathrm{i}\lambda_2 E_j \tag{3.22a}$$

$$\left(\frac{\partial}{\partial t}+\frac{1}{T_1}\right)N_{12}=\mathrm{i}\lambda_3(E_1P_2^*+E_2P_1^*)+\text{c.c.} \tag{3.22b}$$

with $j=1, 2$. The λ's contain material parameters, e.g. the electric transition dipole moment. Equations (3.21, 22) refer to the low intensity case with small relative population changes $N_{12}\ll 1$ and to the slowly varying amplitude approximation. One interesting point in the above equations is to realize that the population grating is not produced by the intensity grating of the interfering pump pulses ($\propto E_1E_2^*+$ c.c.). Equation (3.22b) shows that the interference of the polarization P_j induced by one excitation pulse with the field amplitude of the other, and vice versa, produces the population change. This fact is important in experiments where an intensity grating is missing, but a population grating may be generated nevertheless, e.g. for perpendicular polarization of the pump pulses or for a large time delay between the pump pulses.

The various dynamical processes which cause probe scattering off induced gratings have been extensively discussed in the literature for dye solutions and semiconductors. The reader is referred to the other chapters of this volume, to the review article by von Jena and Lessing (1979) and to the 1986 August issue of the IEEE Journal of Quantum Electronics with contributions by Fayer, Myers and Hochstrasser, Smirl et al., Trebino et al., Wherrett et al. and others. Of the variety of spectroscopic results, just one example is given here in Fig. 3.21. The lowest electric-dipole allowed transition of polydiacetylene (PTS) was investigated. 300 fs pulses of a synchronously coupled dye laser were used in a fully degenerate interaction. The result for the population lifetime of the excited electronic state is 1.8 ps (Carter et al. 1986).

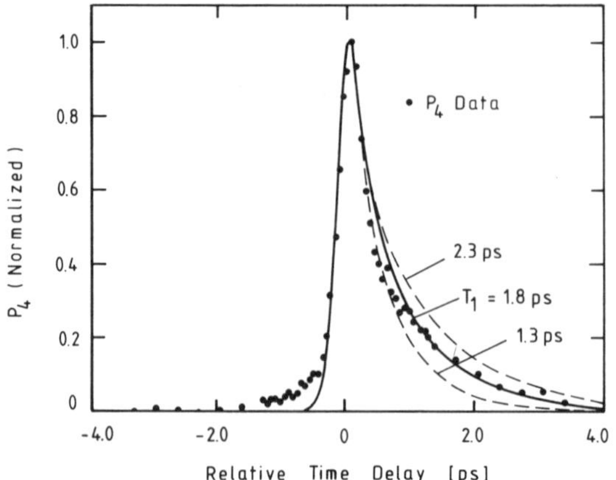

Fig. 3.21. Transient DFWM representing scattering off an induced population grating in polydiacetylene (PTS). Normalized power of the scattered beam vs relative time delay between pump and probe pulses; experimental points and calculated curves; excitation wavelength 652 nm; sample length $\simeq 1\ \mu m$. The population lifetime is measured to be 1.8 ± 0.5 ps (after Carter et al. 1986)

As well as temporal relaxation, spatial effects, e.g. energy transport were also investigated (Dlott et al. 1978). A recent application of the grating technique is the generation of high frequency acoustic phonons (Nelson and Fayer 1980; Nelson et al. 1982; Robinson et al, 1984). Since the phonon wave vector equals the grating vector k_g, the phonon frequency can be adjusted over a large range from GHz to THz with the help of the beam intersection angle θ (Fig. 3.20). The excitation mechanism is termed impulsive stimulated Brillouin scattering.

The k-vector diagram of the ordinary four-wave interaction represents first order light scattering. A sinusoidal grating only contains the fundamental wave vector k_g and higher order scattering vanishes. For an arbitrary grating pattern, high order components in the Fourier series, i.e. multiples of k_g are present as well. In fact, higher order light scattering from induced index gratings was observed by Göbel and Saito (1986) in thin crystalline CdS platelets at 170 K. The distortion of the originally sinusoidal grating could be attributed to an exciton–free-carrier transition.

The theoretical analysis (3.22) showed that the population changes were generated by the combination of the linear polarization induced by one pulse with the electric field amplitude of the second one. As a consequence, dephasing processes can also be studied with the grating technique. When the two pulses do not overlap in time a grating and consequently a probe scattering signal can still be generated via the phase memory of the medium (dephasing time T_2). To monitor the decay of the induced polarization the two excitation pulses are delayed with respect to one another while the probe delay is kept constant with respect to the second pump component (Weiner and Ippen 1984). Alternatively a two-beam grating may be used (Fig. 3.20b) where one pump component also acts as the probing pulse. This effect is called self-diffraction; the process is fully frequency degenerate and non-phase-matched. Small angles θ and short samples are required to achieve sufficiently small values of Δkl.

An early investigation of this kind was performed by Yajima et al. (1979, 1980) who studied the dephasing of dye molecules on the subpicosecond time scale with dye laser pulses of 0.2–0.3 ps. An upper limit of $T_2 < 0.1$ ps was stated by the experimental data. These authors call the two-beam grating experiment the "transient spatial parametric effect in resonant media". A recent three-beam investigation by Weiner et al. (1984) with 70 fs dye laser pulses gave evidence for inhomogeneous broadening in solid polymer solutions of cresyl violet at 15 K.

Smirl et al. (1982) studied self-diffraction of 8 ps pulses of the Nd:glass laser at 1.06 μm in silicon at $\simeq 8\,\mathrm{GW/cm^2}$. The different signal transients for parallel and perpendicular polarization of the two light pulses was taken as evidence for a laser-induced orientational grating in the semiconductor. In a related experiment in thin germanium samples Smirl et al. (1980) demonstrated phase conjugation of the forward-scattered signal.

Very recently several groups have recognized the value of broadband, non-transform-limited light sources for ultrafast transient spectroscopy; for instance, they investigated degenerate four-wave mixing in the resonant case. Instead of the short pulse duration t_p, the short correlation time τ_c of incoherent radiation provides the desired time resolution (Asaka et al. 1984; Yajima et al. 1984; Beach and Hartmann 1984, 1985; Golub and Mossberg 1986). The versatility of the technique for the measurement of dephasing times and population lifetimes was demonstrated. For example, the excitation lifetime of brilliant green in ethylene glycol was measured to be $T_1 = 11$ ps with broadband dye laser pulses ($t_p \simeq 1$ ns, $\tau_c \simeq 6$ ps). The signal transients were shown to depend on the statistical properties of the incoherent light. Careful control of the coherence and modulation properties of the light source is required to obtain reliable results.

Finally, the close relationship between the population grating method discussed above and the resonant Rayleigh-type mixing spectroscopy introduced by Yajima and coworkers (1975, 1978; Souma et al. 1980) should be mentioned. The latter technique may be interpreted as the frequency-domain counterpart of the former one. The dynamical information contained in resonant Rayleigh-type mixing is extracted from the frequency dependence of the scattered signal under quasi-stationary conditions. One input beam is tuned to the absorption resonance while the second beam is slightly shifted off-resonance by a small offset ω_d. While $\omega_d = 0$ produces a spatially fixed intensity profile with maximum grating amplitude, $\omega_d \neq 0$ leads to an oscillating local intensity and consequently smaller population amplitude. For $\omega_d T_1 \geqslant 1$, the signal decreases notably. A further decrease occurs for $\omega_d T_2 \geqslant 1$. The nonlinear spectroscopy was carried out with tunable picosecond pulses for liquid and solid solutions of dye molecules (Souma et al. 1980). The short pulses allow one to work at a high intensity level without damaging the sample. Time constants in the range 20 fs to 2 ps were reported for several systems. This indirect method requires a careful theoretical analysis in order to extract the proper time constants from the measured frequency dependence. A major problem is the proof of the validity of the theoretical approach.

Induced population gratings have found an important application in distributed feedback lasers for the generation of short light pulses. This point is discussed in Chap. 2 of this volume on pulse generation.

b) The Coherence Peak

When pump-probe experiments are performed with identical pulses, it is necessary to consider the nonlinear interaction of the two pulses with particular care since additional features occur for small delay time. In fact, in various experiments to study transient bleaching, photo-induced absorption, induced dichroism and population gratings, a significant signal overshoot was noted and was termed the coherence peak or coherent coupling artifact (Shank and Auston 1975; Ippen and Shank 1977; Vardeny and Tauc 1981; Reiser and Laubereau 1982; Oudar 1983; Eichler et al. 1984). Very recently, the effect was also observed in time-resolved CARS (Kohles and Laubereau 1986, 1988). An example is shown in Fig. 3.22 for a transient absorption experiment. In some cases the signal enhancement is not clearly resolved but a shift of the signal maximum towards $t_D = 0$ results (Ippen and Shank 1977). The phenomenon has been theoretically discussed by several authors, e.g. Wilhelmi and Herrmann (1980).

The basic mechanism is an interchange of the role of the pump and probe pulses which is possible during the temporal overlap of the two pulses. In this way an additional signal is generated. The amount of signal enhancement depends on the specific situation. If pump and probe possess the same frequency and polarization, the same coupling mechanism is available as for the normal process observed for large t_D. A signal enhancement of the order of two results. In other cases the signal enhancement may be of minor importance.

An example of a k-vector diagram for the three-beam grating technique is shown in Fig. 3.23. The population grating of the normal excitation process (Fig. 3.23a) and of the coherent coupling artifact (Fig. 3.23b) are readily distinguished. Interchanging the k-vectors of pump and probe does not alter the k-matching nor the emission direction. As an example we will consider the ensemble of two-level molecules that was discussed in part a) of this section for the off-axis geometry of Fig. 3.23. Including the pump-probe coupling artifact into (3.21) and (3.22) one finds in the quasi-stationary case, $T_2 \ll t_p$:

$$E_{sc}(t, t_D) = \text{const}\left\{E_{pr}(t - t_D) \int_{-\infty}^{t} dt' e^{(t'-t)/T_1} E_1(t') E_2^*(t') + E_1(t) \int_{-\infty}^{t} dt' e^{(t'-t)/T_1} E_{pr}(t' - t) E_2^*(t')\right\} . \tag{3.23}$$

The second term on the right hand side of (3.23) represents the additional contribution of the pump-probe coupling artifact, which is present only for a

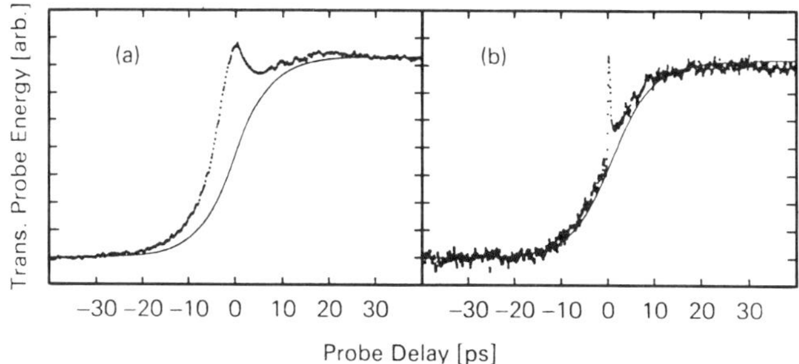

Fig. 3.22a, b. Coherence peak as observed in transient probe transmission measurements of cresyl violet with resonant 7 ps pulses (590 nm) and collinear beam geometry; (**a**) nearly transform-limited pulses; (**b**) pulses spectrally broadened by self-phase-modulation in an optical fiber. The short coherence time causes a sharp coherence spike at $t_D = 0$ (after Heinz et al. 1984)

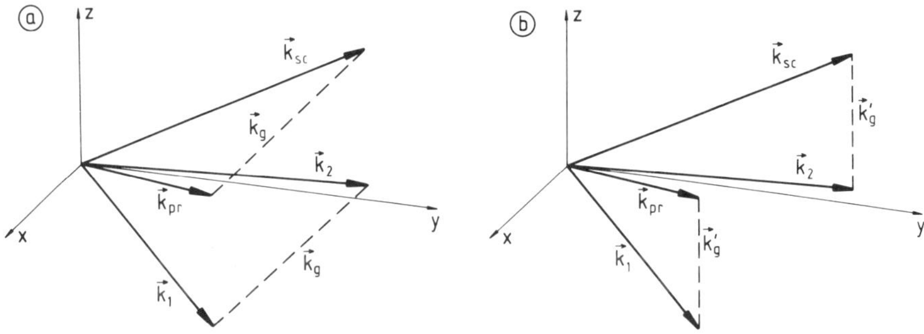

Fig. 3.23a, b. Pump-probe coupling artifact for the three-beam grating experiment with frequency degeneracy and three-dimensional phase-matching; (**a**) normal process with population grating vector k_g generated by the two pump beams (k_1, k_2) and interrogated by the probe pulse (k_{pr}); (**b**) additional interaction for temporally overlapping pulses with population grating vector k'_g excited by pump and probe pulses (k_1 and k_{pr}, respectively) and interrogated by the second pump (k_2). After Troeger et al. (1985)

small delay. Equation (3.23) also provides an explanation of the sharp coherence peak observed in Fig. 3.22b for frequency modulated pulses, which is shorter than the pulse duration. For the normal interaction [first term on the rhs. of (3.23)], the frequency modulation cancels in the integrand $(E_1 = E_2)$ and does not change the t_D-dependence. The second term, however, responsible for the coherent artifact, is strongly affected by time-dependent phases. For a slow response of the medium, $T_1 \gg t_p$, the second integral on the rhs. of (3.23) represents the electric field autocorrelation function which is characterized by the coherence time t_c of the field (Eichler et al. 1984).

The situation for the common two-beam experiment is illustrated in Fig. 3.20b. It is important to note that phase-matched self-diffraction of the pump pulse (k_1) occurs off the grating (k'_g) generated by the combined action of the two incident pulses; the scattering signal travels precisely in the direction of the transmitted probe beam. For a collinear geometry representing the limiting case $k_g \to 0$ the artifact is also present (Heinz et al. 1984). The phenomenon was also observed for crossed polarization of the pump and probe pulses (Wherrett et al. 1983). Coupling via the nonresonant part of $\chi^{(3)}$ may also contribute to the coherent artifact.

In conclusion it should be emphasized that a careful analysis of the coherence peak is crucial when the rapid material response is comparable to the pulse duration (Reiser and Laubereau 1982; Weiner and Ippen 1985).

c) Continuum Generation

For various spectroscopic applications it is highly desirable to generate and control ultrashort radiation with a wide spectral bandwidth. Besides the stimulated amplification of broadband emission by parametric three- or four-wave interactions already discussed above (Sects. 3.2.4d and 3.3.3a), self-phase modulation via the nonlinear refractive index n_2 is also an important mechanism to produce broadband emission. The effect is often combined with drastic changes of the beam geometry, i.e. optical self-focusing (also via n_2). In spite of the difficulties of proper control and reproducibility in the presence of self-focusing, many experimentalists observed and applied picosecond continua under these complex experimental conditions. The importance of self-phase-modulation via self-focusing as a source of superbroadening of short pulses was emphasized by Bloembergen and coworkers (Smith et al. 1977). These authors pointed to the drastic index changes expected in dielectric breakdown and observed under self-focusing conditions in H_2O and D_2O; the fundamental and frequency-doubled pulses of a mode-locked Nd:YAG laser were used in these studies. A theoretical discussion of superbroadening in the absence of self-focusing was recently given by Yang and Shen (1984). These authors considered phase modulation, including four-wave mixing and pulse deformation, but omitted linear dispersion. An asymmetry of Stokes–anti-Stokes broadening was found, in fair agreement with experimental data. Manassah et al. (1986) performed a similar theoretical investigation, again disregarding self-focusing and dispersion. The range of the experimentally observed continua was correctly predicted without recourse to additional assumptions. Computational results for nonlinear Kerr media with a finite response time of the nonlinear refractive index were reported by Mestdagh and Haelterman (1987).

The bandwidth generated by the phase modulation mechanism increases with decreasing pulse duration, since the nonlinear frequency shift is proportional to $1/t_p$. In a recent experiment Fork et al. (1983) observed superbroadening on the fs time-scale. Amplified pulses of a CPM dye laser (80 fs) propagated at an estimated intensity level of 10^{13}–10^{14} W/cm^2 through a 500 μm jet of ethylene glycole yielding drastic spectral changes. The broad spectral range from the uv to the ir (190–1600 nm), the stable and repeatable pulses with powers extending

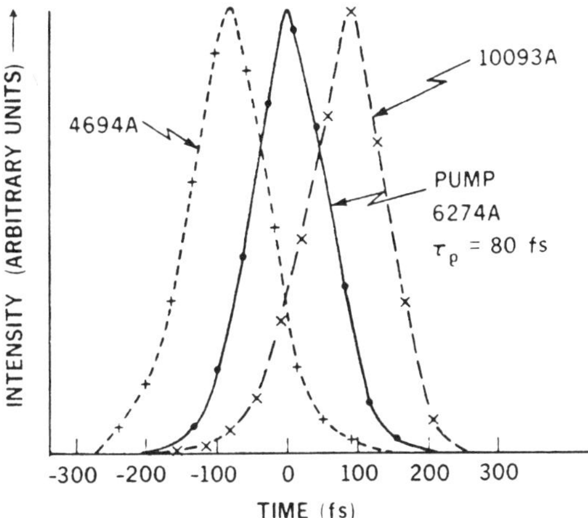

Fig. 3.24. Time evolution of the femtosecond continuum generated in a 500 µm jet of ethylene glycole; the cross-correlation traces are displayed for representative blue (+) and infrared (×) portions of the continuum. An autocorrelation trace of the pump pulse (•) at low intensity is also shown (after Fork et al. 1983)

to the gigawatt range make the femtosecond continuum a powerful tool for time-resolved spectroscopy. Self-focusing and optical breakdown are presumably absent in this work. The time characteristics of the emission are consistent with self-phase-modulation. Cross correlation results between the pump pulse and an infrared or blue portion of the continuum are depicted in Fig. 3.24. The duration of the continuum emission is close to 80 fs i.e. the value of the pump pulse. The infrared component coincides with the leading edge of the excitation pulse while the blue part of the continuum occurs $\simeq 170$ fs later.

Corkum et al. (1985) observed large spectral broadening of picosecond pulses in the infrared which they call supercontinuum, covering the range 3–14 µm. Gigawatt picosecond CO_2 laser pulses (2.5 and 8 ps) at 9.3 µm were used as pump source attaining 10^{11} W/cm^2 in several dielectrics and semiconductors, e.g. GaAs, CdS, ZnSe and AgBr. The data provide qualitative evidence for the importance of the nonlinear refractive index which is known to exceed n_2 of water by two orders of magnitude for the semiconductor materials (e.g. GaAs). Supercontinuum generation was also reported for pico- and femtosecond pulses in several gases at medium pressure (Corkum et al. 1986; Glownia et al. 1986). Induced spectral broadening of a weak picosecond pulse (530 nm) in long BK-7 glass samples by an intense picosecond pulse (1060 nm) was observed by Alfano et al. (1986). The dominant mechanism for the continuum generation was an induced phase modulation process; in other words the spectral broadening of the intense pulse was transferred to the weak one via the rapid nonlinear refractive index changes.

d) Ultrafast Bistability and Hysteresis

An exciting and rapidly expanding field of nonlinear optics is the investigation of bistable optical devices (BOD). The effect was first demonstrated for a nonlinear Fabry Perot resonator where the cavity contains a nonlinear absorber (Szöke et al. 1969). Alternatively, a dispersive mechanism, i.e. the nonlinear refractive index change detuning the cavity modes has been considered (McCall 1974; Gibbs et al. 1976; Felber and Marburger 1976). Later work showed that an artifical nonlinearity is also possible, for example with an electrooptic crystal and an electronic feedback system (Smith et al. 1977, 1978). The first observation of dispersive bistability by Gibbs et al. (1976) stimulated further extensive theoretical and experimental work (Bonifacio and Lugiato 1976, 1978; Gronchi et al. 1981). It was subsequently realized that a resonant cavity is not necessary for optical bistability (Garmire et al. 1978). Optical bistability was shown to be a more general phenomenon which may, for example, occur in degenerate four-wave mixing (Flytzanis and Tang 1980; Winful and Marburger 1980; Goldstone and Garmire 1984). Since the stationary behavior of BOD's is widely discussed in the literature (for a review see Goldstone 1985), the present chapter concentrates on the dynamic features of fast bistable systems.

The transient properties of a nonlinear resonator originate from the finite photon lifetime in the cavity and the memory time of the nonlinear medium. The delayed response increases the switching time of the BOD and can cause ringing and overshoot of the transmitted intensity as the system changes from one transmission value to another. Dynamic properties were discussed by Bonifacio and Meystre (1978), Bischofsberger and Shen (1978, 1979), Bonifacio et al. (1979), Hopf et al. (1979), Goldstone and Garmire (1981) and others.

Several dynamic properties of bistabile devices under transient conditions have been observed on relatively long time scales (μs to seconds) only. Regenerative pulsation of a dispersive nonlinear Fabry-Perot device was demonstrated experimentally by McCall (1978) and theoretically discussed by Lugiato (1980) and Benza et al. (1980). Related phenomena including monostable pulse generation, overshoot switching by incident pulses and subharmonic generation were observed shortly after (Okada and Takizawa 1980; Goldstone et al. 1980); these phenomena await applications on ultrashort time scales. A fast bistable device with switching times less than 1 ns was first described by Grischkowsky (1978).

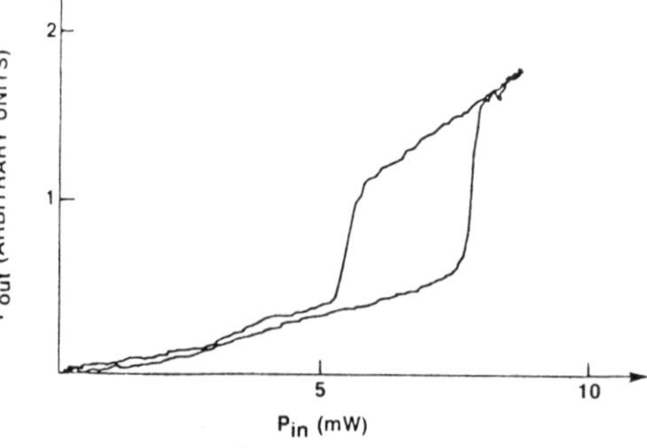

Fig. 3.25. Optical bistability and hysteresis in cadmium sulfide. Single shot measurement of transmitted power P_{out} vs P_{in} with the laser detuned by less than $0.3\,cm^{-1}$ from the I_2 bound exciton. The deconvoluted switch-on and switch-off times are measured to be less than 1 and 2 ns, respectively (after Dagenais and Sharfin 1985)

Figure 3.25 shows a quasi-steady-state bistability in cadmium sulfide on the nanosecond time scale (Dagenais and Sharfin 1985). The hysteresis of the transmission characteristics is a necessary requirement for bistability. Detector-limited switching times of less than 1 and 2 ns were reported for ON and OFF switching, respectively (upper and lower curves in Fig. 3.25). To observe bistability, the laser had to be tuned in close proximity ($<0.3\,\mathrm{cm}^{-1}$) to the I_2 bound exciton of the semiconductor material which was responsible for the large nonlinearity of the device. A cavity of length 25 μm accomplishes a short cavity lifetime of a few picoseconds. A lower limit to the switching times is determined by the radiative lifetime of the bound exciton of approximately 500 ps. At higher input intensities further hysteresis loops appear; i.e. the system displays multistability as expected theoretically.

Optical bistability and hysteresis of a colliding pulse mode-locked fs dye laser was recently reported by Jacobovitz et al. (1986). These authors observed two lasing states, one at 612 nm with fs pulse formation and a second one with cw output at 570 nm depending on the input power of the cw pump laser. The spectral shift could be explained by the wavelength dependence of the small signal gain. An ultrafast BOD was recently proposed by Otsuka et al. (1985) utilizing the nonresonant $\chi^{(3)}$ of a Kerr-like medium. Compression of ultrashort pulses by a nonlinear Fabry-Perot cavity synchronously coupled to an incident pulse train was discussed by Piché and Quellette (1986). On the basis of computer calculations, compression factors of 10 to 10^3 were predicted.

Ultrafast switching of (monostable) nonlinear Fabry-Perot cavities has been demonstrated very recently. Very thin etalons of $\simeq 1.5$ μm and the large nonlinearity of GaAs-AlGaAs multiple-quantum-well structures combine in these devices. An earlier experiment (Migus et al. 1985) utilized the nonlinear refractive index change which originates from a strong variation of the excitonic absorption line at room temperature for resonant excitation pulses. A switching time of $\simeq 1$ ps was measured for a NOR gate operation by time-resolved absorption spectroscopy with a $\simeq 100$ fs white-light continuum derived from a fs dye laser. With an improved device Hulin et al. (1986b) subsequently observed subpicosecond ON and OFF switching times. The etalon consisted of one hundred alternating layers of 100 Å GaAs and 25 Å AlGaAs grown by molecular beam epitaxy on a GaAs substrate. Gold layers of 200 Å on the surfaces provided high reflectivity (90% at 800 nm). A nonresonantly tuned fs pump pulse induced an immediate transient shift of the exciton resonance during the irradation of the sample; this optical Stark effect was exploited for the gate operation. The switching speed was limited only by the cavity lifetime of $\simeq 250$ fs.

Bistable optical devices display many modes of operation analoguous to electronic circuits. Thus these systems might constitute the basic elements for optical computers and communication systems. Further progress in materials science towards larger nonlinearities, improved minaturization, and a better understanding of dynamic effects will open up many new possibilities in the field of picosecond and subpicosecond bistability.

3.4 Short Pulse Nonlinearities with Special Boundary Conditions

3.4.1 Nonlinear Effects at Interfaces

The growing interest in surface science has stimulated the demand for new experimental methods to study surfaces and solid-gas or solid-liquid interfaces. Ultrashort nonlinear optical techniques were introduced in this field recently. The value of short pulses in this context is the higher intensity level, improving the detection sensitivity, and of course the superior time resolution.

Second harmonic generation by nonlinear reflection of laser light from surfaces was studied as early as two decades ago, both theoretically and experimentally, by Bloembergen and coworkers (1962, 1966, 1968). Of special interest are centrosymmetric media, where the bulk effect (Sect. 3.2.1) is absent in the electric dipole approximation. The lack of inversion symmetry at the solid interface gives rise to a non-vanishing second order susceptibility $\chi^{(2)}$ which initiates surface SHG. The angle and polarization dependence of SH reflection, and the physical origin of the nonlinearity, were first studied on Si, Ge and Ag samples.

Application of the nonlinear reflectivity to measure the autocorrelation function of the pulse was suggested (Tomov 1974). Because of the absence of group velocity dispersion at the surface, this proposal may gain practical importance for the analysis of fs pulses in the future. Third harmonic generation by nonlinear reflection of picosecond pulses from metal and semiconductor surfaces was studied by Bloembergen et al. (1969). The use of the pulse train of a mode-locked Nd:glass laser provided high fundamental intensities without damaging the surface of the material. The nonlinear susceptibility $\chi^{(3)}$ was determined for silicon, germanium, silver and gold samples; values of $\chi_{1111} = 5.6 \times 10^{-12}$ esu (Ag) to 5.1×10^{-11} esu (Si) were found from an experimental comparison with LiF. The angle and polarization dependences agreed well with theory. Bulk absorption or color dispersion do not hinder surface harmonic generation.

The distinct anisotropy of the $\chi^{(2)}$ tensor at crystal surfaces is an important feature for spectroscopic studies. Exploiting this property, SHG studies allowed the determination of the symmetry properties of surfaces (Guidotti et al. 1983). The 2×1 and 7×7 reconstruction of clean Si(111) surfaces was recently observed with this technique (Heinz et al. 1985). Of particular interest is the study of the solid-liquid phase transition accompanying laser-heating of the surface. Strong evidence for ultrafast melting on the subpicosecond time scale was gained by Shank et al. (1983) for silicon surfaces. The angular and delay-time dependence of the SHG of weak probe pulses was analyzed for different power levels of the 90 fs pump pulses. Related results were obtained by Malvezzi et al. (1984) and Akhmanov et al. (1985) studying GaAs surfaces. Malvezzi et al. applied frequency-doubled Nd:YAG laser pulses of 12.5 ps for a detailed analysis of the SH signals which established an upper limit of 2 ps for the structural transition associated with the melting of the surface. A similar conclusion was reached by

Akhmanov et al. (1985) in their work with 300 ps pulses. For details the reader is referred to Chap. 4 of this volume on solid state spectroscopy.

Soon after the discovery of surface enhanced Raman spectroscopy, an important break-through in optical surface spectroscopy (Fleischmann et al. 1974; Jeanmaire and van Duyne 1977), other related nonlinear phenomena were found. Chen et al. (1981a) observed surface-enhanced SH generation for adsorbed pyridine and AgCl molecules on roughened silver surfaces in an electrolytic solution. The experimental sensitivity allowed the detection of a molecular monolayer on the metal surface. A similar effect was found for the silver–air interface (Chen et al. 1981b). A large enhancement of up to a factor of 10^4 due to surface roughness was measured and attributed to the local field enhancement. The importance of surface-plasmon and surface-polariton excitations was pointed out in several theoretical studies (Agarwal and Jha 1982; Farias and Maradudin 1984; Deck and Grygier 1984). Further experimental work demonstrated the potential of surface SHG for the spectroscopy of adsorbed molecules and Langmuir-Blodgett films even without enhancement by surface roughening (see references in Nguyen et al. 1986). The sensitivity achieved with nanosecond pulses suffices to detect a small fraction of a monolayer.

Information on the orientation of solute molecules at the surface of a liquid solution was recently obtained via the SHG technique by Rasing et al. (1985) and extended to the picosecond time domain by Hicks et al. (1986). In the latter investigation, aqueous phenol solutions were studied. The solute molecules possess a strong nonlinear response, primarily along one axis, so that the measured polarization of the SH emission is directly related to the molecular orientation at the surface. The molecules were found to be oriented with their long axes tilted 50° from the surface normal independent of concentration. Using short pulses, SH signals from molecules at surface coverages as small as 5% could be detected without resonance enhancement of $\chi^{(2)}$.

Surface vibrational spectroscopy plays an important role for in situ identification of adsorbates. Besides the picosecond Raman gain method (Heritage and Allara 1980) a new vibrational technique was demonstrated very recently with ultrashort time resolution: vibrationally-resonant sum-frequency generation (SFG) (Hunt et al. 1987). The basic idea of SFG is to convert the resonant infrared signal, which probes the surface, to the visible, enabling it to be more easily detected. As a three-wave process based on $\chi^{(2)}$, the up-conversion process is as highly surface specific as SHG for centrosymmetric bulk material. Surface SFG of electronic resonances was observed with nanosecond pulses by Akhmanov et al. (1985) and Nguyen et al. (1986) and theoretically discussed by Dick et al. (1985). These investigations represent a direct generalization of the previous SH studies. In the work of Nguyen et al., adsorbed monolayers of Rhodamine 6G on glass were investigated using the electronic transition frequencies $S_0 \rightarrow S_1$ and $S_0 \rightarrow S_2$ of the adsorbant for a double-resonant enhancement of $\chi^{(2)}$.

The first observation of the vibrational spectrum of a molecular monolayer by infrared-visible sum-frequency was reported by Zhu et al. (1987). Adsorbed

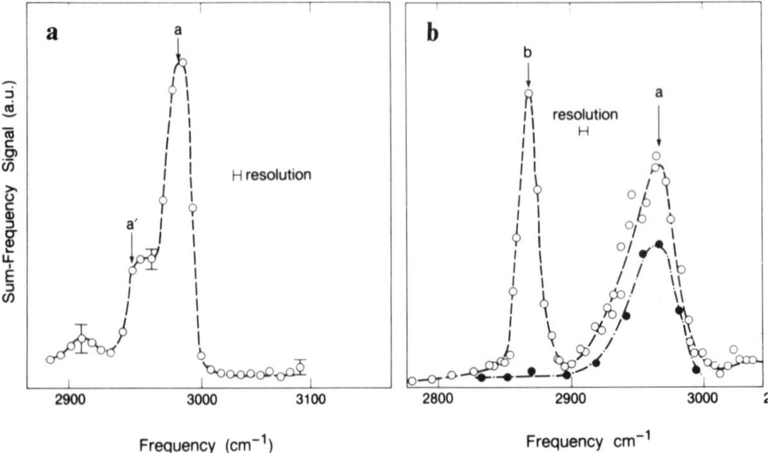

Fig. 3.26a, b. Vibrational sum-frequency spectrum of adsorbed molecules; (**a**) Methanol on glass; the open circles are the experimental points. The dotted line is a guide to the eye. Arrows a and a′ indicate the CH_3 resonances of methanol in the liquid phase. (**b**) Same as in (**a**) but for pentadecanoic acid. The filled-in circles represent the solvent background. Arrows a and b indicate the *d*- and *s*-stretch frequencies of the methyl group in PDA (after Hunt et al. 1987)

coumarin 504 molecules on a substrate of fused quartz were studied with the help of resonant enhancement via the visible transition. The feasibility of the technique for smaller molecules was demonstrated very recently by the same group (Hunt et al. 1987) for monolayers of methanol and pentadecanoic acid adsorbed on glass and water. Tunable ir pulses (15 ps) of a stimulated parametric generator around 3000 cm^{-1} and the second harmonic pulses of a mode-locked Nd:YAG laser were mixed at the air/solid or air/liquid interface. Some spectroscopic data are depicted in Fig. 3.26 for the frequency range of the CH-stretching modes. An electronic resonance of $\chi^{(2)}$ does not occur for the molecules and pulse frequencies used here. The observed SF spectrum is accounted for by vibrational resonances of the second order susceptibility. The authors point out that their picosecond technique could become a most powerful analytical tool for studies of surface dynamics and reactions.

Finally, a few recent studies of surface nonlinearities should be briefly mentioned. A thermal grating which leads to a periodic array of microscopic bubbles has been generated at liquid glass interfaces with 100 ps pulses. Bragg scattering of delayed probing pulses yielded information on the bubble dynamics at the interface (Eyring and Fayer 1983). Several authors have investigated the nonlinear reflection of light at grazing incidence to a liquid/solid interface via nonlinear refractive index changes (Boyko et al. 1975; Kaplan 1976, 1977; Rozanov 1977, 1978; Smith et al. 1979, 1981). As in the case of degenerate four-wave mixing, here too, the possibility of optical bistability was pointed out. In fact, a temporal (nanosecond) hysteresis effect was noted in some of the investigations of the interface glass/CS_2 with nanosecond pulses and taken as evidence for bistability since the material response was believed to be fast, e.g. the fast decay of the orientational Kerr effect in CS_2. New light was shed on these results by careful measurements of Altshuller et al. (1985) of the interface crown-glass K8/benzene alcohol. Using a train of picosecond pulses (15–20 ps) clear evidence was obtained for a slow component of the nonlinear refractive index which dominated the index change via the optical Kerr effect.

3.4.2 Quantum Size Effects in Nonlinear Optics

The search for new materials with higher nonlinear susceptibilities is an expanding field in material science. For instance, some classes of bond-conjugated organic molecules possess an anomalously high second order polarizability. Very large values of the third order nonlinear susceptibility of the order of 10^{-9} esu were also found in THG measurements of polydiacetylenes.

For semiconductors, new highly nonlinear materials with layered or composite miniature structure have been developed; these are known as quantum-well heterostructures and quantum dots respectively. In these artificial structures the bulk properties of the crystal are modified. There is a transition region between normal semiconductor and molecular properties. For bulk material, the diameter of the lowest 1S exciton represents a characteristic length. As the crystallite approaches this size, the electron and hole interactions with the crystallite surface become important since the charges are confined to the semiconductor. The material displays various linear and also nonlinear optical properties that can be tailored by a proper choice of the sample dimensions. The localization of the electrons and holes in one dimension in thin layers (several 10 Å), i.e. in quantum-well structures, enhances the exciton binding energy whereas the spectral broadening due to interaction with thermal phonons is hardly affected. As a result, one finds strong excitonic absorption resonances at room temperature in GaAs/AlGaAs systems. These materials were shown to have great potential in nonlinear devices for optical switching and signal-processing. Some examples of multiple quantum-wells are cited in Sect. 3.3.4d. For further information, see Chap. 4 of this volume.

As an alternative approach, composite materials, e.g. metal inclusions in glasses or semiconductor-doped glasses, have also been shown to display quantum size effects (Brus 1984). Here the electronic motion is confined in microcrystallites in three dimensions. The nonlinear properties of such a composite material were first studied by Jain and Lind (1983) who measured DFWM of nanosecond pulses in commercially available bandpass filters containing CdS_xSe_{1-x} inclusions of dimensions $\simeq 100$ Å. A high nonlinearity of $\chi^{(3)} \simeq 10^{-9} - 10^{-8}$ esu was reported. Subsequent investigations with picosecond pulses gave contradictory results on the response time which were very recently reconciled (Roussignol et al. 1987 and references therein); the decay time of the nonlinearity for intensely illuminated samples is 50–100 ps. Time-resolved absorption spectra on the subpicosecond time scale with 200 fs pulses were measured by Olbright et al. (1986) who observed a blue shift and spectral broadening which were maximum after 400 fs. Evidence for plasma screening, band gap renormalization and band-filling was obtained by a comparison with model calculations. Proposals for further enhancement of the nonlinear properties of semiconductor micro-crystallites have been made (Leung 1986; Chemla and Miller 1986) and await experimental verification.

The unusual linear properties of metal colloids have been known for decades and were explained by surface plasma resonances. These systems consist of small metal spheres in a host dielectric. The nonlinear properties were recently studied by Ricard et al. (1985) using DFWM of 28 ps pulses. These authors showed that the metal-sphere nonlinearity is greatly enhanced near the surface plasma resonance. An impressive value of $\chi^{(3)} = 1.5 \times 10^{-8}$ esu at 0.53 μm was reported for gold spheres ($\simeq$ 100 Å) in water and related to the local field correction in the dielectric. Quantum size effects were treated theoretically by Hache et al. (1986) who also showed experimentally that the nonlinear response is faster than the experimental time resolution of $\simeq$ 5 ps.

3.4.3 Nonlinear Pulse Propagation in Optical Fibers

Propagation effects of guided optical waves have been the subject of numerous investigators in recent years. The motivation for these came both from basic research and from the important potential applications in high speed communication systems. Exciting nonlinear phenomena were discovered, most of which are intimately related to ultrashort laser pulses. Some highlights will be briefly reviewed in this final part of the present chapter.

Even with low power lasers, light propagation in an optical wave-guide rapidly enters the nonlinear optical regime because of long interaction lengths and small mode cross sections. A whole series of nonlinear effects has been studied, e.g. four-wave mixing (Stolen and Ippen 1973; Stolen et al. 1974), continuum generation (Lin et al. 1978), and stimulated Raman scattering (Hill et al. 1976; Lin et al. 1977). Large stimulated Raman gain of $\simeq$ 35 dB was demonstrated for a 100 m long, D_2-doped silica fiber for 120 ps Nd:YAG laser pulses with 130 W peak power. The D_2 concentration in the fiber was 4×10^{20} cm^{-3} (Chraplyvy et al. 1983, 1984). The effect was studied in a stimulated Raman oscillator termed a fiber Raman laser, operating at 1.56 μm. Synchronous pumping with the cw mode-locked YAG laser achieved pulse shortening to 15 ps at 20 W peak power. A single pass of the same laser pulses along a 260 m fiber gave somewhat shorter pulses with 10% conversion efficiency (Chraplyvy and Stone 1985). The stimulated Raman gain of the silica fiber material had to be suppressed in these experiments by exploiting the larger group delay at the SiO_2 Stokes frequency.

With the high peak powers easily attainable with picosecond laser pulses, weak nonlinearities were also observed in optical films, e.g. second harmonic generation and (three-wave) sum-frequency generation (Fuji et al. 1980; Ohmori and Sasaki 1982; Nakazawa et al. 1984; Österberg and Margulis 1987). Phase-matching is achieved by compensating for the bulk dispersion by the waveguide dispersion. The second order susceptibility of the silica fiber possibly originates from a stress induced anisotropy. Following along similar lines, stimulated four-wave parametric amplification has attracted some attention (Stolen and Bjorkholm 1982). Phase-matching of this process is possible in multimode fibers when Stokes and anti-Stokes wave propagate in different modes or in birefringent fibers using the different phase velocities of orthogonal fundamental modes. For phase-matching in single-mode fibers, specific dispersion properties are required (Garth and Pask 1986).

Of central importance for optical communication systems is the transmission of optical signals over long distances. Impressive progress has been achieved in lowering the linear propagation losses to a level significantly below 1 dB/km. Apart from simple absorption, the propagation of short, intense pulses in a medium is governed by at least two factors: color dispersion and the nonlinear refractive index. Including both effects in the nonlinear wave equation yields (with plane waves and slowly varying amplitudes) the expression

$$\frac{\partial E}{\partial z} + \frac{1}{v}\frac{E}{t} = \frac{\mathrm{i}}{2}\frac{\partial^2 k}{\partial \omega^2}\frac{\partial^2 E}{\partial t^2} - \frac{\mathrm{i}\pi}{\lambda n} n_2 |E|^2 E \ . \tag{3.24}$$

The first term on the right hand side of (3.24) leads to the well-known diffusion-like spread of a wave packet due to dispersion while the second term accounts for the nonlinear phase shift in the medium. A simple transformation to dimensionless units yields the nonlinear Schrödinger equation. Zakharov and Shabat (1972) showed that stable, solitary solutions of (3.24) exist for $\partial^2 k/\partial\omega^2 > 0$ in terms of wave-packet solitons. The relevance of these results for picosecond pulse propagation in optical fibers was recognized by Hasegawa and Tappert (1973). For pulses of sech^2-shape representing the fundamental $N = 1$ soliton they predicted stable pulse propagation in a fiber with anomalous dispersion with neither deterioration by group delay nor phase modulation; the two effects cancel on the rhs. of (3.23) for this situation.

First experimental evidence of soliton propagation in optical fibers was reported by Mollenauer et al. (1980). Narrowing and splitting of 7 ps pulses from a mode-locked color-center laser in a 700 m long, single-mode silica-glass fiber was observed. This fiber displays a large negative group velocity dispersion (i.e. $\partial^2 k/\partial\omega^2 > 0$) at the experimental wave-length of 1.55 μm. The observed behavior at certain critical power levels was characteristic of solitons. Some results are displayed in Fig. 3.27. For the soliton experiment it is essential to work with low-loss fibers (0.2 dB/km) of high-quality and with narrow-band picosecond pulses (close to the Fourier-transform limit) in the relevant spectral region of anomalous group velocity dispersion.

Extreme narrowing of picosecond pulses in single-mode fibers by means of the soliton effect was subsequently observed by Mollenauer et al. (1983) and

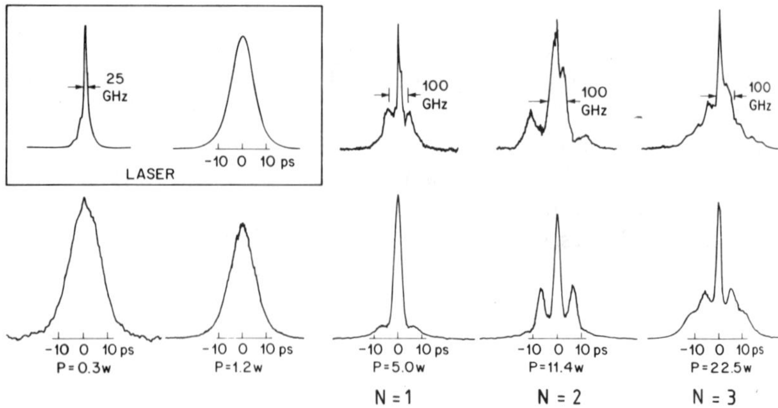

Fig. 3.27. Soliton propagation in a silica-glass fiber at 1.55 μm. Lower traces: autocorrelation of the fiber output as a function of power. Upper traces: corresponding frequency spectra. Inset: Similar data for the direct laser output. The various curves are roughly normalized to a common height. Going from low to high power, the laser pulse widths were 7.2, 7.0, 6.1, 6.8, and 6.2 ps, respectively. Self-compression of the propagating pulse leads to stable pulse shapes at characteristic peak power levels representing the fundamental ($N = 1$) and higher order solitons (after Mollenauer et al. 1980)

shown to be in semiquantitative agreement with predictions from the nonlinear Schrödinger equation. Under conditions of anomalous group dispersion the fiber acts itself as a distributed compressor (and modulator) different to the normal compression scheme discussed in Chap. 2 of this volume. A pulse shortening from an initial 7 ps to 260 fs was reported. Similar results were also obtained by Tai and Tomita (1986). The soliton-like compression scheme was also incorporated in a synchronously-coupled mode-locked color-center laser (Mollenauer and Stolen 1984) yielding fs pulses.

The validity of the nonlinear Schrödinger equation (3.24) representing the lowest-order approximation for nonlinear pulse propagation in fibers, was successfully tested by experiments of Nakatsuka et al. (1981) and of Tomlinson et al. (1985) in the normal dispersion region. In a series of theoretical investigations a generalization of this equation was considered including higher order terms for the nonlinear refractive index and/or color dispersion (Kumar et al. 1986; Wai et al. 1986; Bourkoft et al. 1987). The underlying slowly varying envelope approximation of (3.24) was challenged by Christodoulides and Joseph (1985). For the fundamental ($N = 1$) soliton and for pulse durations $\geqslant 10$ fs, no significant changes were found. Blow et al. (1987) discussed polarization instabilities for solitons in birefringent fibers. Fiber pulse propagation including linear losses was also treated by several investigators (Blow and Doran 1985; Anderson and Lisak 1985; Tajima 1987).

For possible applications in optical communication systems, the compensation of linear loss and the amplification and reshaping of optical solitons in fibers is important. Several schemes have been proposed to solve this problem

(Hasegawa and Kodama 1982, 1984; Menyuk et al. 1985). The stimulated Raman interaction appears to be particularly attractive for this purpose. The effect was demonstrated by Mollenauer et al. (1985) for the propagation of 10 ps pulses at 1.56 μm in a 10 km long single-mode fiber. A counterpropagating cw Raman pump beam at an appropriately chosen shorter wavelength $\lambda_p = 1.46\,\mu m$ provided distributed gain along the fiber. The wavelength difference corresponds to the broad Raman band of SiO_2 (shift $\simeq 440\,cm^{-1}$). The fiber loss of 0.18 dB/km was compensated by a pump intensity of $\simeq 125$ mW. Physical limitations set by spontaneous Raman scattering were pointed out by Gordon and Haus (1986).

Very recently a self-frequency shift of subpicosecond solitons was observed (Mitschke and Mollenauer 1986). According to Gordon (1986), in the Raman effect energy transferred from the higher to the lower frequencies is the main cause of the frequency shift. Since the frequency shift increases with t_p^{-4}, a subpicosecond soliton-based telecommunication system does not appear to be feasible at the present time. As a further application of the soliton concept, a Raman fiber soliton laser was proposed by Dianov et al. (1986). In a series of investigations the generation and modulation of soliton trains was investigated; for details see Tai et al. (1986), Shukla and Rasmussen (1986) and references cited there.

3.5 Conclusions

The emergence of ultrashort nonlinear phenomena has added a new dimension to nonlinear optics. A deeper insight into the light-matter interaction is gained by the higher intensities applicable on the shorter time scales. The present chapter does not attempt a comprehensive review of this field. A few topics were deliberately omitted here because they are already well established in nonlinear optics or because they are treated in other chapters of this volume in the context of spectroscopic applications or mode-locking techniques. The discussion here concentrated on important recent developments intimately connected with ultrashort pulses. Background information on new achievements in related areas, e.g. longer time scales are only briefly mentioned.

The investigations outlined above are mainly concerned with the time scale 10^{-12} s. Relatively little work has been performed as far with pulses of, or shorter than, 10^{-13} s. This fact is not surprising since the techniques to generate and amplify picosecond pulses have themselves only been developed within the last few years. The handling of sufficiently powerful femtosecond pulses is still restricted to a limited number of laboratories. Here a dramatic change is to be expected in the near future. Moderately sized laser systems generating pulses of a few 10^{-14} s and peak powers in the hundred gigawatt range will become more readily available. With these light sources, higher order nonlinearities (and the competition between such effects) will rapidly gain in importance and nonlinear optics may arrive at the stage where the perturbative approach, i.e. expansion of the nonlinear material response in power series will have to be replaced by more refined theoretical tools. Unforeseen discoveries may also be expected in the continuing search for highly nonlinear materials. The manipulation of the light-matter interaction using special geometrical conditions for the electromagnetic or the quantum mechanical wavefunctions could lead to completely new results. In short, the field will become even more active and exciting than in recent years.

References

Adrain, R.S., Arthurs, E.G., Sibbett, W. (1975): Opt. Commun. **15**, 290

Agarwal, G.S., Jha, S.S. (1982): Phys. Rev. B **26**, 482

Agrawal, G.P. (1981): Phys. Rev. A **23**, 1863

Akhmanov, S.A., Khokhlov, R.V., Kovrigin, A.I., Piskarskas, V.I., Sukhorukov, A.P. (1968): IEEE J. QE-**4**, 829

Akhmanov, S.A., Chirkin, A.S., Drabovich, K.N., Kovrigin, A.I., Khokhlov, R.V., Sukhorukov, A.P. (1968): IEEE J. QE-**4**, 598

Akhmanov, S.A., Kovrigin, A.I., Sukharoukov, A.P., Khokhlov, R.V., Chirkin, A.A. (1968): JETP Lett. **7**, 182

Akhmanov, S.A., Sukhorukov, A.P., Chirkin, A.S. (1969): Sov. Phys. JETP **28**, 748

Akhmanov, S.A. (1969): Mat. Res. Bull. **4**, 455

Akhmanov, S.A., Bol'shov, M.A., Drabovich, K.N., Sukhorukov, A.P. (1970): JETP Lett. **12**, 388

Akhmanov, S.A., Drabovich, K.N., Sukhorukov, A.P., Chirkin, A.S. (1971): Sov. Phys. JETP **32**, 266

Akhmanov, S.A., Drabovich, K.N., Sukhorukov, A.P., Shchednova, A.K. (1972): Sov. Phys. JETP **35**, 279

Akhmanov, S.A., Martynov, V.A., Saltiel, S.M., Tunin, V.G. (1975): JETP Lett. **22**, 65

Akhmanov, S.A., Koroteev, N.J., Paitian, G.A., Shumay, I.L. (1985): J. Opt. Soc. Am. B **2**, 284

Albrecht, G., Antonetti, A., Mourou, G. (1981): Opt. Commun. **40**, 59

Albrecht, G.F. (1982): Opt. Commun. **41**, 287

Alexiewicz, W., Buchert, J., Kielich, S. (1978): In *Picosecond Phenomena*, Springer Ser. Chem. Phys., Vol. 4, ed. by C.V. Shank, E.P. Ippen, S.L. Shapiro (Springer, Berlin, Heidelberg)

Alfano, R.R., Shapiro, S.L. (1970): Phys. Rev. A **2**, 2376; (1970): Phys. Rev. Lett. **24**, 584

Alfano, R.R., Shapiro, S.L. (1971): Phys. Rev. Lett. **26**, 1247

Alfano, R.R., Li, Q.X., Jimbo, T., Manassah, J.T., Ho, P.P. (1986): Opt. Lett. **11**, 626

Altshuller, G.B., Ermolaev, V.S., Krylov, K.I., Makarov, M.A. (1985): Opt. Commun. **56**, 131

Anderson, D.B., Boyd, J.T. (1971): Appl. Phys. Lett. **19**, 266

Anderson, D., Lisak, M. (1985): Opt. Lett. **10**, 390

Apanasevich, P.A., Zaporozhchenko, R.G., Zaporozhchenko, V.A., Zakharova, I.S., Kachinskii, A.V. (1982): Sov. J. Quant. Electron. **12**, 861

Apanasevich, P.A., Zaporozhchenko, V.A., Zaporozhchenko, R.G., Kachinskii, A.V. (1984): Sov. J. Quant. Electron. **14**, 609

Arakelyan, S.A., Gyuzalyan, R.N., Sogomonyan, S.B. (1981): Sov. J. Quant. Electron. **11**, 949

Arakelyan, S.A., Gyuzalyan, R.N., Sogomonyan, S.B. (1982): Opt. Commun. **44**, 67

Arecchi, F.T., Schulz-Dubois, E.O. (eds.) (1972): *Laser Handbook* (North-Holland, Amsterdam)

Armstrong, J.A., Bloembergen, N., Ducuing, J., Pershan, P.S. (1962): Phys. Rev. **127**, 1918

Armstrong, J.A. (1967): Appl. Phys. Lett. **10**, 16

Asaka, S., Nakatsuka, H., Fujiwara, M., Matsuoka, M. (1984): Phys. Rev. A **29**, 2286

Asaka, S., Nakatsuka, H., Fujiwara, M., Matsuoka, M. (1984): Opt. Commun. **52**, 150

Atanesyan, V.T., Grigoryan, V.S., Karmenyan, K.V. (1976): Sov. J. Quant. Electron. **6**, 1161
Attwood, D.T., Pierce, E.L., Coleman, L.W. (1975): Opt. Commun. **15**, 10
Auston, D.H. (1971): App. Phys. Lett. **18**, 249
Auston, D. (1971): Opt. Commun. **3**, 272
Babin, A.A., Belyaev, Yu. N., Verevkin, Yu. K., Freidman, G.I. (1979): Sov. J. Quant. Electron. **9**, 728
Bareika, B., Dikchyus, G., Piskarskas, A., Sirutkaitis, V. (1980): Sov. J. Quant. Electron. **10**, 1277
Bareika, B., Dikchyus, G., Piskarskas, A., Yasevichyute, Ya. (1983): Sov. J. Quant. Electron. **13**, 1507
Bareika, B., Birmontas, A., Dikchyus, G., Piskarskas, A., Sirutkaitis, V., Stabinis, A. (1983): Sov. J. Quant. Electron. **12**, 1654
Beach, R., Hartmann, S.R. (1984): Phys. Rev. Lett. **53**, 663
Beach, R., DeBeer, D., Hartmann, S.R. (1985): Phys. Rev. A **32**, 3467
Becker, M.F., Kuizenga, D.J., Phillion, D.W., Siegman, A.E. (1974): J. Appl. Phys. **45**, 3996
Beddard, G.S., Doust, T., Windsor, M.W. (1980): In *Picosecond Phenomena II*, Springer Ser. Chem. Phys., Vol. 14, ed. by R.M. Hochstrasser, W. Kaiser, C.V. Shank (Springer, Berlin, Heidelberg)
Beddard, G.S., Doust, T., Porter, G. (1981): Chem. Phys. **61**, 17
Benza, V., Lugiato, L.A., Meystre, P. (1980): Opt. Commun. **33**, 113
Berg, M., Harris, A.L., Brown, J.K., Harris, C.B. (1984): Opt. Lett. **9**, 50
Berg, A., Harris, C.B., Kenny, J.W., Richards, P.L. (1985): Appl. Phys. Lett. **47**, 206 (1985)
Betin, A.A., Pasmanik, G.A., Piskunova, L.V. (1975): Sov. J. Quant. Electron. **5**, 1309
Bey, P.P., Giuliani, J.F., Rabin, H. (1968): Phys. Rev. Lett. **26A**, 128
Bischofsberger, T., Shen, Y.R. (1978): Appl. Phys. Lett. **32**, 156
Bischofsberger, T. (1979): Phys. Rev. A **19**, 1169
Bloembergen, N., Pershan, P.S. (1962): Phys. Rev. **128**, 606
Bloembergen, N. (1966): Opt. Acta **13**, 311
Bloembergen, N. (1967): Am. J. Phys. **35**, 989
Bloembergen, N., Chang, R.K., Jha, S.S., Lee, C.H. (1968): Phys. Rev. **174**, 813
Bloembergen, N. Burns, W.K., Matsuoka, M. (1969): Opt. Commun. **1**, 195
Bloom, D.M., Bekkers, G.W., Young, J.F., Harris, S.E. (1975): Appl. Phys. Lett. **26**, 687
Bloom, D.M., Bjorklund, G.E. (1977): Appl. Phys. Lett. **31**, 592
Bloom, D.M., Shank, C.V., Fork, R.L., Teschke, O. (1978): In *Picosecond Phenomena*, Springer Ser. Chem. Phys., Vol. 4, ed. by C.V. Shank, E.P. Ippen, S.L. Shapiro (Springer, Berlin, Heidelberg)
Blow, K.J., Doran, N.J. (1985): Opt. Commun. **52**, 367
Blow, K.J., Doran, N.J., Wood, D. (1987): Opt. Lett. **12**, 202
Boichenko, V.L., Zasavitskii, I.I., Kosichkin, Yu. V., Tarasevich, A.P., Tunkin, V.G., Shotov, A.P. (1984): Sov. J. Quant. Electron. **14**, 141
Bonafacio, R., Lugiato, L.A. (1976): Opt. Commun. **19**, 172
Bonafacio, R., Lugiato, L.A. (1979): Nuovo Cimento B **53**, 311
Bonifacio, R., Meystre, P. (1978): Opt. Commun. **27**, 147
Bonifacio, R., Gronchi, M., Lugiato, L.A. (1979): Opt. Commun. **30**, 129
Bourkoff, E., Zhao, W., Joseph, R.J. (1987): Opt. Lett. **12**, 272
Boyd, G.D., Kleinman, D.A. (1968): J. Appl. Phys. **39**, 3597
Boyd, R.W., Raymer, M.G., Narum, P., Harter, D.J. (1981): Phys. Rev. A **24**, 411
Boyko, B.B., Jillavdary, I., Petrov, N.S. (1975): Sov. Phys. J. Appl. Spectr. **23**, 888
Bradley, D.J. (1977): In *Ultrashort Light Pulses*, Topics Appl. Phys., Vol. 18, ed. by S.L. Shapiro (Springer, Berlin, Heidelberg)
Brekhovskikh, G.L., Sokolovskaya, A.I., Ferrier, C., Wu, Z., Rivoire, G., Okladnikov, N.V., Kudryavtseva, A.D. (1983): Bull. Acad. Sci. USSR; Phys. Ser. (USA) **47**, 165
Bret, C.G., Weber, H.P. (1968): IEEE J. QE-**4**, 807
Brus, L.E. (1984): J. Chem. Phys. **80**, 4404
Buchert, J., Dorsinville, R., Delfyett, P., Krimchansky, S., Alfano, R.R. (1985): Opt. Commun. **52**, 433
Burneika, K.P., Ignatavichus, M.V., Kabelka, V.I., Piskarskas, A.S., Stabnis, Yu. A. (1972): JETP Lett. **16**, 257

Byer, R.L. (1975): In Quantum Electronics, Vol. I, Part B, Chap. 9, ed. by R. Rabin, C.L. Tang (Academic, New York)

Campillo, A.J., Hyer, R.C., Shapiro, S.L. (1979a): Opt. Lett. **4**, 325

Campillo, A.J., Hyer, R.C., Shapiro, S.L. (1979b): Opt. Lett. **4**, 357

Carman, R.L., Mack, M.E., Shimizu, F., Bloembergen, N. (1969): Phys. Rev. Lett. **23**, 1327

Carman, R.L., Shimizu, F., Wang, C.S., Bloembergen, N. (1970): Phys. Rev. A **2**, 60

Carman, R.L., Mack, M.E. (1972): Phys. Rev. A. **5**, 341

Carman, R.L., Lowdermilk, W.H. (1974): Phys. Rev. Lett. **33**, 190

Carter, G.M., Hryniewicz, J.V., Thakur, M.K., Chen, Y.J., Meyler, S.E. (1986): Appl. Phys. Lett. **49**, 998

Chatelet, M., Oksengorn, B. (1975): Chem. Phys. Lett. **36**, 73

Chemla, D.S., Miller, D.A.B. (1986): Opt. Lett. **11**, 522

Chen, C.K., Heinz, T.F., Ricard, R., Shen, Y.R. (1981a): Chem. Phys. Lett. **83**, 455

Chen, C.K., de Castro, A.R.B., Shen, Y.R. (1981b): Phys. Rev. Lett. **46**, 145

Chen, C.K., Heinz, T.F., Ricard, D., Shen, Y.R. (1983): Phys. Rev. B **27**, 1965

Chiao, R.Y., Kelley, P.L., Garmire, E. (1966): Phys. Rev. Lett. **17**, 1158

Choi, K.J., Hallidy, L.A., Topp, M.R. (1980): In *Picosecond Phenomena II*, Springer Ser. Chem. Phys. Vol. 14, ed. by R.M. Hochstrasser, W. Kaiser, C.V. Shank (Springer, Berlin, Heidelberg)

Chraplyvy, A.R., Stone, J., Burns, C.A. (1983): Opt. Lett. **8**, 415

Chraplyvy, A.R., Stone, J. (1984): Opt. Lett. **9**, 241

Chraplyvy, A.R., Stone, J. (1985): Opt. Lett. **10**, 344

Christodoulides, D.N., Joseph, R.J. (1985): Appl. Phys. Lett. **47**, 76

Colles, M.J. (1969): Opt. Commun. **1**, 169

Colles, M.J. (1971): Appl. Phys. Lett. **19**, 23

Comly, J., Garmire, E. (1968): Appl. Phys. Lett. **12**, 7

Corkum, P.B., Ho, P.P., Alfano, R.R., Manassah, J.T. (1985): Opt. Lett. **10**, 624

Corkum, P.B., Rolland, C., Srinivasan-Rao, T. (1986): In *Ultrafast Phenomena V*, Springer Ser. Chem. Phys., Vol. 46, ed. by G.R. Fleming, A.E. Siegman (Springer, Berlin, Heidelberg)

Cotter, D., White, K.I. (1984): Opt. Commun. **49**, 205

Craxton, R.S. (1980): Opt. Commun. **34**, 474

Cronin-Golomb, M., Paslaski, J., Yariv, A. (1985): Appl. Phys. Lett. **47**, 1131

Dagenais, M., Sharfin, W.F. (1985): Appl. Phys. Lett. **46**, 230

Danelyus, R., Dikchyus, G., Kabelka, V., Piskarskas, A., Stabinis, A., Yasevichyute, Ya. (1977): Sov. J. Quant. Electron. **7**, 1360

Danelyus, R., Kabelka, V., Piskarskas, A., Smil'gyavichyus, V. (1978): Sov. J. Quant. Electron. **8**, 398

Darée, K., Kaiser, W. (1974): Opt. Commun. **10**, 63

Darée, K. (1975): Opt. and Quant. Electron. **7**, 263

Deck, R.T., Grygier, R.K. (1984): Appl. Opt. **23**, 3203

DeMaria, A.J., Stetser, D.A., Heynau, H. (1966): Appl. Phys. Lett. **8**, 174

DeMartini, F., Ducuing, J. (1966): Phys. Rev. Lett. **17**, 117

Dennis, W.M., Blau, W., Bradley, D.J. (1985): Appl. Phys. Lett. **47**, 200

Dennis, W.M., Blau, W. (1986): Opt. Commun. **57**, 371

DeSilvestri, S., Fujimoto, J.G., Ippen, E.P., Gamble Jr., E.B., Williams, L.R., Nelson, K.A. (1985): Chem. Phys. Lett. **116**, 146

Dianov, E.M., Prokhorov, A.M., Serkin, V.N. (1986): Opt. Lett. **11**, 168

Dick, B., Gierulski, A., Marowsky, G. (1985): Appl. Phys. **38**, 107

Diels, J.-C.M., Fontaine, J.J., McMichael, I.C., Simone, F. (1985): Appl. Opt. **24**, 1270

Dlott, D.D., Fayer, M.D., Salcedo, J., Siegman A.E. (1978): In *Picosecond Phenomena*, Springer Ser. Chem. Phys., Vol. 4, ed. by C.V. Shank, E.P. Ippen, S.L. Shapiro (Springer, Berlin, Heidelberg); (1978): Phys. Rev. Lett. **41**, 131

Drabovich, K.N., Metchkov, D.I., Mitev, V.M., Pavlov, L.I., Stamenov, K.V. (1977): Opt. Commun. **20**, 350

Duguay, M.A., Hansen, J.W. (1969): Appl. Phys. Lett. **15**, 19

Duncan, M.D., Mahon, R., Reintjes, J., Tankersky, L.L. (1986): Opt. Lett. **11**, 803

Dutta, N.K. (1980): J. Appl. Phys. **51**, 84

Dychyus, G.A., Kabelka, V.I., Piskarskas, A.S., Stabnis, Yu. A. (1975): Sov. J. Quant. Electron. **4**, 1402

Eckardt, R.C., Lee, C.H. (1969): Appl. Phys. Lett. **15**, 425

Eichler, H.J., Langhans, D., Massmann, F. (1984): Opt. Commun. **50**, 117

Elgin, J.N., New, G.H.C., Smith, P.R.C. (1980): J. Phys. B **13**, 1663

Elsaesser, T., Seilmeier, A., Kaiser, W. (1983): Opt. Commun. **44**, 293

Elsaesser, T., Seilmeier, A., Kaiser, W. (1984): Appl. Phys. Lett. **44**, 383

Elsaesser, T., Lobentanzer, H., Seilmeier, A. (1985): Opt. Commun. **52**, 355

Etchepare, J., Grillon, G., Antonetti, A., Orszag, A. (1980): In *Picosecond Phenomena II*, Springer Ser. Chem. Phys., Vol. 14, ed. by R.M. Hochstrasser, W. Kaiser, C.V. Shank (Springer, Berlin, Heidelberg)

Etchepare, J., Grillon, G., Astier, R., Martin, J.L., Bruneau, C., Antonetti, A. (1982): In *Picosecond Phenomena III*, Springer Ser. Chem. Phys., Vol. 23, ed. by K.B. Eisenthal, R.M. Hochstrasser, W. Kaiser, A. Laubereau (Springer, Berlin, Heidelberg)

Etchepare, J., Grillon, G., Thomazeau, I., Hamoniaux, G., Orszag, A. (1986): In *Ultrafast Phenomena V*, Springer Ser. Chem. Phys., Vol. 46, ed. by G.R. Fleming, A.E. Siegman (Springer, Berlin, Heidelberg)

Eyring, G., Fayer, M.D. (1983): Chem. Phys. Lett. **98**, 428

Fabricius, N., Nattermann, K., von der Linde, D. (1984): Phys. Rev. Lett. **52**, 113

Farias, G.A., Maradudin, A.A. (1984): Phys. Rev. **30**, 3002

Fayer, M.D. (1986): IEEE J. of Quant. Electron. QE-**22**, 1437

Felber, F.S., Marburger, J.H. (1976): Appl. Phys. Lett. **28**, 731

Fendt, A., Kranitzky, W., Laubereau, A., Kaiser, W. (1979): Opt. Commun. **28**, 142

Ferguson, A.T., Arthurs, E.G. (1976): Phys. Lett. **58A**, 298

Ferrier, J.L., Wu, Z., Gazengel, J., Phu Xuan, N., Rivoire, G. (1982): Opt. Commun. **41**, 135

Ferrier, J.L., Gazengel, J., Nguyen Phu, X., Rivoire, G. (1984): Opt. Commun. **51**, 285

Ferrier, J.L., Gazengel, J., Nguyen Phu, X., Rivoire, G. (1986): Opt. Commun. **58**, 343

Fisher, R.A. (ed.) (1983): *Optical Phase Conjugation* (Academic, New York)

Fleischmann, M., Hendra, P.J., McQuillan, A.J. (1974): Chem. Phys. Lett. **26**, 163

Flytzanis, C. (1975): in *Quantum Electronics*, Vol. I, Part A, Chap. 2, ed. by H. Rabin, C.L. Tang, (Academic, New York)

Flytzanis, C., Tang, C.L. (1980): Phys. Rev. Lett. **45**, 443

Fork, R.L., Shank, C.V., Hirliman, C., Yen, R.T. (1983): Opt. Lett. **8**, 1

Franken, P., Hill, A., Peters, C., Weinreich, G. (1961): Phys. Rev. Lett. **7**, 118

Freund, I. (1968): Phys. Rev. Lett. **21**, 1404

Fuji, Y., Kawasaki, B.S., Hill, K.O., Johnson, D.C. (1980): Opt. Lett. **5**, 48

Gandelman, G.M., Iskovich, Yu. O., Kondratenko, P.S., Sobolev, S.S., Stepanov, B.M., Chalkin, S.F. (1984): Sov. J. Quant. Electron. **14**, 911

Garmire, E., Marburger, J.H., Allen, S.D. (1978): Appl. Phys. Lett. **32**, 320

Garth, S.J., Pask, C. (1986): Opt. Lett. **11**, 380

Gibbs, H.M., McCall, L.L., Venkatesan, T.N.C. (1976): Phys. Rev. Lett. **36**, 1135

Giordmaine, J.A. (1962): Phys. Rev. Lett. **8**, 19

Giordmaine, J.A., Miller, R.C. (1965): Phys. Rev. Lett. **14**, 973

Giordmaine, J.A., Kaiser, W. (1966): Phys. Rev. **144**, 676

Giordmaine, J.A., Rentzepis, P.M., Shapiro, S.L., Wecht, K.W. (1967): Appl. Phys. Lett. **11**, 216

Glass, A.M., Auston, D.H. (1972): Opt. Commun. **5**, 45

Glenn, W.H. (1967): Appl. Phys. Lett. **11**, 333

Glenn, W.H. (1969): IEEE J. QE-**5**, 284

Glownia, J.H., Arjavalingam, G., Sorokin, P.P., Rothenberg, J.E. (1986): Opt. Lett. **11**, 79 (1986)

Glownia, J.H., Misewich, J., Sorokin, P.P. (1987): Opt. Lett. **12**, 19

Göbel E.O., Saito, H. (1986): In *Ultrafast Phenomena V*, Springer Ser. Chem. Phys. Vol. 46, ed. by G.R. Fleming, A.E. Siegman (Springer, Berlin, Heidelberg)
Goldstone, J.A., Ho, P.T., Garmire, E. (1980): Appl. Phys. Lett. **37**, 126
Goldstone, J.A., Garmire, E.M. (1981): IEEE J. Quantum Electr. QE-**17**, 366
Goldstone, J.A., Garmire, E. (1984): Phys. Rev. Lett. **53**, 911
Goldstone, J.A. (1985): In *Fourth Laser Handbook*, ed. by M. Bass, M.L. Stitch (North-Holland, Amsterdam)
Golub, J.E., Mossberg, T.W. (1986): Opt. Soc. of Am. **11**, 431
Gordon, J.P., Leite, R.C.C., Moore, R.S., Porto, G.P.S., Whinnery, J.R. (1965): J. Appl. Phys. **36**, 3
Gordon, J.P. (1986): Opt. Lett. **11**, 662
Gordon, J.P., Haus, H.A. (1986): Opt. Lett. **11**, 665
Graener, H., Laubereau, A. (1982): Appl. Phys. B **29**, 213
Greene, B.I., Farrow, R.C. (1982): J. Chem. Phys. **70**, 4779
Greene, B.I., Farrow, R.C. (1983): Chem. Phys. Lett. **98**, 273
Grischkowsky, D. (1978): J. Opt. Soc. Am. **68**, 641
Gronchi, M., Benza, V., Lugiato, L.A., Meystre, P., Sargent III, M. (1981): Phys. Rev. A **24**, 1419
Guha, S., Falk, J. (1980): J. Appl. Phys. **51**, 50
Guidotti, D., Driscoll, T.A., Gerritsen, H.J. (1983): Solid State Commun. **46**, 337
Gulamov, A.A., Ibrahim, E.A., Redkorechev, V.I., Usmanov, T. (1983): Sov. J. Quant. Electron. **13**, 844
Gyuzalyan, R.N., Sarkisyan, D.G., Ter-Mikaelyan, M.L. (1977): Sov. J. Quant. Electron. **7**, 645
Gyuzalyan, R.N., Sogomonian, S.B., Horvath, Z.Gy. (1979): Opt. Commun. **29**, 239
Hache, F., Ricard, D., Flytzanis, C. (1986): J. Opt. Soc. Am B **3**, 1648
Hagen, W.F., Magnante, P.C. (1969): J. Appl. Phys. **40**, 219
Halbout, J.M., Tang, C.L. (1982): Appl. Phys. Lett. **40**, 765
Hallidy, L.A., Topp, M.R. (1977): Chem. Phys. Lett. **46**, 8
Harris, A.L., Berg, M., Brown, J.K., Harris, C.B. (1984): In *Ultrafast Phenomena IV*, Springer Ser. Chem. Phys., Vol. 38, ed. by D.H. Auston, K.B. Eisenthal (Springer, Berlin, Heidelberg)
Harris, S.E., Miles, R.B. (1971): Appl. Phys. Lett. **19**, 385
Hartmann, H.-J., Laubereau, A. (1981): Appl. Opt. **20**, 4259
Hartmann, H.-J., Laubereau, A. (1983): Opt. Commun. **47**, 117
Hartmann, H.-J., Laubereau, A. (1984): J. Chem. Phys. **80**, 4663
Hartmann, H.-J., Bratengeier, K., Laubereau, A. (1984): Chem. Phys. Lett. **108**, 555
Hartmann, H.-J., Schleicher, H., Laubereau, A. (1985): Chem. Phys. Lett. **116**, 392
Hasegawa, A., Tappert, F. (1973): Appl. Phys. Lett. **23**, 142
Hasegawa, A., Kodama, Y. (1982): Opt. Lett. **7**, 285
Hasegawa, A. (1984): Appl. Opt. **23**, 3302
Heinz, T.F., Palfrey, S.L., Eisenthal, K.B. (1984): Opt. Lett. **9**, 359
Heinz, T.F., Loy, M.M.T., Thompson, W.A. (1985): Phys. Rev. Lett. **54**, 63
Hellwarth, R.W. (1970): J. Chem. Phys. **52**, 2128
Hellwarth, R.W. (1977): J. Opt. Soc. Am. **67**, 1; (1977): Prog. Quant. Electr. **5**, 1
Heritage, J.P. (1979): Appl. Phys. Lett. **34**, 470
Heritage, J.P., Allara, D.L. (1980): Chem. Phys. Lett. **74**, 507
Herman, J.P. (1974): Opt. Commun. **12**, 102
Herrmann, J. (1975): Sov. J. Quant. Electron. **5**, 207
Hicks, J.M., Kemnitz, K., Eisenthal, K.B., Heinz, T.F. (1986): J. Phys. Chem. **90**, 561
Hill, K.O., Kawasaki, B.S., Johnson, D.C. (1976): Appl. Phys. Lett. **29**, 181
Hopf, F.A., Meystre, P., Drummond, P.D., Walls, D.F. (1979): Opt. Commun. **31**, 245
Hoshimiya, T., Inaba, H., Tan-no, N. (1977): Opt. and Quant. Electr. **9**, 448
Hu, P., Geschwind, S., Jedju, T.M. (1976): Phys. Rev. Lett. **37**, 1357
Hulin, D., Migus, A., Antonetti, A., Ledoux, I., Badan, J., Oudar, J.L., Zyss, J. (1986a): Appl. Phys. Lett. **49**, 761
Hulin, D., Mysyrowicz, A., Antonetti, A., Migus, A., Masselink, W.T., Morkoç, H., Gibbs, H.M., Peyghambarian, N. (1986b): Appl. Phys. Lett. **49**, 749

Hunt, J.H., Guyot-Sionnest, P., Shen, Y.R. (1987): Chem. Phys. Lett. **133**, 189
Hutchinson, M.H.R., Manning, R.J. (1985): Opt. Commun. **55**, 55
Ippen, E.P., Shank, C.V. (1975): Appl. Phys. Lett. **26**, 92
Ippen, E.P., Shank, C.V. (1976): Opt. Commun. **18**, 26
Ippen, E.P., Shank, C.V. (1977): In *Ultrashort Light Pulses*, Topics Appl. Phys., Vol. 18, ed. by S.L. Shapiro (Springer, Berlin, Heidelberg)
Ivanova, Z.I., Kabelka, V., Magnitskii, S.A., Piskarskas, A., Smil'gyavichus, V., Rubinia, N.M., Tumkin, V.G. (1977): Sov. J. Quant. Electron. **7**, 1414
Jacobovitz, G.R., Brito Cruz, C.H., Mansur, N.P., Scarparo, M.A. (1986): Opt. Commun. **59**, 233
Jain, R.K., Lind, R.C. (1983): J. Opt. Soc. Am. **73**, 648
Janszky, J., Corradi, G., Gyuzalian, R.N. (1977): Opt. Commun. **23**, 293
Janszky, J., Corradi, G. (1986): Opt. Commun. **60**, 251
Jeanmaire, D.C., Van Duyne, R.P. (1977): J. Electroanal. Chem. **84**, 1
von Jena, A., Lessing, H.E. (1979): Opt. and Quant. Electron. **11**, 419
Jensen, S.L., Hellwarth, R.W. (1978): Appl. Phys. Lett. **32**, 166
Johnson, L.M., Pratt, Jr., G.W. (1978): Appl. Phys. Lett. **33**, 1002
Johnson, A.M., Simpson, W.M. (1983): Opt. Lett. **8**, 554
Johnson, A.M., Simpson, W.M. (1986): IEEE J. Quant. Electr. QE-**22**, 133
Kabelka, V., Kutka, A., Piskarskas, A., Smil'gyavichyus, V., Yasevichyute, Ya. (1979): Sov. J. Quant. Electron. **9**, 1022
Kachen, G.I., Lowdermilk, W.H. (1976): Phys. Rev. **14**, 1472
Kaplan, A.E. (1976): JETP Lett. **24**, 132
Kaplan, A.E. (1977): Sov. Phys. JETP **72**, 1710
Karamzin, Yu. N., Sukhorukov, A.P. (1975): Sov. J. Quant. Electron. **5**, 496
Kenney-Wallace, G.A. (1984): In *Applications of Picosecond Spectroscopy to Chemistry*, ed. by K.B. Eisenthal (Reidel, Dordrecht)
Kohles, N., Laubereau, A. (1985): In *Time-Resolved Vibrational Spectroscopy*, Springer Proc. Phys., Vol. 4, ed. by A. Laubereau, M. Stockburger (Springer, Berlin, Heidelberg) p. 35
Kohles, N., Laubereau, A. (1986): Appl. Phys. B **39**, 141
Kohles, N., Laubereau, A. (1988): Opt. Commun. (to be published)
Kolmeder, C., Zinth, W., Kaiser, W. (1979): Opt. Commun. **30**, 453
Kopainsky, B., Kaiser, W. (1978): Opt. Commun. **26**, 219
Kosic, T.I., Cline, Jr., R.E., Dlott, D.D. (1983): Chem. Phys. Lett. **103**, 109
Kranitzky, W., Ding, K., Seilmeier, A., Kaiser, W. (1980): Opt. Commun. **34**, 483
Krivoshtshekov, G.V., Stroganov, V.I. (1970): In *Nelineynie processi v optike* (Nauka, Novosibirsk)
Krylov, V.N., Papernyi, S.B. (1980): Sov. Phys. Techn. Phys. **25**, 268
Kryukov, P.G., Matveets, Yu. A., Nikogosyan, D.N., Sharkov, A.V., Gordeev, E.M., Fanchenko, S.D. (1977): Sov. J. Quant. Electron. **7**, 127
Kuhl, J., von der Linde, D. (1982): In *Picosecond Phenomena III*, Springer Ser. Chem. Phys., Vol. 23, ed. by K.B. Eisenthal, R.M. Hochstrasser, W. Kaiser, A. Laubereau (Springer, Berlin, Heidelberg)
Kumar, A., Sarkar, S.N., Ghatak, A.K. (1986): Opt. Lett. **11**, 322
Kung, A.H., Young, J.F., Bjorklund, G.C., Harris, S.E. (1972): Phys. Rev. Lett. **29**, 985
Kung, A.H., Young, J.F., Harris, S.E. (1973): Appl. Phys. Lett. **22**, 301
Kung, A.H. (1974): Appl. Phys. Lett. **25**, 653
Kurobori, T., Cho, Y., Matsuo, Y. (1981): Opt. Commun. **40**, 156
Kwok, H.S., Chiu, P.H. (1985): Opt. Lett. **10**, 28
Laubereau, A., von der Linde, D., Kaiser, W. (1971): Rev. Lett. **27**, 802
Laubereau, A., von der Linde, D., Kaiser, W. (1972): Chem. Rev. Lett. **28**, 1162
Laubereau, A., von der Linde, D., Kaiser, W. (1973): Opt. Commun. **1**, 173
Laubereau, A., Kirschner, L., Kaiser, W. (1973): Opt. Commun. **9**, 182
Laubereau, A. (1974): Chem. Phys. Lett. **27**, 600
Laubereau, A., Greiter, L., Kaiser, W. (1974): Appl. Phys. Lett. **25**, 87

Laubereau, A., Kaiser, W. (1978): Rev. Mod. Phys. **50**, 607
Laubereau, A., Fendt, A., Seilmeier, A., Kaiser, W. (1978): In *Picosecond Phenomena*, Springer Ser. Chem. Phys. Vol. 4, ed. by C.V. Shank, E.P. Ippen (Springer, Berlin, Heidelberg)
Lavoine, J.P., Villaeys, A.A. (1986): Opt. Commun. **59**, 160
Ledoux, I., Zyss, J., Migus, A., Etchepare, J., Grillon, G., Antonetti, A. (1986): Appl. Phys. Lett. **48**, 1564
Lehmeier, H.J., Leupacher, W., Penzkofer, A. (1985): Opt. Commun. **56**, 67
Leung, K.M. (1986): Phys. Rev. A **33**, 2461
Leupacher, W., Penzkofer, A. (1985): Appl. Phys. **36**, 25
Lin, C., Stolen, R.H., Cohen, L.G. (1977): Appl. Phys. Lett. **31**, 96
Lin, C., Nguyen, V.T., French, W.G. (1978): Electron. Lett. **14**, 822
von der Linde, D., Maier, M., Kaiser, W. (1969): Phys. Rev. **178**, 11
von der Linde, D., Laubereau, A. (1971): Opt. Commun. **3**, 279
von der Linde, D., Laubereau, A., Kaiser, W. (1971): Phys. Rev. Lett. **26**, 954
Liu, P.L. (1979): Appl. Opt. **18**, 3545
Loree, I.R., Cantrell, C.D., Barker, D.L. (1976): Opt. Commun. **17**, 160
Lowdermilk, W.H., Kachen, G.I. (1975): Appl. Phys. Lett. **27**, 133
Luigiato, L.A. (1980): Opt. Commun. **33**, 108
Luk, T.S., McPherson, A., Jara, H., Johann, U., McIntyre, I.A., Schwarzenbach, A.P., Boyer, K., Rhodes, C.K. (1986): In *Ultrafast Phenomena V*, Springer Ser. Chem. Phys., Vol. 23, ed. by G.R. Fleming, A.E. Siegman (Springer, Berlin, Heidelberg)
Maaswinkel, A.G.M. (1980): Opt. Commun. **35**, 236
Mack, M.E., Carman, R.L., Reintjes, J., Bloembergen, N. (1970): Appl. Phys. Lett. **16**, 209
Madden, P.A. (1984): In *Ultrafast Phenomena IV*, Springer Ser. Chem. Phys. Vol. 38, ed. by D.H. Auston, K.B. Eisenthal (Springer, Berlin, Heidelberg)
Magnitskii, S.A., Pryalkin, V.I., Tunkin, V.G., Kholodnykh, A.I. (1982): Sov. J. Quant. Electron. **12**, 900
Magnitskii, S.A., Malachova, V.I., Tarasevich, A.P., Tunkin, V.G., Yabubovich, S.D. (1986): Opt. Lett. **11**, 18
Mahr, H., Hirsch, M.D. (1975): Opt. Commun. **13**, 96
Maier, M., Kaiser, W., Giordmaine, J.A. (1966): Phys. Rev. Lett. **17**, 1275; (1969): Phys. Rev. **177**, 580
Maker, P.D., Terhune, R.W., Nisenoff, M., Savage, C.M. (1962): Phys. Rev. Lett. **8**, 21
Maker, P.D., Terhune, R.W., Savage, C.M. (1964): Phys. Rev. Lett. **12**, 507
Malvezzi, A.M., Liu, J.M., Bloembergen, N. (1984): Appl. Phys. Lett. **45**, 1019
Manassah, J.T., Mustafa, M.A., Alfano, R.R., Ho, P.P. (1986): IEEE J. Quant. Electron. QE-**22**, 197
Maple, J.R., Knudtson, J.T. (1978): Chem. Phys. Lett. **56**, 241
Marburger, J.H. (1978): Appl. Phys. Lett. **32**, 372
Massey, G.A., Johnson, J.C., Elliot, R.A. (1976): IEEE J. QE-**12**, 143
Matsuoka, M., Nakatsuka, H., Okada, J. (1975): Phys. Rev. A **12**, 1062
McCall, S.L. (1974): Phys. Rev. A **9**, 1515
McCall, S.L. (1978): Appl. Phys. Lett. **32**, 284
McDonald, D.B. (1984): In *Ultrafast Phenomena IV*, Springer Ser. Chem. Phys. Vol. 38, ed. by D.H. Auston, K.B. Eisenthal (Springer, Berlin, Heidelberg)
Medvedev, B.A. (1971): Sov. Phys. JETP **33**, 19
Medvedev, B.A., Parshkov, O.M., Gorshenin, V.A., Dmitriev, A.E. (1974): Sov. Phys. JETP **40**, 36
Menyuk, C.R., Chen, H.H., Lee, Y.C. (1985): Opt. Lett. **10**, 451
Mestdagh, D., Haelterman, M. (1987): Opt. Commun. **61**, 291
Metchkov, D.J., Mitev, V.M., Pavlov, L.I., Stamenov, K.V. (1977): Phys. Lett. **61A**, 449
Midwinter, J.E., Warner, J. (1967): J. Appl. Phys. **38**, 519
Migus, A., Antonetti, A., Hulin, D., Mysyrowicz, A., Gibbs, H.M., Peyghambarian, N., Jewell, J.L. (1985): Appl. Phys. Lett. **46**, 70
Miles, R.B., Harris, S.E. (1973): IEEE J. QE-**9**, 470
Miller, R.C. (1968): Phys. Lett. **26A**, 177

Mironov, G.V., Filonenko, N.N. (1982): Sov. J. Quant. Electron. **12**, 723
Mitschke, F.M., Mollenauer, L.F. (1986): Opt. Lett. **11**, 659
Miyazaki, K., Sato, T., Kashiwagi, H. (1981): Phys. Rev. **23**, 1358
Mokhtari, A., Fini, L., Chesnoy, J. (1987): Opt. Commun. **61**, 421
Mollenauer, L.F., Stolen, R.H., Gordon, J.P. (1980): Phys. Rev. Lett. **45**, 1095
Mollenauer, L.F., Stolen, R.H., Gordon, J.P., Tomlinson, W.J. (1983): Opt. Lett. **8**, 289
Mollenauer, L.F., Stolen, R.H. (1984): Opt. Lett. **9**, 13
Mollenauer, L.F., Stolen, R.H., Islam, M.N. (1985): Opt. Lett. **10**, 229
Moore, C.A., Goldberg, L.S. (1976): Opt. Commun. **16**, 21
Mostowski, J., Raymer, M.G. (1981): Opt. Commun. **36**, 237
Myers, A.B., Hochstrasser, R.M. (1986): IEEE J. of Quant. Electron. QE-**22**, 1482
Nakatsuka, H., Grischkowsky, D., Balant, A.C. (1981): Phys. Rev. Lett. **47**, 910
Nakazawa, M., Nakashima, T., Seikai, S. (1984): Appl. Phys. Lett. **45**, 823
Narayana Rao, D., Swiatkiewicz, J., Chopra, P., Ghoshal, S.K., Prasad, P.H. (1986): Appl. Phys. Lett. **48**, 1187
Neef, V.E. (1975): Ann. Phys. **32**, 191
Nelson, K.A., Fayer, M.D. (1980): J. Chem. Phys. **72**, 5202
Nelson, K.A., Miller, R.J.D., Lutz, D.R., Fayer, M.D. (1982): J. Appl. Phys. **53**, 1144
New, G.H.C. (1976): Opt. Commun. **19**, 177
Nguyen, D.C., Muenchhausen, R.E., Keller, R.A., Nogar, N.S. (1986): Opt. Commun. **60**, 111
Nosach, O.Y., Popovichev, V.I., Ragulskii, V.V., Faisullov, F.S. (1972): Sov. Phys. JETP **16**, 435
Österberg, U., Margulis, W. (1987): Opt. Lett. **12**, 57
Ohmori, Y., Sasaki, Y. (1982): IEEE J. Quant. Electron. QE-**18**, 758
Okada, M., Takizawa, K. (1980): IEEE J. Quant. Electron. QE-**16**, 772
Olbright, G.R., Fluegel, B.D., Koch, S.W., Peyghambarian, N. (1986): In *Ultrafast Phenomena V*, Springer Ser. Chem. Phys. Vol. 46, ed. by G.R. Fleming, A.E. Siegman (Springer, Heidelberg, Berlin)
Onishchukov, G.I., Fomichev, A.A., Kholodnykh, A.I. (1983): Sov. J. Quant. Electron. **13**, 1001
Onodera, N., Ito, H., Inaba, H. (1983): Appl. Phys. Lett. **43**, 720
Orlov, R.U., Usmanov, T., Chrikin, A.S. (1970): Sov. Phys. JETP **30**, 584; (1970): Rws: **57**, 1069
Otsuka, K., Yumoto, J., Song, J.J. (1985): Opt. Lett. **10**, 508
Oudar, J.L., Shen, Y.R. (1980): Phys. Rev. **22**, 1141
Oudar, J.L. (1983): IEEE J. Quant. Electron. QE-**19**, 713
Palfrey, S.L., Heinz, T.F., Eisenthal, K.B. (1980): Opt. Lett. **9**, 359
Papernyl, S.B., Petrov, V.F., Serebryakov, V.A., Startsev, V.R. (1982): Sov. J. Quant. Electron. **12**, 584
Penzkofer, A., Laubereau, A., Kaiser, W. (1973a): Phys. Rev. Lett. **14**, 863
Penzkofer, A., Seilmeier, A., Kaiser, W. (1973b): Opt. Commun. **14**, 363
Penzkofer, A., Kaiser, W. (1977): Opt. Quant. Electr. **9**, 315
Penzkofer, A., Laubereau, A., Kaiser, W. (1979): Prog. Quant. Electron. **6**, 55
Penzkofer, A., Kraus, J., Sperka, J. (1981): Opt. Commun. **37**, 437
Penzkofer, A., Schmailzl, J., Glas, H. (1982): Appl. Phys. B **29**, 37
Piché, M., Quellette, F. (1986): Opt. Lett. **11**, 15
Pochon, E., Bourene, M. (1976): J. Chem. Phys. **65**, 2056
Pokhsraryan, K.M. (1985): Opt. Commun. **55**, 439
Prasad, P.N., Rao, D.N., Swiatkiewicz, J., Chopra, P., Ghoshal, S.K. (1986): In *Ultrafast Phenomena V*, Springer Ser. Chem. Phys., Vol. 46, ed. by G.R. Fleming, A.E. Siegman (Springer, Heidelberg, Berlin)
Puell, H., Scheingraber, H., Vidal, C.R. (1980): Phys. Rev. **22**, 1165
Rabson, T.A., Ruiz, H.J., Shah, P.L., Tittel, F.K. (1972): Appl. Phys. Lett. **20**, 282; (1972): Appl. Phys. Lett. **21**, 129
Rahn, O., Maier, M., Kaiser, W. (1969): Opt. Commun. **1**, 109
Rao, D.N., Swiatkiewicz, J., Chopra, P., Ghoshal, S.K., Prasad, P.N. (1986): Appl. Phys. Lett. **48**, 1187
Rasing, Th., Shen, Y.R., Kim, M.W., Valint Jr. P., Bock, J. (1985): Phys. Rev. A **31**, 537

Reintjes, J.F., Carman, R.L., Shimizu, F. (1973): Phys. Rev. A **8**, 1486
Reintjes, J.F., Eckardt, R.C. (1977): Appl. Phys. Lett. **30**, 91
Reintjes, J.F., She, C.Y., Eckardt, R.C. (1978): IEEE J. Quant. Electron. QE-**14**
Reintjes, J.F. (1984): In *Nonlinear Optical Parametric Processes in Liquids and Gases* (Academic, Orlando)
Reiser, D., Laubereau, A. (1982): Opt. Commun. **42**, 329
Ricard, D., Roussignol, P., Flytzanis, C. (1985): Opt. Lett. **10**, 511
Robinson, M.M., Yan, Y.-X., Gamble Jr. E.B., Williams, L.R., Meth, Y.S., Nelson, K.A. (1984): Chem. Phys. Lett. **112**, 491
Rothenberg, J.E., Grischkowsky, D., Balant, A.C. (1984): Phys. Rev. Lett. **53**, 552
Rothenberg, J.E., Grischkowsky, D. (1987): Opt. Lett. **12**, 99
Roussignol, P., Ricard, D., Rustagi, K.C., Flytzanis, C. (1985): Opt. Commun. **55**, 143
Roussignol, P., Ricard, D., Lukasik, J., Flytzanis, C. (1987): J. Opt. Soc. Am. B **4**, 6
Rozanov, N.N. (1977): JETP Lett. **3**, 583
Rozanov, N.N. (1978): JETP Lett. **4**, 74
Rzazewski, K., Lewenstein, M., Raymer, M.G. (1982): Opt. Commun. **43**, 451
Sala, K.L., Richardson, M.C. (1975): Phys. Rev. A **12**, 1030
Sala, K.L., Kenney-Wallace, G., Hall, G.E. (1980): IEEE J. Quant. Electron. QE-**16**, 990
Saltiel, S.J., Savov, S.D., Tomov, I.V., Telegin, C.S. (1981): Opt. Commun. **38**, 443
Sapondzhyan, S.O., Sarkisyan, D.G. (1983): Sov. J. Quant. Electron. **13**, 1062
Sarkisyan, D.G. (1978): Sov. J. Electron. **8**, 535
Schwartz, C.A., Oudar, J.L., Batifol, E.M. (1975): IEEE J. Quant. Electron. QE-**11**, 616
Seilmeier, A., Spanner, K., Laubereau, A., Kaiser, W. (1978): Opt. Commun. **24**, 234
Seilmeier, A., Kaiser, W. (1980): Appl. Phys. **23**, 113
Seka, W., Jacobs, S.D., Rizzo, J.E., Boni, R., Craxton, R.S. (1980): Opt. Commun. **34**, 469
Shank, C.V., Auston, H.H. (1975): Phys. Rev. Lett. **34**, 479
Shank, C.V., Ippen, E.P., Teschke, O. (1977): Chem. Phys. Lett. **45**, 291
Shank, C.V., Yen, R., Hirliman, C. (1983): Phys. Rev. Lett. **51**, 900
Shapiro, S.L., Giordmaine, J.A., Wecht, K.W. (1967): Phys. Rev. Lett. **19**, 1093
Shapiro, S.L. (1968): Appl. Phys. Lett. **13**, 19
Sharma, D.K., Yip, R.W. (1979): Opt. Commun. **30**, 113
Shelton, J.W., Shen, Y.R. (1970): Phys. Rev. Lett. **25**, 23
Shelton, J.W., Shen, Y.R. (1971): Phys. Rev. Lett. **26**, 538
Shih, C.C. (1986): Opt. Lett. **11**, 641
Shimoda, K. (1970): Z. Phys. **234**, 293
Shukla, P.K., Juul Rasmussen, J. (1986): Opt. Lett. **11**, 171
Smirl, A.L., Boggess, T.F., Hopf, F.A. (1980): Opt. Commun. **34**, 463
Smirl, A.L., Boggess, I.I., Wherrett, B.S., Perryman, G.P., Miller, A. (1982): In *Picosecond Phenomena III*, Springer Ser. Chem. Phys. Vol. 23, ed. by K.B. Eisenthal, R.M. Hochstrasser, W. Kaiser, A. Laubereau (Springer, Berlin, Heidelberg)
Smirl, A.L., Boggess, I.I., Wherrett, B.S., Perryman, G.P., Miller, A. (1983): IEEE J. Quant. Electro. QE-**19**, 690
Smith, P.W., Turner, E.H. (1977): Appl. Phys. Lett. **30**, 281
Smith, P.W., Turner, E.H., Mumford, B.B. (1978): Opt. Lett. **2**, 55
Smith, P.W., Herman, I.P., Tomlinson, W.I., Maloney, P.I. (1979): Appl. Phys. Lett. **35**, 846
Smith, P.W., Herman, I.P., Tomlinson, W.I., Maloney, P.I. (1981): IEEE J. Quant. Electron. QE-**17**, 340
Smith, W.L., Liu, P., Bloembergen, N. (1977): Phys. Rev. A **15**, 2396
Sorokin, P.P., Lankard, J.R. (1973): IEEE J. QE-**9**, 227
Souma, H., Yajima, T., Taira, Y. (1980): J. Phys. Soc. Japan **48**, 2040
Srinivasan, T., Boyer, K., Egger, H., Luk, T.S., Muller, D.F., Pummer, H., Rhodes, C.K. (1982): In *Picosecond Phenomena II*, Springer Ser. Chem. Phys., Vol. 23, ed. by K.B. Eisenthal, R.M. Hochstrasser, W. Kaiser, A. Laubereau (Springer, Berlin, Heidelberg)

Stolen, R.H., Ippen, E.P. (1973): Appl. Phys. Lett. **22**, 276
Stolen, R.H., Bjorkholm, J.E., Ashkin, A. (1974): Appl. Phys. Lett. **24**, 308
Stolen, R.H., Bjorkholm, J.E. (1982): IEEE J. Quant. Electron. QE-**18**, 1062
Szabó, G., Bor, Zs. (1983): Appl. Phys. **31**, 1
Szöke, A., Daneau, V., Goldhar, J., Kurnit, N.A. (1969): Appl. Phys. Lett. **15**, 376
Tai, K., Tomita, A. (1986): Appl. Phys. Lett. **48**, 1033
Tai, K., Tomita, A., Jewell, J.L., Hasegawa, A. (1986): Appl. Phys. Lett. **49**, 236
Tajima, K. (1987): Opt. Lett. **12**, 54
Takagi, Y., Sumitani, M., Nakashima, N., O'Connor, D.V., Yoshihara, K. (1983): Appl. Phys. Lett. **42**, 490
Tanaka, Y., Kushida, T., Shionoya, S. (1978): Opt. Commun. **25**, 273
Tanaka, Y., Kuroda, H., Shionoya, S. (1982): Opt. Commun. **41**, 434
Tang, C.L. (1975): In *Quantum Electronics*, Vol. I, Part B, Chap. 6, ed. by R. Rabin, C.L. Tang (Academic, New York)
Tan-no, N., Shirahata, T., Yokoto, K. (1975): Phys. Rev. A **12**, 159
Taylor, J.R. (1976): Opt. Commun. **18**, 504
Telegin, L.S., Chirkin, A.S. (1982): Sov. J. Quant. Electron. **12**, 1354
Telle, H.R., Laubereau, A. (1980): Opt. Commun. **34**, 287
Telle, H.R., Laubereau, A. (1983): Chem. Phys. Lett. **94**, 467
Telle, H.R., Laubereau, A. (1984): Appl. Phys. B **34**, 43
Thalhammer, M., Penzkofer, A. (1983): Appl. Phys. **32**, 137
Tocho, J.O., Sibbett, W., Bradley, D.J. (1980): Opt. Commun. **34**, 122
Tocho, J.O., Sibbett, W., Bradley, D.J. (1981): Opt. Commun. **37**, 67
Tomlinson, W.J., Stolen, R.H., Johnson, A.M. (1985): Opt. Lett. **10**, 457
Tomov, I.V. (1974): Opt. Commun. **10**, 154
Tomov, I.V., Fedosejevs, R., Offenberger, A.A. (1982): IEEE J. Quant. Electron. QE-**18**, 2048
Treacy, E.B. (1969): Appl. Phys. Lett. **14**, 112
Trebino, R., Barker, C.E., Siegman, A.E. (1986): IEEE J. Quant. Electron. QE-**22**, 1413
Trippenbach, M., Rzazewski, K., Raymer, M.G. (1984): J. Opt. Soc. Am. B **1**, 671
Tröger, P., Liu, C.H., Laubereau, A. (1985): In *Time-Resolved Vibrational Spectroscopy*, Springer Proc. Phys., Vol. 4, ed. by A. Laubereau, M. Stockburger (Springer, Berlin, Heidelberg)
Valat, P., Tourbez, H., Reiss, C., Gex, J.P., Schelev, M. (1978): Opt. Commun. **25**, 407
Van Laak, J., Giuliani, J.F., Lee, Chi.H. (1980): Appl. Opt. **19**, 1844
Vardeny, Z., Tauc, J. (1981): Opt. Commun. **39**, 396
Volosov, V.D., Kalintsev, A.G., Krylov, V.N. (1976): Sov. J. Quant. Electron. **6**, 1163
Vorobiev, N.S., Ruddock, I.S., Illingworth, R. (1982): Opt. Commun. **41**, 216
Wai, P.K.A., Menyuk, C.R., Lee, Y.C., Chen, H.H. (1986): Opt. Lett. **11**, 464
Weber, H.P. (1967): J. Appl. Phys. **38**, 2231
Weiner, A.M., DeSilvestri, S., Ippen, E.P. (1984): In *Ultrafast Phenomena IV*, Springer Ser. Chem. Phys., Vol. 38, ed. by D.H. Auston, K.B. Eisenthal (Springer, Berlin, Heidelberg)
Weiner, A.M., Ippen, E.P. (1984): Opt. Lett. **9**, 53
Weiner, A.M., Ippen, E.P. (1985): Chem. Phys. Lett. **114**, 456
Weisman, R.B., Rice, S.A. (1976): Opt. Commun. **19**, 28
Welford, D., Sibbett, W., Taylor, J.R. (1980): Opt. Commun. **35**, 283
Wherrett, B.S., Smirl, A.L., Boggess, T.F. (1983): IEEE J. Quant. Electron. QE-**19**, 680
Wilhelmi, B., Herrmann, J. (1980): Sov. J. Quant. Electron. **10**, 1082
Winful, H.G., Marburger, J.H. (1980): Appl. Phys. Lett. **36**, 613
Wondrazek, F., Seilmeier, A., Kaiser, W. (1983): Appl. Phys. B **32**, 39
Wu, C.K., Agostini, P., Petite, G., Fabre, F. (1983): Opt. Lett. **8**, 67
Wyatt, R., Cotter, D. (1980): Opt. Commun. **32**, 481
Wyatt, R., Cotter, D. (1981): Opt. Commun. **37**, 421
Wyatt, R., Marinero, E.E. (1981): Appl. Phys. **25**, 297
Yang, G., Shen, Y.R. (1984): Opt. Lett. **9**, 511

Yajima, T. (1975): Opt. Commun. **14**, 378
Yajima, T., Souma, H., Ishida, Y. (1978): Phys. Rev. A **17**, 324
Yajima, T., Taira, Y. (1979): J. Phys. Soc. Japan **47**, 1620
Yajima, T., Ishida, Y., Taira, Y. (1980): In *Picosecond Phenomena II*, Springer Ser. in Chem. Phys., Vol. 14, ed. by R.M. Hochstrasser, W. Kaiser, C.V. Shank (Springer, Berlin, Heidelberg)
Yajima, T., Morita, N., Ishida, Y. (1984): J. Opt. Soc. Am. B **1**, 526; (1984): Phys. Rev. A **30**, 2525
Yariv, A. (1976a): In *Introduction to Optical Electronics* (Holt, Rinehardt, Winston)
Yariv, A. (1976b): Appl. Phys. Lett. **28**, 88
Yariv, A. (1976c): J. Opt. Soc. Am. **67**, 301
Yariv, A., Pepper, D.M. (1977): Opt. Lett. **1**, 16
Yariv, A. (1978): IEEE J. Quant. Electron. QE-**14**, 680
Ye, P., Shen, Y.R. (1982): Phys. Rev. **25**, 2183
Zakharov, V.E., Shabat, A.B. (1972): Sov. Phys. JETP **34**, 62
Zeldovich, B.Y., Popovichev, V.I., Ragulskii, V.V., Faisullov, F.S. (1972): Sov. Phys. JETP **15**, 109
Zhu, X.D., Suhr, H., Shen, Y.R. (1987): Phys. Rev. B **35**, 3047
Zinth, W., Laubereau, A., Kaiser, W. (1978): Opt. Commun. **26**, 457
Zinth, W., Kaiser, W. (1980): Opt. Commun. **32**, 507

4. Ultrashort Interactions in Solids

Dietrich von der Linde

With 42 Figures

The forerunner of the present volume, *Ultrashort Light Pulses – Picosecond Techniques and Applications*, appeared almost a decade ago [4.1]. At that time the review of the applications of picosecond optical techniques to solid-state phenomena involved roughly a hundred publications and provided a fairly complete coverage of the work on the subject. The field was undoubtedly in its infancy, as were most of the applications of picosecond techniques in other areas. It was also quite clear that a significant expansion of the activities was to be expected in the years to come. However, the development that actually followed almost assumed the proportions and dynamics of an explosion. Today it is quite a task to keep track of the published work on the subject, and the number of people entering the field as well as the number of publications is increasing at a faster pace than ever.

It appears that two of the major driving forces behind this development are the following. Firstly, in addition to the general maturing of the experimental tools and techniques there has been a major breakthrough in ultrafast optical techniques which has opened up the subpicosecond and femtosecond time domain, a regime of great importance in solid-state physics. Secondly, ultrafast solid-state spectroscopy is participating in the current surge of interest in semiconductor physics, which is being fuelled by the never ending quest for faster and more sophisticated microelectronic devices. For example, progress in ultrafast optical techniques and spectroscopy laid the foundations of the newly emerging field of ultrafast electronics, and is beginning to make some impact on certain branches of materials science and laser materials processing, in particular, processing of semiconductor materials.

In writing this chapter the decision on including or omitting specific topics becomes a problem in view of the vast amount of material at hand. It is also not easy to reconcile the sometimes conflicting requirements of scientific systematics, an accurate historic survey, and a timely description of current mainstream developments. The author requests the reader's indulgence if it is felt that the attempt to strike a proper balance has failed.

The chapter is divided into four parts. The first section covers various aspects of ultrafast dynamics of excitons. Dynamics of electrons and holes and of vibrational excitations are discussed in the second and third section. The final part of the chapter deals with ultrafast processes related to phase transformations in solids. An effort is made to provide some background information or brief comments concerning the pertinent basic physics before entering a detailed discussion of the various topics. The ordering of the sections of this chapter is essentially arbitrary; readers with some special interest may proceed directly to the respective section of the article.

4.1 Exciton Dynamics

Excitons represent the lowest intrinsic excited electronic states of semiconductors. An exciton can be visualized as a quasi-particle formed from a negatively charged electron in the conduction band and a positively charged hole in the valence band. The attractive Coulomb interaction which binds the two constituents together reduces the exciton energy below the energy of a free electron-hole pair. There is a rather close physical analogy between the exciton and a hydrogen-like positronium atom.

For excitons formed from electron and hole states connected by direct, allowed optical transitions the interaction with resonant electromagnetic radiation is very strong. A powerful concept to deal with the strong coupling is the introduction of yet another quasi-particle, the exciton-polariton [4.2], which corresponds to a coupled wave of mixed character, partly material and partly electromagnetic. The exciton-polariton has a characteristic dispersion relation $\omega(k)$, where ω and k are the frequency and the wavevector of the exciton polariton, respectively.

Excitons interact in various ways with other intrinsic elementary excitations, with impurities and defects of the semiconductor crystal, and, of course, with other excitons. These interactions cause rapid relaxation, redistribution and recombination phenomena, which are interesting objects of ultrafast spectroscopy.

The covalent forces between two excitons can lead to the formation of an excitonic molecule, or biexciton [4.3], in analogy with the formation of a covalent molecular bond in a hydrogen molecule. Excitonic molecules can be produced either directly via resonant two-photon absorption, or else by collisions of two single excitons.

Interesting changes of the excitonic properties occur when the exciton density becomes high [4.4]. Excitons behave as non-interacting bonson-like particles only in the limit of vanishing number density. However, as the density increases, the fermion character of the basic constituents may become noticable and lead to a variety of interesting phenomena, including non-linear optical effects such as bleaching of the optical absorption.

Other interesting high density effect are related to the screening of the Coulomb interaction. When the exciton density exceeds a critical value, the binding between individual electron-hole pairs is screened out, and the excitonic system undergoes a transition from an electrically non-conducting phase of neutral excitons to a conducting plasma of collectively interacting electrons and holes, in analogy with the insulator-metal Mott transition [4.5]. The Coulomb forces between the electron and the hole can also be screened by the injection of free carriers. Excitons become unstable, if the density of free carriers exceeds the Mott density.

Exciton effects have been observed in a great number of different materials. However, two materials, cuprous chloride (CuCl), and gallium arsenide (GaAs) have played a particularly important role. CuCl is a direct gap material with parabolic conduction and valence bands separated by an energy gap of $E_g =$ 3.416 eV. The basic properties of the excitons in CuCl are very well known. The exciton binding energy of the lowest exciton (Z_3) is quite large, $E_x = 0.212$ eV, and in the optical spectra the excitonic features are well separated. The binding energy of the excitonic molecule is also rather large, $E_m = 33$ meV. A variety of complex excitonic phenomena are readily observable in CuCl, and this material has been very popular for the investigation of exciton physics.

GaAs, of course, is an important material because of its great technological relevance. Exciton research of III-V compound semiconductors has received additional new impact from recent advances in materials science, which have made possible the growth of layered structures of ternary III-V-type semiconductors such as $Ga_xAl_{1-x}As$. Excitons in such layer structures exhibit a number of attractive new properties. For example, it has been shown that the quasi-two-dimensional restriction of the electronic motion leads to a significant increase of the exciton binding energy, typically a factor of 2 to 3 in GaAlAs [4.6]. Some of the unique properties of these quasi-two-dimensional quantum well structures will also be discussed in Sect. 4.2.3.

4.1.1 Exciton-Polariton Dispersion

The frequency–wavevector relation of the exciton-polaritons is a property of fundamental importance in exciton physics. We shall discuss some works in which ultrafast optical techniques have been used to measure exciton-polariton dispersion curves.

A schematic of a typical polariton dispersion curve is depicted in Fig. 4.1. There are two different transverse branches which are called the upper polariton branch (UPB), and the lower polariton branch (LPB), respectively. E_T denotes the energy of the transverse exciton, which can be viewed as the energy proper of the exciton (at rest). E_L is longitudinal exciton energy (the longitudinal exciton branch is not shown in Fig. 4.1). The longitudinal-transverse splitting, E_L-E_T, is a measure of the oscillator strength of the exciton resonance. Another important characteristic of the polariton dispersion curve is the upward curvature of the UPB, which reflects the increase of kinetic energy of the exciton with increasing wavevector. The curvature of the UPB is related to the effective mass of the exciton. Recall also that for $\hbar\omega \gg E_T$ and $\hbar\omega \ll E_T$ the polariton assumes photon-like character with a simple approximately linear $\omega(k)$ relation, whereas near resonance $\omega(k)$ varies quite strongly.

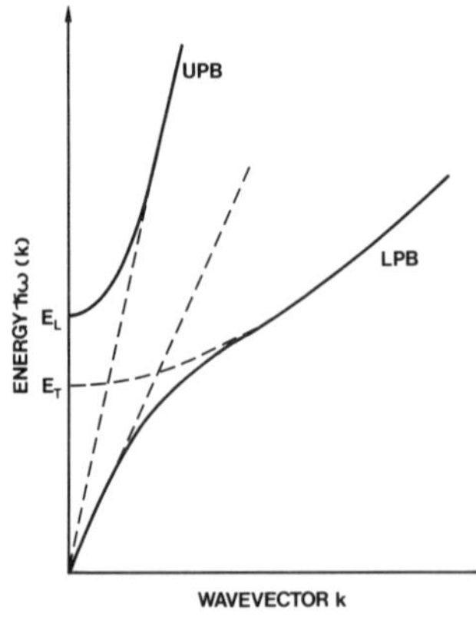

Fig. 4.1. Schematic of the exciton-polariton dispersion relation. Shown is the upper transverse polariton branch (UPB), and the lower transverse polariton branch (LPB). E_T and E_L are the transverse and the longitudinal exciton energy, respectively. The longitudinal branch is not shown. The straight and the curved dashed lines indicate the photon-like and the exciton-like asymptotes

Information about $\omega(k)$ can be obtained from the group velocity, $v_g = d\omega/dk$, which can be obtained from a measurement of the time required for a light pulse to pass through a certain length of material. Near the exciton resonance the material absorbs light very strongly, and therefore thin samples must be used. As a consequence the transit times can be very short, and picosecond optical techniques are called for.

Group velocity measurements of the Z_3 exciton-polariton of CuCl have been performed by *Masumoto* et al. [4.7]. In their experiments exciton-polariton waves were excited in a thin CuCl sample (immersed in liquid helium) with uv light pulses of 20 to 40 ps duration. These pulses could be wavelength-tuned around $\lambda = 387$ nm, which is the wavelength corresponding to the exciton energy, $E_T = 3.2$ eV. The transmitted light pulses were analyzed by means of a CS_2 optical Kerr shutter, which permits measurement of the optical delay time introduced by the sample. The Kerr shutter was driven by an infrared pulse at 1064 nm, providing an open time of about 20 ps.

Results of these measurements are reproduced in Fig. 4.2. The measured group velocity as a function of the photon energy is shown on the right, and the polariton dispersion curve of CuCl on the left. The striking (but not surprising) result apparent from these data is the fact that the velocity of the light pulse is drastically reduced by as much as a factor 20,000 compared with the speed of light in vacuum. Note also the two different sets of data which demonstrate very clearly the two different modes of propagation, i.e. propagation as a lower branch and a upper branch polariton, respectively. The solid and the dashed curves are fits to the theoretical dispersion relation. From the fit of the data the effective mass of the exciton has been determined to be $M = (2.0 \pm 0.1)m_0$, where m_0 is the free electron mass.

Measurements of the group velocity dispersion in CuCl were also reported by *Segawa* et al. [4.8,9]. The same group of authors studied exciton-polariton group velocity dispersion in hexagonal CdS [4.10]. *Masumoto* et al. [4.11] reported group velocity measurements in ZnSe. In this material the valence band

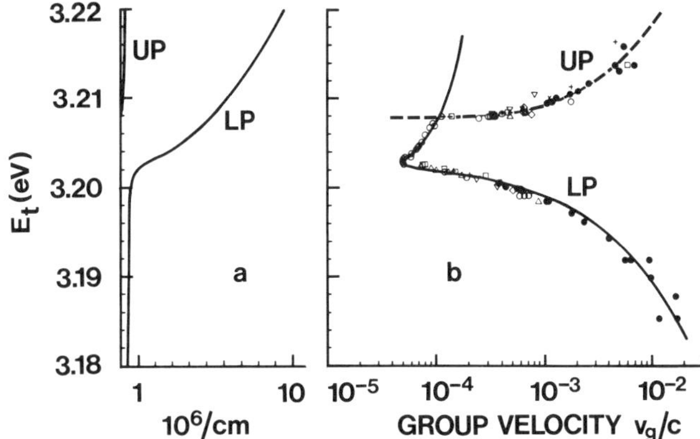

Fig. 4.2. (**a**) Exciton-polariton dispersion curves of CuCl. (**b**) Measured group velocity as a function of energy. The dashed and the solid curves represent the calculated group velocity for the upper and the lower branch, respectively. After [4.7]

is four-fold degenerate at the center of the Brillouin zone, and as a result an intermediate polariton branch is observed.

Ulbrich and *Fehrenbach* [4.12] used a different technique to study the exciton-polariton dispersion in GaAs. In their experiment the output of a synchronously mode-locked cw dye laser operating in the near infrared close to the exciton resonance of GaAs ($E_T = 1.515$ eV) was split into a reference beam with an adjustable time delay, and a probe beam. Only the probe beam was passed through the sample. The two beams were then recombined in a nonlinear crystal for second harmonic generation. By scanning the delay time of the reference beam the cross-correlation function of the reference pulses with the pulses transmitted through the sample can be measured.

Figure 4.3 shows a series of second harmonic signals measured by Fehrenbach and Ulbrich. In these measurements the laser frequency is tuned through the exciton resonance. The distinct increase of the transit time of the pulses when the resonance is approached is quite obvious from these data (maximal increase is about 30 ps). It is interesting to note that the shape of the signals is also changed, which is an indication that the pulses are significantly distorted upon propagation through the crystal. Apparently the dispersion near resonance is so strong that the concept of group velocity is no longer adequate for a description of the pulse propagation. Similar observations have also been made in CuCl [4.7].

The group velocity data of Ulbrich and Fehrenbach are shown in Fig. 4.4 by the open circles. The solid and the dashed line represent the calculated v_g for the lower and the upper branch, respectively. Generally speaking, there is fair agreement, except in the immediate vicinity of the resonance (E_T). This discrepancy may be due to problems related to the smallness of $E_L - E_T = 7.44 \times 10^{-5}$ eV, which is comparable to the frequency width of the picosecond pulses.

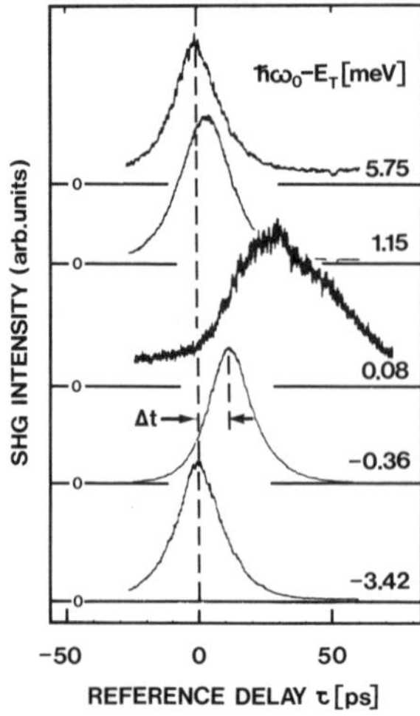

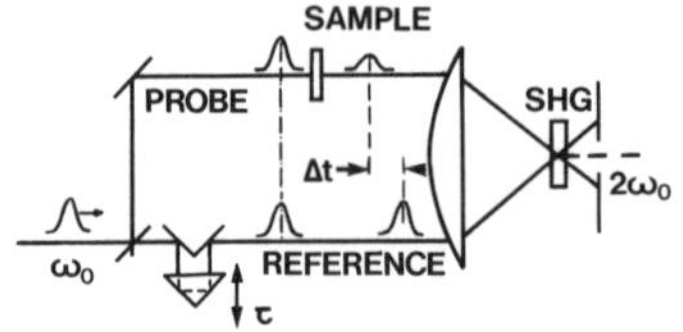

Fig. 4.3. Left: Measured cross-correlation second-harmonic signal for various values of the energy difference $\hbar\omega$-E_T between the incident photon energy and the transverse exciton energy. Right: Schematic of the experimental arrangement. After [4.12]

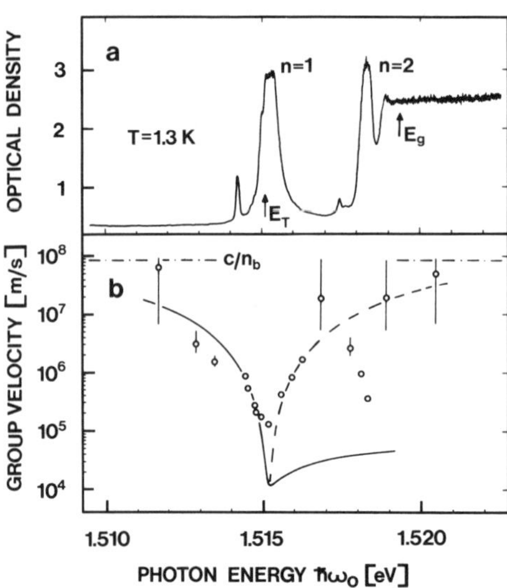

Fig. 4.4. (**a**) Optical absorption spectrum of a 3.7 μm thick GaAs sample. $n = 1$ and $n = 2$ refer to the principal quantum number of the exciton. E_g is the band gap energy. (**b**) Group velocity in GaAs as a function of photon energy. The dots indicate the measured data. The solid and the dashed line are the calculated dispersion relation of the lower and the upper branch, respectively. c/n_b denotes the phase velocity due to the background refractive index. After [4.12]

There is another interesting deviation of the measured data near 1.1518 eV. This feature of the group velocity dispersion has been attributed by Fehrenbach and Ulbrich to the first excited state ($n = 2$) of the exciton, which the authors were able to observe in the absorption spectrum of very thin high quality GaAs samples [4.12] (see top of Fig. 4.4).

An excellent review on the subject of exciton-polariton dispersion and a detailed discussion of the various alternative techniques can be found in the article by *Koteles* [4.13].

4.1.2 Exciton Relaxation

Exciton-polariton relaxation phenomena have been a field of active research for many years. In the earlier work photoluminescence has often been used to obtain information about recombination and relaxation of excitons [4.14,15]. Unfortunately, the extraction of exciton distributions from photoluminescene spectra is somewhat complicated, mainly due to polariton effects. Recall also that, in the light of the polariton concept, the notion of "exciton recombination" or "exciton lifetime" can be misleading and should be treated with some care [4.16]. More recently, nonlinear optical techniques have been used to study the dynamics of exciton-polariton distributions. For example, one can make use of transitions from a populated exciton-polariton branch to biexciton states; this is a form of excited state spectroscopy. For this purpose, CuCl is an ideal material because two step excitation of excitonic molecules in CuCl is particularly well established [4.17].

Masumoto and *Shionoya* [4.18] have reported a detailed investigation of the time evolution of the exciton-polariton distribution in CuCl. In these experiments they used tunable excitation pulses of about 20 ps duration from a frequency-doubled optical parametric oscillator. The photon energy of the excitation pulse was chosen to exceed the longitudinal exciton energy of CuCl, $E_L = 3.208$ eV ($\lambda < 3865$ Å). Time-resolved excited state absorption spectra of CuCl were measured with the help of broad-band probe pulses from a synchronously mode-locked dye laser. From the photon energy of the induced absorption, the energy and the wavevector of the exciton-polariton involved in any given excited-state transition can be unambiguously determined, because the energy-wavevector relation of both the exciton-polariton and the biexciton are precisely known in CuCl.

Figure 4.5 shows examples of the measured induced absorption spectra from the work of *Masumoto* and *Shionoya* [4.18]. The change of the absorption is plotted versus photon energy E_{IA} for four different probe pulse delay times (a: -16.7 ps, b: 0 ps, c: 50 ps, d: 1083.3 ps). The arrows labeled E_t and E_l indicate the probe photon energy corresponding to the transverse exciton energy, $E_T = E_M - E_{IA}$, and the longitudinal photon energy, $E_L = E_M - E_{IA}$, respectively, where $E_M = 6.372$ eV is the energy of the excitonic molecule at rest (zero wavevector). The induced absorption band marked UP1 was interpreted by Masumoto and Shionoya as being due to "initially generated" polaritons, i.e., polaritons in states directly pumped by the excitation pulse ($\hbar\omega = 3.215$ eV). The bands UP2 and LP, on the other hand, were attributed to upper branch and lower branch "relaxed" polaritons.

Quite a lot of detailed information has been extracted from such induced absorption spectra. The UP polaritons disappear quite rapidly, very nearly following the excitation pulse. The build up of the LP polaritons, on the other hand, approximately follows the intergral of the excitation. The subsequent

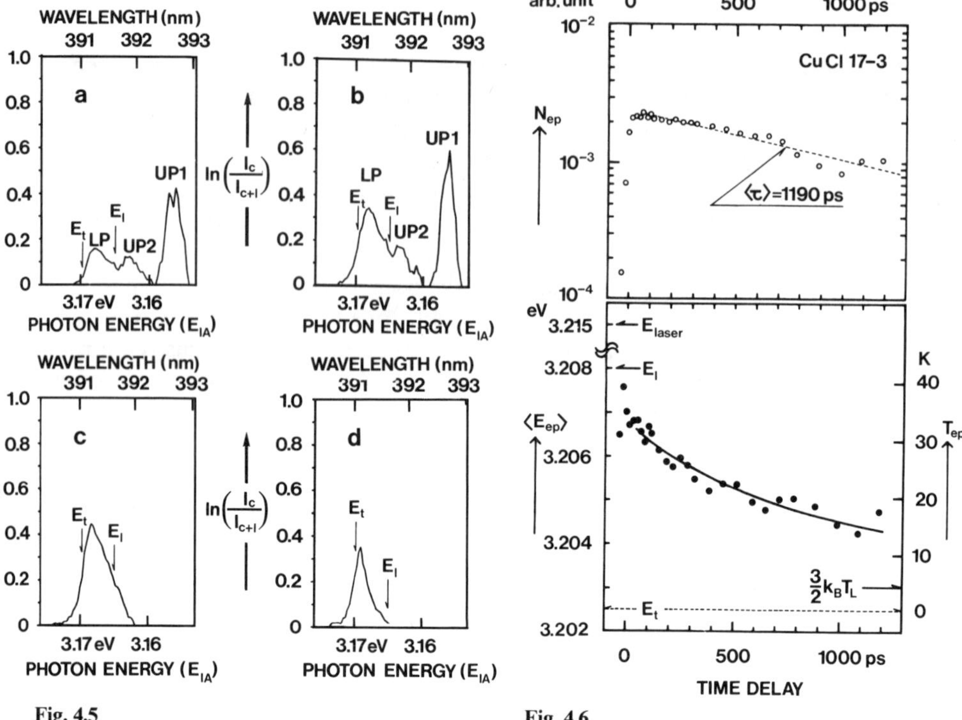

Fig. 4.5

Fig. 4.6

Fig. 4.5a–c. Induced absorption spectra of CuCl pumped slightly above the longitudinal exciton energy ($\hbar\omega_{excite} = 3.215$ eV). E_{IA} is the probe photon energy. E_t and E_e are the transverse and the longitudinal exciton energy. Probe pulse delay times: (**a**) -16.7 ps; (**b**) 0 ps; (**c**) 50 ps; (**d**) 1083.3 ps. I_c/I_{c+e} denotes the ratio of the transmission spectra without and with excitation. After [4.18]

Fig. 4.6. Total exciton number N_{ep} (*upper part*) and average energy $\langle E_{ep} \rangle$ (*lower part*) of lower branch polaritons as a function of time for a CuCl crystal (thickness 24.8 μm). After [4.18]

changes of the shape and the magnitude of the LP band reflect cooling of the distribution and reduction in the number of polaritons as time proceeds.

The behavior of the exciton-polariton population is summarized by Fig. 4.6 from *Masumoto* and *Shionoya* [4.18,19]. The total number of exciton-polaritons in the lower branch (upper half of Fig. 4.6), and the mean exciton-polariton energy averaged over the entire lower branch population (lower half) are plotted as a function of time. The decay time of the exciton-polariton population was observed to increase with the thickness d of the samples, e.g. 750 ps for $d =$ 14.2 μm (not shown in Fig. 4.6), and 1190 ps for $d = 24.8$ μm. This observation is in agreement with the decay being dominated by radiation losses, i.e., losses due to polaritons which propagate to the boundaries of the crystals and escape as free-space photons. The characteristic time for this radiative escape is obviously proportional to the sample thickness.

Figure 4.6 shows that the average energy of the exciton-polariton decreases faster than their number. Since the distribution functions of the LB polaritons are found to be Maxwell-Boltzmann distributions [4.19] the energy relaxation can be described as a cooling process characterized by an exciton-polariton temperature T_{ep}. The data show that T_{ep} decreases from 40 K to about 15 K in approximately a nanosecond, but the equilibrium temperature of the crystal, $T_L = 4.2$ K, is not reached. The observed cooling of the exciton-polariton gas was found to be in good agreement with the calculated decrease of the average energy due to the polariton-LA-phonon scattering (deformation potential scattering).

In related work, *Segawa* et al. [4.20] have investigated relaxation of exciton-polaritons in CuCl using a similar scheme as in [4.18], but with direct, resonant excitation of lower branch exciton-polaritons. Segawa et al. have observed a spectrally narrow induced absorption, with the sum of the photon energies of the induced absorption and the excitation pulses being equal to the energy of the excitonic molecule. They conclude that the induced absorption is caused by polariton states directly pumped by the excitation pulses, i.e., by the unrelaxed polariton population. The bandwidth and the peak of the absorption does not change with delay times up to about 300 ps, the longest delay used in their experiments. The decay time of the induced absorption is measured as a function of the excitation photon energy, $\hbar\omega_{ex}$, over an energy range of approximately $E_T < \hbar\omega_{ex} < E_L$. This time constant, which has been interpreted in [4.20] as polariton lifetime, increases from 150 ps near E_L to about 300 ps near E_T; the increase has been shown to be proportional to the inverse group velocity of the exciton-polariton.

When the results of the two experiments, relaxation of non-resonantly excited [4.18] and of resonantly excited [4.20] lower branch exciton-polaritons in CuCl are compared, interesting differences of the results can be noticed. Apparently the resonantly excited, energetically narrow initial population does not relax to a thermal distribution, whereas for nonresonant excitation [4.18] establishment of quasi-thermal equilibrium distributions is observed for times greater than 50 ps after excitation. Also, the measured decay time of the population in [4.20] is much shorter than in non-resonant experiment [4.18]. Segawa et al. argued that under the conditions of their experiment polariton-polariton scattering is reduced because almost all of the polaritons are generated with the same momentum.

Time-resolved luminescence has been used by *Masumoto* and *Shionoya* [4.21] to study exciton-polariton relaxation in CdSe. The sample is excited via band-to-band transitions using pulses from a synchronously mode-locked Rhodamine 6G dye laser. A synchro-scan streak camera coupled to a spectrometer measures time-resolved photoluminescence spectra (system time resolution of 70 ps). Masumoto and Shionoya study the LO-phonon side band of the *A*-exciton, which – as these authors note – directly reflects the distribution function of the excitons. From time and frequency resolved luminescence measurements they extract the average energy and the total number of exciton-polariton as a function of time. It is observed that the exciton energy relaxes in about 150 ps, much faster than the measured decay of the total number, 2.8 ns. A detailed comparison is made with theoretical energy loss rates due to piezoelectric, deformation potential, and Fröhlich-interaction-type phonon scattering. The measured and the calculated energy relaxation are in reasonably good agreement, but for long times (greater than 300 ps) significant deviations from the theoretical relaxation model have been noted.

Picosecond transient grating experiments have been performed by *Aoyagi* et al. [4.22] to study the diffusion and decay of excitonic molecules in CuCl. The lifetime of the excitonic molecules extracted from these grating decay experiments has been reported to be 280 ps.

The lifetime of excitonic molecules is an interesting problem for picosecond investigations. The oscillator strength for a radiative decay of an excitonic molecule into a photon ("photon-like polariton") and an exciton ("exciton-like polariton") have been predicted to be two to three orders of magnitude greater than the exciton oscillator strength of the single exciton [4.23]. Therefore the radiative lifetime of excitonic molecules is expected to be in the picosecond time domain.

Unuma et al. [4.24] have made measurements of the luminescence of the excitonic molecule in CuBr. They used both band-to-band and resonant two-photon excitation to produce excitonic molecules. Time-resolved luminescence measurements were made with the help of a CS_2 optical Kerr gate with 40 ps resolution. Their data are reproduced in Fig. 4.7. For resonant excitation the rise of the luminescence is observed to follow the integral of the pump pulse as expected. For band-to-band excitation the onset of the luminescence is delayed due to the time it takes for the formation of molecules by fusion of two excitons. The entire data could be explained with a simple rate equation model for the

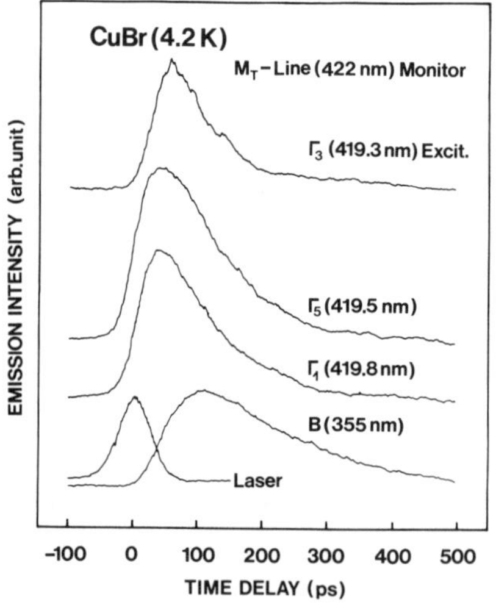

Fig. 4.7. Time-resolved luminescence from excitonic molecules in CuBr due to transitions from biexciton states to transverse exciton-polariton states (M_T line). Γ_1, Γ_5, and Γ_3 correspond to resonant two-photon excitation of the respective biexciton states. Curve B corresponds to non-resonant band-to-band excitation. After [4.24]

population of excitons and excitonic molecules with a molecular lifetime of 60 ps. The authors of [4.24] compare this result with theoretical estimates of the radiative lifetime of biexcitons. For a Coulomb-type and a well-type molecular binding potential they estimate 3 ps and 24 ps, respectively; both values are much less than the observed lifetime of 60 ps.

Results of lifetime measurements of excitonic molecules in CdS have been reported by *Göbel* [4.25]. Excitonic molecules are resonantly excited via two-photon absorption with excitation pulses of 25 ps from a tunable dye laser. The luminescence due to the decay of excitonic molecules is measured using a CS_2 optical Kerr gate. The lifetime of the excitonic molecules in CdS has been determined to be 20 ps.

4.1.3 Exciton Screening

In the following, we are going to discuss experiments on highly excited materials in which the electronic states – and hence the optical properties – are strongly modified by the presence of a high density of photoexcited particles and their mutual interactions. These changes are basically due to the modification of the simple two-body Coulomb interaction in the presence of other charges, or other electrically polarizable particles. This "screening" of the Coulomb interaction is of great importance for the understanding of high excitation phenomena in semiconductors.

As a starting point let us consider the beautiful examples of *low density* exciton optical spectra from Ulbrich and Fehrenbach depicted in the top of Fig. 1.1 and in Fig. 4.9. Shown are the optical absorption spectra of GaAs near the fundamental absorption edge. The spectra exhibit hydrogen-like excitonic absorption peaks corresponding to the principal quantum numbers $n = 1$, and $n = 2$. At the high energy side the exciton series merges with the continuum absorption due to optical transitions corresponding to the formation of unbound electron-hole pairs.

The principal changes of the spectra caused by *high excitation* can be described very briefly as follows. Screening weakens the attractive forces between electrons and holes and impairs the mutual binding. Bound, hydrogen-like states can no longer exist, and the discrete excitonic absorption peaks disappear. The exciton line spectra are replaced by the broad continuum spectra of the electron-hole plasma, which differ, however, from the low excitation continuum by a shift to lower energies. This shift reflects the renormalization of the band gap, which is a characteristic manifestation of the electron-hole interaction of the many-body system. Electron-hole plasmas will be discussed in more detail in Sect. 4.2.1.

A direct illustration of the ultrafast screening of the exciton in GaAs upon the photoinjection of free carriers is given by Fig. 4.8, taken from *Shank* et al. [4.26]. In this experiment a 1.5 μm thick molecular-beam-grown sample of GaAs (sandwiched between suitable GaAlAs layers for support) is photoexcited by a 0.5 ps light pulse at a wavelength of 7500 Å. The excitation pulse generates free

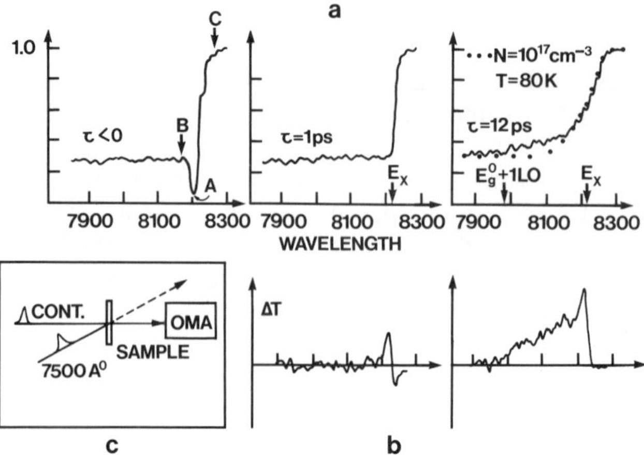

Fig. 4.8a–c. Transmission spectra of GaAs for excitation with subpicosecond pulses at 750 nm. (**a**) Spectra for three different delay times between pump pulse and probe pulse; (**b**) difference between spectra for delayed probe pulses with respect to the spectrum of a leading probe pulse; (**c**) schematic of experimental arrangement. After [4.26]

electron-hole pairs at a density of approximately $10^{17}\,\mathrm{cm}^{-3}$, and an energy of about 100 meV in excess of the band gap energy.

The optical transmission spectrum of the photoexcited sample is measured with a subpicosecond broad band continuum probe pulse at controlled time delays with respect to the excitation (see schematic in Fig. 4.8c). The first spectrum (at the top left) taken at a negative delay time, i.e., prior to photoexcitation, is the low density exciton spectrum. The label A indicates the exciton absorption peak (minimum of the optical transmission); labels B and C mark, respectively, the beginning of the continuum absorption and of the optically transparent region above and below the exciton resonance.

The spectrum of the highly excited sample is shown in the middle for a probe delay of 1 ps after the photoinjection of free electrons and holes. From a comparison with the spectrum on the left it is seen that the sharp exciton resonance has virtually disappeared. The difference spectrum depicted in Fig. 4.8b reveals that in addition to the decrease right at the exciton resonance there is also an increase of the absorption just below the resonance. This observation has been attributed by Shank et al. to band gap shrinkage (energy gap renormalization) resulting from exchange and correlation effects of the photogenerated free carriers (see Sect. 4.2.1).

Shank et al. concluded from their results that the screening of the exciton and the shift of the band edges occurs in less than 0.5 ps. The changes of the spectra at later times, viz. for a delay of 12 ps, were explained by band-filling effects due to the relaxation of the hot carrier distribution.

While clearly demonstrating complete wipe-out of the exciton on a subpicosecond time scale, the experiments of *Shank* et al. [4.26] do not provide any information on how screening of the exciton develops as a function of the density of photoinjected carriers. In fact, the density of free carriers employed in their experiment exceeded the critical density for the stability of the exciton suggested by the Mott criterion [4.27] (the Mott density) by approximately two orders of magnitude.

The details of the dependence of exciton screening on the intensity of the photoexcitation have been explored by *Fehrenbach* et al. [4.28,29]. In one type of experiment they use pulses of 8 ps duration from a near infrared synchronously mode-locked tunable dye laser to *resonantly* excite thin samples of GaAs (thickness 0.5 μm and 4.2 μm) at the center of the 1s exciton resonance (1.515 eV; $\lambda = 8183$ Å). A weak fraction of the pump beam serves to probe the changes in the absorption. The density of photoexcited excitons is determined from measurements of the incident, the reflected, and the transmitted energy of the excitation pulse. The energy balance gives the actual energy deposited in the sample and hence the density of excitons.

In Fig. 4.9 exciton spectra measured by *Fehrenbach* et al. [4.28] are reproduced. The spectra at the bottom and in the middle of Fig. 4.9 show the low density absorption of two GaAs crystals of different thickness. Transient high excitation spectra are shown in the upper part of Fig. 4.9 for three different pair densities at a probe delay time of about 10 ps. From these spectra the following interesting features can be noticed: (i) Up to a pair density as high as $2 \times 10^{16}\,\mathrm{cm}^{-3}$ the exciton spectrum is hardly changed at all; (ii) for higher densities progressive bleaching of the exciton absorption is observed; the peak

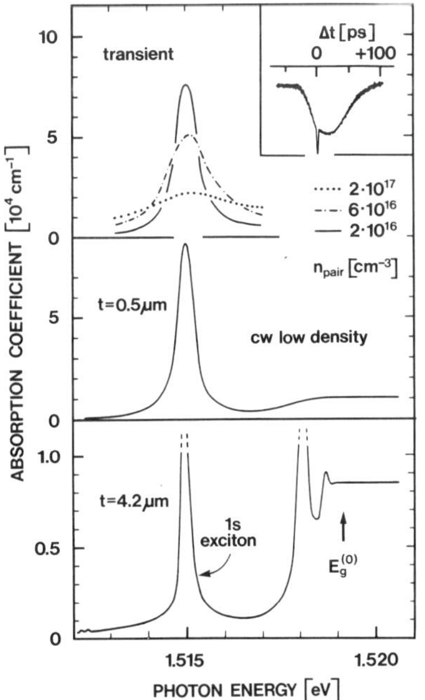

Fig. 4.9. Absorption spectra of GaAs near the fundamental exciton resonance measured with a weak probe pulse with 10 ps time delay with respect to the excitation pulse. The three traces at the top represent different excitation conditions with the indicated densities of photoexcited excitons. The insert shows time-dependence of the probe pulse absorption at 1.515 eV for a pump pulse producing an exciton density of $6 \times 10^{16}\,\mathrm{cm}^{-3}$. The middle and the bottom trace show the low density absorption spectrum for two different GaAs crystals with thicknesses 0.5 and 4.2 μm, respectively. After [4.28]

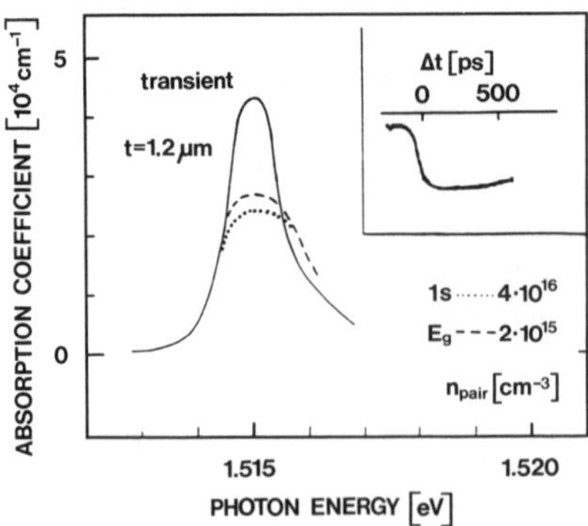

Fig. 4.10. Exciton absorption spectra of GaAs comparing resonant excitation (···) and non-resonant excitation (– – –). For resonant excitation the pump pulse wavelength was tuned into resonance with the 1s-excitation peak. For nonresonant excitation the crystal was pumped 10 meV above the band gap energy. After [4.29]

broadens asymmetrically, and the maximum decreases; (iii) up to the measured maximum density of about $2 \times 10^{17}\,\mathrm{cm}^{-3}$ the energy of the 1s exciton peak remains practically constant. The authors point out that for resonant excitation, distinct exciton resonances persist for densities exceeding the Mott density by approximately a factor of one hundred.

Exciton bleaching experiments have also been performed with non-resonant excitation [4.29]. The excess energy of the photogenerated free carriers was 10 meV. In these experiments two tunable synchronized mode-locked dye lasers were used to enable independent adjustment of the wavelengths of the excitation pulses and of the probe pulses. Qualitatively speaking, the bleaching caused by free carriers in non-resonant pumping was observed to be similar to the resonant pumping case. However, the most significant quantitative difference is the fact that comparable bleaching was observed to occur for a free carrier density roughly ten times lower than the exciton density in resonant bleaching. This behavior is illustrated by Fig. 4.10 taken from *Fehrenbach* et al. [4.29]. The full curve in Fig. 4.10 shows the low excitation 1s exciton resonance of GaAs. The dashed and the dotted curves represent, respectively, the bleached exciton resonance for a free carrier density of $2 \times 10^{15}\,\mathrm{cm}^{-3}$ (non-resonant case) and an exciton density of $4 \times 10^{16}\,\mathrm{cm}^{-3}$ (resonant excitation).

The more effective bleaching caused by the free carriers is quite obvious from these data. *Fehrenbach* et al. have discussed their results in terms of two different screening mechanisms: (i) A relatively weak dielectric screening due to excitons, which is operative during resonant excitation in the absence of free charges, and (ii) a strong metallic plasma screening caused by the free carriers for non-resonant excitation.

Ultrafast dynamics of exciton screening has also been studied in a number of other materials. *Fujimoto* et al. [4.30] have looked at the transient behavior of the optical reflectivity near the excitonic resonance in CdSe after the injection of free carriers by a 70 fs pump pulse. Bleaching of the excitonic resonance in CuCl has been studied by *Hulin* et al. [4.31,32]. Because of the small exciton Bohr radius, $a_B = 10\,\text{Å}$, exciton screening in CuCl is expected to occur at a much higher carrier densities than in GaAs ($a_B \simeq 600\,\text{Å}$).

Recent work on multiple-quantum-well structures (MQW) revealed new interesting aspects of the ultrafast dynamics of exciton screening and exciton bleaching. Striking observations concerning the difference of the screening by free carriers and excitons in GaAlAs/GaAs MQWs were reported by *Knox* et al. [4.33]. They studied the changes of the absorption near the $n = 1$ heavy-hole exciton resonance induced by subpicosecond pump pulses. Two different excitation conditions were compared: (i) nearly resonant situation with a pump wavelength chosen to primarily create excitons, and (ii) a non-resonant situation with pump pulses of shorter wavelength for the generation of free carriers. Suitable pump pulses were produced by selecting and subsequently amplifying the desired pump pulse spectrum from broadband subpicosecond continuum pulses. Another part of the continuum served as probe pulse to measure the changes of the absorption.

Figure 4.11 depicts examples of spectra measured by *Knox* et al. [4.33]. The trace at the bottom (Fig. 4.11d) shows the pump pulse spectra for resonant and non-resonant excitation. Traces a), b) and c) represent the absorption spectrum for the resonant case and probe pulse delay times corresponding to 0, 0.1, and 0.5 ps. The absorption spectrum of the unperturbed sample is shown as a thin line in each trace for comparison. The feature of interest here is the behavior of the heavy-hole exciton peak marked by the dashed vertical line. It is seen that the heavy-hole exciton resonance is bleached rapidly and efficiently. A fast partial recovery is observed in a few hundred femtoseconds after the excitation pulse maximum. For non-resonant excitation, on the other hand, this fast partial recovery is not observed. This point is illustrated in more detail by Fig. 4.12, from *Knox* et al. [4.33]. The full curve and the dashed curve represent, respectively, the time dependence of the absorption at the peak of the heavy-hole exciton for resonant and non-resonant pumping. The fast partial recovery of bleaching for resonant, and the progressive bleaching for non-resonant pumping is quite obvious from these data.

Knox et al. offered the following interpretation of their results. Since the resonant pump pulse primarily increases the density of excitons, the initial bleaching of the absorption must be caused by the excitons themselves. On the other hand, the excitons are short lived because at room temperature they are ionized very rapidly by collisions with thermal phonons, yielding free electron-hole pairs. The observed 300 fs time constant of the partial recovery (see full curve in Fig. 4.12) is interpreted as the exciton ionization time. After the complete fission of the primary excitons, only free electron-hole pairs are left. The system has assumed a state that is the same as the state established by non-resonant pumping. Thus, for times much longer than the exciton ionization time, the spectra are independent of the pump wavelength.

As a test for the proposed role of thermal phonons the resonant pumping experiment was repeated at low temperature $T = 15$ K with negligible concentration of thermal phonons. The result is shown by the dot-dashed curve in Fig. 4.12. The recovery time of the absorption is substantially longer – according to [4.33] – consistent with exciton ionization via phonon collisions being the

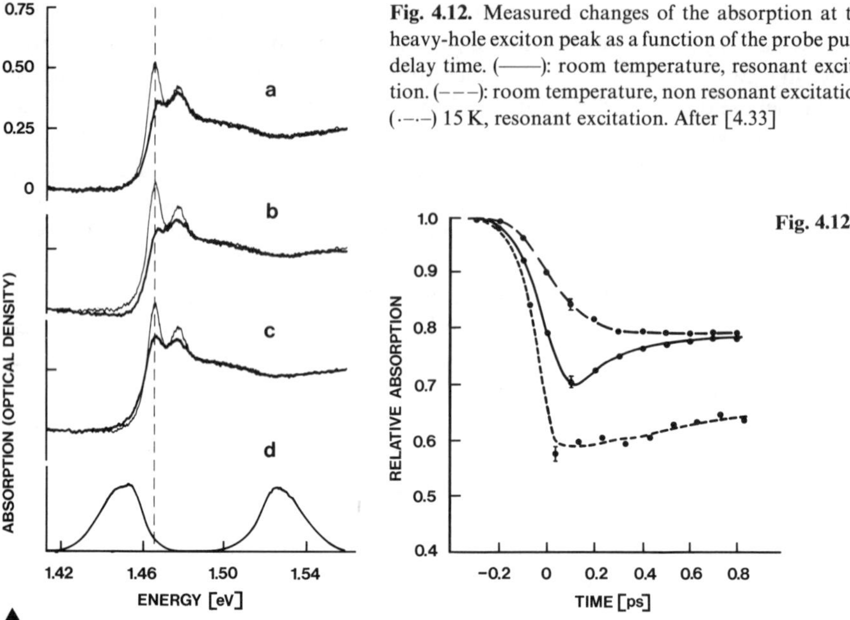

Fig. 4.12. Measured changes of the absorption at the heavy-hole exciton peak as a function of the probe pulse delay time. (——): room temperature, resonant excitation. (– – –): room temperature, non resonant excitation. (·–·–) 15 K, resonant excitation. After [4.33]

Fig. 4.11. Exciton absorption spectra of a GaAs/GaAlAs multiple quantum well sample. The vertical dashed line indicates the position of the heavy-hole exciton resonance. The thin line shows the spectrum without excitation. The heavy lines represent the spectra for different delay times between the continuum probe pulse and the pump pulse; a: 0; b: 0.1 ps; c: 0.5 ps. Trace d shows the pump pulse spectra for resonant and non-resonant excitation. After [4.33]

mechanism of the partial absorption recovery at room temperature. From the interpretation of Knox et al. it follows that in the GaAs/GaAlAs MQW excitons bleach the absorption more effectively than free electron-hole pairs – a conclusion which seems to be in contradiction with the results obtained by *Fehrenbach* et al. [4.29] in their comparison of exciton and free carrier screening in bulk GaAs samples. Knox et al. invoked excitonic phase-space filling as the basic mechanism of the observed exciton bleaching. Excitonic states are built up from single particle states of the system. These states are filled up as more excitons are excited. The authors of [4.33] argue that bleaching caused by exciton phase-space filling is large, whereas band filling and screening are strongly reduced because of the high carrier temperature. It appears that the issue of screening by excitons and free carriers in bulk and in quasi-two-dimensional materials is an interesting open question.

Peyghambarian et al. [4.34] reported a similar study of the effect of non-resonant and resonant photoexcitation on the heavy-hole exciton in GaAs/GaAlAs MQWs. The experimental technique was practically the same as in [4.33]. Peyghambarian et al. observed a small shift of the exciton resonance of about 2 meV to higher energies. The blue shifts have been tentatively attributed to exciton-exciton interactions in a dense exciton gas, which – according to the authors of [4.34] – could lead to a renormalization of the exciton self-energy.

4.2 Electron-Hole Dynamics

Electrons and holes are the basic carriers of electrical charge in semiconductors. The fundamental importance of understanding the dynamics of charge carriers hardly needs to be emphasized. A wealth of useful information about carrier interactions has been obtained in the past from electrical transport measurements [4.35]. The dynamics of the carriers generally involves a great variety of different interactions. Electrical transport experiments usually give some kind of integral information involving averages over the various individual processes. The study of the dynamics of photoexcited carriers, on the other hand, is in principle capable of providing more selective information.

Optical excitation with visible or near infrared light promotes electrons and holes to states with excess energies of typically several hundred meV with respect to the ground state. There is a hierachy of relaxation mechanisms which cause the excited electrons and holes to return to thermal equilibrium. Relaxation involves a number of different steps with rates ranging from 10^{-14} s to about 10^{-9} s.

The fate of the photoexcited carriers may be qualitatively described as follows. The first step is scattering of the carriers out of the primary optically coupled valence and the conduction band states. This process is believed to be dominated by very fast carrier–carrier scattering. A quasi-thermal distribution is eventually established among the electrons and the holes by carrier–carrier collisions and carrier–phonon scattering processes. At this stage the energy distributions of the electrons and the holes are given by Fermi-Dirac distributions characterized by an electron temperature T_e, and a hole temperature T_h. Usually it is assumed that the interaction between electrons and holes is sufficiently strong that the carriers possess a common temperature, $T_c = T_e = T_h$.

The carrier temperature may still be substantially greater than the lattice temperature, $T_c > T_L$. The further evolution can then be described as cooling of the carriers resulting from coupling of the electronic system to the cold lattice. The carriers lose their energy primarily by emission of phonons. The energy loss is usually dominated by emission of optical phonons, mainly because on the average they carry away more energy than acoustic phonons. In semiconductors with polar or partial polar character energy loss due to emission of longitudinal optical phonons (LO-phonons) is likely to be the most efficient energy loss mechanism.

The life of the electrons and holes is finally terminated by carrier recombination processes, which may occur even before the carriers have equilibrated with the lattice temperature. In radiative recombination an electron-hole pair is destroyed and a photon carrying away the pair energy is created. Non-radiative recombination processes may also play a role. For example, at high electron-hole density the recombination may be dominated by a non-radiative Auger process in which the recombination energy of the electron-hole pair is given to a third carrier rather than to a photon.

Much of the picosecond work on electron-hole dynamics has been concerned with strong excitation situations in which the density of photoexcited carriers is so large that the carrier-carrier interactions and many-body effects become important. The properties and the relaxation of hot and dense electron-hole plasmas (EHP) will be the first topic of this section. While the cooling and the recombination of the quasi-thermal EHPs occur on the time scale of a few picoseconds up to perhaps a few hundred picoseconds, the much faster subpicosecond processes which establish quasi-thermal carrier distributions have become accessible to direct observation only recently with the advent of femtosecond optical techniques. Several examples of this category of experiments will be discussed. The remainder of this section will be devoted to electron-hole dynamics in disordered semiconductors and semiconductors with quasi two-dimensional quantum well structures.

4.2.1 Electron-Hole Plasma

The following two characteristic features of the electron-hole plasma (EHP) state should be recalled [4.36]: (i) The density of electron-hole pairs is relatively high and thus screening of the Coulomb interaction and many-body exchange and correlation effects play an important role; (ii) the electrons and holes are assumed to be in thermal equilibrium; the energy distributions of the electrons and holes are given by Fermi-Dirac distributions characterized by an electronic carrier temperature T_c.

Figure 4.13 is a schematic comparison of the low density excitonic optical spectrum with a typical EHP spectrum in direct gap semiconductors. The low

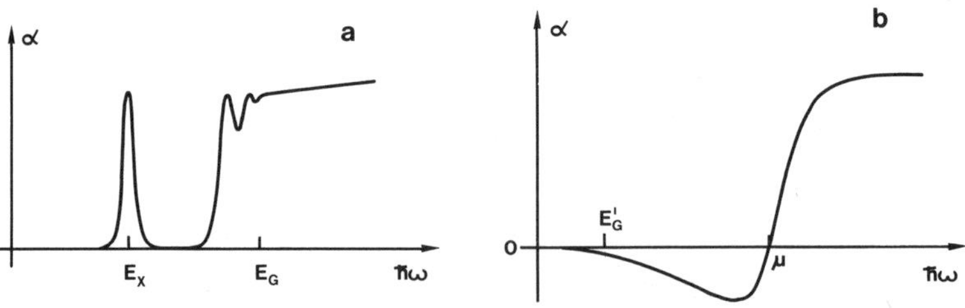

Fig. 4.13a, b. Schematic representation of the absorption spectrum of a direct gap semiconductor near the fundamental band gap. Absorption coefficient α versus photon energy $\hbar\omega$. (**a**) Low excitation excitonic spectrum with the $n = 1$ exciton peak (E_x), some higher exciton peaks, and the band gap energy (E_g). (**b**) High excitation spectrum with the band gap energy shifted to E'_g. μ is the chemical potential of the electron-hole plasma

density spectrum (Fig. 4.13a) is dominated by the fundamental exciton peak ($n = 1$) at E_x, which is followed at higher energies by excited state exciton peaks, and the onset of continuum absorption at the band gap energy E_G.

In the EHP state (Fig. 4.13b) the band gap is shifted to a lower energy $E'_G < E_G$. The shift is a result of a renormalization of the band gap due to exchange and correlation effects [4.36]. The band states just above (below) the renormalized gap are filled with electrons (holes); thus an inversion of the population exists, and optical gain due to stimulated emission occurs for energies $E'_G < \hbar\omega < E'_G + F_e + F_h$, where F_e and F_h are the Fermi energies of the electrons and the holes, respectively. The chemical potential $\mu = E'_G + F_e + F_h$ marks the transition from optical gain to absorption.

The gain and the absorption spectra of EHPs are strongly dependent on the carrier density N_c (density of electron-hole pairs), and on the carrier temperature T_c. Time-resolved measurements of the optical spectra can be used to obtain carrier cooling rates and carrier recombination rates. However, the interpretation of experimental EHP spectra in terms of N_c and T_c is complicated. Different simplified models have been used. *Göbel* [4.37] has pointed out that for energies not too much greater than E'_G the shape of the experimental spectra can be very well accounted for by assuming that wave-vector conservation for the electrons and holes during the interaction with the light can be neglected; thus the interaction is effectively described in terms of indirect optical transitions. *Shah* et al. [4.38] showed that for energies well above E'_G the absorption spectrum of the EHP can be approximated by $\alpha(\omega) = \alpha_0(\omega)(1 + F_e + F_h)$, where $\alpha_0(\omega)$ is the unperturbed continuum absorption coefficient, and F_e and F_h are the Fermi-Dirac distribution functions of the electrons and the holes.

More rigorous treatments of the problem based on calculations of the optical dielectric function from many-body theory have been presented by *Arya* and *Hanke* [4.39], and *Haug* et al. [4.40]. Fortunately, it turned out that the results agreed surprisingly well with the simplified models. For example, in GaAs there is good agreement of the electronic densities and temperatures inferred in both ways from the spectra of quasi-stationary photoexcited EHPs [4.30]. Thus it appears that there is a fairly sound basis for deducing temperatures and densities from measured spectra.

Most of the picosecond time-resolved work has been concerned with the dynamics of carrier relaxation and recombination. Some work has also been done on the dynamics of the Mott transition and the possibility of a condensation of the EHP to droplets of electron-hole liquid.

One of the first observations of EHP dynamics on a time scale of a few picoseconds is due to *Shank* et al. [4.26]. They photoexcited GaAs via band-to-band transitions with subpicosecond pump pulses and measured the induced changes of the optical transmission spectrum with the help of broadband sub-picosecond probe pulses. They were able to observe fast screening of the exciton resonance, shifting of the band gap energy, and energy relaxation of the photo-excited EHP. A time constant of 4 ps was obtained for the cooling of the hot

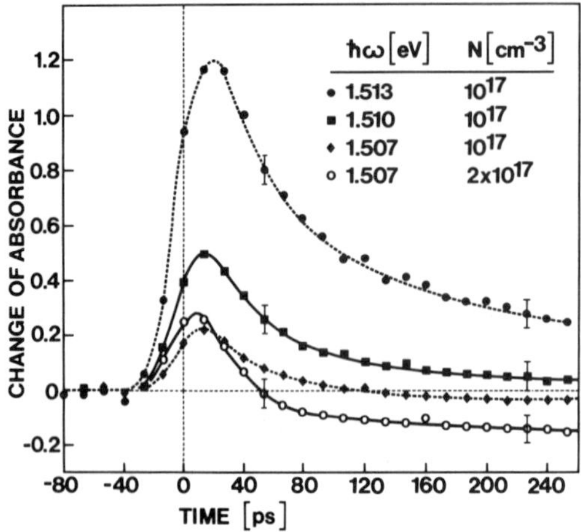

Fig. 4.14. Change of absorbance of a GaAs crystal as a function of time after two-photon band-to-band excitation. $\hbar\omega$ is the photon energy of the probe pulse. N is the density of electron-hole pairs. After [4.41]

EHP to the lattice temperature, which was 80 K in their experiments. The cooling was attributed to the emission of LO-phonons.

The detailed temperature evolution of a photoexcited EHP in GaAs has been obtained by *von der Linde* and *Lambrich* [4.41] from time-resolved measurements of gain and absorption at the low energy edge of the spectrum below the exciton energy, $\hbar\omega < E_x = 1.515\,\mathrm{eV}$. Because the induced optical changes in this spectral region are rather small, a relatively thick GaAs crystal of 8 μm had to be used. To ensure homogeneous photoexcitation, the EHP was generated via two-photon absorption of a 25 ps near-infrared (1064 nm) pump pulse from a mode-locked Nd:YAG laser. The photoexcited EHP was probed using frequency-tunable picosecond pulses from a synchronously mode-locked $LiNbO_3$ parametric oscillator pumped by the second harmonic of the Nd:YAG laser.

Results from von der Linde and Lambrich are reproduced in Fig. 4.14. The observed changes of the absorbance are plotted as a function of time for three different probe photon energies, $\hbar\omega < E_x$. In all cases an increase of the absorption during the pump process is observed, followed by a decrease on a time scale of about 150 ps. Note that for the probe photon energy $\hbar\omega = 1.507\,\mathrm{eV}$ the absorbance changes sign and becomes negative, that is to say, a transition from induced absorption to optical gain is observed. The general features of the absorbance changes follow directly from the properties of the EHP: (i) Induced absorption for $\hbar\omega < E_x$ signifies the shift of the band gap due to renormalization; (ii) the positive sign of the initial induced change is an indication that the plasma is hot. For high electron temperatures the chemical potential is less than the renormalized band gap, $\mu < E'_G$, and the plasma is absorbing for all frequencies; (iii) as the plasma cools down, the chemical potential moves to higher energies. The onset of optical gain at the photon energy $\hbar\omega$ corresponds to the time when $\mu(T_c) = \hbar\omega$.

The detailed temperature evolution of the EHP was obtained from a quantitative evaluation of the data based on the indirect transition model (i.e. neglect

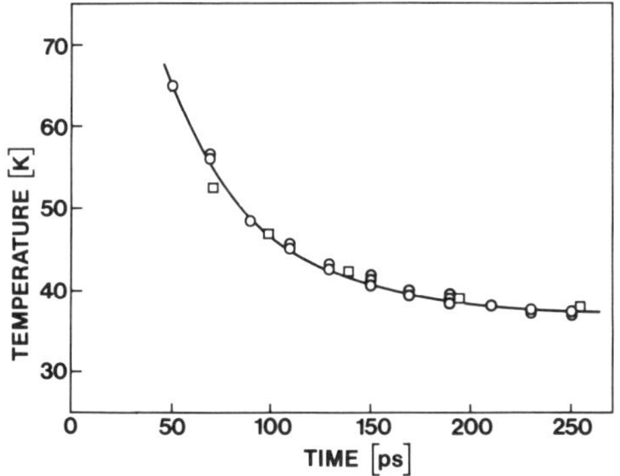

Fig. 4.15. Temperature of the electron-hole plasma as a function of time as inferred from the observed changes of the absorption for two-photon excited GaAs. After [4.41]

of the k-selection rule). The analysis indicated that a maximal plasma temperature of 120 K is reached at 20 ps after the maximum of the excitation pulse. The subsequent cooling of the EHP for times greater than 50 ps is shown in Fig. 4.15. It can be seen that the relatively fast initial temperature drop levels off at a plasma temperature near 40 K, which is still much higher than the lattice temperature, $T_L = 7$ K. Von der Linde and Lambrich showed that the observed cooling of the plasma is consistent with an energy loss mechanism dominated by LO-phonon emission. The solid curve in Fig. 4.15 represents the calculated temperature variation assuming unscreened carrier–LO-phonon scattering processes. The calculation neglects carrier recombination and acoustic phonon emission which are expected to play a role for longer times.

Time-resolved measurements of spontaneous photoluminescence spectra have been used by *Tanaka* et al. [4.42,43] to study the dynamics of EHPs in GaAs. They used excitation pulses of 20 ps from the second harmonic of a Nd:YAG laser and a CS_2 optical Kerr shutter gated by the fundamental pulses (1064 nm). The time evolution of the chemical potential and of the renormalized band gap were deduced from the luminescence data on the basis of the indirect transition model. They inferred a carrier temperature of $\simeq 90$ K at 10 ps after the excitation and found a cooling down to $\simeq 50$ K within 150 ps. For longer times the chemical potential and the gap energy remained approximately constant, indicating that the temperature and the density of the EHP in the late stage of the evolution are slowly varying.

Tanaka et al. discussed the possibility of the formation of electron-hole liquid droplets. Since both the observed carrier temperatures and the estimated density were much greater than the expected critical values for liquid formation, they concluded that in their experiments the formation of an electron-hole liquid is unlikely.

One of the problems of *spontaneous* luminescence spectroscopy in highly excited direct gap semiconductors is to avoid stimulated emission. However, using a suitable experimental geometry *Tanaka* et al. [4.44] were able to obtain time-resolved gain spectra from measurements of the *stimulated* emission in GaAs. From an indirect-transition-model line shape analysis they obtained the time evolution of the temperature (T_c) and the density (N_c) of the EHP. Their results indicate that in 160 ps following the excitation pulse T_c decreases from $\simeq 60$ to 34 K, and N_c decreases from $1.5 \times 10^{17}\,\mathrm{cm}^{-3}$ to $0.7 \times 10^{17}\,\mathrm{cm}^{-3}$. Good agreement of the measured cooling with the LO-phonon emission loss mechanism was noted. The observed time variation of the plasma density could be accounted for by a bimolecular electron-hole recombination process with a coefficient $B = 10^{-7}\,\mathrm{cm}^3/\mathrm{s}$.

From the different types of picosecond measurements a fairly consistent picture arises for the dynamics of EHPs in low temperature GaAs ($T = 5$ to 10 K). For pump pulses providing typically 100 meV or several 100 meV of electronic excess energy and carrier densities of about $10^{17}\,\mathrm{cm}^{-3}$ a maximal electron-hole temperature of roughly 100 K is achieved. The plasma cools down to about 30–40 K in 150–200 ps. The electronic temperature remains much higher than the lattice temperature during the entire lifetime of the plasma. The cooling is dominated by LO-phonon emission, although it appears that during the late stage of the evolution LO-phonon emission becomes ineffective and energy losses due to acoustic phonon interactions begin to play a role.

Picosecond time scale investigations of the dynamics of dense EHPs in II-VI semiconductors have been reported in a series of papers by *Yoshida* et al. [4.45–49]. They used a CS_2 optical Kerr shutter to observe time-resolved spontaneous luminescence spectra. According to the authors great care was exercised to avoid competition of stimulated emission. The time evolution of electronic densities and temperatures was determined from analyses of the spectral line shapes using analogous procedures as for GaAs [4.42].

In wurtzite semiconductors like CdSe and CdS, two different kinds of EHPs can be observed which consist of conduction electrons and A-valence-band holes or B-valence-band holes. The luminescence bands due to the two different components of the plasma can be readily distinguished because of the different polarization properties. With suitable pumping both the A-band and the B-band luminescence appears initially. According to *Yoshida* et al. [4.45] the B-band luminescence decays rapidly and vanishes because the B-holes relax to the higher A valence band. The inter-valence-band hole relaxation time was measured to be 35 ps and 55 ps for CdSe and CdS, respectively. The hole relaxation has been attributed by Yoshida et al. to acoustic phonon scattering in CdSe, and acoustic and optical phonon processes in CdS.

In a different experiment *Yoshida* et al. [4.46] were able to selectively excite the A-type EHP in CdSe using a tunable picosecond dye laser. From time-resolved spontaneous luminescence measurements they deduced the temporal evolution of the density and the temperature of the photoexcited electrons and holes. It was reported for example, that N_c decreases from $8 \times 10^{17}\,\mathrm{cm}^{-3}$ to $3 \times 10^{17}\,\mathrm{cm}^{-3}$, and T_c from 90 to 20 K during the first 150 ps following excitation. Somewhat surprisingly, N_c and T_c seem to remain practically constant during the remainder of the observation time up to 800 ps. Similar results were obtained in CdS [4.47].

While the initial relaxation of the EHPs in CdSe and CdS can be understood in terms of carrier cooling by LO-phonon emission and electron-hole recombination, the explanation of the constant spectra for long times turned out to be more difficult. According to *Yoshida* et al. [4.48] the temperatures inferred from the spectra of the late stage are less than the critical temperatures, and the densities are close to the critical densities for the formation of a condensed electron-hole liquid state. Nevertheless this possibility was ruled out. In fact, experiments performed at temperatures much higher than the critical temperatures showed similar constant spectra for long times [4.49]. More recent experiments of *Unuma* et al. in CdSe [4.50], and of *Saito* and *Göbel* in CdS [4.51] provided further convincing evidence against an interpretation of the late stage of the luminscence spectra in terms of electron-hole liquid formation. These authors presented data suggesting that as time proceeds the EHP state is transformed into an excitonic state. In other words, a reverse Mott transition takes place, and the spectra at late times cannot be interpreted in terms of temperature and density of an EHP. Rather, this emission is more likely to be due to excitonic scattering processes and/or biexciton recombination. In fact, similar observations on the transformation of the initial photoexcited EHP into biexcitons have been reported by *Hulin* et al. [4.52] for CuCl, where the situation is less complicated, because the EHP luminescence band and the biexciton band are quite well seperated in energy.

Let us now continue with the discussion of carrier cooling in EHPs. Experiments in GaAs have indicated that for carrier densities up to about $10^{17}\,\mathrm{cm}^{-3}$ the observed cooling rates can be explained by energy losses due to the emission of LO-phonons, with the Fröhlich-interaction as the dominant carrier–phonon coupling mechanism. However, for higher plasma densities the energy relaxation is slowed down below the rates expected for unscreened Fröhlich-type carrier–phonon scattering. For example, in an extension of the work of *Shank* et al. [4.26] *Leheny* et al. [4.53] have studied the dependence of the cooling rate on the density of photoexcited carriers. To explain the measured cooling curve in GaAs for $N_c = 5 \times 10^{17}\,\mathrm{cm}^{-3}$ Leheny et al. had to assume a decrease of the carrier–phonon scattering rate by a factor of 5.

Shank et al. [4.54] compared hot carrier relaxation in GaAs/GaAlAs multiple-quantum-well (MQW) structures and in bulk GaAs. They found that the cooling rates in the quasi-two-dimensional system and in bulk material are approximately the same; slowing down of the energy relaxation with increasing carrier density was also observed in the MQW samples. *Graudszus* and *Göbel* [4.55] obtained the EHP temperature in GaAs from time-resolved measurements of the spontaneous luminescence for densities of several times $10^{18}\,\mathrm{cm}^{-3}$. Their data indicated much higher maximal EHP temperatures than the values expected for unscreened carrier–phonon interaction. Related observations on "anomalously" slow energy relaxation at high density have been reported by *Seymour* et al. [4.56] for GaAs, by *Xu* and *Tang* [4.57] for GaAs/GaAlAs MQW structures, and also for CuCl by *Antonetti* et al. [4.58].

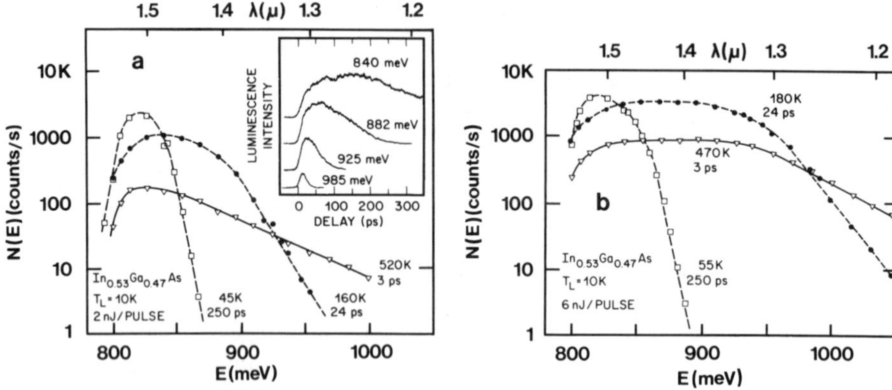

Fig. 4.16a, b. Photoluminescence spectra of InGaAs after excitation with an 8 ps laser pulse at 610 nm. Excitation pulse energy: (a) 2 nJ; (b) 6 nJ. Spectra are shown for three different delay times: 3 ps, 24 ps, and 250 ps. After [4.60]

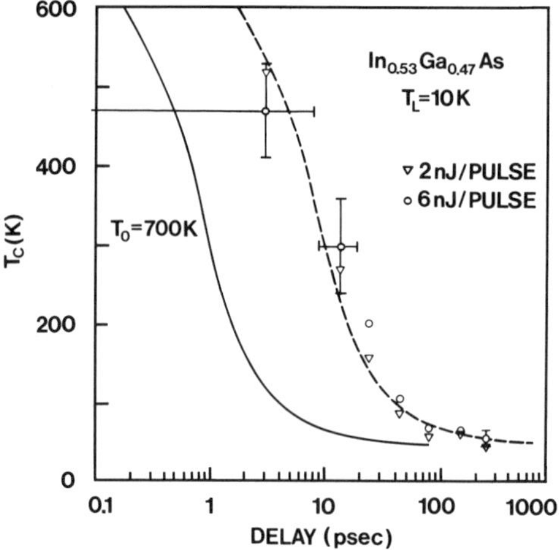

Fig. 4.17. Carrier temperature as a function of time. Solid line: Theoretical curve for uncreened carrier–LO-phonon interaction, for an initial carrier temperature of $T_0 = 700$ K. The dashed line was obtained by shifting the solid line to longer times by a factor of 10. After [4.60]

A detailed study of EHP cooling for several alloy III-V semiconductors and GaAs MQWs has been presented by *Kash* et al. [4.59]. They used an up-conversion technique to observe picosecond time-resolved near-infrared photoluminescence spectra. Figure 4.16 is an example taken from *Kash* and *Shah* [4.60] which depicts a series of InGaAs luminescence spectra corresponding to different delay times after photoexcitation. Results for two excitation energies are shown: a) 2 nJ, and b) 6 nJ. Note the approximately exponential shapes of the high energy edge of the luminescence spectra which signify the Maxwellian tails of the thermalized carrier distribution. The excellent dynamic range of about three orders of magnitude permits the carrier temperature to be directly extracted from the shape of the high energy edge of the spectra. As shown in Fig. 4.17 Kash and

Shah were able to follow the temperature evolution of the photoexcited EHP from 500 K at 3 ps down to about 50 K at 250 ps.

In Fig. 4.17 the experimental temperature curve (dashed line) is compared with the expected cooling curve for unscreened carrier–LO-phonon interaction (solid curve). The comparison shows that the measured cooling is much slower. Another interesting observation is that an increase of the photoexcitation by a factor of three does not change the cooling curve within experimental accuracy. According to Kash and Shah this result is inconsistent with the suggestion that the reduced cooling is due to screening of the carrier–LO-phonon interaction.

A number of authors have pointed out the significance of hot phonon effects for understanding the energy relaxation of EHPs at high densities [4.61–64]. The key point is that during cooling of the plasma by LO-phonon emission a substantial non-equilibrium population of LO-phonons is built up in the receiving LO-modes. It is clear that the transfer of energy from the EHP to the LO-lattice modes is strongly reduced as the effective phonon temperature approaches the electronic temperature. In fact, *Pötz* and *Kocevar* [4.63] showed that in GaAs the rapid build-up of a non-equilibrium LO-phonon population may actually lead to a *reversal* of the energy transfer, that is to say, a reabsorption of LO-phonons by the plasma. This situation is possible since the cooling of the EHP by other phonon emission processes, e.g. emission of transverse optical phonons by holes, can be faster than the relaxation of the nonequilibrium LO-phonons.

Notwithstanding the role of nonequilibrium phonons it is clear that screening of the Fröhlich interactions by high density electrons and holes must also be present. However, the interplay between LO-phonon screening and population effects can be quite complicated. The presence of screening, on the one hand, decreases the cooling rate of the EHP because the emission rate of LO-phonons is decreased. Reduced LO-phonon emission rate, on the other hand, reduces the non-equilibrium phonon population, leaving the EHP cooling rate unaffected by hot phonon effects. Thus it appears that for a complete understanding of the cooling of dense EHPs a consistent treatment of the coupled carrier–phonon system is required.

4.2.2 Carrier Thermalization

Progress in ultrafast laser pulse techniques has provided opportunities to perform optical experiments in the subpicosecond time regime, with a resolution now reaching 10 fs. This is the appropriate time scale for the study of the primary relaxation process that immediately follow photoexcitation of electrons and holes. As discussed in the previous section, the dynamics of *thermalized* carriers is the domain of *picosecond* experiments. *Femtosecond* techniques, on the other hand, enable investigation of the earlier stage, the study of the actual process by which thermal equilibrium among the photoexcited carriers is established.

As a first example of the direct observation of the initial relaxation of photoexcited carriers let us discuss experiments reported by *Tang* and coworkers

[4.65–69]. They used femtosecond pulses to study the saturation of the optical absorption in GaAs and GaAlAs semiconductors and related multiple-quantum-well structures. In these experiments the output pulse train from a colliding pulse, passively mode-locked dye laser with pulses of about 90 fs at 615 nm is used. The experiments are based on a variant of the equal-pulse-correlation technique [4.70]. Two equally intense, collinear, orthogonally polarized pulses with an adjustable delay interact with the semiconductor sample. The first pulse produces a population of carriers in certain energy states of the band structure. After some delay time the second pulse interacts with the very same band states. In the experiment, the combined, time-averaged transmission is measured as a function of the delay time between the two beams. If the delay time is much longer than both the pulse duration and the lifetime of the population in the optically coupled band states, the transmission is independent of the delay. However, a delay-dependent enhancement of the transmission will be observed when the second pulse arrives earlier, e.g., at a time when the population of the electronic states produced by the first pulse is still appreciable. From the time-correlated transmission measurement information about the carrier relaxation times can be obtained, if the pulse duration is shorter than the relaxation times. If not, the time-correlated transmission simply reproduces the autocorrelation function of the pulses.

In experiments with overlapping pulses of the same frequency some care must be exercised to avoid complications related to the so-called coherent artifact. To some extent the problem can be minimized with the use of orthogonal polarization. A detailed discussion of the relevance of coherent artifacts in the time-correlated transmission experiments has been given by *Taylor* et al. [4.69].

An example of a time-correlated transmission measurement on a GaAs/GaAlAs multiple-quantum-well sample [4.66] is given by the solid curve in Fig. 4.18a. The data show a small enhancement of the transmission for a delay time of a few tenths of a picosecond. The signal maximum at zero delay corresponds to an increase of the total transmitted light flux of about 0.1%. Comparison with

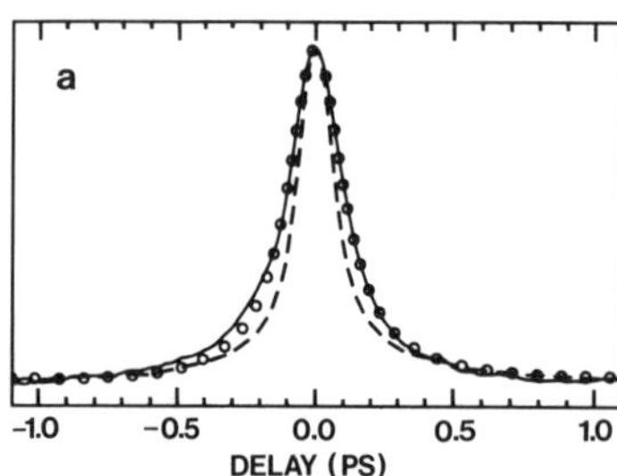

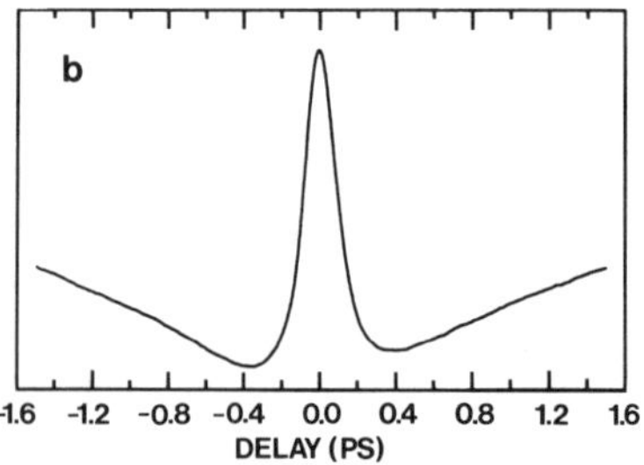

Fig. 4.18. (**a**) Time-correlated transmission measurement for a GaAs/-GaAlAs MQW sample (*solid line*). The dashed line is the autocorrelation function of the laser pulses. The open circles represent a fit of the data with an exponential response function with a time constant $t_r = 44$ fs. (**b**) Time correlated transmission measurement at higher laser power (not discussed here). After [4.66]

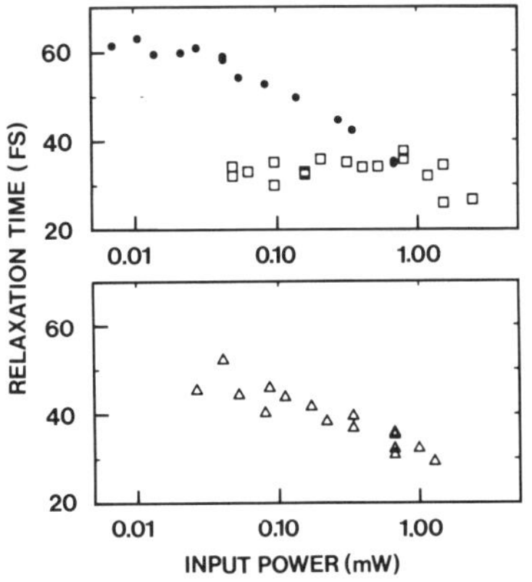

Fig. 4.19. Measured relaxation times as a function of laser input power for various samples. Top: bulk GaAs (●) and GaAlAs (□). Bottom: GaAlAs MQW sample (△). 1 W corresponds to an estimated carrier density of 1.8×10^{19} cm^{-3} (●), $4 \times 10^{19} cm^{-3}$ (□), and $6 \times 10^{19} cm^{-3}$ (△). After [4.66]

the pulse autocorrelation function (dashed curve) indicates a small, but distinct broadening of the time-correlated transmission peak. It has been shown that under suitable circumstances the shape of the transmission peak is given by the convolution of the autocorrelation of the excitation pulse with the population decay function of the excited states [4.65,68,69]. The circles in Fig. 4.18a represent a fit of the data by a convolution with an exponential decay function. An excellent fit is obtained with an excited state lifetime of $t_r = 44$ fs.

Figure 4.19 from *Erskine* et al. [4.66] shows the measured relaxation times as a function of the input power for three different materials. The carrier densities corresponding to the input powers in Fig. 4.19 have been estimated to range from about $10^{17} cm^{-3}$ to $5 \times 10^{19} cm^{-3}$. In all cases the relaxation times t_r are very short, well below 100 fs. A distinct decrease of t_r with carrier density is observed for GaAlAs (circles in the top part of Fig. 4.19), whereas in GaAs t_r appears to be independent of the density of photoexcited carriers. It is also interesting to note that the measured initial relaxation times in bulk material and in multiple quantum wells are comparable, i.e., the two-dimensional substructure of the quantum-well samples apparently does not strongly effect the primary relaxation processes.

Tang and coworkers have discussed the possible physical mechanisms of the initial relaxation of the excited states. They came to the conclusion that the dominant rclaxation mechanism in their experiments is intervalley electron-phonon scattering.

More information about the detailed changes of the initial photoexcited carrier distribution in GaAs/GaAlAs multiple-quantum-well samples has been obtained by *Knox* et al. [4.71]. They were able not only to observe the population and depopulation of the initally excited band states, but also to follow the intermediate steps during the evolution of the carrier distribution to a thermalized distribution. To perform these measurements pulses of 50 fs duration from a colliding pulse passively mode-locked dye laser are amplified to energies of a few microjoules. These energetic pulses can be used to generate an intense white-light

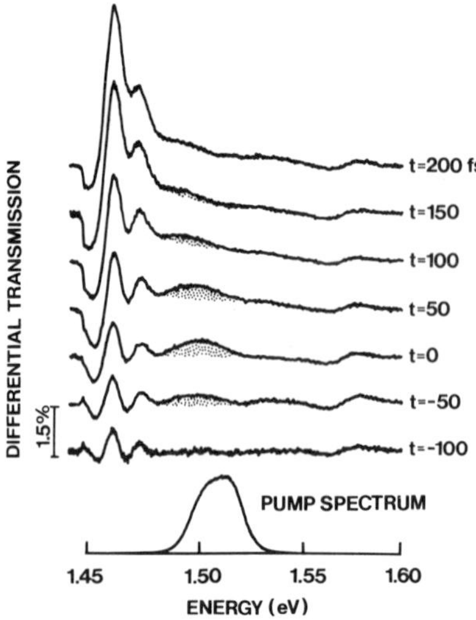

Fig. 4.20. Differential transmission spectra of a GaAs/GaAlAs MQW sample for various delay times of the continuum probe pulse. The bottom trace shows the spectrum of the pump pulse. The dotted area indicates the bleached feature discussed in the text. After [4.71]

continuum of about 50 fs duration. A small fraction of the continuum is used as a broad-band probe pulse to measure changes of the optical absorption spectrum of the sample. The main portion of the continuum is spectrally filtered to a bandwidth of about 10 nm, and serves as a frequency-tunable pump pulse.

Knox et al. measured the changes of the transmission of a GaAs/GaAlAs MQW sample induced by the femtosecond pump pulses with photon energies relatively close to the lowest exciton (excess energies of several tens of meV). An example of their results is depicted in Fig. 4.20, which shows the measured differential transmission spectra for different probe pulse delay times. The pump pulse spectrum is shown at the bottom of Fig. 4.20. The spectral features of interest here are marked by the dotted areas. For example, the spectrum corresponding to a delay of −50 fs exhibits an area of increased transmission, which appears to be slightly broadened and shifted to lower energies in comparison with the pump pulse spectrum. This bleaching of the absorption is caused be the initial non-equilibrium population of the primary optically coupled states. It appears, however, that the broadening and the shifts already signify some effect of relaxation at this very early stage. As time proceeds the transmission peak shifts to lower energies and continues to broaden, merging into a continuum after about 200 fs. Knox et al. noted that after 200 fs the spectra remain practically unchanged, indicating that by this time the thermalization of the carriers is almost complete. It was inferred from the spectral data that the average kinetic energy of the carriers is conserved during the thermalization, suggesting that thermalization mechanism is dominated by carrier–carrier scattering, while LO-phonon scattering does not seem to play an important role at this stage.

Some interesting conclusions about the mechanisms of exciton bleaching could also be derived from the data. The strong peaks of the differential spectra near 1.46 eV and 1.475 eV are due to bleaching of the heavy-hole and the light-hole exciton, respectively. It can be seen that it takes about 150 fs for the bleaching to fully develop. According to Knox et al. this observation indicates a dominance of exclusion principle effects over the screening of the Coulomb interaction in the bleaching mechanism of the lowest excitons.

In related work *Oudar* et al. [4.72] have reported the observation of dynamic spectral hole burning in bulk GaAs at low temperature (15 K). They used pump pulses of 0.5 ps duration from a suitably filtered and amplified white-light-continuum pulse to populate states near the band gap. The pump pulse photon energy ($\hbar\omega = 1.54$ eV) is chosen such that the carrier excess energy is less than the LO-phonon energy. The transmission changes of the bulk GaAs sample induced by the pump pulse were measured with a weak 100 fs broad-band-continuum pulse. Oudar et al. observed a bleached feature centered around the frequency of the pump pulse. The spectral width of the "burned hole" was about three times larger than the width of the pump pulse spectrum. According to Oudar et al. this bleached feature disappears together with the excitation pulse, leaving a broad, continuous, bleached transmission spectrum characteristic of a thermalized distribution of hot carriers.

To describe the transition from the initial non-equilibrium carrier distribution to a quasi-thermal distribution a simple model was used which treats the band-to-band transition as an assembly of homogeneously broadened two-level systems. According to this model the absorption spectrum of the photoexcited material exhibits a quasi-continuous bleached feature due to the thermalized carriers plus a spectral hole in the form of a power-broadened Lorentzian. The latter reflects the non-equilibrium carrier population in the primary optically coupled band states. From a lineshape analysis in the framework of the two-level model Oudar et al. derived the value of the dephasing time, $T_2 = 0.3$ ps. The dephasing time is related to the intrinsic lifetime of the Bloch states.

In some earlier work *Oudar* et al. [4.73] dealt with another interesting aspect of the initial evolution of the photoexcited carriers. It is known that for photo-excitation with linearly, polarized light the momentum distribution of electrons and holes may be anisotropic. For example, in GaAs the carrier momenta (wave vectors) for transitions between the heavy-hole valence band and the conduction band are preferentially oriented perpendicular to the polarization of the light. The randomization of this initial anisotropic momentum distribution is sometimes called orientational relaxation. The orientational relaxation time is of course closely related to the momentum relaxation time of the carriers.

Effects due to transient anisotropic carrier momentum distributions had already been observed previously by *Smirl* et al. [4.74] in picosecond transient grating experiments in Ge, but at that time the available time resolution was not sufficient to resolve the orientational relaxation.

For a measurement of the orientational relaxation time Oudar et al. made use of the transient dichroism which is associated with anisotropic state filling. In their experiment GaAs was pumped near the band gap. The bleaching of the

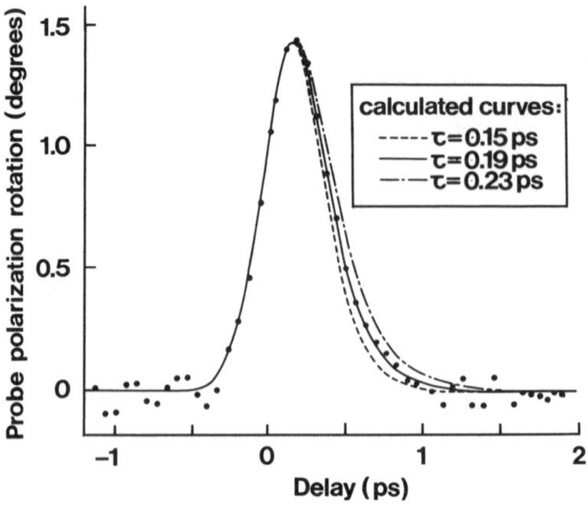

Fig. 4.21. Measurement of the relaxation of the anisotropic momentum distribution of photoexcited carriers in GaAs. The dots indicate the measured rotation angle of the probe pulse as a function of time. The three curves represent fits to the experimental data with different values of the orientational relaxation time. After [4.73]

absorption seen by a probe pulse of the same frequency as the pump is different for parallel and perpendicular probe polarization. As a result, the polarization direction of a probe pulse is rotated by a certain angle. For small rotation this angle is proportional to the anisotropic part of the carrier distribution function. Figure 4.21 from *Oudar* et al. [4.73] shows the measured rotation angle as a function of delay time. The experimental data are given by the full circles. The curves represent the calculated response (convolutions of the autocorrelation function of the pump pulse with a single exponential decay process) for three different time constants. From a fit of the experimental data the orientational relaxation time is determined to be 190 fs. Oudar et al. suggest that the dominant mechanism for the orientational relaxation in their experiment is carrier–carrier scattering.

4.2.3 Electronic Relaxation in Disordered Materials and Quantum Wells

So far, electronic carrier dynamics has been discussed basically in terms of the properties of ideal crystals. Let us now briefly look at some aspects of electronic relaxation related to deviations from the ideal crystal structure. Specifically, we shall discuss characteristic features of carrier dynamics which are caused by material disorder, and by carrier confinement in quasi-two-dimensional structures, such as "artificial" quantum wells. An interesting aspect of these structural effects is that by changing the structure one may be able to control relaxation properties and eventually tailor materials for specific device applications.

Let us recall some basic semiconductor properties that are changed by the introduction of disorder. In strongly disordered materials such as amorphous semiconductors there is no well-defined energy gap separating valence and conduction bands [4.75]. Instead, the density of states between the valence band and the conduction band varies continuously. However, an important feature of amorphous semiconductors is that the electronic states between the bands are localized. As a matter of fact, the gap region may be defined as the approximate energy range of localized states. States close to the band edges (mobility edges) can act as shallow traps of carriers, whereas the states closer to the gap center behave as deep traps.

It is clear that structural disorder can have a profound influence on the dynamics of photoexcited carriers. A particularly illustrative demonstration of the effect of disorder is due to *Göbel* and *Graudzus* [4.76] who studied picosecond relaxation of optically excited carriers in heavily doped, *p*-type GaAs. Strong doping introduces a continuum of quasi-localized electronic states below the band gap, similar to the situation in amorphous semiconductors. These states arise from the pertubation of the band edges of the undoped material by randomly distributed ionized dopant atoms. By varying the concentration of the dopants, the degree of disorder can be changed continuously from an ideal crystal situation all the way to a highly disordered system, and it is possible to study the concomitant changes of the material properties.

Göbel and Graudzus used picosecond time-resolved luminescence spectroscopy to observe the relaxation of photoexcited carriers as a function of dopant concentration. Figure 4.22 shows results from this work; the luminescence spectra of a nominally undoped GaAs sample and that of heavily doped GaAs are compared. It can be seen that, as time increases, the spectra of the undoped material (top of Fig. 4.22) shift to *higher* energies and become narrower. According to Göbel and Graudzus the observed behavior can be explained in terms of the usual dynamics of electron-hole plasma cooling and recombination in an ideal crystal. On the other hand, the time evolution of the spectra of the heavily doped sample shown in the lower part of Fig. 4.22 is strikingly different. In particular, the spectra shift in the opposite direction, i.e. towards *lower* energies, as time increases.

From the detailed time and temperature dependence of the luminescence spectra Göbel and Graudzus concluded that the carrier relaxation mechanism

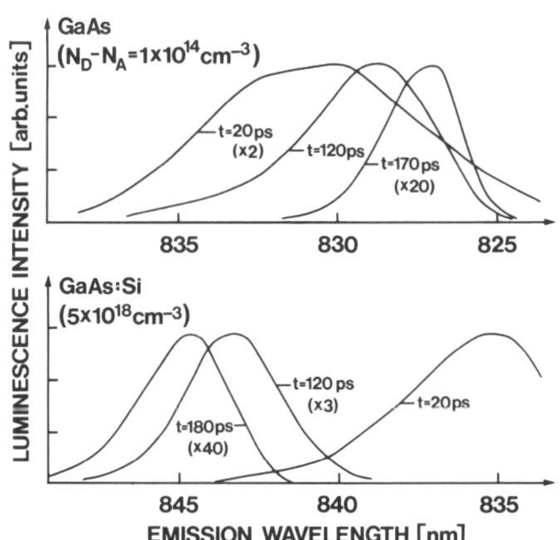

Fig. 4.22. Time resolved photoluminescence spectra of undoped GaAs (*top*) and heavily Si-doped GaAs (*bottom*). After [4.76]

causing the low energy shift in heavily doped GaAs is directly connected with the presence of disorder, being dominated by a thermally activated multiple trapping process of the carriers involving localized states below the gap. They also observed a saturation of the trapping process at high temperature and small defect concentration.

Tauc and coworkers [4.77–81] have performed a series of picosecond pump-probe experiments on amorphous semiconductor materials. They used pump and probe pulses at the same wavelength of somewhat less than a picosecond in duration from a passively mode-locked Rh 6G dye laser. It has been observed in these experiments that in amorphous materials the absorption at the excitation wavelength is usually increased by pumping, whereas in pure crystalline semiconductors bleaching of the band-to-band transitions is more typical. The different behavior is belived to be related to the relaxation of the wave-vector selection rules in disordered materials. In the absence of strict k-selection rules absorption from the excited states is likely to dominate over bleaching because the excited state absorption cross sections are determined by the density of states, which is greater at higher energies.

Tauc and coworkers studied a variety of materials and conditions. The observed time evolutions of the induced absorption have been interpreted in terms of hot carrier relaxation [4.77], carrier trapping [4.78], and recombination [4.79]. Detailed reviews of this work have been published by *Vardeny* and *Tauc* [4.80], and *Tauc* [4.81].

Electronic relaxation in quantum wells has been discussed several times during this article. Here we shall address specific features of the electron-hole dynamics which are distinct from the situation in bulk material. So-called quantum-well structures are fabricated by epitaxially growing crystalline semiconductor layers of varying composition on top of a suitable substrate. For example, in GaAs/GaAlAs heterostructures abrupt changes of the Ga/Al ratio within one or a very few layers can be produced [4.82]. Since the local band gap energy of the layers is a function of composition, sharp transitions of the band edge energies occur between the layers. Thus the layer structure introduces potential wells which for sufficiently small well thicknesses can confine the carriers in a quasi-two-dimensional structure. The composition of the layers and the layer thicknesses can be readily controlled during crystal growth. These additional degrees of freedom offer unprecedented means to shape the electronic properties of semiconductor materials.

The possible influence of the two-dimensional carrier confinement on the carrier dynamics is a question of great current interest. For example, it has been observed by *Ryan* et al. [4.83] that electronic cooling of two-dimensionally confined carriers in GaAs/GaAlAs heterostructures is much slower than the cooling predicted for the bulk material. They discuss the possibility that the reduction of the cooling rate may be due to the reduced dimensionality, an interpretation which has recently been challenged by *Shah* et al. [4.84], who have attributed the effect to the presence of non-equilibrium hot phonons.

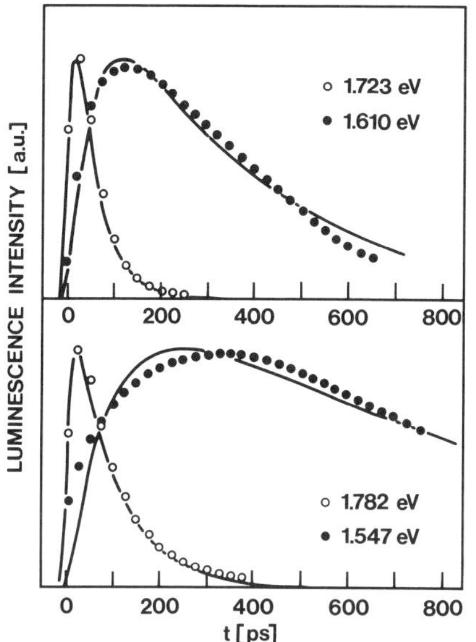

Fig. 4.23. Photoluminescence as a function of time for two different GaAs/GaAlAs quantum well samples. Top: single well of 5 nm thickness. Bottom: Double well with 14 nm well thickness. (o): Luminescence due to GaAlAs. (•): Luminescence from the quantum well levels. After [4.85]

While most of the differences between the electron-hole dynamics of bulk semiconductors and quasi-two-dimensional materials are still quite far from being understood, some of the unique effects of carrier confinement in two dimensions on excitonic recombination have been well documented. For instance, using time-resolved photoluminescence *Göbel* et al. [4.85] were able to demonstrate an enhancement of the exciton recombination rate in quantum wells. In these experiments pulses of 4.7 ps duration at 575 nm from a synchronously mode-locked dye laser were used for photoexcitation of GaAs/GaAlAs quantum-well samples. The rise and decay of the photoluminescence at different luminescence wavelengths was detected by a synchronously scanned streak camera coupled to a spectrometer; the time resolution of the system was 25 ps. An example of the results is depicted in Fig. 4.23 which shows the measured photoluminescence as a function of time for two different samples. The top and the bottom of Fig. 4.23 represent, respectively, results for a single quantum well with a well thickness $L_z = 5$ nm, and a double quantum well with $L_z = 14$ nm. In both cases the wells are sandwiched between GaAlAs cladding layers. The open circles correspond to the photoluminescence due to carrier recombination in the GaAlAs cladding. In bulk GaAlAs the spontaneous luminescence decay time would be expected to be about 1 ns. Here, much shorter decay times are observed, viz. 50 ps ($L_z = 5$ nm) and 100 ps ($L_z = 14$ nm). The shortening of the decay time is caused by trapping of the carriers in the quantum wells. These results demonstrate quite convincingly that the wells represent very efficient traps of free carriers.

The interpretation of the observed shortening of the decay rate of the luminescence of the cladding material in terms of carrier trapping by the wells is consistent with the observed rise time of the luminescence from quantum-well transitions. The full circles in Fig. 4.23 the data represent the photoluminescence due to transitions between the lowest quantum well subbands. It can be seen that the rise time of the quantum-well luminescence is in approximate agreement with the decay time of the free carrier luminescence from the cladding material.

The data of Göbel et al. [4.85] demonstrate yet another interesting aspect of electronic dynamics in quantum-well structures. It can be seen from Fig. 4.23 that the luminescence decay time for the two samples are different, e.g., 1 ns and 330 ps for $L_z = 14$ nm and $L_z = 5$ nm, respectively. Göbel et al. show that there is a distinct *decrease* of the exciton recombination time with decreasing well thickness. This reduction of the recombination lifetime in the quantum wells has been attributed to the increase of overlap of the wavefunctions of the electrons and the holes that is associated with a decrease of the well thickness. Thus the change of the exciton lifetime is a direct consequence of the carrier localization caused by the well structure.

A drastic *increase* of the exciton lifetime time in GaAlAs/GaAs quantum wells has been observed by *Polland* et al. [4.86] when electric fields of 10^4 to 10^5 V/cm are applied perpendicular to the plane of the wells. The change of the lifetime has also been attributed to a change of the overlap between the electron and the hole wave functions. In this case the overlap is decreased (and the lifetime increased) because the electric field pulls the electron and the hole in the well in opposite directions. It should be recalled that the effect of an electric field on exitons in bulk material is quite different; for rather modest field strengths the exciton becomes unstable and is ionized. In quantum wells, on the other hand, the well potential confines the carriers and prevents ionization. The electric field induces a shift of the sublevel transitions to lower energies; one speaks of a quantum-confined Stark effect. Polland et al. have observed that the quantum-confined Stark shifts are accompanied by an increase of the recombination lifetime. The quantum-confined Stark effect is more pronounced in quantum wells of relatively large well widths. For example, no shift and no lifetime enhancement has been observed in a 5 nm well, whereas in a 20 nm well the lifetime increases by as much as a factor of 100.

4.3 Phonon Dynamics

The fundamental vibrational excitations in solids are usually described in terms of propagating lattice waves [4.87]. If the vibrational field is quantized, the quanta in the various lattice modes are called phonons.

The plane wave phonon modes can be characterized by three parameters: (i) the phonon frequency ω, (ii) the phonon wavevector k, and (iii) a vector describing the type of polarization of the vibrational wave. The characteristic phonon dispersion relations $\omega(k)$ are of fundamental importance for the understanding of vibrational processes in crystals.

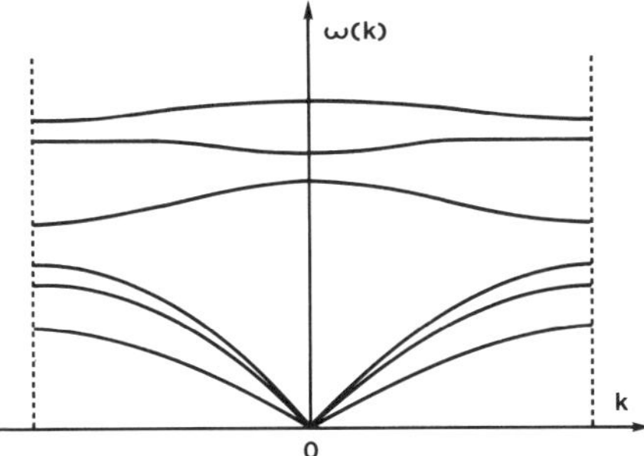

Fig. 4.24. Schematic representation of typical phonon dispersion curves along a certain general direction in k-space. The vertical dashed lines mark the boundaries of the first Brillouin zone

As shown schematically in Fig. 4.24, the typical phonon dispersion curves consist of several different branches. The three $\omega(k)$ curves starting from the origin are called acoustic branches [4.87]; they are related to the center of mass motion of the unit cell. The dispersion curves with a non-zero value of the frequency at $k = 0$ are referred to as optical branches; they are related to the internal relative motion of the atoms in the unit cell. The number of optical branches varies, depending on the number of atoms in the unit cell. Recall also that the maximal length of phonon wavevectors is of the order of π/a, where a is the lattice constant of the crystal.

Lattice vibrations may be associated with a change of the electric dipole moment or with a modulation of the electronic polarizability [4.88]. In the former case the vibrational modes interact with the light by direct processes of first order in both the vibrational and the electric fields, whereas in the latter case the interaction energy is quadratic in the electric fields. The former and the latter type of interaction are, respectively, the basis of single phonon infrared and first order Raman vibrational spectroscopy.

For any type of phonon-photon interaction, energy and momentum must be conserved. An important consequence of momentum conservation is that single phonon infrared spectroscopy and first order Raman spectroscopy give access only to long wavelength zone center phonons, because the photon momentum corresponds to less than one percent of the width of the Brillouin zone. This limitation may be overcome by multi-phonon spectroscopy. When more than one phonon is involved in the interaction with the photons, phonons with large momenta can be arranged such that their momenta cancel in the total momentum balance.

The zone center acoustic phonon modes are ordinary sound or hypersound waves, and the time scale of relaxation of these low frequency modes is nanoseconds or longer. The vibrational frequencies of the zone center optical phonons, on the other hand, range from about $100\,\mathrm{cm}^{-1}$ to perhaps $1000\,\mathrm{cm}^{-1}$. Thus the duration of a typical vibrational cycle in a solid is of the order of 100 fs. The damping times are typically ten to hundred times the duration of a cycle. For the relaxation of optical phonons we are therefore in the proper time regime for picosecond spectroscopy.

For a discussion of phonon lifetimes in solids, a comparison with the energy relaxation time T_1 and the dephasing time T_2 may be useful. T_1 and T_2 are commonly used in quantum optics to describe, respectively, the excited state lifetime and the dephasing of the induced dipole moments in an ensemble of two-level atoms [4.89]. The phonon lifetime τ describes the decay of a single lattice mode of frequency ω and wavevector k. In such a state of vibrational excitation there are well-defined phase relations between the individual vibrating atoms which are described by $\exp(ikr_i)$, r_i being a vector pointing to the site of the ith atom. Any interaction process which terminates the life of a phonon by annihilating a quantum in that mode contributes to the destruction of the phase-coherent state of vibrational excitation. Thus the phonon lifetime τ is associated with a dephasing process which is closely related to the dephasing process in an ensemble of two-level atoms described by the dephasing time T_2 [4.89].

On the other hand, if we consider the total phonon population in a number of adjacent modes of a certain volume element of the Brillouin zone, the lifetime of that population is in some sense equivalent to the energy relaxation time T_1. The population in the considered k-space volume element can obviously only be diminished by scattering of phonons *out* of the volume element, whereas scattering process which merely redistribute the energy within the volume element conserve the energy, but lead to a dephasing of the vibrational field.

As we shall discuss below, picosecond time-resolved spectroscopy has been used successfully to determine the lifetime of optical phonons by direct observation of the decay of the vibrational field in the time domain. In many cases, however, the phonon lifetime can also be obtained from line broadening data of the spontaneous Raman spectra or infrared absorption spectra. For example, in the absence of inhomogenous broadening, the phonon lifetime is given by $\tau = 1/2\pi\Delta f$, where Δf is the frequency width (full width at half maximum, FWHM) of the line, say, the spontaneous Raman line. However, because the phonon lifetime involves contributions from all possible phonon scattering processes, it may be rather difficult if not impossible to unravel individual scattering mechanisms on the basis of spectral linewidth data. Direct observations in the time domain, on the other hand, often provide an opportunity for pinning down specific details of a complex relaxation process, which cannot be obtained from the integral information of the spectra.

In the first demonstration of direct, picosecond time-domain vibrational lifetime measurements [4.90,91] the vibrational excitation was generated via the stimulated Raman effect. The advantage of this simple excitation scheme is that only one pump pulse is required, and that the pump frequency is not important. There is a drawback, however, in that only the vibrational mode with the largest Raman gain can be excited. More advanced schemes also use Raman-type interaction but with two pump pulses of different frequency, the frequency difference being equal to the vibrational frequency of the desired phonon mode. Once the sample has been prepared in some vibrationally excited state, the subsequent evolution of the system can be monitored with a delayed probe pulse by measuring the intensity of the anti-Stokes (or Stokes) Raman scattering of the probe, which is a measure of the degree of vibrational excitation. An excellent and detailed account of the Raman technique has been published by *Laubereau* and *Kaiser* [4.92].

In addition to single phonon Raman processes there are also a number of different interaction schemes which can be used for ultrafast generation and detection of phonons. Examples include schemes based on the interaction of conduction electrons and holes with phonons [4.93], two-phonon Raman processes [4.94], vibronic side-band spectroscopy [4.95], infra-red spectroscopy [4.96]. Some of these methods will also play a role in the experiments on ultrafast phonon dynamics which will be discussed in the following sections.

4.3.1 Optical Phonon Lifetimes

Direct measurements of the decay of coherently excited optical phonon modes in the time domain were among the first applications of picosecond spectroscopy more than a decade ago. This early work which has been reviewed in several articles [4.97,98] will be mentioned only very briefly. In 1971 *Alfano* and *Shapiro* [4.91] reported the first direct measurement of the decay of an optical phonon mode, the A_{1g} mode at 1086 cm^{-1} of calcite. Shortly thereafter, *Laubereau* et al. [4.99] studied the lifetime of the threefold-degenerate zone-center optical mode of diamond. Pure exponential decay of the vibrational excitation was observed, and the phonon lifetime was measured to be (3.4 ± 0.3) ps and (2.9 ± 0.3) ps at 77 K and at room temperature, respectively. The directly measured phonon lifetimes were in good agreement with spontaneous Raman linewidth data. Comparison of directly measured phonon lifetimes and Raman linewidth data of the A_{1g} mode of calcite by *Laubereau* et al. [4.100] also gave very good agreement.

More recently a detailed study of the temperature dependence of the lifetime of the A_1 mode at 465 cm^{-1} of α-quartz has been reported by *Gale* and *Laubereau* [4.101]. In their experiment the optical phonon mode is driven slightly off resonance by a pair of temporally coincident pulses. They used the second harmonic of a single pulse of about 5 ps from Nd-phosphate-glass laser together with a second pulse frequency-shifted by 460 cm^{-1} via stimulated Raman scattering in carbon tetracloride. Figure 4.25 depicts decay curves measured by Gale and Laubereau, the anti-Stokes scattered probe pulse as a function of delay time for a number of different temperatures. Exponential decay is observed for all temperatures, and the phonon lifetime was measured to be (3.4 ± 0.5) ps, (1.65 ± 0.15) ps, and (0.8 ± 0.08) ps, at 23 K, 120 K, and room temperature, respectively. Note the excellent, subpicosecond time resolution of the experiment. The dashed curve in Fig. 4.25 represents the cross correlation between the two simultaneous pump pulses and the probe pulse (the system response), which was measured by non-resonant, coherent anti-Stokes scattering in cyclohexane (expected to have an instantaneous response).

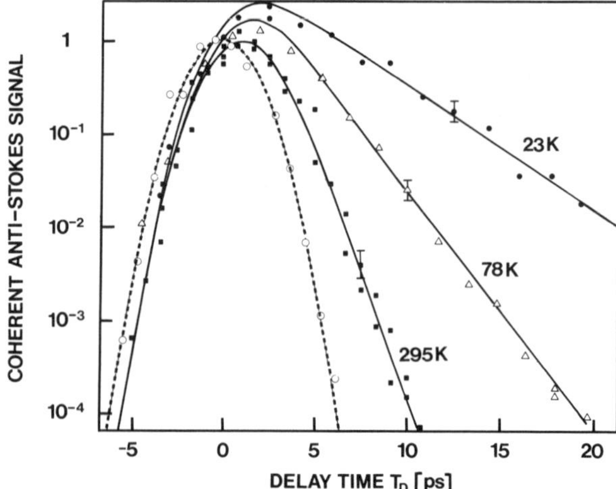

Fig. 4.25. Coherent anti-Stokes scattering of the probe pulse from the A_1-mode at $465\,\mathrm{cm}^{-1}$ of quartz as a function of delay time. Results from three different crystal temperatures are shown. The open circles represent the measured cross-correlation of the pump pulses and the probe pulse. After [4.101]

The detailed measurements of the temperature dependence of the phonon life-time provided new interesting information about the decay mechanism of the $465\,\mathrm{cm}^{-1}$ A_1 mode of α-quartz. Gale and Laubereau proposed a three phonon decay process in which the primary optical phonon decays into a low frequency acoustic phonon at $51\,\mathrm{cm}^{-1}$, and an optical phonon at $414\,\mathrm{cm}^{-1}$ with large, but opposite wave vectors near the border of the Brillouin zone. This rather asymmetric break-up of the primary phonon is consistent with the observed linear variation with temperature of the inverse phonon lifetime down to very low temperature, which is an indication that one of the decay products involves a low frequency mode (low activation energy). Gale and Laubereau also noted the good agreement of their phonon lifetimes with Raman linewidth data.

Progress in picosecond pulse technology and new pulse generation schemes which became available around 1980 have greatly extended the opportunities for doing time-resolved vibrational spectroscopy. *Kuhl* and *von der Linde* [4.102] have demonstrated the utility of frequency tunable cw mode-locked dye lasers for the purpose of vibrational spectroscopy. They used a dual laser system with two coupled, synchronously mode-locked cw dye lasers which produced picosecond pulses at two independently tunable frequencies. The frequency difference between the two sets of pulses can be readily tuned to match the frequency of any desired vibrational mode. This system is one example of several new schemes ideally suited for coherent excitation of vibrations via Raman-type interactions which do not suffer from the wavelength limitations of the earlier schemes.

Kuhl and *von der Linde* have studied the fundamental optical lattice modes of GaP [4.102]. For excitation they used the two pulse trains from the dual dye laser system with an average power of about 20 mW in each beam, and a pulse duration of 2.6 ps. The light paths were adjusted such that the pulses from the two dye lasers arrive at the sample at the same time. By chosing suitable polarizations to satisfy the respective Raman selection rules, and by adjusting the frequencies

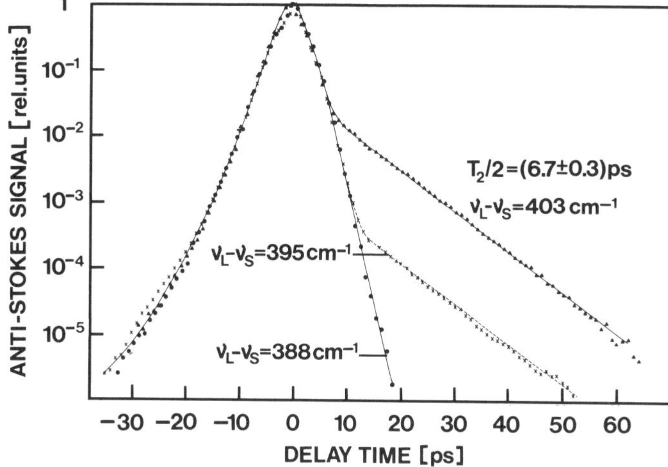

Fig. 4.26. Coherent anti-Stokes Raman scattering of the probe pulse from the LO-phonon mode of GaP (room temperature) as a function of delay time. $\nu_L - \nu_s$ is the wavenumber difference of the two pump pulses which are used to excite the LO-mode. $T_2/2 = 6.7$ ps is the lifetime of the LO-phonon mode. After [4.102]

ω_L and ω_S of the two dye lasers to match the desired vibrational frequency $\omega_L - \omega_S = \omega_v$ either LO-phonon modes or TO-phonon-polariton modes could be coherently excited. The generated vibrational excitation was probed using properly phase-matched, delayed probe pulses, in this case at the frequency ω_L of one of the pump pulses. Figure 4.26 shows an example of the results for an experimental geometry arranged for the generation and detection of LO-phonons. The measured anti-Stokes signal is plotted versus probe pulse delay time for three different values of the pump pulse frequency difference $\omega_L - \omega_S$ (the numbers in Fig. 4.26 actually indicate the wave-number difference). It is quite remarkable that the variation of the anti-Stokes signal can be followed over a dynamic range of one million.

For probe delays less than 10 ps the signal is independent of the driving frequency $\omega_L - \omega_S$; at these short times the signal is dominated by a four-wave-mixing process of the pump and probe pulses due to the non-resonant part of the third order non-linear susceptibility $\chi^{(3)}$, which produces a signal at $\omega_{AS} = 2\omega_L - \omega_S$. The interesting phonon dynamics is revealed by the exponential parts of the anti-Stokes signal for delays greater than 10 ps. From these data a value of (6.7 ± 0.3) ps is obtained as the LO-phonon lifetime in GaP at room temperature. When the driving frequency is detuned by more than 15 cm^{-1} from the LO-phonon resonance (403 cm^{-1}) the signal due to LO-phonons is lost, and one is left with the non-resonant four-wave-mixing signal.

Kuhl and *Bron* [4.103] have extended this work and reported detailed measurements of the temperature dependence of the LO-phonon lifetime in GaP for temperatures 5 K $> T >$ 300 K. These authors compared directly measured phonon lifetimes with Raman linewidth data; again, a very good agreement betweem time domain and frequency domain measurements has been established. Kuhl and Bron have also discussed the temperature dependence of the phonon lifetime. They showed that the experimental data can be accounted for by a phonon process in which the LO-phonon decays into two LA-phonons of about half the LO-phonon frequency with equal, but oppositely directed wave vectors. A detailed account of transient phonon dynamics in GaP also including some new data on ZnSe has been published by *Bron* et al. [4.104].

Direct time-domain observations have also been used to study picosecond vibrational dephasing in some organic molecular crystals. For example, *Duppen* et al. [4.105] measured the decay of delocalized low frequency librations ("librons") in naphtalene crystals. At $T = 1.5$ K the decay times of the A_g and B_g librons at 69 cm^{-1} and 57.5 cm^{-1} were measured to be 225 ps and 141 ps, respectively. From the observed temperature dependence of the linewidth of the libron modes the authors concluded that cubic anharmonic interactions play an important role in the decay mechanism. Two different processes are considered: (i) break-up of the primary librons into two accoustic phonons each with half the energy and with opposite wave vectors, and (ii) an up-conversion-type process in which the libron and a thermal accoustic phonon are annihilated, and an optical phonon is generated. The temperature-dependent linewidth data suggest that the second process (up-conversion) is more efficient.

Dephasing of several high frequency vibrational modes of naphtalene was investigated by *Dlott* and coworkers [4.106]. The temperature dependence of the dephasing times could be accounted for by a single "activation" energy for each mode, indicating a decay route dominated by interaction with a single frequency mode (100 to 175 cm^{-1}).

Kosic et al. [4.107] have reported dephasing measurements in an amino acid crystal, l-alanine. They studied a number of low frequency libron modes (40 to 137 cm^{-1}). A rapid increase of the dephasing rate with the mode frequency ω has been noticed, $\Gamma \sim \omega^{4.1}$. The temperature dependence of the dephasing rate was interpreted in terms of up- and down-conversion anharmonic processes.

4.3.2 Relaxation of Incoherent Phonons

The phonon lifetime measurements discussed in the preceding section showed that in general there is very good agreement between direct time-resolved phonon decay measurements and spectral linewidth data. Thus, in these cases time-domain and frequency-domain experiments provide essentially the same kind of information. Now we are going to discuss a different type of phonon relaxation phenomena which are much more difficult or even impossible to observe from measurements of frequency spectra. Such phenomena offer unique opportunities for time-resolved spectroscopy.

As a first example let us consider an experiment of *von der Linde* et al. [4.108,109] in which the relaxation of a non-equilibrium population of LO-phonons in GaAs is measured. Unlike the examples considered before where the phonons are generated via coherent Raman processes, in the present case the LO-phonons are produced by electron-phonon scattering as follows. The GaAs sample is photoexcited by pulses of 2.5 ps duration from a synchronously mode-locked cw dye laser at 575 nm. These pulses are strongly absorbed and free electron-hole pairs with an excess energy of about 0.6 eV (with respect to the band gap energy of 1.52 eV) are generated. In a semiconductor with polar or partially polar character the hot carriers lose their excess energy very rapidly, primarily by emission of LO-phonons. With $\hbar\omega_{LO} = 36.5$ meV one estimates that in GaAs each electron-hole pair generates 17 LO-phonons.

Note that a very large number of modes will be populated during the relaxation of the hot carriers. The corresponding volume of the Brillouin zone is given by energy and momentum conservation of the electron-phonon interaction. Because the number of modes is large and the modes are randomly phased one can speak of an *incoherent* phonon population, in contrast to the *coherent* phonons in a mode with a well-defined wave vector which are generated with coherent Raman processes.

In the experiment of von der Linde et al. the photoexcited non-equilibrium incoherent phonons were detected by spontaneous (incoherent) anti-Stokes Raman scattering of a delayed probe pulse, which was obtained by splitting off a fraction of the pump pulse. Actually, the experimental system differed from a classical laser Raman experiment only in the type of laser light used to illuminate the sample: a train of picosecond pump and probe pulses from a mode-locked laser. Although pump and probe are at the same wavelength, it is nevertheless possible to discriminate between Raman scattering from the pump and the probe pulse by exploiting the Raman selection rules of the material, and by using suitable orientations for the polarizations of pump and probe. Thus the rise and the decay of the non-equilibrium LO-phonons generated by the pump pulse can be monitored simply by measuring the anti-Stokes intensity (which is proportional to the number of phonons) as a function of the probe pulse delay time.

A result from *von der Linde* et al. [4.108] is shown in Fig. 4.27. From the measured anti-Stokes signal it can be seen that the number of LO-phonons rises to a maximum in about 3 ps. However, the observed rise time in this case is probably limited by the duration of the light pulses. Nevertheless, noting that 17 LO-phonons are emitted during the relaxation of each electron-hole pair, an upper limit of an average carrier–phonon scattering time of 180 fs can be given. For delay times greater than 5 ps an exponential decay of the LO-phonon number with a time constant $\tau' = (7 \pm 1)$ ps was observed (crystal temperature 77 K).

It should be noted that τ' is *not* the LO-phonon lifetime. Rather, it is the lifetime of the incoherent phonon population representing an entire range of phonon wave vectors of a k-space volume which is defined by the observation geometry of the probe scattering, e.g. the solid angle for the detection of the anti-Stokes spontaneous Raman light. It is interesting to compare the relaxation

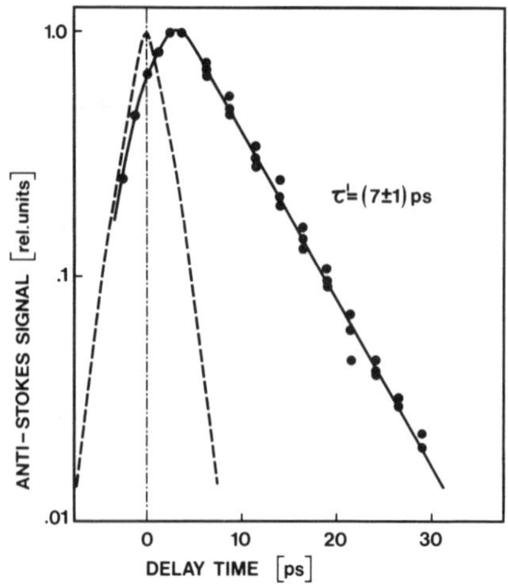

Fig. 4.27. Incoherent anti-Stokes Raman scattering of the probe pulse from photoexcited GaAs (77 K) as a function of delay time. The dashed line is the autocorrelation function of the laser pulses. After [4.108]

time τ' of the distribution with the LO-phonon lifetime proper (τ). From Raman linewidth data (direct LO-phonon lifetime measurements were not available) *von der Linde* et al. [4.108] obtained a phonon lifetime $\tau = (6.3 \pm 0.7)$ ps. They concluded that within the experimental accuracy the lifetime τ' of the population and the LO-phonon lifetime τ are the same. This result suggests that intrabranch LO-phonon scattering processes are not important and that both τ' and τ are determined by familiar three phonon interbranch decay processes due to the cubic anharmonicity terms of the lattice potential. However, *Pötz* and *Kocevar* [4.63] have pointed out that in the presence of a large density of photoexcited electrons and holes the interpretation of the observed relaxation time of the LO-phonon distribution solely in terms of intrinsic phonon processes may be questioned. They emphasized the necessity of a consistent treatment of the coupled non-equilibrium LO-phonon–carrier system.

The dynamics of LO-phonons in GaAs was re-examined with better time resolution by *Kash* et al. [4.110], with the same method for time-resolved spontaneous Raman scattering. Temporally compressed pulses from a synchronously mode-locked cw dye laser with a duration of 0.64 ps were used for photoexcitation and Raman probing. These authors obtained a rise time of the photoexcited LO-phonon population of 2–3 ps. Noting that about 12 LO-phonons are involved in building up the phonon population, an electron–phonon scattering time of one-twelfth of the observed rise time, i.e. 165 fs, is deduced. This number is consistent with the theoretical predictions for LO-phonon–carrier scattering via Fröhlich interaction.

When the power density of the excitation pulse was increased such that the estimated density of photoexcited carriers exceeded $10^{18}\,\mathrm{cm}^{-3}$, Kash et al. observed changes of the emission spectra of the GaAs sample near the LO-phonon Raman resonance, e.g., a broad continuum-like contribution which disappears

3 to 4 ps after the excitation pulse. Similar observations have been reported earlier, and parts of the continuum could be attributed to luminescence due to recombination of hot carriers [4.69]. The time-resolved Raman spectra of Kash et al. indicated a decreased LO-phonon peak at a delay time of 10 ps. The authors speak of a loss of phonon oscillator strength which they attributed to screening of the LO-phonons by relaxed free carriers. Kash et al. also believe that they have observed a time-dependent change of the LO-phonon oscillation frequency which they explained as being due to transients associated with the short electron–phonon scattering times.

An interesting application involving generation and detection of LO-phonons in GaAs has been reported by *Collins* and *Yu* [4.111, 112]. They utilized the non-equilibrium LO-phonons generated by relaxation of photoexcited carriers as an internal probe of intervalley scattering of hot carriers. In their experiments they photoexcited GaAs with pulses of 4 ps duration from a synchronously mode-locked cw dye laser which was operated with different laser dyes to give a wavelength-tuning range from 575 nm to about 720 nm. The key point is that by tuning the dye laser the excess energy of the electron-hole pairs could be varied over rather a wide range. The non-equilibrium LO-phonons generated by relaxation of the energetic carriers were probed by spontaneous Raman scattering. The occupation number of the LO-phonon modes could be obtained from the measured anti-Stokes-to-Stokes ratio of LO-phonon Raman scattering. Figure 4.28 shows an example of the results from the work of Collins and Yu. The measured phonon occupation number N_q is plotted as a function of the photon energy of the dye laser pulse. The marks on the energy axis indicate the number

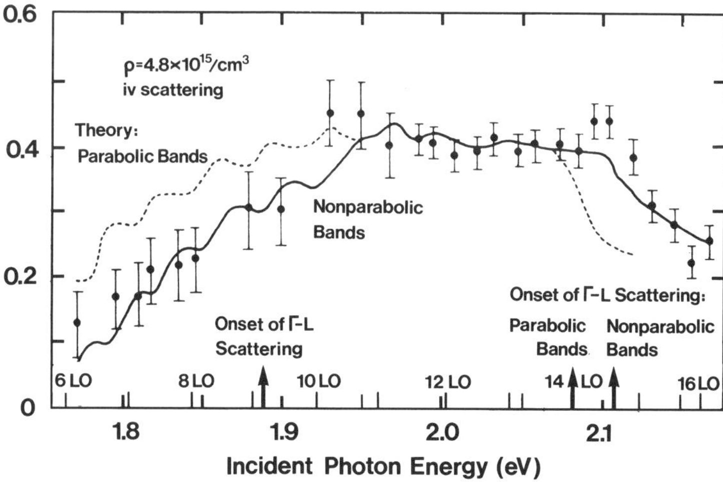

Fig. 4.28. LO-phonon occupation number in GaAs from incoherent anti-Stokes-Raman measurements (crystal temperature 10 K) as a function of the photon energy. (•): Experimental data. (– – –) and (——): Theoretical results based on a parabolic and a non-parabolic band model. After [4.112]

of LO-phonons which correspond to the initial excess energy of the photoexcited electrons. Note that the maximum phonon occupation number is $N_q \simeq 0.4$, which is approximately a factor of 10^{18} greater than the equilibrium phonon concentration corresponding to the sample temperature of 10 K.

Collins and *Yu* presented a thorough theoretical analysis of the phonon generation process [4.112]. The N_q calculated by these authors is given by the solid line in Fig. 4.28. An excellent agreement between the theory and the experiments can be noted. It was shown that the experimental results (N_q vs photon energy) could not be explained with a simple model involving only intravalley carrier relaxation in parabolic bands. Rather, the measured distribution of phonon occupation numbers reflects a number of interesting band-structure effects. For example, it was necessary to take into account the non-parabolicity of the conduction band and valence band warping to produce the remarkably good agreement with the measured data. In addition, there is clear evidence of the importance of intervalley scattering, both between the $\Gamma - X$ valleys and the $\Gamma - L$ valleys. For example, the decrease of the phonon occupation number N_q for photon energies greater that 2.1 eV could be shown to be due to the onset of efficient $\Gamma - X$ scattering of the photoexcited conduction electrons. This process drains off electrons from the Γ-valley very efficiently, thereby reducing the generation of (detectable) LO-phonons in the Γ-valley.

From the fit of the experimental data with the theory Collins and Yu have obtained the values of the intervalley electron–phonon deformation potentials for the $\Gamma - X$, $\Gamma - L$, and $X - L$ processes. This work demonstrates quite convincingly that the mechanism of non-equilibrium LO-phonon generation in GaAs is well understood and that the phonon distribution is very sensitive to intervalley electron–phonon scattering and band-structure effects.

4.3.3 Multiphonon States

Picosecond time-resolved vibrational spectroscopy has been extended to the investigation of overtones and multiphonon states. Before discussing some examples of this work a few remarks concerning vibrational overtones of molecules and bound two-phonon states of a solids may be useful.

One speaks of the excitation of a molecule to the first overtone of some vibrational mode v when a transition is made from the ground state to the second excited state. The v-mode is then occupied with two quanta. Typically, the overtone excitation energy of the molecule is less than twice the vibrational energy of the first excited state, $\hbar\omega_{2v} < 2\hbar\omega_v$. This is a manifestation of the anharmonicity of the molecular potential. While the overtone excitation represents an *intramolecular* two-quantum excitation, a different, *intermolecular* type of two-quantum state may occur when *two* interacting molecules are simultaneous excited, each by one vibrational quantum. In this case the total energy may also be different from the energy of two fundamental quanta of isolated molecules, as a result of the intermolecular interaction.

Let us now consider a typical solid, say, a crystal with ionic, covalent, or metallic chemical binding. For such crystals the interaction between atoms in neighboring unit cells is usually quite strong. As a result of the strong intermolecular forces, the vibrational modes of the crystal are delocalized and exhibit strong dispersion, the well-known phonon dispersion with acoustic and optical phonon branches being typical examples. Anharmonicity of the lattice potential gives rise – among other effects – to a continuum of delocalized multiphonon states. The width of the multiphonon band is essentially given by the sum of the widths of the underlying single phonon bands. While the continuum of free two-phonon states can be regarded as the analogy of the two-quantum intermolecule excitations, the solid state analogy of the molecular overtone is usually not observed, because in general the intermolecular interaction is much stronger than intramolecular anharmonicities. However, in some solids, in particular molecular crystals, the situation may be different. The Van-der-Waals-type intermolecular interaction in typical molecular crystals is relatively weak and may be comparable with the intramolecular anharmonicity. It has been shown that in this case, bound two-phonon states may exist [4.113] which are *localized* two-phonon excitations, the analogues of the overtones of a molecule.

In the work of *Geirnaert* et al. [4.114] the vibrational dephasing times of the Raman active first overtone $2\nu_2$ (at $795\,\mathrm{cm}^{-1}$) of the fundamental ν_2-mode of liquid CS_2 was measured and compared with measurements of the corresponding bound two-phonon state at $801\,\mathrm{cm}^{-1}$ of solid (crystalline) CS_2. The linewidth of the bound two-phonon band is relatively narrow, and Geirnaert et al. pointed out that the two-phonon states can be coherently driven with high coupling efficiency by suitable two-pulse laser excitation. In fact, their experimental technique for the excitation and subsequent probing of bound two-phonon states is quite similar to the familiar schemes used in one-phonon experiments. For excitation they used a simultaneous pair of pulses (~ 5 ps) with a difference in frequencies equal to the $2\nu_2$ overtone frequency ($795\,\mathrm{cm}^{-1}$) or the bound two-phonon frequency ($801\,\mathrm{cm}^{-1}$). A delayed portion of one of the pump pulses was suitably phase matched and served as a probe pulse. The coherent two-phonon excitation was detected by measuring anti-Stokes scattering of the probe pulse, which is due to a second-order, coherent anti-Stokes Raman interaction.

The experimental results for the ν_2 overtone of liquid CS_2 ($T = 165$ K), and the bound two-phonon state ("bi-phonon") of solid CS_2 ($T = 160$ K) are shown in Fig. 4.29 and Fig. 4.30, respectively. In both cases a pure exponential decay is observed with decay times ($T_2/2$, half the dephasing time) of 0.9 ps (liquid) and 7 ps (solid). According to Geirnaert et al. the bi-phonon decay is due to coupling with the overlapping continuum of *free* two-phonon states, i.e., the bound biphonon decays into two free, delocalized phonons. The much faster decay of the $2\nu_2$ overtone of liquid CS_2 has been tentatively attributed to additional line broadening mechanisms resulting from the random fluctuations of the environment of a molecule in the liquid.

Another interesting example of the vibrational dynamics of a system with strong anharmonic effects has been represented by *Gale* et al. [4.115]. They studied

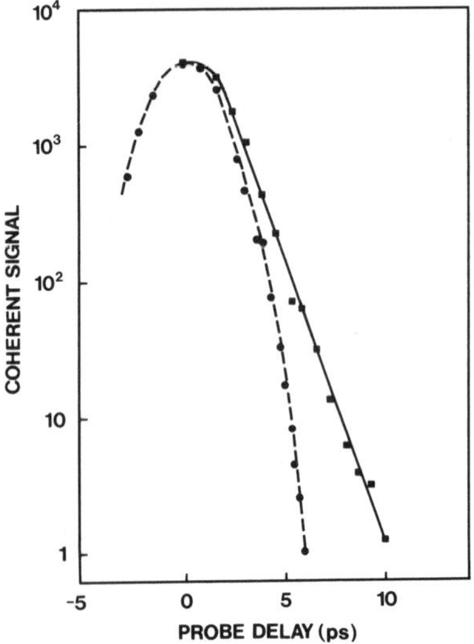

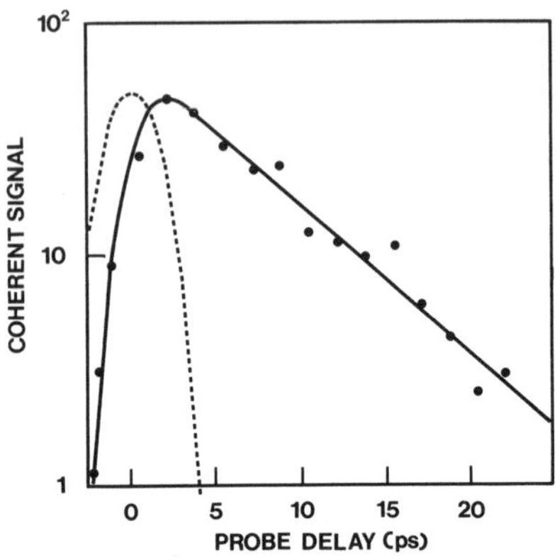

Fig. 4.29. Coherent anti-Stokes signal of the probe pulse from the $2\nu_2$ mode of liquid CS_2 (165 K) as a function of delay time. The dashed line is the system response function. The measured dephasing time is $T_2 = 1.8$ ps. After [4.114]

Fig. 4.30. Coherent anti-Stokes signal of the probe pulse from the two-phonon bound state in solid CS_2 (160 K) as a function of delay time. The dashed line is the system response. The measured dephasing time is $T_2 = 14$ ps. After [4.114]

a Fermi doublet in solid CO_2 (a molecular crystal with four CO_2 molecules per unit cell) which arises from the Fermi resonance between the strongly Raman-active symmetric ν_1 stretching mode and the first overtone $2\nu_2$ of the infrared active ν_2 bending modes of CO_2. The strong coupling between the $2\nu_2$ overtone and the ν_1 fundamental leads to two new hybrid states of mixed $2\nu_2/\nu_1$ character. These hybrid modes are both Raman active because they share the Raman oscillator strength of the fundamental. The two hybrid lines appear on the high frequency side and the low frequency side of the continuum of free two-phonon states at frequencies ω_+ and ω_- corresponding to $1383\,\mathrm{cm}^{-1}$ and $1276\,\mathrm{cm}^{-1}$, respectively. In the experiments of Gale et al. the hybrid modes were coherently pumped by two simultaneous pulses at frequencies ω_{S1} and ω_{S2}, with $\omega_{S1} - \omega_{S2} = \omega_+$ or ω_-. These excitation pulses were generated by stimulated Raman scattering in two different Stokes generators pumped by the second harmonic of a 5 ps pulse from a Nd-phosphate glass laser. A properly phase-matched portion of the second harmonic served as a probe pulse.

Measurements of the decay time were carried out over a temperature range from 9 K up to the melting point of solid CO_2 at 217 K. In Fig. 4.31 the measured decay rates are plotted as a function of temperature, according to *Gale* et al. [4.115]. It is seen that the high frequency mode ω_+ is more heavily damped than the ω_- mode. However, the increase of the decay rates with temperature is very

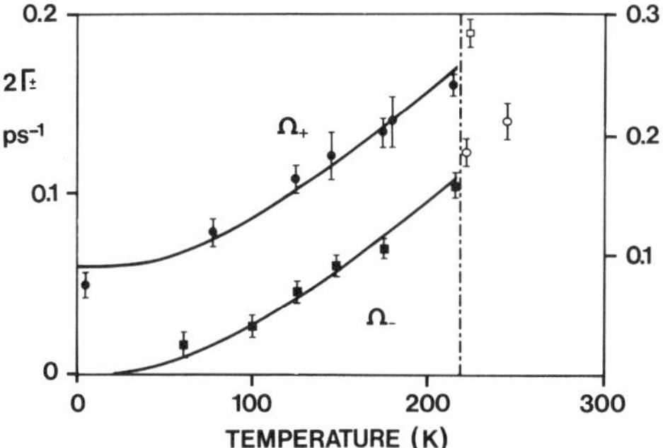

Fig. 4.31. Measured relaxation rates Γ_+ and Γ_- of the Fermi doubled of CO_2 as a function of temperature. The dash-dotted line marks the solid-liquid phase transition. Full symbols: solid, left vertical scale. Open symbols: liquid, right vertical scale. After [4.115]

similar for both modes; apparently it is just the values of the temperature-independent part of the rates which are different for the two modes. Gale et al. suggested that this temperature-independent contribution to the decay rate may be due to the mixing with the continuum of the free two-phonon states, similar to the relaxation mechanism discussed above in connection with the dephasing of the $2\nu_2$ overtone of CS_2. On the other hand, the temperature-dependent contributions to the relaxation rates point to an additional mechanism. The authors discussed parametric modulation of the *intramolecular* frequencies, i.e. $\omega(\nu_1)$ and $\omega(2\nu_2)$ resulting from *intermolecular* low frequency modes. The temperature dependence predicted by such a model was found to be in good agreement with the experimental data (solid curve in Fig. 4.31).

Another example of bi-phonon dynamics in a similar system, crystalline N_2O, with somewhat weaker Fermi interaction has been reported by *Valée* et al. [4.116].

4.4 Ultrafast Phase Transformations

Over the last decade lasers have become increasingly important for the processing of materials. Many processing applications involve some kind of phase transition, e.g. melting, evaporation, resolidification, and so on. When the materials are treated with extremely short pulses such phase transitions can occur under rather unusual conditions. The possibility of heating materials very rapidly is immediately obvious. For example, with today's laser-pulse technology, sufficient energy for melting and vaporization can be deposited in less than a hundred femtoseconds (10^{-13} s). On the other hand, cooling can also be very fast. Extremely steep temperature gradients can be set up in strongly absorbing materials by ultrashort laser pulses. Under these conditions heat conduction becomes so effective that the times for cooling by thermal conduction are very short [4.117].

Phase transformations driven by very fast rates of heating and cooling lead to interesting phenomena such as the formation of new metastable phases. An example is the formation of metallic glasses of "unusual" composition [4.118]. Materials can also be driven into very highly supercooled or superheated transient phases with extremely strong deviations from the equilibrium situation. Although, until now, ultrashort laser pulses have not played any significant role in industrial processes there is nonetheless a wealth of interesting ultrafast phenomena related to the fundamentals of materials processing.

A major incentive in much of the work reviewed in this section has been to unravel the fundamental physical mechanism of pulsed laser annealing of radiation damage in ion-implanted semiconductors [4.119]. A few remarks explaining this background may be appropriate. In ion-implantation, which is an important doping technique for semiconductors, the ionized dopants are fired onto the semiconductor crystal with kinetic energies typically of the order of 10^5 eV. Implantation with particles of such a high energy causes very strong lattice damage. In fact, the material eventually becomes so strongly disordered that the doped layer is left practically amorphous. It was demonstrated almost a decade ago that this radiation damage can be annealed very effectively by pulsed laser radiation [4.120].

The central issue that has attracted a great deal of attention concerns the nature of the transformation of the strongly disordered, amorphous semiconductor layer into a highly perfect crystal. Two fundamentally different theoretical models have been offered. The thermal model [4.121], on the one hand, assumes that the laser energy, after being absorbed by the semiconductor via strong electronic transitions, is converted into heat almost instantaneously, causing the temperature of the material to rise up to the melting point. The surface of the solid is melted down to a depth somewhat greater than the thickness of the disordered, amorphous layer, such that the meltfront contacts the unperturbed crystalline substrate. During the subsequent cooling an epitaxial recrystallization takes place as the solid-liquid interface retreats back to the surface. Thus in the thermal model annealing is explained in terms of conventional thermal processes such as melting and epitaxial crystallization. The plasma annealing model [4.122], on the other hand, emphasizes the role of the hot, dense electron-hole plasma which is photoexcited in the primary absorption process of the laser light. In contrast to the thermal model, it is assumed that the transfer of the electronic energy to the lattice – which is usually believed to be a very fast, efficient process – is somehow inhibited. Therefore, lattice heating, which is the precondition of the normal thermal melting, is greatly delayed or reduced. The plasma annealing model assumes that photoexcitation of a large number of electrons, speaking in chemical terms, from bonding states (valence band) to anti-bonding states (conduction band) weakens the shear forces between the lattice atoms to such an extent that the atoms rearrange into annealed positions without the lattice being significantly heated up.

Obviously, the thermal model and the plasma model represent rather antagonistic ideas, and, as a matter of fact, the strongly opposing views fueled many heated debates between proponents and opponents. Today it appears that there is almost overwhelming experimental evidence in favor of the thermal mechanism of pulsed laser annealing [4.123]. However, a non-thermal mechanism has recently been invoked in connection with far uv pulsed laser processing of organic materials [4.124]. Also, it is quite obvious that departures from a strictly thermal description are expected for shorter and shorter processing times. For instance, if we consider "heating" of a material with subpicosecond laser pulses, say pulses of 100 fs, the pulse duration is comparable to, or even shorter than, the various time constants which govern the redistribution of the absorbed energy among the various electronic and vibrational states of the system. Thus a material could undergo a phase transformation in a situation in which the thermalization of the energy is incomplete and the concept of temperature is inadequate. The issue of how fast the energy thermalizes and the details of the relaxation and redistribution routes still pose many open fundamental questions for the understanding of ultrafast phase transformations. These questions reach far beyond the initial laser annealing issues.

So far the bulk of the work on ultrafast phase transitions has centered on just a few materials, mainly semiconductors such as Si, Ge, and GaAs. In the following a selection of experiments will be reviewed which attempt to shed light on the mechanism and the dynamics of ultrafast phase transitions in these semiconductors.

4.4.1 Amorphization

Annealing of radiation damage implies a transformation of a highly disordered material back to a nearly defect-free crystalline structure. Here we start by discussing a situation which can be regarded as the opposite process: the transformation of a surface layer of a crystalline solid into a disordered, amorphous film by irradiation with suitable laser pulses. The thermal model of pulsed laser annealing explains the disorder-order transition as being due to expitaxial regrowth of the laser-melted liquid layer. However, if the resolidification velocity in this process exceeds some critical value, a liquid-to-amorphous transition may be observed rather than expitaxial crystal growth. *Tsu* et al. [4.125] demonstrated that in silicon such an order-disorder transition can be induced with 10 ns uv light pulses at 266 nm. This liquid-to-amorphous transition is observed only for laser fluences close to the melting threshold where the depth of the molten layer is still relatively small.

Thompson and *Galvin* [4.126] used a technique based on measuring the electrical resistance of the laser-melted surface film of metallic liquid silicon to measure the resolidification velocity. They found that for (100) silicon surfaces amorphization is observed if the regrowth velocity exceeds 15 m s^{-1}. According to one model [4.126] the velocities of both epitaxial and amorphous resolidification are strongly dependent on the degree of supercooling of the solid–liquid interface. If the interface is supercooled sufficiently below the equilibrium melting point of the amorphous phase, the amorphous regrowth velocity exceeds

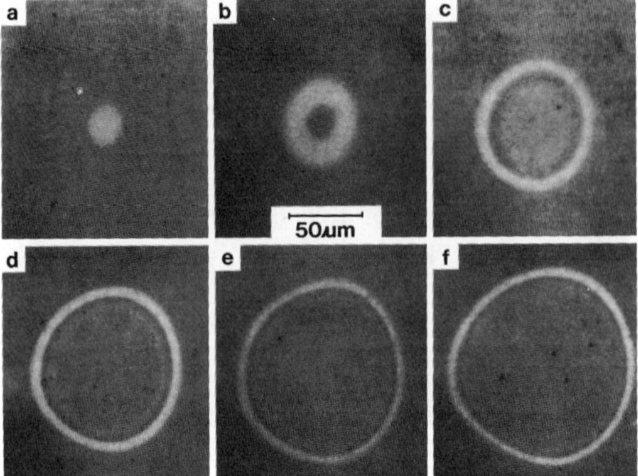

Fig. 4.32a–f. Optical micrographs of the surface of a crystalline silicon wafer after exposure to a single 20 ps laser pulse at 532 nm. The bright areas indicate an amorphous surface layer. Energy fluence in the center of the laser beam: (**a**) 0.2J cm^{-2}; (**b**) 0.26 J cm^{-2}; (**c**) 0.5 J cm^{-2}; (**d**) 0.85 J cm^{-2}; (**e**) 1.125 J cm^{-2}; (**f**) 1.75 J cm^{-2}. After [4.128]

the velocity for epitaxial crystal growth. As a result the liquid resolidifies as amorphous material. Thus very fast cooling is the key point in the liquid-to-amorphous transitions. Since cooling occurs via thermal conduction, steep temperature gradients are essential to obtain high cooling rates. Experimentally, these conditions can be met when laser heating pulses with short optical absorption depth and/or very short duration are used.

Amorphization of (111) and (100) surfaces of crystalline silicon after irradiation with visible and uv picosecond light pulses has been demonstrated by *Liu* et al. [4.127,128]. Figure 4.32 depicts a series of micrographs of the surface of a crystalline silicon sample [(111) orientation, 532 nm pulses]. These pictures clearly demonstrate the permanent changes of the surface caused by the laser pulses. At the lowest fluence corresponding to 0.2 J/cm^2 (in the center of the Gaussian beam profile) a uniform bright spot is obtained. As the fluence is increased the spot develops into a ring pattern with increasing diameter. The optical reflectivity of the bright areas is identical to the reflectivity of amorphous silicon [4.127]. More detailed electron diffraction studies of the microstructure of the irradiated surface areas confirmed the presence of amorphous layers. Transient optical reflectivity measurements indicated that the threshold fluences for amorphization and surface melting are the same. The observed changes of the surface morphology have been interpreted as follows: Amorphization occurs only for local laser fluences close to the threshold value, i.e. 0.2–0.26 J/cm^2. With increasing fluence the amorphous area moves together with the threshold contour of the beam profile. For higher local fluences in the center of the beam the thickness of the molten layer is greater, as is the cooling time. Thus the cooling rate in the center of the spot is slowed down sufficiently that epitaxial crystal growth rather than amorphous regrowth takes place.

More recently, *Boyd* et al. [4.129] have shown that amorphization of (111) crystalline silicon surfaces can also be achieved with infrared laser pulses, provided that the pulse duration is less than 10 ps. Formation of uniform amorphous spots has been observed for energy fluences near the threshold, which was measured to be 0.6 J/cm^2 for pulses of 7 ps at 1054 nm (Nd-phosphate glass laser). With increasing fluence the amorphous spots transform into an annular pattern, very similar to the observations of *Liu* et al. [4.127] for picosecond pulses at 532 nm. At first sight these results of Boyd et al. appear surprising. The optical absorption of silicon at 1054 nm is very weak, and heating with ir laser pulses is expected to be rather uniform, whereas in fact, very steep temperature gradients are produced with 532 nm and 266 nm laser pulses. Thus it appears that the conditions for rapid cooling are not met.

Smirl et al. [4.130] have discussed the situation in more detail and point out the strong non-linear increase of the optical absorption with fluence in silicon around 1000 nm. The non-linearity of the absorption is due to efficient free carrier absorption and an enhancement of the indirect band-to-band absorption caused by band gap narrowing with increasing temperature. Smirl et al. discuss a "thermal runaway" process in which lattice heating by free carrier absorption leads to an enhancement of the band-to-band absorption, which in turn causes more efficient free carrier generation and so on.

Detailed information about the velocities of the liquid–solid interface during ultrafast melting and resolidification of silicon has been obtained from the experiments of *Bucksbaum* and *Bokor* [4.131]. They used 15 ps uv pulses at 248 nm for heating and melting of (100) and (111) silicon surfaces. By measuring reflectivity and transmission of a delayed infrared probe pulse (1640 nm) the thickness of the liquid layer as a function of time could be determined with a resolution of about 20 ps. Bucksbaum and Bokor find that the time required to melt the silicon to the maximum depth is independent of the energy fluence of the uv heating pulse (melt-depth 2–50 nm). A maximum melt-in velocity as high as 750 m s^{-1} has been measured. On the other hand, a nearly constant resolidification velocity of 25 m s^{-1} practically independent of the laser fluence has been observed. Smaller velocities occur at the beginning and the end of the resolidification process.

Comparing the measured melt-depth vs time profiles with thermal model calculations, Bucksbaum and Bokor were unable to fit their data, when supercooling is neglected, i.e., when the resolidification temperature is assumed to be the equilibrium melting temperature. They modified the model to include supercooling of the melt. The results of *Thompson* et al. [4.126] who showed that the crossover of the velocities for crystalline and amorphous regrowth occurs at about 15 m s^{-1} were taken into account in these calculations. Good agreement with the experimental data is obtained with supercooling up to 700 K. Thus the results of Bucksbaum and Bokor have provided strong evidence that in picosecond experiments the liquid–solid interface is very strongly undercooled during resolidification.

4.4.2 Surface Melting

Laser-induced surface melting plays a central role in the understanding of pulsed laser annealing and related phenomena such as formation of amorphous surface films. In the following we shall discuss examples of experiments which provide some insight into solid–liquid phase transitions driven by ultrashort laser pulses.

The liquid phase of silicon, germanium, gallium arsenide and other related semiconductors is metallic. The solid-to-liquid transition is accompanied by drastic changes of the optical properties. For example, in the visible part of the spectrum the optical reflectivity of liquid silicon is roughly a factor of two greater than the reflectivity of crystalline or amorphous silicon. Thus phase transitions between the liquid and the solid state can be readily detected by optical measurements. The first quantitative measurements of the changes of the optical reflectivity and transmission of silicon during pulsed laser annealing were performed by *Auston* et al. [4.132,133]. Abrupt reflectivity changes were observed and it was shown that these changes are consistent with melting of the material followed by recrystallization. Time-resolved reflectivity measurements have become an important standard tool in phase transition studies.

Picosecond measurements of *Liu* et al. [4.128] and *Kim* et al. [4.134] provided the first experimental evidence that changes of the reflectivity take place on a picosecond time scale. Kim et al. used 25 ps annealing pulses at 532 nm and delayed infrared probe pulses for picosecond time-resolved measurements of the reflectivity changes during annealing of ion-implanted silicon. These experiments showed that the reflectivity rise occurred in about 30 ps – limited by the duration of the pulses used in the experiment – and remained at the high value over the entire time span of the experiment of about 1 ns. *Liu* et al. [4.128] used crystalline silicon samples and measured the reflected excitation pulse (25 ps, 532 nm) as a function of energy fluence. For fluences exceeding some critical value a strong increase of the self-reflectivity was measured. These observations demonstrated that a significant increase of the reflectivity occurs even during the pulse.

Picosecond excite-and-probe experiments similar to the measurements of *Kim* et al. [4.134] but with much better spatial resolution have been carried out by *Liu* et al. [4.135] and *von der Linde* and *Fabricius* [4.136]. These improved measurements provided more detailed insight into the role of the photoexcited electron-hole plasma and gave further compelling evidence that surface melting takes place during the laser heating pulse. Examples of the results of *von der Linde* and *Fabricius* [3.136] are given in Fig. 4.33, which shows the measured reflectivity of a 1064 nm probe pulse as a function of delay time. Let us compare the two reflectivity curves in Fig. 4.33, which correspond to different energy fluences, a) 0.11 J/cm^2, and b) 0.35 J/cm^2. At the lower fluence a small transient decrease of the reflectivity is measured which recovers almost completely during 50 ps. Very

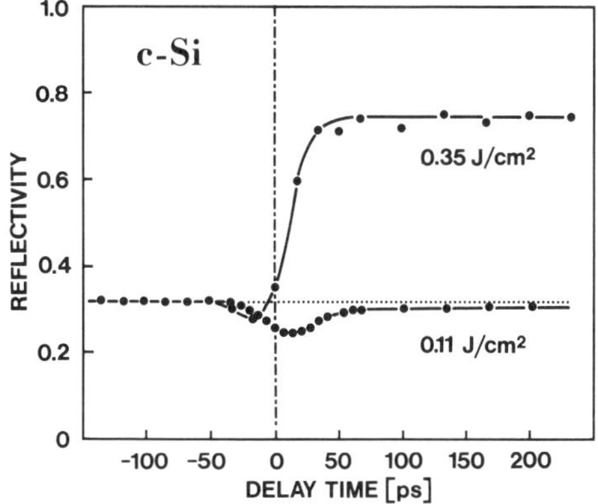

Fig. 4.33. Optical reflectivity of crystalline silicon measured with a 25 ps probe pulse at 1064 nm, as a function of the delay time with respect to the 25 ps pump pulse (532 nm). The probe pulse reflectivity is shown for two different pump pulse fluences, 0.11 J cm^{-2} (below the melting threshold) and 0.35 J cm^{-2} (above the melting threshold). After [4.136]

different behavior is observed at the higher fluence. Following a small initial decrease there is a distinct jump of the reflectivity to a value of 76% with a rise time of about 25 ps (limited by the experimental time resolution). Using a continuous 1064 nm probe beam it was shown that the reflectivity remains at a high value all the way into the nanosecond time regime. These results are in excellent quantitative agreement with the work of *Liu* et al. [4.135].

The observed small decrease of the reflectivity is the characteristic signature of a plasma of photoexcited electrons and holes with a plasma frequency less than the frequency of the probe pulse. The rapid recovery of the reflectivity is in agreement with the expected electron-hole plasma lifetime which is limited by non-radiative Auger recombination. A more detailed discussion of the density and temperature of the photoexcited electron-hole plasma will be given later.

There is now almost general agreement that the reflectivity behavior at higher fluence (curve *b*) in Fig. 4.33) cannot be explained in terms of a photoexcited electron-hole plasma. To explain the long-lived high reflectivity phase one would be compelled to assume that at the threshold the plasma lifetime increases discontinuously by several orders of magnitude, and that Auger recombination becomes ineffective. On the other hand, the reflection coefficient during the high reflectivity phase is independent of the fluence and has a constant value of $R = (76 \pm 1)\%$ at 1064 nm, which is in excellent agreement with the reflectivity of liquid silicon [4.137]. These considerations suggest that the transition to the high reflectivity phase does indeed signify the onset of surface melting and the formation of an overlayer of metallic liquid silicon which persists for many nanoseconds depending on the amount of energy deposited in the material.

In the experiments with heating pulses of a few tens of picoseconds the ultimate speed of phase transition was not resolved. The time-resolution limitation was overcome in experiments of *Shank* et al. [4.138] who measured the reflectivity changes of silicon following excitations with intense laser pulses of 90 fs duration at 620 nm. In this case the energy deposition is clearly separated from the subsequent response of the system and the evolution of the phase transition.

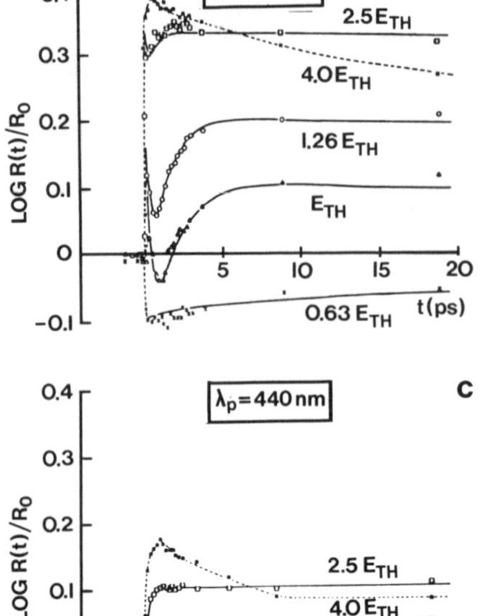

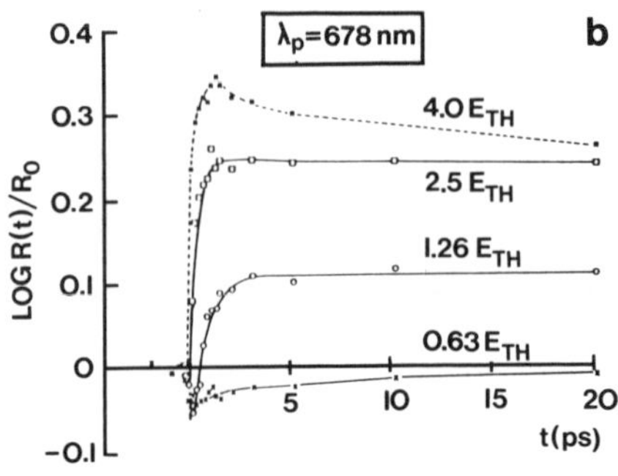

Fig. 4.34a–c. Change of the optical reflectivity of crystalline silicon as a function of time after excitation with a 90 fs excitation pulse at 620 nm. Probe pulse wavelengths: (**a**) 1000 nm; (**b**) 678 nm; (**c**) 440 nm. The different energy fluences of the excitation pulses are normalized to the threshold fluence, $E_{TH} = 0.1\,J\,cm^{-2}$. The solid lines are calculated according to the model discussed in the text. After [4.131]

The results of the femtosecond excite-and-probe reflectivity measurements of Shank et al. are reproduced in Fig. 4.34. The changes of the reflectivity are plotted as a function of time for three different probe wavelengths and different laser fluences. Note that on the time scale of Fig. 4.34 the excitation pulse would be represented by a thin vertical line. There is a wealth of useful information in these femtosecond data. Very briefly, two time regimes can be distinguished: For times less than 1 ps rapid variations of the reflectivity are observed. For times greater than 1 ps and pump fluences greater than the threshold (E_{TH}) the reflectivity assumes a constant, fluence-dependent value. Shank et al. have explained their data by assuming that a molten layer of liquid silicon propagates from the surface into the bulk. As long as the thickness of the layer is less than the absorption length of the probe pulse there is interference between light reflected from the air–liquid interface and the moving liquid–solid interface. This interference gives rise to the observed oscillations of the reflectivity. On the other hand, when the melt-front has penetrated to a depth greater than the absorption length and bulk reflectivity of liquid silicon is measured.

Shank et al. found that this model permitted a consistent fit of the data for all probe wavelengths with a single melt-front velocity for each value of the fluence. One of the principle conclusions of this work is that a surface layer corresponding to the absorption length of the probe light of about 200 Å can be completely melted in a time less than one picosecond. These observations suggest

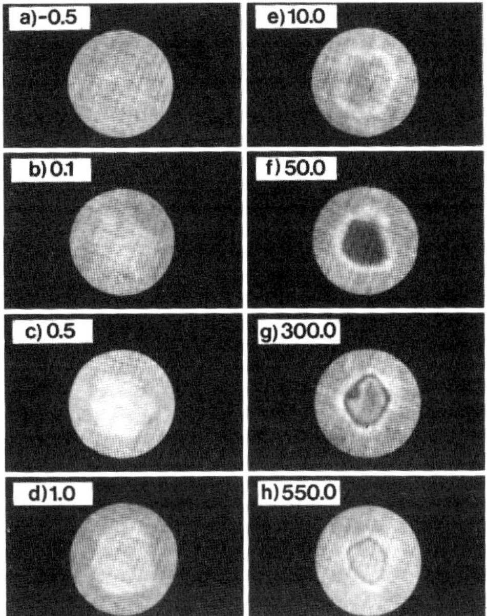

Fig. 4.35a–h. Time-resolved optical micrographs of the surface of crystalline silicon after excitation with a 80 fs pulse at 620 nm (energy fluence $0.5\,\mathrm{J\,cm^{-2}}$). The pictures were taken with a white light sub-picosecond probe pulses for illumination of the crystal surface. The delay time of the probe pulse (in ps) is indicated in the upper left corner of each picture. After [4.139]

melt-front velocities which greatly exceed the speed of sound. On the other hand, the sound velocity is expected to represent an upper limit to the propagation of the liquid-solid phase boundary. This contradiction has led to speculations that a different melting mechanism may be operative when femtosecond pulses are used, and that in this case bulk nucleation may play a role.

In a further extension of the femtosecond experiments *Downer* et al. [4.139] were able to obtain "flash photography" pictures of the silicon surface following excitation with intense 80 fs laser heating pulses. Exposure times of 100 fs have been achieved by using delayed femtosecond probe pulses to illuminate the scene. In these pictures (Fig. 4.35) the onset melting shows up as a bright spot in the center of the excited surface area. At high fluences ($E > 2.5\ E_{\mathrm{TH}}$) a dark region begins to obscure the center after a delay time of 5 to 10 ps. Downer et al. have attributed these observations to the development and subsequent dissipation of a cloud of droplets of liquid silicon ejected from the overheated molten surface layer. Quantitative transmittance spectra of the cloud could be fitted assuming scattering and absorption of light according to a Mie mechanism.

Boyd et al. [4.140] have also used an imaging technique to re-examine the reflectivity changes of silicon after irradiation with picosecond infrared pulses. Earlier reports had indicated rather long (nanosecond) reflectivity rise times [4.141], in contrast to the observations with visible and uv laser heating pulses. Boyd et al. showed that their spatially well-resolved data are consistent with rapid melting in a few picoseconds. They pointed out that with ir heating pulses, melting is always accompanied by violent evaporation. The reason is the large increase of the infrared optical absorption of silicon upon melting. Just a very small increase of the fluence above the melting threshold is sufficient to cause very strong overheating of the melt and substantial evaporation. It is argued that the earlier rise-time measurements were obscured by the strong optical losses due to the evaporated material.

A principal disadvantage of techniques measuring changes of the optical properties is that very little or no direct information about the microscopic structural changes during a phase transition can be obtained. Recent EXAFS (Extended x-ray Absorption Fine Structure) measurements with nanosecond resolution [4.142] and picosecond electron diffraction experiments [4.143] (see below) are examples of non-optical methods which are not subject to such limitations. However, higher order optical processes can also give more detailed structural information, for example, Raman scattering and various non-linear optical effects. *Shank* et al. [4.144] used femtosecond time-resolved second-harmonic generation to observe the structural dynamics in silicon. In a crystal with a center of inversion such as silicon (cubic diamond structure) bulk dipolar second-harmonic generation is forbidden. Second-harmonic generation can nevertheless result from surface electric dipole as well as bulk magnetic dipole and electric quadrupole non-linear polarizations [4.145]. It is well known that these second-harmonic processes can be used to obtain detailed structural information [4.146].

In the experiments of *Shank* et al. [4.144] the (111) surface of a crystalline silicon sample is optically excited with a 90 fs pulse at 620 nm, and the second harmonic of a delayed probe pulse is measured as a function of the angle of rotation about the (111) surface normal. The second-harmonic experiment probes a thin surface layer given by the optical absorption depth at 310 nm. Polar plots of the measured second harmonic are reproduced in Fig. 4.36. Part a) of the figure shows the data for an excitation pulse corresponding to 50% of the threshold fluence (E_{TH}). Two points should noted: (i) The second harmonic clearly exhibits the threefold rotational symmetry of the (111) surface; (ii) the

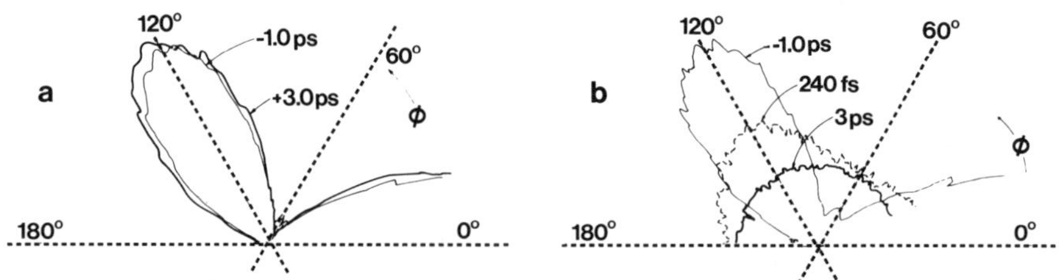

Fig. 4.36a, b. Second harmonic of a probe pulse generated at the surface of a silicon sample after excitation with a 90 fs excitation pulse at 620 nm. The data are presented as a polar plot: The second harmonic is plotted as a function of the rotation angle about the surface normal of the crystal [(111) direction]. The various delay times of the probe pulse are indicated in the figure. Energy fluence of the excitation pulses: (**a**) 0.5 E_{TH}; (**b**) 2.0 E_{TH}. After [4.144]

subthreshold excitation pulse does not significantly alter the second-harmonic pattern, although the density of photoexcited carriers is already very high, approximately $10^{21}\,cm^{-3}$. For pump pulse fluences greater than E_{TH} drastic changes of the second-harmonic pattern with time are observed. Figure 4.36b shows that the second harmonic has become more isotropic after only 240 fs, and that the distribution is completely isotropic after 3 ps. The change of the threefold rotational symmetry of the second-harmonic pattern to an isotropic one is precisely what is expected during the transformation of a (111) crystalline silicon surface into an isotropic disordered liquid. The time for this change to occur has been measured to be about 1 ps, in excellent agreement with the conclusions from the femtosecond reflectivity measurements [4.139].

Rotational anisotropy of the second harmonic from (111) and (100) silicon surfaces was also observed in earlier picosecond experiments by *Guidotti* et al. [4.147]. However, they fitted their data with a dipolar second-harmonic tensor of D_{2d} symmetry and concluded that the O_h centrosymmetric symmetry of silicon is changed under intense laser excitation. They suggested the formation of localized tightly bound electron-hole pairs (Frenkel exciton) [4.148] which would lower the symmetry to non-centrosymmetric T_d. This mechanism would be closely linked to the presence of a very high density of electron-hole pairs. On the other hand, the results of *Shank* et al. [4.144] (Fig. 4.36a) show that up to electron-hole pair densities as high as $10^{21}\,cm^{-3}$ the symmetry of the second harmonic is not changed and that there is no indication of change of the crystal symmetry before melting. An explanation of this apparent discrepancy was suggested by *Litwin* et al. [4.149] who re-examined the situation both theoretically and experimentally. They came to the conclusion that the data of *Guidotti* et al. [4.147] and *Driscoll* and *Guidotti* [4.148] are consistent with surface-dipolar/bulk-quadrupolar second-harmonic generation if the wavelength dispersion of the pertinent non-linear optical coefficients is taken into account.

4.4.3 Density of Electrons and Holes

For visible and ultraviolet pulsed laser processing of semiconductors the primary light–matter interaction involves creation of electron-hole pairs by absorption of photons, and the formation of an electron-hole plasma. The density and temperature of the electron-hole plasma can be very high when intense short laser pulses are used. Recall that the presence of an electronic plasma of very high density is the key point in the proposed non-thermal mechanism of the plasma laser annealing model [4.122]. Note also that the photoexcited solid state electron-hole plasma is replaced by the electronic plasma of the metallic liquid phase, when the solid-to-liquid transition occurs. Thus it is clear that the dynamics of electronic plasmas play a fundamental role in the understanding of the nature of the phase transformations induced by laser pulses.

Information about electronic plasmas can be obtained from the optical spectra. Usually it is assumed that the optical response of the free carriers can be described by a simple Drude-Zener formalism [4.150], which involves just two characteristic parameters, the damping time τ of the plasma oscillations, and the plasma frequency, $\omega_p^2 = N/m^* \varepsilon_0 \varepsilon_\infty$. Here N is the density of electron-hole pairs, and m^* is the reduced electronic effective mass. ε_∞ is the high frequency dielectric constant ($\varepsilon_\infty = 11.8$ for silicon) and the other symbols have their usual meaning. Recall the typical features of a Drude reflectivity spectrum: There is a transparency regime of relatively low reflectivity for frequencies $\omega > \omega_p$, and an opaque regime of high reflectivity for $\omega < \omega_p$. The plasma frequency and thus the ratio N/m^* can be determined from the position of the characteristic reflectivity minimum, which occurs near $\omega = \omega_p$. The damping constant, on the other hand, determines the depth of the reflectivity minimum.

As discussed in the previous section, the presence of a photoexcited electron-hole plasma in picosecond laser melting experiments has been demonstrated by *Liu* et al. [4.135] and *von der Linde* and *Fabricius* [4.136], who detected a slight decrease of the reflected probe pulse (1064 nm) during picosecond laser heating (see e.g., Fig. 4.33). However, it was quite clear that in these experiments the actual reflectivity minimum occurred at a wavelength longer than the employed probe wavelength (1064 nm). A crude upper limit of the plasma density can be obtained from these measurements by setting the frequency of the minimum ω_0 equal to the probe pulse frequency. With an effective mass ratio $m^*/m_0 = 0.15$ the upper limit of the plasma density is calculated to be $1.6 \times 10^{21}\,\mathrm{cm}^{-3}$. Somewhat higher densities can be estimated from the femtosecond reflectivity measurements of *Shank* et al. [4.138]. For example, for a pump fluence of 1.26 E_{TH} they see an initial increase of the reflectivity with a 1000 nm probe pulse, but a decrease with 677.8 nm, suggesting that the plasma frequency is between these two wavelengths. With $m^*/m_0 = 0.15$, the plasma density falls between 2.1×10^{21} and $4.3 \times 10^{21}\,\mathrm{cm}^{-3}$. Using the free electron mass, $m^* = m_0$, Shank et al. gave a density estimate of $5 \times 10^{21}\,\mathrm{cm}^{-3}$.

Van Driel et al. [4.151] used longer wavelength infrared probe pulses and were able to fully resolve the plasma resonance. Figure 4.37 depicts results from their work showing the measured reflectivity and transmission as a function of time for a silicon crystal excited by a picosecond pump pulse at 532 nm. The two different infrared probe pulse wavelengths used in these experiments were generated by stimulated Raman scattering of the fundamental pulse (1064 nm) in compressed hydrogen gas (1900 nm, first Raman Stokes component), and methane (2900 nm, second Raman Stokes component). At 1900 nm van Driel et al. observed a distinct reflectivity minimum, with a reflection coefficient as low as 12%. The pump fluence in this case corresponds to the maximum fluence for the occurrence of just a single reflectivity minimum (Fig. 4.37a). Under these conditions the frequency of the probe pulse is equal to the critical frequency of the reflectivity minimum ω_0. In this way they determine $N/m^* = 3.6 \times 10^{48}\,\mathrm{g}^{-1}\mathrm{cm}^{-3}$.

The results of reflectivity and transmission for the longer probe wavelength (2800 nm) with otherwise the same conditions, are shown in Fig. 4.37b. Now there are two reflectivity minima separated by a transient reflectivity maximum. These

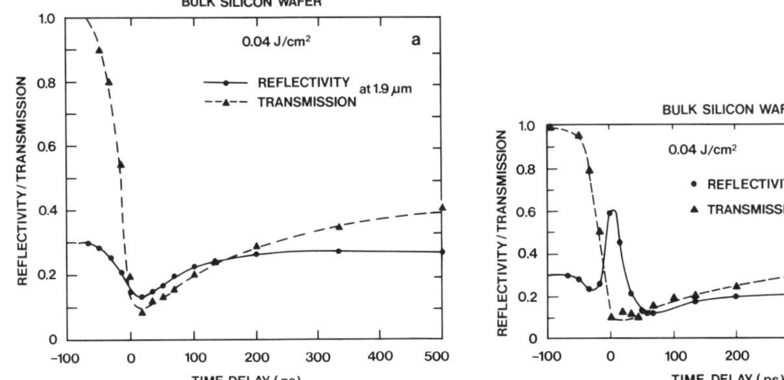

Fig. 4.37a, b. Reflectivity and transmission of crystalline silicon after excitation with a 20 ps laser pulse at 532 nm (energy fluence 0.04 J cm^{-2}). Reflectivity and transmission are measured with delayed probe pulses at 1.9 μm (**a**), and 2.8 μm (**b**). After [4.151]

results represent clear evidence that the critical plasma density corresponding to minimal reflectivity occurs twice, a first time during the build-up, and a second time during the decay of the electron-hole plasma. Between the two minima the system passes through a high reflectivity phase corresponding to the condition $\omega_{\text{probe}} < \omega_p$. *Van Driel* et al. [4.151] have estimated an instantaneous peak reflectivity of more than 90%. Using an effective mass $m^* = 0.123\ m_0$ and the measured $N/m^* = 3.6 \times 10^{48}\ \text{g}^{-1}\text{cm}^{-3}$ they calculate a peak electron-hole pair density of $4 \times 10^{20}\ \text{cm}^{-3}$ at a pump fluence of 40 mJ/cm^2 (about 20% of E_{TH}).

Even if one assumes the validity of the Drude model – a detailed critical assessment of this assumption is still lacking – considerable uncertainty remains in the determination of the plasma density because of the uncertainties in the correct value of the effective mass. Various groups have used values ranging from $m^*/m_0 = 0.123$ [4.154] up to $m^*/m_0 = 1$ [4.135,136,152].

The effective mass parameter m^* of the Drude formula is the reduced conductivity effective mass of the ambipolar plasma, $1/m^* = 1/m_e + 1/m_h$. The electron and hole effective masses m_e and m_h must be calculated from the band structure. Different procedures have been applied [4.153,154] in the calculation of m^* in various works. The case is further complicated by the fact that the effective mass parameter taken at the band extrema may not be adequate to describe the situation of a very dense and hot electron-hole plasma. For high density and high temperature the electrons and holes occupy states of higher energy. The non-parabolicity of the band structure comes into play tending to increase the effective mass.

Van Driel [4.155] presented a calculation of the temperature and density dependence of the effective mass for silicon. According to this work, the effective mass deviates significantly from the low density room temperature value $m^* = 0.16\ m_0$ as soon as the density exceeds about $10^{20}\ \text{cm}^{-3}$. Van Driel concludes that for the density and temperature range of interest the effective mass could increase by as much as a factor two. *Yang* and *Bloembergen* [4.156] have also published theoretical calculations of the temperature and density dependence of m^* in silicon. Their low density room temperature value is $m^*/m_0 = 0.17$. This model predicts a weaker increase of m^* of only about 30%. Experimental data have been reported that indicate a stability of the effective mass up to densities around $10^{21}\,cm^{-3}$ [4.157]. However, it appears from the work of *Lompre* et al. [4.158] that a 30% increase would also be consistent with the experimental data.

Summarizing the discussion of the plasma density, experimental data and theoretical work indicate that with visible picosecond excitation of silicon an electron-hole plasma density of about $10^{21}\,cm^{-3}$ can achieved before the phase transition occurs. The uncertainty in the density appears to be less than a factor of two. Thus the density is still lower than the values envisaged for the plasma-assisted non-thermal phase transition mechanisms [4.122,4.154]. Somewhat higher densities may be achieved with femtosecond excitation.

4.4.4 Electron and Lattice Temperature

It is interesting to compare melting and resolidification under highly transient conditions with normal melting close to thermal equilibrium. For example, normal thermal melting implies that the energy distributions of electrons (holes) and phonons obey Fermi-Dirac or Bose-Einstein distribution, respectively, and that the lattice and the electron temperature are the same. Clearly the assumption of thermal conditions becomes questionable when picosecond or even femtosecond laser pulses are involved. Determination of the energy distribution over the electronic and vibrational states of the system and measurement of the electron and lattice temperature are very important for the understanding of phase transitions under highly transient conditions.

In a series of papers *Liu* et al. [4.159] and *Malvezzi* et al. [4.160,161] reported on measurements of charged particle emission during picosecond laser heating of silicon. These experiments provide qualitative information about the electron and the lattice temperature. For example, if the energy transfer from electrons and holes to the lattice is slow compared with the heating rate, the photoexcited electron-hole plasma is extremely hot at the beginning and gives rise to intense thermionic emission of electrons. On the other hand, if the energy transfer to the lattice is very fast, the electron temperature is not expected to exceed the lattice temperature a great deal, and lattice heating is practically instantaneous. Evaporation of atoms and ions is a clear indication of a hot lattice.

Liu et al. and Malvezzi et al. used a collection electrode with either positive or negative bias voltage for the detection of electrons or positive ions. An example from the work of *Malvezzi* et al. [4.161] is given in Fig. 4.38, which shows the total time-integrated charge of electrons and ions as a function of fluence for uv pump pulses (266 nm). Three characteristic regimes can be distinguished. The

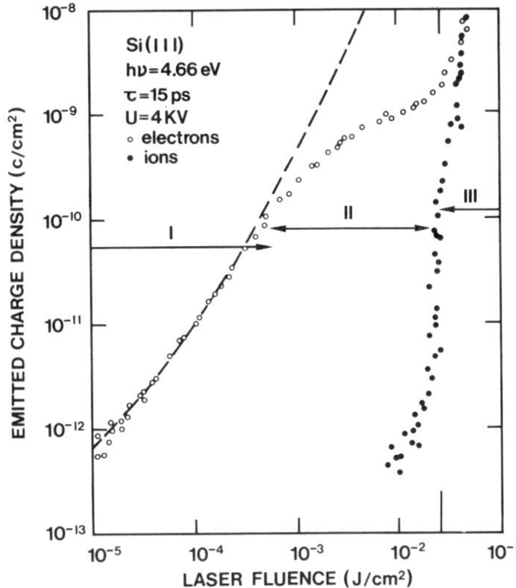

Fig. 4.38. Electronic and ionic charge emitted from a silicon crystalline surface after exposure to a 30 ps laser pulse at 266 nm, plotted as a function of laser fluence. The melting threshold corresponds to 25 mJ cm^{-2}. The dashed line represents the calculated electronic charge for linear and quadratic photo-electric emission. After [4.161]

electron emission for fluences less than 5×10^{-4} J/cm^2 (region I) has been identified as being due to a superposition of linear and quadratic photoelectric emission. No contribution from thermionic emission was found. Application of the Richardson-Dushman equation gave an upper limit of the electron temperature of 2200 K [4.160]. In region II the electron emission is space-charge limited and depends on the bias voltage. No ion emission can be detected in region I and II. Positive particle emission sets in when the laser fluence approaches the threshold fluence (25 mJ/cm^2) which is marked by the onset of surface amorphization. In region III (above threshold) equal amounts of electronic and ionic charge are detected, because the positive particles neutralize the electronic charge and lift the space charge limitation.

Liu et al. and Malvezzi et al. have drawn the following conclusions from their charged particle measurements. The absence of thermionic emission puts an upper limit of about 3000 K on the average electron temperature during the phase transition. The onset of positive ion emission is interpreted as evidence that the lattice is indeed getting hot. The charged particle measurements are consistent with a rapid energy transfer from electrons to phonons on a time scale of 1 ps, much faster than the duration of the excitation pulse, suggesting that the phase transformation can be described as a thermal process.

More detailed information about the lattice temperature rise during picosecond irradiation of silicon has been obtained by *Lompré* et al. [4.162,163]. They took advantage of the precise knowledge of the temperature dependence of the optical constants in silicon [4.164] which permits the use of thermally induced changes of the optical properties as a "thermometer". Lompre et al. measured the changes of the reflectivity and transmission of silicon-on-sapphire (SOS) samples induced by 25 ps pulses at 532 nm. An advantage of using SOS structures

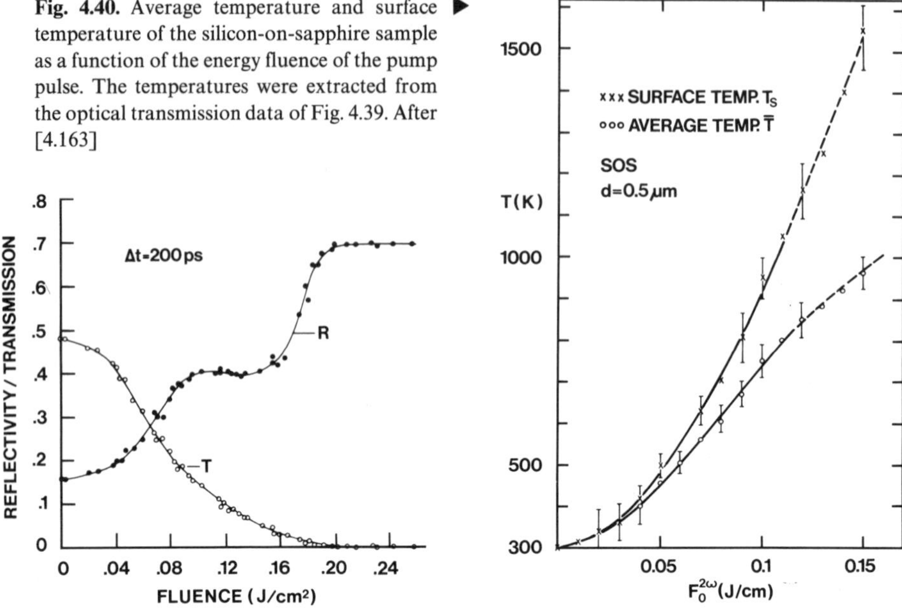

Fig. 4.40. Average temperature and surface temperature of the silicon-on-sapphire sample as a function of the energy fluence of the pump pulse. The temperatures were extracted from the optical transmission data of Fig. 4.39. After [4.163]

Fig. 4.39. Reflectivity and transmission of a silicon-on-sapphire sample (0.5 μm thick) after excitation with a picosecond pulse at 532 nm, probed after a delay of 200 ps with a pulse at the same wavelength. Reflectivity and transmission are plotted as a function of the pump pulse energy fluence. Melting starts at 0.16 J cm^{-2}. After [4.163]

is that the induced optical changes are enhanced by Fabry-Perot-type interference effects. A similar technique has also been used earlier by *Murakami* et al. [4.165] in nanosecond laser heating experiments. An example of the results of *Lompré* et al. [4.163] is shown in Fig. 4.39. It is seen that at a constant probe delay of 200 ps the reflectivity increases and the transmission decreases as the laser fluence is increased. The pronounced structure of the reflectivity curve signifies the Fabry-Perot interference effects. For fluences greater than 0.16 J/cm^2 the reflection coefficient reaches a fluence-independent maximal value of 69%, which corresponds to the metallic reflectivity of liquid silicon.

Temperature information is extracted from these measurements by a fitting procedure in which thermal-model calculations, including the temperature dependence of the optical and the various thermal parameters, are fitted to the experimental data. Figure 4.40 shows the spatially averaged temperature and the surface temperature obtained from such fits. The data indicate a rapid increase of the lattice temperature with fluence. When the temperature curves are extrapolated to higher fluences (dashed parts in Fig. 4.40), it is seen that the melting temperature of silicon (T_m = 1685 K) is reached at 0.16 J/cm^2 which is the fluence level for the onset of metallic reflectivity (see Fig. 4.39).

Fabricius et al. [4.166,167] have compared laser surface heating of gallium arsenide (GaAs) for nanosecond and picosecond pulses. In these experiments the surface temperature is inferred from the velocity distributions of the atoms which evaporate from the laser-heated surface. A time-of-flight method, with a quadrupole mass spectrometer as particle detector, was used to determine the atomic velocities. The same technique was employed earlier by *Stritzker* et al. [4.168, 169] to measure the surface temperature of silicon and GaAs during laser heating with nanosecond ruby laser pulses. Fabricius et al. found that the velocity distributions of the evaporated atoms can be represented by Maxwell distributions, both for nanosecond and picosecond laser heating, and that the temperature obtained from gallium and arsenic atoms are in excellent agreement. The fact that Maxwellian distributions are observed is somewhat surprising because in these experiments spatially and temporally averaged velocity distributions are measured, corresponding to the spatial and temporal variation of the surface temperature. In general such averaged distributions are expected to be different from simple Maxwell distributions. However, Fabricius et al. argue that the particle emission rates are expected to be strongly increasing functions of the local temperature. The highest temperatures therefore carry much more weight, and the averaged distributions are likely to be close to the distributions corresponding to the maximum temperature.

The results of Fabricius et al. for laser heating of GaAs with 10 ns laser pulses are shown in Fig. 4.41. The measured temperature is plotted as a function of fluence. The dashed vertical line marks the threshold fluence E_{TH} for the onset of the high reflectivity phase, established from independent time-resolved measurements of the optical reflectivity. The following two points should be noted: (i) The measured temperature at the threshold fluence is in excellent agreement with the normal melting point of GaAs, $T_m = 1511$ K (upper horizontal dashed line); (ii) there is a distinct step or plateau at the temperature $T = T_m$. Such a plateau at T_m would be expected for a normal melting process in which the phase transformation proceeds close to the equilibrium melting point. The stagnation of the temperature signifies the consumption of the supplied energy as latent heat of melting. The temperature rise is resumed at higher fluences when the entire heated surface layer is completely melted and the liquid film is overheated by the excess energy of the laser pulse. The temperature curve for heating with laser pulses of 25 ps duration is shown in Fig. 4.42. Note the striking difference between the nanosecond and the picosecond data: the absence of a step at $T = T_m$ in the picosecond temperature curve. Figure 4.42 shows that the temperature passes smoothly through $T = T_m$. Note also that the fluence necessary to heat the surface to the melting point (35 mJ/cm^2) is less than the threshold fluence E_{TH} for the onset of the high reflectivity phase (45 mJ/cm^2). The measured temperature at E_{TH} is approximately 2000 K, much greater than T_m. It appears from Fig. 4.42 that there is some temperature step near E_{TH}, somewhat reminescent of the temperature plateau at T_m of the nanosecond data in Fig. 4.41.

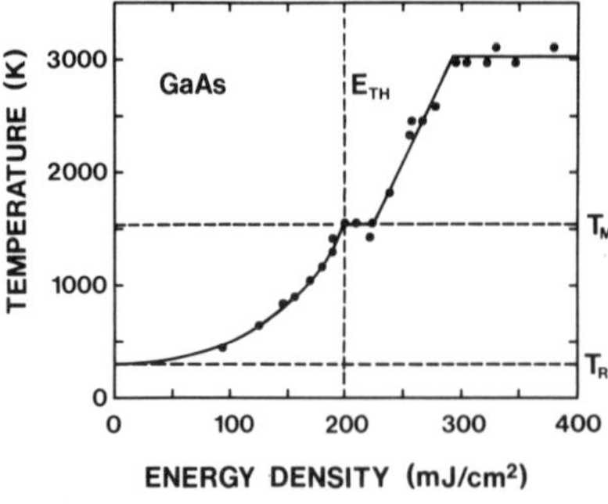

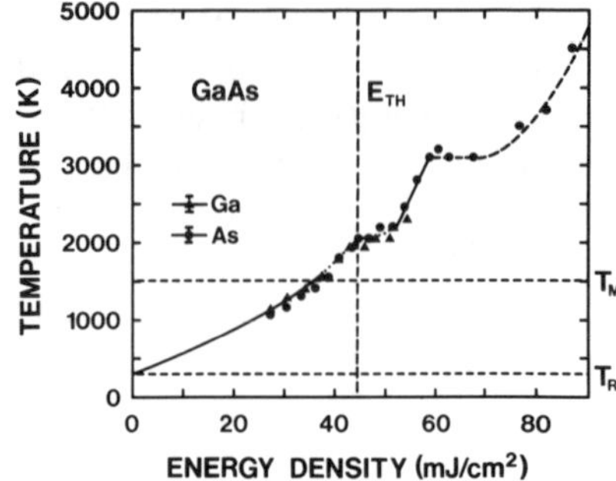

▲

Fig. 4.41. Temperature deduced from the velocity distribution of Ga atoms evaporated from the surface of a GaAs crystal during heating with laser pulses of 10 ns duration at 532 nm. The upper dashed line marks the melting point of GaAs, 1511 K. $E_{TH} = 200\,\mathrm{mJ\,cm^{-2}}$ is the threshold fluence for the onset of metallic reflectivity for nanosecond heating. After [4.167]

Fig. 4.42. Temperature deduced from the velocity distributions of Ga and As atoms evaporated from the surface of a GaAs crystal during heating with laser pulses of 25 ps duration at 532 nm. For picosecond heating the reflectivity rise starts at $E_{TH} = 45\,\mathrm{mJ\,cm^{-2}}$. After [4.167]

Fabricius et al. have interpreted the picosecond temperature data as evidence of superheating of the solid. Superheating occurs [4.170] when the heating rate is so high that the evolution of the phase transformation is no longer limited by the heat supply, which would be the "normal" situation. Instead, the progress of the phase transformation is limited by the kinetics of the atomic rearrangement at the solid–liquid interface. In such a situation the solid may be driven into a metastable state with the temperature greatly exceeding the equilibrium melting temperature. Fabricius et al. concluded that the picosecond temperature data indicate superheating of solid GaAs by several hundred Kelvin.

Very large superheating of the order of 1000 K has been reported by *Williamson* et al. [4.171] who studied picosecond laser melting of a very thin aluminium film. The metal film is melted by irradiation with an intense near-infrared picosecond pulse (1064 nm). A remarkable feature of this experiment is the use of a picosecond electron probe pulse for monitoring the phase transition, a technique pioneered by *Mourou* and coworkers [4.143]. The picosecond electron pulse is generated by shining a suitably delayed pulse at the fourth harmonic of the heating pulse onto the photocathode of a modified streak camera, which is operated as a photon-electron converter in a static mode without sweeping the electron beam. The 25 keV electron pulse is passed through the polycrystalline aluminium film. Diffracted electrons forming a ring pattern characteristic of Debye-Scherrer-type electron diffraction are recorded on a phosphorescent screen. The rise of the lattice temperature during laser heating can be inferred from the measured changes of the ring diameters in the electron diffraction pattern, using the known changes of the lattice parameters with temperature. Disappearance of the diffraction rings signifies the collapse of the crystal lattice when melting occurs. The data of *Williamson* et al. [4.171] indicated very strong superheating and a time delay between the heating pulse and the complete disappearance of the diffraction rings, which varied from several nanoseconds near the threshold energy for melting down to about 20 ps for higher energies of the heating pulse.

4.5 Concluding Remarks

It is hoped that the topics and examples discussed in this article have served to provide an impression of the great variety and breadth of past and current applications of ultrashort laser pulses to the study of very fast solid state phenomena. A few concluding remarks may be appropriate.

If the various activities in ultrafast solid state spectroscopy are compared, it appears that during the last few years the work on electronic relaxation phenomena in semiconductors has really dominated the field, to a greater extent perhaps than reflected by the selection and number of topics considered in this article. Due to the necessity to limit the scope of the review, many important topics had to be omitted, most notably, for instance, transient carrier transport phenomena [4.172].

While investigations of electronic relaxation processes in semiconductors have been particularly stimulated by the maturing of femtosecond optical technology, studies of vibrational phenomena have expanded to a somewhat lesser degree. Very recently, however, it has been demonstrated that femtosecond laser pulses with durations comparable or less than a vibrational cycle can be used for impulsive excitation of high frequency vibrational modes [4.173–175], and that the direct observation in the time domain of the atomic displacements associated with vibrational lattice mode is feasable. These exciting developments may well stimulate new directions in the study of high frequency phonons.

The fascinating past achievements and the exciting current developments have created a great deal of enthusiasm as regards the application of ultrafast measuring techniques to semiconductor physics. The opinion of long-time experts in the semiconductor field may be quite helpful for a proper assessment of the impact of these developments. Many experts do indeed share the optimism and agree that ultrafast spectroscopy has become a useful item in the arsenal of complex scientific tools of semiconductor physics. It is generally acknowledged that pico- and femtosecond experiments have provided new means of detailed and thorough testing of the existing theories of electronic relaxation in semiconductors. On the other hand, the complexity of exciton screening phenomena discovered in recent time-resolved experiments serves as an example to show that ultrafast spectroscopy is reaching beyond the limits of present expectations. Some observers feel that the true impact of ultrafast techniques on semiconductor physics and technology is still to be felt in the future.

References

4.1 S.L. Shapiro: (ed.) *Ultrashort Light Pulses*, Topics Appl. Phys. Vol. 18 (Springer, Berlin, Heidelberg 1977)

4.2 J.J. Hopfield: Phys. Rev. **112**, 1555 (1958)

4.3 M.A. Lampert: Phys. Rev. Lett. **1**, 450 (1958); S.A. Moskalenko: Optics and Spectroscopy **5**, 147 (1958)

4.4 See, e.g. *Excitons at High Densities*, ed. by H. Haken, S. Nikitine, Springer Tracts Mod. Phys. Vol. 73 (Springer, Berlin, Heidelberg 1975)

4.5 N.F. Mott: *Metal-Insulator Transitions* (Taylor and Francis, London 1974)

4.6 For an overview of the properties of excitons in quantum well structures, see, e.g., D.S. Chemla, D.B.A. Miller: J. Opt. Soc. Am. B **2**, 1155 (1985)

4.7 Y. Masumoto, Y. Unuma, Y. Tanaka, S. Shionoya: J. Phys. Soc. Jap. **47**, 1844 (1979)

4.8 Y. Segawa, Y. Aoyagi, K. Azuma, S. Namba: Sol. State Commun. **28**, 853 (1978)

4.9 Y. Segawa, Y. Aoyagi, S. Namba: Solid State Commun. **32**, 229 (1979)

4.10 Y. Segawa, Y. Aoyagi, S. Namba: J. Phys. Soc. Japan **52**, 3664 (1983)

4.11 Y. Masumoto, Y. Unuma, S. Shionoya: J. Phys. Soc. Japan Suppl. A **49**, 393 (1980)

4.12 R.G. Ulbrich, G.W. Fehrenbach: Phys. Rev. Lett. **43**, 963 (1979)

4.13 See e.g., E.S. Koteles: In *Excitons*, ed. by E.I. Rashba, M.D. Sturge, (North Holland, New York 1982)

4.14 U. Heim, P. Wiesner: Phys. Rev. Lett. **30**, 1205 (1973)

4.15 P. Wiesner, U. Heim: Phys. Rev. B **11**, 3071 (1975)

4.16 See, e.g., D. von der Linde: in *Ultrashort Light Pulses*, ed. by S.L. Shapiro, Topics Appl. Phys. Vol. 18 (Springer, Berlin, Heidelberg 1977) p. 254

4.17 R. Levy, B. Hönerlage, J.B. Grun: Solid State Commun, **29**, 103 (1979)

4.18 Y. Masumoto, S. Shionoya: J. Phys. Soc. Japan **51**, 181 (1982)

4.19 Y. Masumoto, S. Shionoya: J. Lumin. **24/25**, 421 (1981)

4.20 Y. Segawa, Y. Aoyagi, S. Namba: Solid State Commun. **39**, 535 (1981)

4.21 Y. Masumoto, S. Shionoya: Phys. Rev. B **30**, 1076 (1984)

4.22 Y. Aoyagi, Y. Segawa, S. Namba: Phys. Rev. B **25**, 1453 (1982)

4.23 A.A. Gogolin, E.I. Rashba: JETP Lett. **17**, 478 (1973)

4.24 Y. Unuma, Y. Masumoto, S. Shionoya: J. Phys. Soc. Japan **51**, 1200 (1982)

4.25 E.O. Göbel: In *Nonlinear Optics: Processes and Devices*, ed. by C. Flytzanis, J.L. Oudar, Springer Proc. Phys., Vol. 4 (Springer, Berlin, Heidelberg 1986) p. 104

4.26 C.V. Shank, R.L. Fork, R.F. Leheny, J. Shah: Phys. Rev. Lett. **42**, 112 (1972)

4.27 N.F. Mott: Philos. Mag. **6**, 287 (1961)

4.28 G.W. Fehrenbach, W. Schäfer, J. Treusch, R.G. Ulbrich: Phys. Rev. Lett. **49**, 1281 (1982)

4.29 G.W. Fehrenbach, W. Schäfer, R.G. Ulbrich: J. Lumin. **30**, 154 (1985)

4.30 J.G. Fujimoto, S.G. Shevel, E.P. Ippen: Solid State Commun. **49**, 605 (1984)

4.31 D. Hulin, A. Antonetti, L.L. Chase, J.L. Martin, A. Migus, A. Mysyrowicz, J.P. Löwenau, S. Schmitt-Rink, H. Haug: Phys. Rev. Lett. **52**, 779 (1984)

4.32 A. Antonetti, D. Hulin, A, Migus, A. Mysyrowicz, L.L. Chase: J. Opt. Soc. Am. B **2**, 1197 (1985)

4.33 W.H. Knox, R.L. Fork, M.C. Downer, D.A.B. Miller, D.S. Chemla, C.V. Shank, A.C. Gossard, W. Wiegmann: Phys. Rev. Lett. **54**, 1306 (1985)

4.34 N. Peyghambarian, H.M. Gibbs, J.L. Jewell, A. Antonetti, A. Migus, D. Hulin, A. Mysyrowicz: Phys. Rev. Lett. **53**, 2433 (1984)

4.35 See, e.g., B.R. Nag: *Theory of Electrical Transport in Semiconductors* (Pergamon, Oxford 1972)

4.36 For a review see, e.g., C. Klingshirn, H. Haug: Phys. Reports, **70**, 317 (1981)

4.37 E. Göbel: Appl. Phys. Lett. **24**, 492 (1974)

4.38 J. Shah, R.F. Leheny, W. Wiegmann: Phys. Rev. B **26**, 1577 (1977)

4.39 K. Arya, W. Hanke: Solid State Commun. **33**, 739 (1980)

4.40 H. Haug, D.B. Tran Thoai: Phys. Status Solidi B **98**, 581 (1981)

4.41 D. von der Linde, R. Lambrich: Phys. Rev. Lett. **42**, 1090 (1979)
4.42 S. Tanaka, H. Kobayashi, H. Saito, S. Shionoya: Solid State Commun. **33**, 167 (1980)
4.43 S. Tanaka, H. Koabyashi, H. Saito, S. Shionoya: J. Phys. Soc. Japan **49**, 1051 (1080)
4.44 S. Tanaka, T. Kuwata, T. Hokimoto, H. Kobayashi, H. Saito: J. Phys. Soc. Japan **52**, 677 (1983)
4.45 H. Yoshida, H. Saito, S. Shionoya: J. Phys. Soc. Japan **50**, 881 (1981)
4.46 H. Yoshida, H. Saito, S. Shionoya: Phys. Status Solidi B **104**, 331 (1981)
4.47 H. Yoshida, H. Saito, S. Shionoya, V.B. Timofev: Solid State Comm. **33**, 161 (1980)
4.48 S. Tanaka, H. Yoshida, H. Saito, S. Shionoya: In *Semiconductors-Probed by Ultrafast Spectroscopy*, Vol. I, ed. by R.R. Alfano (Academic, New York 1984) p. 171
4.49 H. Yoshida, S. Shionoya: Phys. Status Solidi B **115**, 203 (1983)
4.50 Y. Unuma, Y. Abe, Y. Masumoto, S. Shionoya: Phys. Status Solidi B **125**, 735 (1984)
4.51 H. Saito, E.O. Göbel: Phys. Rev. B **31**, 2360 (1985)
4.52 D. Hulin, A. Mysyrowicz, A. Migus, A. Antonetti: J. Lumin. **30**, 290 (1985)
4.53 R.F. Leheny, J. Shah, R.L. Fork, C.V. Shank, A. Migus: Solid State Comm. **31** 809 (1979)
4.54 C.V. Shank, R.L. Fork, R. Yen, J. Shah, B.I. Green, A.C. Gossard, C. Weisbuch: Solid State Comm. **47**, 981 (1983)
4.55 W. Graudszus, E.O. Göbel: Physica **117B** & **118B**, 555 (1983)
4.56 R.J. Seymour, M.R. Junnarkar, R.R. Alfano: Solid State Comm. **41**, 657 (1982)
4.57 Z.Y. Xu, C.L. Tang: Appl. Phys. Lett. **44**, 492 (1984)
4.58 A. Antonetti, D. Hulin, A. Migus, A. Mysyrowicz, L.L. Chase: J. Opt. Soc. Am. B **2**, 1197 (1985)
4.59 K. Kash, J. Shah, D. Block, A.C. Gossard, and W. Wiegmann, Physica **134 B**, 189 (1985)
4.60 K. Kash, J. Shah: Appl. Phys. Lett. **45**, 401 (1984)
4.61 J. Collet, A. Cornet, M. Pugnet, T. Amand: Solid State Comm. **42**, 883 (1982)
4.62 R. Luzzi, A.R. Vasconcellos: In *Semiconductors Probed by Ultrafast Spectroscopy*, Vol. I, ed. by R.R. Alfano (Academic, New York, 1984) p. 135
4.63 W. Pötz, P. Kocevar: Phys. Rev. B **28**, 7040 (1983)
4.64 P. Kocevar: Physica **134 B**, 155 (1985)
4.65 C.L. Tang, D.J. Erskine: Phys. Rev. Lett. **51**, 840 (1983)
4.66 D.J. Erskine, A.J. Taylor, C.L. Tary: Appl. Phys. Lett. **45**, 54 (1984)
4.67 C.L. Tang: In *Nonlinear Optics: Processes and Devices*, ed. by C. Flytzanis, J.C. Oudar, Springer Proc. Phys., Vol. 4 (Springer, Berlin, Heidelberg 1986) p. 80
4.68 D.J. Erskine, A.J. Taylor, C.L. Tang: Appl. Phys. Lett. **43**, 989 (1983)
4.69 A.J. Taylor, D.J. Erskine, C.L. Tang: J. Opt. Soc. Am. B **2**, 663 (1985)
4.70 D. von der Linde, J. Kuhl, E. Rosengart: J. Lumin. **24/25**, 675 (1981)
4.71 W.M. Knox, C. Hirlimann, D.A.B. Miller. J. Shah, D.S. Chemla, C.V. Shank: Phys. Rev. Lett. **56**, 1191 (1986)
4.72 J.L. Oudar, D. Hulin, A. Migus, A. Antonetti, F. Alexandre: Phys. Rev. Lett. **55**, 2074 (1985)
4.73 J.L. Oudar, A. Migus, D. Hulin, G. Grillon, J. Etchepare, A. Antonetti: Phys. Rev. Lett. **53**, 384 (1984)
4.74 A.L. Smirl, T.F. Boggers, B.S. Wherrett, G.P. Perryman, A. Miller: Phys. Rev. Lett. **49**, 933 (1982)
4.75 See, e.g., N.F. Mott, E.A. Davies: *Electronic Processes in Non-Crystalline Materials* (Clarendon, London 1979)
4.76 E.O. Göbel, W. Graudzus: Phys. Rev. Lett. **48**, 1277 (1982)
4.77 Z. Vardeny, J. Tauc: Phys. Rev. Lett. **46**, 1223 (1981)
4.78 Z. Vardeny, J. Strait, D. Pfost, J. Tauc, B. Abeles: Phys. Rev. Lett. **48**, 1132 (1982)
4.79 Z. Vardeny, J. Tauc: J. Phys. C **7**, 477 (1981)
4.80 Z. Vardeny, J. Tauc: In *Semiconductors Probed by Ultrafast Spectroscopy*, Vol. II, ed. by R.R. Alfano (Academic, Orlando 1984)
4.81 J. Tauc: In *Semiconductors and Semimetals*, Vol. 21, Part B (Academic, New York 1984) p. 299
4.82 See. e.g., A.C. Gossard: In *Thin Film Preparation and Properties*, ed. by K.N. Tu, R. Rosenberg (Academic, New York 1983)

4.83 J.F. Ryan, R.A. Taylor, A.J. Turberfield, A. Maciel, J.M. Worlock, A.C. Gossard, W. Wiegmann: Phys. Rev. Lett. **53**, 1841 (1984)
4.84 J. Shah, A. Pinczuk, A.C. Gossard, W. Wiegmann: Phys. Rev. Lett. **54**, 2045 (1985)
4.85 E.O. Göbel, H. Jung, J. Kuhl, K. Ploog: Phys. Rev. Lett. **51**, 1588 (1983)
4.86 H.-J. Polland, L. Schultheis, J. Kuhl, E.O. Göbel, C.W. Tu: Phys. Rev. Lett. **55**, 2610 (1985)
4.87 See, e.g., N.W. Ashcroft, N.D. Mermin: *Solid State Physics* (Holt Saunders, Philadelphia 1976), or any other textbook on solid state physics
4.88 See, e.g., M. Born, K. Huang: *Dynamical Theory of Crystal Lattices* (Clarendon, Oxford 1961)
4.89 See, e.g., Y.R. Shen: *The Principles of Nonlinear Optics*, (Wiley, New York 1984)
4.90 D. von der Linde, A. Laubereau, W. Kaiser: Phys. Rev. Lett. **26**, 954 (1971)
4.91 R.R. Alfano, S.L. Shapiro: Phys. Rev. Lett. **26**, 1247 (1971)
4.92 A. Laubereau, W. Kaiser: Rev. Mod. Phys. **50**, 608 (1978)
4.93 J. Shah, R.C.C. Leite, J.F. Scott: Solid State Comm. **8**, 1089 (1970)
4.94 K.T. Tsen, D.A. Abramson, R. Bray: Phys. Rev. B **26**, 4770 (1982)
4.95 See, e.g., W.E. Bron: Rep. Prog. Phys. **43**, 303 (1980)
4.96 H.J. Hartmann, A. Laubereau: J. Chem. Phys. **80**, 4663 (1984)
4.97 See, e.g., D. von der Linde: In *Ultrashort Light Pulses*, ed. by S.L. Shapiro (Springer, Berlin, Heidelberg 1977)
4.98 A. Laubereau: In *Semiconductors Probed by Ultrafast Laser Spectroscopy*, Vol. I, ed. by R.R. Alfano (Academic, New York 1984) p. 275
4.99 A. Laubereau, D. von der Linde, W. Kaiser: Phys. Rev. Lett. **27**, 802 (1971)
4.100 A. Laubereau, G. Wochner, W. Kaiser: Opt. Commun. **14**, 75 (1975)
4.101 G.M. Gale, A. Laubereau: Opt. Commun. **44**, 273 (1983)
4.102 J. Kuhl, D. von der Linde: In *Picosecond Phenomena III*, Springer Ser. Chem. Phys. Vol. 23, ed. by K.B. Eisenthal, R.M. Hochstrasser, W. Kaiser, A. Laubereau (Springer, Berlin, Heidelberg 1982) p. 201
4.103 J. Kuhl, W.E. Bron: Solid State Commun. **49**, 935 (1984)
4.104 W.E. Bron, J. Kuhl, B.K. Rhee: Phys. Rev. **34**, 6961 (1986)
4.105 K. Duppen, B.M.M. Hesp, D.A. Wiersma: Chem. Phys. Lett. **79**, 399 (1981)
4.106 D.D. Dlott, C.L. Schlosser, E.L. Chronister: Chem. Phys. Lett. **90**, 386 (1982)
4.107 T.J. Kosic, R.E. Cline, D.D. Dlott: Chem. Phys. Lett. **103**, 109 (1983)
4.108 D. von der Linde, J. Kuhl, H. Klingenberg: Phys. Rev. Lett. **44**, 1505 (1980)
4.109 D. von der Linde, J. Kuhl, R. Lambrich: In *Picosecond Phenomena II*, Springer Ser. Chem. Phys., Vol. 14, ed. by R.M. Hochstrasser, W. Kaiser, C.V. Shank (Springer, Berlin, Heidelberg, 1980) p. 336
4.110 J.A. Kash, J.C. Tsang, J.M. Hvam: Phys. Rev. Lett. **54**, 2151 (1985)
4.111 C.L. Collins, P.Y. Yu: Phys. Rev. B **27**, 2602 (1983)
4.112 C.L. Collins, P.Y. Yu: Phys. Rev. B **30**, 4501 (1984)
4.113 F. Bogani: J. Phys. C **11**, 1283 (1978)
4.114 M. L. Geirnaert, G.M. Gale, C. Flytzanis: Phys. Rev. Lett. **52**, 815 (1984)
4.115 G.M. Gale, P. Guyot-Sionnest, W.Q. Zheng, C. Flytzanis: Phys. Rev. Lett. **54**, 823 (1985)
4.116 F. Vallée, G.M. Gale, C. Flytzanis: Chem. Phys. Lett. **124**, 216
4.117 N. Bloembergen: In *Laser Solid Interactions and Laser Processing*, ed. by S.D. Ferris, M.J. Leamy, J.M. Poate (American Institute of Physics Conference Proceedings No. 50, New York 1979), p. 1
4.118 Chieng-Jung Lin, F. Spaepen: Appl. Phys. Lett. **41**, 721 (1982)
4.119 For a general reference, see e.g., *Laser Annealing of Solids*, ed. by J.M. Poate, J.W. Meyer (Academic, New York 1982)
4.120 I.B. Khaibullin, E.I. Shtyrkov, M.M. Zaripov, R.M. Bayazitov, M.F. Galjantdinov: Radiat. Eff. **36**, 225 (1978)
4.121 See, e.g., P. Baeri, S.U. Campisano, G. Foti, E. Rimini: J. Appl. Phys. **50**, 788 (1978)
4.122 J.A. Van Vechten, R. Tsu, F.W. Saris: Phys. Lett. A **74**, 4222 (1979)
4.123 W.L. Brown: Mater. Res. Soc. Symp. Proc. **23**, 9 (1984)

4.124 R. Srinivasan, V. Mayne-Banton: Appl. Phys. Lett. **41**, 576 (1982)
4.125 R. Tsu, R.T. Hodgson, T.Y. Tan, J.E. Baglin: Phys. Rev. Lett. **42**, 1356 (1979)
4.126 M.O. Thompson, G.J. Galvin: Mater. Res. Soc. Symp. Proc. **13**, 57 (1983), and references therein
4.127 P.L. Liu, R. Yen, N. Bloembergen, R.T. Hodgson: Appl. Phys. Lett. **34** 864 (1979)
4.128 J.M. Liu, R. Yen, H. Kurz, N. Bloembergen: Appl. Phys. Lett. **39**, 755 (1981)
4.129 I.W. Boyd, S.C. Moss, T.F. Bogges, A.L. Smirl: Appl. Phys. Lett. **45**, 80 (1984)
4.130 A.L. Smirl, T.F. Bogges, S.C. Moss, I.W. Boyd: J. Lumin. **30**, 272 (1985)
4.131 P.H. Bucksbaum, J. Bokor: Phys. Rev. Lett. **53**, 182 (1984)
4.132 D.H. Auston, C.M. Surko, T.N.C. Venkatesan, R.E. Slusher, J.A. Golovchenko: Appl. Phys. Lett. **33**, 437 (1978)
4.133 D.H. Auston, J.A. Golovchenko, A.L. Simons, C.M. Surko, T.N.C. Venkatesan: Appl. Phys. Lett. **34**, 777 (1979)
4.134 D.M. Kim, R.R. Shah, D. von der Linde, D.L. Crosthwait: Mater. Res. Soc. Symp. Proc. **4**, 85 (1982)
4.135 J.M. Lui, H. Kurz, N. Bloembergen: Appl. Phys. Lett. **41**, 643 (1982)
4.136 D. von der Linde, N. Fabricius: Appl. Phys. Lett. **41**, 991 (1982)
4.137 K.M. Sharev, B.A. Baum, P.V. Gel'd: Sov. Phys. Solid State **16**, 2111 (1975)
4.138 C.V. Shank, R. Yen, C. Hirlimann: Phys. Rev. Lett. **50**, 454 (1983)
4.139 M.C. Downer, R.L. Fork, C.V. Shank: J. Opt. Soc. Am. B **2**, 595 (1985)
4.140 I.W. Boyd, S.C. Moss, T.F. Bogges, A.L. Smirl: Appl. Phys. Lett. **46**, 366 (1985)
4.141 K. Gamo, K. Murakami, M. Kawabe, S. Namba, A. Aoyagi: Mater. Res. Soc. Symp. Proc. **1**, 97 (1981)
4.142 K. Murakami, H.C. Gerritsen, H. Van Brug, F. Bijkerk, F.W. Saris, M.J. van der Wiel: Phys. Rev. Lett. **56**, 655 (1986)
4.143 S. Williamson, G. Mourou, J.C.M. Li: Phys. Rev. Lett. **52**, 2364 (1984)
4.144 C.V. Shank, R. Yen, C. Hirlimann: Phys. Rev. Lett. **51**, 900 (1983)
4.145 H.W.K. Tom, T.F. Heinz, Y.R. Shen: Phys. Rev. Lett. **51**, 1983 (1983)
4.146 N. Bloembergen, R.K. Chang, S.S. Jha, C.M. Lee: Phys. Rev. Lett. **174**, 813 (1968)
4.147 D. Guidotti, T.A. Driscoll, H.J. Gerritson: Solid State Comm. **46**, 337 (1983)
4.148 T.A. Driscoll, D. Guidotti: Phys. Rev. B **28**, 1171 (1983)
4.149 J.A. Litwin, J.E. Sipe, H.M. van Driel: Phys. Rev. B **31**, 5543 (1985)
4.150 See, e.g., M. Born, E. Wolf: *Principles of Optics* (Pergamon, Oxford 1975)
4.151 H.M. van Driel, L.A. Lompre, N. Bloembergen: Appl. Phys. Lett. **44**, 285 (1984)
4.152 N. Bloembergen, H. Kurz, J.M. Liu, R. Yen: Mater. Res. Soc. Symp. Proc. **4**, 3 (1982)
4.153 M. Combescot, J. Bok: Phys. Rev. Lett. **51**, 519 (1983)
4.154 J.A. Van Vechten, In *Semiconductors Probed by Ultrafast Laser-Spectroscopy*, Vol. II, ed. by R.R. Alfano (Academic, New York 1984) p. 95
4.155 H.M. van Driel: Appl. Phys. Lett. **44**, 617 (1984)
4.156 G.Z. Yang, N. Bloembergen: IEEE QE-**22**, 195 (1986)
4.157 H. Kurz, N. Bloembergen: Mater. Res. Soc. Symp. Proc. **35**, 3 (1985)
4.158 L.A. Lompre, J.M. Liu, H. Kurz, N. Bloembergen: Appl. Phys. Lett. **44**, 3 (1984)
4.159 J.M. Liu, R. Yen, H. Kurz, N. Bloembergen: Mater. Res. Soc. Symp. Proc. **4**, 29 (1982)
4.160 A.M. Malvezzi, J.M. Liu, N. Bloembergen: Mater. Res. Soc. Symp. Proc. **23**, 135 (1984)
4.161 A.M. Malvezzi, H. Kurz, N. Bloembergen: Mater. Res. Soc. Symp. Proc. **35**, 75 (1985)
4.162 L.A. Lompre, J.M. Liu, H. Kurz, N. Bloembergen: Appl. Phys. Lett. **43**, 168 (1983)
4.163 L.A. Lompre, J.M. Liu, H. Kurz, N. Bloembergen: Mater. Res. Soc. Symp. Proc. **23**, 57 (1984)
4.164 G.E. Jellison, F.A. Modine: Appl. Phys. Lett. **41**, 180 (1982) G.A. Jellison, F.A. Modine: Phys. Rev. B **27**, 7466 (1983)
4.165 K. Murakami, K. Takita, K. Masuda: Jap. J. Appl. Phys. **20**, L 867 (1981)
4.166 N. Fabricius, P. Hermes, D. von der Linde, A. Pospieszczyk, B. Stritzker: Solid State Comm. **58**, 239 (1986)

4.167 N. Fabricius, P. Hermes, D. von der Linde, A. Pospieszczyk, B. Stritzker: Mater. Res. Soc. Symp. Proc. **51**, 219 (1986)
4.168 B. Stritzker, A. Pospieszczyk, J. Tagle: Phys. Rev. Lett. **47**, 356 (1981)
4.169 A. Pospieszczyk, M. Abdel Harith, B. Stritzker: J. Appl. Phys. **54**, 3176 (1983)
4.170 F. Spaepen, D. Turnbull: In *Laser Annealing of Semiconductors*, ed. by J.M. Poate, J.W. Mayer (Academic, New York 1982) p. 15
4.171 S. Williamson, G. Mourou, J.C. Lee: Mater. Res. Soc. Symp. Proc. **35**, 876 (1985)
4.172 See, e.g., D.K. Ferry, H.L. Grubin, G.J. Iafrate: In *Semiconductors Probed by Ultrafast Spectroscopy*, Vol. I, ed. by R.R. Alfano (Academic, New York 1984) p. 413
4.173 M.M. Robinson, Y.X. Yan, E.B. Gamble, L.R. Williams, J.S. Meth, K.A. Nelson: Chem. Phys. Lett. **112**, 491 (1984)
4.174 S. DeSilvestri, J.G. Fujimoto, E.P. Ippen, E.G. Gamble, L.R. Williams, K.A. Nelson: Chem. Phys. Lett. **116**, 146 (1985)
4.175 K.P. Cheung, D.H. Auston: Phys. Rev. Lett. **55**, 2152 (1985)

5. Ultrafast Optoelectronics

David H. Auston

With 26 Figures

This chapter reviews recent progress in high speed optoelectronics and its applications to the measurement of high speed electronic devices and materials. Basic device concepts using photoconducting and electro-optic materials are described and their implementation in high speed measurement systems is illustrated with specific applications to discrete device and integrated circuit testing, high power switching, and radio frequency generation and detection. The chapter concludes with a discussion of current challenges and possible future trends.

5.1 Introduction

The other chapters in this book vividly illustrate the remarkable progress in the development of high speed optics technology and the science of using optics to measure extremely rapid events. A parallel development which has occurred at an equally rapid pace is the steady progress in high speed electronics technology. Spurred on by new materials capabilities such as molecular beam epitaxy, and metal-organic chemical vapor deposition, semiconductor electronic devices have now demonstrated switching speeds as fast as 6 picoseconds [5.1]. The evolution of these two technologies is plotted in Fig. 5.1. Clearly, optics still leads electronics in raw speed capability. There is a substantial gap of one to three orders of magnitude between optical and electronic measurement capabilities. It is in this gap between the two technologies that optoelectronics resides and plays its unique role in combining the speed and flexibility of optics to develop new and faster electronic devices and measurement systems

There is an important distinction between the concept of ultrafast optoelectronics described in this chapter, and the more conventional descriptions given in texts on optical communications. In the latter, the emphasis is on modulation and demodulation of an optical signal, with the light beam being regarded as the carrier of information. Our interest is to use optical pulses as sources of power and timing in addition to their use as information carriers. Although the distinction is a subtle one, it is important since it provides enormously greater flexibility in the application of high speed optics technology and has enabled an entirely new class of devices and approaches to electronic measurements to be developed. This has evolved from the somewhat unconventional use of optics to control and

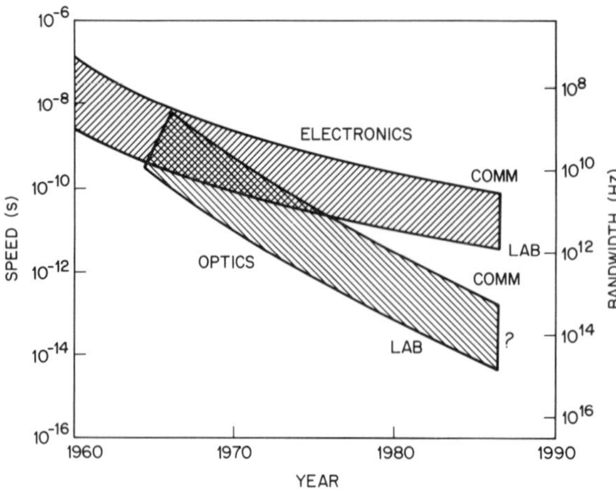

Fig. 5.1. General trend of increasing speed in electronics and optics technology plotted vs year. The spread of these curves is intended to represent the difference between commercially available devices ("comm") and what can be achieved in the laboratory ("lab"). The left hand vertical scale is a measure of speed in units of seconds and the right hand scale is the equivalent base bandwidth

measure electrical signals, rather than the reverse, which is usually done in conventional optoelectronics. Consequently, this review does not include that part of optoelectronics that relates to optical communications. The latter subject has a large literature which has been extensively reviewed.

In this chapter, we review the recent research in ultrafast optoelectronics, with particular emphasis on novel device and measurement concepts. We begin with a brief overview of basic device concepts to illustrate the wide range of novel applications of ultrafast optoelectronics for the generation and measurement of ultrafast electrical transients, including microwave, millimeter-wave, and far-infrared generation and detection. This is then followed by a discussion of materials for ultrafast optoelectronics, emphasizing the properties of photoconductors and electro-optical materials. The specific details of these applications are described in subsequent sections. Many of these, such as the use of non-invasive electro-optic sampling for integrated circuit measurements, high power switching photoconductors, and time-domain far-infrared measurements of material properties, are advancing at a rapid rate. For this reason, it is possible that our description of many of these topics will be outdated by the time this article appears. We conclude with a brief discussion of current challenges and future trends.

The scope of this review is not intended to be exhaustive and all-inclusive, but merely to highlight the key concepts of ultrafast optoelectronics and to illustrate them with specific examples of measurements and devices. This will require some omissions which the author regrets. Extensive references to the literature cited in the text are listed at the end of this chapter. For additional information the reader is referred to the articles in *Picosecond Optoelectronic Devices* [5.2] the Proceedings of the 1985 and 1987 Picosecond Electronics and Optoelectronics Conferences [5.3,4], and the 1984 and 1986 Ultrafast Phenomena Conferences [5.5,6].

The reader will find little resemblance between this chapter and the equivalent chapter on this subject in the first edition of this book [5.7], which was written almost ten years ago, a clear indication of the rapid pace of developments in this field.

5.2 Optoelectronic Device Concepts

In this section we give a brief overview of the basic device concepts which use ultrafast optical pulses to generate and detect high speed electrical signals. In subsequent sections we will give the details of specific materials, devices and applications including complete references to the literature.

Clearly, a nonlinear interaction is required to provide the coupling between the optical and electrical signals. Two specific classes of materials have been used extensively for this purpose. These are photoconducting and electro-optic materials.

Some of the optoelectronic devices based on the use of photoconducting materials are illustrated in Figs. 5.2–4. Although each performs a different electronic function, they are all based on variations of the central concept of a light pulse producing conductivity modulation by electron-hole injection in a semiconductor. With moderate optical pulse energies it is possible to produce a photoresistance that is relatively low compared to the characteristic impedance of the transmission line. This results in a switching action which permits a fast optical pulse to initiate a high speed electrical signal. Figure 5.2a is a schematic

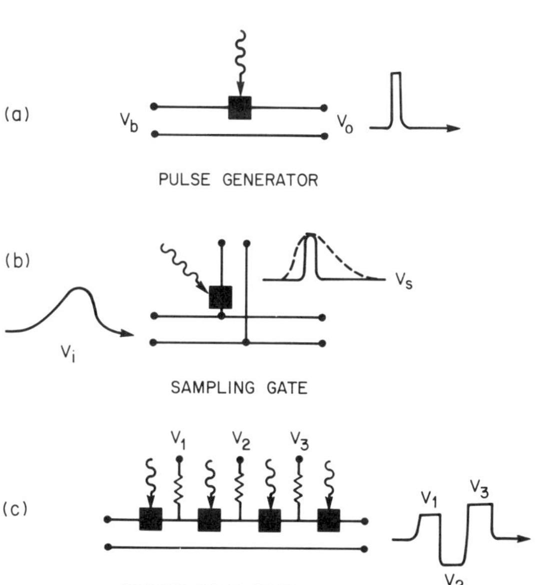

Fig. 5.2a–c. Schematic illustration of basic optoelectronic device concepts: (**a**) photoconducting electrical pulse generator, (**b**) photoconducting electrical sampling gate, and (**c**) "frozen wave" generator. In each case an ultrashort optical pulse illuminates a photoconductor to produce a fast conducting transient in a transmission line

illustration of the basic photoconducting electronic pulse generator. It consists of a photoconducting material mounted in a high speed transmission line. As we shall discuss in detail in Sect. 5.4, the amplitude and shape of the electrical pulse depends on the details of the device geometry and the materials as well as on the optical pulse. With proper choice of these parameters extremely fast electrical pulses having relatively large amplitudes can be generated by this technique. Figure 5.2b illustrates a sampling gate. In this case the input electrical signal is a time-varying waveform which is sampled by the photoconductor by diverting a small portion of the signal to a sampling electrode. By varying the relative timing between the incoming electrical waveform and the optical pulse, the amplitude, v_s, of the sampled pulse gives a stroboscopic replica of the desired waveform. As in other sampling measurement systems, it is not necessary to time-resolve the sampled signal v_s, so that highly sensitive low frequency electronics can be used. A novel variation on the pulse generator is the use of multiple switches to form a frozen wave generator as shown in Fig. 5.2c. Simultaneous illumination of the photoconductors releases the "frozen" waveform which can have an arbitrary shape determined by the number of photoconductors and the *dc* bias voltages applied to them. In this case, a long photoconductivity fall-time is required to enable each section to discharge in tandem. For the same reason, a relatively high optical pulse energy is required to ensure a low resistance for each photoconductor. The amplitudes and durations of each segment of the waveform can in principle be adjusted arbitrarily by changing the bias voltages and lengths of transmission line separating each photoconductor.

Figure 5.3a is an application to a radio frequency mixer in which the input signal is a high frequency sine wave whose amplitude is modulated by the optical

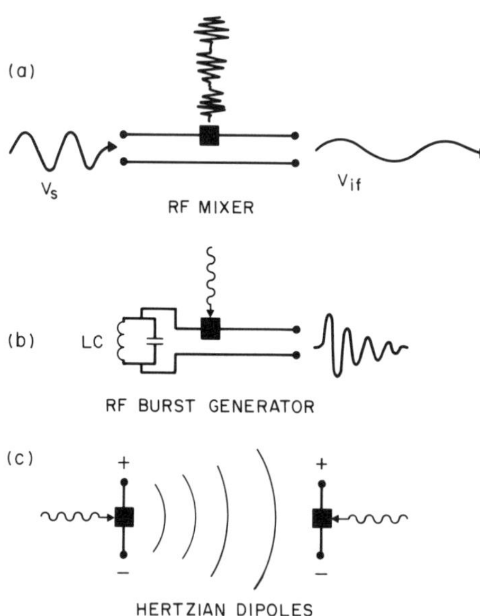

Fig. 5.3a–c. Other ultrafast optoelectronic devices which use high speed photoconductors: (**a**) a radio frequency mixer, which mixes a local oscillator signal, V_s, with an envelope encoded optical signal, which illuminates a high speed photoconductor, to produce an intermediate frequency, V_{if}; (**b**) a radio frequency burst generator which produces a damped oscillation at the frequency of the tank circuit in which the photoconductor is mounted; (**c**) photodonductors in which a rapid current transient radiates into free space, and is detected by a second receiving photoconducting antenna

pulse. The optical signal has an envelope modulation at a frequency equal to the local oscillator so that the output signal at the intermediate frequency is proportional to the product of the electrical and optical input signals.

Figure 5.3b illustrates the use of photoconductors for producing short bursts of radio frequencies. In this case, a resonant circuit controls the frequency of the generated waveform. The coupling of the circuit to an external transmission system causes the signal to decay rapidly, resulting in short bursts of radio frequency energy. With the use of high bias voltages and large optical pulse energies, this method can produce relatively high power rf signals.

When the rise-time of the photocurrent is extremely short, the photoconductor can directly radiate an electromagnetic signal into free space as illustrated in Fig. 5.3c. Photoconductors can also be used as receiving antennas by sampling an rf pulse when illuminated by an optical pulse. The combination of optically triggered transmitting and receiving antennas forms a measurement system which is phase coherent and has extremely good time resolution.

Electro-optic materials have also been extensively used for high speed optoelectronics. The electro-optic property of these materials has been used both for measuring high speed electrical signals using the Pockel's effect and also for generating short electromagnetic pulses by optical rectification. Figure 5.4a–c illustrates some of these approaches. Figure 5.4a shows a traveling-wave Pockel's cell for measuring fast electrical signals. This approach uses the small change in optical birefringence of an electro-optic material (Pockels effect) to measure the electrical waveform. In this example, the electro-optic material is used for the insulating substrate of the transmission line. The measurement does not perturb the electrical signal and is extremely fast. An important variation on this approach is illustrated in Fig. 5.4b, in which picosecond electrical signals in GaAs integrated circuits are probed by using the electro-optic property of the substrate.

Figure 5.4c illustrates the use of electro-optic materials for generating extremely short electro-magnetic transients. This approach uses the optical rectification effect to produce a nonlinear polarization in an electro-optic material. This polarization follows the intensity envelope of the optical pulse and can have an extremely short duration. The polarization can be converted to an electrical signal by using it to charge a capacitor as illustrated in Fig. 5.4c. When femto-

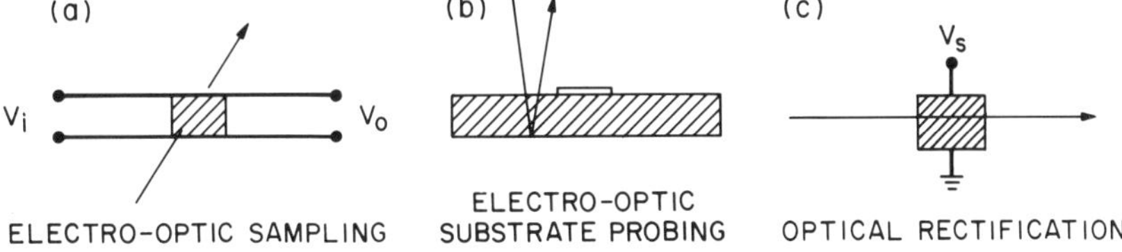

Fig. 5.4a–c. Ultrafast optoelectronic devices which use electro-optic materials: (**a**) electro-optic sampling of a fast electrical signal by probing the induced birefringence in a traveling-wave Pockels cell; (**b**) electro-optic measurement of the local electric field in a circuit for which the substrate material is electro-optic; (**c**) the generation of an electrical pulse in an electro-optic material by rectification of an optical pulse

second optical pulses are used, the polarization can be used as a radiative source and produce short electrical pulses analagous to the Cherenkov radiation from a supra-relativistic charged particle.

5.3 Materials for Ultrafast Optoelectronics

5.3.1 Photoconducting Materials

A wide variety of materials have been used for picosecond photoconductors. A summary of the key properties of some of the more important materials is given in Table 5.1. All of the materials listed in this table are semiconductors. They can be divided into the following classes: intrinsic, impurity-dominated, radiation damaged, polycrystalline, and amorphous semiconductors (note, in Table 5.1: d: radiation damaged; p: polycrystalline; and a: amorphous). Before discussing their relative merits, we will first summarize some of the more important intrinsic properties of semiconductor materials that are of interest to us for applications

Table 5.1. Materials for picosecond photoconductors

Material	Band Gap E_g[eV]	Resistivity $\rho[\Omega\,\mathrm{cm}]$	Mobility $\mu[\mathrm{cm}^2/\mathrm{vs}]$	Decay Time τ_c[ps]	Reference(s)
Si	1.12	4×10^4	1950	$10^7 - 10^3$	[5.56]
GaAs:Cr	1.43	$10^6 - 10^7$	~2000	300	[5.151]
InP:Fe	1.29	2×10^8	2200	150–1000	[5.42]
$CdS_{0.5}Se_{0.5}$	2.0	10^7	400	2×10^4	[5.152]
GaP	2.24	10^8	240	60–500	[5.153]
Diamond	5.5	$10^{13} - 10^{15}$	1800	50–300	[5.154]
d-SOS	1.12	10^5	~10 to 100	1 to 300	[5.62]
d-GaAs	1.43	10^7	~100	<5	[5.66]
d-InP	1.29	10^6	100–1000	1–100	[5.41,155]
d-InGaAs	0.75	~1	800–4000	40–800	[5.156,158]
p-Si	~1.1	10^5		2–50	[5.117]
p-Ge	0.85		~3	~50	[5.44]
p-CdTe	~1.5	3×10^7	60	4	[5.157]
a-Si	~1.4	$10^5 10^7$	~10	330	[5.61]

to picosecond photoconductors (for a review of the basic properties of photoconductors, the reader is referred to the book by *Rose* [5.8]).

a) Intrinsic Speed of Response of Semiconductors

The absorption of a photon by a semiconductor and the subsequent generation of an electron-hole pair is an intrinsically fast process, being limited by the uncertainty principle and the requirement that the frequency spectrum of the optical pulse falls within the absorption bands corresponding to electronic transitions from bound to free states. Since the width of these bands are a few eV, this time can in principle be as short as 10^{-15} s or one optical cycle, and consequently does not limit the rise-time of the photocurrent.

The quantum efficiency of the initial photocurrent is determined by the probability that the electron-hole pair will escape its mutual coulomb field. In most high mobility semiconductors this probability is essentially 100%. In some low mobility materials, however, such as amorphous selenium, the situation is quite different, and "geminate" recombination of electron-hole pairs can produce a significantly lower quantum efficiency.

Although the onset of photoconduction is very rapid, there are a number of effects that can influence the subsequent evolution of the current following excitation by a very short optical pulse. These are illustrated schematically in Fig. 5.5, where the photocurrent due to an optical pulse of infinitesimal duration is plotted vs time on a logarithmic scale. The initial rise-time is limited by the elastic scattering rate of the photo-generated electrons and holes (momentum relaxation rate). If the photon energy exceeds the threshold for photoconduction (the band gap in an intrinsic semiconductor) by an amount greater than the thermal energy, kT, the excess energy is imparted to the electron-hole pair in the form of kinetic energy. Since most scattering mechanisms [5.9,10] are stronger at higher energies (an important exception being ionized impurity scattering),

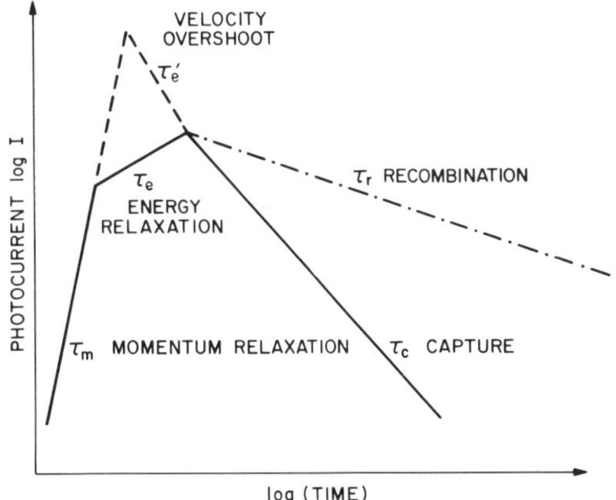

Fig. 5.5. Schematic illustration of the temporal response of a semiconductor to a fast optical pulse. The initial rise time is limited by the elastic scattering of the free electrons and holes. Energy relaxation then occurs and can either enhance or slow the response depending on the electric field and photon energy. The decay of the photocurrent is dominated by recombination, but may be speeded up by the introduction of defects to produce a rapid trapping of free carriers

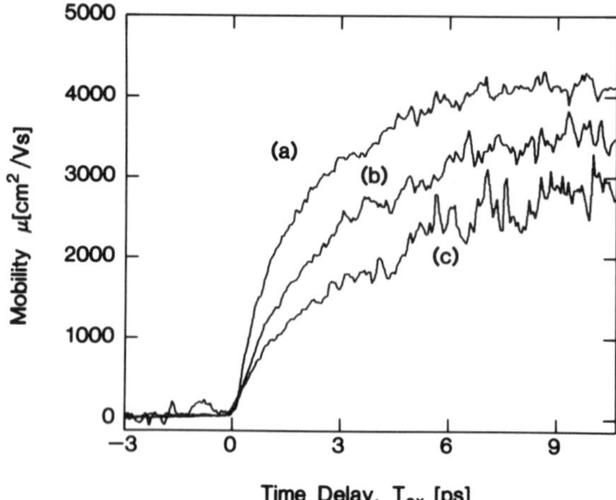

Fig. 5.6. The rise of the low field mobility of a sample of GaAs due to the optical injection of carriers at 2 eV by an ultrashort optical pulse. The three curves illustrate the dependence on injection density and are for densities of (*a*) $5 \times 10^{17}\,cm^{-3}$, (*b*) $5 \times 10^{18}\,cm^{-3}$, and (*c*) $1.2 \times 10^{19}\,cm^{-3}$ [5.16]

the initial mobility of the free carriers will be smaller than under equilibrium conditions. The transfer of this excess energy to the lattice by phonon emission produces an additional mobility transient with a rise-time that may range from picoseconds to nanoseconds depending on the material, the excitation density, the amount of excess energy, and the lattice temperature. These effects have been studied extensively using picosecond and femtosecond optical techniques [5.11–15]. At high excitation densities, these rates can be influenced by screening of the electron-phonon interaction, which tends to slow the cooling of the electrons and holes. Recent measurements in GaAs [5.16], show that these effects produce a mobility transient that rises from an initial value of approximately 500 cm^2/Vs to 4000 cm^2/Vs in a time which varies from 2 to 5 picoseconds depending on the excitation density. This result is illustrated in Fig. 5.6. The photon energy in this experiment (2 eV) was sufficient to cause an appreciable population of the satellite valleys of GaAs, accounting for the relatively "slow" rise-time of the mobility. The value of the mobility at longer times depended on the excitation density, an effect that is expected from electron-hole scattering. The density and temperature dependence of e-h scattering is relatively complicated due to the screening of the coulomb cross-section at high densities. This tends to produce a minimum in the carrier mobility vs density dependence. Theoretical calculations of e-h scattering have been done by *McLean and Paige* [5.17,18], *Paige* [5.19], *Appel* [5.20,21] and *Meyer* and *Glicksman* [5.22]. Measurements of *e-h* scattering have been made by *Vaitkus* et al. [5.23], *Auston* and *Johnson* [5.24], and *Grivitskas* et al. [5.25]. In the experiment of Auston and Johnson, the effective mobility ($\mu_n + \mu_h$) of silicon was measured in a picosecond photoconductor over a wide range of density and two different temperatures. At room temperature, they found that the mobility was depressed by a factor of approximately two due to e-h scattering at densities in the range of $10^{19}\,cm^{-3}$. At 80 K, however, the mobility was depressed almost an order of magnitude.

b) High Electric Field Effects

If the bias voltage applied to the photoconductor is very large, the photogenerated carriers will be heated by the electric field and produce a non-equilibrium distribution which can produce a number of important effects that strongly influence the dynamical and steady-state properties of photoconductors [5.26]. On a short time scale (less than 10 ps) this can produce a transient in the current waveform due to an effect known as velocity overshoot [5.26,27]. A number of experiments have been directed at measuring this effect [5.28–31]. Although many of these measurements demonstrate that velocity overshoot can influence the transient photocurrent under high field conditions, the detailed features of velocity overshoot remain to be elucidated. The situation is particularly complicated when electron heating is caused by both optical injection and high electric fields, as in the case of photoconductors with large bias voltages.

On longer time scales, once an equilibrium carrier distribution has been established, high electric fields can produce a substantially lower mobility due to saturation of the drift velocity. For example, in silicon, at fields in excess of 10^4 V/cm the electron and hole drift velocities saturate at approximately 10^7 cm/s. In GaAs and other materials having "satellite" valleys in their conduction bands, the steady-state drift velocity reaches a maximum at a field of approximately 3×10^3 V/cm, and then decreases at higher fields. This negative resistance property can lead to electric field instabilities due to the Gunn effect, and can limit the maximum *dc* bias that can be placed across a photoconductor.

c) Trapping and Recombination

In an intrinsic semiconductor, recombination is determined by radiative transitions. These are relatively slow, producing a long-lasting current waveform which is a major problem for applications to picosecond photoconductors. Direct-gap semiconductors have radiative lifetimes that are a few nanoseconds. Indirect-gap materials can have lifetimes that are much longer, extending into the millisecond range in some cases. Clearly, alternative materials and techniques must be used to produce the fast current decays that are necessary for picosecond applications.

In p-n junction diodes, carrier sweep-out determines the speed of response, which can be substantially faster than the intrinsic recombination time. Maximum sweep-out rates, however, are limited by drift velocity saturation (e.g. 10 ps for a saturation velocity of 10^7 cm/s in a 1 μm path length). As a result, sweep-out is primarily used in semiconductor devices where extremely high speeds are not required. Also the photovoltaic behavior of p-n junctions and the need for a large reverse bias make them unsuitable for some picosecond devices such as sampling gates.

An effective method of reducing the free carrier lifetime is to introduce a moderate density of defects into the semiconductor which act as traps and recombination centers. This can be achieved by radiation damage, impurities, or the use of materials with large densities of naturally occurring defects such as polycrystalline and amorphous semiconductors. In Table 5.1, some examples of materials in these classes are listed. The much faster decay times relative to the intrinsic materials are evident.

The capture time, τ_c can be estimated from the expression:

$$\tau_c = \frac{1}{N_t \sigma_c \langle v_{th} \rangle}, \tag{5.1}$$

where N_t is the trap density, σ_c the capture cross-section, and $\langle v_{th} \rangle$ the mean thermal carrier velocity. According to (5.1) trap densities of 10^{18} to $10^{20}\,cm^{-3}$ are sufficient to produce free carrier lifetimes of approximately one picosecond if the capture cross-sections are between 10^{-13} and $10^{-15}\,cm^2$, respectively.

d) Radiation Damage

The specific nature of the defects produced by radiation damage is a subject that has been extensively studied (for a review see [5.32]). The type, density, and stability of the defects depend on the material, type of radiation, background impurities, and temperature. In the case of damage produced by light- and medium-weight ions, it is thought that clusters of defects, possibly as large as 10–20Å are also produced [5.33] as well as elementary displacements such as Frenkel pairs (vacancy plus interstitial). Capture cross-sections have been measured by deep level transient spectroscopy (DLTS) for a variety of materials and defects. Although there exists a wide range of levels, each having different energies and capture cross-sections, certain general trends can be identified. For example, in silicon, a commonly occurring defect is the electron trap at -0.53 eV having a capture cross-section of approximately $2 \times 10^{-15}\,cm^2$ at room temperature [5.34]. A similar confluence of hole traps occurs at an energy of 0.30 eV above the valence band with a cross-section of approximately $10^{-15}\,cm^2$. In GaAs, there exist many deep levels having capture cross-sections in the range of 10^{-14} to $10^{-13}\,cm^2$ [5.35,36]. For example, bombardment of GaAs produces electron and hole traps at -0.71 eV with cross-sections of 1.3×10^{-13} and $2.3 \times 10^{-13}\,cm^2$ [5.37].

Experiments with picosecond photoconductors have shown that extremely fast photocurrent decay times can be produced with moderate levels of radiation damage (see Table 5.1 for specific references). In Fig. 5.7 the carrier lifetime of a radiation-damaged silicon-on-sapphire film is plotted as a function of radiation dose [5.38]. This measurement used femtosecond optical pulses to probe the free carrier absorption following excitation. The maximum decay rate of 0.7 ps was produced with damage levels that are close to the point of rendering the material completely amorphous.

The introduction of moderately large densities of defects into crystalline semiconductors has a number of additional effects, some of which are advantageous and others that are not. If the traps lie deep within the band gap, free

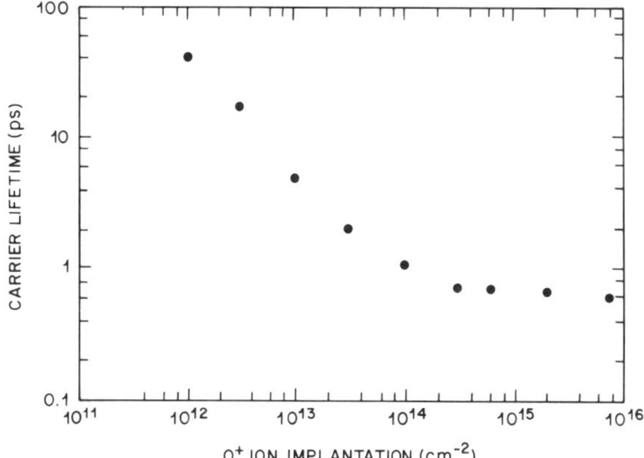

Fig. 5.7. The relaxation of optically injected carriers in radiation-damaged silicon-on-sapphire, as measured by time-resolved optical reflectivity with femtosecond optical pulses [5.38]

carriers due to dopants are removed from the transport bands, moving the Fermi level close to mid-gap and greatly increasing the resistivity. For most applications to photoconductors this is a desirable property since it reduces the dark current. A second feature of high defect densities is an enhanced optical absorption in the spectral region below the edge for direct transitions due to the introduction of new states and the relaxation of selection rules for indirect transitions. This effect is more pronounced in indirect-gap materials such as silicon.

Another advantage of the use of high defect density materials is the ease of fabrication of ohmic contacts. Semi-insulating materials are generally very difficult to contact with low resistance, non-rectifying contacts. A Schottky barrier usually forms at the metal semiconductor interface which produces non-ohmic behavior along with an internal electrostatic field which results in a photovoltaic response in addition to a photoconductive signal [5.39]. The problem of contacts is made even more difficult for cases where the injected carrier density exceeds the background carrier density, as usually occurs in picosecond applications. It is important in this situation to have contacts which are ohmic for both electrons and holes, a condition that is extremely difficult to achieve. Although not fully understood, an empirical property of metallic contacts on semiconductors with moderate to high defect densities is that they tend to exhibit ohmic behavior without the need for special processing. The most probable explanation for this phenomenon is the extremely short depletion layers in these materials which permit efficient tunneling currents to pass through the contacts, analogous to heavily doped materials.

e) Carrier Mobilities

The major disadvantage of the use of high defect densities is the reduction in carrier mobilities due to increased elastic scattering from the defects. The lack of a strong temperature dependence of the mobilities suggests also that most of the defects are neutral rather than charged if relatively high doses of radiation damage are used. Estimates of the influence of elastic scattering from neutral defects have been made by *Erginsoy* [5.40]. He estimates a mobility:

$$\mu = \frac{1.4 \times 10^{22}}{N_t} \frac{m^*}{m_0} \frac{\varepsilon}{\varepsilon_0} [\mathrm{cm^2/Vs}] , \tag{5.2}$$

where the defect density, N_t, is in units of $\mathrm{cm^{-3}}$. Using the previously estimated numbers for N_t required to produce capture times of 1 ps in silicon and GaAs, the corresponding mobilities from (5.2) would be 350 $\mathrm{cm^2/Vs}$ for GaAs and 5 $\mathrm{cm^2/Vs}$ for silicon (using $\sigma = 10^{-13}\,\mathrm{cm^2}$ for GaAs and $10^{-15}\,\mathrm{cm^2}$ for Si). The clear advantage of GaAs relative to silicon arises primarily from the apparently larger capture cross-sections observed in GaAs. This conclusion appears to be borne out by experiments.

Although the introduction of relatively high densities of defects has a dramatic effect on the free carrier lifetime, it is also important that recombination paths be provided to prevent thermal emission of trapped carriers which can produce sustained photoconduction in the form of long tails on the current waveforms. If the defect density is sufficiently high, trapped carriers can tunnel between defect sites, and recombination will occur without reemission. For this reason it is usually best to use a defect density somewhat higher than is necessary to achieve short capture times. This problem is usually exacerbated by the simultaneous presence of shallow defects which can have relatively fast emission rates. If the emission times are shorter than the measurement interval, their effect is equivalent to a reduction in mobility due to the inelastic capture and subsequent fast release of free carriers.

In general, a particular method of introducing defects results in a variety of traps, each having different energies and capture cross-sections. Some of these will be deep levels with fast capture times and slow emission rates, but others may be too shallow to be useful for picosecond applications. It is important that different combinations of materials and methods of introducing defects be explored to optimize this approach. The use of annealing to remove shallow defects, but retain deep traps, is a potentially useful technique for improving the mobility without loss of speed [5.41].

f) Other Photoconducting Materials

Other methods of achieving fast photocurrent decay times are the use of compensating impurities such as Fe in InP [5.42,43], polycrystalline materials such as p-Ge 5.44 and amorphous semiconductors [5.45,46]. In the latter case, the mobilities tend to be very small (1–10 $\mathrm{cm^2/Vs}$) and although the response times can be very fast, radiation damage appears to provide a better compromise between speed and sensitivity than the use of completely disordered (i.e. amorphous) materials.

5.3.2 Electro-optic Materials

Electro-optic materials have been used extensively for high speed optoelectronics. The key property of interest is the Pockels effect [5.47] whereby the optical indices of refraction can be perturbed by a low frequency electric field. This property is described explicitly by the relationship [5.48]:

$$\delta\left(\frac{1}{\varepsilon_{ij}}\right) = r_{ijk}E_k \ , \tag{5.3}$$

where the reciprocal dielectric tensor (sometimes referred to as the impermeability tensor) is defined as

$$\left(\frac{1}{\varepsilon_{ij}}\right) = \varepsilon_0 \frac{\partial E_i}{\partial D_j} \ , \tag{5.4}$$

and E_k is the low frequency electric field and r_{ijk} is the electro-optic coefficient. In general, r_{ijk} depends on the polarization direction of the low frequency and optical electric fields and is a third rank tensor having the symmetry of the point-group of the particular crystal. Electro-optic crystals do not have center-of-inversion symmetry and are the same class of materials as those which have piezoelectric and second harmonic generation properties (for a review, see [5.49]).

To calculate the indices of refraction of an electro-optic material in the presence of an electric field requires the solution of the equation of the index ellipsoid (for details see [5.48]). The particular electro-optic coefficients used depend on the symmetry of the crystal and its orientation relative to the light beams and applied electric field. For example, in the case of lithium tantalate, a material widely used for picosecond optoelectronics, the electric field is usually applied along the optic axis to produce a birefringence equal to:

$$n_z - n_x = (n_e - n_o) - \tfrac{1}{2}(n_e^3 r_{333} - n_o^3 r_{113})E_3 \ , \tag{5.5}$$

where n_o and n_e the ordinary and extraordinary indices of refraction. The first term in (5.5) is the static birefringence which is usually canceled by introducing an equal and opposite birefringence with an optical compensator or a second crystal rotated 90 degrees. The second term due to the electro-optic effect causes a relative phase retardation between the x and z polarization components of a light beam propagating in the y direction (transverse geometry). Interference between these two components can be converted to an amplitude variation using standard polarizing optics.

The conventional use of the Pockels effect is for phase and amplitude modulation of optical signals (for a recent review of high speed electro-optic modulation for optical communications see [5.50]. Two frequently used applications are coding of signals for optical communications and Q-switching of lasers. In both cases, the electric field is treated as a known quantity and is used as a driving field to achieve the desired result. In picosecond opto-electronics, the electric field is usually an unknown quantity and the Pockel's effect is used to determine it by measuring the induced birefringence produced on a probing optical beam. This factor along with the emphasis on speed, suggest that the criteria for the selection of electro-optic materials and geometric configurations will differ from the more conventional applications. In the remainder of this section, the specific properties of electro-optic materials will be discussed with this in mind.

a) Selection of Electro-optic Materials for Picosecond Optoelectronics

Table 5.2 summarizes the properties of some of the more frequently used electro-optic materials (from [5.49]). The magnitudes of the electro-optic coefficients are generally very small, being measured in units of 10^{-12} m/V. There is, nevertheless, a wide variation in the magnitudes of these coefficients between materials. These large differences are deceptive, however, and do not necessarily produce better results. This point requires some elaboration. The physical mechanism responsible for the electro-optic effect is the change in optical polarizability due to the coupling of optically active electronic states to a low frequency electric-field-induced polarization. Both ionic lattice displacements and electronic displacements contribute to this low frequency polarization. Those materials having the largest effects are usually dominated by ionic lattice displacements. A further enhancement occurs in materials in which an unstable or "soft" lattice mode plays a role. These modes have extremely large low frequency polarizations and produce substantial electro-optic effects. For example, $BaTiO_3$ and $Sr_{75}Ba_{25}Nb_2O_6$ (SBN) have electro-optic coefficients that are almost 1000

Table 5.2. Electro-optic materials

Material	Sym-metry	Electro-optic coefficients $r_{jk}[10^{-12}\,m/V]$ @ $\lambda[\mu m]$	Refractive Indices	Dielectric constants
$LiTaO_3$	3m	$r_{13} = 6.2$ @ 0.63 μm $r_{33} = 28.5$ @ 0.63 μm	$n_o = 2.1834$ @ 0.6 μm $n_e = 2.1878$ @ 0.6 μm	$\varepsilon_1 = 41$ $\varepsilon_3 = 43$
$LiNbO_3$	3m	$r_{13} = 4.5$ @ 0.63 μm $r_{33} = 28.8$ @ 0.63 μm	$n_o = 2.2716$ @ 0.7 μm $n_e = 2.1874$ @ 0.7 μm	$\varepsilon_1 = 43$ $\varepsilon_3 = 28$
$BaTiO_3$	4mm	$r_{51} = 820$ @ 0.55 μm	$n_o = 2.46$ @ 0.55 μm	$\varepsilon_1 = 2300$
$Sr_{0.75}Ba_{0.25}Nb_2O_6$	4mm	$r_c = 1090$ @ 0.63 μm	$n_e = 2.2987$ @ 0.63 μm	$\varepsilon_3 = 3400$
ZnO	6mm	$r_{33} = 2.6$ @ 0.63 μm	$n_e = 2.123$ @ 0.63 μm	$\varepsilon_3 = 8$
SiO_2(Quartz)	32	$r_{11} = 0.174$ @ 0.63 μm	$n_o = 1.546$ @ 0.63 μm	$\varepsilon_1 = 3.78$
GaAs	$\bar{4}3$m	$r_{41} = 1.2$ @ 1.0 μm	$n_o = n_e = 3.6$ @ 1.0 μm	$\varepsilon_1 = \varepsilon_3 = 12.3$

times larger than GaAs. In the latter material, electronic rather than ionic displacements are dominant. In spite of their large r_{ijk} coefficients, $BaTiO_3$ and SBN are not necessarily preferable to GaAs. This is due to the fact that the presence of the soft lattice mode also increases the low frequency dielectric constant and microwave absorption. In general, the magnitude of the electro-optic coefficients scale with the static dielectric constant (Miller's rule). A static dielectric constant of a few thousand presents a serious problem for the design of high speed electro-optic devices for picosecond applications. The high capacitance of these materials make it difficult to make high speed structures even when distributed or traveling wave geometries are employed. More serious is the strong microwave absorption due to the presence of the soft mode whose resonant frequency may be so low that it falls in the microwave and millimeter-wave region. In addition, the electro-optic and dielectric properties of $BaTiO_3$ and SBN are strongly temperature sensitive.

The most frequently used material for picosecond optoelectronics is $LiTaO_3$. Although its electro-optic coefficients are not the largest, it has some very desirable properties which account for its popularity. Most important is that it is readily available in single crystals of excellent optical quality. It also has a very small static birefringence. This makes it easy to compensate and less sensitive to temperature variations. The dielectric constant also has a small anisotropy. The microwave and millimeter-wave absorption is also relatively low, as indicated in Fig. 5.8. These plots of transmission vs frequency are based on measurements of the far-infrared absorption in this material using coherent time domain FIR spectroscopy [5.51]. The experimental data was fitted to the expression for the complex dielectric permittivity due to a single infrared-active lattice mode:

$$\varepsilon(\omega) = \varepsilon_{\text{op}} + \frac{(\varepsilon_{\text{dc}} - \varepsilon_{\text{op}})\omega_{\text{v}}^2}{\omega_{\text{v}}^2 - \omega^2 - \mathrm{i}\omega\Gamma} , \tag{5.6}$$

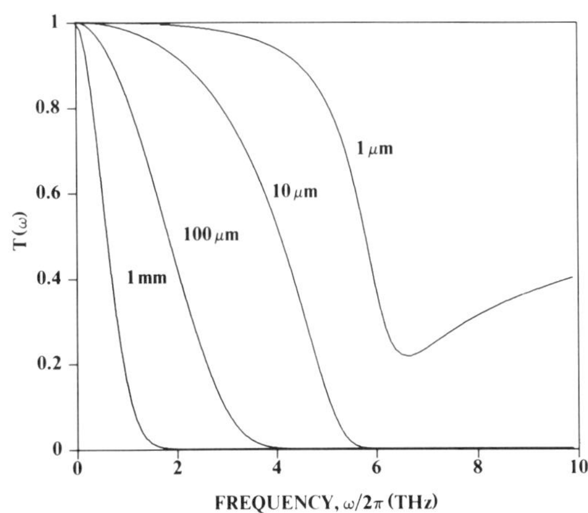

Fig. 5.8. Transmission of lithium tantalate in the millimeter-wave and far-infrared region of the spectrum. The strong lattice resonance at 6.3 THz is responsible for the increasing absorption at high frequencies

where ε_{op} and ε_{dc} are the optical and dc dielectric constants, ω_v is the lattice vibrational resonance frequency, and Γ is the damping rate of this mode. The experimentally determined values are: $\omega_v = 6.3$ THz and $\Gamma = 1.1$ THz. The resonance behavior of the dielectric response clearly has an important influence on the millimeter-wave absorption. For sub-picosecond applications where bandwidths up to one Terahertz are required, the interaction length will be strongly limited by lattice absorption. The resonance also influences the electro-optic effect, producing an enhancement of approximately a factor of 5 at ω_v relative to the dc value. Since the absorption is extremely strong at the resonant frequency this enhancement is of little practical value. The limiting speed of response of this material is estimated to be approximately 300 fs, and is primarily determined by the damping rate, Γ [5.52]. Materials in which electronic displacements dominate, such as gallium arsenide, have potentially a much greater speed of response. Far-infrared lattice absorption, however, will still limit their useful speed of response.

5.4 Generation and Detection of Ultrashort Electrical Pulses

5.4.1 Picosecond Photoconductors

Many different materials and geometric configurations have been used for picosecond photoconductors (for a review, see [5.53]). Most, however, have in common a design which optimizes the speed of response by directly coupling the photocurrent to a high speed transmission line structure. This is usually accomplished by making the electrodes of the photoconductor an integral part of the transmission line. A typical example is illustrated in Fig. 5.9. It consists of a photoconducting film on an insulating substrate on which metallic electrodes have been deposited in the configuration of a microstrip transmission line. A picosecond light pulse is focused on the active region of the photoconductor

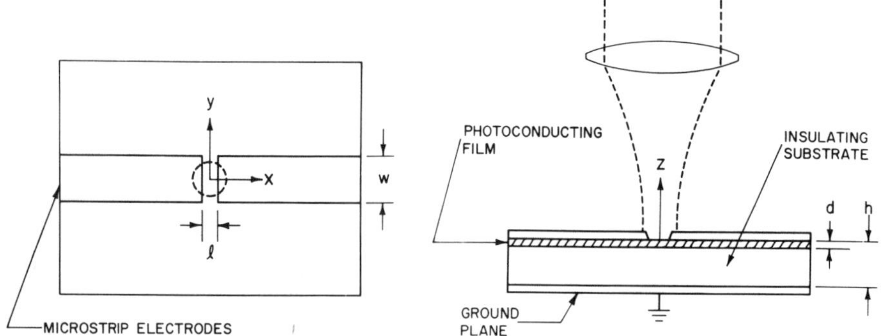

Fig. 5.9. Schematic diagram of a thin-film photoconductor as an integral component of a high-speed microstrip transmission line. The active region is the gap in the left electrode at which the light is focused

which consists of a small gap in the top microstrip electrode. Other transmission line configurations such as coaxial and coplanar waveguides have also be used (a more complete discussion of different transmission line structures will be given in Sect. 5.4.3). Also the photoconductor may be used as the substrate material if it is semi-insulating. A rigorous analysis of the voltage waveform produced in Fig. 5.9 is a difficult task which requires the solution of the time-varying electromagnetic equations for the electric field produced by the radiating currents and charges in the gap. A more expedient method is to neglect retardation effects and represent the photoconductor by time-varying macroscopic circuit elements. These approximations should be valid if the dimensions of the gap and the transmission line cross-section are small relative to the distance an electromagnetic signal travels in the shortest time interval of interest.

Following this approach the photoconductor is represented as a time-varying conductance $G(t)$, in parallel with a capacitance C_g embedded in a transmission line. Theoretical estimates of the static capacitance across this gap have been made for microwave applications, and have typical values of 10–100 fF for the specific geometries of interest to us here [5.54]. The capacitance increases only logarithmically as the gap becomes smaller, and as we shall see, the sensitivity increases as l^{-2}, so that it is possible to use very small gap lengths, l, to make sensitive photoconductors without appreciable loss of speed. Although a small shunt capacitance may exist in some configurations, its value is usually small and will be omitted here.

A general expression for the time-varying conductance $G(t)$ of a photoconductor can be derived from the rate of dissipation of electrical energy:

$$GV_g^2 = \int_v \boldsymbol{E} \cdot \boldsymbol{J} dx \ , \tag{5.7}$$

where V_g is the voltage across the gap, $\boldsymbol{E}$ is the electric field, and $\boldsymbol{J}$ is the current density, and the integration extends over the entire active volume of the photoconductor. When Ohm's law can be used to relate the current density and electric field, the conductance is

$$G(t) = \frac{1}{V_g^2} \int_v dx \sigma |\boldsymbol{E}|^2 \ , \tag{5.8}$$

where σ is the conductivity. G can be separated into two components, one of which is a constant G_0, representing the dark conductivity, and a second time-varying term, $g(t)$, which is due to the photoconductance:

$$g(t) = \frac{1}{V_g^2} \int_v dx (ne\mu_n + pe\mu_p) |\boldsymbol{E}|^2 \ , \tag{5.9}$$

where n, p and μ_n, μ_p are the electron and hole densities and mobilities, respectively. Although $n = p$ initially, their time evolution may differ due to different capture rates by defect sites, and drift. Assuming the optical pulse to be negligibly short in duration, the initial carrier densities are

$$n(t=0^+) = p(0^+) = \frac{(1-R)\alpha E(x,y)\mathrm{e}^{-\alpha z}}{\hbar\omega} \ , \tag{5.10}$$

where R is the reflectivity of the photoconductor surface, α is the optical absorption constant, E is the incident optical pulse energy per unit area, and $\hbar\omega$ is the energy of one photon.

The electric field distribution in the gap, $\boldsymbol{E}(\boldsymbol{x})$ depends both on the geometrical configuration of the photoconductor and also on the distribution of the photo-excited carriers. The latter effect is particularly important in the high injection regime which is typical of picosecond photoconductors. If the electrical contacts to the photoconductor are not ohmic, they can also influence the electric field distribution. For the radiation-damaged and amorphous semiconductors, however, good contacts can usually be made using relatively simple methods.

a) Response of an Ideal Photoconductor in a Transmission Line

The incorporation of a high speed photoconductor into a transmission line has some unique properties which will now be considered in detail [5.55]. The most important of these is that the bias voltage can be a fast electrical transient and need not be a dc voltage. The effective load seen by the photoconductor is the characteristic impedance Z_0, of the transmission line. Also, due to the traveling-wave nature of the electrical signals it is necessary to describe the response of the photoconductor in terms of incident, reflected, and transmitted waves. To determine the response for an arbitrary incident signal, $v_\mathrm{i}(t)$, and conductance, $G(t)$, it is convenient to first focus attention on the instantaneous voltage, $v_\mathrm{g}(t)$, across the gap. By straight-forward application of circuit laws, the following equation for $v_\mathrm{g}(t)$ is derived:

$$Z_0 C_\mathrm{g}\frac{dv_\mathrm{g}}{dt} + \frac{1}{2}[1 + 2Z_0 G(t)]v_\mathrm{g}(t) = v_\mathrm{i}(t) \ , \tag{5.11}$$

from which the reflected and transmitted waves can be determined:

$$v_\mathrm{r}(t) = \tfrac{1}{2}v_\mathrm{g}(t) \ , \tag{5.12}$$

$$v_\mathrm{t}(t) = v_\mathrm{i}(t) - \tfrac{1}{2}v_\mathrm{g}(t) \ . \tag{5.13}$$

b) Photoconducting Electrical Pulse Generators

To illustrate some of the basic features of the response of an ideal photoconductor in a transmission line, we first consider the case of a step-function conductance with a dc bias:
i.e.

$$G(t) = \begin{cases} 0; & t < 0 \\ G_1; & t \geqslant 0 \end{cases} \quad \text{and} \quad v_\mathrm{i}(t) = \tfrac{1}{2}V_\mathrm{b} = \text{constant} \ .$$

From (5.11,12), the transmitted signal is:

$$v_t(t) = \left(\frac{V_b}{2}\right)\left[\frac{2Z_0G_1}{1 + 2Z_0G_1}\right]\left\{1 - \exp\left[-\left(\frac{1}{2Z_0C_g} + \frac{G_1}{C_g}\right)t\right]\right\}. \tag{5.14}$$

It is important to distinguish two different regimes, depending on the magnitude of the photoconductance, G_1. For small values of G_1, $v_t(t)$ has a rise-time of $2Z_0C_g$ and a steady-state value of $Z_0G_1V_b$ which is linearly proportional to G_1. For large values of G_1 (i.e. $Z_0G_1 \gg 1$), the rise-time decreases and the steady-state transmitted signal saturates at one-half the dc bias voltage. For example, if $Z_0G_1 = 5$, the rise-time is 11 times faster than the small signal case and the incident wave is transmitted with approximately 90% efficiency. The response is thus very different in the small signal or linear response regime, and the large signal or saturated regime.

Picosecond photoconductors were first demonstrated by *Auston* [5.56] and *Lawton* and *Scavannec* [5.57]. When the optical illumination is large, picosecond photoconductors are more appropriately described as light-activated electric switches, as demonstrated experimentally in crystalline semiconductors by *Auston* [5.56]. In this case, illustrated in Fig. 5.10, optical pulses from a mode-locked Nd:glass laser were used to inject high densities of free carriers into a silicon

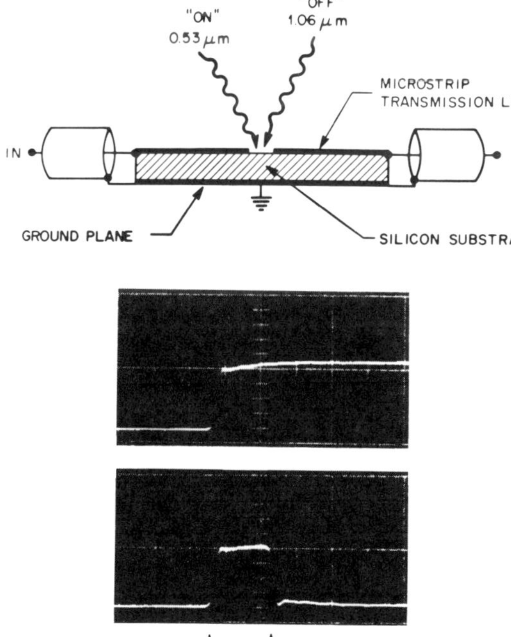

Fig. 5.10. A picosecond photoconductor which uses a visible (0.53 μm) picosecond pulse of approximately 5 μJ to initiate conduction and an infrared pulse of approximately 5 μJ to terminate the photocurrent by introducing a short circuit between top electrodes and ground plane. The oscilloscope photograph (2 ns/div.) illustrates the variable aperture feature of the device which can be controlled by the time delay between the two optical pulses. Peak electrical transmission of the gate was 95% [5.56]

microstrip photoconductor. A visible pulse at 0.53 μm obtained by second harmonic generation was used to initiate the photocurrent. Since the material used was a high quality crystal, the free carrier lifetime produced a current which persisted for a relatively long time. It could be terminated, however, by absorbing a second optical pulse at the fundamental wavelength of 1.06 μm which penetrated through to the ground plane, making a short circuit between with the top electrode. This turned "off" the switch by reflecting all further incoming signals. By varying the delay between the two pulses, the duration of the current pulse could be continuously varied from approximately 15 ps to many nanoseconds. The same device also functioned as a sampling gate when a pulse was used as a bias rather than a dc voltage.

c) Photoconducting Electronic Sampling

Picosecond photoconductors can also be used as electronic gates to measure the waveform of a high speed electrical signal [5.56,58,59]. In this case, the incident bias signal on the photoconductor is the pulsed waveform rather than a dc signal. The active photoconducting area is usually a gap between a main transmission line and a secondary line. When a light pulse strikes the sampling gap, a small sample of charge is transferred from the signal on the main transmission line to the sampling line. The response of this sampling gate can be analyzed in a manner similar to the single gap in a main line considered in the last section. Since the speed of the sampling gate comes from the response time of the photoconductor, it is sufficient to measure the total charge generated in the sampling transmission line without the necessity of time-resolving the sampled voltage waveform. This procedure also has to advantage that it does not contain a contribution from the signal which is capacitively coupled across the sampling gap. This can be written in the form of a convolution of a sampling function, $f_s(t)$ with the incident signal $v_i(t)$:

$$Q_s(\tau) = Q_0 + \int_{-\infty}^{+\infty} f_s(t' - \tau) v_i(t') dt' \ , \tag{5.15}$$

where

$$f_s(t) = \frac{2}{3Z_0 C_s} \left(\frac{1}{1 + \frac{3}{2} Z_0 G_0} \right) \int_t^{\infty} g(t') e^{-\gamma(t'-t)} dt' \ , \tag{5.16}$$

and

$$\gamma = \frac{2}{3Z_0 C_s} + \frac{G_0}{C_s} \ ,$$

where C_s is the capacitance of the sampling gap.

The sensitivity and signal-to-noise ratio of photoconductive sampling is extremely good, permitting fast electrical signals having amplitudes of approximately 1 μV to be measured with realistic integration times. The limiting noise properties of photoconductive sampling gates have not been studied in detail, nor have they been optimized. In the ideal case, the limiting noise level would be determined by generation-recombination noise and Johnson noise [5.60]. For typical sampling gates this would predict a limiting sensitivity of a few tens of $nV/\sqrt{Hz}$. In practice, however, imperfect ohmic contacts give rise to a photovoltaic signal which tends to dominate the noise properties. This produces a noise background signal due to laser amplitude fluctuations. Although the quantum efficiency of the photovoltaic signal is extremely small, it is often sufficient to dominate the noise properties and results in sensitivity limits which are more typically a few $\mu V/\sqrt{Hz}$. Further work to make better ohmic contacts would greatly improve the minimum detectable signal that can be measured by photoconductive sampling.

d) Electronic Autocorrelation Measurements

A simple and effective technique for evaluating the speed of response of photoconducting pulse generators and sampling gates is to directly connect them together so that the output of the generator is the input to the sampling gate [5.56,46]. This approach, which we call an electronic autocorrelation measurement gives an accurate measurement of the system response and has important applications for characterizing specific photoconducting materials, electrode geometries, transmission line structures, and mounting configurations. The system response is approximately equal to the convolution of the individual responses of the generator and sampling gate (for a detailed discussion see [5.55]).

This approach has been used to measure the photocurrent decay rates in amorphous silicon [5.46,61], silicon-on-sapphire [5.25], radiation-damaged silicon-on-sapphire [5.62,63], radiation-damaged indium phosphide [5.64–66], gallium arsenide [5.67], radiation-damaged gallium arsenide [5.64,66] and polycrystalline silicon [5.66]. In many cases, the capacitance of the electrode geometry limits the speed of response. For example, in microstrip structures having widths of 0.2 mm, gaps of 25 μm and substrate thicknesses of 0.25 mm, the limiting circuit response has a fullwidth at half maximum of approximately 10 ps. Recently, *Ketchen* and coworkers [5.63] have scaled down an autocorrelation circuit using 5 μm coplanar strip transmission lines and obtained a response with a FWHM of only 1.6 ps, as illustrated in Fig. 5.11. They used a novel "sliding contact" technique which involved the illumination of regions between the conductors. They have applied this technique to the measurement of dispersion on coplanar transmission lines by sampling the generated electrical pulse farther down the coplanar line. By using superconducting coplanar lines they have recently been able to observe the contribution to the propagation constant due to the finite superconducting bandwidth [5.68].

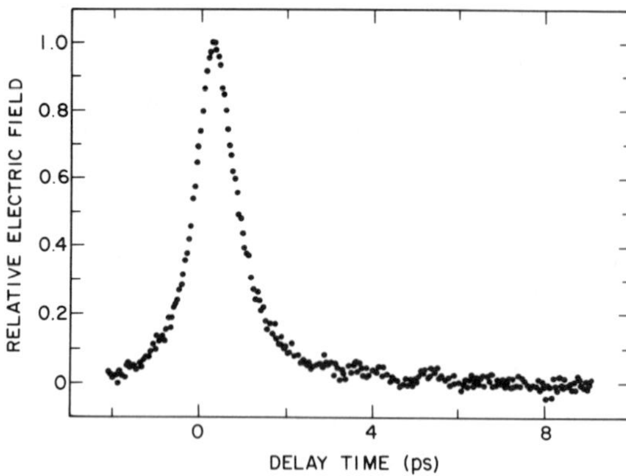

Fig. 5.11. The picosecond electronic autocorrelation of two radiation-damaged silicon-on-sapphire photoconductors in a coplanar microstrip transmission line [5.63]

5.4.2 Electro-optic Devices

a) Electro-optic Sampling

As discussed in Sect. 5.2.2, electro-optic materials have been widely used for the measurement of high speed electrical signals by the Pockel's effect (for recent reviews, see [5.69,70]). To ensure good speed of response, a traveling-wave Pockel's cell geometry is usually employed. This distributes the capacitance of the crystal, and permits the electrical waveform to be measured without perturbing it. This approach was used to measure single picosecond electrical transients on a lithium tantalate traveling-wave Pockel's cell having amplitudes in the range of hundreds of volts to kilovolts [5.71,72]. With the use of high repetition rate lasers it has been possible to greatly improve the sensitivity with signal averaging, making possible the measurement of signals below one millivolt [5.73]. The use of subpicosecond optical pulses and velocity matching has resulted in subpicosecond electrical response times [5.74].

Figure 5.12 is an illustration of an electro-optic sampling system. The trigger beam is used to excite a picosecond photoconductor to produce a fast electrical transient which is coupled to a traveling-wave Pockels cell. This electrical signal is then measured by the probe which passes through the cell. The change in polarization ellipticity which this beam encounters is detected with polarizing optics consisting of a polarizer, compensator, and analyzer. A slow conventional photodiode is sufficient for detecting the probe beam. The compensator is used to cancel the static birefringence of the Pockels cell and to set the operating point of the cell for optimum signal-to-noise ratio. With femtosecond pulses it is important to compensate the cell in zeroth order. To accomplish this, a second identical electro-optic crystal is often placed in the path of the probe beam in a 90°-rotated position. This also improves the temperature stability. The operating

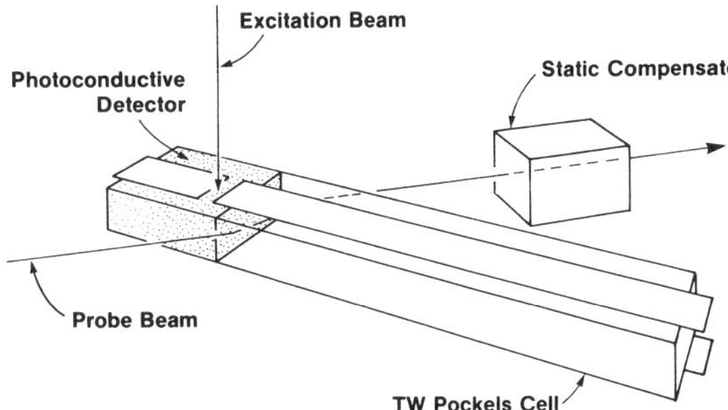

Fig. 5.12. Schematic diagram of a traveling wave Pockels cell coupled to a high speed photoconductor. The electrical pulse generated in the photoconductor is measured by sampling the induced birefringence in the electro-optic Pockels cell as it travels along the structure

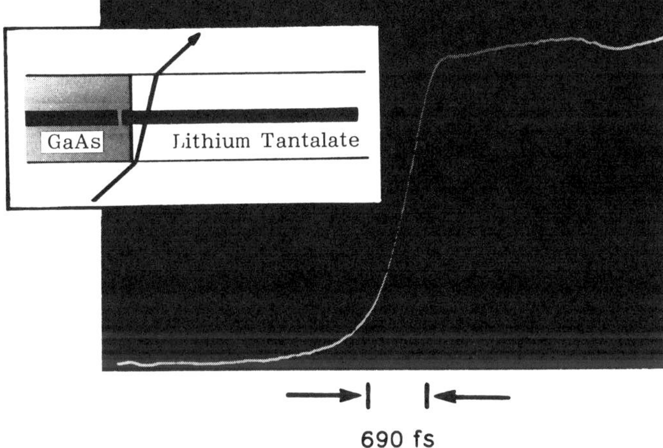

Fig. 5.13. The experimental response of the electro-optic sampling device illustrated in Fig. 5.12. The probing optical pulse passed through the cell at an angle to match the longitudinal velocities of the optical and electrical pulses. The measured rise time of 690 fs illustrates the extremely fast speed of response this approach [5.74]

point is usually set at the 50% transmitting point by introducing an additional $\lambda/4$ retardation with the compensator. This biases the cell at a point where the incremental change in transmission of the cell is linearly proportional to the electric field of the test signal.

Figure 5.13 is an example of a measurement of a fast electrical transient made with the electro-optic sampling system illustrated in Fig. 5.12 [5.73,74]. The electrical signal was produced by illuminating a semi-insulating GaAs:Cr photoconductor with a 100 fs optical pulse. For optimum temporal resolution, the probe beam was transmitted through the lithium tantalate traveling-wave Pockels cell at an angle of 17° with respect to the normal of the crystal face. This matched the velocity of the probe with the electrical pulse velocity in the direction along the traveling-wave Pockels cell. The probe beam was also focused to approximately 10 μm to minimize the time smear due to integration along the cell. The rise-time of the signal is 690 fs. This signal was measured with the probe beam passing as close to the GaAs:Cr pulse generator as possible. When the probe is positioned farther along the cell, a substantial broadening of the signal is observed due to waveguide dispersion in the lithium tantalate traveling-wave Pockels cell. This effect has been numerically modeled by *Haisnan* et al. [5.75] and agrees well with the observed results of *Valdmanis* et al. [5.73].

The transmission of the probe beam through the lithium tantalate Pockel's cell biased at the quarter-wave point is

$$T(\tau) = \frac{1 + \sin \Gamma(\tau)}{2}, \tag{5.17}$$

where Γ is the retardation of the probe beam due to the local electric field, $E_z(\boldsymbol{r}, t)$, in the cell. Assuming the electro-optic effect is instantaneous, Γ is given by the expression [5.70]:

$$\Gamma(\tau) = \frac{\pi}{\lambda}\frac{1}{2}[n_e^3 r_{333} - n_o^3 r_{113}]\frac{\int dt \int d\boldsymbol{r} E_k(\boldsymbol{r}, t) I(\boldsymbol{r}, t + \tau)}{\int dt \int d\boldsymbol{r} I(\boldsymbol{r}, t)}. \tag{5.18}$$

This expression shows that the measurement is a convolution of the local electric field with the intensity profile, $I(\boldsymbol{r}, t)$ of the probing optical pulse. In a typical case, the amplitude of the electric field is in the range of $1-10^3$ V/cm and the corresponding relative change in probe transmission is only $10^{-6}-10^{-3}$. Extensive signal averaging is necessary to bring signals of this magnitude above the noise level. Fortunately this can be readily accomplished with high repetition rate lasers operating at 100 MHz or higher. The main source of noise is intensity fluctuations in the laser. The use of high frequency synchronous detection can reduce this noise component to a negligible level, leaving only shot noise and Johnson noise in the probe detector. *Kolner* and *Bloom* [5.70] have made a detailed analysis of the sources of noise in electro-optic sampling measurements. They estimate the minimum detectable voltage in a 50 Ω lithium tantalate microstrip traveling-wave Pockel's cell to be 188 $\mu V/\sqrt{Hz}$ for an average detector current of 10 mA and a probe wavelength of 1.06 μm.

The convolution involved in the measurement as illustrated by expression (5.18) shows that it is important to use a tightly focused geometry to avoid a loss of time resolution. This requires the dimensions of the traveling-wave Pockel's cell to be as small as possible. In Sect. 5.5, where applications of electro-optic sampling are summarized, some alternatives to the traveling-wave geometries will be described.

b) Optical Rectification

Electro-optic materials can also be used to generate fast electrical waveforms as well as to measure them. Optical rectification is the property of electro-optic crystals that permits one to produce an electrical waveform which follows the envelope of an optical pulse. This effect was first observed by *Bass* et al. [5.76] using millisecond pulses from a solid-state laser. The effect was extremely weak and required very energetic optical pulses to produce an observable electrical pulse. With the picosecond and femtosecond lasers available today, however, it is possible to generate relatively large amplitude electrical signals with extremely short durations. For this reason, it is an attractive method of producing electrical transients for the measurement of high speed devices and materials.

The basic physical mechanism of optical rectification comes from a second-order nonlinear polarization which can be written as:

$$P_i = \varepsilon_0 \chi_{ijk} E_j E_k \ . \tag{5.19}$$

The nonlinear susceptibility, χ_{ijk}, is related to the electro-optic tensor, r_{kij}, by the expression:

$$\chi_{kij} = -\tfrac{1}{4} n_i^2 n_j^2 r_{ijk} \ . \tag{5.20}$$

If the optical pulse has a rapidly varying envelope as in the case of picosecond and femtosecond pulses, the nonlinear polarization, $P(t)$, will also be rapidly varying and will have a very broad frequency spectrum which can extend from dc up through microwaves, millimeter waves and even into the far infrared.

When an electro-optic crystal is used as a voltage source, as in the first experiments of *Bass* et al. [5.76], the signal is relatively weak and difficult to couple to a high speed circuit. A more useful configuration is to use the electro-optic crystal as a current source. In this case, the crystal is mounted on a low impedance electrical transmission line, or may even form a part of the line. This provides a tight coupling between the crystal and the line, and current flows readily down the line to produce the signal. The use of optical rectification for generating short current waveforms from picosecond optical pulses has been demonstrated by *Auston* et al. [5.77]. Assuming a roughly cubic shaped crystal of dimensions a^3, the current waveform is given by the expression:

$$i(t) = \frac{1}{4c} a^3 r_{33} \frac{\partial W}{\partial t} \ , \tag{5.21}$$

where $W(t)$ is the instantaneous optical power in the electro-optic crystal. The current waveform is proportional to the first derivative of the envelope of the optical pulse. This produces a bipolar waveform having positive and negative lobes. Another important property is that the peak current is inversely proportional to the square of the optical pulse duration for a given pulse energy. Hence the efficiency of optical rectification improves greatly with the use of very short pulses. For example, a 1 μJ pulse of 1 ps duration will produce a peak current of approximately 0.3 A in lithium tantalate.

c) Electro-optic Cherenkov Radiation

When the pulse duration is less than one picosecond, it becomes more difficult to design circuits which can handle the extremely large bandwidths associated with these pulses. A solution to this problem which we have recently demonstrated [5.78] is to let the polarization current radiate directly into a uniform dielectric medium, rather than attempt to couple it into a transmission line. In this case, the size of the crystal is not critical and it can have dimensions that are substantially larger to provide an extended interaction. The polarization source is not stationary, however, but moves with the group velocity of the optical pulse. If the optical pulse is tightly focused, its spatial extent can be only a few microns so that is has the appearance of a moving dipole moment. Since it has rapidly varying temporal components, it will radiate in a broad range of frequencies from radio through microwave, millimeter waves, and into the far-infrared. Due to the additional contribution to the low frequency dielectric response from the infrared lattice vibrations, the velocity of the source exceeds the radiation velocity. This produces a characteristic cone of radiation in the form of a shock wave, analogous to the classical Cherenkov radiation from a charged particle moving at a velocity that exceeds the radiation velocity, as illustrated in Fig. 5.14. The temporal profile of the radiation pulse is approximately a single cycle having a period comparable to the envelope duration of the optical pulse. This technique produces electrical pulses that are shorter than any other method. Effective durations as short as 300 fs have been measured. Due to the reciprocal nature of the electro-optic effect and optical rectification, it is possible to measure the electrical waveforms produced by optical rectification in the same crystal by electro-optic sampling. Perfect velocity synchronism between the generating and detecting optical beams is also possible, permitting very fast time resolution.

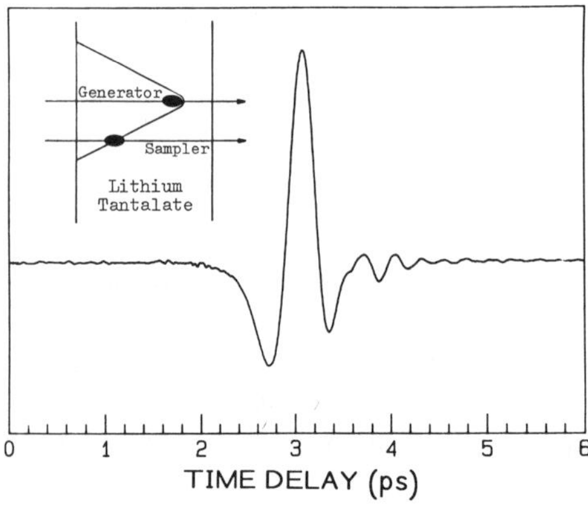

Fig. 5.14. The electrical signal produced by the rectification of a femtosecond optical pulse in the electro-optic material, lithium tantalate. The ringing on the trailing edge of the pulse is due to the resonance of the nonlinearity of the material [5.78]

Figure 5.14 illustrates this experimental approach and shows a waveform of an electrical signal produced by the rectification of femtosecond optical pulses.

Recently, it has been possible to substantially improve the amplitude of the electro-optic Cherenkov pulses by using optical pulses which have been amplified to an energy of approximately 0.1 μJ. This has resulted in electric field strengths as high as 7 kV/cm, which corresponds to a peak power of approximately 1 MW/cm^2.

An interesting aspect of the electro-optic Cherenkov radiation technique is that the peak electric field depends on the inverse 5/2 power of the optical pulse duration (the peak power depends on the inverse 5th power of τ_p). Consequently the efficiency is such that femtosecond optical pulses are essential.

The spectral content of these pulses extends as high as 5THz, making them valuable sources of coherent radiation in a region of the electromagnetic spectrum where tunable coherent sources are not available. When combined with phase-sensitive coherent detection, a complete far-infrared spectroscopy system results which has been used to measure the momentum relaxation times of free carriers in semiconductors [5.79], the dynamics of lattice vibrations [5.51], and the mobility transient of a semiconductor arising from optical carrier injection [5.16].

5.4.3 Ultrafast Electrical Transmission Lines

Ultrafast electrical pulses require transmission lines which are extremely broadband, have low dispersion and loss, and which can be readily interconnected to optoelectronic and other devices. Unlike microwave and millimeter-wave signals, the frequency spectrum of ultrafast electrical pulses covers a range from dc to many hundreds of gigahertz, and in some cases even terahertz. For this reason, waveguides are unsuitable due to their large dispersion and low frequency cut-off. Some transmission line structures that have been used for ultrafast optoelectronics are: coaxial, microstrip, coplanar, and coplanar microstrip. For details of the electrical properties of these structures, the reader is referred to the standard texts on this subject such as [5.80–82]. Coaxial lines have good bandwidth and low losses and dispersion. They have been used extensively for high power optoelectronic switching with photoconductors mounted in a gap in the center conductor. They tend to be used less for high speed low power optoelectronic applications since they are difficult to interconnect to other structures such as microstrip lines and discrete devices, although schemes for accomplishing this have been developed. The microstrip, coplanar waveguide, and coplanar microstrip structures are all compatible with semiconductor microelectronic fabrication technology. Consequently, optoelectronic devices such as photoconductors can be integrated into the transmission line being placed at a gap in an electrode or between two electrodes. Electro-optic sampling can also be readily accomplished by constructing a transmission line on an electro-optic substrate. The ability to use microelectronic processing technology also means

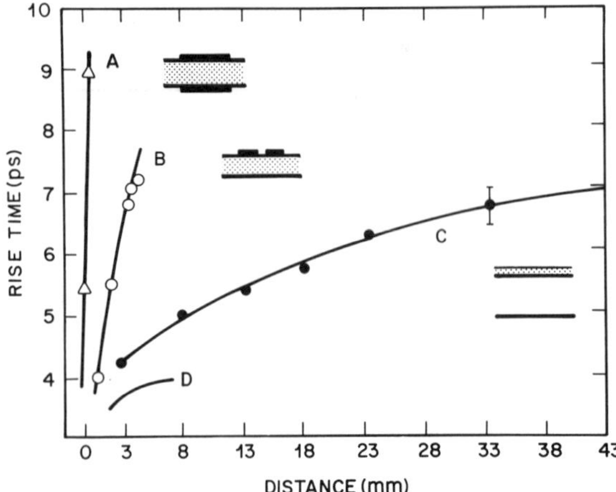

Fig. 5.15. Measured pulse dispersion on different transmission line structures: (*A*) balanced stripline on $LiTaO_3$ with 500 μm electrode separation; (*B*) coplanar strip on $LiTaO_3$ with 50 μm electrode separation (*C*) inverted air-spaced stripline – the top electrode is supported by a thin glass superstrate 500 μm above the ground plane; (*D*) superconducting coplanar transmission line with line dimensions of 50 μm [5.87]

that these structures can be scaled to extremely small sizes, thereby increasing the useful bandwidth.

Dispersion is a major problem in transmission lines for ultrafast electrical pulses since it can easily produce a loss of speed due to pulse broadening. Theoretical calculations of the dispersive broadening of pulses on microstrip and coplanar lines have been made by *Li* et al. [5.83] and *Hasnain* et al. [5.75]. They find good agreement with experimental measurements of *Mourou* and *Meyer* [5.84]. Measurements of pulse dispersion on microstrip lines by *Cooper* [5.85] indicates that microstrip is much less dispersive than expected by theory. *Gooson* and *Hammond* [5.86], have made time-domain calculations of the dispersion and loss of electrical pulses on microstrip lines on silicon substrates. Their results show the strong influence of substrate losses and demonstrate the need for high resistivity substrates. *Mourou* [5.87] has made extensive measurements of pulse dispersion on different lines. His results, which are illustrated in Fig. 5.15 demonstrate that ultrashort pulses can propagate on suitably designed lines with relatively little broadening due to dispersion.

The principal advantage of the coplanar geometries is that the size scale can be reduced to very small dimensions. Although microstrip can also be scaled down, it requires the use of very thin substrates which is often impractical.

The geometrical properties of various transmission line structures have been exploited for pulse shaping applications. For example, *Buck* et al. [5.88] have shown how a stub microstrip transmission line can be used to shape the electrical pulse from a photoconductor. *Margulis* and *Persson* [5.89], have used a coaxial differentiator to accomplish similar results. *Chang* et al. [5.90] have mounted a picosecond photoconductor in a coaxial Blumlein structure for pulse shaping applications. *Li* et al. [5.83] have reviewed the topic of pulse forming with optoelectronic switches.

The properties of nonlinear transmission lines have been examined theoretically by a number of workers. For example, *Landauer* [5.91] and *Khokhlov*

[5.92] have shown that shock waves can be expected to develop for high voltage pulses propagating on transmission lines with nonlinear dielectrics. *Paulus* et al. [5.93] have suggested that soliton-like behavior should be possible in transmission lines having a quadratic nonlinearity and a cubic dispersion. The implementation of nonlinear transmission lines to the picosecond time domain could result in the development of electrical pulse compression techniques analogous to the soliton-like techniques that have been developed for optical pulses.

5.5 Optoelectronic Measurement Systems and Their Applications

5.5.1 Optoelectronic Measurement Systems

As mentioned in the previous sections, the use of picosecond and femtosecond optical pulses for control and measurement of electronic devices and circuits can produce a substantial improvement in both performance and flexibility relative to conventional electronic measurement systems such as sampling oscilloscopes. In addition, an entirely new class of devices has evolved from the use of this approach having properties and applications that go beyond conventional electronic devices.

In the previous section we briefly described a number of different optoelectronic devices using photoconducting or electro-optic materials. These are basically of two types: sources of electrical transients and electrical sampling gates. When one of these optoelectronic signal generators is combined with an optoelectronic sampling gate as illustrated in Fig. 5.16, a flexible high speed measurement system results which can be used to measure a wide range of electronic devices and materials. An important feature of this system is the use of the same optical pulse for triggering both the optoelectronic pulse generator and sampling gate [5.94]. An optical beam splitter is used to split this pulse into two pulses, one of which goes to the generator and the other which goes to the sampling gate after passing through a variable path-length delay line. The variation of this delay permits accurate stroboscopic measurements with negligible jitter.

Clearly, speed is the single most important advantage of using optoelectronics, and it provides the main incentive for the extensive proliferation of activity

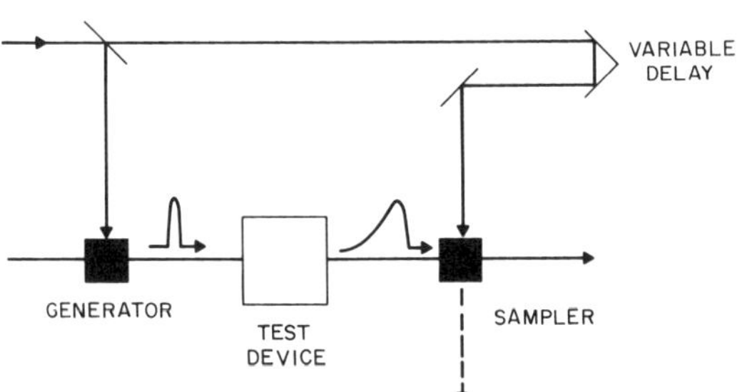

Fig. 5.16. An ultrafast optoelectronics measurement system for measuring the response of high speed devices and materials. Both the generator and sampler can be photoconducting or electro-optic devices. The stroboscopic measurement is accomplished by varying the optical delay between the two optical pulses used to time the generator and sampler

in this field. As illustrated by Fig. 5.1, this speed advantage comes from the availability of short pulse lasers. In addition to speed, there are a number of equally important advantages that derive from the use of optics, which we will now summarize and illustrate with examples in the remaining part of this section.

a) Sensitivity

The high sensitivity of optoelectronic measurement systems is primarily due to the high repetition rate of pulsed lasers which permits extensive signal averaging. For a single event these measurement techniques are relatively insensitive. The signal-to-noise can be readily enhanced however by averaging multiple events at high repetition rates. At 100 MHz rates, this results in an improvement of $10^4:1$ in S/N for a one second integration time making possible microvolt sensitivity for photoconducting sampling gates and millivolt sensitivity for electro-optic sampling.

b) Timing Accuracy

The absence of trigger jitter is an extremely important feature. Without it, it would not be possible to obtain high timing precision. In a conventional sampling oscilloscope, the jitter between the triggering of the sampling gate and the measured signal can limit the useful time resolution. With optoelectronic sampling the timing of both the test electrical pulse and the sampling is controlled by the same optical pulse, as previously described (Fig. 5.16).

c) Non-invasive Probing

Non-invasive probing is a novel capability and an extremely important feature of optoelectronic measurements. It can be achieved in a number of ways. One of these is possible when the substrate material of the particular device or circuit under test is an electro-optic material, such as gallium arsenide. This permits direct measurement of the signal waveforms at any point in the circuit without requiring physical contact. In another approach, a small electro-optic crystal is brought close to a circuit to sample the fringing electric fields. Picosecond photoelectric emmission can also be used for non-invasive measurements. These approaches will be described in more detail in Sect. 5.5.3.

d) Low Temperature Environments

The basic optoelectronic devices described in the previous sections are relatively rugged and perform well at low temperature as well as room temperature. Picosecond photoconductors have been used at 125 K [5.61,95] and at 4 K

[5.38]. Electro-optic sampling has been accomplished at temperatures less than 2 K using Indium superconducting coplanar strip transmission line structures [5.96]. A potential problem with the use of electro-optic materials at low temperatures is the increased sensitivity to photorefractive damage. This primarily effects pulse generation applications using optical rectification which generally requires higher optical intensities than electro-optic sampling.

e) High Power

The high power scaling of photoconductors is an extremely important property for generating multi-kilovolt electrical transients. As will be discussed in Sect. 5.5.5, it is possible to switch electrical signals which have peak powers that exceed the incoming optical pulse.

f) Simultaneous Time and Frequency Measurements

The bandwidth of picosecond optoelectronic measurement systems must be extremely broad to avoid loss of speed due to pulse broadening. For picosecond systems, this means that a flat baseband response is required from dc to hundreds of gigahertz. This places some demanding constraints on the design of transmission structures and interconnections between devices. Once achieved, however, an important benefit results from the broad bandwidth. Since the measurement system is phase-coherent (i.e. amplitudes are measured, not intensities), the spectral response of a device can be inferred directly from its temporal response. This concept of time-domain metrology is equivalent to variable frequency measurements of both phase and amplitude over bandwidth of hundreds of GHz. In the femtosecond time domain the equivalent bandwidths extend as high as 5 THz, making possible coherent time-domain spectroscopy in the far-infrared [5.79].

g) Electrical Isolation

The use of optical pulses for triggering optoelectronic devices results in excellent isolation between the control signal (optical) and the generated or measured signal (electrical). The ability to accurately focus and position optical beams also results in good isolation between different optoelectronic devices. Since the triggering and sampling information is conveyed on optical beams, the device or material under test can also be placed in remote environments such as low temperature cryostats or high magnetic fields, and can be interfaced to an external environment with optical beams and perhaps one or two dc or low frequency electrical connections. This avoids the problems of requiring high speed coax or other electrical transmission systems to bring out (or send in) electrical transients.

5.5.2 Characterization of High Speed Discrete Devices

a) Photoconductive Sampling of High Speed Photodetectors

Radiation damaged GaAs photoconductors have been used as sampling gates to measure the response of a high speed silicon p-i-n photodiode [5.95]. The photodiode was a p-i-n structure having an intrinsic drift region of 3 μm and an active area of 25 μm. The diode chip, which measured 0.010″ square, was mounted on a microstrip in a configuration that permitted the photocurrent pulses to be injected into a 50 Ω transmission line that was terminated at each end. The sampling photoconductor was a small chip of semi-insulating GaAs which had been irradiated with 3×10^{15} protons/cm^2 at 300 keV. An electronic autocorrelation measurement of the sampling gate with an identical GaAs photoconductor used as a pulse generator gave a response time of approximately 12 ps for the effective sampling aperture. When used with the photodiode, the FWHM was observed to be 36 ps at full bias at room temperature. A comparable measurement of the same photodiode with a sampling oscilloscope showed the photoconductive sampling technique to be superior with regard to time resolution, trigger jitter, signal-to-noise level, and ringing. Measurements of the photodiode were also made at 125 K and with different bias voltages. Since the dominant contribution to the speed of response was due to the electron transit time in the intrinsic layer of the photodiode, it was possible to estimate the electron drift velocities from the results. It was found that the saturated drift velocity of electrons in silicon was 8.8×10^6 cm/s and 1.0×10^7 cm/s at 125 K and 293 K respectively.

b) Electro-optic Sampling of High Speed Photodetectors

An example which illustrates the application of electro-optic sampling to the measurement of discrete high speed devices is the characterization of a GaAs Schottky photodiode [5.97,98]. The photodiode was made by molecular beam epitaxy and had an active area only 10 μm^2. It was carefully mounted in a special coaxial configuration and connected to a lithium tantalate Pockel's cell. Optical pulses of 5 ps duration from a synchronously pumped mode-locked dye laser were used for excitation of the photodiode and for electro-optic sampling. A careful deconvolution of the measurement produced a measured response time for the photodiode of only 5.4 ps FWHM. A Fourier analysis of this response gave an equivalent frequency response which was flat from dc to approximately 100 GHz. The use of radio frequency modulation of the trigger beam and synchronous detection of the probe with a high frequency lock-in amplifier gave a shot-noise limited sensitivity of 11 $\mu V/\sqrt{Hz}$.

c) Optical Mixing in Photodetectors

A very different approach to characterizing the response of high speed photodetectors has recently been demonstrated by *Carruthers* and *Weller* [5.99]. They have mixed two picosecond optical pulses in a GaAs Schottky barrier photodiode and observed the variation in the average detector current as a function of the delay between the optical pulses. A nonlinearity with respect to the excitation amplitude produces an autocorrelation signal that can be used to extract information about the intrinsic speed of response of the detector. They suggest the most probable physical mechanism for the nonlinearity is the perturbation of the electric field by the injected carriers and the consequent change in carrier collection efficiency.

d) Impulse Response of High Speed Transistors

An early example of the application of a picosecond optoelectronic instrumentation system is the measurement of the impulse response of a GaAs FET [5.94].

The configuration used for these measurements is illustrated in Fig. 5.17. Picosecond photoconductors were used both for electrical pulse generators and for sampling gates. The circuit consists of two radiation-damaged silicon-on-sapphire wafers with microstrip transmission lines in the pulse-injection and sampling geometries previously described. The use of a "floating" main transmission line onto which pulses are injected and sampled has the advantage of providing flexibility for making many different types of measurements. In addition, the dc bias for the gate and drain can be supplied through the lines to set the operating point of the FET without disturbing the photoconducting sampling gates and pulse generators. The use of two photoconductors on the input (left) side allows for calibration of the input signal by using one as a pulse generator

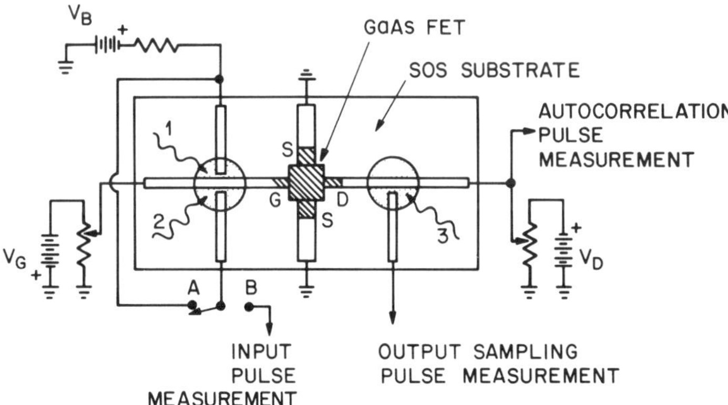

Fig. 5.17. Schematic of picosecond optical electronic circuit used to measure electronic response of GaAs FET. Picosecond optical pulses, indicated by wavy lines 1, 2, and 3 are focused on gaps in the microstrip transmission lines where very high speed photoconductors have been made by radiation damage to the silicon-on-sapphire wafers. Biased photoconductors (V_B) act as electrical pulse generators by injecting charge onto the main transmission line. When the electrical pulses of interest (e.g. the output drain current pulse) are used as bias signals, the photoconductors act as sampling gates (e.g. photoconductor #3). DC gate-source and drain-source bias voltages are applied through the main microstrip line [5.94]

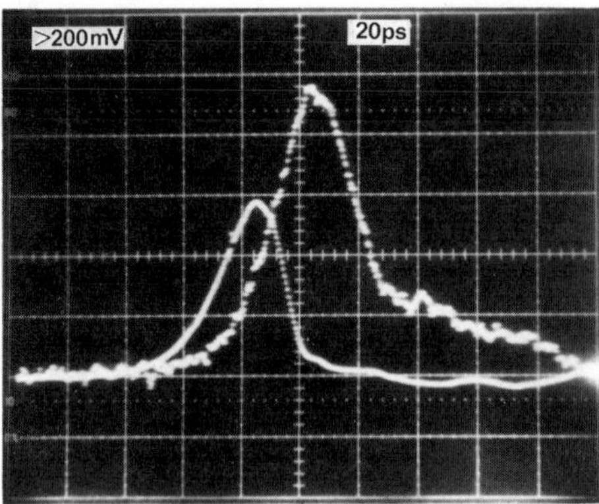

Fig. 5.18. Electronic response of GaAs FET using circuit in Fig. 5.17. The larger amplitude signal is the impulse response of the FET using photoconductor #1 as a pulse generator and #3 as a sampling gate under bias conditions for optimum gain (+3.7 db). The lower amplitude curve is the response with the FET replaced by 50 Ω transmission line (i.e. the system response). The delay and broadening of the signal are clearly evident [5.94]

(with dc bias) and the second as a sampling gate to make an autocorrelation measurement as described in Sect. 5.2.4. This same configuration can also be used in a time-domain reflectometer mode if the optical decay on pulse #2 is lengthened to observe the reflected electrical signal returning from the gate. A third type of measurement which provides information about the nonlinear response of the device is to use both input photoconductors as pulse generators and to observe the total average current in the drain (without sampling pulse #3) as a function of the relative delay between optical pulses #1 and #2.

Figure 5.18 shows a result of the measurement of the GaAs FET. Two signals are illustrated. The first (smaller) is the response of the system without any FET in place using photoconductor #1 as a pulse generator and photoconductor #3 as a sampling gate. This gives an accurate measure of the zero time reference and shows also the time resolution of the system. A certain amount of broadening and asymmetry of the pulse is clearly evident due to the dispersion of the 2.5 cm length of microstrip between photoconductors 1 and 3. The second larger signal shows the response with the FET in place and biased for optimum gain. The larger signal amplitude shows a gain of approximately 4 dB, a delay of 19 ps, a broadening to approximately 40 ps, and a long tail which extends out to 75 ps. The fact that the system is virtually jitter-free provides a very accurate measure of the delay between gate and drain signals.

Optoelectronic measurements have been used by *Cooper* and *Moss* [5.100, 101] to determine the high frequency scattering parameters of a GaAs FET. By Fourier analysis of time-domain waveforms of the reflected and transmitted drain and gate signals, they were able to determine the scattering parameters to frequencies greater than 60 GHz. They have also used direct optical stimulation of FETs [5.101].

Meyer et al. [5.102] have applied electro-optic sampling to the measurement of the impulse response of a modulation-doped gallium arsenide field effect transistor, otherwise known as a MODFET or TEGFET. The measurement approach is similar to that employed by *Smith* et al. [5.94], except that a lithium

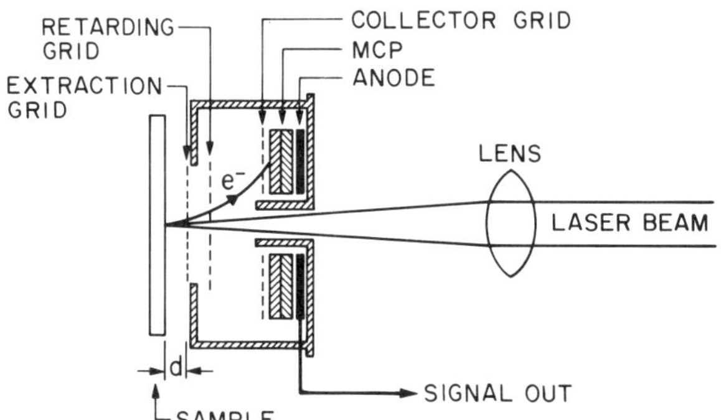

Fig. 5.19. Schematic diagram of an experimant used to measure the response of picosecond devices and circuits by photoemissive sampling [5.103]

tantalate electro-optic crystal was used to measure the drain signal. The measured rise time of the TEGFET was 16 ps which agreed well with the calculated cutoff frequency of 23 GHz. Similar measurements by *Dykaar* et al. [5.96] of the response of a permeable base transistor (PBT) gave a rise-time of 5.3 ps.

e) Picosecond Photoemissive Sampling

Photoelectric emission due to the direct illumination by picosecond optical pulses of high speed devices and circuits in vacuum has recently been demonstrated [5.103,104]. This approach, which is illustrated in Fig. 5.19 uses picosecond pulses to illuminate metallic electrodes in the circuit. Multi-photon photoemission produces electrons that are accelerated by the electric field between the sample and the extraction grid. Voltage signals on the sample modulate the accelerating field producing a shift in the energy of the electrons as they pass through the extraction grid. An electron analyzer consisting of a retarding grid and microchannel plate detector is used to measure this small energy shift which is proportional to the signal voltage on the sample. In general, the time resolution of this technique is limited by the transit time of the electrons from the sample surface to the extraction grid. In preliminary experiments, speeds of approximately 40 ps have been demonstrated. Recently, *Weiner* et al. [5.105] have shown that a substantial improvement in speed can result from measurements of relative potential between two electrodes for which the electric field lines are confined to a region close to the surface of the circuit, thereby reducing the effective transit time.

5.5.3 Optoelectronic Measurements of Integrated Circuits

An important application of picosecond optoelectronic techniques is for the measurement of high speed integrated circuits. Conventional wafer probers, and network analyzers have limited speed and bandwidths and cannot probe internal nodes of a complex circuit without perturbing its performance. A number of

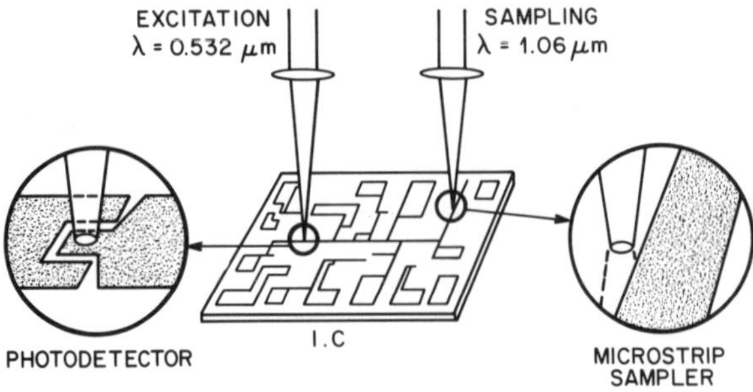

Fig. 5.20. Schematic diagram of an experimantal approach to the noninvasive measurement of high speed electrical signals at internal nodes of an integrated circuit by optical probing of an electro-optic substrate [5.108]

approaches which use picosecond optoelectronics have recently been developed and will be summarized in this section.

a) Substrate Probing of GaAs Integrated Circuits

Gallium arsenide, unlike silicon, is an electro-optic material. The substrate can be used to directly probe the local electrical signals without requiring interconnections to external measurement devices [5.70,106–108]. This non-invasive probing technique has excellent speed and sensitivity and is rapidly being implemented for the characterization of a wide range of circuits and devices. Fig. 5.20 illustrates one of the basic measurement approaches used by *Bloom* and his coworkers [5.106–108]. The circuit is stimulated by an optical pulse striking a photoconductor and the resulting electrical response of the circuit is measured with the probe beam which passes through the GaAs substrate in close proximity to an electrode and is reflected from the bottom of the substrate. This probe beam is detected using polarization optics to measure the small retardation arising from the electro-optic effect in the substrate. In some cases the probe pulse enters from the backside of the substrate and is reflected from electrodes on the top surface. This gives a signal proportional to the integral of the electric field over the probe path length which is equal to the voltage on the specific electrode. In another configuration, a synchronized electrical signal is used to trigger the circuit instead of a photoconductor. This rf signal is carefully synchronized to the laser repetition rate clock and can have a small frequency offset, making possible variable rate stroboscopic measurements.

Substrate electro-optic probing of integrated circuits has recently been demonstrated with the use of a semiconductor diode laser as the source of optical pulses [5.109]. Temporal resolutions of 12 ps and a sensitivity of $2\,\mathrm{mV}/\sqrt{\mathrm{Hz}}$ were obtained with an InGaAsP injection laser.

b) A Picosecond Electro-optic Wafer Prober

Valdmanis et al. [5.110] have recently developed a non-invasive electro-optic wafer prober for testing high speed integrated circuits. The technique uses a small

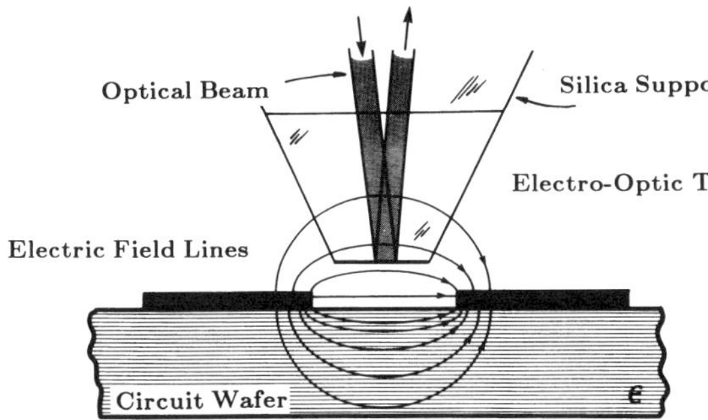

Fig. 5.21. An electro-optic wafer prober which uses a small electro-optic crystal to probe the fringing electric fields above an integrated circuit [5.110]

(40 μm) crystal of lithium tantalate to probe the fringing electric fields above an integrated circuit as illustrated in Fig. 5.21. A focused (5 μm) optical beam detects the small electric field induced birefringence in the crystal using polarizing optics as described previously in the section on electro-optic sampling. The probe tip can be positioned over any location in the circuit permitting non-invasive measurement of high speed signals with excellent speed and sensitivity. The small loading of the circuit due to the capacitive coupling to the electro-optic crystal is estimated to limit the speed of response to approximately 2 ps. The approach is very flexible and can be used for probing silicon as well as GaAs circuits. A similar approach which uses a GaAs injection laser as a source of picosecond pulses has been demonstrated by *Nees* and *Mourou* [5.111].

c) Charge Density Probing in Integrated Circuits

The optical detection of charge density variations in silicon devices has recently been demonstrated by *Heinrich* et al. [5.112]. This measurement technique is based on the detection of the small change in optical index of refraction due to the presence of free charge. The sensitivity is estimated to be capable of detecting sheet charge densities as low as 2.5×10^8 electrons/cm^2 using an optical probe of approximately 1 mW average power and a short-noise limited detection system. The technique has been used to measure internal node voltages in a silicon IC, and to measure μmV signal levels in a forward biased diode and digital signals in an NMOS inverter.

d) Direct Optical Stimulation of Integrated Circuits

Another approach to characterize high speed integrated circuits developed by *Jain* and coworkers [5.113–116] is to directly illuminate logic elements in an IC with picosecond optical pulses to change their logic state. By directly injecting optical pulses into the channel of a GaAs FET they have demonstrated fast logic level switching. The output of the stimulated FET initiates a sequence of logic operations in subsequent connected elements. To obtain accurate relative timing,

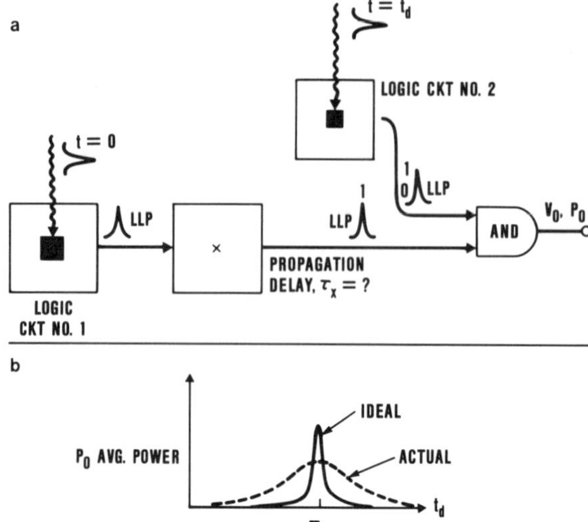

Fig. 5.22a, b. Measurement of the speed of response of integrated circuits by the direct optical stimulation of logic gates [5.113–115]

two or more stimulating optical pulses are used to trigger different logic gates in the circuit. A NOR or AND circuit can be used to determine the coincidence of logic signals produced by different optical pulses. The basic concept of this approach is illustrated in Fig. 5.22. In this case the propagation delay of an intervening set of logic elements can be measured with high precision. An advantage of this approach is that it is non-invasive since it does not require physical contact to the IC and can in principle be used with silicon ICs as well as gallium arsenide ICs. Experiments on conventional GaAs D-flip-flop logic circuits have demonstrated latching times of 475 ps and transition times < 100 ps [5.115].

More recent measurements in higher speed GaAs ICs have demonstrated switching speeds less than 10 picoseconds [5.116].

e) Integrated Photoconductors

Hammond and coworkers [5.66,117,118] have developed an approach to characterizing silicon integrated circuits which involves the fabrication of polysilicon photoconductors on the IC for pulse generators and sampling gates. The polysilicon photoconductors are made with standard silicon integrated circuit processing techniques to insure full compatibility with standard VLSI processes. Photoresist-masked, ion beam irradiation was used to generate trapping sites in the polysilicon to give fast response times. Test measurements of dispersion in silicon-based stripline show response time of the photoconductors to be less than 3 ps FWHM.

5.5.4 Microwave, Millimeter-Wave, and Far-Infrared Applications

The ability to generate and measure extremely fast electrical signals by optoelectronic techniques has extremely important applications for the generation and detection of high frequency electromagnetic waves. The frequency range spanned by optoelectronic devices is very large. For example, an electrical signal having a rise-time of 1 ps has a base bandwidth which extends from dc to approximately 300 GHz. Faster electrical transients such as those produced by optical rectification [5.78] have spectral components as high as 5 THz. The reciprocal property of photoconducting and electro-optic devices makes possible the phase-coherent detection of these high frequency signals with good sensitivity.

a) Photoconductive Switching and Gating

The use of photoconductive switches to gate cw microwave sources is an effective method of producing short bursts of radiation with precise timing and variable durations. Since photoconductive switches are designed to have large bandwidths, it is usually possible to apply a high frequency microwave bias to the input instead of a dc bias voltage. When two optical pulses are used to control the opening and closing times of the switch, a variable duration burst of microwaves is produced [5.119–122]. A disadvantage of this approach is the poor on-off ratio of the switch due to capacitive coupling of the microwave bias across the unilluminated switch. Also the peak microwave power is limited by the cw microwave source. *Platte* [5.120] has reviewed the use of photoconductive gating of microwave signals.

A second method of producing short bursts of microwaves is the impulse excitation of dc-biased photoconductors which are mounted in tuned circuits or waveguides. This approach has the potential for producing much higher peak powers. The basic concept is to use the tuned circuit or waveguide to act as a frequency selective load so that a particular frequency component of the current pulse is extracted. In the experiment of *Mourou* et al. [5.123,124] the current pulse from a semi-insulating GaAs photoconductor was coupled into an X-band waveguide to produce short bursts of X-band radiation. They measured the duration of these pulses to be approximately 50 ps by observing the time resolved reflection of the microwave signals from an optically pumped sample of germanium. The technique was also applied to a radar ranging experiment in which microwave echoes from targets were resolved within a few centimeters a distance of 4 m.

Auston and *Smith* [5.125] have used small radiation damaged silicon-on-sapphire photoconductors which were resonant at 55 GHz to produce short bursts of millimeter waves. In this experiment, the reciprocal property of the photoconductors was used so that phase-coherent detection of the signals was possible with excellent signal-to-noise ratio. *Lee* et al. [5.126] have used impulse excitation of GaAs:Cr and InP:Fe photoconductors in coaxial resonators to produce bursts of rf energy of 300 MHz. The dc to rf energy conversion efficiency was as high as 50%.

Proud and *Norman* [5.127] have applied the concept of a frozen-wave generator to picosecond optoelectronics. They used a sequence of photoconductive switches mounted in tandem in a microstrip line. The line segments between the switches were independently biased. When the photoconductors were simultaneously illuminated, the frozen wave determined by the bias conditions was then launched down the line. The result is a short rf burst whose waveshape was determined by the static voltage profile established in the device by the bias voltages. *Lee* et al [5.126] have extended this approach to high powers by generating an rf pulse of two and one-half cycles in duration at 250 MHz having a peak-to-peak amplitude of 850 Volts.

Mooradian [5.128] has demonstrated a novel form of microwave generation by using multiple pulses to illuminate either a GaAs avalanche photodiode or an InP : Fe photoconductive switch. An optical circuit consisting of beam splitters and multiple path length delays was used to produce an optical pulse having a variable pulse repetition rate, so that a microwave signal of variable frequency could be produced.

The use of optical pulses to modulate and control the operation of conventional semiconductor devices has been explored by *Kiehl* [5.129–131], and *Carruthers* [5.132,133]. *Kiehl* [5.130] used optical excitation of an avalanche photodiode to gate rf signals with excellent on/off ratios and reverse isolation. He also used optical injection to quench the output of an IMPATT oscillator to produce short bursts of microwaves [5.131]. Phase control of a TRAPATT oscillator was also achieved by injection of a modulated optical signal [5.129,131]. *Carruthers* [5.132,133] has made extensive measurements of the effects of injecting picosecond optical pulses into transfered electron devices.

b) Optical Rectification

Optical rectification of femtosecond pulses in electro-optic materials has been used to produce broad spectral radiation which extends into the far-infrared [5.78]. The approach uses the electro-optic Cherenkov effect described in Sect. 5.3.2. The frequency spectrum of these electromagnetic pulses is given by the expression:

$$E(\omega) \approx \omega^{3/2} \exp\left(\frac{-\omega^2\tau^2}{4}\right) , \tag{5.22}$$

where τ is a pulse-width parameter which depends on the (Gaussian) optical beam waist, w_0, the optical pulse duration, τ_p (1/e half-width of a Gaussian), and the Cherenkov angle, θ_c:

$$\tau = \left\{\tau_p^2 + \frac{w_0^2}{v^2}\tan^2\theta_c\right\}^{1/2} , \tag{5.23}$$

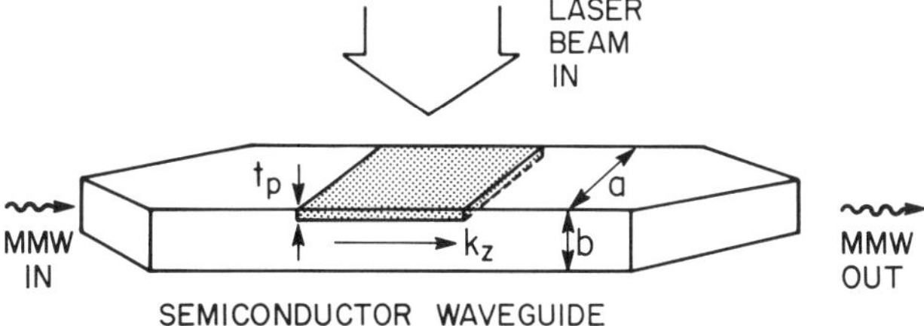

Fig. 5.23. A photoconducting microwave phase modulator. The phase velocity of the dielectric waveguide is modified by the optical injection of free carriers [5.150]

where v = optical group velocity. Spectral peaks as high as 3 THz and useful spectral components as high as 5 THz have been produced and measured in $LiTaO_3$. When combined with coherent electro-optic detection, this technique has been developed into a complete measurement system for millimeter wave and far-infrared spectroscopy [5.79].

c) Phase Modulation of Microwave and Millimeter Waves

Phase modulation of microwave and millimeter-wave signals has been accomplished by *Lee* et al. [5.134] using optically injected electron-hole plasmas in dielectric waveguides. The basic concept of their millimeter-wave phase modulator is illustrated in Fig. 5.23. The phase shift arises from the contribution to the real part of the dielectric response due to the free electrons and holes and can be estimated from the expression:

$$\varepsilon(\omega) = (n + \mathrm{i}k)^2 = \varepsilon_s - \sum_\alpha \frac{\omega_{p\alpha}^2}{\omega^2 + \gamma_\alpha^2} + \mathrm{i}\sum \frac{\gamma\alpha}{\omega} \frac{\omega_p^2\alpha}{\omega^2 + \gamma_\alpha^2}\,, \tag{5.24}$$

where the plasma frequencies, $\omega_p\alpha$ are (MKS):

$$\omega_{p\alpha}^2 = \frac{n_\alpha e^2}{m_\alpha}\,, \tag{5.25}$$

and the summation is over all species of free carriers (i.e. electrons + light and heavy holes). The static dielectric constant is ε_s, and n_α, γ_α and m_α are the carrier densities, damping rates, and effective masses. Useful phase shifts at 94 GHz have been obtained with this device without incurring excessively large attenuations due to the losses arising from the imaginary part of the dielectric response.

d) Radio Frequency Mixing

Foyt et al. [5.135] have demonstrated a radio frequency mixer which used an interdigitated InP optoelectronic switch [5.42]. This device, which is illustrated in Fig. 5.24, uses a semiconductor laser diode as a light source. It is modulated

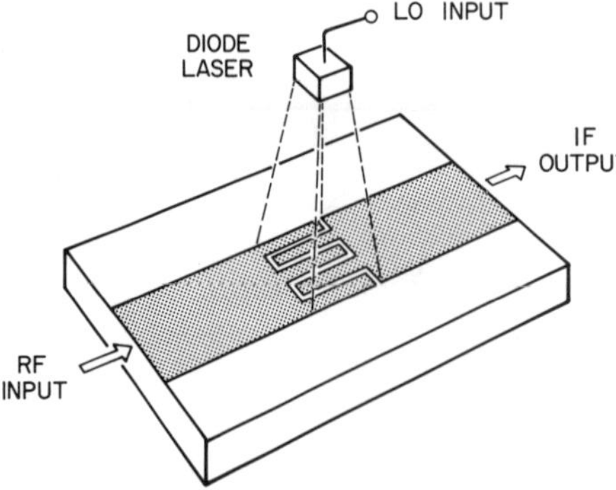

Fig. 5.24. A radio frequency mixer utilizing a photoconductor illuminated by a semiconductor laser diode [5.135]

at the local oscillator frequency and the switch is operated in a bilinear mode so that the output signal at the intermediate frequency is proportional to the product of the input rf signal and the light signal. This gives an intermediate frequency signal of excellent quality with a very low third-order intermodulation component. This device and related applications of InP optoelectronic devices have been summarized in a review article by *Foyt* and *Leonberger* [5.136].

e) Photoconducting Antennas

The concept of a photoconducting antenna is based on the radiation of a photocurrent into free space as opposed to being coupled into a transmission line or other guiding structure. This approach has the advantage of potentially greater speeds of response by overcoming limitations imposed by dispersion and losses of conventional transmission systems.

Photoconductors have been used to drive a dipole antenna by *Mourou* et al. [5.123,124]. A GaAs:Cr photoconductor was illuminated with subpicosecond pulses to produce a short current pulse which was then fed to a dipole antenna. The duration of the radiated signal was measured by the gated transmission through a thin slab of germanium and was estimated to be approximately 3 ps. In a similar experiment, *Heidemann* et al. [5.137,138] have used a photoconductive switch to drive an exponentially tapered slot-line antenna. They produced a radiated pulse of a single cycle having a duration of approximately 1 ns.

Auston and *Cheung* [5.139] have used a geometry in which the photoconductor itself forms the antenna. In this case the photocurrent radiates directly into space without requiring a separate antenna structure. The photocurrent signal produces a time-varying dipole moment whose radiation field, $E_\theta(\theta)$ at a distance r, and angle θ, can be estimated by the classical field of a Hertzian dipole:

$$E_\theta(r,\theta) = \frac{1}{4\pi\varepsilon}\left\{\frac{p}{r^3} + \frac{n}{cr^2}\frac{\partial p}{\partial t} + \frac{n^2}{c^2 r}\frac{\partial^2 p}{\partial t^2}\right\}\sin\theta \ , \tag{5.26}$$

where the dipole moment $p(t)$ is related to the photocurrent, $i(t)$, by the expression $i(t) = \dot{p}/l$, where l is the length of the radiating current (assumed small relative to the shortest radiated wavelength). The three terms in the expression for the electric field can be identified as the static, near, and far field components, varying respectively as r^{-3}, r^{-2} and r^{-1}. Each of these has a different temporal variation. The far field term is proportional to the second derivative of the photocurrent. Hence a short unipolar current pulse will radiate a far field which is bipolar. For short pulses, the distinction between the far and near fields is given by the simple relation $r \gg \tau_p c/n$, where τ_p is the pulse duration.

Photoconductors can also be used as receiving antennas. In this case, the bias signal is induced by the radiation field rather than a dc signal. By measuring the average current at the receiving dipole as a function of the delay between the optical pulses illuminating the transmitting and receiving dipoles, autocorrelation measurements of the system response can be made [5.78]. Response times of approximately 1.6 ps were measured for radiation damaged silicon-on-sapphire Hertzian dipoles. *Karin* et al. [5.65] have observed response times less than 1 ps using He^+ ion bombarded InP:Fe Hertzian dipoles. Their experimental result is shown in Fig. 5.25. This result is the fastest autocorrelation measurement of a photoconducting device and indicates that the intrinsic response time of photoconductors can be subpicosecond. They used a geometry in which the photocurrent was orthogonal to the photoconductor electrodes to suppress the slower radiation signal arising from currents in the electrodes.

Smith and *Auston* [5.140] have recently made resonant half-wave photoconducting dipole antennas with measured frequency responses up to 2 THz.

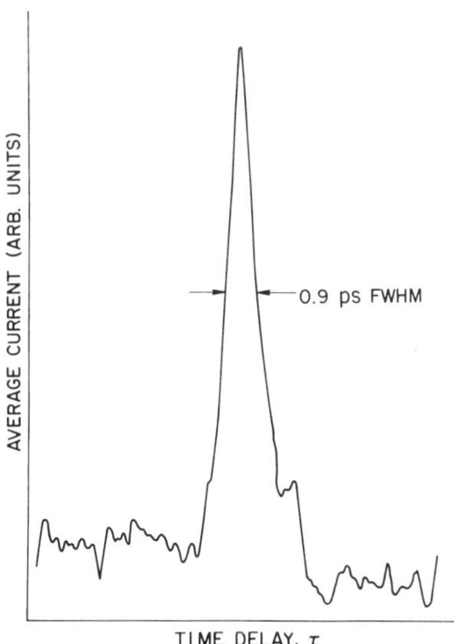

Fig. 5.25. Electronic autocorrelation of two photoconducting antennas made from radiation damaged InP [5.65]

5.5.5 High Power Optoelectronic Switches

Photoconductors use the bulk properties of materials and consequently can be readily scaled to switch high voltages and currents. This has led to a number of important applications where extremely fast *dv/dt* transients are required for switching high voltage instruments. In principal it is possible to have power gain in a photoconductor, i.e. to switch an electrical power which exceeds the optical power.

The design of photoconductors for high power applications requires a substantially different approach than for low powers. Electric field breakdown and thermal dissipation are two effects that greatly influence the choice of materials and geometry. Since the size of high power photoconductors scales up with the switching voltage, careful attention to geometry and mounting is also important to optimize the speed vs power trade-off.

As discussed in Sect. 5.3, electronic transport in semiconductors under high electric field conditions differs substantially from that at low fields. In silicon, for example, the electron velocity saturates at fields above 3–5 kV/cm at a value of approximately 10^7 cm/s. The behavior of holes is similar. In GaAs and related III–V direct-gap semiconductors the electron velocity reaches a maximum at a few kV/cm and then decreases at higher fields producing a negative differential mobility which can lead to instabilities of the Gunn type. These departures from linear ohmic behavior can influence the efficiency and stability of high power photoconductors. Additional effects such as inductive [5.141] and electromagnetic [5.142] transit times, may also play a role in determining the speed of response of high power photoconductors.

High power photoconductors use higher light intensities and consequently the carriers densities can be extremely high, and can produce results which differ from low injection conditions. For example, in silicon at room temperature, electron densities above $10^{19}\,\mathrm{cm}^{-3}$ will produce a degenerate Fermi distribution for which the conductivity is determined by elastic scattering at the Fermi energy, and generally results in a low effective mobility. An additional effect which can occur at high carrier densities is a reduction of the mobility due to electron-hole scattering (see Sect. 5.3).

The size of a high power picosecond photoconductor is constrained by high field breakdown. This determines the gap size which scales linearly with voltage. Since the photoconductance is proportional to the optical energy and inversely proportional to the square of the gap length, it follows that the required optical energy increases as the square of the operating voltage.

Thermal effects are also important considerations in high power photoconductors. Thermal runaway can limit the hold-off voltage. This effect occurs when the temperature rise due to the dark current is sufficient to create additional carriers by thermal generation of e-h pairs. Solutions to this problem that have been employed are to use high resistivity materials, low temperatures, and pulsed bias voltages. For a more detailed discussion of these and related aspects of high power photoconductors the reader is referred to the review articles by *Mourou* et al. [5.143] and *Nunnally* and *Hammond* [5.141].

a) Applications of High Power Photoconductors

Picosecond photoconductive switching above 1 kV was first demonstrated by *LeFur* and *Auston* [5.72]. They used a silicon microstrip photoconductor with a pulsed bias. The output voltage was used to drive a traveling-wave Pockel's cell of $LiTaO_3$. Transverse probing of the Pockel's cell by a second light pulse showed the voltage wave had a rise time less than 25 ps.

Antonetti et al. [5.144] used a silicon photoconductor mounted in a coaxial transmission line to switch voltages up to 10 kV. These were used to drive a traveling-wave Kerr cell and a fast commercial Pockel's cell. Rise times less than 50 ps were reported.

A GaAs photoconductive switch was used by *Agostinelli* et al. [5.145] to generate electrical pulses of 3 kV. They used them to drive a Pockel's cell and measured rise and fall times as fast as 40 ps. A streak camera was used to observe the transmitted optical signals which were sliced out of a long optical pulse by the Pockel's cell. Improved switching of high voltages has been reported by *Mourou* and *Knox* [5.146], *Stavola* et al. [5.147], and *Koo* et al. [5.148].

The highest voltage photoconductor reported to date was demonstrated by *Nunnally* and *Hammond* [5.141] to produce 1.8 kA into a 25 Ω load for a peak voltage of 45 kV and a peak power of 80 MW. Although the rise time was

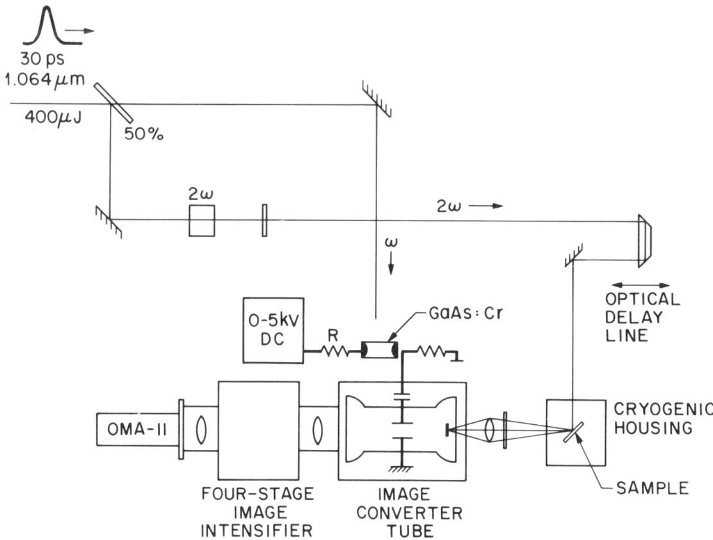

Fig. 5.26. Optoelectronic synchronization of a streak camera by a high voltage photoconductor [5.149]

relatively slow, 5 ns, it is expected that with further improvements in high power switching technology, similar results will be obtained on the subnanosecond time scale.

Mourou and *Knox* [5.149] have used a high voltage silicon photoconductor to synchronize a streak camera to optical pulses from a mode-locked Nd:YAG laser. Their experiment, which is illustrated in Fig. 5.26, used the output of the silicon photoconductor to drive the deflection plate of the streak camera. Greatly reduced jitter enabled extensive shot-to-shot averaging for improved time resolution and signal-to-noise ratios.

5.6 Discussion

It is clear from the diversity and volume of work summarized in this article that ultrafast optoelectronics is a well developed and rapidly growing field of research. The technology of using optical pulses to generate and detect picosecond and subpicosecond electrical pulses is now relatively well developed. We can expect to see substantial further improvements, however, both with regard to speed and sensitivity. Optical pulses are now available with durations as short as 6 fs. The application of these extremely short pulses to optoelectronics presents an exciting challenge for future work. The available bandwidth extends as high as 10^{14} Hz, well into the far-infrared region of the spectrum. If techniques can be developed which can utilize most of this bandwidth, it will make possible measurements over an enormous range of the electromagnetic spectrum. This will essentially fill in the "gaps" in the spectrum where coherent sources do not exist today. In this realm, the distinction between optics and electronics disappears.

New materials for optoelectronics will undoubtedly have an important influence on future trends. Although the traditional bulk properties of the better known semiconductors and electro-optic materials have served to establish the foundation for picosecond optoelectronic device concepts, new materials are required to go beyond the limits of what can be done today. For example, semiconductors grown by molecular beam epitaxy, have already proven to have valuable properties for high speed optical modulators and bistable optical logic elements. The "self-electro-optic" effect and the quantum-confined Stark effect will probably find applications in high speed optoelectronic measurement systems as well. Organic nonlinear optical materials also have the potential for new applications in optoelectronics.

It is in the area of applications that ultrafast optoelectronics is now experiencing its most rapid growth. As a precision tool for testing high speed discrete devices and integrated circuits, optoelectronic measurement systems are now being widely implemented and are at the stage where useful design information can be obtained about device performance. The non-invasive probing techniques described in Sect. 5.5 are giving circuit and device engineers information that a few years ago was not considered possible. Measurements of novel high speed electronics devices such as resonant tunneling transistors and "ballistic" transistors will be an important guide to research and development of new high speed technologies.

In the very high speed area, the trend toward applications to materials characterization is expected to continue. For example, measurements of the detailed kinetics of electronic transport on the subpicosecond time scale over a wide range of density, electric field, and temperature, are vitally important to the development and understanding of electronic and optical materials. Picosecond optoelectronic tools have recently been developed which will make this possible. The recent measurement of subpicosecond mobility transients by *Nuss* and *Auston* [5.16] is an example of what can be accomplished with these capabilities.

The continued development of techniques for generating and detecting large amplitude electrical pulses will have important applications for the study of the transient nonlinear response of materials and devices. For example, we might expect to observe electrical solitons on nonlinear transmission lines or in nonlinear dielectric materials, analogous to the optical solitons that have been produced in optical fibers. This could lead to pulse compression and result in even shorter electrical pulses than can be produced today.

References

5.1 N.J. Shah, S.S. Pei, C.W. Tu, R.C. Tiberio: IEEE Trans. Electr. Dev., ED-**33**, 543–547 (1986)
5.2 C.H. Lee (ed.): *Picosecond Optoelectronic Devices* (Academic Press, New York 1984)
5.3 G.A. Mourou, D.M. Bloom, C.H. Lee (eds.): *Picosecond Electronics and Optoelectronics*, Springer Ser. Electrophys., Vol. 21 (Springer, Berlin, Heidelberg 1985)
5.4 F.J. Leonberger (ed.): *Picosecond Electronics and Optoelectronics II*, Springer Ser. in Electronics and Photonics, Vol. 24 (Springer, Berlin, Heidelberg 1987)
5.5 D.H. Auston, K.B. Eisenthal (eds.): *Ultrafast Phenomena IV*, Springer Ser. Chem. Phys., Vol. 38 (Springer, Berlin, Heidelberg 1984)
5.6 G.R. Fleming, A.E. Siegman: In *Ultrafast Phenomena IV*, Springer Ser. Chem. Phys., Vol. 38 (Springer, Berlin, Heidelberg 1986)
5.7 S. Shapiro (ed.): *Ultrashort Light Pulses*, Topics Appl. Phys., Vol. 18 (Springer, Berlin, Heidelberg 1977)
5.8 A. Rose: *Concepts in Photoconductivity and Allied Problems* (Kreiger, New York 1963)
5.9 B.R. Nag: *Electronic Transport in Compound Semiconductors* (Springer, Berlin, Heidelberg 1980)
5.10 K. Seeger: *Semiconductor Physics*, Springer Ser. Solid-State Sci., Vol. 44 (Springer, Berlin, Heidelberg 1982)
5.11 J. Shah: J. Phys., **42**, Suppl. 10, C7445-C7462 (1981)
5.12 J. Shah: IEEE J. Quant. E1. QE-**22**, 1728 (1986)
5.13 J.L. Oudar, D. Hulin, A. Migus, A. Antonetti, F. Alexandre: Phys. Rev. Lett. **55**, 1191 (1985)
5.14 W.H. Knox, C. Hirliman, D.A.B. Miller, J. Shah, D.S. Chemla, C.V. Shank: Phys. Rev. Lett. **56**, 1191 (1986)
5.15 D. Block, J. Shah, A.C. Gossard: Solid State Commun. **59**, 527 (1986)
5.16 M.C. Nuss, D.H. Auston: Phys. Rev. Lett. **58**, 2355 (1987)

5.17 J.P. McLean, E.G.S. Paige: J. Phys. Chem. Sol. **16**, 220 (1960)
5.18 J.P. McLean, E.G.S. Paige: J. Phys. Chem. Sol. **18**, 139 (1961)
5.19 E.G.S. Paige: J. Phys. Chem. Sol. **16**, 207 (1960)
5.20 J. Appel: Phys. Rev. **122**, 1760 (1961)
5.21 J. Appel: Phys. Rev. **125**, 1815 (1962)
5.22 J.R. Meyer, M. Glicksman: Phys. Rev. B **17**, 3227–3238 (1978)
5.23 J. Vaitkus, V. Grivitskas, J. Storasta: Sov. Phys. Semicond. **9**, 883 (1976)
5.24 D.H. Auston, A.M. Johnson: In *Ultrashort Light Pulses*, ed. by S.L. Shapiro, Topics Appl. Phys., Vol. 18 (Springer, Berlin, Heidelberg 1977)
5.25 V. Grivitskas, M. Willander, J. Vaitkus: Sol. St. Electr. **27**, 565–572 (1984)
5.26 L. Reggiani (ed.): *Hot-Electron Transport in Semiconductors*, Springer Ser. Solid-State Sci., Vol. 58 (Springer, Berlin, Heidelberg 1985)
5.27 H.D. Rees: J. Phys. Chem. Sol. **30**, 643–655 (1969)
5.28 C.V. Shank, R.L. Fork, B.I. Greene: Appl. Phys. Lett. **38**, 104 (1981)
5.29 S. Laval, C. Bru, C. Arnodo, R. Castagne: Digest of the 1980 IEDM Conference, IEEE Press, New York (1980) p. 626
5.30 R.B. Hammond: Proceedings of the 1987 International Conference on Hot Electrons, ed. by E. Gornick, Innsbruck, Austria (1985)
5.31 G.A. Mourou: Digest of The Second Topical Meeting on Picosecond Electronics and Optoelectronics, Optical Society of America, Washington DC (1987) pp. 186–187
5.32 V.S. Vavilov, N.A. Ukhin: *Radiation Effects in Semiconductors and Semiconductor Devices* (Plenum, New York 1977)
5.33 A.M. Stoneham: Physics Today, January 1980, pp. 34–42
5.34 J.W. Chen, A.G. Milnes: Sol. St. Electr. **22**, 684–686 (1977)
5.35 G.M. Martin, A. Mitonneau, A. Mircea: Electr. Lett. **13**, 191–193 (1977)
5.36 A. Mitonneau, G.M. Martin, A. Mircea: Electr. Lett. **13**, 666–667 (1977)
5.37 S.S. Li, W.L. Wang, P.W. Lai, R.T. Owen: J. Electr. Mat. **9**, 335–354 (1980)
5.38 F.E. Doany, D. Grischkowsky, C.C. Chi: Digest of the Second Topical Conference on Picosecond Electronics and Optoelectronics, Optical Society of America, Washington, DC (1987)
5.39 E.H. Rhoderick: *Metal-Semiconductor Contacts* (Oxford Univ. Press, Oxford 1978)
5.40 C. Erginsoy: Phys. Rev. **79**, 1013–1014 (1950)
5.41 A.G. Foyt, F.J. Leonberger, R.C. Williamson: Appl. Phys. Lett. **40**, 447–449 (1982)
5.42 F.J. Leonberger, P.F. Moulton: Appl. Phys. Lett. **35**, 712–714 (1979)
5.43 R.B. Hammond, N.G. Paulter, A.E. Iverson, R.C. Smith: Tech. Dig. 1981 Int. Electr. Dev. Mtg., 157–159 (1981)
5.44 A.P. DeFonzo: Appl. Phys. Lett. **39**, 480–481 (1983)
5.45 D.H. Auston, P. Lavallard, N. Sol, D. Kaplan: Appl. Phys. Lett. **35**, 66–68 (1980a)
5.46 D.H. Auston, A.M. Johnson, P.R. Smith, J.C. Bean: Appl. Phys. Lett. **37**, 371–373 (1980b)
5.47 F. Pockels: Annal. D. Physik **37**, 158 (1889)
5.48 A. Yariv, P. Yeh: *Optical Waves in Crystals* (Wiley, New York 1984)
5.49 I. Kaminow: In *Handbook of Lasers*, ed. by R.J. Pressley (Chemical Rubber Co., Cleveland 1971)
5.50 R.C. Alferness: Science **234**, 825–829 (1986)
5.51 K.P. Cheung, D.H. Auston: Phys. Rev. Lett. **55**, 2152 (1986a)
5.52 K.P. Cheung, D.H. Auston: Infrared Phys. **26**, 23–27 (1986b)
5.53 D.H. Auston: In *Picosecond Optoelectronic Devices*, ed. by C.H. Lee (Academic Press, New York 1984)
5.54 M. Maeda: IEEE Trans., MTT-**20**, 390–396 (1972)
5.55 D.H. Auston: IEEE J. Quant. E1., QE-**19**, 639–648 (1983a)
5.56 D.H. Auston: Appl. Phys. Lett. **26**, 101–103 (1975)
5.57 R.A. Lawton, A. Scavannec: Electr. Lett. **11**, 74–75 (1975)
5.58 R.A. Lawton, J.R. Andrews: Electr. Lett. **11**, 138 (1975)
5.59 R.A. Lawton, J.R. Andrews: IEEE Trans. Instr. Meas. **25**, 56–60 (1976)

5.60 S.R. Forrest: IEEE J. Lightwave Tech., LT-**3**, 347 (1985)
5.61 A.M. Johnson, D.H. Auston, P.R. Smith, J.C. Bean, J.P. Harbison, A.C. Adams: Phys. Rev. B **23**, 6816–6819 (1981)
5.62 P.R. Smith, D.H. Auston, A.M. Johnson, W.M. Augustyniak: Appl. Phys. Lett. **38**, 47–50 (1981a)
5.63 M.B. Ketchen, D. Grischkowsky, T.C. Chen, C.C. Chi, I.N. Duling, N.J. Halas: Appl. Phys. Lett. **48**, 751–753 (1986)
5.64 P.M. Downey, D.H. Auston, P.R. Smith: Appl. Phys. Lett. **42**, 215–217 (1983)
5.65 J.R. Karin, P.M. Downey, R.J. Martin: IEEE J. Quant. E1. QE-**27**, 677 (1986)
5.66 R.B. Hammond, N.G. Paulter, R.S. Wagner, W.R. Eisenstadt: Appl. Phys. Lett. **45**, 404 (1984)
5.67 C.H. Lee, A. Antonetti, G.A. Mourou: Optics Comm. **21**, 158 (1977)
5.68 W.J. Gallagher, C.C. Chi, I.N. Duling, D. Grischkowsky, N.J. Halas: Appl. Phys. Lett. **50**, 350 (1987)
5.69 J.A. Valdmanis, G. Mourou: IEEE J. Quant. Elect. QE-**22**, 69–78 (1986)
5.70 B.H. Kolner, D.M. Bloom: IEEE J. Quant. Elect. QE-**22**, 79 (1986)
5.71 D.H. Auston, A.M. Glass: Appl. Phys. Lett. **20**, 398–399 (1972)
5.72 P. LeFur, D.H. Auston: Appl. Phys. Lett. **28**, 21–23 (1976)
5.73 J.A. Valdmanis, G. Mourou, C.W. Gabel: Appl. Phys. Lett. **41**, 211–213 (1982)
5.74 J.A. Valdmanis, G. Mourou, C.W. Gabel: IEEE J. Quant. Elect. QE-**19**, 664–667 (1983)
5.75 G. Haisnan, G. Arjavalingam, A. Dienes, J.R. Whinnery: Proc. SPIE Conf. on Picosecond Electro-optics, San Diego (1983)
5.76 M. Bass, P.A. Franken, J.F. Ward, G. Weinreich: Phys. Rev. Lett. **9**, 446–448 (1962)
5.77 D.H. Auston, A.M. Glass, A.A. Ballman: Phys. Rev. Lett. **28**, 897–900 (1972)
5.78 D.H. Auston, K.P. Cheung, J.A. Valdmanis, D.A. Kleinman: Phys. Rev. Lett. **53**, 1555–1557 (1984)
5.79 D.H. Auston, K.P. Cheung: J. Opt. Soc. Amer. B **2**, 606–612 (1985)
5.80 S. Ramo, J.R. Whinnery, T. van Duzer: *Fields and Waves in Communication Electronics* (Wiley, New York 1984)
5.81 K.C. Gupta, R. Garg, I.J. Bahl: *Microstrip Lines and Slotlines* (Artech House 1979)
5.82 T.C. Edwards: *Foundations for Microstrip Circuit Design* (Wiley, New York 1981)
5.83 K.K. Li, J.R. Whinnery, A. Dienes: In *Picosecond Optoelectronic Devices*, ed. by C.H. Lee (Academic, New York 1984)
5.84 G.A. Mourou, K.E. Meyer: Appl Phys. Lett. **45**, 492–494 (1984b)
5.85 D.E. Cooper: Appl. Phys. Lett. **47**, 33–35 (1985)
5.86 K.W. Goosen, R.B. Hammond: IEEE Trans. Micr. Th. Tech. MTT-**33** (1985)
5.87 G.A. Mourou: *High Speed Electronics*, ed. by B. Kallback, H. Beneking, Springer Ser. Electronics and Photonics, Vol. 22 (Springer, Berlin, Heidelberg 1986) p. 191
5.88 J.A. Buck, K.K. Li, J.R. Whinnery: J. Appl. Phys. **51**, 769 (1980)
5.89 W. Margulis, R. Persson: Rev. Sci. Instr. **56**, 1586–1588 (1986)
5.90 C.S. Chang, V.K. Mathur, M.J. Rhee, C.H. Lee: Appl. Phys. Lett. **41**, 392–394 (1982)
5.91 R. Landauer: IBM J. Res. Dev. **4**, 391–401 (1960)
5.92 R.V. Khokhlov: Radio Eng. Electr. Phys. (USSR) **6**, 817–824 (1961)
5.93 P. Paulus, B. Wedding, A. Gasch, D. Jager: Phys. Lett. **102A**, 89–92 (1984)
5.94 P.R. Smith, D.H. Auston, W.M. Augustyniak: Appl. Phys. Lett. **39**, 739–741 (1981)
5.95 D.H. Auston, P.R. Smith: Appl. Phys. Lett. **41**, 599–601 (1982)
5.96 D.R. Dykaar, R. Sobelewski, J.F. Whitaker, T.Y. Hsiang, G.A. Mourou, M.A. Hollis, B.J. Clifton, K.B. Nichols, C.O. Bozler, R.A. Murphy: *Ultrafast Phenomena V*, ed. by G.R. Fleming, A.E. Siegman, Springer Ser. Chem. Phys., Vol. 46 (Springer, Berlin, Heidelberg 1986) p. 103
5.97 S.Y. Wang, D.M. Bloom, D.M. Collins: Appl. Phys. Lett. **42**, 190–192 (1983)
5.98 B.H. Kolner, D.M. Bloom, P.S. Cross: Electr. Lett. **19**, 574–575 (1983)
5.99 T.F. Carruthers, J.F. Weller: Appl. Phys. Lett. **48**, 460–462 (1986)
5.100 D.E. Cooper, S.C. Moss: IEEE J. Quant. E1., QE-**22**, 94–100 (1986)
5.101 D.E. Cooper, S.C. Moss: *Ultrafast Phenomena V*, ed. by G.R. Siegman, A. Fleming, Springer Ser. Chem. Phys., Vol. 46 (Springer, Berlin, Heidelberg 1986)

5.102 K.E. Meyer, D.R. Dykaar, G.A. Mourou: In *Picosecond Electronics and Optoelectronics*, ed. by G.A. Mourou, D.M. Bloom, C.-M. Lee, Springer Ser. in Electrophys., Vol. 21 (Springer, Berlin, Heidelberg 1985)
5.103 J. Bokor, A.M. Johnson, R.H. Storz, W.M. Simpson: Appl. Phys. Lett. **49**, 226–228 (1986)
5.104 R.B. Marcus, A.M. Weiner, J.H. Abeles, P.S.D. Lin: Appl. Phys. Lett. **49**, 357–359 (1986)
5.105 A.M. Weiner, R.B. Marcus, P.S.D. Lin, J.H. Abeles: In *Picosecond Electronics and Optoelectronics II*, ed. by F.J. Leonberger, Springer Ser. in Electronics and Photonics, Vol. 24 (Springer, Berlin, Heidelberg 1987)
5.106 B.H. Kolner, D.M. Bloom: Electr. Lett. **20**, 818–819 (1984)
5.107 J.L. Freeman, S.K. Diamond, H. Fong, D.M. Bloom: Appl. Phys. Lett. **47**, 1083–1084 (1985)
5.108 K.I. Weingarten, M.J.W. Rodwell, H.K. Heinrich, B.H. Kolner, D.M. Bloom: Electr. Lett. **21**, 765–766 (1985)
5.109 A.J. Taylor, J.M. Wiesenfeld, G. Eisenstein, J.R. Talman, U. Koren: Electr. Lett. **21**, 765 (1986)
5.110 J.A. Valdmanis, S.S. Pei: Technical Digest of the Second Topical Conference on Picosecond Electronics and Optoelectronics, Optical Society of America, Washington, DC (1987) pp. 4–6
5.111 J. Nees, G.A. Mourou: Electr. Lett. **22**, 913 (1986)
5.112 H.K. Heinrich, D.M. Bloom, B.R. Hemenway: Appl. Phys. Lett. **48**, 1066–1068 (1986)
5.113 R.K. Jain, D.E. Snyder: IEEE J. Quant. E1., QE-**19**, 658 (1983a)
5.114 R.K. Jain, D.E. Snyder: Opt. Lett. **8**, 85 (1983b)
5.115 R.K. Jain, D.E. Snyder, K. Stenersen: IEEE Electr. Dev. Lett. EDL-**5**, 371 (1984)
5.116 X.-C. Zhang, R. Jain: Electr. Lett. **22**, 264 (1986)
5.117 D.R. Bowman, R.B. Hammond, R.W. Dutton: IEEE Electr. Dev. Lett. EDL-**6**, 502–504 (1985)
5.118 W.R. Eisenstad, R.B. Hammond, R.W. Dutton: IEEE Trans. Electr. Dev., ED-**32**, 364–369 (1985)
5.119 A.M. Johnson, D.H. Auston: IEEE J. Quant. E1., QE-**11**, 283–87 (1975)
5.120 W. Platte: Optics and Laser Technology, Feb. 1978, 40–42 (1978)
5.121 W. Platte, G. Appelhaus: Electr. Lett. **12**, 270 (1976)
5.122 W. Platte: Electr. Lett. **12**, 437–438 (1977)
5.123 G. Mourou, C.V. Stancampiano, D. Blumenthal: Appl. Phys. Lett. **38**, 470 (1981a)
5.124 G. Mourou, C.V. Stancampiano, A. Antonetti, A. Orszag: Appl. Phys. Lett. **39**, 295 (1981b)
5.125 D.H. Auston, P.R. Smith: Appl. Phys. Lett. **43**, 631–633 (1983b)
5.126 C.H. Lee, M.G. Li, C.S. Chang, A.M. Yurek, M.J. Rhee, E. Chauchard, R.P. Fischer, A. Rosen, H. Davis: Proceedings of IEEE-MTT-S Intl. Micr. Symp., St. Louis (1985) pp. 178–191
5.127 J.M. Jr. Proud, S.L. Norman: IEEE Trans. Micr. Th. Tech., MTT-**26**, 137 (1978)
5.128 A. Mooradian: Appl. Phys. Lett. **45**, 494 (1984)
5.129 R.A. Kiehl: IEEE Trans. Electr. Dev., ED-**25**, 703–710 (1978)
5.130 R.A. Kiehl: IEEE Trans. Micr. Th. Tech., MTT-**27**, 533–539 (1979)
5.131 R.A. Kiehl: IEEE Trans. Micr. Th. Tech., MTT-**28**, 409–413 (1980)
5.132 T.F. Carruthers, J.F. Weller, H.F. Taylor, T.G. Mills: Appl. Phys. Lett. **38**, 202–204 (1981)
5.133 T.F. Carruthers: In *Picosecond Optoelectronic Devices*, ed. by C.H. Lee (Academic, New York 1984) pp. 339–371
5.134 C.H. Lee, S. Mak, A.P. DeFonzo: Electr. Lett. **14**, 733 (1978)
5.135 A.G. Foyt, F.J. Leonberger, R.C. Williamson: Proc. SPIE **269**, 109–114 (1981)
5.136 A.G. Foyt, F.J. Leonberger: In *Picosecond Optoelectronic Devices*, ed. by C.H. Lee (Academic, New York 1984) pp. 271–311
5.137 R. Heidemann, Th. Pfeiffer, D. Jager: Electr. Lett. **18**, 783–784 (1982)
5.138 R. Heidemann, Th. Pfeiffer, D. Jager: Electr. Lett. **19**, 316–317 (1983)
5.139 D.H. Auston, K.P. Cheung: Appl. Phys. Lett. **45**, 284–286 (1984)
5.140 P.R. Smith, D.H. Auston: unpublished
5.141 W.C. Nunnally, R.B. Hammond: In *Picosecond Optoelectronic Devices*, ed. by C.H. Lee (Academic, New York 1984) pp. 373–398
5.142 F. Kerston, M. Schubert: Optical and Quant. E1., **16**, 477–486 (1984)

5.143 G. Mourou, W.H. Knox, S. Williamson: In *Picosecond Optoelectronic Devices*, ed. by C.H. Lee (Academic, New York 1984)
5.144 A. Antonetti, M.M. Malley, G. Mourou, A. Orszag: Opt. Commun. **23**, 435–438 (1977)
5.145 J. Agostinelli, G. Mourou, C.W. Gavel: Appl. Phys. Lett. **35**, 731–733 (1979)
5.146 G. Mourou, W. Knox: Appl. Phys. Lett. **35**, 492–495 (1979)
5.147 M. Stavola, J.A. Agostinelli, M.G. Sceats: Appl. Opt. **18**, 4101–4105 (1979)
5.148 J.C. Koo, G.M. McWright, M.D. Pocha, R.B. Wilcox: Appl. Phys. Lett. **45**, 1130–1131 (1984)
5.149 G. Mourou, W. Knox: Appl. Phys. Lett. **36**, 623–626 (1980)
5.150 C.H. Lee, P.S. Mak, A.P. De Fonzo: IEEE J. Quant. Electr. QE-**16**, 217–288 (1980)
5.151 C.H. Lee: Appl. Phys. Lett. **30**, 84–86 (1977)
5.152 P.S. Mak, V.K. Mathur, C.H. Lett: Opt. Commun. **32**, 485–488 (1980)
5.153 W. Margulis, W. Sibbett: Opt. Commun. **37**, 224–228 (1981)
5.154 L.A. Vermeulen, J.F. Young, M.I.A. Gallant, H.M. von Driel: Solid State Commun. **38**, 1223–1225 (1981)
5.155 P.M. Downey, B. Tell: J. Appl. Phys. **56**, 2672–2674 (1984)
5.156 P.M. Downey, B. Schwartz: Appl. Phys. Lett. **44**, 207 (1984)
5.157 A.M. Johnson, D.W. Kisker, W.M. Simpson, R.D. Feldman: In *Picosecond Electronics and Optoelectronics*, ed. by G.A. Mourou, D.M. Bloom, C.H. Lee, Springer Ser. in Electrophys., Vol. 21 (Springer, Berlin, Heidelberg 1985) p. 188
5.158 P.M. Downey, R.J. Martin, R.E. Nahory, O.G. Lorimer: Appl. Phys. Lett. **46**, 396–398 (1985)

6. Ultrafast Coherent Spectroscopy

Wolfgang Zinth and Wolfgang Kaiser

With 20 Figures

Coherent spectroscopy, initially started in the mid-sixties, is intimately conncected with the development of intense coherent light sources. The generation of ultrashort light pulses enabled coherent spectroscopy to be extended to real-time measurements of rapid dynamic processes on the time scale of 10^{-12} s. The early measurements of picosecond time-resolved coherent Raman scattering in 1971 were followed by numerous time-resolved techniques adapted to a variety of ultrafast phenomena. Most investigations are concerned with rapid dynamic processes in the condensed phases. The experimental progress stimulated theoretical investigations, improving our understanding of the interactions in liquids and solids. In recent years, a number of dynamic relaxation processes have been elucidated.

The title of this chapter, *Ultrafast Coherent Spectroscopy*, requires some explanation since it implies investigations in the time *and* frequency domain. Actually, it adequately characterizes the present situation. Coherent techniques give information on dynamic processes in the time domain as well as on linewidths and line positions in the frequency domain.

Numerous ultrafast investigations are concerned with coherent Raman scattering by molecular levels in the electronic ground state. The experimental techniques are well established on the picosecond time scale. Currently, these experiments are being extended to the study of even faster processes in the femtosecond time domain. A substantial fraction of coherent experiments deal with the study of vibrational levels in excited electronic states. Here, echo and induced-grating experiments have been successfully applied to reveal the relaxation dynamics.

The present chapter is organized as follows: In Sect. 6.6.1 the theoretical background of ultrafast coherent spectroscopy is given together with a short description of dynamical processes in condensed matter. Section 6.2 reports on time-resolved investigations using a single excitation process. Various time resolved coherent Raman and infrared techniques are discussed. Finally, in Sect. 6.3 echo-type experiments, i.e. experimental techniques with multiple excitation are presented.

6.1 Theory of Time-Resolved Coherent Spectroscopy

This section deals with the principles of time-resolved coherent experiments in a semi-classical approach. The electromagnetic light field is treated classically, the molecular system quantum-mechanically by its density matrix. Dynamic interaction processes between molecules have two important relaxation times – the phase relaxation time T_2 and the energy relaxation time T_1. We shall repeatedly refer to the relationship between time-domain coherent experiments and frequency-domain steady-state measurements.

6.1.1 The Theoretical Model

Coherent spectroscopy is presented here as a tool to investigate transitions between different energy levels or – more precisely – to study the transition frequencies and the relaxation processes related to these energy states. The essential features of the theoretical model are:

i) Of the many molecular transitions we treat only two levels, the ground state $|b\rangle$ and the excited state $|a\rangle$ separated by the energy $E = \hbar\omega_0$. All other states are assumed not to interact with the electromagnetic excitation and probing fields. This approach is well justified for most molecular systems [6.1].

ii) We assume that each molecular two-level system is only weakly interacting with the other molecules which are assumed to act as a fluctuating bath [6.2–6]. The Hamiltonian H of the molecule is written as a sum of an unperturbed Hamiltonian H_0 and two interaction Hamiltonians, the latter are: H_{im} which gives the interaction between the molecules and H_{ie} characterizing the interaction of the molecules with the electromagnetic fields [6.7,8]. The molecular interaction consists of a static contribution which leads to a constant frequency shift, and of a fluctuating contribution which modulates the resonance frequency and leads to a broadening of the transition band. The microscopic nature of the dynamic interaction may be understood in liquids in terms of "collisions" between individual molecules [6.2,4]. The collision process will be treated in detail in the context of Sect. 6.2.1. A major interaction in solids is caused by the thermal distribution of the acoustic phonons. In the macroscopic ensemble of two-level systems the dynamic interaction leads to relaxation time constants in the equation of motion (see below).

iii) The ensemble of two-level systems is described by the density matrix ρ with the diagonal elements ρ_{aa} and ρ_{bb} and the off-diagonal elements $\rho_{ab} = \rho_{ba}^*$. The density matrix follows the equation of motion

$$\frac{\partial\rho}{\partial t} = \frac{\mathrm{i}}{\hbar}[\rho, H] \ . \tag{6.1}$$

iv) The electromagnetic field is treated classically, since the number of photons is quite large in most coherent experiments. We work with plane waves, i.e.

the fields have the form: $E = E_0 \cos(\Omega t + kr)$ and consider electric dipole interactions with the Hamiltonian $H_{\text{ie}} = \hat{\mu}E(t)$. $\hat{\mu}$ is the component of the dipole operator along the direction of the electric field E with $\mu_{aa} = \mu_{bb} = 0$ and $\mu_{ba} = \mu_{ab} = \mu$.

The evolution of the electromagnetic fields is described by the wave equation

$$\Delta E - \frac{n^2}{c^2}\frac{d^2 E}{dt^2} = \frac{1}{\varepsilon_0 c^2}\frac{d^2 P^{\text{NL}}}{dt^2} \ . \tag{6.2}$$

The source term on the right-hand side of (6.2) describes the generation of light fields. The nonlinear polarization is proportional to the macroscopic dipole moment, $P = N\langle\hat{\mu}\rangle = N \times \text{Tr}\{\hat{\mu}\rho\} = N\mu(\rho_{ab} + \rho_{ba})$, where N is the number of molecules per unit volume. Thus the radiation intensity observed in the coherent experiment is proportional to $|N\,\text{Tr}\{\hat{\mu}\rho(t)\}|^2$.

6.1.2 Equations of Motion

Here we discuss the equations for the different components of the density matrix. The system of equations can be solved in a formal way by simple multiplication of matrices. In this way it is possible to readily see the salient features of the coherent spectroscopy. The discussion presented here follows a treatment given by *Hesselink* and *Wiersma* [6.9].

The equation of motion (6.1) is rewritten in (6.3–5) for the different density matrix elements. The electric field is coupled to the molecules via the interaction Hamiltonian $H_{\text{ie}} = -\hat{\mu}E(t)$. It is convenient to introduce $\tilde{\rho}_{ba} = \rho_{ba}\exp(\text{i}\Omega t)$ and to apply the rotating frame approximation.

$$\dot{\rho}_{aa} = \frac{\text{i}\mu E_0}{2\hbar}(\tilde{\rho}_{ba}\text{e}^{-\text{i}kr} - \tilde{\rho}_{ab}\text{e}^{\text{i}kr}) - \frac{\rho_{aa}}{T_1} \ , \tag{6.3}$$

$$\dot{\tilde{\rho}}_{ba} = \frac{\text{i}\mu E_0}{2\hbar}(\rho_{aa} - \rho_{bb})\text{e}^{-\text{i}kr} + \left(\text{i}\Delta - \frac{1}{T_2}\right)\tilde{\rho}_{ba} \ , \tag{6.4}$$

$$\dot{\rho}_{bb} = 1 - \rho_{aa} \ . \tag{6.5}$$

Δ determines the detuning between the driving frequency Ω and the molecular resonance frequency ω_0, i.e. $\Delta = (\omega_0 - \Omega)$. It is assumed in (6.3–5) that the molecular system consists only of two levels $|a\rangle$ and $|b\rangle$. The equations may be modified to describe other situations, e.g. when relaxation to a third level is relevant. In the latter case one has to extend (6.5).

The significance of the relaxation times T_1 and T_2 is readily seen, when the electric field is turned off, i.e. for $E_0 = 0$. The energy relaxation time T_1 describes the relaxation of the population of level $|a\rangle$ with $\rho_{aa}(t) = \rho_{aa}(0)\exp(-t/T_1)$. The phase relaxation time T_2 determines the decay of the dipole moment $P = N\mu(\rho_{ba} + \rho_{ab})$ with $P(t) = P(0)\exp(-t/T_2)$. There are three contributions which determine the dephasing of the system. (i) The macroscopic dipole moment P

may decay by orientational motion of the individual molecules (orientational relaxation time T_R); (ii) the molecules may "get out of step", i.e. lose their phase relation by interaction with the surroundings (pure dephasing time T_2^*); (iii) the population of the upper level may decay (energy relaxation time T_1). The observed effective dephasing time T_2 is related to the individual time constants as follows: $1/T_2 = 1/T_R + 1/T_2^* + 1/(2T_1)$.

The applied electric field excites the system to the upper level $|a\rangle$ via (6.3) and generates a polarization via (6.4). For the following discussion it is important to introduce the pulse area $A = \int \mu E/\hbar \, dt$ [6.10]. The magnitude of A frequently determines whether a certain nonlinear coherent process occurs or not, e.g. whether a photon echo may be observed. When the applied electric field is strong and the pulse area is large, $A \gg 2\pi$, repeated excitation and de-excitation processes occur, causing the population ρ_{bb} to oscillate with the Rabi frequency $\chi = \mu E/\hbar$. In most of the following investigations with short pulses one has $A < 1$.

Analytical solutions of (6.3–5) can be found for two limiting conditions:

i) Without an applied electric field the density matrix evolves freely. Only the relaxation processes influence $\rho(t)$. The four components of the density matrix may be asigned to a vector $\boldsymbol{\rho}$, as listed in the Appendix [Eq. (6.20)]. The density matrix at time t is obtained by a linear transformation from the initial density matrix $\boldsymbol{\rho}(0)$:

$$\rho_j(t) = \sum_{k=1}^{4} Y_{jk}(t)\rho_k(0) \quad \text{or} \quad \boldsymbol{\rho}(t) = \underline{\boldsymbol{Y}}(t)\boldsymbol{\rho}(0) \ . \tag{6.6a}$$

The transformation matrix contains exponentials of t/T_2 and t/T_1. It is given explicitly in the Appendix [Eq. (6.21)].

ii) When a resonant electric field of pulse area A is applied, which is short compared to the relaxation times T_1 and T_2, the density matrix at the end of the pulse is calculated by a linear transformation $\underline{\boldsymbol{X}}(A)$,

$$\boldsymbol{\rho}(A) = \underline{\boldsymbol{X}}(A)\boldsymbol{\rho}(0) \ . \tag{6.6b}$$

The elements of the transformation matrix $\underline{\boldsymbol{X}}(A)$ depend on the pulse area and the wave vector of the electric field. They are given in the Appendix, [Eq. (6.22)]. The important aspects of many coherent experiments can be seen from the analytical solutions. Experimentally one has periods of short excitations with pulse areas A_i, $i = a, b, \ldots$ and free evolution periods of durations $t_1, t_2, \ldots$ At the time of observation, t, after the final excitation pulse with area A_f, the density matrix is calculated as a product of the individual excitation and evolution matrices $\underline{\boldsymbol{X}}(A_i)$, $\underline{\boldsymbol{Y}}(t_i)$:

$$\boldsymbol{\rho}(t) = \underline{\boldsymbol{Y}}(t)\underline{\boldsymbol{X}}(A_f) \ldots \underline{\boldsymbol{Y}}(t_2)\underline{\boldsymbol{X}}(A_b)\underline{\boldsymbol{Y}}(t_1)\underline{\boldsymbol{X}}(A_a)\boldsymbol{\rho}(0) \ . \tag{6.7}$$

Detailed examples for the solution of (6.7) are given below in Sect. 6.1.3. Here we refer briefly to the situation depicted in Fig. 6.1c, where three excitation fields are applied. The density matrix at time t is as follows:

$$\rho(t) = \underline{Y}(t)\underline{X}(A_c)\underline{Y}(t_2)\underline{X}(A_b)\underline{Y}(t_1)\underline{X}(A_a)\rho(0) \ .$$

The treatment given above in (6.6,7) is well justified when the applied light fields are separated in time and the radiating polarization of the sample is small at all times t_i. In the more general case reference is made to the literature cited in Sect. 6.3.2.

6.1.3 Ultrafast Coherent Techniques

A short overview of coherent spectroscopy is given here. Two types of molecular systems are treated: System I contains molecules with a single resonance frequency ω_0 (homogenous broadening), while System II shows a distribution of resonance frequencies $f(\omega_0)$ (inhomogeneous broadening).

a) One Excitation Pulse

In the most elementary coherent experiment one short laser pulse traverses the sample. After the excitation of the system by the pulse with area A_a and wave vector k_a radiation is emitted from the sample in the direction k_a. From (6.7) one derives the polarization $P(t)$ observed at the time t after the excitation pulse $P(t) = \mu N[\rho_{ab}(t) + \rho_{ba}(t)]$:

$$P(t) \propto \int d\omega\, f(\omega) \mathrm{e}^{-\mathrm{i}\omega t} \mathrm{e}^{-t/T_2} \sin(A_a) \mathrm{e}^{-\mathrm{i}k_a r} \ . \tag{6.8}$$

A schematic of the decay of $P(t)$ is shown in Fig. 6.1a. The generated polarization has an amplitude proportional to sin (A_a); it has the same wave vector as the exciting light pulse. For the molecular system I with $f(\omega) = \delta(\omega - \omega_0)$

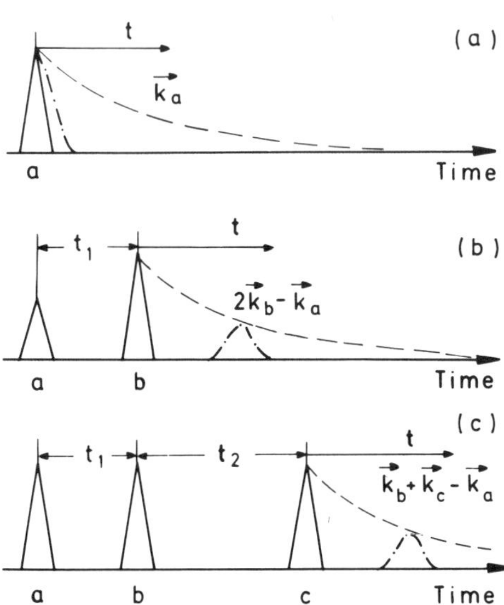

Fig. 6.1a–c. Schematic of various coherent techniques. The electric field pulses (*solid triangles*) excite the sample. The resulting coherent signal is shown for homogeneously broadened transitions (*dashed curves*), and for inhomogeneously broadened transitions (*dash-dotted curves*). **(a)** A single excitation pulse with wave vector k_a. Homogeneously broadened transitions lead to an exponential decay with dephasing time T_2, while inhomogeneous broadening induces a faster decay. **(b)** Two excitation pulses a and b with wave vectors $\boldsymbol{k}_a$ and $\boldsymbol{k}_b$ give echos in the direction $2\boldsymbol{k}_b - \boldsymbol{k}_a$. For inhomogeneously broadened transitions the echo is emitted at time $t = t_1$. The echo amplitude measured as a function of t_1 allows determination of the dephasing time T_2. **(c)** When three excitation pulses are used the coherent signal emitted in the direction $\boldsymbol{k}_b + \boldsymbol{k}_c - \boldsymbol{k}_a$ allows determination of the dephasing time T_2 and the energy relaxation time T_1

(broken line in Fig. 6.1a), the polarization decays exponentially with the dephasing time T_2. For a distribution of resonance frequencies $f(\omega_0)$ the decay of polarization starts exponentially, but accelerates due to the interference of the different resonance frequencies (dash-dotted line in Fig. 6.1a). With increasing width of the frequency distribution, the decay becomes more rapid and it is difficult – often impossible – to determine the dephasing time T_2. The explicit mathematical solution for the signal decay is readily obtained from (6.8). We note that the polarization is proportional to the product of the Fourier transform $g(t)$ of the distribution function $f(\omega_0)$ of the resonance frequencies times the exponential decay with $\exp(-t/T_2)$.

Investigations with single-pulse excitation are well established in ultrafast coherent spectroscopy. A series of interesting experiments are discussed in Sect. 6.2.

b) Two Excitation Pulses

When two excitation pulses of areas A_a and A_b separated by the evolution time t_1 (see Fig. 6.1b) are applied to the (molecular) system, a more complex time dependence of the coherent signal or of the density matrix is found. At time t after the second pulse, the radiating polarization consists of three contributions: two parts are independently produced by the two exciting pulses, radiating in the corresponding directions k_a and k_b. The two polarizations are described by (6.8). Of special interest is a third term, where the polarization evolves as follows:

$$P_{2\text{PE}}(t) \propto \int d\omega\, \underline{f(\omega)\exp\left[-\mathrm{i}\omega(t-t_1)\right.} - \left.\frac{t+t_1}{T_2}\right] \times \sin A_a \sin^2\frac{A_b}{2}\exp[-\mathrm{i}(2k_b-k_a)r] \; . \tag{6.9}$$

The signal is emitted with the wave vector $2k_b - k_a$ and may readily be separated from the exciting beams. For a homogeneously broadened transition (System I), the peak of the polarization is proportional to $\exp(-t_1/T_2)$. The polarization decays at later times with $\exp(t/T_2)$ (see Fig. 6.1b, broken line). Quite different is the situation for strong inhomogeneous broadening (System II). Now the (underlined) interference term in (6.9) is removed at one delay time, $t = t_1$, and a delayed signal appears, i.e. one finds a photon echo with an amplitude proportional to $\exp(-2t_1/T_2)$ [6.11–13]. Variation of t_1, i.e. of the distance between the two exciting pulses, therefore enables one to deduce the dephasing time T_2 even for inhomogeneously broadened transitions [6.13]. (For the intermediate case with small inhomogeneous broadening see [6.14]).

An important aspect for the application of the photon echo technique is the dependence of the signal on the area of the excitation pulses. Good echo signals are generated for $A_a \simeq \pi/2$ and $A_b \simeq \pi$. For small areas A_a, $A_b \ll \pi$ the signal amplitude drops with $A_a A_b^2$ or the signal intensity with $(A_a A_b^2)^2$. We recall that in the case of a single excitation pulse the emitted signal intensity is proportional to A_a^2. Consequently, coherent echo experiments become more difficult when the transitions are weak and the pulses and relaxation times are very short.

c) Three Excitation Pulses

When three excitation pulses with areas A_a, A_b, and A_c separated in time by t_1 and t_2 are applied to the molecular system, a variety of coherent signals is produced: single excitation signals, two-pulse echos from each pair of the three pulses, and so-called three-pulse echos at the time $t = t_1$ after the third pulse, the latter echos are emitted in the direction $k_b + k_c - k_a$ (see Fig. 6.1c). The introduction of the third pulse allows one to measure the dephasing time T_2 of the molecular transition *and* to determine the energy relaxation time T_1. The application of three pulses gives considerable flexibility to the design of experiments. Similar limitations regarding the signal strength exist as in two-pulse experiments; the three-pulse echo intensities are proportional to $(\sin A_a \sin A_b \sin A_c)^2$. For small area pulse the echo signal decreases proportion to $A_1^2 A_2^2 A_3^2$. Experiments with ultrafast three-pulse echos have been performed for a number of electronic transitions. These are discussed in Sect. 6.3.2.

6.1.4 Electric Dipole and Raman-Type Transitions

The interaction between the system of interest and the light field depends on the specific type of interaction Hamiltonian H_{ie}. We address here two situations, the electric dipole and the Raman-type transitions. The common aspect of the (molecular) excitation is best demonstrated by introducing the operator $\hat{q}$ for the excitation coordinate (of the molecules). In the case of a vibrational transition, $\hat{q}$ represents the coordinate of the nuclear motion of the vibrating molecules. For an electronic transition, $\hat{q}$ is related to the motion of the electrons. For an ensemble of two-level systems the expectation value $\langle \hat{q} \rangle$, called the coherent or collective amplitude, is given by:

$$\langle \hat{q} \rangle = \mathrm{Tr}\{\rho \hat{q}\} = q(\rho_{ab} + \rho_{ba}) \ , \tag{6.10}$$

where $q_{ab} = q = \langle a | \hat{q} | b \rangle$ is the transition matrix element.

a) Electric Dipole Transitions

Electric dipole transitions are commonly investigated with light frequencies close to a resonance frequency. In this case the interaction cross sections are large and the polarizations resulting from the excitation process generate new electromagnetic waves.

The electric dipole moment $\boldsymbol{\mu}$ related to the transition between the levels $|a\rangle$ and $|b\rangle$ (of one molecule) is a vector; its direction (unit vector $\boldsymbol{u}$) is fixed in the reference frame of the molecule and its amplitude is proportional to the amplitude of the coordinate q: $\boldsymbol{\mu} = \boldsymbol{u}\, d\mu/dq\, q$[6.15,16]. The interaction of the dipole moment with an electric field is given by the Hamiltonian $H_{ie} = -\boldsymbol{\mu}\boldsymbol{E} = d\mu/dq\, qE \cos\theta$, where θ is the angle between the dipole moment and the electric field. For an ensemble of molecules the relaxation of the total dipole moment $\langle\boldsymbol{\mu}\rangle$ as a function of time depends on (i) the orientational relaxation of the vector $\boldsymbol{u}$ (irrelevant in most solids) and (ii) the relaxation of the absolute value of the coherent amplitude $\langle\hat{q}\rangle$. This topic is the subject of intensive investigations discussed in this chapter.

It is of interest to compare the quantities measured in time-resolved coherent investigations and in steady-state spontaneous spectroscopy, i.e. the time dependence of the expectation values $\langle\hat{q}\rangle$ or $\langle\boldsymbol{\mu}\rangle$ with the absorption spectra [6.3,15]. The latter can be calculated from (6.3–5), when the left-hand side of the equations, i.e. the time-dependent terms, are equal to zero. One readily finds the absorption line to be of Lorentzian shape with a peak absorption cross section $\sigma = (d\mu/dq)^2 \omega T_2/(\varepsilon_0 cn\hbar)$ and a linewidth (FWHM) of $\Delta\nu = 1/\pi T_2$. Here ε_0 is the permittivity, n the refractive index, c the velocity of light, and $\hbar$ Planck's constant.

The absorption spectrum $J(\omega)$ may be related to the time-dependent dipole moment in a general way without referring to the special model introduced above. Line-shape theory gives [6.2–6,15,16]:

$$J(\omega) \propto \int_{-\infty}^{+\infty} e^{i\omega t}\phi(t)\, dt = \int_{-\infty}^{+\infty} e^{i\omega t}\langle\langle\mu(t)\mu(0)\rangle\rangle\, dt \ , \tag{6.11}$$

where $\langle\langle\cdots\rangle\rangle$ denotes the equilibrium ensemble average. According to (6.11) the spectral line shape is the Fourier transform of the correlation function $\phi(t)$, i.e. of the time dependence of $\mu(t)$, which is measured in straightforward coherent experiments [see Sect. 6.1.3(a)]. Introducing an exponential decay with time constant T_2 for the dipole moment in (6.11) one obtains the same Lorentzian line shape as determined above directly from (6.3–5).

b) Raman-Type Transitions

Raman type transitions have the advantage that excitation is possible with a wide range of available electromagnetic fields, i.e. resonance frequencies are not required. On the other hand, the Raman interaction is a weak, second-order process; the pulse area and the degree of excitation are small. In spite of this difficulty, ultrafast Raman spectroscopy has provided valuable new information. Direct observation of several rapid dynamic processes has become possible and superior spectral resolution was demonstrated in congested spectral regions.

According to *Placzek* [6.17], the interaction between a given vibrational mode and the electric field can be described by the interaction Hamiltonian $H_{\mathrm{ie}} = (\partial\underline{\boldsymbol{\alpha}}/\partial q)q\boldsymbol{E}\boldsymbol{E}/2$, where $d\underline{\boldsymbol{\alpha}}/dq$ stands for the change of the electric polarizability tensor with the vibrational coordinate q. It leads to the excitation of the molecular mode (frequency ω_0) under the action of the light field E, provided that components i, j of the electric field exist, where the product $E_i E_j^*$ oscillates at the frequency ω_0. The Raman process may again be described by (6.3–5), if we replace the quantity μE_0 by $(\partial \underline{\boldsymbol{\alpha}}/\partial q) q_{ab} E_i E_j/2$. In a similar way, the radiating polarization is obtained as

$$P_i = N\mu_i = N \frac{\partial \boldsymbol{\alpha}_{ij}}{\partial q} E_j \langle \hat{q} \rangle \ . \tag{6.12}$$

The polarization is proportional to the electric field $\boldsymbol{E}$, i.e. the molecular excitation $\langle q \rangle$ leads to a scattered light field when monitored by a probing field. Without a probing field the coherent amplitude $\langle \hat{q} \rangle$ does not produce electromagnetic radiation. When the transition is electric dipole and Raman active, some light emission may occur, provided the molecules are in a crystal without inversion symmetry. This case will not be treated here. For isotropic Raman transitions the tensor $\partial \underline{\boldsymbol{\alpha}}/\partial q$ is reduced to a scalar quantity $\partial \alpha/\partial q$. In this case, rotational motion of the molecules does not effect the evolution of the coherent signal and the decay of the coherent amplitude $\langle q \rangle$. The more complex case of an anisotropic Raman tensor has been treated in the literature by *Laubereau* and *Kaiser* [6.7], by *Kohles* and *Laubereau* [6.18], and by *Dick* [6.19]; it will not be discussed here in detail.

Comparing time-resolved coherent and spontaneous Raman scattering, one obtains very similar equations as in the case of electric dipole transitions. The spontaneous Raman line shape can again be calculated from (6.3–5) or from the line-shape theory [6.2–6]. The spectral shape is the Fourier transform of a molecular correlation function; for isotropic Raman lines it is proportional to the Fourier transform of $\langle\langle q(t) q(0) \rangle\rangle$.

c) Pulse Area and Population Changes in Ultrafast Coherent Experiments

For practical experiments it is important to know how strongly the (molecular) system is excited under the action of a light pulse. The strength of the excitation determines which type of coherent experiment can be performed with given light pulses. The important quantity is the pulse area A which is directly related to the changes of the population of the various levels [see (6.6,22)]; e.g. for a system initially in the ground state a short pulse with area A changes the ground-state population by $\Delta\rho_{bb} = (1 - \cos A)/2$. Using an exciting pulse with peak intensity I, duration t_p, and absorption cross section σ one estimates the pulse area to be $A = \mu E_0 t_\mathrm{p}/\hbar = (2n^2 I T_2 \sigma/\hbar\omega_0)^{1/2}$. We assume here that the duration of the pulses is equal to the dephasing time. We now give numbers for the pulse area for three relevant examples using the following parameters: dephasing time $T_2 = 10^{-12}$ s, intensity $I = 10^{10}$ W/cm^2, refractive index $n = 1.5$.

i) For an infrared-active vibrational CH-stretching mode at $\lambda = 3\,\mu$m with $\sigma = 10^{-19}$ cm^2 one obtains $A = 2.6 \times 10^{-1}$ and $\Delta\rho_{bb} = 1.6 \times 10^{-2}$.

ii) For an electronic transition at $\lambda = 0.5\,\mu$m with $\sigma = 10^{-16}$ cm^2 one calculates $A = 3.4$ or $\Delta\rho_{bb} = 0.99$. The two examples show that electronic transitions may lead to large pulse areas making coherent experiments quite easy, while for vibrational transitions the observation of two- and three-pulse echos with echo amplitudes proportional to A^6 is difficult.

iii) For Raman active transitions the pulse area $A = 1/2(\partial\alpha/\partial q)q_{ab}t_p E^2/\hbar$ may be calculated using the definitions of $\partial\alpha/\partial q$ and q_{ab} from the literature [6.8]. One obtains $A = In\lambda^2 t_p \sqrt{\delta\sigma}/(\pi c\hbar)$, where $\delta\sigma$ is the Raman cross section integrated over one Steradian. Taking $\delta\sigma = 2.5 \times 10^{-29}$ from benzene [6.10] (which has a rather large Raman cross section for a molecular vibration) and $\lambda = 0.5\,\mu\text{m}$ we compute an area A of $A = 1.9 \times 10^{-2}$ or a population change $\Delta\rho_{bb} \simeq 10^{-4}$. These numbers are even smaller than the values calculated for resonant infrared absorption, see (i). It is evident that under the present conditions, ultrafast Raman-echo experiments appear to be unrealistic, while single-pulse coherent Raman experiments are readily possible. It should be noted that considerably higher cross sections are found for resonant Raman transitions. Only under these conditions might ultrafast Raman-echo experiments become possible.

In this section a set of equations was introduced which allows one to treat time-resolved coherent spectroscopy within the framework of the model presented in Sect. 6.1.1. We have briefly outlined and compared different experimental methods for which applications are discussed in the following sections.

6.2 Coherent Spectroscopy Using a Single Excitation Process

We focus here on investigations of time-resolved coherent spectroscopy using single excitation processes. It is the aim of this section to demonstrate the different time-resolved techniques and to show the advantages of coherent techniques over spontaneous frequency-domain methods.

6.2.1 Time-Resolved Coherent Raman Spectroscopy of a Single Homogeneously Broadened Transition

The experiments of time-resolved coherent Raman spectroscopy are performed in two steps. First, the sample is excited by a pair of light pulses via transient stimulated Raman scattering (see Fig. 6.2, upper part) [6.20–23]. Subsequently, the degree of excitation is interrogated by a probe pulse, which is delayed in time relative to the excitation pulses. The probe pulse monitors the coherent amplitude by Raman scattering [6.1,7,8]. The experiments are described by two different sets of equations: the material equation governing the generation of the excitation under the action of the light fields [see (6.3–5)] and the wave equation (6.2) determining the light field due to the interaction with the material excitation. We recall that in coherent Raman processes the pulse areas are quite small [see Sect. 6.1.4(c)]. For this reason the changes of the diagonal elements of the density matrix are small and the off-diagonal elements may be calculated by using only

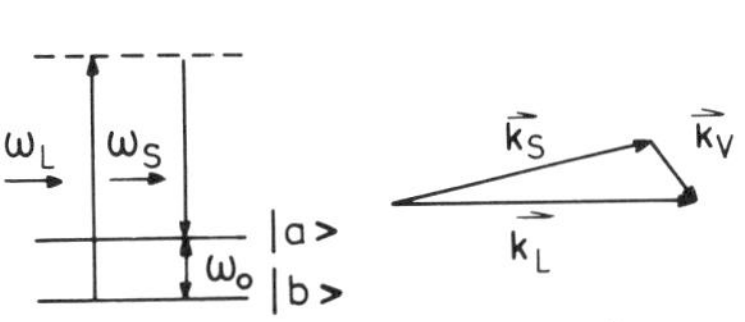

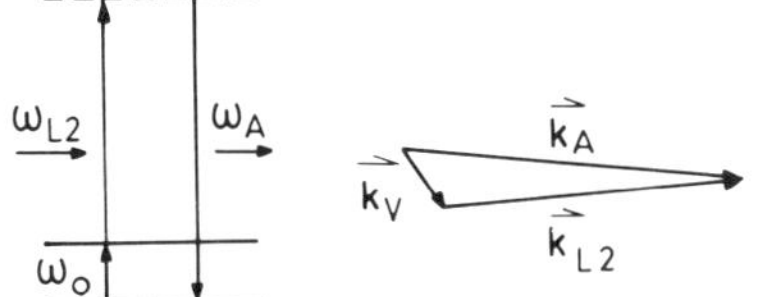

Fig. 6.2. Schematic of a time-resolved coherent Raman experiment. The excitation of the upper level $|a\rangle$ is accomplished via stimulated Raman scattering by the laser and Stokes pulses at frequencies ω_L and ω_s, respectively. The wave vectors of the two laser fields determine the wave vector of the coherent excitation, $\boldsymbol{k}_v = \boldsymbol{k}_L - \boldsymbol{k}_s$ (*upper part*). At a later time the coherent Raman probing process (*lower part*) with a second laser pulse E_{L2} $(\omega_{L2}, \boldsymbol{k}_{L2})$ monitors the coherent excitation. The coherent anti-Stokes signal emitted under phase matching conditions $\boldsymbol{k}_a = \boldsymbol{k}_{L2} - \boldsymbol{k}_v$ is a measure of the coherent amplitude at the probing time

(6.4). For Raman-type interactions (6.4) may be solved by direct integration. The electric field producing excitation consists of two components:

$$E = E_L \cos(-\omega_L t + k_L x) + E_s \cos(-\omega_s t + k_s x) \ ,$$

and for the coherent material excitation one makes the ansatz of a plane wave

$$\langle q \rangle = \frac{\mathrm{i}}{2} Q \exp(-\mathrm{i}\omega_0 t + k_v x) + \text{c.c.} = q_{ab}(\rho_{ab} + \rho_{ba}) \ .$$

With these expressions (6.4) yields the amplitude Q after one integration [6.1,7,8,24]:

$$Q(x,t) = \frac{\mathrm{i} q_{ab}^2}{2\hbar} \frac{\partial \alpha}{\partial q} \int_{-\infty}^{t} E_L(t',x) E_s^*(t',x) \exp\left[\frac{1}{T_2}(t'-t) + \mathrm{i}\Delta t'\right] dt' \ . \tag{6.13}$$

The excitation process is shown schematically in Fig. 6.2, upper part. The two components of the electric field at frequency ω_L and ω_s drive the vibration with frequency $\Omega = \omega_L - \omega_s$ close to the resonance frequency ω_0. The wave vector of the coherent amplitude k_v is determined by $k_v = k_L - k_s$. Q shows the properties discussed previously: a rapid rise with the driving force $E_L E_s^*$ and an exponential decay with $\exp(-t/T_2)$ of the freely vibrating system. The 'back' reaction of the coherent amplitude on the light fields is governed by the polarization $\boldsymbol{P}$:

$$\boldsymbol{P} = N \frac{\partial \alpha}{\partial q} \langle q \rangle \boldsymbol{E} + \underline{\chi}_{\mathrm{NR}}^{(3)} \varepsilon_0 \boldsymbol{EEE} \ . \tag{6.14}$$

The coupling of the wave equation and the material equation via the polarization leads to stimulated Raman scattering or Stokes/anti-Stokes coupling, which have been discussed in detail elsewhere [6.8,20–23,25–28]. The nonresonant nonlinear polarization $\chi_{\text{NR}}^{(3)}$ is due to distant electronic transitions. It leads to self-phase-modulation and nonresonant scattering [6.10,28].

For time-resolved coherent Raman spectroscopy a probing pulse is necessary to obtain information on the time dependence of the coherent amplitude (see Fig. 6.2, lower part). The third light pulse $E_{\text{L2}} = E_{\text{L2}}(t)\cos(-\omega_{\text{L2}}t + k_{\text{L2}}x)$ generates a radiative polarization according to (6.14). A short probing pulse at variable time delay t_{D} after the excitation pulses allows one to map out the time dependence of the coherent amplitude. The scattered signal at the anti-Stokes frequency $\omega_{\text{A}} = \omega_{\text{L2}} + \omega_0$ or at the Stokes frequency $\omega_{\text{s2}} = \omega_{\text{L2}} - \omega_0$ is observed. Efficient generation of the coherent signal requires phase matching, i.e. the wave vectors of the material excitation $\boldsymbol{k}_{\text{v}}$, the probing field $\boldsymbol{k}_{\text{L2}}$, and generated light, e.g. $\boldsymbol{k}_{\text{A}}$, have to satisfy the condition $\boldsymbol{k}_{\text{v}} + \boldsymbol{k}_{\text{L2}} = \boldsymbol{k}_{\text{A}}$.

a) Experimental Arrangements

A typical experimental system is depicted schematically in Fig. 6.3. A pumping laser source generates exciting and probing light fields at the frequencies ω_{L}, ω_{s}, and ω_{L2} in several frequency converters. In early experiments a simple optical system was used: the Stokes pulse was generated via stimulated Raman scattering of the strong laser pulse in the sample itself [6.29,30]. More recently, frequency conversion by dye lasers produced tunable Stokes pulses [6.31–34]. The exciting

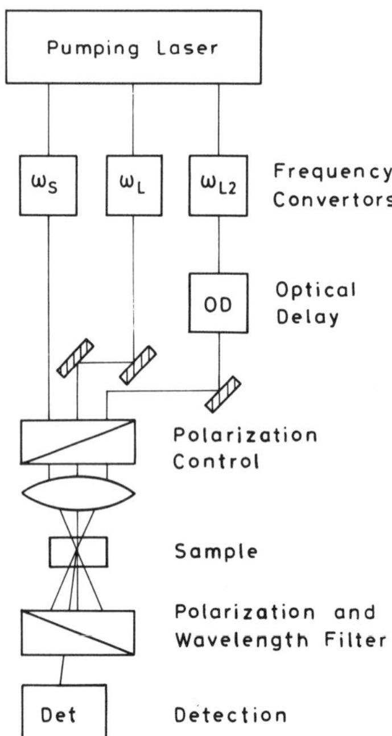

Fig. 6.3. Experimental system used for time-resolved coherent Raman scattering. Synchronized pulses at frequencies ω_{L}, ω_{s}, and ω_{L2} are generated by different frequency converters from one pumping laser source. The pulses at ω_{L} and ω_{s} excite the sample via stimulated Raman scattering. The coherent excitation is monitored by the delayed (optical delay OD) probing pulses at frequency ω_{L2}. The different beams cross in the sample at phase-matching angles. Polarization and wavelength filters separate the coherent signal from the excitation light

Stokes and laser pulses are focused at a specific angle into the sample generating the coherent material excitation. The probing light pulse at frequency ω_{L2} is delayed relative to the excitation process in the optical delay system OD and monitors the excited volume of the sample. The geometry of the exciting and probing beams in the sample have to be adjusted to obtain phase matching for the coherent probe-scattering process. In most experiments the signal intensities are much smaller than the intensities of the excitation pulses. Consequently, the coherently scattered light must be separated carefully from the incoming light beams by diaphragms and spectral and/or polarization filtering. The detection system is a highly sensitive photo-detector. Frequently, the spectra of the coherently scattered light are analyzed with the help of a spectrograph.

b) Dephasing in Liquids

A straightforward application of time-resolved coherent Raman spectroscopy on the picosecond time scale is the investigation of single Raman active vibrational transitions. Such an experiment was performed for the first time by *von der Linde* et al. [6.29] in 1971. The authors observed the coherent anti-Stokes signal for ethanol (C_2H_5OH) and carbontetrachloride (CCl_4) on the picosecond time scale.

A large number of other molecules and transitions have been studied in the meantime [6.32–43]. The state of the art of coherent measurements with high time resolution is demonstrated in two recent publications by *Gale* et al. [6.42] and by *Zinth* et al. [6.44]. In the first experiment, a Nd:glass laser system supplied light pulses with relatively long duration ($t_p = 5$ ps). The high time resolution was achieved by a specially designed excitation and probing system that allows the signal decay to be measured over 9 orders of magnitude. With the nearly Gaussian-shaped pulses from the Nd:glass laser, the decay of the response function (see broken curve in Fig. 6.4) accelerates with decreasing signal at later delay times. As a consequence, the favorably high signal-to-noise ratio of the experiment allowed the measurement of decay times considerably shorter than the pulse duration. In Fig. 6.4a the solid curve corresponding to the ν_1 mode of acetonitrile at $\nu/c = 2943\,\mathrm{cm}^{-1}$ shows an exponential decay over the entire observation range. This fact clearly demonstrates that the vibrational transition is homogeneously broadened with a dephasing time $T_2 = 1.63 \pm 0.07$ ps.

Superior time resolution is achieved with very short light pulses. *Zinth* et al. [6.44] introduced an experimental system with femtosecond pulses ($t_p \simeq 80$ fs) from a ring dye laser. A recent result is presented in Figs. 6.4b and 6.5. In Fig. 6.4b the same acetonitrile mode was studied as in Fig. 6.4a. Using femtosecond pulses the coherent signal is well separated from the autocorrelation trace, just 100 fs after time zero. Consequently, the decay time is determined with very high precision to be $T_2 = 1.70 \pm 0.02$ ps. In Fig. 6.5a and b the coherent signal decays are shown for the ν_2 and ν_7 modes of liquid acetone. In both cases one finds for delay times larger than 200 fs a clear exponential decay. The time constants are 510 ± 30 fs (ν_7 mode) and 305 ± 10 fs (ν_2 mode); i.e. one obtains

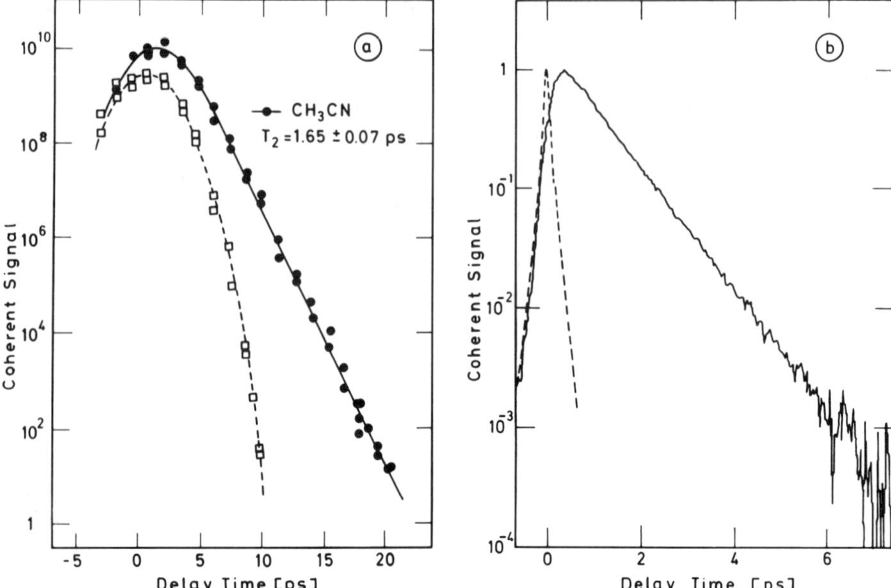

Fig. 6.4a, b. Coherent anti-Stokes signal for the ν_1 mode of acetonitrile plotted as a function of the time delay of the probing pulses. (**a**) Laser pulses of 5 ps duration were used. The signal is recorded over nine orders of magnitude. •: Coherent response of acetonitrile CH_3CN. From the exponential decay a dephasing time $T_2 = 1.65 \pm 0.07$ ps is determined. o: Response function of the experimental system measured by coherent scattering in C_2H_5OH. (Redrawn from *Gale* et al. [6.42]). (**b**) Laser pulses of 80 fs duration were used. A ready separation between coherent signal and autocorrelation trace starts within 100 fs. The dephasing time is determined to be $T_2 = 1.70 \pm 0.02$ ps [6.44b]

dephasing times of $T_2 = 1020 \pm 60$ fs and 610 ± 20 fs. At shorter delay times the signal deviates from the pure exponential decay. In Fig. 6.5a this effect is mainly due to the nonresonant susceptibility. There may also be some residual influence from collisional processes (see below) which are not included in (6.3–5).

The experiments on time-resolved coherent Raman scattering have initiated a number of theoretical investigations treating dephasing and energy relaxation of vibrational levels in molecular liquids. *Fischer* and *Laubereau* [6.2] have estimated relaxation times assuming a binary collision between the molecules with exponential intermolecular interaction potential. Despite its several simplifying assumptions this model is capable of predicting dephasing times in liquids over a range of parameters such as temperature, concentration, and viscosity. A number of other theoretical approaches, e.g. molecular dynamics simulations, correlation function modelling, or hydrodynamic models have been discussed in the literature [6.45–60]. For a review see *Oxtoby* [6.3].

In many coherent experiments the initial signal is strongly affected by the nonresonant susceptibility $\chi_{NR}^{(3)}$ deviating substantially from the single exponential decay discussed theoretically (for an example see Fig. 6.5 or 6.6). The influence

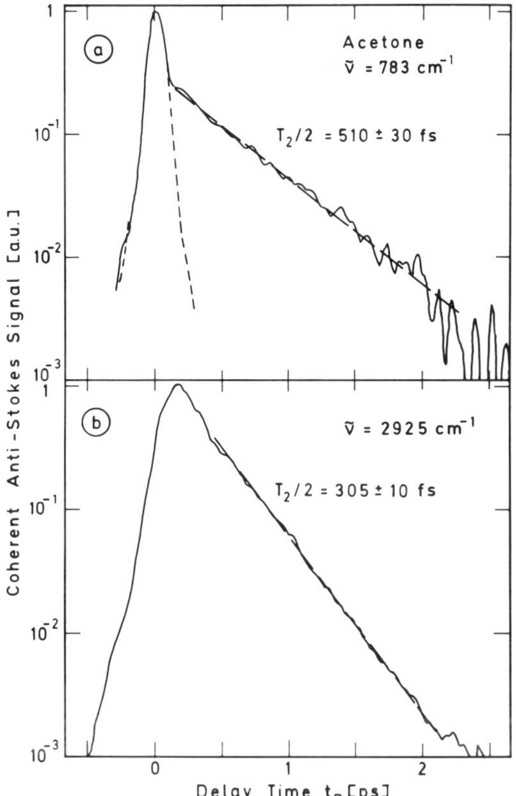

Fig. 6.5a, b. Coherent signal from aceton measured with femtosecond laser pulses. The ν_7 (**a**) and ν_2 (**b**) modes are studied. Exponential decay with time constants $T_2/2 = 510 \pm 30$ fs (ν_7) and 305 ± 10 fs (ν_2) are found at later delay times. The signal shape in (**a**) around time zero is influenced by the nonresonant susceptibility $\chi^{(3)}_{NR}$. The dashed curve shows the response function of the experiment measured with the help of sum-frequency generation in a nonlinear KDP crystal. The narrow response function clearly demonstrates the power of the experimental system to measure very short decay times [6.44]

of the nonresonant susceptibility $\chi^{(3)}_{NR}$ on time-domain coherent Raman experiments was investigated for the first time by *Zinth* et al. [6.61] in mixtures, where molecular modes of highly diluted molecules were studied. Figure 6.6 shows the data for the ν_1 mode of liquid CCl_4 (9 vol.%) dissolved in cyclohexane, C_6H_{12}, (experimental system Nd:glass laser with $t_p = 7$ ps). The peak of the scattered Stokes signal is close to $t_D = 0$, where the driving force ($E_L E_s^*$) overlaps with the probing light pulse. At first the signal decays very rapidly by a factor of 50; it subsequently decays more slowly with a time constant of $T_2/2 = 3.6$ ps (broken line). This time dependence is readily understood by taking into account the nonresonant susceptibility $\chi^{(3)}_{NR}$ of the system [see (6.14)]. The nonresonant susceptibility induces an additional signal close to time zero due to a polarization proportional to $\chi^{(3)}_{NR} E_L E_s E_{L2}$. This part of the signal has the time dependence of the experimental response function. At later delay times, the slower decay due to the resonantly excited solute molecules (CCl_4) takes over and determines the decay curve. Under the existing experimental conditions the later decay is close to an exponential (further details are given together with the discussion of results on pure CCl_4 in Sect. 6.2.2). Comparing the experimental signal curves with the

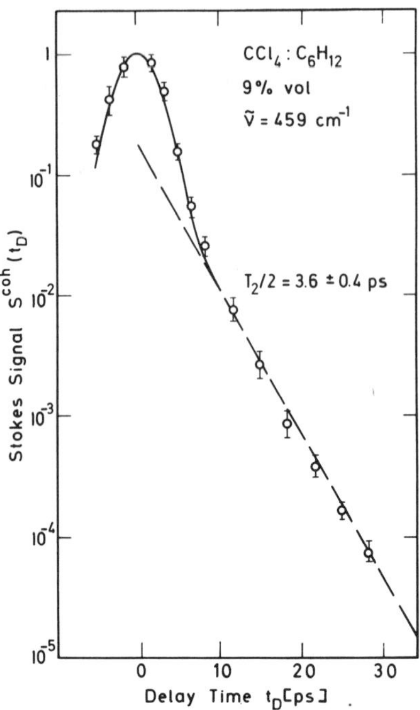

Fig. 6.6. Coherent signal from a mixture of CCl_4 and C_6H_{12}. Here, the effect of the nonresonant susceptibility $\chi_{NR}^{(3)}$ in a time resolved Raman experiments was demonstrated for the first time. While the exponential decay at later delay times is due to the resonant coherent excitation of the CCl_4 molecules, the signal peak at time zero originates from the nonresonant susceptibility $\chi_{NR}^{(3)}$ of the solvent molecules C_6H_{12} [6.61]

theory of (6.3–5) and (6.13,14) allows one to deduce the ratio of the nonresonant to the resonant susceptibility in the mixture.

The experimental results show that the resonant and the nonresonant part of the nonlinear susceptibility may be separated in a time-domain experiment. While at time zero, both the nonresonant and the resonant part contribute to the signal, the measurement at later delay time gives access to the resonant part of the susceptibility. This separation allows frequency-domain coherent anti-Stokes Raman experiments (CARS) without the disturbing influences of $\chi_{NR}^{(3)}$; picosecond light pulses and delayed probing with tuning of the frequency difference $\omega_L - \omega_s$ are required (see [6.61–64]).

In the experiment of Fig. 6.6 the strong Raman band of CCl_4 at $\tilde{\nu} = 459\,\mathrm{cm}^{-1}$ was investigated and the influence of the nonresonant susceptibility of the solvent cyclohexane was evident because of the high dilution of the CCl_4 molecules, which reduced the resonant contribution to the signal. There are many experiments, where $\chi_{NR}^{(3)}$ contributes to the probe signal even in concentrated samples. This situation occurs, when (i) weak Raman transitions are investigated, (ii) very strong nonresonant susceptibilities exist, as e.g. in semiconductors, or (iii) the excitation process is very transient, i.e. when the duration of the excitation pulses is much shorter than the dephasing time T_2. The latter case was realized under the experimental conditions of Fig. 6.5a, where the excitation pulses of $t_p = 70\,\mathrm{fs}$ were considerably shorter than the dephasing time $T_2 = 1\,\mathrm{ps}$.

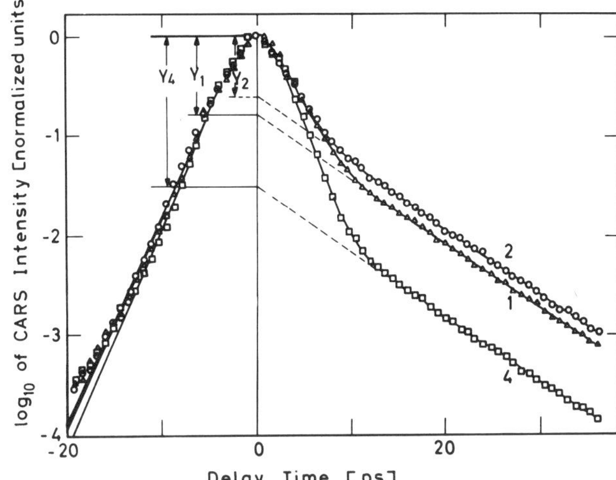

Fig. 6.7. Time-resolved CARS signal for LO-phonons in GaP measured for different polarization conditions showing the influence of the nonresonant susceptibility $\chi^{(3)}_{\mathrm{NR}}$ (around time zero) and the exponential decay of the resonant signal at later delay times [6.65]

An interesting example, where a time-domain experiment allows determination of the components of the nonresonant susceptibility tensor $\chi^{(3)}_{\mathrm{NR}}$ in crystalline gallium phosphate, GaP, is presented in Fig. 6.7 [6.65]. The coherent anti-Stokes signal was recorded for three different polarization conditions of excitation and probing fields (curves 1, 2, and 4 in Fig. 6.7). The slow exponential decay at later times corresponds to the LO-Raman resonance of GaP at $403\,\mathrm{cm}^{-1}$. Each configuration of polarization of excitation and probing light fields yields the same dephasing time for the LO-phonons, but a different ratio between nonresonant and resonant part of the signal (represented by the values Y_i obtained after extrapolation of the resonant part back to time zero). A complete analysis of the data of Fig. 6.7 together with measurements at other polarization configurations has provided numbers for the components of the nonlinear susceptibility tensor $\chi^{(3)}_{\mathrm{NR}}$.

c) Vibrational Transitions in Solids

Time-resolved coherent Raman scattering supplies valuable dynamical information on vibrations (phonons) in the solid phase at low temperatures. Here the relaxation times are often long, e.g. $T_2 \simeq 10^{-10}$ s, and the study of the corresponding line shapes (width $\Delta\nu/c \sim 0.1\,\mathrm{cm}^{-1}$) via spontaneous Raman scattering becomes difficult. On the other hand, time-resolved coherent experiments are readily made in this time domain. Due to the well-defined dispersion relations $\omega(k)$ of the phonons in crystalline solids, the time-resolved experiments require special care to properly adjust the polarization and propagation conditions to the symmetry of the investigated modes (see e.g. *Giordmaine* and *Kaiser* [6.1] and *Velsko* et al. [6.66]).

The relaxation time T_2 of vibrational modes in crystals has been treated theoretically in a number of publications [6.9,56,67–79]. It could be shown that the pure dephasing times increase with decreasing temperature proportional to $(T_D/T)^7$ for $T \ll T_D$, where T_D is the Debye temperature [6.67]. On the other hand, the energy relaxation time T_1 approaches a finite value at low temperatures and therefore determines the total phase relaxation $1/T_2 = 1/(2T_1)$ for $T \to 0$.

The equivalence of the coherent relaxation time τ to the energy relaxation times T_1 at low temperatures has been shown experimentally for molecular crystals, e.g. by echo experiments for pentacene in naphthalene and by measuring time-resolved spontaneous anti-Stokes Raman scattering for non-thermal LO-phonons in GaAs at 77 K [6.9,80,81].

Time-domain coherent Raman investigations exist for a number of crystals. In an early investigation the TO lattice mode of diamond was found to have a relaxation time $T_2/2$ of 3.4 ps and 2.9 ps at 77 K and 295 K, respectively [6.82]. Another example is given in Fig. 6.7, where GaP is investigated. The slow exponential decay of the coherent signal allows one to deduce the lifetime of the LO-phonon mode of GaP at 403 cm^{-1} A value of $T_2/2 = 6.7 \pm 0.3$ ps was reported [6.33]. In a recent paper, *Bron* et al. [6.83] studied the temperature dependence of T_2 of the LO-phonons of GaP and ZnSe by time-resolved coherent Raman scattering. The authors interpret their data as follows: At low temperatures the LO-phonons with wave vector $|k| \simeq 0$ decay into two acoustic phonons with half the energy and with equal wave vectors q_i of opposite sign, $q_1 = -q_2$. Impurity scattering was not relevant in the presence of nitrogen impurities of the order of 10^{16} cm^{-3} in the GaP crystal. For elevated temperatures $T > 150$ K, higher-order phonon-phonon interactions begin to play a role.

The internal A_1 mode of α-quartz at $\nu/c = 465$ cm^{-1} was investigated over a wide temperature range by *Gale* and *Laubereau* [6.84]. The experiments yielded a pronounced temperature dependence of the relaxation times, e.g. time constants of $T_2/2 = 0.8$ and 3.4 ps were found at 295 and 23 K, respectively. The phonon relaxation in α-quartz was discussed in terms of a three-phonon process, where the population of the excited 465 cm^{-1} phonon decays, generating an acoustic phonon at $\nu_1 = 51$ cm^{-1} and an optical phonon at $\nu_2 = 414$ cm^{-1} The temperature dependence of the decay time T_2 was successfully described by the relation $2/T_2 = \gamma[1 + n_1(T) + n_2(T)]$, where n_1 and n_2 are the thermal occupation numbers of the created phonons, γ is the low-temperature relaxation rate and T the temperature.

Besides the examples given above numerous publications have focused on other inorganic crystals. They have treated, for example, the dephasing of one-phonon states in calcite [6.30,85,86], of polaritons in ammonium chloride [6.87], or of two-phonon states [6.88–90].

A number of papers have addressed vibrational modes in organic molecular crystals [6.91–102]. In these materials the transition frequencies are often similar to those in molecular liquids. When several molecules are in the elementary cell, a splitting of the transition lines may occur due to the different local symmetry

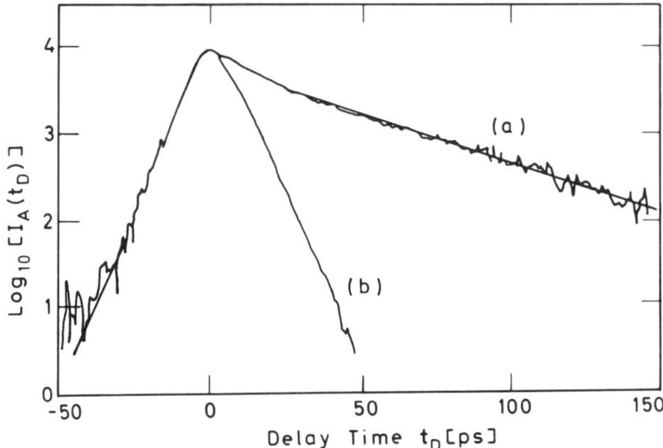

Fig. 6.8. Time-resolved CARS signal decay for the ν_1 mode, A_g factor group of crystallized benzene of natural isotope composition at 1.6 K. (*a*) Measured signal and best-fit curve. From the exponential decay a time constant of $T_2/2 = 40 \pm 2$ ps is deduced. (*b*) Instrumental response function [6.97]

of the molecules (factor group splitting). Extensive investigations of the relaxation processes in crystalline benzene were carried out by *Hochstrasser*'s group [6.96–101].

An example is given in Fig. 6.8, curve *a*, where the coherent anti-Stokes signal at 1.6 K is plotted as a function of the time delay t_D for the A_{1g} factor group of the ν_1 mode ($\nu/c = 991\ \text{cm}^{-1}$) of benzene with natural carbon isotopic composition (curve *b* gives the instrumental response) [6.97]. The signal decays exponentially with a time constant of $\tau = T_2/2 = 40 \pm 2$ ps at later delay times. Extrapolating the exponential slope back to time zero one finds a slight contribution from the nonresonant susceptibility $\chi_{NR}^{(3)}$. A number of additional experiments were carried out with benzene at low temperatures in order to deduce the relevant relaxation mechanism. Measurements of different factor groups indicated faster decay times of $\tau = 35.6$ ps for the B_{2g} component [6.99]. Mixed crystals containing C_6D_6 molecules gave increased decay rates [6.97]. Important for the understanding of the relaxation mechanism is the finding that the decay rates depend on the isotopic composition of carbon in the benzene molecule. It should be noted that natural benzene contains 1.1% of ^{13}C carbon. Experiments have been performed with neat $^{12}C_6H_6$ benzene crystals giving considerably longer decay times of $T_2/2 = 61.7$ ps (A_{1g}) and $T_2/2 = 56$ ps (B_{2g}). Strong isotope effects have also been found for a number of other vibrational modes of benzene crystals [6.99].

The low-temperature relaxation rates of the isotopically pure benzene crystals show mode specific energy relaxation. The ^{13}C "impurities" in the molecules cause increased relaxation rates due to impurity phonon scattering or near resonant ($\Delta E \sim 9\ \text{cm}^{-1}$) energy trapping. The experiments with benzene crystals demonstrate that the relaxation rates measured via time-resolved coherent Raman scattering give a lower value for the energy relaxation time T_1 in the isotopically mixed samples, while they give T_1 in the pure crystals. These observations provide convincing evidence that the isotopic composition of the molecules has to be taken into account when interpreting the phonon relaxation.

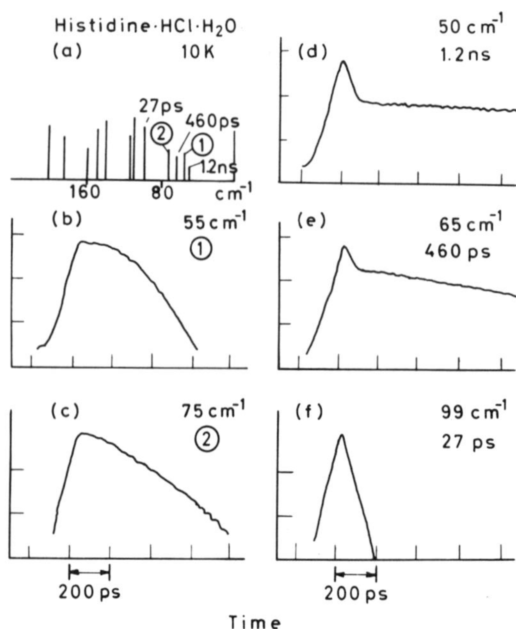

Fig. 6.9a–f. Time-resolved coherent Raman data for 1-(t)-histidine. $HCl \cdot H_2O$ crystals at 10 K (semi-log plots). (**a**) Positions of the low-frequency modes. (**b**)–(**f**) Time-resolved data for the five low-frequency modes shown in (**a**). Exponential decays are found for the totally symmetric modes at 99 cm^{-1}, 65 cm^{-1}, and 50 cm^{-1}, whereas the two modes at 55 cm^{-1} (**b**) and 75 cm^{-1} (**c**) suggest inhomogeneous line-broadening [6.102]

Of special interest are time-resolved coherent Raman investigations in large molecules which are of biological relevance. Extensive investigations of low-frequency vibrational modes (librons) in crystals of aminoacids and peptides were reported by *Dlott* [6.79]. An example is presented in Fig. 6.9, where crystals of *L*-histidine hydrochloride monohydrate were investigated at 10 K [6.102]. The frequency positions, the intensities, and the observed lifetimes are summarized schematically in Fig. 6.9a. A common feature of the relaxation properties of the vibrations in the aminoacid crystals is the following: Modes with frequencies higher than 150 cm^{-1} show lifetimes $\tau = T_1$ shorter than 10 ps. For the symmetric modes the lifetimes increase with decreasing frequency ω_0. The relaxation data for the totally symmetric 99, 65, and 50 cm^{-1} lattice modes are depicted in Fig. 6.9d–f, where the lifetimes extend up to $\tau = 1.2$ ns. From the frequency dependence of T_1 the authors concluded that the decay times are determined by spontaneous decay of librons into two counterpropagating acoustic phonons or into one libron and one acoustic phonon. The non-totally symmetric local modes at 55 and 75 cm^{-1} show non-exponential (Gaussian) decay (see Fig. 6.9b,c). The inhomogeneous broadening of these local modes is believed to be due to dipolar interactions or small variations in the crystal packing.

A short summary should be given here concerning the coherent relaxation times and the relaxation mechanisms found in molecular crystals:

i) Homogeneously broadened lines are frequently found. The inhomogeneity which one would expect due to the site-specific frequency shifts appears in most systems as factor group splitting. It is commonly accepted that imperfections of the crystals are less effective due to the delocalisation of the phonons (motional narrowing); but they may strongly affect the relaxation processes.

ii) The frequency widths or time constants at elevated temperatures are determined by energy relaxation, dephasing or intraband relaxation, and impurity scattering. In a number of cases the temperature dependence of the time constant allows one to determine the relaxation mechanism.

iii) At very low temperatures the relaxation rates become constant. They are determined by energy relaxation and impurity scattering. The decay times measured in coherent experiments give directly the energy relaxation time T_1. Care has to be taken when impurities influence the coherent decay times; in this case only a lower limit for T_1 may be estimated.

d) The Collision Time

We return now to molecular modes in liquids. The theoretical model of Section 6.1 treated the relaxation processes in terms of two time constants, T_1 and T_2. At very early times, $t_D \ll T_2$, however, the time-dependent interaction during the collisions of the molecules becomes important. As a consequence, the correlation function $\langle\langle q(0)q(t)\rangle\rangle$ is no longer a single exponential, but has the form [6.3,4,50,103]:

$$\langle\langle q(0)q(t)\rangle\rangle \propto \exp\left\{-\frac{t}{T_2}-\frac{\tau_c}{T_2}\left[\exp\left(-\frac{t}{\tau_c}\right)-1\right]\right\} . \tag{6.15}$$

Equation (6.15) was derived by Kubo for an exponential decay of the frequency correlation function, $\langle\delta\omega(t)\delta\omega(0)\rangle = \exp(-t/\tau_c)$. The collision time τ_c is the time for which a certain frequency shift $\delta\omega(t) = \omega(t) - \omega_0$ is maintained.

At later times (6.15) gives the exponential decay of the coherent amplitude with $\exp(-t/T_2)$, while at short times, $t \to 0$, (6.15) implies a Gaussian time dependence $\exp[-t^2/(T_2\tau_c)]$. For short collision times $\tau_c \ll T_2$ the collision process leads to small corrections of the coherent signal. Nevertheless, time-resolved coherent scattering may be used to determine the collision time τ_c or can enable one to estimate an upper limit for τ_c. For an example we refer to Figs. 6.4b and 6.5. The decay of the coherent signal was found to be exponential for times larger than 0.3 ps. Taking into account the accuracy of the experimental data and the nonresonant susceptibility $\chi_{NR}^{(3)}$, one may estimate an upper limit for the collision time of both acetone modes and for the acetonitrile mode of $\tau_c < 0.3$ ps.

For sufficiently long collision times $\tau_c \geqslant 0.5$ ps, picosecond light pulses from a Nd:glass laser may be used for the direct determination of τ_c. *Telle* and *Laubereau* have performed coherent experiments with a carefully controlled time resolution of the experiment [6.104]. They simultaneously measured the temporal response of their system (by coherent scattering of a purely nonresonant susceptibility) and the coherent signal of the investigated transition (liquid CH_2BrCl). From the experimental curves and numerical calculations based on the (measured) temporal system response the authors estimated a collision time of

$$t_c = 0.4\begin{pmatrix}+0.4\\-0.1\end{pmatrix}\text{ps} , \qquad \text{i.e. } 0.3 - 0.8\text{ ps} .$$

In a recent paper *Chesnoy* investigated the kinetics of vibrational dephasing of nitrogen under supercritical conditions [6.41]. He demonstrated that, after some initial delay, the coherent signal decays exponentially with time. Comparing the time-resolved coherent Raman data with spontaneous Raman spectra enabled a collision time of $\tau_c \simeq 10$ ps to be deduced; this is of the same order as the measured dephasing time of $T_2 \simeq 20$ ps. The long collision time τ_c indicates the importance of density fluctuations as a source of frequency modulation close to the critical point.

6.2.2 Time-Resolved Coherent Raman Scattering of a Distribution of Resonance Frequencies

a) Time-Domain Experiments

In the previous sections we discussed time-resolved coherent investigations, where dynamic processes such as dephasing and energy relaxation were of major relevance. Here we address new spectroscopic studies which allow accurate measurement of frequency differences between vibrational modes and the resolution of transitions within congested spectral regions.

The following time-resolved coherent experiments are described by (6.1–13) discussed above. For each vibrational mode one introduces individual coherent amplitudes Q_j and vibrational resonance frequencies ω_j. The coherently scattered field is a sum of field components scattered from the individual molecular modes Q_j [6.7,37,105].

$$E_A(t, t_D) = \sum_j E_{Aj}(t, t_D) \propto \sum_j Q_j(t) E_{L2}(t - t_D) \exp(-i\omega_j t) . \qquad (6.16)$$

The observed coherent signal has the form $S(t_D) \propto \int dt |E_{AS}(t, t_D)|^2$. It can be shown that in the limit of short excitation and probing pulses the signal $S(t_D)$ at $t_D > 0$ is proportional to the absolute square of the Fourier transform of the spontaneous Raman spectrum $R(\omega)$, i.e. $S(t_D) \propto |\int d\omega \exp(i\omega t_D) R(\omega)|^2$ [6.3,105]. In the time domain – according to (6.16) – simultaneously excited modes lead to a beating of the coherent signal at the frequency differences $\delta\omega_{ij}$ between the various transition frequencies ω_i and ω_j, with $\delta\omega_{ij} = \omega_i - \omega_j$. These "quantum" beats of the coherent signal were found for the first time by *Laubereau* et al. [6.36] in a number of tetrahalides. Later, similar beating phenomena were reported in several publications [6.37,44,94,98,102,106–108].

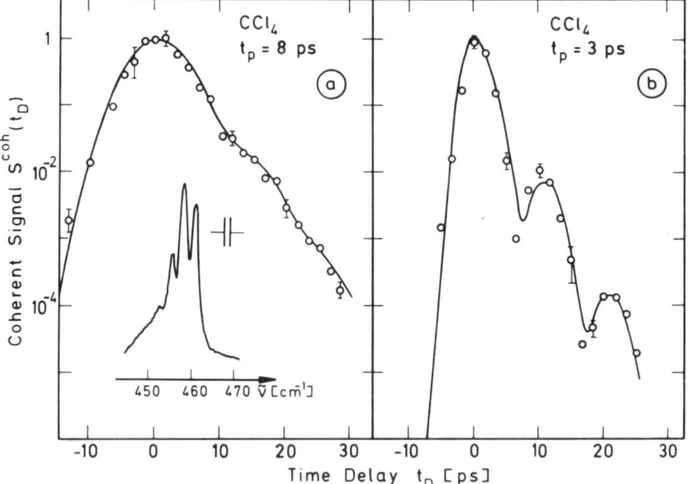

Fig. 6.10a, b. Coherent Stokes signal from the ν_1 mode of liquid CCl_4. The isotope composition of CCl_4 leads to a splitting of the ν_1 band in the spontaneous Raman spectrum – see insert in (**a**). The different resonance frequencies lead to the beating phenomenon which is clearly resolved in (**b**) where exciting and probing pulses of 3 ps duration were used. For longer ($t_p = 8$ ps) pulses the beating structure is smeared out. Nevertheless, the dephasing time T_2 is readily determined in both experiments [6.57]

Figure 6.10 presents the results of a study of the ν_1 mode of liquid carbontetrachloride ($\nu/c \simeq 459\,\mathrm{cm}^{-1}$) where the influence of the pulse duration on the observed beating was investigated [6.37]. Due to the natural isotopic composition of the CCl_4 molecules (the two chlorine isotopes ^{37}Cl and ^{35}Cl have abundances of 24% and 76%, respectively), the transition is split in five equidistant components (distance $\Delta\nu/c = 3\,\mathrm{cm}^{-1}$). Four of the more abundant components are readily resolved in the spontaneous Raman spectrum as shown in the insert of Fig. 6.10a. In the time-resolved coherent experiment one expects a beating phenomenon with maxima separated by $\Delta t = 1/\Delta\nu \simeq 11$ ps. This beating is indeed observed in Fig. 6.10b, where short excitation and probing pulses with $t_p = 3$ ps were used. In addition, one obtains information on the dephasing time T_2. The solid curve was calculated using the measured pulse parameters ($t_p =$ 3 ps, Gaussian shape), a difference of the neighboring frequency components of $3.0\,\mathrm{cm}^{-1}$, a dephasing time $T_2 = 6$ ps (assumed to be the same for all components) and the nonresonant susceptibility $\chi^{(3)}_{NR}$. Figure 6.10a shows the signal curve for longer excitation and probing pulses ($t_p \simeq 8$ ps). Here the beating phenomenon is smeared out and the decay, with the time constant $T_2 = 6$ ps, appears quite clearly. The longer pulses influence the signal curve in two ways: (i) During the longer excitation process the most intense transition of the Raman spectrum is preferentially excited, and (ii) the long duration of the probing pulse gives a levelling of the modulation. The two experiments of Fig. 6.10 show convincingly

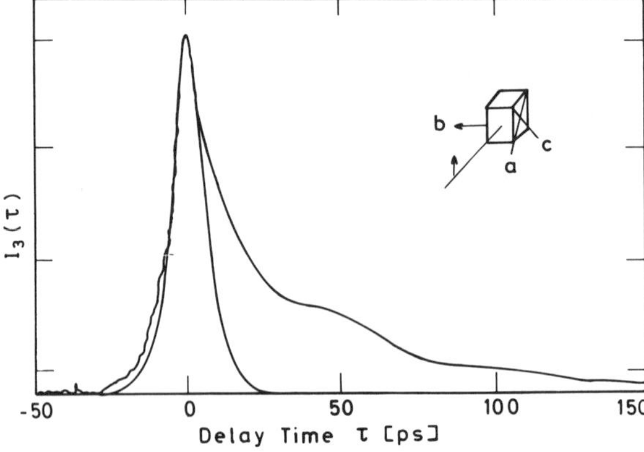

Fig. 6.11. Time-resolved coherent Raman data for a special phase-matching condition, where the A_g and the B_{2g} factor group components of the 991 cm^{-1} vibron of crystallized benzene are probed simultaneously. The modulation of the signal is due to the frequency shift of 0.64 cm^{-1} between the two interfering lines. Also shown is the experimental response function [6.98]

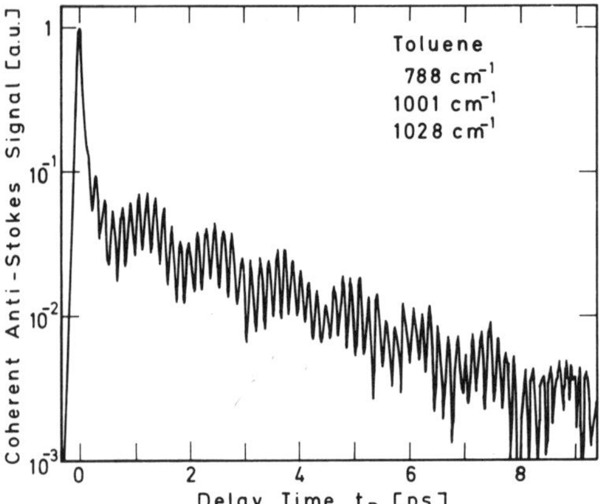

Fig. 6.12. Terahertz beats observed in time-resolved coherent Raman scattering from liquid toluene using femtosecond light pulses. The interference of the three simultaneously excited toluene modes at 788, 1001, and 1028 cm^{-1} generates a rich structure with beat frequencies up to 7.2 THz [6.106]

that time-resolved coherent experiments are well suited to investigate frequency differences and decay times in spectral regions consisting of discrete frequency components.

Beating between transitions of different factor groups was studied by *Velsko* et al. [6.98] in crystallized benzene at low temperature (1.6 K). Some results are shown in Fig. 6.11. The geometry of the excitation and probing light field was selected in such a way that coherent signals were obtained from the A_g and B_{2g} states in the factor groups of the 991 cm^{-1} transition of benzene. From the beating of the signal (i.e. the time between the peaks) and from its magnitude the authors determined the frequency difference to be $\Delta\nu = 19$ GHz, the relative amplitudes to be 20:1, and the decay times of the two modes to be 40 and 34 ps, respectively.

While the experiment on crystallized benzene gave beating frequencies in the 20 GHz range, the advent of femtosecond light pulses allows one to measure beating frequencies far into the terahertz regime. Beat frequencies of more than 10 THz have been reported [6.44]. Figure 6.12 shows very recent data from a coherent Raman experiment on liquid toluene [6.106]. In the wide frequency range of 700–1100 cm^{-1} three vibrational modes are excited simultaneously by the (broadband) femtosecond light pulses. The scattered coherent anti-Stokes signal exhibits an amazing beating structure with beat frequencies of up to 7.2 THz. The frequency components of the signal curve are evaluated by Fourier transformation or other more specialized numerical methods in order to obtain the spectral information. This procedure called "Fourier transform Raman spectroscopy" was first demonstrated by Graener and Laubereau [6.107]. The authors studied coherent beating in gaseous methane-argon mixtures on a much longer time scale of many picoseconds to nanoseconds.

It should be noted here that the case of continuously distributed resonance frequencies (e.g. an inhomogeneously broadened transition) leads to a more difficult situation. There is no recurrence of the signal but only destructive interference with a rapid signal decay. An early attempt to measure T_2 in inhomogeneously broadened frequency distributions by using a "selectivity in k-space" had to be reconsidered [6.36,109–114]. *Zinth* et al. showed experimentally that with the existing dispersion of the samples, a selectivity for a molecular subgroup is not possible [6.37]. In addition, general theoretical arguments prove that the information from a single-pulse coherent experiment and from a spontaneous Raman spectrum is equivalent [6.103,115,116]; i.e. T_2 cannot be obtained in a coherent Raman experiment, where the spontaneous Raman line is strongly inhomogeneously broadened.

Under certain conditions, however, time-resolved coherent Raman techniques may give information on the line-broadening mechanism. Two processes cause similar non-Lorentzian spontaneous Raman bands: (i) Long collision times $\tau_c \simeq T_2$ and (ii) inhomogeneous broadening (cross-relaxation times much longer than T_2).

Gale et al. [6.42] pointed out that coherent experiments give new information on the broadening mechanism: (i) If τ_c is relatively long, $\tau_c \simeq T_2$, the coherent signal decays non-exponentially around time zero but turns to an exponential decay at later times. (ii) For a small inhomogeneous contribution an initial exponential decay with time constant $T_2/2$ (close to time zero) is followed by an increasingly rapid decay at later times [6.105]. If the inhomogeneous width adds 10% to the homogeneous linewidth, the slope of the signal decay (on a semilog plot) gives $T_2/2$ around time zero. This slope increases with time. For example, four orders of magnitude below the signal peak, the slope is larger than at time zero by more than 50%. While the influence of weak inhomogeneous broadening [case (ii)] and of long collision times τ_c [case (i)] yield very similar spontaneous Raman spectra, the time-resolved coherent data are different and may be used to decide which line-broadening mechanism exists.

b) Transient Frequency-Domain Experiments

In the previous section we discussed investigations where the spectral information on the molecular system was deduced from time-resolved coherent data followed by a numerical analysis. Now we present a technique which avoids the Fourier transformation by directly measuring the spectrum of the coherent signal generated by properly shaped probing pulses [6.64,117–121].

The basic idea of the technique is readily seen by considering the spectrum of the scattered coherent anti-Stokes signal of a single vibrational transition

$$I_A(\omega) \propto \left| \int dt \exp(i\omega t) E_A(t) \right|^2 \tag{6.17a}$$

$$I_A(\omega) \propto \left| \int dt \exp(i\omega t) E_{L2}(t) \exp\left(-\frac{t}{T_2} \right) \right|^2 \tag{6.17b}$$

$$I_A(\omega) \propto \left| \int dt \exp(i\omega t) \exp\left[-\left(\frac{t}{t_0} \right)^2 - \frac{t}{T_2} \right] \right|^2 \tag{6.17c}$$

$$I_A(\omega) \propto \left| \int dt' \exp(i\omega t') \exp\left[-\left(\frac{t'}{t_0} \right)^2 \right] \right|^2 . \tag{6.17d}$$

In these equations it is assumed that the excitation and the initial effects of the collision process are terminated. Equation (6.17a) represents the Fourier transform of the anti-Stokes signal. The freely relaxing coherent amplitude $Q = Q_0 \exp(-t/T_2)$ is monitored by a Gaussian shaped laser field $E_{L2}(t) \propto \exp(-t^2/t_0^2)$ [see (6.17b,c)]. By introducing a new time t', one eliminates the dephasing time. According to (6.17d), the anti-Stokes spectrum has the same frequency shape, i.e. the same frequency bandwidth, as the probing laser pulse E_{L2}. With the phase factors included in (6.17) one finds the anti-Stokes band centered at the anti-Stokes frequency $\omega_{AS} = \omega_{L2} + \omega_0$. When sufficiently long ($t_p > 1.5\, T_2$) probing pulses of Gaussian shape are used, the spectral width $\Delta\nu$ of the time-resolved coherent anti-Stokes band, $\Delta\nu = 0.44/t_p$, is narrower than the spontaneous Raman line $\Delta\nu = 1/\pi T_2$. In practical applications it is advantageous to use a short exciting force and a prolonged Gaussian shaped interrogation pulse. As an acronym the spectroscopic technique is known as SEPI spectroscopy. A more detailed description of the theory is given in the literature by *Zinth* et al. [6.119] and *Collins* et al. [6.122].

An experimental example of the line-narrowing SEPI spectroscopy is given in Fig. 6.13. The ν_1 mode of liquid CCl_4 was investigated using a Nd:glass laser system supplying 20 ps probing pulses of Gaussian shape. In Fig. 6.13a the spontaneous Raman spectrum of the ν_1 mode is shown as measured with a high-resolution Raman spectrometer; the isotopic structure has been discussed above. The SEPI spectrum (Fig. 6.13b) was taken by using a high-resolution spectrograph in conjunction with an optical spectrum analyser. The entire coherently scattered spectrum is recorded immediately for each probe pulse. When the excitation force (i.e. $E_L E_s^*$) is centered at $460\,\mathrm{cm}^{-1}$ (see Fig. 6.13b), the two high-frequency components of the isotopically split Raman band appear very narrow and thus well resolved. The SEPI spectrum clearly separates the two

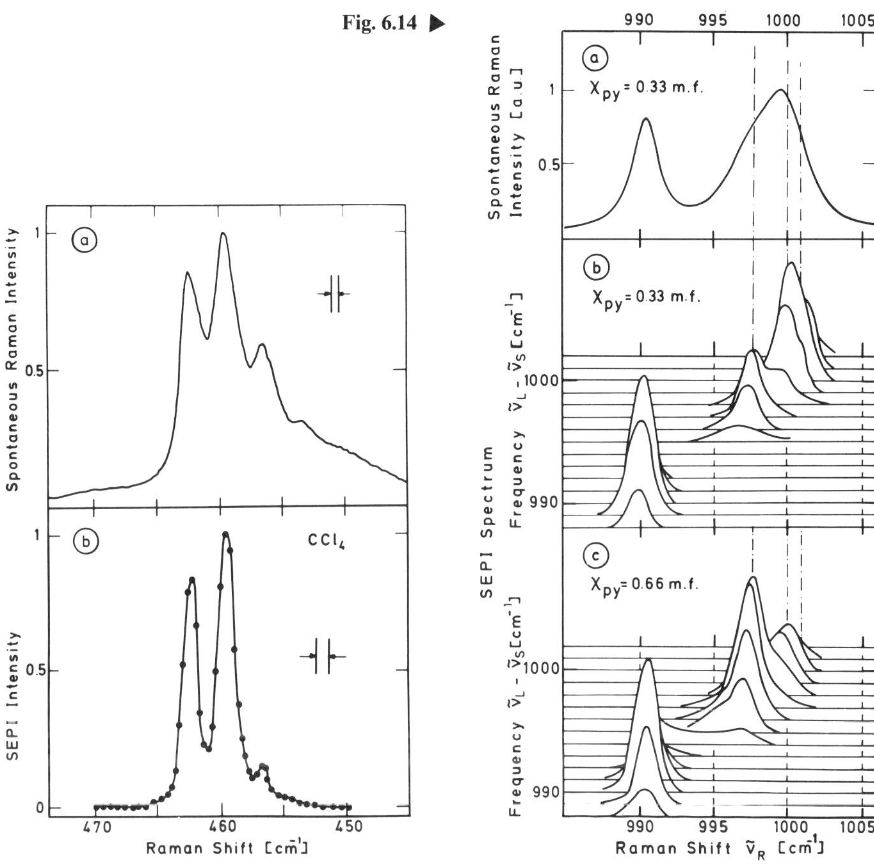

▲

Fig. 6.13a, b. Line-narrowing coherent Raman spectroscopy using short excitation and prolonged coherent interrogation (SEPI) of the ν_1 mode of liquid CCl_4. Broad overlapping lines of the spontaneous Raman spectrum (**a**) are considerably narrowed by the coherent SEPI technique (**b**) [6.105]

Fig. 6.14a–c. Raman spectra of pyridine: methanol mixtures. (**a**) Polarized spontaneous Raman spectrum measured at a pyridine concentration of 0.33 molar fraction taken with a standard Raman spectrometer of spectral resolution of $0.5\,\mathrm{cm}^{-1}$. (**b**) and (**c**) Short excitation and prolonged interrogation (SEPI) spectra obtained for a set of excitation frequencies, $\nu_L - \nu_s$, at two pyridine concentrations, $\chi_{py} = 0.33$ molar fraction (**b**) and $\chi_{py} = 0.66$ molar fraction (**c**). The spectra taken at excitation frequencies $\nu_L - \nu_s < 994\,\mathrm{cm}^{-1}$ are drawn with reduced amplitudes. The SEPI spectra resolve three Raman lines of different hydrogen-bonded aggregates hidden under the broad band of the spontaneous Raman spectrum (see dash-dotted lines) [6.120]

bands and allows a very accurate determination of the frequency difference of $\delta\nu/c = 3.0 \pm 0.08\,\mathrm{cm}^{-1}$.

An interesting application of the line-narrowing SEPI spectroscopy is presented in Fig. 6.14 [6.120]. The molecular system pyridine: methanol was investigated, which exhibits a broad featureless spontaneous Raman band around

1000 cm^{-1} (see Fig. 6.14a). The position and amplitude of this band changes with varying methanol concentration. With the help of a picosecond dye laser system, the frequency of the excitation force was tuned over the frequency range of interest. SEPI spectra taken at fourteen excitation frequencies are shown in Fig. 6.14b, c for two different methanol concentrations. In both cases the band at 1000 cm^{-1} can be resolved in three components at 997.3, 1000.0, and 1001 cm^{-1}. The positions of the three bands remain constant while the amplitudes depend on the methanol concentration. From these data new information on hydrogen-bonded complexes between pyridine and methanol were obtained: there exist at least three distinct molecular complexes of pyridine and methanol with frequency positions remaining constant with methanol concentrations. The apparent frequency shift of the Raman band observed in spontaneous spectroscopy results from the changing abundance of the complexes as a function of the methanol concentration.

The following points concerning the line-narrowing transient Raman spectroscopy are of interest:

i) The SEPI method investigates the freely relaxing molecules.
ii) Influences of $\chi_{NR}^{(3)}$ are eliminated as the probing process is performed at late delay times.
iii) We emphasize that the amplitudes of the SEPI bands depend on the Raman cross sections and also on the dephasing times T_2 of the individual modes. This fact provides increased spectral resolution in congested spectral regions when fast decaying bands disappear at late delay times.
iv) The SEPI experiments may be performed with coherent anti-Stokes or Stokes scattering. The two spectra give the same information.

6.2.3 Resonant Pulse Propagation in the Infrared

In the previous sections the discussion concentrated on Raman-type interactions. The absence of a dipole moment required a pair of light pulses for the excitation of the molecules and a separate probing pulse to study the time evolution of the material excitation. In this section we focus on infrared-active transitions where the material is excited by resonant ir light and where the oscillating molecules emit radiation. We discuss the case of a single small area excitation pulse propagating resonantly through the absorbing medium [6.7,123–125].

It has been shown in [6.7] that small area pulse propagation may be treated by a single equation containing the dephasing time T_2 and a modified Bessel function. The discussion may be further simplified for the case of optically thin samples, i.e. where the absorbance $a = \alpha l =$ (absorption coefficient × sample length) is small, $a \ll 1$. The transmitted electric field may then be written as follows [6.125]:

$$E_{tr}(t) = E_0(t) - \frac{\alpha l}{2T_2} \int_{-\infty}^{t} dt' E_0(t') \exp\left[\left(t' - t\right)\left(\frac{1}{T_2} - \mathrm{i}\Delta\right)\right] . \tag{6.18}$$

As previously, Δ stands for the difference between resonance and driving frequency. According to (6.18) the transmitted electric field consists of the incoming light field minus a second term which decays with the dephasing time T_2. Thus the dephasing time T_2 can be directly measured from the slope of the trailing part of the transmitted, i.e. coherently emitted, light field. In the general case of $\alpha l \gtrsim 1$ the trailing part of the signal decays more rapidly. For values $\alpha l = 1$ a simple correction formula may be used in order to deduce T_2 from the measured decay [6.126].

An elegant experimental system to measure the small area pulse propagation in the infrared was developed by *Hartmann* and *Laubereau* [6.124]. The main difficulty which had to be overcome was the slow time resolution of the photodetectors. Existing infrared detectors are not able to measure the transmitted pulse with picosecond time resolution. The detection of the pulse intensity as a function of time is accomplished by gated ultrafast infrared up-conversion. This technique provides the convolution of the instantaneous infrared intensity with a short probing pulse. Tuning the time delay between the exciting infrared pulse and the probing gate pulse allowed the evolution of the coherently radiated infrared signal to be followed.

In a recent experiment, the coherent interaction of an infrared pulse with rotational transitions of HCl:Ar gas mixtures at medium pressure was studied [6.127]. HCl was used with natural isotope abundance of the ^{35}Cl and ^{37}Cl isotopes. Transitions of the $R(J)$ branch of HCl around 3000 cm^{-1} were investigated for different values of the rotational quantum number J of the $J \rightarrow J+1$ rotation-vibrational transition. An example is presented in Fig. 6.15. The infrared

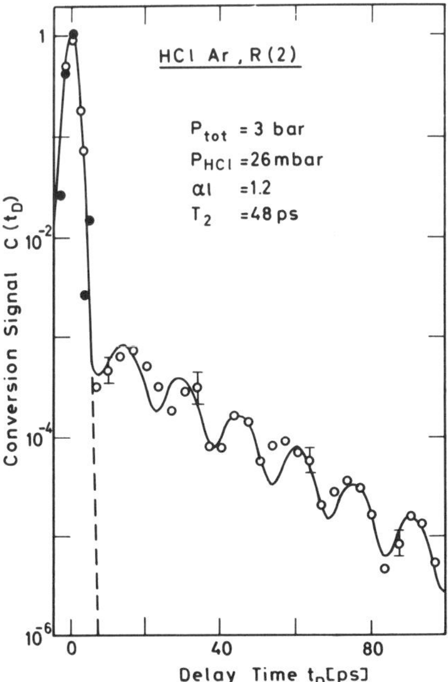

Fig. 6.15. Resonant propagation of an infrared pulse through a HCl:Ar mixture. The conversion signal representing the transmitted and reemitted infrared intensity is plotted versus delay time t_D. Note the beating in the reemitted part at $t_D > 10$ ps which is due to the interference from the emission from the $R(2)$ transition of HCl containing the two chlorine isotopes ^{35}Cl and ^{37}Cl. From the decay of the signal a dephasing time $T_2 = 48 \pm 5$ ps was determined [6.127]

pulse is tuned to the $R(2)$ transition of $H\,^{35}Cl$. The related $R(2)$ transition of $H\,^{37}Cl$ is shifted by $\simeq 2\,\mathrm{cm}^{-1}$ to smaller frequencies but also interacts with the incident pulse. [For the numerical analysis of the experimental data the second transition has to be incorporated in (6.18)]. The time integrated conversion signal $C(t_D)$ is plotted in Fig. 6.15 on a logarithmic scale as a function of the delay time of the probing pulse. Around $t_D = 0$ the conversion signal of the emitted pulse (open circles, solid line) follows the input pulse (full circles, broken line). For longer delay times, $t_D > 10\,\mathrm{ps}$, the emitted pulse decays more slowly. A beating phenomenon (beat time 15.3 ps) is found which is superimposed on an approximately exponential decay ($\tau = 19.5\,\mathrm{ps}$). The analysis of the data – taking into account the optical thickness of the sample, $\alpha l = 1.2$ – yielded a frequency distance between the isotope components of $2.18\,\mathrm{cm}^{-1}$ and the dephasing time of $T_2 = 48 \pm 5\,\mathrm{ps}$. Measurements at different partial pressures of HCl and Ar gave the following results: HCl–HCl collisions are much more efficient for the dephasing process than HCl–Ar collisions; the relaxational data for the rotational-vibrational transitions of HCl agree well with the relaxation data for pure rotational transitions, suggesting that vibrational relaxation contributes little to the observed dephasing process. The authors proposed that total dephasing is due to pure rotational dephasing (quasi-elastic collisions) and the population decay of the rotational levels (inelastic collisions).

6.3 Coherent Spectroscopy Using Multiple Excitation Processes

In the preceding sections we have discussed various applications of coherent spectroscopy using a single excitation process. One major result was the determination of the dephasing time T_2 in a variety of systems. On the other hand, these techniques do not allow measurement of the dephasing time T_2 for wide inhomogeneously broadened lines and give no information on the energy relaxation time T_1 (except in solids at low temperature).

In the theoretical part (Sect. 6.1.3) it was shown that multiple excitation experiments provide additional information. These experiments, frequently called echo experiments, have their analogues in magnetic resonance. They have been demonstrated for optical transitions, initially on a nanosecond or even longer time scale [6.11–13]. More recently, picosecond and femtosecond measurements have been reported. Successful echo experiments require a pulse area A of approximately one. For small pulse areas, the echo signal is drastically reduced, since the signal intensity is proportional to A^6. For this reason ultrafast echo experiments are only possible when the relaxation times T_1 and T_2 are sufficiently long and when the transition dipole moment μ is large. Since 1975 ultrafast echo experiments have been performed on electronic transitions in dye solutions [6.128–130], in low-temperature solids (e.g. excitons in semiconductors [6.131, 132] and molecular crystals [6.9,133–139] on a picosecond time scale or, more recently, in dye solutions with femtosecond pulses [6.140–142]. A number of theoretical papers treat the different echo techniques [6.143–147].

In the literature different – and in some cases misleading – names have appeared for the same coherent experiment. We briefly discuss (i) the echo experiments and (ii) the induced grating experiments, two aspects of the same coherent investigation.

i) In a coherent experiment with multiple excitation pulses, the system is excited (at time zero) by a first pulse (wave vector $\boldsymbol{k}_a$). After an evolution time t_1 a second excitation pulse (wave vector $\boldsymbol{k}_b$) passes through the sample and interacts with the excited molecules. Rephasing of the excited molecule leads (in the case of inhomogeneous broadening) to the formation of a delayed radiation – the echo. The echo is emitted at a time $t = t_1$ after the second pulse under the well-defined phase-matching condition with wave vector $\boldsymbol{k}_e = 2\boldsymbol{k}_b - \boldsymbol{k}_a$.

ii) In the excitation process, the polarization produced by a first excitation pulse and part of the second pulse, form an excitation grating in the sample (the induced grating) with the wave vector $\boldsymbol{k}_g = \boldsymbol{k}_b - \boldsymbol{k}_a$. The trailing part of the second pulse or a third pulse (see Sect. 6.3.2) interacts with the grating, generating radiation at the wave vector $\boldsymbol{k}_s = \boldsymbol{k}_b + \boldsymbol{k}_g = 2\boldsymbol{k}_b - \boldsymbol{k}_a$. This radiation – which forms the echo – is emitted with a time delay of $t = t_1$ in the case of inhomogeneously broadened transitions.

6.3.1 Ultrafast Two-Pulse Echo Spectroscopy

In the theoretical part [Eq. (6.7) in Sect. 6.1.2] a formal solution for the evolution of a two-level system was derived for the limiting case of light pulses with duration t_p much shorter than the system's time constants T_1 and T_2. In the case of a two-pulse echo, two pulses of area A_a and A_b are applied to the sample with a time separation of t_1. The resulting echo polarization is given by (6.9). The signal amplitude is proportional to $\sin A_a \sin^2(A_b/2)$ and decays with the dephasing time T_2. The emission direction is determined by the wave vector $k_e = 2k_b - k_a$. An important point emerges from the underlined part of (6.9): in the case of a wide inhomogeneously broadened transition the signal is emitted close to time $t = t_1$, i.e. the emission occurs delayed as a photon echo.

In a standard two-pulse echo experiment the echo energy, i.e. the time integral of the absolute square of the echo amplitude, is measured as a function of the separation t_1 between the two excitation pulses. In the case of a broad inhomogeneous line the signal is emitted only close to $t = t_1$ and its intensity decays with $\exp(-4t_1/T_2)$. For homogeneously broadened transitions the signal is emitted starting at $t = 0$. The decay (as a function of t_1) follows $\exp(-2t_1/T_2)$. (For intermediate line broadening see [6.14]). In order to obtain the dephasing time independently of the nature of the line broadening, only the polarisation at the time $t_e = t_1$ should be measured (this may be done by optical gating techniques). Under these conditions the signal depends on t_1 as $\exp(-4t_1/T_2)$ and is independent of the line shape.

The preceding discussion shows that the two-pulse echo technique allows determination of the dephasing time T_2 even for inhomogeneously broadened transitions, where experiments with single excitation pulses fail.

A number of two-pulse echo experiments have been performed on exciton bands of molecular crystals at low temperature. For instance, the relaxation of the 0–0 transition of pentacene in naphthalene (and in p-terphenyl) was studied as a function of temperature by *Hesselink* and *Wiersma* [6.9,133]. Here the energy relaxation times T_1 are much longer ($\simeq 20$ ns) than the dephasing times T_2 and the pure dephasing times T_2^* could be readily determined from the relation $1/T_2 = 1/T_2^* + 1/(2T_1) \simeq 1/T_2^*$. The temperature dependence of T_2^* was fitted with a relaxation theory involving low-energy ($\nu/c \simeq 16\,\mathrm{cm}^{-1}$) pseudo-local phonons coupled to the pentacene transition. The experiment clearly demonstrated the importance of this relaxation channel for the dephasing of the excitonic transitions.

Recently, the two-pulse echo technique has been applied to two-dimensional excitons in GaAs-AlGaAs quantum wells by *Schultheis* et al. [6.132]. The authors used a synchronously pumped dye laser emitting pulses (with 12.6 ps autocorrelation width) tunable in the range of the exciton absorption band between $\lambda = 820$ nm and $\lambda = 750$ nm. A pair of pulses at the temporal separation t_1 with wave vectors $\boldsymbol{k}_a$ and $\boldsymbol{k}_b$ was focussed into a sample consisting of 78 alternating layers of GaAs and AlGaAs (thickness 102 Å and 200 Å, respectively), grown by molecular beam epitaxy. The sample was kept at 2 K. The transmission spectrum of the inhomogeneously broadened exciton band is shown in the insert of Fig. 6.16 (broken curve). The echo signal was detected in the phasematching direction $2\boldsymbol{k}_b - \boldsymbol{k}_a$ with the help of a lock-in system. A series of data is shown in Fig. 6.16. The echo signals are plotted as a function of delay time t_1 for three different energies of the excitation photons. Curve *a* investigates the exciton at the low-energy side of the absorption band, curve *b* was measured at somewhat higher energies, and curve *c* was taken close to the peak of the exciton absorption.

The relaxation times of the sample do not fulfill the limit of $T_2 \ll t_p$ as assumed in the theoretical discussion of (6.19). A more detailed analysis was required in order to derive the relaxation time T_2 from the experimental curve. Together with T_1-data from hole-burning experiments, the analysis yielded interesting numbers. Close to the peak of the exciton band (curve *c*) the relaxation times are fast; $T_2 \simeq 4$ ps and $T_1 = 8$ ps. Consequently, the diffracted signal intensity is small and the signal decay is very rapid. At the low-energy side of the exciton band the energy relaxation time T_1 is much longer (100 ps), while the dephasing increases to 22 ps. There is no change of the time constants T_1 and T_2 at the frequency positions *a* and *b* in the low-energy tail of the exciton band. The results are interpreted as follows: Near the peak of the exciton band, there are several contributions to the coherent decay: scattering with acoustic phonons, impurities and well thicknesses (spectral diffusion). Below the absorption peak, excitation is more localized and spectral diffusion contributes less to the dephasing pro-

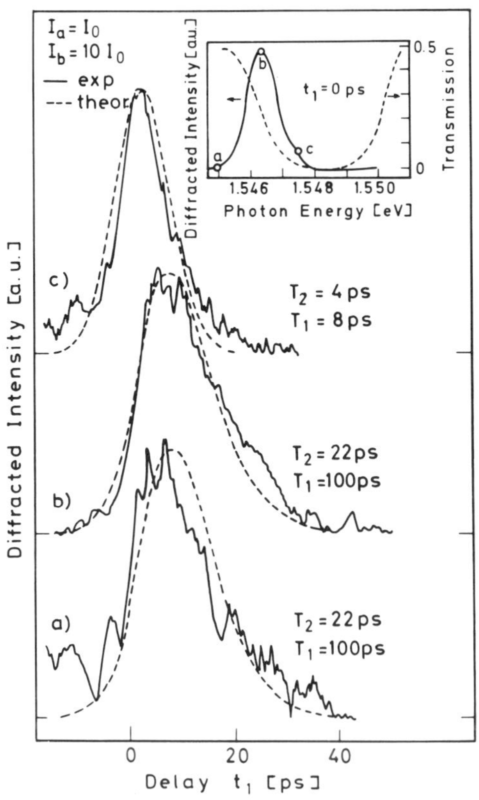

Fig. 6.16. Two-pulse photon echo experiments from two-dimensional excitons in GaAs-AlGaAs quantum wells. Experimental (——) and theoretical (– – –) scattered intensity as a function of the delay time t_1 for three different photon energies. The data show different relaxation times T_2 depending on the spectral position of the excitation pulse within the heavy-hole exciton transition. The insert gives the scattered intensity at zero delay (*solid curve*) together with the transmission spectrum (*dashed curve*) [6.132]

cess. Very recent investigations of high-quality single quantum wells show a homogeneously broadened exciton band. Under these conditions acoustic phonon scattering is found to be the dominant broadening mechanism of the 2*D* excitons [6.148].

6.3.2 Three-Pulse Echos

Experiments with three incident pulses constitute an important extension of the preceding coherent techniques. A schematic of this method was shown in Fig. 6.1c. Adding a third excitation pulse – which enters the sample at a time t_2 after the second pulse – gives new possibilities for studying molecular systems. For a two-level system, the three-pulse technique allows one to determine T_1 in a coherent experiment by varying the delay time t_2. When other molecular levels are involved in the relaxation process, a third pulse at a wavelength differing from that of the first two excitation pulses gives information on the path of the energy decay.

The three-pulse echo is calculated for a two-level system by solving (6.7) for the appropriate pulse sequence. The solution for the polarization $P(t)$ contains a number of different terms, which describe (i) the single-pulse excitation due to pulses a, b, and c, radiating in the directions $\boldsymbol{k}_a$, $\boldsymbol{k}_b$, and $\boldsymbol{k}_c$, respectively, (ii) two-pulse echos originating from pairs of excitation pulses radiating in the direction $2\boldsymbol{k}_i - \boldsymbol{k}_j$ for the various pulses $i, j = a, b, c$, (iii) higher-order terms of the order A^5, and finally (iv) the three-pulse echo:

$$P_{3PE}(t) \propto \int d\omega f(\omega) \exp[-i\omega(t - t_1) - (t_1 + t)/T_2] G(T_1, t_2) \times \sin A_a \sin A_b \sin A_c \exp[-i(\boldsymbol{k}_b + \boldsymbol{k}_c - \boldsymbol{k}_a)r] \ , \tag{6.19}$$

where $G(T_1, t_2)$ depends on the level scheme involved. For a closed system consisting of only two states $|a\rangle$ and $|b\rangle$, $G(T_1, t_2)$ has the form $G^c(T_1, t_2) = \exp(-t_2/T_1)$. For an open system, where the upper level decays to a long-lived $(\tau \gg t_1, t_2, t)$ state, which is not probed, $G(T_1, t_2)$ is $G^0(T_1, t_2) = 1 + \exp(-t_2/T_1)$. Equation (6.19) suggests a variety of possibilities for applying the three-pulse echo technique.

1) Variation of the excitation time t_1 allows the determination of the dephasing time T_2 in the case of inhomogeneously and homogeneously broadened transitions (in the same way as in the two-pulse echo experiment).

2) Variation of the time t_2 gives access to the population decay of the upper state. The experimentally observed kinetics depend on the details of the molecular system. If a closed two-level system is investigated, the coherent signal intensity decays with $\exp(-2t_2/T_1)$. A more complex time dependence is observed for an open system where the decay of the signal is proportional to $[1 + \exp(-t_2/T_1)]^2$; i.e. one observes a biexponential decay with decay times $T_1/2$ and T_1, and the signal shows a constant background. The repopulation of the lower level $|b\rangle$ may also be deduced from the later kinetics.

3) The constant background, which is found due to the long-lived intermediate state in open systems allows accumulation of the echo by a sequence of excitation pulse pairs with a repetition rate faster than the spectral cross-correlation time or the repopulation of the ground state. The accumulation of the grating may become so efficient that even weak cw mode-locked lasers can be used to induce a grating and to stimulate an echo.

4) Other aspects of the three-pulse echo technique may be inferred from the grating picture of the three-pulse echo experiment. The first two exciting pulses induce a population grating for the two levels coupled by the two pulses. The populations of the ground and excited state are changed. Energy transfer of either of the two grating states can be monitored. Using the third pulse at a new wavelength, it becomes possible to selectively monitor the relaxation processes. An example of such a two-color echo will be given below.

The laser systems for three-pulse or induced-grating experiments are cw mode-locked dye lasers with a repetition rate of ~ 100 MHz and a mean power of 30 mW. In some cases, amplified pulses from cw mode-locked dye lasers at repetition rates of 10–100 Hz were applied. For high repetition rate lasers it is advantageous to use perpendicular polarizations for the excitation and probing process in order to remove artifacts due to temperature changes.

a) Measurements of Energy Relaxation Times

A number of interesting grating experiments were performed by *Ippen*'s group using femtosecond light pulses [6.140–142]. In Fig. 6.17 the three-pulse echo experiment was done with amplified 75 ps pulses ($\lambda = 620$ nm) from a colliding pulse mode-locked ring dye laser (repetition rate of the amplified pulses 10 Hz) [6.147]. The first two excitation pulses with wave vectors k_a and k_b entered the sample simultaneously, i.e. $t_1 = 0$. The scattered energy induced by a third pulse is plotted in Fig. 6.17 as a function of the delay time t_2. Two different samples, the dyes cresyl violet and oxazine 720, were investigated and the results are shown in Fig. 6.17a and b, respectively. Both dyes were embedded in a thin film of a polymethylmethacrylate (PMMA) polymer and held at 15 K. The excitation laser pulses populate vibronic levels in S_1, the first excited electronic state of the dyes. The first decay of the signal occurs on a time scale of several 100 fs. It results from an energy redistribution within the electronic state of the dye molecules. Similar rapid relaxation processes were found for dye molecules in solutions at room temperature [6.149]. At later times t_2, longer than 2 ps, a constant level of the signal is approached which decays within nanoseconds with the S_1 lifetime. The different amplitudes of the slower component found for the two dyes suggest complicated relaxation processes. They cannot be explained by simple closed or open two-level systems.

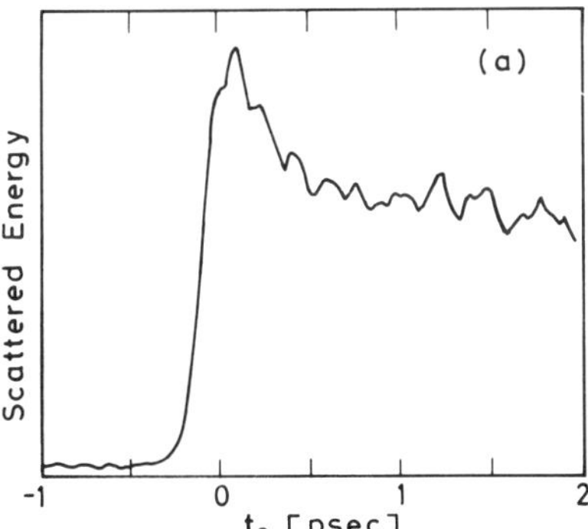

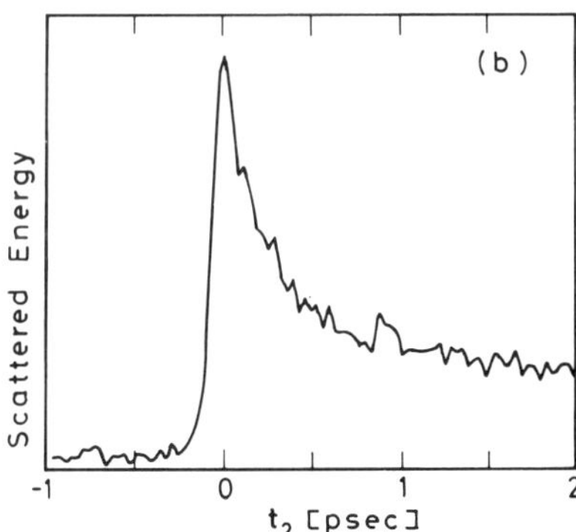

Fig. 6.17. Three-pulse scattering experiment from the organic dye molecules cresyl violet (**a**) and oxazine 720 (**b**) at 15 K using femtosecond light pulses at 620 nm. The scattered energy is plotted as a function of time t_2 for zero relative delay t_1. The signal decay with a time constant of several hundred femtoseconds corresponds to the energy redistribution within the first excited electronic states of the dye molecules [6.142]

b) Measurements of Dephasing Times

Ippen et al. have extended their experiments to measure the short dephasing times in dye molecules. The dephasing processes were determined by varying the time t_1, while keeping t_2 constant at 1.3 ps. The signal scattered by the third pulse was detected in two directions $k_4 = k_c + k_b - k_a$ and $k_5 = k_c + k_a - k_b$. It has been shown in [6.142] that inhomogeneously broadened transitions with long dephasing times $T_2 > t_p$ give scattered signal curves which do not peak at time zero.

Figure 6.18 shows the results for cresylviolet in PMMA at 15 K (Fig. 6.18a) and at 290 K (Fig. 6.18b). The asymmetry is detected for the low-temperature sample. One observes a 60 fs separation of the peaks of the signal curves which demonstrates the presence of inhomogeneous line broadening. With increasing temperature the splitting of the peaks decreases. At room temperature (see Fig. 6.18b) the coincidence of the two signal curves suggests that homogeneous broadening dominates. The same experiments have been performed with Nile blue and oxazine 720. For both dyes the scattered curves coincide over the entire temperature range (15–300 K). These results have been explained as follows: For Nile blue and oxazine 720, the excitation photons at 620 nm have energies of several hundred wavenumbers in excess of the 0–0 transition energy. At large excess energies, background states of high densities exist and a rapid decay of the initially prepared states into these background states can occur. In cresyl violet, on the other hand, low-lying vibronic states, interacting with a small number of background states, were populated; these decay relatively slowly at low temperatures.

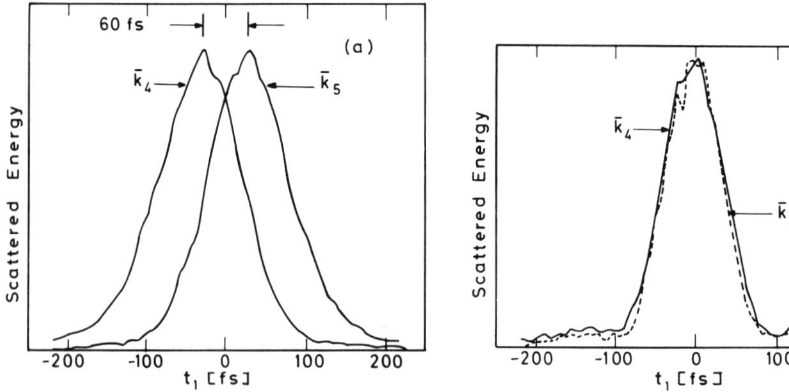

Fig. 6.18a, b. Three-pulse echo experiments for the study of dephasing processes in large dye molecules. For a constant pulse separation, $t_2 = 1.3$ ps, the first delay time t_1 was varied and the three-pulse echo energy was detected in the two directions $\boldsymbol{k}_4 = \boldsymbol{k}_c + \boldsymbol{k}_b - \boldsymbol{k}_a$ and $\boldsymbol{k}_5 = \boldsymbol{k}_c + \boldsymbol{k}_a - \boldsymbol{k}_b$. The asymmetry found in **(a)** – cresyl violet in PMMA at 15 K – indicates the presence of inhomogeneous line broadening. At higher temperatures the asymmetry disappears. At room temperature **(b)** the two curves coincide. Here, homogeneous line broadening is dominant [6.142]

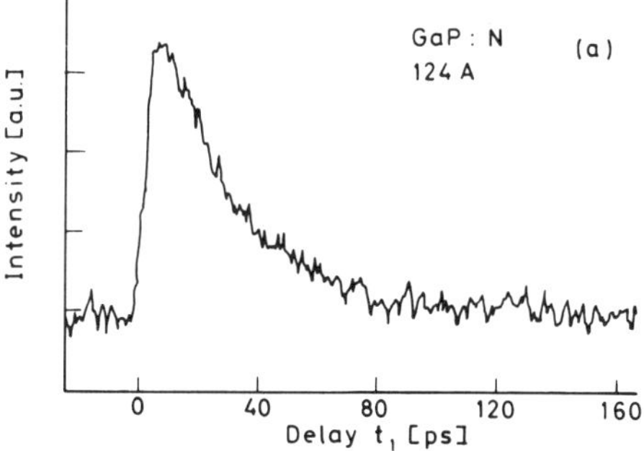

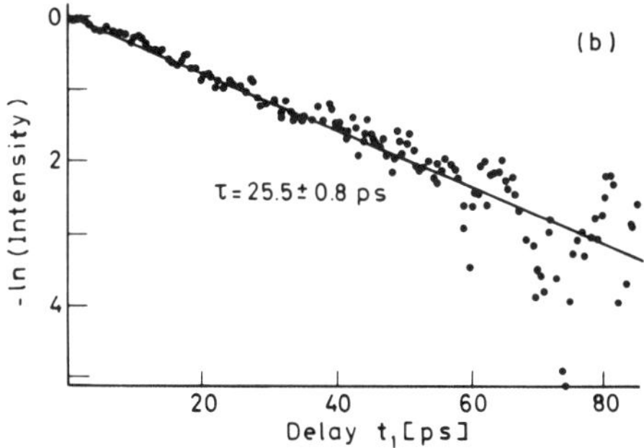

Fig. 6.19a, b. Accumulated photon echo from the nitrogen bond exciton in GaP:N at 1.5 K ($\lambda =$ 534.95 nm). The echo intensity is plotted in a linear scale (**a**) and in a logarithmic scale (**b**) versus delay time t_1. The low-temperature decay time of 25.5 ps is interpreted as the exciton relaxation time out of the initially excited A state [6.131]

c) Accumulated Three-Pulse Echos

Accumulated three-pulse echos have been observed in a number of molecular crystals and semiconductors where the existence of a long-lived bottleneck state allowed the accumulation of the population grating [6.9,134,136–138]. The experiments discussed here were performed by *Wiersma*'s group [6.131]. The authors investigated nitrogen-doped GaP with a N-concentration of $1.3 \times 10^{16}\,\mathrm{cm}^{-3}$. The nitrogen bound exciton at 534.95 nm was investigated at 1.5 K. In optical absorption only the A-state ($J = 1$) can be reached from the ground state ($J = 0$) while the energetically lower B-state ($J = 2$) is forbidden and is only seen in emission. The long lifetime of 4 μs makes the B-state a bottleneck state. In the experiment a synchronously pumped picosecond dye laser was used at the high repetition rate of 80 MHz without amplification. The time dependence of the accumulated echo intensity as a function of the delay time t_1 is plotted on a linear scale and logarithmic scale in Fig. 6.19a and b, respectively. From Fig. 6.19b, a low-temperature decay time of 25.5 ps was deduced. This decay time is interpreted as the exciton relaxation time from the A- to the B-state. The rapid transfer process explains the absence of the A-line in the low-temperature emission spectrum. The temperature dependence of the decay times gives two activation energies. It appears that two distinct interaction processes determine the exciton lifetime.

d) Two-Color Echos

The three-pulse echo proves to be a versatile tool for the study of vibrational deactivation processes when different laser frequencies are applied for the excitation and probing processes. In the upper part of Fig. 6.20 a four-level system (A) is depicted, where the vibrational states $|a\rangle$ and $|b\rangle$ belong to the electronic

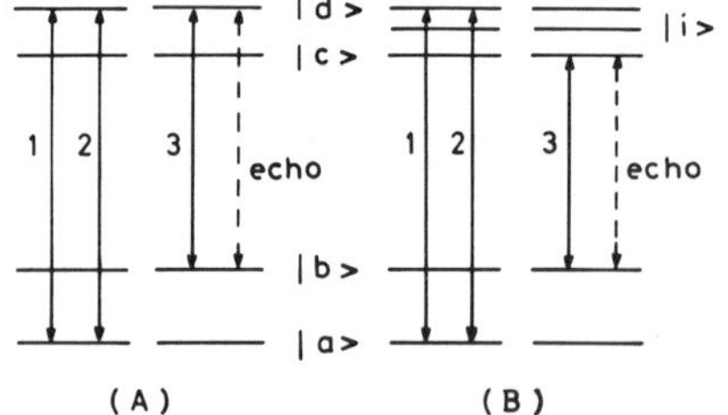

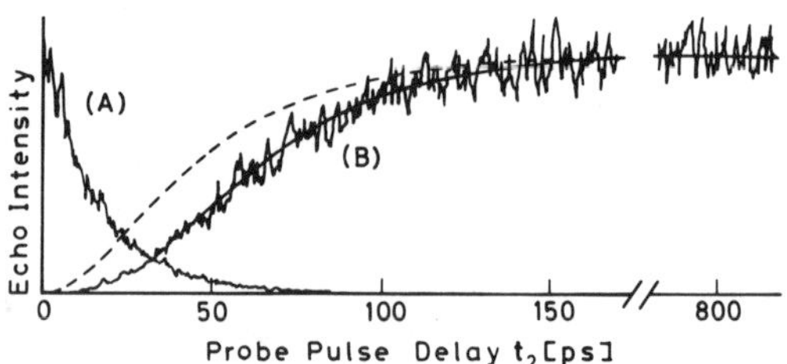

Fig. 6.20. Two-color stimulated photon echos applied to the study of energy relaxation pathways. Left part: Level scheme of the two-color experiment. In case A the first two excitation pulses 1 and 2 generate the population grating in states $|a\rangle$ and $|d\rangle$. Pulses 3 has a wavelength probing only the grating of levels $|b\rangle$ and $|d\rangle$. Varying the delay time t_2 between pulses 2 and 3 allows one to measure the decay of the population of state $|d\rangle$ (curve A). In case B the wavelength of the probing pulse is selected in order to probe states $|b\rangle$ and $|c\rangle$. An echo found with this probing wavelength allows one to follow the population grating of state $|c\rangle$ or $|b\rangle$ built up in the course of relaxation out of state $|d\rangle$. Curve B in the right part of the figure shows a delayed build-up of this echo suggesting the existence of additional levels $|i\rangle$ populated during the relaxation process [6.131]

ground state, and $|c\rangle$ and $|d\rangle$ to the excited electronic state. With two excitation pulses 1 and 2, a population grating is generated in the levels $|a\rangle$ and $|d\rangle$; the probe pulse 3 connects levels $|b\rangle$ and $|d\rangle$. When the inhomogeneous broadening in the different levels is correlated, the third pulse generates an echo signal. The echo amplitude measured as a function of the time t_2 monitors the population decay of state $|d\rangle$. Data from a related experiment are shown in curve A of the right part of Fig. 6.20. The system pentacene in naphthalene was investigated. The first excitation pulse at 17 335 cm^{-1} prepares a vibrational level at 747 cm^{-1} above the absorption origin. The probing pulse at $\nu_3 = 16\,579\,\mathrm{cm}^{-1}$ monitors the occupation of state $|d\rangle$; curve A may be fitted by an exponential decay with a time constant of 33 ps. With the smaller probing frequency $\nu_3/c = 15\,832\,\mathrm{cm}^{-1}$ transitions between levels $|b\rangle$ and $|c\rangle$ are monitored [see Fig. 6.20, scheme (B)]. According to the experimental curve B in the right part of Fig. 6.20 no signal is detected at time zero, i.e. no population is found in the two levels $|b\rangle$ and $|c\rangle$. The echo intensity starts to increase delayed. Within 100 ps it reaches a plateau, where the signal stays constant for the observation time of 800 ps (the lifetime of state $|c\rangle$ is 20 ns). The echo-curve B indicates a population transfer from level $|d\rangle$ to level $|c\rangle$. A more careful inspection of the experimental data suggests a complex transfer process. The experimental echo-curve B deviates from the broken curve which is calculated for a direct population transfer from state $|d\rangle$ to state $|c\rangle$. As a consequence, the authors conclude that there exist intermediate levels $|i\rangle$ with short lifetimes of the order of 20 ps involved in the relaxational processes from level $|d\rangle$ to level $|c\rangle$.

This section clearly demonstrates the power of time-resolved multiple-pulse coherent techniques. The possibility of measuring dephasing times T_2 for inhomogeneously broadened transitions, and also of studying relaxational path ways is very promising. We mention corresponding experiments in the frequency domain: The techniques of spectral hole-burning have become well established. Holes are first "burnt" in an inhomogeneously broadened transition and the widths are subsequently monitored by properly delayed pulses. From the frequency width, the coherent dephasing time may be determined, while the decay of the holes gives information on the energy relaxation T_1 due to cross correlation within the inhomogeneously broadened line or due to the energy decay to other levels. Time- and frequency-domain methods are intimately related. Optimum information is gained by a combination of both techniques.

6.4 Summary

In this chapter we have discussed more than thirty investigations of time-resolved coherent spectroscopy. Many more experiments may be found in the literature. From the various time-resolved coherent techniques we can obtain the following information:

i) whether a transition is homogeneously broadened,
ii) the dephasing time of homogeneously or inhomogeneously broadened transitions,
iii) dominant line broadening mechanisms,
iv) pure dephasing times,
v) energy relaxation times,
vi) mechanisms and pathways of energy relaxation,
vii) collision times,
viii) nonresonant susceptibilities,
ix) transition frequencies within congested spectral regions,
x) precise frequency differences of vibrational modes separated by up to 10 THz.

6.5 Appendix

For the analytical solution of (6.3–5) we introduced in Sect. 6.1.2 the vectorial form $\boldsymbol{\rho}(t)$ of the density matrix and the transformation matrices $\underline{\boldsymbol{X}}(A)$ and $\underline{\boldsymbol{Y}}(t)$ for the excitation process and the free-evolution process, respectively. In detail they are given as follows:

$$\rho_1(t) = \rho_{bb}\ ; \qquad \rho_2(t) = \rho_{ba}\ ; \qquad \rho_3(t) = \rho_{ab}\ ; \qquad \rho_4(t) = \rho_{aa} \tag{6.20}$$

$$Y(t) = \begin{pmatrix} 1 & 0 & 0 & Y_{14} \\ 0 & e^{(i\Delta - 1/T_2)t} & 0 & 0 \\ 0 & 0 & e^{(-i\Delta - 1/T_2)t} & 0 \\ 0 & 0 & 0 & e^{-t/T_1} \end{pmatrix} \tag{6.21}$$

$$X(A_j) = \frac{1}{2} \begin{pmatrix} 1 + \cos A_j & -i \sin A_j e^{ik_j r} & i \sin A_j e^{-ik_j r} & 1 - \cos A_j \\ -i \sin A_j e^{-ik_j r} & 1 + \cos A_j & (1 - \cos A_j) e^{-2ik_j r} & i \sin A_j e^{-ik_j r} \\ i \sin A_j e^{ik_j r} & (1 - \cos A_j) e^{2ik_j r} & 1 + \cos A_j & -i \sin A_j e^{ik_j r} \\ 1 - \cos A_j & i \sin A_j e^{ik_j r} & -i \sin A_j e^{-ik_j r} & 1 + \cos A_j \end{pmatrix} \tag{6.22}$$

where A_j is the pulse area of the excitation pulse, k_j is the wave vector. The matrix element Y_{14} depends on the type of two-level system used. When a closed two-level system is treated, where the system is only in level $|a\rangle$ or $|b\rangle$, Y_{14} becomes $Y_{14} = 1 - \exp(-t/T_1)$. For an open system, where level $|a\rangle$ decays into a long-lived state $|c\rangle$, Y_{14} vanishes: $Y_{14} = 0$.

References

6.1 J.A. Giordmaine, W. Kaiser: Phys. Rev. **144**, 676 (1966)
6.2 S.F. Fischer, A. Laubereau: Chem. Phys. Lett. **35**, 6 (1975)
6.3 D.W. Oxtoby: Adv. Chem. Phys. **40**, 1 (1979)
6.4 R. Kubo: In *Fluctuations, Relaxation and Resonance in Magnetic Systems*, ed. by D. ter Haar (Plenum, New York 1962) p. 101
6.5 R.G. Gordon: Adv. Magn. Reson. **3**, 1 (1968)
6.6 R.G. Gordon: J. Chem. Phys. **39**, 2788 (1963)
6.7 A. Laubereau, W. Kaiser: Rev. Mod. Phys. **50**, 607 (1978)
6.8 A. Penzkofer, A. Laubereau, W. Kaiser: Progr. Quant. Electron. **6**, 55 (1979)
6.9 W.H. Hesselink, D.A. Wiersma: J. Chem. Phys. **73**, 648 (1980)
6.10 A. Yariv: *Quantum Electronics* (Wiley, New York 1975) p. 372
6.11 N.A. Kurnit, I.D. Abella, S.R. Hartmann: Phys. Rev. Lett. **13**, 567 (1964)
6.12 I.D. Abella, N.A. Kurnit, S.R. Hartmann: Phys. Rev. **141**, 391 (1966)
6.13 S.R. Hartmann: IEEE J. Quant. Electron. QE-**4**, 802 (1968)
6.14 R.F. Loring, S. Mukamel: Chem. Phys. Lett. **114**, 426 (1985)
6.15 R.G. Gordon: J. Chem. Phys. **43**, 1307 (1965)
6.16 R.G. Gordon: J. Chem. Phys. **44**, 1830 (1966)
6.17 G. Placzek: *Handbuch der Radiologie*, ed. by E. Marx (Akad. Verlagsges., Leipzig 1934) p. 1
6.18 N. Kohles, A. Laubereau: Appl. Phys. B **39**, 141 (1986)
6.19 B. Dick: Chem. Phys. **113**, 131 (1987)
6.20 N. Bloembergen: Am J. Phys. **35**, 989 (1967)
6.21 W. Kaiser, M. Maier: In *Laser Handbook*, Vol. 2, ed. by F.T. Arecchi, E.O. Schulz-Dubois (North-Holland, Amsterdam 1972) p. 1077
6.22 S.A. Akhmanov: Mater. Res. Bull. **4**, 455 (1969)
6.23 R.L. Carman, F. Shimizu, C.S. Wang, N. Bloembergen: Phys. Rev. A **2**, 60 (1970)
6.24 M. Maier, W. Kaiser, J.A. Giordmaine: Phys. Rev. **177**, 580 (1969)

6.25 Y.R. Shen, N. Bloembergen: Phys. Rev. **137A**, 1786 (1965)
6.26 N. Bloembergen: *Nonlinear Optics* (Benjamin, New York 1965)
6.27 P.D. Maker, R.W. Terhune: Phys. Rev. **137A**, 801 (1965)
6.28 A. Penzkofer, W. Kaiser: Opt. Quant. Electron. **9**, 315 (1977)
6.29 D. von der Linde, A. Laubereau, W. Kaiser: Phys. Rev. Lett. **26**, 954 (1971)
6.30 R.R. Alfano, S.L. Shapiro: Phys. Rev. Lett. **26**, 1247 (1971)
6.31 C.H. Lee, D. Richard: Appl. Phys. Lett. **32**, 168 (1978)
6.32 J.P. Heritage: Appl. Phys. Lett. **34**, 470 (1979)
6.33 J. Kuhl, D. von der Linde: In *Ultrafast Phenomena III*, Proc. Topical Meeting, Garmisch-Partenkirchen, Germany, June 1982, Springer Ser. Chem. Phys., Vol. 23, ed. by K.B. Eisenthal, R.M. Hochstrasser, W. Kaiser (Springer, Berlin, Heidelberg 1982) p. 201
6.34 R. Leonhardt, W. Holzapfel, W. Zinth, W. Kaiser: Chem. Phys. Lett. **133**, 373 (1987)
6.35 A. Laubereau: Chem. Phys. Lett. **27**, 600 (1974)
6.36 A. Laubereau, G. Wochner, W. Kaiser: Opt. Commun. **17**, 91 (1976)
6.37 W. Zinth, H.J. Polland, A. Laubereau, W. Kaiser: Appl. Phys. B **26**, 77 (1981)
6.38 H. Graener, A. Laubereau, J.W. Nibler: Opt. Lett. **9**, 165 (1984)
6.39 M.L. Geirnaert, G.M. Gale, C. Flytzanis: Phys. Rev. Lett. **52**, 815 (1984)
6.40 M.L. Geirnaert, G.M. Gale: Chem. Phys. **86**, 205 (1984)
6.41 J. Chesnoy: Chem. Phys. Lett. **125**, 267 (1986)
6.42 G.M. Gale, P. Guyot-Sionnest, W.Q. Zheng: Opt. Commun. **58**, 395 (1986)
6.43 M. van Exter, A. Lagendijk: Opt. Commun. **56**, 191 (1985)
6.44a W. Zinth, R. Leonhardt, W. Holzapfel, W. Kaiser: J. Quant. Electron. QE-**24**, 455 (1988)
6.44b R. Leonhardt, W. Holzapfel, W. Zinth, W. Kaiser: private communication
6.45 D.W. Oxtoby, D. Levesque, J.J. Weis: J. Chem. Phys. **68**, 5528 (1978)
6.46 J.P. Riehl, D.J. Diestler: J. Chem. Phys. **64**, 2593 (1976)
6.47 D.J. Diestler, R.S. Wilson: J. Chem. Phys. **62**, 1572 (1975)
6.48a P.A. Madden, R.M. Lynden-Bell: Chem. Phys. Lett. **38**, 163 (1976)
6.48b R.M. Lynden-Bell, G.C. Tabisz: Chem. Phys. Lett. **46**, 175 (1977)
6.49 S. Bratos, E. Marechal: Phys. Rev. A **4**, 1078 (1971)
6.50 W.G. Rothschild: J. Chem. Phys. **65**, 455 (1976)
6.51 W.G. Rothschild: J. Chem. Phys. **65**, 2958 (1976)
6.52 D.J. Diestler, J. Manz: Mol. Phys. **33**, 227 (1977)
6.53 R. Wertheimer: Mol. Phys. **35**, 257 (1978)
6.54 R. Wertheimer: Chem. Phys. Lett. **52**, 224 (1977)
6.55 D. Oxtoby: J. Chem. Phys. **70**, 2605 (1979)
6.56 R.M. Shelby, C.B. Harris, P.A. Cornelius: J. Chem. Phys. **70**, 34 (1979)
6.57 S.F. Fischer, A. Laubereau: Chem. Phys. Lett. **55**, 189 (1978)
6.58 S. Mukamel: Phys. Rev. A **26**, 617 (1982)
6.59 S. Mukamel: Phys. Rev. A **28**, 3480 (1983)
6.60 W.G. Rothschild, M. Perrot, F. Guillaume: Chem. Phys. Lett. **128**, 591 (1986)
6.61 W. Zinth, A. Laubereau, W. Kaiser: Opt. Commun. **26**, 457 (1978)
6.62 F.M. Kamga, M.G. Sceats: Opt. Lett. **5**, 126 (1980)
6.63 M.G. Sceats, F. Kamga, D. Podolski: In *Picosecond Phenomena II*, Proc. Topical Meeting, Cape Cod, Mass., USA, June 1980, Springer Ser. Chem. Phys., Vol. 14, ed. by R.M. Hochstrasser, W. Kaiser, C.V. Shank (Springer, Berlin, Heidelberg 1980) p. 348
6.64 W. Zinth: Opt. Commun. **34**, 479 (1980)
6.65 B.K. Rhee, W.E. Bron, J. Kuhl: Phys. Rev. B **30**, 7358 (1984)
6.66 S. Velsko, J. Trout, R.M. Hochstrasser: J. Chem. Phys. **79**, 2114 (1983)
6.67 D.E. McCumber, M.D. Sturge: J. Appl. Phys. **34**, 1682 (1963)
6.68 R. Orbach: IEEE Trans. Sonics Ultrason. **SU14**, 140 (1967)
6.69 R. Orbach, L.A. Vredevoe: Physics **1**, 91 (1964)
6.70 P.G. Klemens: Phys. Rev. **148**, 845 (1966)

6.71 I.I. Abram, R.M. Hochstrasser: J. Chem. Phys. **72**, 3617 (1980)
6.72 S. Velsko, R.M. Hochstrasser: J. Chem. Phys. **82**, 2180 (1985)
6.73 G.J. Small: Chem. Phys. Lett. **57**, 501 (1978)
6.74 K.E. Jones, A.H. Zewail: In *Advances in Laser Chemistry*, ed. by A.H. Zewail (Springer, New York 1978) p. 196
6.75 C.B. Harris, R.M. Shelby, P.A. Cornelius: Phys. Rev. Lett. **38**, 1415 (1977)
6.76 C.B. Harris J. Chem. Phys. **67**, 5607 (1977)
6.77 P. de Bree, D.A. Wiersma: J. Chem. Phys. **70**, 790 (1978)
6.78 R.G. Delle Valle, P.F. Fracassi, R. Righini, S. Califano: Chem. Phys. **74**, 179 (1983)
6.79 D.D. Dlott: Ann. Rev. Phys. Chem. **37**, 157 (1986)
6.80 W.H. Hesselink, D.A. Wiersma: Chem. Phys. Lett. **65**, 300 (1979)
6.81 D. von der Linde, J. Kuhl, H. Klingenberg: Phys. Rev. Lett. **44**, 1505 (1980)
6.82 A. Laubereau, D. von der Linde, W. Kaiser: Phys. Rev. Lett. **27**, 802 (1971)
6.83a W.E. Bron, J. Kuhl, B.K. Rhee: Phys. Rev. B **34**, 6961 (1986)
6.83b J. Kuhl, W.E. Bron: Solid State Commun. **49**, 935 (1984)
6.84 G.M. Gale, A. Laubereau: Opt. Commun. **44**, 273 (1983)
6.85 A. Laubereau, G. Wochner, W. Kaiser: Opt. Commun. **14**, 75 (1975)
6.86 A. Laubereau: In *Semiconductors Probed by Ultrafast Laser Spectroscopy*, Vol. 1, ed. by R.R. Alfano (Academic, Orlando 1984) p. 275
6.87 G.M. Gale, F. Vallee, C. Flytzanis: Phys. Rev. Lett. **57**, 1867 (1986)
6.88 G.M. Gale, P. Guyot-Sionnest, W.Q. Zheng, C. Flytzanis: Phys. Rev. Lett. **54**, 823 (1985)
6.89 G.M. Gale, F. Vallee, C. Flytzanis: In *Time Resolved Vibrational Spectroscopy*, Proc. Conf. June 1985, Springer Proc. Phys. IV, ed. by A. Laubereau, M. Stockburger (Springer, Berlin, Heidelberg 1985) p. 117
6.90 C. Flytzanis, G.M. Gale, M.L. Geirnaert: In *Applications of Picosecond Spectroscopy to Chemistry*, ed. by K.B. Eisenthal (Reidel, Amsterdam 1984) p. 205
6.91 P.L. Decola, R.M. Hochstrasser, H.P. Frommsdorff: Chem. Phys. Lett. **72**, 1 (1980)
6.92 B.H. Hesp. D.A. Wiersma: Chem. Phys. Lett. **75**, 423 (1980)
6.93 K. Duppen, B.H. Hesp, D.A. Wiersma: Chem. Phys. Lett. **79**, 399 (1981)
6.94 K. Duppen, D.P. Weitekamp, D.A. Wiersma: J. Chem. Phys. **79**, 5835 (1983)
6.95 C.L. Schosser, D.D. Dlott: J. Chem. Phys. **80**, 1394 (1984)
6.96 F. Ho, W.S. Tsay, J. Trout, R.M. Hochstrasser: Chem. Phys. Lett. **83**, 5 (1981)
6.97 F. Ho, W.S. Tsay, S. Velsko, R.M. Hochstrasser: Chem. Phys. Lett. **97**, 141 (1983)
6.98 S. Velsko, J. Trout, R.M. Hochstrasser, J. Chem. Phys. **79**, 2114 (1983)
6.99 T.J. Trout, S. Velsko, R. Bozio, P.L. Decola, R.M. Hochstrasser: J. Chem. Phys. **81**, 4746 (1984)
6.100 S. Velsko, R.M. Hochstrasser: J. Chem. Phys. **89**, 2240 (1985)
6.101 R.M. Hochstrasser: In *Time Resolved Vibrational Spectroscopy*, Proc. Conf. June 1985, Springer Proc. Phys. IV, ed. by A. Laubereau, M. Stockburger (Springer, Berlin, Heidelberg 1985) p. 96
6.102 T.J. Kosic, R.E. Cline Jr., D.D. Dlott: J. Chem. Phys. **81**, 4932 (1984)
6.103 R.F. Loring, S. Mukamel: In *Time Resolved Vibrational Spectroscopy*, Proc. Conf. June 1985, Springer Proc. Phys. IV, ed. by A. Laubereau, M. Stockburger (Springer, Berlin, Heidelberg 1985) p. 293
6.104 H.R. Telle, A. Laubereau: Chem. Phys. Lett. **94**, 467 (1983)
6.105 W. Zinth, W. Kaiser: In *Organic Molecular Aggregates*, Springer Ser. Solid-State Sci., Vol. 49, ed. by P. Reinecker, H. Haken, H.C. Wolf (Springer, Berlin, Heidelberg 1983) p. 124
6.106 R. Leonhardt, W. Holzapfel, W. Zinth, W. Kaiser: private communication
6.107a H. Graener, A. Laubereau: Opt. Commun. **54**, 141 (1985)
6.107b H. Graener, A. Laubereau: In *Time Resolved Vibrational Spectroscopy*, Proc. Conf. June 1985, Springer Proc. Phys. IV, ed. by A. Laubereau, M. Stockburger (Springer, Berlin, Heidelberg 1985) p. 11

6.108 S. De Silvestri, J.G. Fujimoto, E.P. Ippen, E.B. Gamble Jr., L.R. Williams, K.A. Nelson: Chem. Phys. Lett. **116**, 146 (1985)
6.109 A. Laubereau, G. Wochner, W. Kaiser: Chem. Phys. **28**, 363 (1978)
6.110 A. Laubereau, G. Wochner, W. Kaiser: Phys. Rev. A **13**, 2212 (1976)
6.111 C.B. Harris, H. Auweter, S.M. George: Phys. Rev. Lett. **44**, 737 (1980)
6.112 S.M. George, H. Auweter, C.B. Harris: J. Chem. Phys. **73**, 5573 (1980)
6.113 S.M. George, C.B. Harris: Phys. Rev. A **28**, 863 (1983)
6.114 S.M. George, A.L. Harris, M. Berg, C.B. Harris: J. Chem. Phys. **80**, 83 (1984)
6.115 J.C. Diels: IEEE J. Quant. Electron. QE-**16**, 1020 (1980)
6.116 R.F. Loring, S. Mukamel: J. Chem. Phys. **83**, 2116 (1985)
6.117 W. Zinth, W. Kaiser: In *Lecture Notes in Physics*, Vol. 182, ed. by J.D. Harvey, D.F. Walls (Springer, New York 1983) p. 152
6.118 W. Zinth, M.C. Nuss, W. Kaiser: Chem. Phys. Lett. **88**, 257 (1982)
6.119 W. Zinth, M.C. Nuss, W. Kaiser: Opt. Commun. **44**, 262 (1983)
6.120 W. Zinth, M.C. Nuss, W. Kaiser: Phys. Rev. A **30**, 1139 (1984)
6.121 M.C. Nuss, W. Zinth, W. Kaiser: J. Opt. Soc. Am. B **2**, 322 (1985)
6.122 M.A. Collins, P.A. Madden, A.D. Buckingham: Chem. Phys. **94**, 291 (1985)
6.123 M.D. Crisp: Phys. Rev. A **1**, 1604 (1970)
6.124 H.J. Hartmann, A. Laubereau: Opt. Commun. **47**, 117 (1983)
6.125 H.J. Hartmann, A. Laubereau: J. Chem. Phys. **80**, 4663 (1984)
6.126 H.J. Hartmann, K. Bratengeier, A. Laubereau: Chem. Phys. Lett. **108**, 555 (1984)
6.127 H.J. Hartmann, H. Schleicher, A. Laubereau: Chem. Phys. Lett. **116**, 392 (1985)
6.128 D.W. Phillion, D.J. Kuizenga, A.E. Siegman: Appl. Phys. Lett. **27**, 85 (1975)
6.129 A.E. Siegman: Appl. Phys. Lett. **30**, 21 (1977)
6.130 R. Trebino, A.E. Siegman: J. Chem. Phys. **79**, 3621 (1983)
6.131 K. Duppen, L.W. Molenkamp, D.A. Wiersma: Physica **127B**, 349 (1984)
6.132 L. Schultheis, M.D. Sturge, J. Hegarty: Appl. Phys. Lett. **47**, 995 (1985)
6.133 W.H. Hesselink, D.A. Wiersma: Chem. Phys. Lett. **56**, 227 (1978)
6.134 W.H. Hesselink, D.A. Wiersma: Phys. Rev. Lett. **43**, 1991 (1979)
6.135 W.H. Hesselink, D.A. Wiersma: J. Chem. Phys. **74**, 886 (1981)
6.136 L.W. Molenkamp, D.A. Wiersma: J. Chem. Phys. **80**, 3054 (1984)
6.137 K. Duppen, D.P. Weitekamp, D.A. Wiersma: Chem. Phys. Lett. **108**, 551 (1984)
6.138 K. Duppen, D.P. Weitekamp, D.A. Wiersma: Chem. Phys. Lett. **106**, 147 (1984)
6.139 L.W. Molenkamp, D.P. Weitekamp, D.A. Wiersma: Chem. Phys. Lett. **99**, 382 (1983)
6.140 A.M. Weiner, E.P. Ippen: Opt. Lett. **9**, 53 (1984)
6.141 S. De Silvestri, A.M. Weiner, J.G. Fujimoto, E.P. Ippen: Chem. Phys. Lett. **112**, 195 (1984)
6.142 A.M. Weiner, S. De Silvestri, E.P. Ippen: J. Opt. Soc. Am. B **2**, 654 (1985)
6.143 J.L. Skinner, H.C. Andersen, M.D. Fayer: J. Chem. Phys. **75**, 3195 (1981)
6.144 W.H. Hesselink, D.A. Wiersma, J. Chem. Phys. **75**, 4192 (1981)
6.145 D.P. Weitekamp, K. Duppen, D.A. Wierma: Phys. Rev. A **27**, 3089 (1983)
6.146 H. Paerschke, K.E. Süsse, B. Wilhelmi: Opt. Quant. Electron. **15**, 41 (1983)
6.147 J.G. Fujimoto, T.K. Yee: Appl. Phys. B **34**, 55 (1984)
6.148 L. Schultheis, A. Honold, J. Kuhl, K. Köhler, C.W. Tu: Phys. Rev. B **34**, 9027 (1986)
6.149 A.M. Weiner, E.P. Ippen: Chem. Phys. Lett. **114**, 456 (1985)

7. Ultrashort Intramolecular and Intermolecular Vibrational Energy Transfer of Polyatomic Molecules in Liquids

Alois Seilmeier and Wolfgang Kaiser

With 22 Figures

In this chapter the fundamental problem of transfer and dissipation of vibrational energy in liquids is investigated on a microscopic level. Vibrational excess energy may appear locally by chemical reactions or via optical excitations. The subsequent redistribution within the excited molecule and the flow of energy to the surroundings occurs very rapidly. Picosecond pulses are necessary to study the various relaxation processes which finally lead to thermal equilibrium.

Of particular interest is the population lifetime of individual vibrational modes (Sect. 7.2). An understanding of this important vibrational parameter has been gained only recently. Prior to the development of ultrashort laser pulses, our knowledge of the population lifetime was almost nonexistent. In early papers a "vibrational lifetime" was frequently deduced from the spectral linewidth of the corresponding vibrational mode. In 1972 a clear distinction was established between the population lifetime T_1 and the vibrational dephasing time T_2 [7.1]. The latter frequently determines the line width of the mode (see Chap. 6). Other line-broadening effects, such as inhomogeneous broadening or isotope multiplicity, may also contribute to the measured line width. In most cases one finds $T_1 > T_2$, i.e. the energy remains longer in the vibrational mode than the phase of the vibration in the coherently excited medium.

The population lifetime of a vibrational mode is of paramount importance for the question of state-selective chemistry. This topic has intrigued chemists for many years. The T_1 values observed so far are too small ($T_1 < 10^{-9}$ s) to give state-selective chemistry a chance in the liquid state.

In larger organic molecules, energies of larger than $1000\,\mathrm{cm}^{-1}$ are – in most cases – very rapidly (<1 ps) redistributed over the vibrational manifold of the electronic ground state. Different experimental methods give evidence of the vibrational redistribution in the electronic ground state. It has been shown that one may speak of a transient internal temperature of the excited molecules. This fact is important for radiationless transitions, where large amounts of energy are transferred from electronic states to high-lying vibrational levels of the electronic ground state. We note that radiationless transitions are most common, while fluorescence transitions occur less frequently. In spite of this fact very little consideration has been given to the question of what happens to an individual molecule after internal conversion, where an energy of several eV or 100 kcal/mol is suddenly transferred to the molecule's ground state. This subject will be discussed in Sect. 7.3.

The question now arises of how long the excess energy (or temperature) stays within an individual molecule in the liquid phase. A series of experimental studies show quite convincingly that intermolecular energy dissipation occurs fast, within the order of 10^{-11} s but specific to the degree of excitation and to the solvent molecules. We note that in large molecules (e.g. dyes) intramolecular energy redistribution is faster than intermolecular energy transfer. In many cases one may readily separate the two processes (Sect. 7.4).

The equalization of energy can be studied in organic liquids with the help of a very rapid molecular thermometer. Individual vibrational modes of the solvent molecules are excited (in the electronic ground state) and the build-up and decay of the temperature in the medium is followed on a picosecond time scale. It is possible to observe intramolecular and intermolecular energy transfer processes within the time for equalization of energy of the order of 10^{-11} s (Sect. 7.5).

Large molecules (e.g. dyes) excited to high vibronic states, substantially above the 0-0 transition, exhibit fast intramolecular relaxation to the bottom of the S_1 state. Vibrational redistribution of energy in S_1 and S_0 appears to be equally rapid, of the order of 500 fs (Sect. 7.6).

In this chapter we concentrate on the liquid phase; reference to gaseous data is made when appropriate. The dynamics of vibrational relaxation of gases [7.2] and of low-temperature solids [7.3] proceed on a much slower time scale and standard electronic detection systems may be applied.

7.1 Experimental Techniques

The ultrafast intramolecular and intermolecular vibrational relaxation processes of polyatomic molecules in liquids require new experimental methods for the direct observation of the relevant time constants. Photocells or photodiodes are not fast enough to measure times of 10 ps and less.

Generally, pump-probe techniques are used. A first ultrashort light pulse excites the molecules and a second, delayed pulse probes the momentary population of the vibrational states. There exist different excitation and probing processes adapted to different problems and to various molecular systems.

7.1.1 Excitation

In Fig. 7.1 three different excitation processes are depicted schematically. They have been used successfully in a variety of investigations.

i) By stimulated Raman excitation an excess population may be generated in a vibrational state (Fig. 7.1a). The method is experimentally simple; only one intense laser pulse of any fixed frequency smaller than the electronic transition frequencies is required for the excitation of a defined vibrational mode. Only one

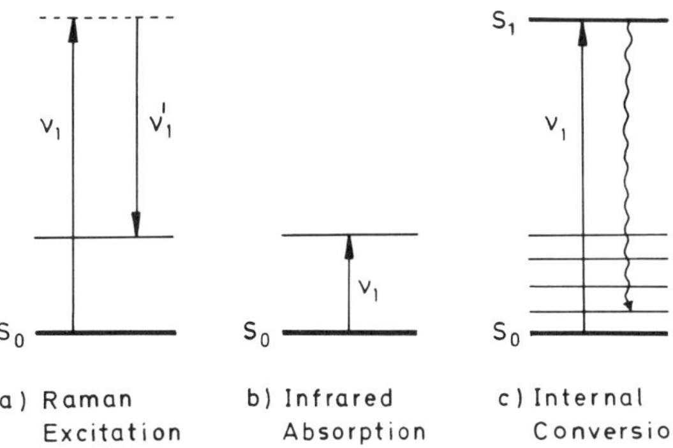

Fig. 7.1a–c. Three techniques for the vibrational excitation in the electronic ground state S_0. A single vibrational state is populated via (**a**) Raman excitation or (**b**) resonant infrared absorption. (**c**) A larger amount of energy is supplied to the vibrational manifold via electronic excitation and subsequent rapid internal conversion

specific vibration, the mode with the largest Raman scattering cross-section, is excited. Excess populations of the order of 10^{-3} are achieved in neat liquids and in highly concentrated systems [7.4]. In a more elaborate system, the liquid is excited by two ultrashort pulses [7.5,6] at the frequencies ν_L and ν_S (the Stokes frequency $\nu_S = \nu_L - \nu_{vib}$). In this way excess populations as high as 3×10^{-2} have been reported.

ii) More flexibility in the vibrational excitation is provided by the infrared absorption technique (Fig. 7.1b). An infrared light pulse of frequency ν_1 resonantly excites a vibrational mode of the molecule in the electronic ground state [7.7–9]. Any infrared-active mode may be populated when the corresponding infrared pulses are available. Population numbers exceeding 10^{-1} have been reported for transitions with favorably large absorption cross-sections [7.10].

These numbers have to be compared with the thermal population at room temperature. For vibrational states around $3000\,\mathrm{cm}^{-1}$ and $1000\,\mathrm{cm}^{-1}$ one calculates thermal populations of approximately 10^{-6} and 10^{-2}, respectively, i.e. substantial excess populations can readily be generated for higher vibrational states with picosecond pulses.

iii) Large vibrational excitation with total energies of several $10^4\,\mathrm{cm}^{-1}$ or more may be supplied to a molecule by internal conversion of electronic energy (Fig. 7.1c). With a first visible or UV pulse, transitions are made to the first excited state S_1. In molecules with rapid internal conversion (radiationless transitions) the electronic energy is transferred to the vibrational manifold of the ground state. In this way large excess vibrational populations (corresponding to a high transient internal temperature) may be generated [7.11,12].

The excess population of the primarily excited vibrational modes in techniques (i) and (ii) relaxes with a specific lifetime T_1 and lower lying states are populated subsequently. The energy transferred to the vibrations of the ground states by method (iii) is redistributed within the molecule. For an analysis of the momentary populations, fast probing techniques are necessary. Ideally, these monitor the instantanous excitation of different vibrational modes ν_i.

7.1.2 Probing

Four different probing techniques developed in recent years, are shown schematically in Fig. 7.2. In the figure the molecules are assumed to be vibrationally excited at frequency ν_1, e.g. by an infrared pulse. Other excitation techniques may be applied as well.

i) The anti-Stokes Raman probing technique is shown schematically in Fig. 7.2a. The momentary population of different vibrational modes ν_i is proportional to the magnitude of the spontaneous anti-Stokes Raman signal at the frequency $\nu_{AS} = \nu_2 + \nu_i$ produced by a probe pulse of fixed frequency ν_2. In a number of cases ν_2 is the second harmonic frequency at $18\,940\,cm^{-1}$ of a Nd:glass system. Time-resolved data are obtained by proper delay of the monitoring pulse relative to the excitation pulse [7.1,5,13,22]. A geometric pulse delay of 3×10^{-2} cm gives a time delay of one picosecond.

The anti-Stokes technique represents a powerful method, since the anti-Stokes spectrum gives detailed information on the degree of population, on the lifetimes of different vibrational states and on the energy flow between the different levels. Unfortunately, the technique is experimentally difficult on account of the small Raman scattering cross-sections of vibrational modes [7.14]. There are also upper limits to the intensities of the probing pulse. At very high intensities spurious signals are generated by various nonlinear optical processes, e.g. by dielectric breakdown, multiphoton transitions or self–phase modulation. Careful experimentation is required to obtain reliable data. The method is restricted to highly concentrated systems or pure liquids.

The following numbers are relevant for an estimate of the magnitude of the expected Raman signal. In liquids the number density is approximately 5×10^{21}

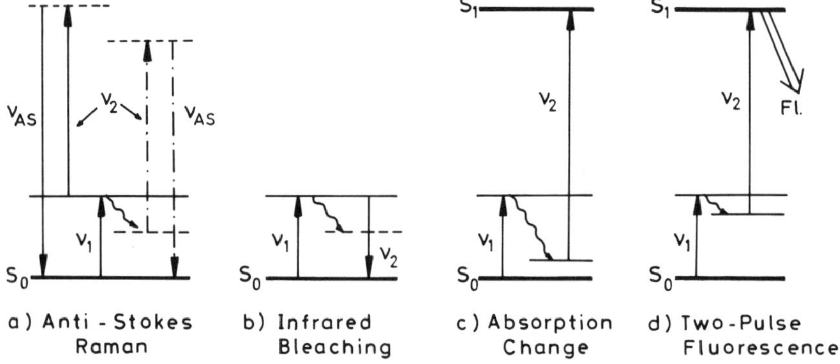

Fig. 7.2a–d. Methods for probing the instantaneous population of vibrational state: The population of a well-defined vibrational level is monitored (**a**) by spontaneous anti-Stokes Raman scattering or (**b**) via changes of the sample transmission at the corresponding infrared frequency. The vibrational population may be measured via transitions to the electronic S_1 state. Changes in the absorption (**c**) or the subsequently emitted fluorescence signal (**d**) are proportional to the momentary excess population of the monitored vibrational state

molecules/cm^3. For a relative vibrational population of 10^{-2}, an interaction length of 0.1 cm, an acceptance angle of 1 sr, a Raman scattering cross-section of 10^{-30} cm^2/sr, and a probe pulse of 10^{16} photons ($\sim$1 mJ) one calculates 10^4 accepted photons. These numbers show that the relaxation of the excess population can be measured over one to two orders of magnitude.

ii) In Fig. 7.2b the infrared bleaching technique is depicted schematically. A vibrational transition is partially saturated by a first intense infrared excitation pulse at the resonance frequency ν_1. A second, properly delayed probe pulse of the same frequency monitors the time dependence of the change in transmission (bleaching of the sample). The return to the initial transmission gives the population lifetime T_1 of the excited vibration [7.15,16].

In order to observe bleaching of a transition, the excitation intensity has to be of the order of the saturation intensity I_s. One has $I_s = h\nu_i/\sigma_i t_p$ if the population lifetime T_1 of the mode i is longer than the duration of the excitation pulse t_p of several 10^{-12} s. $h\nu_i$ is the energy and σ_i the absorption cross-section of the transition. Since σ_i is very small for vibrational modes in the liquid state (e.g. 10^{-20} cm^2 for a CH mode), very high infrared intensities of 10^{11} to 10^{12} W/cm^2 are calculated. The bleaching technique has been used for vibrational transitions with favorably large values of σ_i ($\sim 10^{-18}$ cm^2) and relatively long T_1 times exceeding 10^{-11} s [7.10]. It should be noted that saturation of a vibrational transition requires a certain degree of anharmonicity of the mode. The frequency of the first overtone has to be smaller (or larger) than twice the fundamental frequency.

Special care has to be taken when the dephasing time of the investigated vibrational state is longer than, or comparable to, the pulse duration. In this case coherent effects have to be considered which make the analysis of the bleaching data more complicated [7.17]. The full Maxwell-Bloch equations have to be applied instead of the simple rate equations.

iii) In large molecules the vibrational excess energy is rapidly distributed over the vibrational manifold of the molecule (see Sect. 7.3). A pulse with photon energy $h\nu_2$ smaller than the 0-0 transition monitors the change of absorption in the long-wavelength tail of the electronic absorption (see Fig. 7.2c). Here, transitions to the S_1 state have to start from populated vibrational states in S_0. Excess population of low-lying modes gives an enhanced absorption [7.18,19]. For higher vibrational modes the thermal population is small and the absorption is barely measurable. For small excitations the absorption change is measured with greatest accuracy when the sample transmission is approximately 1/e. For a probing frequency ν_2, vibrational states at the frequency $\nu_i \gtrsim \nu_{00} - \nu_2$ are monitored (ν_{00} is the frequency of the electronic $S_0 \to S_1$ transition).

iv) A sensitive technique to detect the change in population of higher vibrational states is the two-pulse fluorescence probing technique (Fig. 7.2d). Actually, one measures a fluorescence activation spectrum of the molecule and substracts the signals of the sample in thermal equilibrium. The fluorescence signal is directly proportional to the population of the probed vibrational state. Large changes in the fluorescence signal are observed when high lying vibrational modes with a very small thermal population are investigated. The method is quite sensitive; highly diluted systems ($c \sim 10^{-5}$ M) may be studied in this way [7.7,20,21].

v) Independent data on the T_1 lifetime of vibrational modes may be obtained from the energy equalization in solutions (see Sect. 7.5). For molecules with long T_1 times the equalization of energy, i.e. the energy transfer to a neighboring molecular detector, is time retarded. Delay times of the order of several 10^{-11} s have been observed [7.18,23].

The excitation and probing techniques introduced here are applied in various experiments discussed in this chapter. For example, excitation via internal conversion of electronic energy (Fig. 7.1c) and probing by the absorption changes (Fig. 7.2c) have been used for the investigation of the dynamics of vibrationally very hot molecules in solutions (see Sect. 7.4.2).

7.1.3 Experimental System

Several of the experimental techniques outlined above require tuning of the infrared pulses at frequency ν_1 and of the probing pulses at frequency ν_2. Both infrared and probing pulses, should have a duration of a few picoseconds with a time jitter between them of less than 1 ps. As an example we show in Fig. 7.3

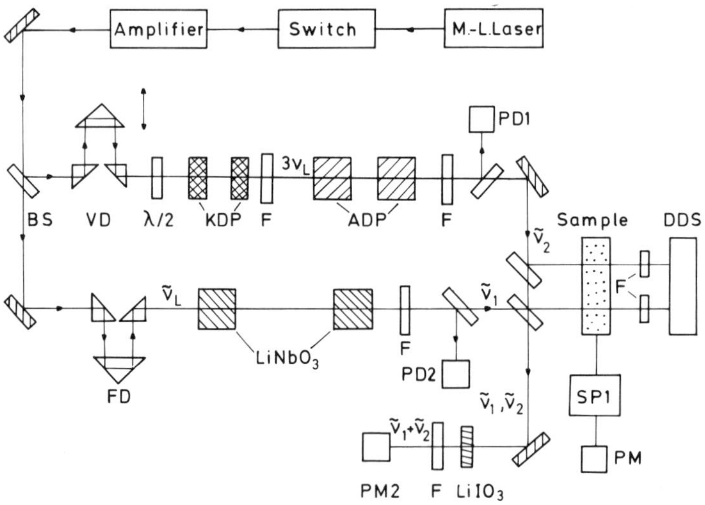

Fig. 7.3. Schematic of an experimental system for the investigation of vibrational relaxation in liquids. An infrared excitation pulse (ν_1) is generated in a parametric system consisting of two $LiNbO_3$ crystals. The visible probe pulse (ν_2) is produced by a parametric process in two ADP crystals pumped by the third harmonic $3\nu_L$ of a mode-locked Nd: laser. The cross-correlation curve is measured simultaneously with the help of a thin $LiIO_3$ crystal. The two pulses at frequencies ν_1 and ν_2 travel through the sample with variable delay times. Absorption changes are measured with a differential detector system DDS. The fluorescence signal is monitored by a spectrometer SP1 and a photomultiplier PM [7.21]

an experimental system that fulfills the preceeding requirements [7.24]. A mode-locked Nd:glass laser provides a train of near infrared pulses of 5 ps duration at $\tilde{\nu}_L = 9500\,\mathrm{cm}^{-1}$. A single, nearly bandwidth-limited pulse is cut from the leading edge of the pulse train and is amplified in two glass amplifiers to a pulse energy of approximately 50 mJ. The beam splitter (BS) generates two pulses of equal intensity. In the upper beam the infrared pulse is first tripled to the third harmonic frequency, and subsequently passes through a parametric frequency converter, in the present case through two ADP crystals. Rotation of the crystals provides signal and idler pulses of the appropriate frequency ν_2 tunable over a wide frequency range from the UV to the near infrared (see Fig. 3.6) [7.24]. The second pulse at ν_L traverses two $LiNbO_3$ crystals generating infrared pulses of frequency ν_1 tunable between $2500\,\mathrm{cm}^{-1}$ and $4000\,\mathrm{cm}^{-1}$ [7.8,15,25,26]. The two pulses, ν_1 and ν_2, are advanced or delayed relative to each other by the fixed and variable delay lines FD and VD, respectively. The two pulses travel collinearly through the sample, where Raman signals, absorption or fluorescence changes may be measured.

In Fig. 7.3 transmission changes of the sample are monitored employing a differential detector system, DDS, consisting of two photodiodes. The photodiodes measure the energy of the two transmitted pulses of frequency ν_2 at two sample positions, namely with and without infrared excitation. The differential detector system determines the energy difference of the two pulses optoelectronically [7.27]. With this arrangement, absorption changes may be measured with an accuracy of $\simeq 10^{-4}$. The spectrometer SP1 and a photomultiplier PM are used to measure the fluorescence signal when high lying modes are investigated.

Of considerable importance is the autocorrelation measurement depicted at the lower right of the figure. A small part of the two pulses at frequencies ν_1 and ν_2 passes through a thin $LiIO_3$ crystal and the sum frequency at $\nu_1 + \nu_2$ is registered while the delay position between the two pulses is varied. The measured cross-correlation curve gives the zero point of the time scale (when the peaks of both pulses overlap) and indicates the time resolution of the experimental system [7.28]. In most of the following experiments the cross-correlation curve is measured simultaneously with the main investigations. In this way, the precision of the data is substantially improved and the limit of time resolution is readily seen.

7.2 Vibrational Population Lifetimes T_1 of Small Molecules in the Electronic Ground State S_0

In recent years a considerable amount of experimental and theoretical information has been accumulated on the relaxation of vibrational energy in small molecules. Small molecules have – in most cases – well-separated vibrational states. For this reason it is possible to study the population dynamics of well-defined vibrational modes. The small number of vibrational states has the additional advantage of making the experimental relaxation data more readily amenable to theoretical interpretations. In a number of favorable cases (e.g. in acetylene [7.29]) the flow of vibrational energy from a high energetic state to a lower mode has been experimentally observed and theoretically interpreted.

In the following a review of the molecules investigated so far is presented. A theoretical discussion of the relaxation mechanisms follows. A formula is discussed which allows one to estimate the vibrational population lifetimes from independent spectroscopic information.

7.2.1 Experimental Data

As a first example, the transient population and depopulation of the symmetric CH_2-stretching mode in two halogenated hydrocarbons is discussed [7.8]. Figure 7.4 shows the scattered anti-Stokes Raman signal plotted as a function of the delay time for $(CH_2Cl)_2$ (circles) and for CH_2BrCl (solid points). The dashed line represents the cross-correlation function between infrared pump and visible probe pulse. It determines the zero point on the time scale and the time resolution of the experimental system. The rise of the anti-Stokes signal reflects the growing excess population of the monitored vibrational state. The excess population decays with time constants of 6.5 ps and 13 ps for $(CH_2Cl)_2$ and CH_2BrCl, respectively. The signal is large enough to follow the relaxation process over a factor of approximately 100. The error bars in Fig. 7.4 represent the statistics of the photon counting system. The observed time constants are specific to the vibrational states and the molecules investigated here (see Sect. 7.2.2).

An exceptionally long population lifetime for a hydrocarbon molecule is found in the next example (see Fig. 7.5). Here the vibrational relaxation in acetylene C_2H_2 dissolved in CCl_4 is investigated [7.29]. At first, the ν_3 mode at $3265\,cm^{-1}$, – a CH-stretching mode – is excited by a resonant infrared picosecond pulse. The excess energy is transferred from the ν_3 mode to the symmetric CC-stretching mode ν_2 within a time shorter than 2 ps. The fast rise of the signal in Fig. 7.5 represents the population of the ν_2 mode due to the rapid intramolecular energy transfer. For the depopulation of the ν_2 mode at $1968\,cm^{-1}$ a time constant of 240 ps is measured. The observed time constants are explained by a theoretical analysis of the energy transfer rates between vibrational states. For acetylene, the frequencies of the fundamental vibrations and of numerous overtones and combination modes are available. There also exists information on the anharmonicities of the potential from rotational spectra in the gas phase. Based on these parameters the transfer rates were calculated [7.30]. The theoretical analysis shows that the excitation of the ν_3 mode is rapidly distributed via anharmonic intramolecular interactions among combination tones, all of which contain the ν_2 mode. In this way a rapid population of the ν_2 mode is achieved as seen experimentally. Quite different is the situation for the relaxation of the ν_2 mode. Because of symmetry relations, the ν_2 mode can only decay to lower

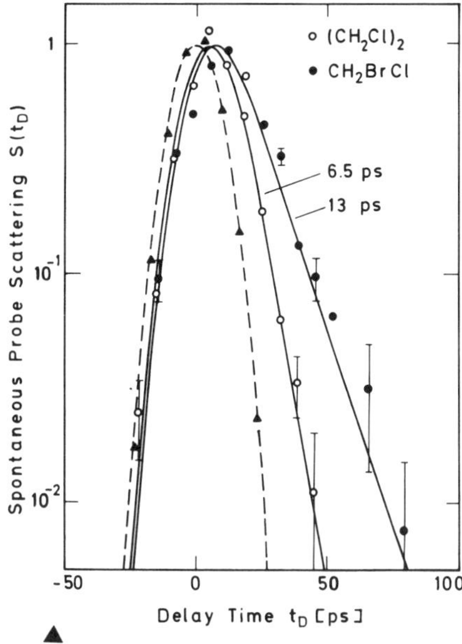

Fig. 7.5. Time dependence of the population of the C≡C mode of acetylene in liquid CCl_4 following excitation of the ν_3 CH-stretching mode at 3265 cm^{-1}. After a rapid rise the population decays slowly with a decay time of 240 ps [7.29]

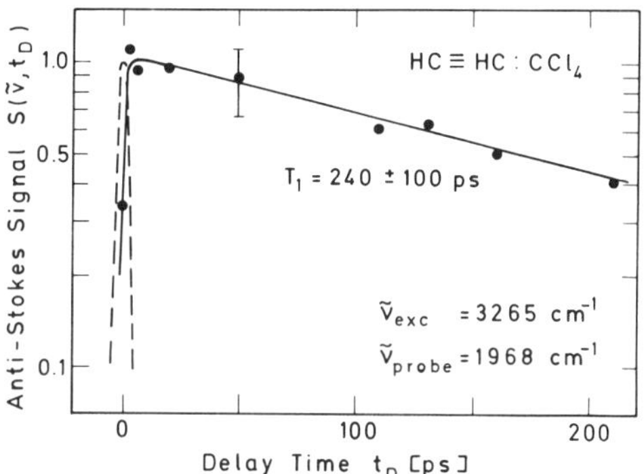

▲

Fig. 7.4. Spontaneous anti-Stokes scattering signal $S(t_D)$ of the probe pulse versus delay time t_D for the symmetric CH_2-stretching modes of $(CH_2Cl)_2$ (*circles*) and CH_2BrCl (*points*); calculated solid curves. Broken line and triangles represent the instrumental response of the measurement [7.8]

lying bending modes, which are separated in energy from the ν_2 vibration by several hundred wave numbers. A considerable amount of energy has to be transferred to translational and rotational degrees of freedom via intermolecular processes, making the relaxation of the ν_2 mode in acetylene rather long.

Experimental data for a large number of molecules are summarized in Table 7.1. The molecules are listed in order of decreasing number of atoms. The investigated vibrational modes are given in the second column. The population lifetimes determined experimentally, T_1(exp), are compared with theoretical values, T_1(theor). Calculations of T_1(theor) are discussed in Sect. 7.2.2.

The data of Table 7.1 represent our present state of knowledge on population lifetimes of polyatomic (and several diatomic) molecules. Reported values of T_1, which are questionable on account of experimental uncertainties or sample purity (unknown presence of stabilizers), are not listed here.

According to Table 7.1, vibrational frequencies around 3000 cm^{-1} are most frequently investigated; i.e., extensive information exists on CH-, CH_2-, and CH_3-stretching modes. One reason for the extensive studies in this frequency range is the availability of intense and tunable picosecond pulses around 3000 cm^{-1}.

Inspection of Table 7.1 shows that for large molecules, consisting of more than ten atoms, the time constants are very short, close to or equal to the time resolution of the experimental system, which is 2 ps in most cases. This result may be rationalized as follows. At vibrational energies of 3000 cm^{-1} the density

Table 7.1. Population lifetimes T_1 of individual vibrational states of small molecules after excitation of CH- and OH-stretching modes. The second column gives the vibrational modes investigated. The experimental time constants (column 3) are compared with calculated values (column 4). In the lower part of the table several time constants for diatomic molecules are listed for comparison

Molecule	Vibrational frequency [cm^{-1}]	Time T_1(exp) [ps]	Time T_1(theor) [ps]	Ref.
Excitation via CH-*stretch*				
$C_{10}H_8$ in CCl_4	3050 (ν_1CH)	2 ± 0.5		[7.13]
$(CH_2ClCH_2)_2O$ in CCl_4	~2964 ($\nu_1 CH_2^s$)	2 ± 2		[7.8]
$C_6H_5CH_3$	2920 ($\nu_4 CH_3$)	2	6	[7.34]
	3054 (ν_3CH)	1		
$C_6H_5-C_2H$ in CCl_4	2111 (νCC)	15 ± 7		[7.29]
C_6H_6	3063 (ν_2CH)	1 ± 0.5	<1	[7.34,35]
$C_6D_3H_3$	3055 (ν_2CH)	5 ± 2	6	[7.35]
C_6H_5Cl	3068 (ν_2CH)	1		[7.34]
C_5NH_5	3057 (ν_2CH)	1		[7.34]
CH_3CH_2OH (in CCl_4)	2928 (νCH_3^s)	22 ± 5	18 ± 9	[7.1,22,36,37]
$(CH_2Cl)_2$ in CCl_4	2950 ($\nu_1 CH_2^s$)	6.5 ± 1		[7.8]
CH_3CCl_3 (in CCl_4)	2939 ($\nu_1 CH_3^s$)	7.5 ± 1	11	[7.1,38]
	3007 ($\nu_7 CH_3^{as}$)	7.5 ± 1	11	
CH_2-CCl_2	3036 ($\nu_1 CH_2^s$)	3 ± 1	2	[7.35]
$CHCl-CHCl$	3073 ($\nu_1 CH_2^s$)	10 ± 2	13	[7.35]
CH_3OH (in CCl_4)	2935 ($\nu_2 CH_3^{as}$)	1.5	1.6 ± 0.8	[7.37]
$CHCl_3$ in CCl_4	3020 (ν_1CH)	1.6	2.8	[7.37,39]
$CHBr_3$ (in CCl_4)	3020 (ν_1CH $v=1$)	40 ± 4		[7.40]
	5920 (ν_1CH $v=2$)	5 ± 1	~9	
CH_2Cl_2 (in CCl_4)	2989 ($\nu_1 CH_2^s$)	12 ± 2	23 ± 7	[7.8,37]
CH_2Br_2 (in CCl_4)	2987 ($\nu_1 CH_2^s$)	7 ± 1	35 ± 12	[7.8]
CH_2I_2 in CCl_4	2967 ($\nu_1 CH_2^s$)	45 ± 5	150 ± 70	[7.8]
CH_2ClBr in CCl_4	2987 ($\nu_1 CH_2^s$)	13 ± 2	34 ± 8	[7.8]
CH_2ClI in CCl_4	2979 ($\nu_1 CH_2^s$)	14 ± 2	36 ± 9	[7.8]
CH_3I in CCl_4	2950 ($\nu_1 CH_3^s$)	1		[7.39,41]
C_2H_2 in CCl_4	3265 (ν_3CH)	<2	1.6	[7.29]
	1968 (ν_2CC)	240 ± 100	150	
Excitation via OH-*stretch*				
CH_3OH neat	3350 (OH)	<2		[7.23]
in CCl_4	3641 (OH)	$15-30$		[7.10]
CH_3OD in CCl_4	2685 (OD)	52 ± 17		[7.10]
CD_3OH in CCl_4	3642 (OH)	73 ± 7		[7.10]
CD_3OD in CCl_4	2690 (OD)	79 ± 17		[7.10]
CH_3CH_2OH neat	(OH)	<5		[7.18]
in CCl_4	3625 (OH)	70 ± 10		[7.10]
C_6H_5OH in CCl_4	3610 (OH)	$5-20$		[7.10]
C_6H_5OD in CCl_4	2665 (OD)	$15-25$		[7.10]
$(CH_3)_3SiOH$ in CCl_4	3690 (OH)	205 ± 21		[7.10]
$(C_2H_5)_3SiOH$ in CCl_4	3689 (OH)	185 ± 19		[7.10]
Diatomic Molecules				
CN^- in H_2O ($c = 2.3$ M)	2080	6.7 ± 1.0		[7.5]
HCl [173 K]	2783	2100		[7.42]
	5463	25		
CO [70 K]	2147	7×10^9		[7.43]
O_2 [65 K]	1556	3.1×10^9		[7.44]
N_2 [77 K]	2350	7.7×10^{12}		[7.45]

of states per frequency interval ρ is quite large for molecules with more than ten atoms. For the bandwidth of the excited mode Γ a product $\rho\Gamma \gg 1$ is calculated. This value indicates that numerous channels for intramolecular energy redistribution exist, which lead to a rapid depopulation of the initially excited vibrational mode. In fact, very fast transient quasi-thermal redistribution of vibrational excess energy has been observed in large molecules. These findings will be discussed in Sect. 7.3.

In molecules with less than 10 atoms the observed lifetimes differ drastically from molecule to molecule and for different vibrations of the same molecule. Time constants between less than 1 and 240 ps are found. Here the density of vibrational states is considerably smaller. The energy relaxation has to be treated in the small ($\rho\Gamma \ll 1$) or intermediate ($\rho\Gamma \sim 1$) molecular limit. Individual transfer rates between specific vibrational states have to be considered. The depopulation time depends critically on the individual vibrational frequency and on the structure of the molecule. We return to this important situation below.

Recently, the population lifetimes of the OH and OD stretching vibrations of numerous alcohols and silanols in highly diluted solutions ($<10^{-2}$ per volume) of CCl_4 have been investigated by vibrational saturation [7.10] (see Fig. 7.2b). For alcohols, the T_1 values are smaller than 80 ps and vary strongly between the different molecules (see Table 7.1). An example is presented in Fig. 7.6, where three calculated curves are fitted to experimental points for the OD-stretching vibration of C_6H_5OD. The curves are based on the parameters of the infrared pulses, the molecular absorption cross-section, and the assumed T_1 values of 10,

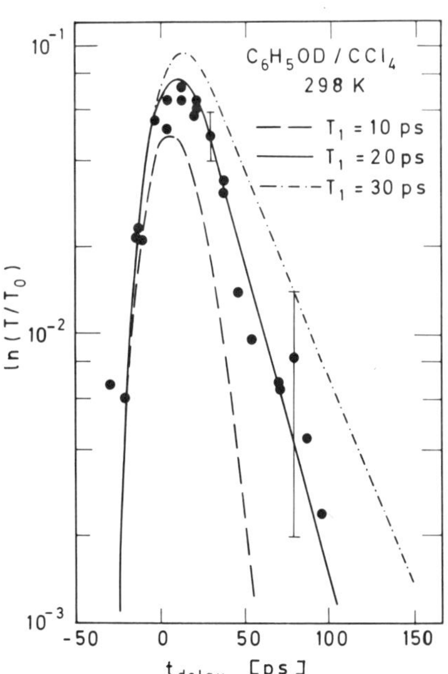

Fig. 7.6. Absorption recovery after excitation of the OD-stretching vibration ($\tilde{\nu} = 2660\,cm^{-1}$) in deuterated phenol ($C_6H_5OD$) in CCl_4 ($c = 10^{-2}$ M). The experimental data are compared with calculated curves for $T_1 = 10$ ps, 20 ps, and 30 ps. A time constant of $T_1 = 20 \pm 5$ ps is estimated [7.10]

20, and 30 ps. A value of $T_1 = 20 \pm 5$ ps is deduced from the computer simulation [7.10].

It should be emphasized that the T_1 values of the OH- and OD-stretching modes of various alcohols are measured in dilute solutions in CCl_4. In neat alcohols the T_1 times of the OH-stretching mode are much shorter. From the temperature equalization of neat methanol and ethanol solutions one estimates $T_1 < 2$ ps for the OH-stretching mode (see Sect. 7.5). It is quite likely that the association of alcohol molecules via hydrogen bridges – known to occur in neat and highly concentrated solutions – is a major factor for the short T_1 times in such systems.

Quite surprising is the constancy of T_1 for different silanols (see Table 7.1). Moreover, T_1 is found to be very similar ($T_1 = 190 \pm 90$ ps) for the OH vibration in all SiOH containing systems, in liquid silanols, in solids (quartz) [7.31], and on surfaces [7.15]. The results for OH groups attached to Si are not understood at present. The effect of a heavy atom (i.e. silicon) in intramolecular vibrational energy transfer has been repeatedly discussed [7.32]. It is quite likely that this effect is important in explaining the large T_1 values of the silanols.

At the end of Table 7.1 a few time constants of diatomic cryogenic liquids are added for comparison. Here the population relaxation times are longer by many orders of magnitude. In diatomic molecules with one fundamental frequency an intramolecular energy transfer does not exist. Relaxation occurs via slow radiative processes or via very slow intermolecular energy transfer to translations and rotations. The vibrational relaxation of diatomic molecules has been reviewed elsewhere [7.33]. We concentrate here on the discussion of organic polyatomic molecules.

7.2.2 Theoretical Interpretation and Examples

As pointed out above, anharmonic intramolecular coupling between neighboring states, the Fermi resonance, represents a major mechanism for energy relaxation in small molecules. The anharmonicities of potential surfaces are known for a few small molecules only, e.g. for C_2H_2. In general, coupling coefficients are estimated from absorption and Raman data which are experimentally accessible.

In the theoretical model discussed here we consider the initially excited states, the final states and a coupling mechanism [7.37]. In most practical cases the initial state is a vibrational eigenstate of the molecule and the final states are overtones or combination modes of lower lying states. The interaction with the medium is treated as a small perturbation. In liquids, intramolecular transitions are collision induced, but most of the original vibrational energy flows to other vibrational states of the same molecule. One speaks of collision-induced intramolecular energy transfer processes. The transfer rate is calculated with the help of the interaction Hamiltonian $H_{\text{int}} = V(Q_1, \ldots, Q_n)U(t)$, where the operator V depends on the normal coordinates Q_j of the molecule, and the time-dependent factor $U(t)$ describes the interaction with the surrounding molecules. The transition rate k_{if} between an initial state ψ_i and a final state ψ_f is given by [7.46]:

$$k_{if} = \left| \frac{\langle \psi_i V(Q_1, \ldots, Q_n) \psi_f \rangle}{\hbar} \right|^2 \cdot \int_{-\infty}^{\infty} dt \langle\langle U(t) U(0) \rangle\rangle e^{i\omega t} \; . \tag{7.1}$$

The first factor represents the vibrational transition-matrix element and the second factor describes the dynamics of the time-dependent interaction. The frequency ω is determined by the energy difference between the initial and the final eigenstates $\hbar\omega = E_i - E_f$. The brackets $\langle\langle \cdots \rangle\rangle$ indicate the average over initial velocities and angular momenta.

The vibrational transition-matrix element may be determined empirically from the Fermi resonance-mixing between neighboring states of the same symmetry. In Fermi resonance the final states borrow oscillator strength from the excited vibrational state as a result of anharmonic mixing. The mixing coefficient may be estimated from the intensity ratio of the two states, $R = I_f/I_i$, which is taken from infrared or Raman spectra [7.47].

The correlation function of the intermolecular force field $\langle\langle U(t) U(0) \rangle\rangle$ can be estimated by comparing the depopulation rate k_{if} with the dephasing rate T_2^{-1} for the final state $|f\rangle$. The dephasing rate is given by

$$T_2^{-1}(f) = \left| \frac{\langle \psi_f V \psi_f \rangle}{\hbar} \right|^2 \int_{-\infty}^{\infty} dt \langle\langle U(t) U(0) \rangle\rangle \; . \tag{7.2}$$

The ω-dependence of the integral in (7.1) has to be estimated in order to introduce T_2 in (7.1). Taking the binary collision model as a guide, the main ω-dependence is of the form [7.48]

$$k_{if} \propto \exp[-(\omega/\Omega)^{2/3}] \; . \tag{7.3}$$

In principle, Ω can be determined within the framework of the SSH theory, provided $\hbar\omega$ is larger than kT. However, in the relaxation processes discussed here, ω is comparable to or smaller than kT. For smaller values of ω, detailed calculations for small molecules show that Ω is always close to $100\,\mathrm{cm}^{-1}$ [7.49]. The dephasing time T_2 may be estimated from the Raman linewidth of the final state or, if this is not available, from the corresponding infrared linewidth. From the preceding equations, (7.1–3), one obtains the following formula for the energy relaxation time T_1 due to the anharmonic coupling:

$$T_1 = (k_{if})^{-1} = N(1 + R^2) R^{-1} \exp(\omega/\Omega)^{2/3} T_2(f) \; , \tag{7.4}$$

where N stands for the number of states initially populated. For instance, the two stretching modes of the CH_2 group exchange their energy within the time of excitation and both are depopulated through one decay channel.

The theoretical values T_1(theor) given in Table 7.1 are calculated according to (7.4). One finds fair agreement with the experimental values for most of the molecules.

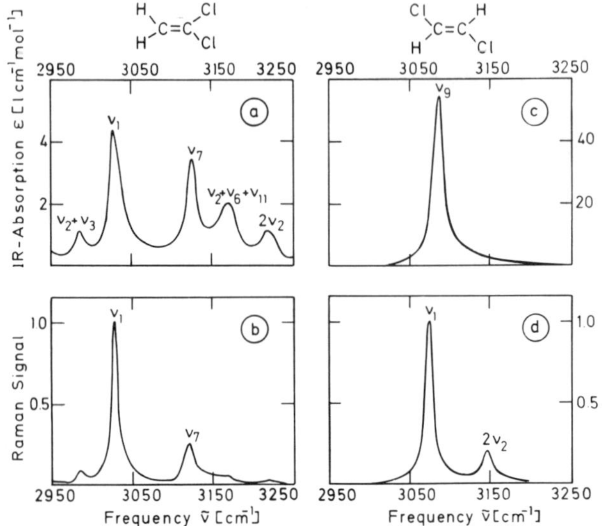

Fig. 7.7. Infrared absorption (**a**) and Raman (**b**) spectra of CH_2CCl_2 between $2950\,cm^{-1}$ and $3250\,cm^{-1}$. Two combination tones are in strong Fermi resonance with the two CH-stretching modes at ν_1 and ν_7. Infrared absorption (**c**) and Raman (**d**) spectra of trans-CHClCHCl. There is less Fermi resonance mixing than in CH_2CCl_2 [7.35]

As an example, the population lifetime T_1 of two similar molecules should be discussed in more detail [7.35]. Dichloroethylene is available in two forms, $CH_2{=}CCl_2$ and *trans*-CHCl=CHCl. In these molecules only two atoms exchange their position, while the basic structure and the type and number of atoms are preserved. The small changes in the molecular structure give rise to a large change in the Fermi-resonance mixing which in turn has a profound effect on the population lifetime. The infrared and Raman spectra between 2950 and $3250\,cm^{-1}$ of both molecules are depicted in Fig. 7.7. The infrared spectra, Fig. 7.7a and c, differ drastically, i.e. there are substantial differences in the anharmonic coupling of the CH-stretching modes of the two molecules.

a) $CH_2{=}CCl_2$

In Fig. 7.7a one finds strong Fermi resonance between the fundamental ν_1 and the combination tone $\nu_2 + \nu_3$, both of A_1 symmetry, and between the fundamental ν_7 and the combination mode $\nu_2 + \nu_6 + \nu_{11}$, both of B_1 symmetry [7.50]. This observation indicates that one has to consider at least two decay channels for the CH_2-stretching modes. For the decay $\nu_1 \rightarrow \nu_2 + \nu_3$ we estimate an intensity ratio $R = 0.2 \pm 0.05$ and calculate a dephasing time $T_2(\mathrm{f}) = (2\pi c \Delta\nu)^{-1} = 0.3\,\mathrm{ps}$ from the Raman linewidth of $\Delta\nu = 17\,cm^{-1}$. With $N = 1$ and $\omega = 45\,cm^{-1}$ a value of $T_1 = 4 \pm 2\,\mathrm{ps}$ is obtained for one decay channel.

According to Fig. 7.7a and b the symmetric (ν_1) and the asymmetric (ν_7) CH_2-stretching vibrations are separated by $100\,cm^{-1}$. There exists experimental evidence that collision-induced interaction leads to a very rapid exchange of vibrational energy between similar fundamental vibrations that are close in

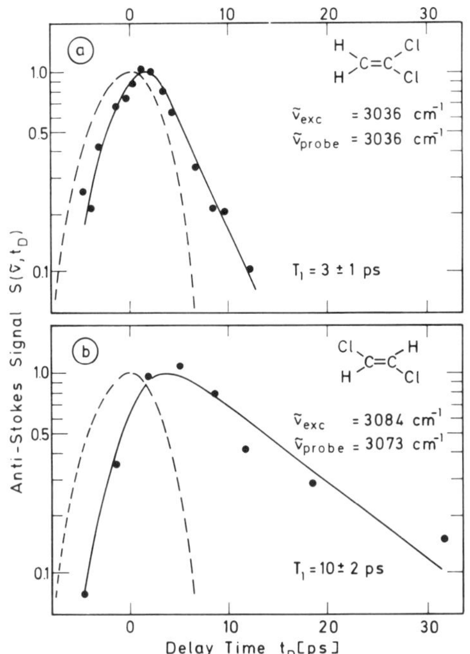

Fig. 7.8a, b. Anti-Stokes scattering signal versus delay time of the probing pulse. (**a**) CH_2CCl_2 in CCl_4. The decay of the CH_2-stretching mode at $3036\,cm^{-1}$ is shown. (**b**) trans-CHClCHCl in CCl_4. The CH-stretching mode at $3084\,cm^{-1}$ is excited and the mode at $3073\,cm^{-1}$ is monitored. The broken curves are the cross-correlation functions of the IR exciting and green probing pulses [7.35]

energy. The energy transfer time between the similar CH_2-stretching modes is frequently $\leqslant 1$ ps. Here, the relatively large separation of $100\,cm^{-1}$ between ν_1 and ν_7 slows down the transfer rate to approximately 2 ps [7.35].

Using (7.4) one calculates a time constant of $T_1 = 1.5$ ps for the second decay channel $\nu_7 \to \nu_2 + \nu_6 + \nu_{11}$. Since the ν_1 mode is excited and interrogated, one has to add the two decay rates of $\nu_1 \to \nu_7$ and $\nu_1 \to \nu_2 + \nu_3$ and arrives at a lifetime of $T_1(\nu_1) = 2$ ps.

In Fig. 7.8a we present experimentally determined T_1 values for the CH_2-stretching mode (ν_1). The scattered anti-Stokes Raman signal rises quickly to a slightly delayed maximum during the excitation process and decays with a relaxation time of $T_1(\text{exp}) = 3 \pm 1$ ps. This experimental finding is in fair agreement with the calculated value $T_1(\text{theor}) = 2$ ps.

b) *trans*-CHCl=CHCl

In Fig. 7.7c one sees the infrared-active CH-stretching mode ν_9 and in Fig. 7.7d the Raman-active symmetric ν_1 mode. For this molecule one finds a considerably smaller Fermi resonance. The Raman spectrum of Fig. 7.7d suggests some Fermi resonance between ν_1 and an overtone $2\nu_2$, both of A_g symmetry. With the values $R = 0.15 \pm 0.02$, $T_2 = 0.3$ ps, $N = 2$, and $\omega = 80\,cm^{-1}$ we calculate $T_1(\text{theor}) =$

13 ps. It should be noted that additional weak Fermi resonances between the ν_9 mode and higher combination modes (e.g. $\nu_2 + \nu_5 + \nu_{10}$) might be buried under the high-frequency tail of the ν_9 fundamental. These additional decay channels may somewhat reduce the estimated T_1 value.

The time dependence of the CH-stretching mode ν_1 is depicted in Fig. 7.8b. The molecule is excited via the ν_9 mode at 3084 cm^{-1} and the transient population of the ν_1 mode at 3073 cm^{-1} is monitored by anti-Stokes Raman scattering. The rapid rise of the signal, i.e. the fast population of the ν_1 mode, gives clear evidence of the fast energy exchange between the two CH-fundamentals ν_1 and ν_9. The decay of the signal curve suggests a relatively long lifetime of the two CH-stretching modes of $T_1(\exp) = 10 \pm 2$ ps. This number is in good agreement with the value estimated above. The small intramolecular coupling with overtones and combination modes gives rise to the longer vibrational lifetime.

The examples discussed here convincingly show that the population lifetimes of vibrational modes in small molecules depend – in a sensitive way – on the molecular structure and the frequencies of the vibrational states. Fermi resonances observed in infrared and Raman spectra serve as a valuable guide for estimating the population lifetime T_1. Vibrational modes with strong anharmonic interaction are readily recognized by their Fermi resonances. Weak mode coupling, however, is not easily seen in standard absorption and Raman spectra. As a result, the decay channels of long-lived modes (with $T_1 > 50$ ps) are more difficult to discern.

7.3 Vibrational Redistribution in Large Molecules in the Ground State S_0

In the preceding section the flow of vibrational energy from well-defined initial states to specific final states was discussed for small molecules.

The situation is quite different in larger molecules. It is well known that the number of vibrational quantum states per energy interval $\rho(E)$ rises strongly with the number of atoms which make up the polyatomic molecule, and with the vibrational energy E (see Fig. 7.9) [7.51]. For instance, one estimates densities of states of $\rho(E) = 5 \times 10^2$ and 10^8 per cm^{-1} for naphthalene (18 atoms) and coumarin 6 (43 atoms), respectively, at $E = 3000\ cm^{-1}$ (the position of the CH-stretching modes). As a result of the large mode density, the energy supplied to a specific vibrational state in large molecules may flow very rapidly to numerous isoenergetic overtones and combination modes. In this way, the excess energy is quickly redistributed over the vibrational manifold of the molecule. In the following we shall present data on the redistribution time, Sect. 7.3.1, give evidence of vibrational redistribution, Sect. 7.3.2, and discuss the concept of an internal temperature, Sect. 7.3.3.

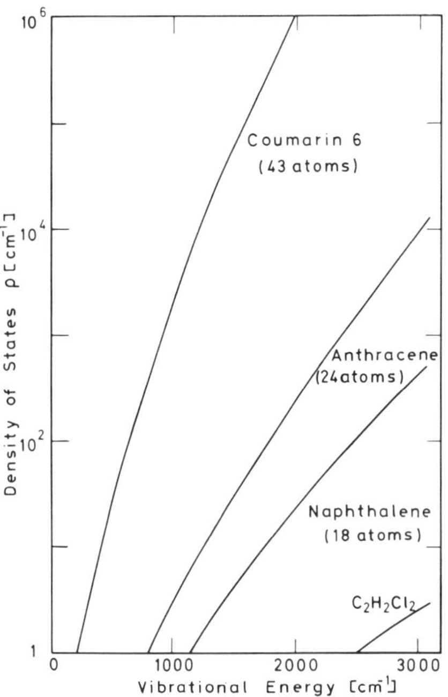

Fig. 7.9. Averaged density of states ρ as a function of the vibrational energy for four molecules. The strong increase of the density of states with the vibrational energy and the number of atoms of the molecule is clearly shown

7.3.1 The Redistribution Time

In Fig. 7.10 the redistribution of vibrational energy in naphthalene is depicted [7.13]. The infrared pulses of a frequency $\tilde{\nu} = 3055\,\mathrm{cm}^{-1}$ are absorbed by two overlapping bands at $\tilde{\nu} = 3064\,\mathrm{cm}^{-1}$ and $\tilde{\nu} = 3048\,\mathrm{cm}^{-1}$. The transient change of the population of two strong Raman transitions at $\tilde{\nu}_1 = 3058\,\mathrm{cm}^{-1}$ (CH-stretching mode) and $\tilde{\nu}_5 = 1380\,\mathrm{cm}^{-1}$ (ring deformation mode) [7.50] is seen in Fig. 7.10a and b, respectively. The strongly rising Raman signal in part a indicates a rapid energy transfer from the pumped IR-active CH-stretching modes to the monitored totally symmetric CH-stretching mode $\tilde{\nu}_1$. The maximum of the Raman signal delayed by 1.5 ps with respect to the maximum of the correlation curve (broken line) and the decay of the Raman signal suggest a lifetime of the coupled set of CH-stretching modes of approximately 2 ps. The Raman signal of the $\tilde{\nu}_5$ mode in Fig. 7.10b rises more slowly than the ν_1 mode suggesting an energy transfer from the CH-stretching modes to the lower-lying ring deformation mode ν_5. The scattered signal of the ν_5 mode decays with a time constant of 9 ± 4 ps, indicative of intermolecular relaxation of energy to the solvent. The solid lines through the data points in Fig. 7.10a and b are calculated from a system of rate equations with the following time constants: energy transfer from the excited modes to the Raman active ν_1 mode, $t_1 \leqslant 0.5$ ps; energy transfer from the CH-

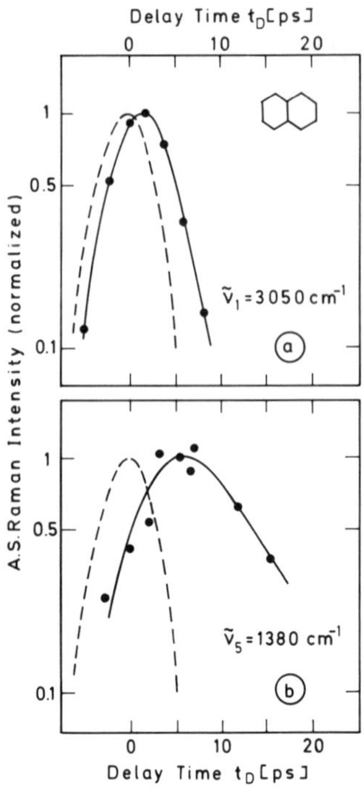

Fig. 7.10a, b. Redistribution of vibrational energy in naphthalene. (**a**) Anti-Stokes Raman signal of the mode ν_1 (3058 cm^{-1}). Rapid transfer of vibrational energy from the initially pumped IR-active CH-stretching mode to the monitored Raman-active CH-stretching mode. The correlation curve is shown for comparison (---). (**b**) Anti-Stokes Raman signal of ν_5 (1380 cm^{-1}). This ring deformation mode is populated by the decay of the CH-stretching modes [7.13]

stretching modes to the ν_5 mode, $t_2 = 2$ ps; and energy dissipation to the surrounding $t_3 = 9$ ps. The model fits the two experimental curves quite well over the whole time range.

Of interest at this point is the discussion of the absolute scattering signal. The values at the maximum of the ν_5 mode (Fig. 7.10b) and of a lower-lying mode $\tilde{\nu}_8 = 765\,\mathrm{cm}^{-1}$ were measured and compared with calculated occupation numbers. In these estimates, a statistical energy randomization over the whole molecule is assumed. This leads to a vibrational temperature of $T = 537$ K. We find good agreement between the experimental and calculated (transient) excess occupation numbers, considering the large experimental uncertainties. Obviously, intramolecular redistribution of vibrational energy is achieved in naphthalene after approximately 5 ps.

In anthracene, a slightly larger molecule with 24 atoms, the vibrational redistribution over the vibrational manifold is even faster when an infrared photon of 3050 cm^{-1} is supplied to the molecule [7.52]. As shown below (Fig. 7.17), the change of absorbance or fluorescence – a measure of the vibrational redistribution – appears very fast, within the time resolution of the experimental system of 2 ps. Similar results were obtained for several larger dye or pigment molecules (e.g. coumarin 7) with approximately 30 to 50 atoms. The redistribution time is less than 1 ps [7.11,21,53].

7.3.2 The Vibrational Distribution

As pointed out in Sect. 7.1, the redistribution of vibrational energy in the electronic ground state affects the shape of the long-wavelength tail of the $S_0 \rightarrow S_1$ transition. Changes of the absorption or fluorescence (the activation spectrum) give direct information on the transient vibrational distribution. Several interesting examples will be discussed here. First, we present data for anthracene in C_2Cl_4 [7.52]. In Fig. 7.11a part of the standard absorption spectrum is presented (curve 1) with the peak absorption of the 0-0 transition at 26 380 cm^{-1}. For smaller energies the absorption decreases steeply; numerous thermally populated levels contribute to the measured absorption. The broad shoulder around 25 000 cm^{-1} is of special interest here; it belongs to a group of vibrations with frequencies of approximately 1400 cm^{-1}. In fact, anthracene has three totally symmetric modes around 1400 cm^{-1} with $\tilde{\nu}_4 = 1557$, $\tilde{\nu}_6 = 1403$, and $\tilde{\nu}_7 = 1255\,cm^{-1}$. The Franck-Condon factors of these three modes are large, they add up to a transition probability similar to that of the 0-0 transition. Since the thermal population of the three modes is approximately 10^{-3} at 300 K, the shoulder appears at an extinction coefficient three orders of magnitude below the peak value (see Fig. 7.11a).

The absorption spectrum 1 of Fig. 7.11a changes drastically within a few picoseconds if an infrared pulse of 5 ps resonantly excites the anthracene molecules via the CH-stretching modes at 3052 cm^{-1}. The transient absorption of the excited molecules – measured 7 ps after the infrared excitation – is shown by full points in Fig. 7.11a, curve 2. A direct comparison with the unexcited molecules is possible, since the degree of excitation – here 20% – is readily estimated from the number of IR-photons, the IR-absorption coefficient, and the beam geometry. Most significant in Fig. 7.11a is the strong increase of the extinction coefficient

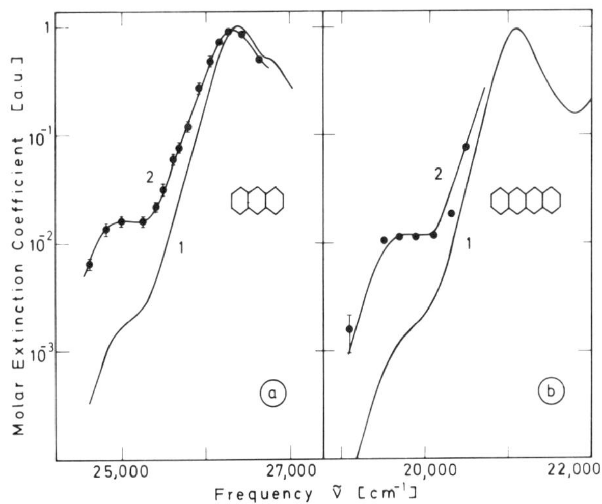

Fig. 7.11. (**a**) Molar extinction coefficient of anthracene in C_2Cl_4. Curve 1: room-temperature extinction coefficient. Curve 2: transient extinction coefficient 7 ps after IR excitation. Note the gain in intensity of the hot band at $\tilde{\nu} = 25\,000\,cm^{-1}$. (**b**) Molar extinction coefficient of tetracene in CCl_4. Curve 1: room-temperature extinction coefficient. Curve 2: transient extinction coefficient 3.5 ps after IR excitation. Note the smaller gain in intensity of the hot band at $\tilde{\nu} = 19\,700\,cm^{-1}$ on account of the reduced heating compared to anthracene [7.54]

at 25 000 cm^{-1} by a factor of ten. This observation suggests a substantial transient increase in population of the three symmetric modes ν_4, ν_6, and ν_7 as a result of the vibrational redistribution. More quantitatively, one estimates a population rise by a factor of ten for a temperature increase of the molecule to 470 K. A similar temperature rise is calculated if the IR quantum of 3050 cm^{-1} is distributed over the vibrational manifold of the molecule, similar to a specific-heat–type argument. The good agreement between the observed and the calculated transient internal temperature suggests a redistribution of energy over very many vibrational modes of the molecule. We note that the Raman data and the transitions with good Franck-Condon factors give information on a small number of modes only. The specific heat is calculated for all modes of the molecule.

The experimental findings for tetracene are similar to the data of anthracene [7.54]. In Fig. 7.11b a shoulder due to several (possibly six) totally symmetric modes with good Franck-Condon factors is again seen in the room temperature spectrum. 3.5 ps after the infrared excitation via CH-stretching modes of $\tilde{\nu} = 3045\,cm^{-1}$ a transient hot band appears which is indicative of a redistribution of vibrational energy with an internal temperature of 450 K. The smaller temperature is consistent with the larger specific heat of the larger molecule.

The next example of vibrational redistribution is concerned with coumarin 7, a large molecule of low symmetry made up of 44 atoms [7.21] (see insert of Fig. 7.12). Curve 1 in Fig. 7.12 shows the absorption edge for frequencies smaller than the 0-0 transition. The absorption rises exponentially over six orders of

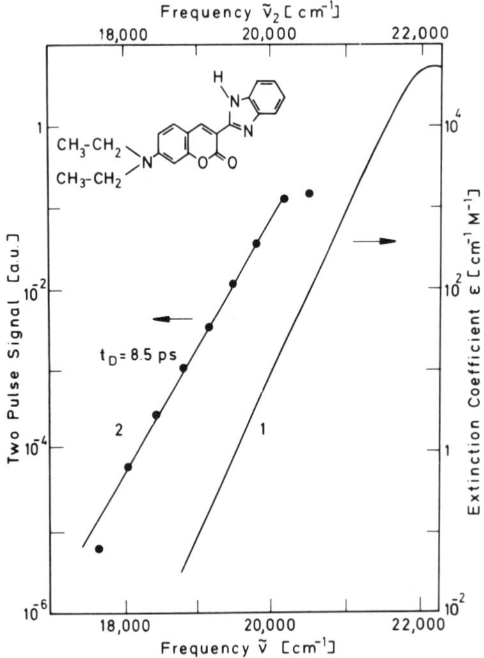

Fig. 7.12. Absorption edge of Coumarin 7 at 300 K (*Curve 1*). Transient fluorescence excitation data. The molecules are excited by infrared pulses with $\tilde{\nu}_1 = 3395\,cm^{-1}$, the frequency of the NH-stretching mode. The straight slope measured after $t_D = 8.5$ ps points to a vibrational distribution with an increased internal temperature of $T^* \sim 400$ K (*Curve 2*) [7.21]

magnitude. The relation $\varepsilon \propto \exp[-(\nu_{00} - \nu)/kT]$ fits the experimental slope quite well with $T = 300$ K. Numerous thermally populated vibrational modes make up the absorption tail. Curve 2 of Fig. 7.12 gives data points of the fluorescence excitation spectrum taken 8.5 ps after excitation of the coumarin molecules by an infrared pulse of $\tilde{\nu}_1 = 3395\,\mathrm{cm}^{-1}$. The latter is strongly resonantly absorbed by the NH-mode. Two findings are important: (i) The data points rise exponentially with frequency over four orders of magnitude suggesting a quasi-thermal distribution; (ii) The new curve 2 is less steep than the room-temperature curve 1 indicating a higher internal temperature of $T^* = 400$ K. These observations provide convincing evidence for a redistribution of vibrational energy within the extremely short time of a few picoseconds. It is interesting to note that the exponential slope of the fluorescence activation spectrum is independent of the excitation process. Pumping the coumarin molecule via NH- or CH-stretching modes gives almost exactly the same slope.

The excess fluorescence measured in the two-pulse technique is directly proportional to the number of molecules excited by the first infrared pulse [7.55]. This fact is clearly demonstrated in Fig. 7.13, where the infrared frequency ν_1 is tuned from $2850\,\mathrm{cm}^{-1}$ to $3250\,\mathrm{cm}^{-1}$, the frequency range where the CH-, CH_2-, and CH_3-stretching modes of the molecule coumarin 6 are located (coumarin 6 has the same structure as coumarin 7, except that the NH-group is replaced by an sulphur atom; see Fig. 7.12). Comparison of the fluorescence spectrum of Fig. 7.13a with the standard infrared absorption spectrum of the same specimen (Fig. 7.13b) shows a marked similarity. Higher IR-absorption leads to more excited molecules and thus to a proportionally higher fluorescence. The fluorescence signal is not specific to the excited vibrational mode. The data of Fig. 7.13a are taken 7 ps after IR excitation at a time when the redistribution of vibrational energy has already occurred.

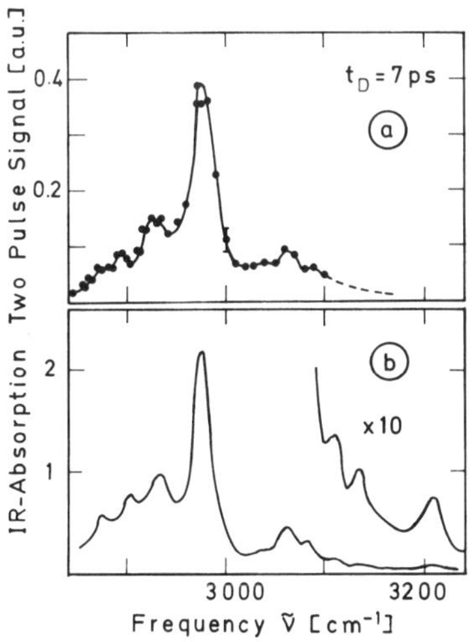

Fig. 7.13a, b. IR-spectra of coumarin 6. (**a**) Two-pulse infrared spectrum measured with a delay time of $t_D = 7$ ps. The fluorescence signal is plotted as a function of the excitation frequency ν_1. (**b**) Standard infrared spectrum showing bands due to CH_3, CH_2, and CH-stretching modes between $2850\,\mathrm{cm}^{-1}$ and $3060\,\mathrm{cm}^{-1}$ [7.55]

In the preceding examples the vibrational redistribution was analyzed after excitation with infrared quanta of approximately $3000\,cm^{-1}$. Excitation with quanta of higher energies should give even higher internal temperatures and thus even more drastic changes of the absorption edge of the excited molecules. Indeed, recent experiments on azulene bear out this prediction [7.11]. Azulene was chosen because of its known very fast internal conversion [7.56]. The lifetime of the S_1 state is very short, approximately 2 ps [7.57,58,59], and the electronic energy flows very rapidly to high lying vibrational states of the electronic ground state S_0. The excess energy in S_0 is quickly redistributed over the vibrational manifold giving rise to large changes of the absorption spectrum at the long-wavelength edge of the 0-0 transition. Figure 7.14 shows some relevant data. The room-temperature absorption spectrum of azulene in CCl_4 is depicted in curve 1. The steep absorption edge contains a small hot band as a shoulder at 13 600 cm^{-1}. Excitation of the azulene molecules with visible pulses at $\tilde{\nu}_1 = 18\,970\,cm^{-1}$ leads to a drastic change of the absorption spectrum. Curve 2 of Fig. 7.14 represents the transient spectrum taken 10 ps after the excitation of the azulene molecules. We emphasize the drastic difference between this result and the room temperature data of curve 1. For instance, at $\tilde{\nu}_2 = 13\,000\,cm^{-1}$ and $13\,600\,cm^{-1}$ the absorption is increased by a factor of one hundred and a factor of ten, respectively. The broken line through the experimental points corresponds to a Boltzmann edge with an internal temperature of 1200 K. This value is consistent

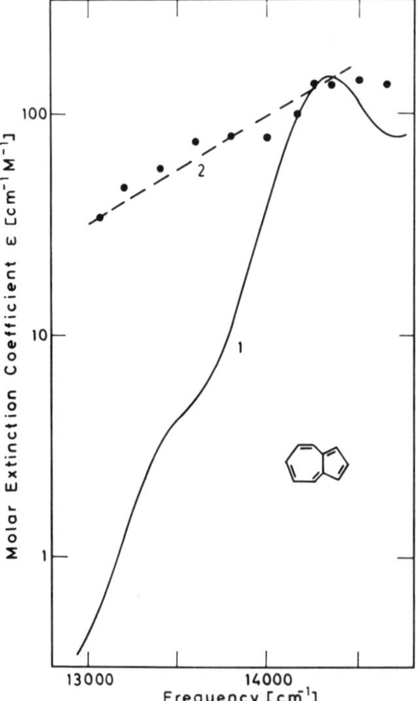

Fig. 7.14. Low-energy absorption tail of azulene at 300 K (*1*) and transient absorption measured 10 ps after excitation by photons of energy $19\,000\,cm^{-1}$ (*2*). The broken line is calculated for a Boltzmann edge of 1200 K [7.11]

with an estimate, in which the temperature-dependent specific heat of azulene was calculated from the known vibrational frequencies, and where an excess energy of 18 970 cm^{-1} was added to the molecule. The cooling, i.e. the return to thermal equilibrium, will be discussed in Sect. 7.4.2.

7.3.3 The Concept of Internal Temperature

Strictly speaking, the temperature of a molecular system is only defined when the whole system is in thermodynamic equilibrium. The transiently hot molecules discussed here are certainly not in thermodynamic equilibrium. On the other hand, the transient absorption increases exponentially with frequency (Boltzmann edge). This behavior is typical for a thermal population of the vibrational states of a molecule. In the following we want to address the question: To what extent we are allowed to talk of an internal temperature after vibrational excitation of the molecules?

In Fig. 7.15 the total energy distribution is presented for azulene. The probability $P(E)$ of finding azulene molecules with a total energy E within an energy interval (here 1 cm^{-1}) is plotted. Curve 1 is calculated from the known density of states of azulene, $\rho(E)$ [7.60], weighted by the Boltzmann factor at 300 K. Most molecules are found to have a total energy of $\lesssim$1000 cm^{-1}.

After excitation of the azulene molecules by a photon of 18 970 cm^{-1}, one has to consider the new probability distribution 2 obtained by shifting curve 1 by 18 970 cm^{-1}. The distribution 2 differs considerably from the thermal equilibrium distribution 3 calculated for hot azulene at 1200 K. Figure 7.15 shows that the distribution of the molecules excited by an ultrashort laser pulse is much narrower than the thermal distribution 3. We recall that in a first approximation, all azulene molecules contain the same total energy of approximately 20 000 cm^{-1}. The distribution is close to a microcanonical distribution.

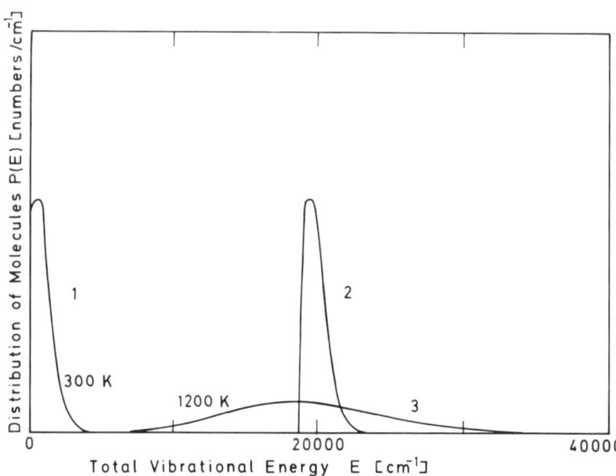

Fig. 7.15. Distribution $P(E)$ of the number of azulene molecules with a total vibrational energy E. Curves 1 and 3: equilibrium distributions for 300 K and 1200 K, respectively. Curve 2: distribution $P(E)$ immediately after excitation and internal conversion; $P(E)$ is determined by the distribution at room temperature, but shifted by the energy of the excitation

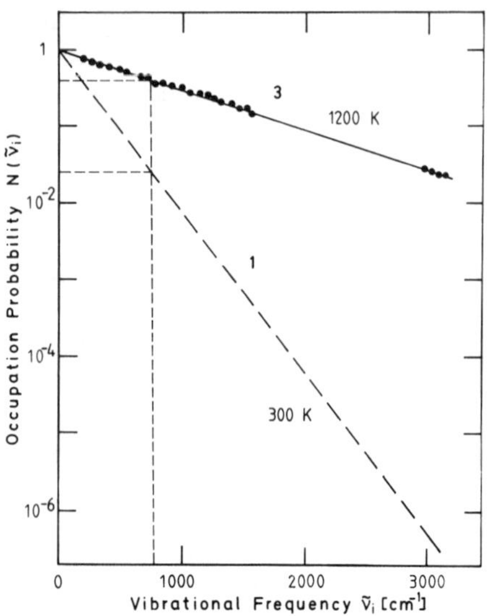

Fig. 7.16. Occupation probability $N(\nu_i)$ of fundamental vibrational modes of energy $h\nu_i$ for two temperatures, 300 K (*1*) and 1200 K (*3*). Values of $N(\nu_i)$, calculated for the distribution curve 2 in Fig. 7.15, are indicated for several modes by full points. Note the close correlation with the line for 1200 K

The energy redistribution and the internal temperature after excitation are more readily discussed with the help of the population probability of vibrational states. Figure 7.16 shows the occupation probability $N(\nu_i)$ of specific vibrational states with energy $h\nu_i$ as a function of ν_i for azulene molecules. The absorption and the two-step fluorescence is directly proportional to the occupation probability $N(\nu_i)$.

In thermal equilibrium the occupation probability $N(\nu_i)$ of a vibrational state decreases exponentially with increasing frequency ν_i according to the Boltzmann factor

$$N(\nu_i) = \exp(-h\nu_i/kT) \ . \tag{7.5}$$

In Fig. 7.16 the curves 1 and 3 represent the population probability for 300 and 1200 K, respectively.

After excitation of a molecule by a picosecond light pulse the vibrational energy is rapidly redistributed over a very large number of isoenergetic combination modes (in azulene $\sim 10^{15}/\mathrm{cm}^{-1}$ at $20\,000\,\mathrm{cm}^{-1}$). Each combination state is assumed to have equal weight. We are interested in the occupation of an individual vibrational state *i*. The probability of finding a molecule with a total energy *E* in a mode *i* where the $v = 1$ level is occupied is given by [7.61]

$$N_E(h\nu_i) = \frac{\rho_i(h\nu_i)\rho(E - h\nu_i)}{\rho(E)} \tag{7.6}$$

with $\rho(e)$ being the density of states at the energy *e* and $\rho_i(h\nu_i) = 1$ is the density of states of the *i*th mode at frequency ν_i. The total occupation probability of the vibrational state of frequency ν_i is given by integration over the total energy distribution $P(E)$

$$N(\nu_i) = \int N_E(\nu_i)P(E)dE = \int \frac{\rho_i(E - h\nu_i)}{\rho(E)} P(E)dE \ . \tag{7.7}$$

First we want to look at (7.7) for a system in thermal equilibrium with a temperature T. The total energy distribution $P(E)$ for molecules in thermal equilibrium has the form

$$P(E) = \rho(E)\exp(-E/kT) \ . \tag{7.8}$$

Inserting (7.8) into (7.7) gives the Boltzmann factor of (7.5).

We now turn to the non-equilibrium distribution obtained after vibrational excitation. For the energy distribution $P(E)$ given in curve 2 of Fig. 7.15 the integral in (7.7) may be solved numerically. Results for the fundamental modes i of azulene are shown as points in Fig. 7.16. Curve 3, the occupation probability for a thermal distribution at 1200 K, closely fits the calculated points. The system behaves as if it had a higher temperature. We call this temperature the internal temperature of the excited molecules.

In a first approximation, the distribution 2 in Fig. 7.15 is a microcanonical distribution. All molecules have the same energy $E_0 \sim 20\,000\,\mathrm{cm}^{-1}$ within a small uncertainty ΔE_0. Each fundamental mode with frequency $\nu_i \lesssim 3000\,\mathrm{cm}^{-1}$ may be considered as a small subsystem of the molecular system of vibrational quantum states. The energy of the quantum state $h\nu_i$ of the subsystem is small compared to the total energy E_0. In this case the occupation probability of the subsystem is given by $N(\nu_i) = \exp(-h\nu_i/kT)$ in a good approximation (see text books for quantum statistics, e.g. [7.62]). For a classical density of states the same expression is derived [7.63]. The numerical result in Fig. 7.16 explains why the Boltzmann distribution for $T = 1200\,\mathrm{K}$ adequately describes the real situation for $h\nu_i \ll E_0$.

We also mention the limitation of the temperature concept. The $N(\nu_i)$ values of higher vibrational states with energies comparable to the total energy of the molecule are not fully accounted for by the internal temperature. Deviations of a thermal distribution are observed. For example, the true $N(\nu_i)$ values are smaller than those estimated from the internal temperature for frequencies ν_i around $3000\,\mathrm{cm}^{-1}$ in anthracene when the molecules are excited by a pulse of $3050\,\mathrm{cm}^{-1}$ energy [7.52].

7.4 Intermolecular Transfer of Vibrational Energy of Large Molecules to the Surrounding Solvents

After absorption of infrared quanta (or visible photons with subsequent internal conversion) polyatomic molecules acquire an excess energy which – as demonstrated in the last section – is rapidly redistributed over the vibrational manifold of the electronic ground state. Transfer of the excess energy from the excited molecule to the liquid surrounding restores the thermal equilibrium. The cooling of the optically pumped molecules is readily inferred from the absorption (or fluorescence) changes which occur in the frequency range of the long-wavelength tail of the absorption spectrum. More accurately, the decay of the excess absorption (or fluorescence) of the hot molecule is a direct measure of the transfer of energy to the solvent.

7.4.1 Molecular Excitation with Infrared Pulses

Intermolecular energy transfer has been studied for a number of molecules under a variety of conditions (see Table 7.2). As an example, experimental data for anthracene in C_2Cl_4 are discussed in more detail here [7.52]. The anthracene molecules are excited by infrared photons of 3050 cm^{-1} via CH-stretching modes. In Fig. 7.17 the temporal evolution of the change of the electronic absorption is presented for two frequencies of the probe pulses at 24 760 cm^{-1} and 23 530 cm^{-1}. The change in fluorescence (which is proportional to the change of absorption) is plotted versus delay time between excitation and probe pulse. The two plots in Fig. 7.17 show a substantially different time dependence for the two probing frequencies. With $\tilde{\nu}_{\text{probe}} = 24\,760\,\text{cm}^{-1}$ in Fig. 7.17a one monitors vibrational energy states of 1600 cm^{-1} and higher, since the 0-0 transition of anthracene is 26 380 cm^{-1}. Many transitions to S_1 are possible and the excess fluorescence

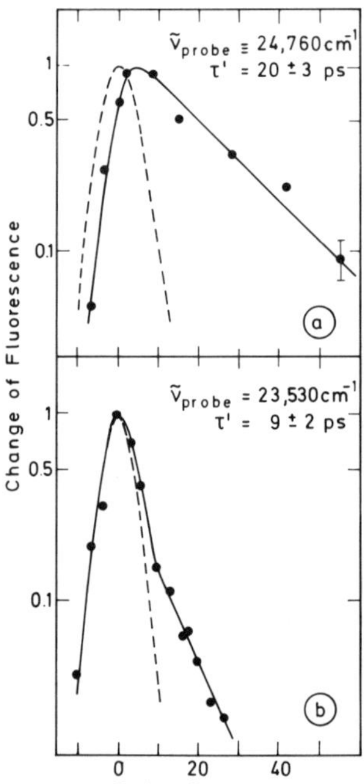

Fig. 7.17a, b. Temporal evolution of the excess fluorescence in anthracene after excitation at $\tilde{\nu}_1 = 3050\,\text{cm}^{-1}$. (**a**) Frequency of the probe pulse $\tilde{\nu}_2 = 24\,760\,\text{cm}^{-1}$. The signal (*solid line*) rises rapidly with IR excitation and decays more slowly. The correlation curve of both pulses is shown by the broken line. (**b**) Frequency of the probe pulse $\tilde{\nu}_2 = 23\,530\,\text{cm}^{-1}$. The signal (*solid line*) shows a fast and a slow component, discussed in the text [7.52]

is large. The excess fluorescence rises rapidly with time due to the fast intramolecular redistribution discussed in Sect. 7.3. The slower decay of the excess fluorescence indicates an intermolecular energy transfer with a time constant of 20 ps. We recall that the probing frequency of 24 760 cm^{-1} monitors the dynamics of the hot band of skeletal modes discussed in Sect. 7.3.2 (see Fig. 7.11a).

With a probe frequency of 23 530 cm^{-1} (Fig. 7.17b) transitions to S_1 from vibrational states with energies of 2850 cm^{-1} and higher are possible. The observed fluorescence signal is smaller and the time evolution shows two physical processes: (i) A very fast component which closely follows the correlation curve (i.e. the time dependence of the excitation pulses folded with the probing pulse). This signal is due to an instantaneous two-photon transition or a two-step process with a real intermediate, short-lived combination state. The transition probabilities are small for both processes at the applied probing intensities. The corresponding fluorescence signals are only visible when the second component is small as well. (ii) For longer delay times, $t_D \geqslant 10$ ps, a second fluorescence component with a time constant of 9 ± 2 ps appears. This longer time corresponds to the intermolecular energy flow to the solvent. As pointed out above, in Fig. 7.17b higher vibrational states are monitored. Their reduced population results in smaller fluorescence signals compared to the probing of the lower states of Fig. 7.17a. The different time constants of 20 and 9 ps, observed in Fig. 7.17a and b may be rationalized as follows. Experimentally, the fluorescence (or absorption) monitors the population of vibrational states, while physically energy is transferred from the hot molecule to the surrounding, according to a certain transfer law. A decrease in excess energy reduces the internal temperature of the molecules and thus the vibrational population. It is important to note that a certain temperature decrease more strongly affects the population of higher vibrational modes; i.e. a faster decay of excess fluorescence is observed for smaller values of ν_{probe} which monitor the higher vibrational states. A more quantitative model will be discussed below.

7.4.2 Molecular Excitation via Rapid Internal Conversion

We return to the discussion of azulene, where the high energy of the absorbed visible light quantum of close to 19 000 cm^{-1} is transferred to the ground state [7.11] within 2 ps by fast internal conversion. The large amount of energy gives rise to a very large internal temperature of approximately 1200 K (see Sects. 7.3.2 and 7.3.3). In Fig. 7.18 the change of absorption is plotted as a function of time. The data are taken with a probing frequency of $\tilde{\nu}_2 = 13\,600$ cm^{-1} in the long-wavelength tail of the S_1 transition (see Fig. 7.14). The signal points rise rapidly with the excitation pulse. Considering the time resolution of the experimental system, one estimates an upper limit for the internal conversion and for the redistribution of vibrational energy of 2 ps, i.e. the redistribution process is faster than 1 ps. The intermolecular energy transfer of vibrational excess energy to the solvent molecules depends upon the surrounding medium. Time constants of

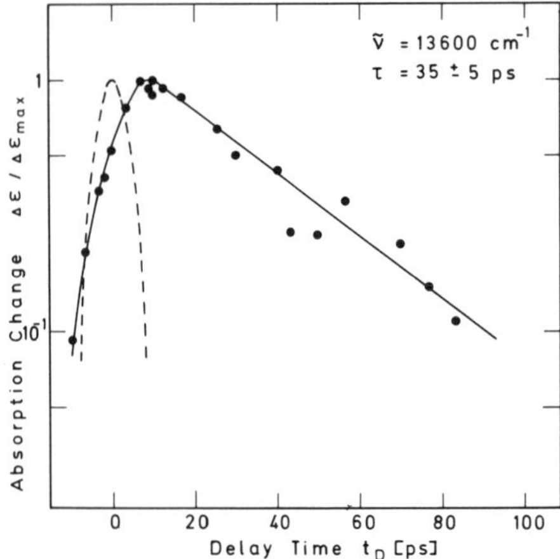

Fig. 7.18. Intermolecular energy dissipation of azulene molecules dissolved in CCl_4. Normalized absorption change at a probe pulse frequency of 13 600 cm^{-1}. A first rise of absorption – due to internal conversion and rapid vibrational redistribution (<2 ps) – is followed by a cooling of the molecules with a time constant of 35 ps [7.65]

Table 7.2 Intermolecular energy transfer of large molecules. The excitation energy E_{exc} leads to a transient internal temperature of T^*_{int}. The time constant τ is observed when vibrational states at the frequency $\nu_{00} - \nu_2$ are monitored. (ν_2 is the frequency of the probing pulse). The energy relaxation time τ_e is calculated from the observed population lifetime τ using (7.9–12)

Molecule	Atoms	Solvent	E_{exc} [cm^{-1}]	T^*_{int} [K]	τ [ps]	τ_e [ps]	$\tilde{\nu}_{00} - \tilde{\nu}_2$ [cm^{-1}]	Ref.
Oxazine 1	63	C_2D_5OD	2975 (CH_3)	370	10	9	1200	[7.18]
					6	9	2000	[7.64]
Coumarin 7	44	C_2Cl_4	3395 (NH)	410	7	10	2430	[7.21]
			2970 (CH_3)	400	7	10	2430	
HRO 5656	55	CCl_4	2960 (CH_3)	390	7	11	2760	[7.55]
Coumarin 6	43	CCl_4	2975 (CH_3)	400	7	10	2560	[7.21,55]
		C_6H_6	2975 (CH_3)	400	7	10	2560	
Tetracene	30	CCl_4	3045 (CH)	450	7	12	2150	[7.54]
Anthracene	24	C_2Cl_4	3050 (CH)	470	9	25	2850	[7.52]
					20	25	1600	
Azulene	18	CCl_4	18970 ($S_0 \rightarrow S_1$)	1200	35	36	700	[7.11,65]
		C_6H_6			15	10		
		CH_3OH			13	9		

35 ps and 13 ps were observed for the solvents CCl_4 and CH_3OH, respectively [7.65].

7.4.3 Discussion and Model for Energy Transfer

In Table 7.2 the observed time constants τ for intermolecular energy dissipation are listed for seven large molecules. The table specifies the solvents, the energy of the exciting photons, the calculated internal temperature of the excited molecule and the energy $h\nu_{00} - h\nu_2$, (the energy of the vibrational states preferentially investigated by the probing process). The data in Table 7.2 provide us with much interesting information: (i) In large molecules (more than 40 atoms) the decay times τ are fairly constant for similar IR excitation, e.g. via CH or NH modes. (ii) The measured value of τ appears to depend on $\nu_{00} - \nu_2$, the vibrational state probed with the second monitoring pulse (see anthracene). (iii) τ varies by as much as a factor of four for different solvents after high excitation (see data of azulene).

We start our discussion with point (ii). In a first approximation, the energy transfer rate is proportional to the instantaneous excess energy E_{exc}: $dE_{\mathrm{exc}}/dt = E_{\mathrm{exc}}/\tau_{\mathrm{e}}$ [7.66]. Integration of this equation gives an exponential energy decay law

$$E_{\mathrm{exc}}(t) = E_{\mathrm{exc}}(0)\mathrm{e}^{-t/\tau_{\mathrm{e}}} \tag{7.9}$$

with an energy relaxation time τ_{e}. The excess energy gives rise to an internal molecular temperature $T(t)$ according to

$$E_{\mathrm{exc}}(t) = \int\limits_{T_0}^{T(t)} C(T')dT' \ , \tag{7.10}$$

where $C(T)$ is the temperature-dependent specific heat of the molecule and T_0 the temperature of the molecule without excitation. The population probability of a vibrational state with energy $h\nu_i = h(\nu_{00} - \nu_2)$ is given by the Boltzmann factor

$$N(\nu_i, t) = \exp\left[\frac{h(\nu_2 - \nu_{00})}{kT(t)}\right] , \tag{7.11}$$

where ν_2 is the frequency of the probing pulse which induces transitions to S_1. Finally, the change of absorption is proportional to the change of population

$$\varDelta\alpha(\nu_2, t) \propto \varDelta\varepsilon(\nu_2, t) \propto N(\nu_i, t) - N(\nu_i, 0) \ . \tag{7.12}$$

Equations (7.9–12) have been solved for the case of anthracene where the temperature-dependent specific heat can be calculated from the known vibrational modes of the molecule [7.50]. In Fig. 7.19 the relative change of the extinction coefficient $\varDelta\varepsilon/\varepsilon$ is plotted versus time for the two vibrational modes at $\tilde{\nu}_i = 1620$ and $2850\,\mathrm{cm}^{-1}$. An energy relaxation time $\tau_{\mathrm{e}} = 25$ ps was assumed in the calculation. The faster decay of the absorption for higher values of ν_i is readily seen from the graph. The decay time of the lower vibrational modes approaches the value of the energy relaxation time τ_{e}. Experimental decay constants of 17 and 10 ps are obtained from the calculations of Fig. 7.19 for the two vibrational modes at $\tilde{\nu}_i = 2850\,\mathrm{cm}^{-1}$ (i.e. $\tilde{\nu}_2 = 23\,530\,\mathrm{cm}^{-1}$) and at $\tilde{\nu}_i = 1620\,\mathrm{cm}^{-1}$ (i.e. $\tilde{\nu}_2 = 24\,760\,\mathrm{cm}^{-1}$), respectively. The agreement with the experimental data of Fig. 7.17, namely 20 and 9 ps, is satisfactory considering the crudeness of the model.

Values of the energy relaxation τ_{e} calculated according to (7.9–12) are presented in Table 7.2. Generally, we find depopulation time constants smaller than the energy relaxation times τ_{e} when vibrational states of an energy of two or

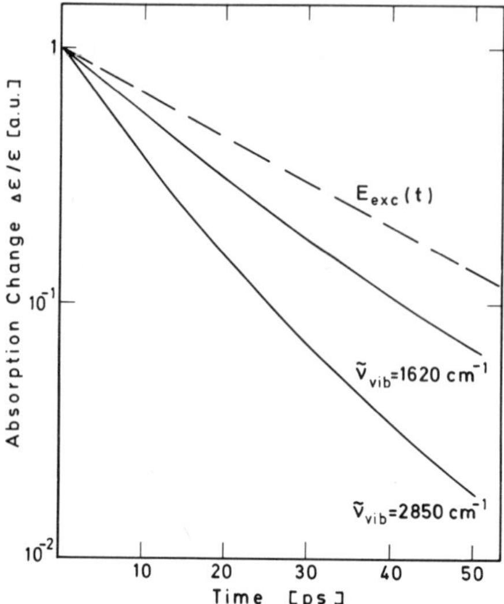

Fig. 7.19. Relative change of the extinction coefficient $\Delta\varepsilon/\varepsilon$ calculated for the two vibrational modes of anthracene at $\tilde{\nu}_i = 1620\,\mathrm{cm}^{-1}$ and $2850\ \mathrm{cm}^{-1}$. An energy relaxation time of $\tau_e = 25\,\mathrm{ps}$ (---) was assumed in the calculation. A faster decay of the absorption for higher values of ν_i is readily seen

three thousand wave numbers are probed. The time constants for depopulation and energy relaxation are very similar for $\tilde{\nu}_i \lesssim 1000\,\mathrm{cm}^{-1}$. An energy relaxation time of $\tau_e \sim 10\,\mathrm{ps}$ is found for all molecules with more than 40 atoms.

To summarize: The measured time constants τ for vibrational depopulation – due to intermolecular energy transfer – may differ from the energy relaxation time τ_e depending on the observation frequency.

Let us now turn to point (iii), the dependence of τ on the solvent at high excitation energies. The following model has been treated [7.67]. First, isolated binary collisions between the excited molecules and solvent molecules are considered. During the collision a partial energy exchange takes place [7.68]. The resulting (exponential) equalization proceeds with time. The extent of equalization is determined by the translational energy of the molecules.

Up to this point the special properties of the liquid state are not taken into account. The first part of the model only describes the situation in the gas phase [7.19,69,70]. We therefore introduce a second step. After the partial energy equalization process the excess energy of the solvent molecule has to be transferred to the rest of the liquid. In our model, this transfer is approximated by macroscopic heat conduction within the solvent.

The model is applied to the azulene data of Table 7.2. Time constants between 13 and 35 ps are observed for the intermolecular energy transfer in different solvents. The dependence of τ on the specific solvent is well reproduced by our model. Very similar time constants are calculated when the corresponding thermal diffusivity of the specific solvent is taken into account.

A comment should be made concerning the collision frequency in our model [7.67]. Generally, collision frequencies are well defined in the gas phase, where the collision partners do not interact between collisions. During the collision the colliding molecule may be captured in the attractive part of the potential. In this case, several collisions with the repulsive part of the potential take place. In the liquid phase, however, the molecules are always within the attractive part of the potential. Here the collision frequency is given by the number of single collisions with the repulsive hard-core potential. This fact has to be taken into account when energy dissipation rates in liquids and gases are compared.

Inspection of the data in Table 7.2 suggests that for low excess energies the transfer time is approximately the same in different solvents [7.21], in contrast to the situation for high excess energies [7.67]. In the present model two parameters depend on the solvent properties: the thermal diffusivity of the solvent and the amount of energy transferred in one collision. For high excess energies a large amount of energy has to be transferred per unit time interval. Consequently, high lying vibrational states like CH modes are important for an efficient relaxation process. Solvents without C-H groups (e.g. CCl_4) are less efficient collision partners and longer decay times are observed. For low excess energies, low lying vibrational modes and translational and rotational motions are dominant in the transfer process; they are similar in different molecules. Thus for this case, the time constants should not exhibit a pronounced dependence on the solvent.

7.5 Equalization of Energy in Liquids Measured with a Molecular Thermometer

7.5.1 The Molecular Thermometer

The change of absorption in the long-wavelength tail of the 0-0 transition can be used to accurately determine local temperature variations [7.18,23]. Of special importance for the following applications is the high speed of the molecular thermometer. In the investigations discussed in Sects. 7.3.1, 7.4 the time resolution was approximately 2 ps for probing pulses of 5 ps. With shorter probing pulses a better time resolution should be possible. A series of investigations were made with the dye oxazine 1, a large molecule consisting of 63 atoms. This molecule has several advantages: high photochemical stability, availability in high purity, convenient spectral position of the absorption edge, favorable solubility of the heptafluorobutyrate salt, and lack of aggregation even at the high concentration of 0.2 M.

The molecular thermometer is readily calibrated by heating the solution to a known temperature and by determining $\Delta\varepsilon/\varepsilon$ at an appropriate wavelength of the absorption tail. For instance, a value of $(\Delta\varepsilon/\varepsilon)/dT = 10^{-2}$ per degree is determined for oxazine 1 in ethanol measured at $14400\,cm^{-1}$. This frequency is approximately $700\,cm^{-1}$ smaller than the 0-0 transition of the dye. Experimentally, it is possible to determine absorption changes with an accuracy of 10^{-4} using a sensitive differential detection system. With such a device temperature changes of 0.1 K may be determined with a temporal resolution of one picosecond [7.23]. We recall that temperature changes of several hundred degrees were measured with different molecules (see Sect. 7.4.2) making the fast molecular thermometer applicable over a wide temperature range.

7.5.2 Energy Transfer Processes

With the help of the molecular thermometer we are able to tackle the following question: How fast is the equalization of energy in organic liquids? To be more specific, when vibrational energy is supplied to a certain vibrational mode of the solvent molecules, how quickly and for how long a time can the excess energy be detected by a solute molecule? The energy flow from the solvent to the solute molecule represents the reverse process to the intermolecular energy transfer discussed in the preceding section.

Several energy transfer processes are involved in the equalization of energy in organic liquids. They are discussed separately with reference to the schematic picture in Fig. 7.20. (i) Solvent molecules absorb infrared photons leading to an excess population of high-lying vibrational modes, e.g. of OH- or CH-stretching vibrations with $\tilde{\nu}_1 = 3400\,cm^{-1}$ or $\tilde{\nu}_1 = 3000\,cm^{-1}$, respectively. According to the results discussed in Sect. 7.2, the lifetimes of vibrational modes strongly depend on the specific mode and on the size and structure of the individual molecule. In

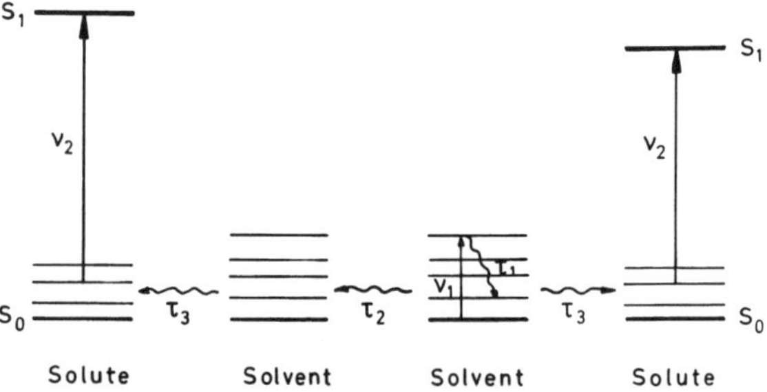

Fig. 7.20. Energy transfer processes involved in the equalization of energy in organic liquids. After infrared excitation of solvent molecules rapid intramolecular redistribution (τ_1) occurs and intermolecular energy transfer to other solvent molecules (τ_2) and to solute molecules (τ_3) takes place. A local equilibrium is established with a small temperature increase ΔT which is monitored by the large solute molecules

some cases the excited CH-modes are relatively long lived (e.g. in ethanol [7.36] or aceton [7.37]); in other molecules lower intermediate vibrational levels have the longest lifetime (e.g. in acetylene). Anharmonic coupling to isoenergetic overtones and combination modes represent a major intramolecular decay mechanism for various states. Lumping together the lifetimes of the different vibrational levels one may introduce an intramolecular redistribution time τ_1 for the solvent molecule. (ii) In the condensed phase, vibrational energy is expected to be rapidly transferred to neighboring, identical molecules. Resonant *VV* transfer and energy transfer via hydrogen bonds represent fast dissipation mechanisms with estimated time constant of picoseconds or even smaller. Interactions with translational and rotational degrees of freedom (*VT* processes) establish a local equalization with a small temperature increase ΔT after a time τ_2. (iii) After a time τ_3 the temperature jump ΔT affects the large solute molecule which now acts as a molecular thermometer. We recall that the redistribution of energy is very fast in large molecules.

7.5.3 Experimental Observations

Several experimental investigations are discussed now. In the first experiment [7.18], the dye oxazine 1 is dissolved in ethanol at a concentration of 0.1 M. For this concentration one calculates an average distance between the "molecular thermometers" of 2.6 nm. The OH-vibration of ethanol has a broad and intense band in the infrared at 3400 cm^{-1} while the oxazine molecule does not absorb at 3400 cm^{-1}. Changes of the absorption tail of oxazine may be monitored at 14 860 cm^{-1} (where ethanol is transparent). In Fig. 7.21a the absorption change $\Delta\varepsilon/\varepsilon$ of oxazine 1 is plotted versus time after exposing the solution to infrared pulses of a frequency of 3400 cm^{-1}. The data points indicate a very fast rise of the temperature of the dye molecules during the passage of the excitation pulse. The cross-correlation curve (dashed line) – taken between infrared and probe pulse – allows one to estimate the time resolution of the system. One obtains an upper limit of 5 ps for the time of energy transfer from the heated solvent molecules to the dye molecules. According to the figure, the absorption change, i.e. the higher temperature, persists for over 200 ps. Actually, the excess temperature decreases within several milliseconds via macroscopic heat conduction. The maximum value of $\Delta\varepsilon/\varepsilon$ corresponds to a temperature jump of $\Delta T = 10\,K$ according to the steady-state calibration discussed above [7.23]. The same temperature rise is calculated for the excited sample volume taking into account the absorbed energy of the infrared pulse and the heat capacity of ethanol.

In the experiment discussed next, the same sample is excited by an infrared pulse of $\tilde{\nu}_1 = 2980\,cm^{-1}$. The result is anticipated to be more involved for two reasons. First at $\tilde{\nu}_1 = 2980\,cm^{-1}$ the solvent *and* the solute absorb via their respective CH_3-modes and, second, the CH_3-mode of ethanol has a relatively

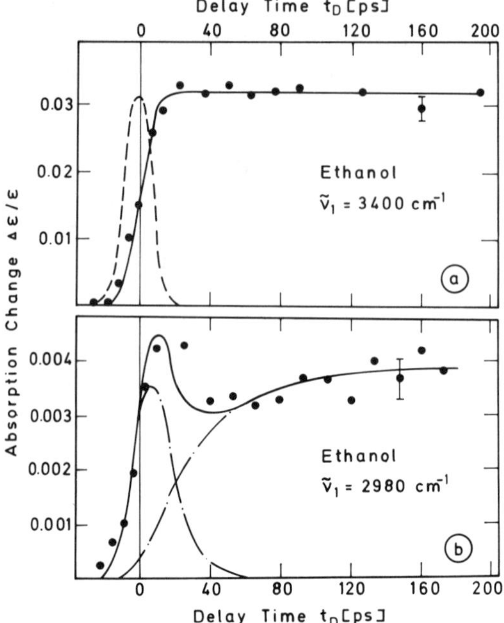

Fig. 7.21a, b. Time dependence of the absorption change of an oxazine 1 solution. (**a**) The OH modes of the solvent ethanol are excited ($\tilde{\nu}_1 = 3400\,\mathrm{cm}^{-1}$). A rapid rise of the absorption (<5 ps) is observed at $\tilde{\nu}_2 = 14\,860\,\mathrm{cm}^{-1}$. The cross correlation of the excitation and monitoring pulses is shown as dashed curve. (**b**) The CH_3 modes of both the dye and the solvent are excited. The result is a superposition of two components (*see dash-dotted curves*): A rapidly decaying direct excitation of the dye and the slow heating of the solvent. The latter is delayed by the slow depopulation of the CH_3 mode in ethanol

long relaxation time T_1, delaying the intramolecular energy redistribution of this molecule (time constant τ_1 in Fig. 7.20).

In Fig. 7.21b the time dependence of the absorption change measured at $\tilde{\nu}_2 = 14\,750\,\mathrm{cm}^{-1}$ is plotted. We find a rapid increase to a first maximum followed by a slow rise to a constant value. The pattern in Fig. 7.21b differs substantially from the data of Fig. 7.21a, where the OH-mode of ethanol was excited.

The time dependence of $\Delta\varepsilon/\varepsilon$ of Fig. 7.21b may be explained as a superposition of two processes with different time behavior. In the first process the dye molecules are directly excited; they relax quickly with a time constant of approximately 10 ps (see Sect. 7.4.1). The second component rises slowly with a time constant of 30 ps (see dash-dotted line). The long relaxation of the excited CH_3-modes slows down the intramolecular energy redistribution in the ethanol molecule; it delays the temperature equalization in the excited volume and thus the temperature rise of the probe molecule.

The observed transient temperature changes are estimated as follows: The absorbed energy from the infrared pulse at $\tilde{\nu}_1 = 2980\,\mathrm{cm}^{-1}$ is $E = 4\,\mu\mathrm{J}$ (i.e. the pulse energy is down by a factor of ten compared to the previous experiment with $\tilde{\nu}_1 = 3400\,\mathrm{cm}^{-1}$). As a result, the temperature of the excited volume rises only by 1 K, and an absorption change $\Delta\varepsilon/\varepsilon \sim 4 \times 10^{-3}$ is expected at late times. At the beginning, approximately 1% of the dye molecules are directly excited and experience a temperature rise of 70 K. The contribution of the absorption change

Table 7.3. Observed time constants of the energy transfer from the solvent molecules to the probe molecules after excitation of OH, CH_3, and C=O ($v = 2$) modes in different solvents [7.23]

Solvent	Excited vibrations OH	CH_3	C=O ($v = 2$)
Methanol	<2 ps	<5 ps	
Ethanol	<5 ps	30 ps	
Ethyleneglycol	<5 ps		
Glycerol	<5 ps		
Acetone		30 ps	<5 ps

of the dye molecules and of the solvent molecules is approximately equal, which is in agreement with Fig. 7.21b.

The experimental results discussed so far are supported by data obtained with other solvents such as methanol, ethyleneglycol, glycerol, and acetone. When the OH-vibration of the alcohols was excited, the time of energy equalization was found to be very short, in the case of methanol even below 2 ps (see Table 7.3). Supplying energy to CH_3-modes gives different results dependent upon the T_1 lifetime of the primary excited mode. In methanol with $T_1 \leqslant 1$ ps (see Table 7.1), the energy equalization is faster than 5 ps while in ethanol and acetone with longer T_1 times the energy equalization is correspondingly slower, approximately 30 ps. Obviously, the study of the dissipation of energy using the molecular thermometer also provides data on intramolecular vibrational relaxation times of the solvent molcculcs.

In this context, the excitation of the overtone ($v = 2$) of the C=0 of acetone should be emphasized. The fast equalization within times <5 ps of this modes at 3410 cm^{-1} suggests a rapid T_1 decay of the overtone of the C=0 vibration. We draw attention to the marked difference of the CH_3-mode of acetone at 2970 cm^{-1}, where a time constant of 30 ps was observed (see Table 7.3).

Very recently, the molecular thermometer has found an interesting new application. With the help of the fluorescence probing technique, the transient temperature changes due to shock loading were studied in a liquid system [7.76].

The findings of this section are summarized as follows:

i) The microscopic equalization of vibrational energy in liquids is very fast. Energy-transfer times to the solute, the molecular detector, may be smaller than 2 ps.

ii) The energy transfer is delayed when the intramolecular energy redistribution of the excited mode and molecule has a longer time constant.

iii) The rapidly attained higher temperature at the end of the pumping process for $t_D > 20$ ps suggests a quasi-thermal equilibrium within the interactive volume.

iv) The temperature increase measured by the molecular thermometer agrees with the calculated temperature rise of the excited sample volume. The supplied energy is distributed over all degrees of freedom of the solvent.

7.6 Intramolecular Vibrational Relaxation of Large Molecules in the First Excited Electronic State S_1

Here the relaxation of vibronic levels in S_1 will be discussed. In a number of investigations two techniques have been applied: (i) With a first intense pump pulse transitions to vibronic levels deplete the ground state and reduce the absorption of the sample (bleaching). The recovery of the change of absorption [7.71,72,73] or the induced dichroism [7.74,75] is monitored with a second probe pulse. When the pump and probe pulse have the same frequency, a coherent artifact appears around time zero, making the interpretation of the early data points more complicated (see below). The analysis of the bleaching data at later times gives time constants below 1 ps, e.g. 0.7 ps for rhodamine 6G in water [7.71] or 0.8 ps for phenoxazone 9 in dioxane [7.74]. The accuracy of the earlier experiments was marginal with an uncertainty in time constants of several tenths of a picosecond. (ii) In a series of measurements fluorescent dyes were first excited to high vibronic levels from which they relax to the bottom of the S_1 state with an intramolecular relaxation time. Using a second probing pulse at a smaller frequency, gain may be observed as a function of delay time between the two pulses. Care has to be taken in the interpretation of these data. Stimulated transitions, i.e. gain for the probe pulse, occur from numerous vibronic levels to higher states in S_0 during the vibrational relaxation and later on from the bottom of the S_1 state to lower vibrational levels in S_0. It is not surprising therefore, that the fluorescence rises with the integral over the correlation curve of the pump and probe pulse. In other words, gain measurements do not give unique information on the vibronic relaxation times.

In a recent advanced investigation, the bleaching dynamics of organic dyes was measured using 70 fs pulses from a colliding pulse mode-locked ring laser operating at a center wavelength of 625 nm [7.73]. With these very short pulses transitions from low-lying levels of the electronic ground state S_0 were made to high vibronic states in S_1 in the dyes nile blue and oxazine 720. The experiment employs the pump and probe technique described in (i) above. The two pulses travel noncollinearly through a small common volume of the sample. The change of transmission induced by the pump pulse is monitored as a function of relative delay time of the probe pulse. Experimental results obtained for nile blue in methanol are shown in Fig. 7.22. Different physical processes contribute to the observed transmission pattern. Around time zero the so-called coherent artifact is quite prominent. This is the result of the coherent interaction of the pump and probe pulse at the same frequency. The width of the coherent artifact is related

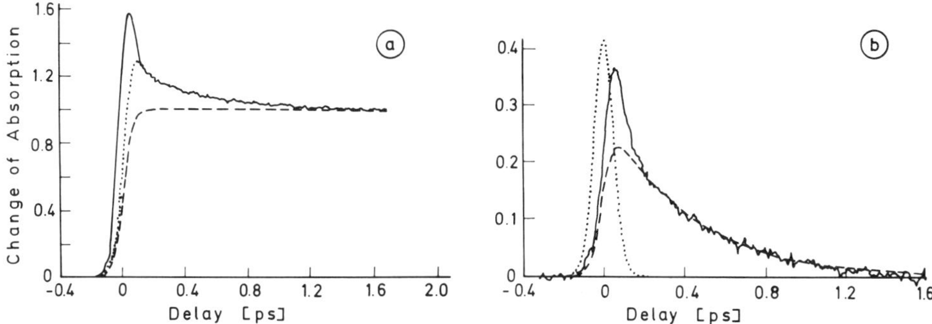

Fig. 7.22a, b. Vibrational relaxation in the S_1 state of nile-blue in methanol. Bleaching is plotted as a function of time. (**a**) The original data (——), the data after removal of the coherent artifact (····), and the integral of the intensity autocorrelation (– – –). (**b**) Portion of the probe signal due to fast molecular dynamics (——); fit assuming a single exponential response with a 410 fs time constant (– – –); intensity autocorrelation curve of the pulses (····) [7.73]

to the coherence time of the pulse and may be estimated from the spectrum of the laser pulses. Subtracting a calculated coherent artifact from the experimental bleaching curve one obtains the dotted curve in Fig. 7.22a. The long-lived population change corresponds to the depletion of the ground state and to the population of S_1 for the duration of the electronic excitation which is of the order of nanoseconds (a constant value on the time scale of Fig. 7.22). The integral over the pulse duration [i.e. the intensity autocorrelation function (dotted in Fig. 7.22b)] allows one to determine the growth of the long-lived population (dashed in Fig. 7.22a). Removal of the latter component finally gives the solid curve in Fig. 7.22b. Of interest here are the data for times longer than 200 fs, where the coherent artifact is negligible. In this time range, the signal decays exponentially with a time constant of 410 fs. Similar data for oxazine 720 in methanol give a value of 480 fs. These time constants for the partial recovery of absorption are assigned to the intramolecular vibronic relaxation times of the two large molecules.

At this point we recall the intramolecular vibrational relaxation times of large molecules in S_0, which are estimated to be $\leqslant 1$ ps (see Sect. 7.3.1). Obviously the intramolecular relaxation and redistribution of vibrational energy of large molecules is extremely rapid in S_0 and in S_1.

References

7.1 A. Laubereau, D. von der Linde, W. Kaiser: Phys. Rev. Lett. **28**, 1162 (1972)

7.2 E. Weitz, G. Flynn: In *Photoselective Chemistry*, Part 2, ed. by J. Jortner, R.D. Levine, S.A. Rice, Adv. Chem. Phys., Vol. XLVII (Wiley, New York 1981), p. 185;
J.T. Yardley: *Introduction to Molecular Energy Transfer* (Academic, New York 1980);

H. Hippler, J. Troe, H.J. Wendelken: J. Chem. Phys. **78**, 5351 (1983);
D.A. Dolson, K.W. Holtzclaw, D.B. Moss, C.S. Parmenter: J. Chem. Phys. **84**, 1119 (1986)
7.3 W.H. Hesselink, D.A. Wiersma: J. Chem. Phys. **74**, 886 (1981);
F. Ho, W.-S. Tsay, J. Trout, R.M. Hochstrasser: Chem. Phys. Lett. **83**, 5 (1981);
H. Dubost: In *Inert Gases*, ed. by M.L. Klein, Springer Ser. Chem. Phys., Vol. 34 (Springer, Berlin, Heidelberg 1984) p. 145
7.4 A. Laubereau, W. Kaiser: Rev. Modern Phys. **50**, 607 (1978)
7.5 E.J. Heilweil, F.E. Doany, R. Moore, R.M. Hochstrasser: J. Chem. Phys. **76**, 5632 (1982)
7.6 A. Laubereau, D. von der Linde, W. Kaiser: Opt. Commun. **7**, 173 (1973)
7.7 A. Laubereau, A. Seilmeier, W. Kaiser: Chem. Phys. Lett. **36**, 232 (1975)
7.8 H. Graener, A. Laubereau: Appl. Phys. B **29**, 213 (1982)
7.9 J. Chesnoy, D. Ricard: Chem. Phys. Lett. **73**, 433 (1980)
7.10 E.J. Heilweil, M.P. Casassa, R.R. Cavanagh, J.C. Stephenson: J. Chem. Phys. **85**, 5004 (1986)
7.11 W. Wild, A. Seilmeier, N.H. Gottfried, W. Kaiser: Chem. Phys. Lett. **119**, 259 (1985)
7.12 For related work in gases see:
V.V. Gruzinskii, N.A. Borisevich: Opt. Spectrosc. **15**, 246 (1963)
L. Brouwer, H. Hippler, L. Lindemann, J. Troe: J. Phys. Chem. **89**, 4608 (1985)
R. Naaman, D.M. Lubman, R.N. Zare: J. Chem. Phys. **71**, 4192 (1979)
J.G. Black, P. Kolodner, M.J. Shultz, E. Yablonovitch, N. Bloembergen: Phys. Rev. **19**, 704 (1979)
J.W. Perry, N.F. Sherer, A.H. Zewail: Chem. Phys. Lett. **103**, 1 (1983)
7.13 N.H. Gottfried, W. Kaiser: Chem. Phys. Lett. **101**, 331 (1983)
7.14 A. Penzkofer, A. Laubereau, W. Kaiser: Progr. Quant. Electron. **6**, 55 (1979)
7.15 E.J. Heilweil, M.P. Casassa, R.R. Cavanagh, J.C. Stephenson: J. Chem. Phys. **82**, 5216 (1985)
7.16 H. Graener, R. Dohlus, A. Laubereau: In *Ultrafast Phenomena V*, ed. by G.R. Fleming, A.E. Siegman, Springer Ser. Chem. Phys., Vol. 46 (Springer, Berlin, Heidelberg 1986) p. 458
7.17 Z. Vardeny, J. Tauc: Opt. Commun. **39**, 396 (1981);
T.F. Heinz, S.L. Palfrey, K.B. Eisenthal: Opt. Lett. **9**, 359 (1984)
7.18 A. Seilmeier, P.O.J. Scherer, W. Kaiser: Chem. Phys. Lett. **105**, 140 (1984)
7.19 H. Hippler, L. Lindemann, J. Troe: J. Chem. Phys. **83**, 3906 (1985)
7.20 J.P. Maier, A. Seilmeier, A. Laubereau, W. Kaiser: Chem. Phys. Lett. **46**, 527 (1977)
7.21 F. Wondrazek, A. Seilmeier, W. Kaiser: Chem. Phys. Lett. **104**, 121 (1984)
7.22 R.R. Alfano, S.L. Shapiro: Phys. Rev. Lett. **29**, 1655 (1972)
7.23 P.O.J. Scherer, A. Seilmeier, W. Kaiser: J. Chem. Phys. **83**, 3948 (1985)
7.24 F. Wondrazek, A. Seilmeier, W. Kaiser: Appl. Phys. B **32**, 39 (1983)
7.25 A. Seilmeier, K. Spanner, A. Laubereau, W. Kaiser: Opt. Commun. **24**, 237 (1978)
7.26 A. Fendt, W. Kranitzky, A. Laubereau, W. Kaiser: Opt. Commun. **28**, 142 (1979)
7.27 H.-J. Polland, W. Zinth: J. Phys. E **18**, 399 (1985)
7.28 E.P. Ippen, C.V. Shank: In *Ultrashort Light Pulses*, ed. by S.L. Shapiro, Topics Appl. Phys., Vol. 18 (Springer, Berlin, Heidelberg 1984) p. 83
7.29 W. Zinth, C. Kolmeder, B. Benna, A. Irgens-Defregger, S.F. Fischer, W. Kaiser: J. Chem. Phys. **78**, 3916 (1983)
7.30 S.F. Fischer, A. Irgens-Defregger: J. Phys. Chem. **87**, 2054 (1983)
7.31 E.J. Heilweil, M.P. Casassa, R.R. Cavanagh, J.C. Stephenson: Chem. Phys. Lett. **117**, 185 (1985)
7.32 K.N. Swamy, W.L. Hase: J. Chem. Phys. **82**, 123 (1985);
V. Lopez, R.A. Marcus: Chem. Phys. Lett. **93**, 232 (1982)
7.33 J. Chesnoy, G.M. Gale: Ann. Phys. Fr. **9**, 893 (1984)
7.34 A. Fendt, S.F. Fischer, W. Kaiser: Chem. Phys. Lett. **82**, 350 (1981)
7.35 C. Kolmeder, W. Zinth, W. Kaiser: Chem. Phys. Lett. **91**, 323 (1982)
7.36 A. Laubereau, G. Kehl, W. Kaiser: Opt. Commun. **11**, 74 (1974)
7.37 A. Fendt, S.F. Fischer, W. Kaiser: Chem. Phys. **57**, 55 (1981)
7.38 W. Zinth, C. Kolmeder, W. Kaiser: unpublished results

7.39 A. Laubereau, S.F. Fischer, K. Spanner, W. Kaiser: Chem. Phys. **31**, 335 (1978)
7.40 H. Graener, A. Laubereau: Chem. Phys. Lett. **102**, 100 (1983)
7.41 K. Spanner, A. Laubereau, W. Kaiser: Chem. Phys. Lett. **44**, 88 (1976)
7.42 J. Chesnoy, D. Ricard: Chem. Phys. Lett. **73**, 433 (1980)
7.43 N. Legay-Sommaire, F. Legay: Chem. Phys. Lett. **52**, 213 (1977)
7.44 R. Protz, M. Maier, Chem. Phys. Lett. **64**, 27 (1979)
7.45 S.R.J. Brueck, R.H. Osgood, Jr.: Chem Phys. Lett. **39**, 568 (1976)
7.46 D.W. Oxtoby: Advan. Chem. Phys. **40**, 1 (1979)
7.47 D.C. McKean: Spectrochim. Acta **29A**, 1559 (1973)
7.48 R.N. Schwarz, Z.I. Slawsky, K. Herzfeld: J. Chem. Phys. **20**, 1591 (1952)
7.49 R. Zygan-Maus: Dissertation, Technische Universität München (1980)
7.50 L.M. Sverdlov, M.A. Korvner, E.P. Krainov: *Vibrational Spectra of Polyatomic Molecules* (Wiley, New York 1974) and references therein
7.51 P.J. Robinson, K.A. Holbrook: *Unimolecular Reactions* (Wiley, Chichester U.K., 1972)
7.52 N.H. Gottfried, A. Seilmeier, W. Kaiser: Chem. Phys. Lett. **111**, 326 (1984)
7.53 T. Robl, A. Seilmeier: to be published in 1988
7.54 F. Wondrazek, A. Seilmeier, W. Kaiser: unpublished results
7.55 J.P. Maier, A. Seilmeier, W. Kaiser: Chem. Phys. Lett. **70**, 591 (1980) (Dye II corresponds to Dye HRO 5656)
7.56 P.M. Rentzepis: Chem. Phys. Lett. **3**, 717 (1969)
7.57 J.P. Heritage, A. Penzkofer: Chem. Phys. Lett. **44**, 76 (1976)
7.58 P. Wirth, S. Schneider, F. Dörr: Chem. Phys. Lett. **42**, 482 (1976)
7.59 C.V. Shank, E.P. Ippen, O. Teschke, R.L. Fork: Chem. Phys. Lett. **57**, 433 (1978)
7.60 The vibrational frequencies originate from:
R.S. Chao, R.K. Khanna: Spectrochim. Acta **33A**, 53 (1977)
7.61 A.J. Chintschin: *Mathematische Grundlagen der Quantenstatistik* (Akademie-Verlag, Berlin 1956)
7.62 L.D. Landau, E.M. Lifschitz: *Course of Theoretical Physics*, Vol. 5, Statistical Physics (Pergamon, Oxford 1958)
7.63 H. Hippler, J. Troe, H.J. Wendelken: J. Chem. Phys. **78**, 5351 (1983)
7.64 H.-J. Hübner, A. Seilmeier, W. Kaiser: unpublished results
7.65 A. Seilmeier, U. Sukowski, W. Kaiser, S.F. Fischer: In *Ultrafast Phenomena V*, ed. by G.R. Fleming, A.E. Siegman, Springer Ser. Chem. Phys., Vol. 46 (Springer, Berlin, Heidelberg 1986)
7.66 K.F. Herzfeld, Th.A. Litowitz (eds.): *Absorption and Dispersion of Ultrasonic Waves* (Academic, New York 1959)
7.67 U. Sukowski, A. Seilmeier, S.F. Fischer, W. Kaiser: to be published
7.68 R. Zygan-Maus, S.F. Fischer: Chem. Phys. **63**, 475 and 485 (1981)
7.69 M.J. Rossi, J.R. Pladziewicz, J.R. Barker: J. Chem. Phys. **78**, 6695 (1983)
7.70 W. Forst, J.R. Barker: J. Chem. Phys. **83**, 124 (1985)
7.71 A. Penzkofer, W. Falkenstein, W. Kaiser: Chem. Phys. Lett. **44**, 82 (1976)
7.72 J.M. Wiesenfeld, E.P. Ippen: Chem. Phys. Lett. **67**, 213 (1979)
7.73 A.M. Weiner, E.P. Ippen: Chem. Phys. Lett. **114**, 456 (1985)
7.74 D. Reiser, A. Laubereau: Appl. Phys. B **27**, 115 (1982)
7.75 D. Reiser, A. Laubereau: Opt. Commun. **42**, 329 (1982)
7.76 B.L. Justus, A.L. Huston, A.J. Campillo: Appl. Phys. Lett. **47**, 1159 (1985)

8. Ultrafast Chemical Reactions in the Liquid State

Kenneth B. Eisenthal

With 16 Figures

8.1 Introduction

The aim of this chapter is to provide some feeling for the impact of picosecond lasers on studies of chemical phenomena in liquids. Although there have been notable advances in the application of picosecond spectroscopy to chemical change in gases and to some extent solids, keeping this chapter to a reasonable length precludes a reasonable discussion of those research areas. For the same reason photophysical processes have not been not included. One might correctly observe that the need to restrict the topics that can be included in a single chapter testifies to the enormous success of ultrafast laser spectroscopy in diverse areas of chemistry. Why then select chemical change in liquids from among these many possibilities? Aside from the author's inclinations, the treatment of "truly" chemical processes extends the scope of this book. Many of the physical relaxation processes are treated, to some extent at least, in the other chapters. Furthermore, most of what we generally recognize as chemistry does take place in solution, and provides a way for both experimentalist and theorist to probe the nature of the liquid state. So much for direction and motivation for this chapter, now for the commentary.

8.2 Fast Chemical Reactions – Solvent Effects

We all recognize that a chemical reaction can be vastly different in a liquid from its gas phase counterpart. Not only the products but also the related kinetics can differ due to the wide ranging effects of interactions of the reactants with the solvent molecules of the liquid. Even for a unimolecular reaction, such as a dissociation or isomerization, the kinetics and channels for the chemical change can be markedly different. In a photochemical process electronic and vibrational energy relaxation can compete with chemical change in the initially excited state and thereby vitiate the change or introduce new channels by relaxation to other states and thus yield different products or distribution of products. The solvent, in acting as a sink for the excess energy of a reacting molecule, can decrease the reaction by reducing the energy available for reaction. On the other hand, it can enhance the reaction by stabilizing the product or intermediate produced, thereby preventing return to the original reactant. Of course, in exchanging energy with reactant molecules the solvent can act as an energy source as well and thus provide the energy necessary to overcome some barrier to the reaction. The interactions of the reactants with the solvent can also serve to "cage" the reactants and thus increase the likelihood of chemical change. By the same token in viewing the opposite reaction the "cage" can reduce the net chemical change by interfering with the escape of the fragments of the reaction. The molecule-solvent interaction alters the energy-structure profile along the chemically active coordinate(s), and thus the dynamics and outcome of a reaction. In a liquid environment the motions of the solvent molecules modulate the energies, geometries and motions of reacting molecules and of the products. Solvent dynamics affects both the intramolecular motions participating in a reaction, e.g. isomerization, as well as the rotational and translational motions involved in such reactions as radical recombinations or charge transfer.

The incentives to study chemical reactions in liquids are based on the fact that most of what we call chemistry takes place in solution. Furthermore, the character of chemical change in solution is strongly connected with the nature of the liquid state and thus reactions can serve as a probe of liquid state properties. The considerable activity in this area of research in recent years derives from the combination of theoretical advances in treating liquid structures and dynamics by analytic, molecular dynamics, and Monte Carlo calculations [8.1], and on the experimental side by the application of laser techniques, in particular ultrashort pulse laser spectroscopy [8.2,3]. The current interplay of theory and experiment makes for considerable excitement, sometimes displayed in the form of heat, but increasingly and hopefully in the form of visible light!

8.3 Photoisomerization

To investigate the nature of chemical reactions in solution, one natural area to focus on is the photoisomerization of molecules. These structural changes are unimolecular processes which thereby simplify the study. By starting on an excited state surface the reaction can be initiated at a defined time by excitation with an ultrashort light pulse having a duration (hopefully) less than the isomerization to be studied. An important idea in the description of chemical reactions is the activation barrier which must be crossed in passing from the reactants to the products. For an isomerization the barrier separates the structural forms along the isomerization reaction coordinate. The photoisomerization process is thus one way to study the role of the solvent in the activation barrier model of chemical reactions.

Perhaps the most studied photoisomerizations are those of substituted ethylenes and polyenes. Using stilbene as an example [8.4], we see that the isomerization connects the *cis* and *trans* forms of the molecule (Scheme 8.1). The various

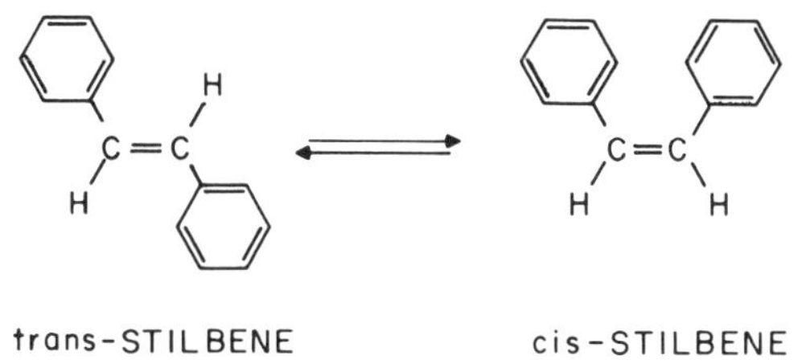

Scheme 8.1

trans-STILBENE ⟷ cis-STILBENE

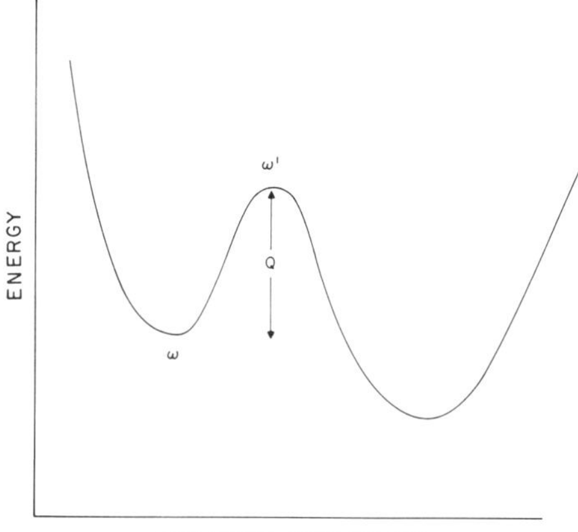

Fig. 8.1. Potential energy diagram for barrier crossing. The frequency of the upper well ω and the barrier ω' are indicated. The barrier height Q is also indicated

models [8.4,6–10] for the photoisomerization of stilbene differ as to whether there is a common excited state intermediate for the excited *cis* and *trans* isomers which precedes decay to the ground-state isomers. However, on the excited *trans* surface, it is generally agreed that there is a barrier which is the rate limiting step in the *trans* to *cis* isomerization. We thus represent the *trans* → *cis* isomerization by a bistable potential with the initial excited singlet state being the *trans* form and the lower energy part remaining undefined for this discussion.

The experiment is initiated by a short pulse of light which promotes the molecule to the excited singlet state surface S_1 shown as the *trans* isomer in Fig. 8.1. The isomerization kinetics are then monitored using a number of approaches, such as decay time of S_1 obtained by fluorescence [8.11–18] $S_1 \rightarrow S_0$, transient absorption [8.19–25] $S_n \leftarrow S_1$, multiphoton ionization [8.26–28], transient Raman spectroscopy [8.29], as well as measuring the kinetics for the appearance of the isomerized form when feasible [8.30,31]. We now turn to an interpretation of the observed dynamics.

8.3.1 Kramers Model of a Chemical Reaction

A historically important model which sought to describe the effects of the solvent on chemical reactions is due to *Kramers* [8.32,33]. In this model the solvent imposes a frictional drag force and random fluctuating force on the solvent motion. These solvent effects together with the intramolecular force defined by the potential, e.g. that shown in Fig. 8.1, describe the Brownian motion of a particle in a force field. It can be expressed in a one-dimensional Langevin equation as a model of the reaction dynamics,

$$F = -\xi v + A(t) - \frac{\partial V}{\partial x} .$$

F is the force on the solute, ξ is the friction, v its velocity along the reaction coordinate, $A(t)$ is the random fluctuating force and V is the intramolecular potential. Kramers obtained the steady-state solution to this equation in the strong coupling limit where the molecules in the reactant well are in thermal equilibrium to the top of the barrier, i.e. vibrational relaxation is rapid compared with the barrier crossing. The Kramers equation for the barrier crossing rate is,

$$k = \frac{\omega}{2\pi\omega'}\left[\left(\frac{\beta^2}{4} + \omega'^2\right)^{1/2} - \frac{\beta}{2}\right] e^{-Q/kT} ,$$

where Q is the height of the barrier, ω and ω' are frequencies associated with the curvatures at the reactant well and at the barrier respectively, and β is the reduced friction coefficient (ξ/I) where I is the moment of inertia.

The comparison of the Kramers and other theoretical models with the experimental data available to date has not been wholly successful [8.34–43]. Several groups have studied the viscosity dependence of the conformational or isomerization dynamics of systems in which bulky molecular groups rotate about single or double bonds. In all these studies the barrier crossing rate was observed to decrease monotonically as the viscosity increased, which is characteristic of the strong coupling regime. Since it seemed reasonable that the solute-solvent coupling in these systems was frictional, these data were compared with Kramers solution of the Langevin model in the strong coupling limit ("Kramers equation"). For diphenylbutadiene [8.35] and *trans*-stilbene [8.38] in alkane solvents and the dye DODCI [8.36] in alcohol solvents, the data showed significant deviation from Kramers equation. One explanation [8.44,45] for this behavior is that for molecules with sharp barriers the effective friction acting on the conformational coordinate is less than the zero frequency (static) friction which appears in the Kramers equation. This view was apparently confirmed by the observation that for diphenylbutadiene and stilbene in alcohol solvents, where the barrier curvature was smaller relative to the alkane solvents, the data were in good agreement with Kramers equation in the Smoluchowski (high viscosity) limit [8.37]. However, for stilbene in alkane solvents [8.38], it was found that agreement with Kramers equation in the Smoluchowski limit could be obtained by "stiffening" the stilbene – that is, restricting the conformational degrees of freedom of the molecule by incorporating the phenyl groups into a fused ring system. Although this agreement in the Smoluchowski limit could be due, in part,

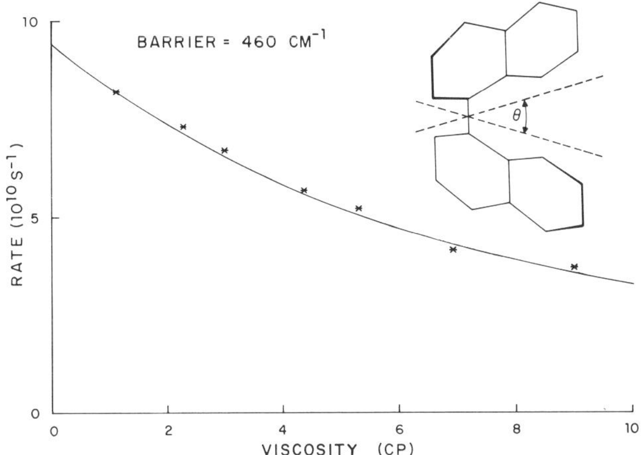

Fig. 8.2. Barrier crossing rate in excited 1,1′- binaphthyl vs. solvent shear viscosity. The * are experimental points and the solid curve is a best fit using the Kramers equation

to a flattening of the barrier, the reduction in the isomerization-related conformational degrees of freedom could also be pertinent [8.38,40,42]. The question is whether the observed deviations from Kramers behavior are due to frequency dependent [8.43–48] solvent friction effects, the participation of more than one conformational coordinate [8.49–53], or other factors [8.43,54–59].

Of direct relevance to this question are the studies on photoisomerization of 1,1′-binaphthyl in solution [8.60,61a]. Excellent agreement [8.34,61] of the Kramers model with experiment, Fig. 8.2, is found from the intermediate to the high friction (Smoluchowski) limit using the static friction, despite the high barrier frequency. For *t*-stilbene, diphenylbutadiene, and DODCI, which showed substantial deviation from the Kramers model it is likely that the isomerization involves more than a single structural coordinate, e.g. twisting of the phenyl group about the carbon-phenyl bond in addition to the isomerization coordinate, and thus leads to a deviation from the Kramers equation which is based on a one-dimensional model. Although the results of 1,1′-binaphthyl show that the assumption that the conformational change involves only a single coordinate is reasonable, it does not establish that the dimensionality is the key or only factor. It does show that deviations from Kramers is not a universal phenomenon in the conformational dynamics of flexible molecules in solution and raises the issue of why a Markovian theory such as Kramers works so well despite the sharp barrier.

8.3.2 Photoisomerization at Low Densities

In our discussion of photoisomerization up to this point we have considered the case where there is strong coupling of the solute to the solvent. This means that energy exchange with the solvent is rapid on the time scale of the reaction, i.e. the rate of isomerization is not limited by the molecule acquiring enough energy to reach the top of the barrier. The strong coupling case is expected to be applicable to liquids and this has been the focus of our discussion.

In a low density gas, or more generally where the coupling of the solute reaction coordinate to the solvent is weak, e.g. at low friction, the rate limiting step is reaching the barrier region. Increasing the coupling increases the reaction rate unlike the strong coupling case where the increasing collisions, friction (viscosity) hinders crossing and causes recrossing of the barrier, and thus yields a slower rate. The transition from the weak coupling or energy controlled regime to the strong coupling or diffusion controlled regime can be examined by varying the solvent density from the "isolated" molecule to the dense liquid. A maximum in the rate will occur at some density as the rate at first increases then eventually decreases at high friction. This behavior is called the Kramers turnover[1] and is of interest in connecting the different kinetic regimes.

Starting at the weakest coupling end of the solute-solvent interactions, namely the isolated molecule case, we can gain important insights into how the solvent affects the reaction dynamics. One way to carry out isolated molecule experiments is to be at densities which are sufficiently low that the photo-excited molecules undergo no collision in the time scale of the reaction. Another powerful way to study the reaction of an isolated molecule is in a jet-cooled molecular beam. Because of the very low vibrational, <40 K, and rotational, <10 K, temperatures of molecules in a cooled jet, a narrow energy distribution of excited molecules can be achieved. In the low density collision-free gas on the other hand a broad thermal distribution of excited molecules is produced at room temperature.

Studies of *t*-stilbene [8.15,43,62–65] and diphenylbutadiene [8.66–68] in cooled jets have shown that there is an energy threshold value E_0 in the isomerization kinetics for these molecules; see Fig. 8.3. Molecules below this threshold energy do not cross the barrier to the isomerized form. Collisions are necessary to provide such molecules below E_0 with necessary energy. Studies on *t*-stilbene in gaseous ethane have shown that the rate increases with pressure [8.41,42,43a]. This indicated that the reaction is in the energy controlled regime. On continuing to increase the pressure the rate is observed eventually to decrease as the reaction enters the strong coupling diffusive regime.

Although the qualitative features of a Kramers turnover are seen in Fig. 8.4, the experimental results do not follow statistical theories which extend from the energy controlled regime to the diffusive regime [8.41]. Among the various possible explanations for the discrepancy are solvent induced shifts in the barrier height as the pressure increases due to solute-solvent complexes [8.43]. There is also incomplete intramolecular vibrational redistribution which can increase or decrease the rate depending on whether the collision populates the reaction mode

---

[1] Evidence for the Kramers turnover has been obtained from NMR studies of cyclohexane in solution [8.61b].

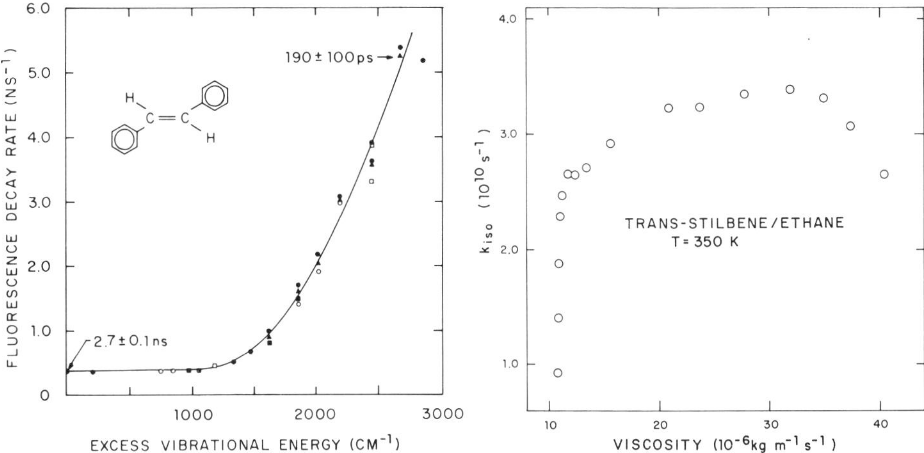

Fig. 8.3. Observed fluorescence decay rates in *t*-stilbene as a function of excess vibrational energy. Figure taken from [8.15]

Fig. 8.4. The isomerization rate of *trans*-stilbene as a function of the experimental viscosity in ethane at 350 K. After [8.41]

or modes strongly coupled to it, or populates modes not strongly coupled to the reactive mode, i.e. breakdown of RRKM on the time scale of reaction [8.41–43]. Entropy effects [8.69] associated with activation volumes in the more dense regions, nonadiabatic effects [8.17], and dimensionality effects as noted earlier can also be important possibilities [8.41–43]. At this time, there is considerable discussion about why the thermally averaged results of the cooled jet are slower than the liquid phase kinetics. One might have expected the thermally averaged results of the jet to be faster than the liquid state kinetics, since the viscous effects in liquids which slow the isomerization are not present. This disagreement is not very surprising in the light of the failure of the statistical theories to describe the kinetics in the Kramers turnover region, since the same factors are operating in both comparisons.

8.3.3 Photoisomerization of Polar Molecules

For a molecule which undergoes a significant charge redistribution on isomerization, it can be expected that the course of the reaction will be sensitive to the polarity of the solvent. The dynamics and even the occurrence of the isomerization can depend on the solute-solvent interactions. A classic example of the effects of charge redistribution on the excited state isomerization is that of *p*-dimethylaminobenzonitrile (DMABN), where a structural change occurs in polar but not nonpolar solvents. From the many studies [8.70–85] of this molecule a general outline of the dynamics has been developed (Fig. 8.5) though important issues remain to be clarified. The general picture [8.71,74,75] involves a rotation of the dimethylamino group about the aminophenyl bond in an excited

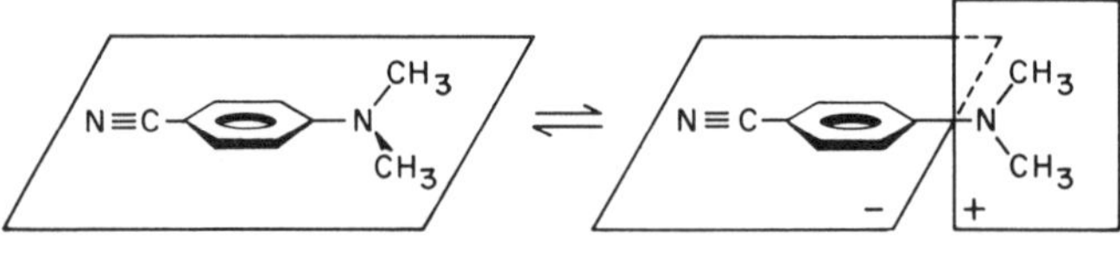

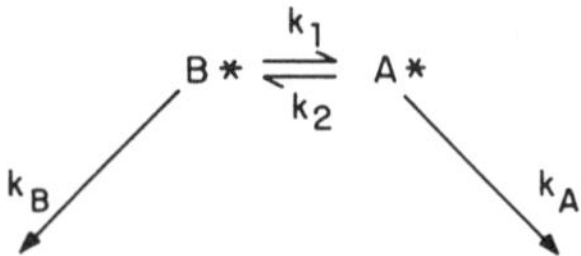

Fig. 8.5. Twisted internal charge transfer in dimethylaminobenzonitrile (DMABN). B^* represents the planar state and A^* represents the twisted charge-transfer state

singlet state. This motion results in a charge separation between the two groups thereby generating a large increase, 10D, in the excited state dipole moment. In polar solvents the twisted, larger dipole form is stabilized. It produces a new emission in the visible which does not appear in nonpolar solvents. Rapid equilibration [8.78] between the twisted and planar forms in polar media yields a dual fluorescence, one due to the planar form in the uv and the other, in the visible, due to the twisted form.

Having established some of the key ideas of the isomerization it becomes possible to investigate how the various solute-solvent interactions contribute to the observed kinetics. A commonly used approach to determine the effects of solution viscosity on the kinetics of a reaction is to carry out measurements in a solvent series, e.g. the linear alcohols, methanol, ethanol, propanol, etc. or the linear nitriles. Following this approach the forward rate constant k at room temperature yielded a viscosity dependence in the nitrile series [8.83] of $k \sim \eta^{-2/3}$, whereas in the alcohol series [8.84] it was roughly η^{-1}. These results could indicate that the frictional effects in alcohols were in the Smoluchowski regime of very strong coupling and perhaps intermediate coupling in the nitriles. However, this interpretation is based on the assumption that it is only the viscosity which is changing as one goes through a solvent series. If a large change in dipole moment is *not* occurring in the isomerization, then the polarity differences within a series are probably of minor importance. For a polar transition, however, this cannot be assumed to be the case. Measurements of the rates for DMABN isomerization carried out at different temperatures so as to have the same viscosity for different liquids of a series, showed that the rates were increasing as temperature decreased [8.85]. If the barrier to isomerization was independent of the temperature and was the same in the different members of a solvent series, as is usually assumed, then the rate would decrease with decreasing temperature due to the Boltzmann factor. To explain these observations it has been proposed [8.83,84] that the barrier is not constant with the temperature but is lowered as temperature goes down. The reason for this is that the polarity of liquid increases with decreasing temperatures. The more polar twisted form of DMABN is stabilized relative to the initially excited less polar, planar form, thereby reducing the barrier height. This decrease in barrier height with lower temperature overcomes the usual temperature effect and is responsible for the increased rate at lower temperatures. Correcting for the change in polarity with temperature results in a "normal" positive activation energy for the isomerization. Further support for the role of solvent polarity on isomerization dynamics in DMABN was obtained by measurements in isoviscous mixtures of a nitrile [8.83] or an alcohol [8.84] with an alkane. The polarity of the solution was controlled by varying the concentration of the polar component, always maintaining the same solution viscosity and temperature. Despite the constancy of temperature and viscosity the isomerization rate was found to depend exponentially on the solvent polarity as measured by the widely used empirical polarity parameter $E_T(30)$. We thus see that the effects of temperature, the effects of variation from one member of a solvent series to another, or the effects of any other variable that changes the polarity of a liquid can play a key role in the dynamics of a chemical reaction. Separating these various parameters, such as viscosity, polarity, activation volume and entropy effects, to name some of the more obvious ones, is an essential step in unravelling the effects of the solvent on solute reaction dynamics.

8.3.4 Photoisomerization with Femtosecond Pulses

One aspect of the photoisomerization of stilbene which was too fast to measure until the development of femtosecond light pulses, was the *cis* → *trans* isomerization. Earlier work [8.30] examined the dynamics of the *cis* → *trans* isomerization using a novel two pulse sequence. As shown in Fig. 8.6, the laser experiment used a single pulse to first excite the *cis* form and then a second pulse was used to probe the dynamics by laser induced fluorescence from the *trans* form. The results established that the process was faster than a few picoseconds.

In the more recent work [8.27] to time resolve the *cis* → *trans*-stilbene isomerization a 190 fs uv pulse was used to excite *cis*-stilbene. A resonantly enhanced two photon ionization of the excited *cis*-stilbene with a visible fs pulse probed the decay of the excited *cis*-stilbene. The amplitude of the current produced as a function of the time separation between the uv and visible pulses yielded the dynamics of the decay. It was 1.2 ps in hexane. Following earlier work [8.86] on

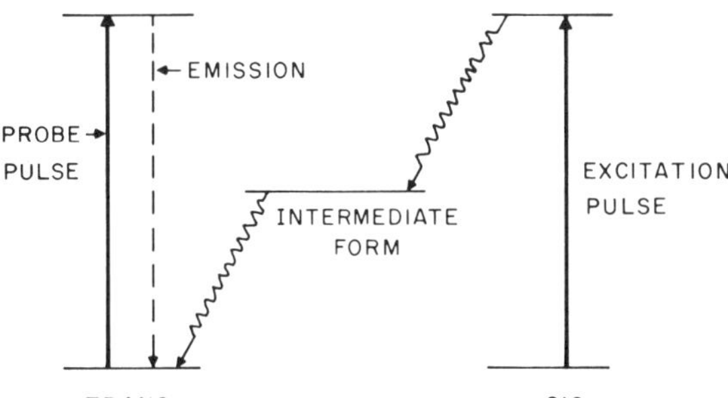

Fig. 8.6. Two pulse sequence used to study the *cis* → *trans* isomerization in stilbene via laser induced fluorescence

the dynamics of tetraphenylethylene isomerization, the multiphoton ionization technique yielded a value of 3.7 ps in hexane which agrees with transient absorption measurements.

A most interesting result was obtained by application of the femtosecond multiphoton ionization technique to the isomerization of *cis*-stilbene in the vapor phase [8.26]. Exciting the *cis*-stilbene with less than 50 cm^{-1} of excess energy in S_1 it was found that the decay time was 320 fs. Theoretical calculations of the torsional frequency of the carbon-carbon double bond yield a time constant of 38 fs [8.87]. It would therefore appear that under the collision-free conditions of this experiment there is at most a very small barrier to the *cis* → *trans* isomerization, in contrast to the *trans* → *cis* isomerization.

As a further point concerning the extensive research on stilbene, and a promising advance in the development of ultrashort pulse methods, is the application of picosecond transient Raman spectroscopy to this problem [8.2,88]. Using a high repetition rate synchronously pumped dye laser the Raman spectrum of the excited *trans*-stilbene was obtained. Based on this data it was proposed that there are various excited state conformers contributing to the spectrum. The interconversion rate of these conformers and their relation to the isomerization dynamics have not been determined.

8.4 Photodissociation and Recombination of Molecular Iodine in Solution

A seemingly simple chemical reaction in solution is the dissociation of the iodine molecule and the subsequent recombination of the iodine atoms. However, just as the spectroscopy of I_2 is highly complicated as seen by frequent theoretical revisits, the dynamics too are similarly complex. An important idea emerging from the observed wavelength and viscosity dependence of the iodine stom dissociation was that of a solvent cage effect [8.89–92]. The idea was that the solvent interferes with the escape of the fragments, the iodine atoms, thereby increasing the likelihood of their recombination and thus decreasing the dissociation yield [8.89–94]. Early time-resolved studies [8.95–100] of the cage effect were initially carried out on I_2 in the solvents CCl_4 and hexadecane, and later extended to other solvents. The picosecond experiments [8.95–102] succeeded in monitoring an important aspect of the solvent role, namely the collision induced predissociation [8.103,104] of the photoexcited I_2 molecule from the bound $^3\Pi_{0^+u}$ state to a repulsive state. This collision induced predissociation, which occurs in less than 10 ps in I_2, had not been seen earlier, even by the usual spectroscopic broadening processes present in solution. The subsequent return of the iodine atoms to the ground electronic state of molecular iodine was measured to be in the range of 50–150 ps, depending upon the solvent [8.95–100]. This latter time constant was interpreted as the time scale of the geminate recombination of the iodine atoms. However, subsequent theoretical calculations [8.105–108] brought this interpretation into question. The various theoretical approaches predicted that the recombination times should be shorter. A new and important proposal was made [8.105]: namely, that the measured relaxation time of about 100 ps was due not to atom recombination but rather to vibrational relaxation in the ground electronic state. Other calculations supported this idea [8.106–108]. The theoretical work was a valuable impetus to new time-resolved experiments [8.101,102,106,108–114]. Picosecond transient absorption experiments in the red and near infrared were consistent with the view that vibrational relaxation was a key part of the observed dynamics.

A new development [8.110–113] in the increasing number of picosecond studies on I_2 was the idea that a fraction of the atoms recombined into the excited electronic states, A and A'. The lifetime of the I_2 molecule in these states was rather long (ns), giving rise to a slow component in the ground-state repopulation kinetics. There is now evidence that the dynamics is not only determined by vibrational relaxation in the ground electronic state but also by the time to get onto the ground electronic surface [8.113,114]. The possibilities of a cage effect and/or the emptying out of I_2 from weakly bound electronic states into the ground state in tens of picoseconds are among the ideas being pursued [8.113–115]. Clearly the effects of the solvent on vibrational relaxation, electronic relaxation e.g. by changes in the position of curve crossing, and cage restrictions, continue to make the iodine problem an interesting one.

8.5 Chemical Intermediates

To fully understand the nature of chemical change, the chemist must study not only the role of the solvent, but at a more elementary level must identify the various chemical species and states involved in the course of a chemical reaction. Most chemical reactions pass through a sequence of energy states, structures and chemical species, in going from reactant to product. To determine the mechanism of a reaction the various steps in the overall change must be elucidated. The short-lived transient species which provide the bridge between reactant and product are often called chemical intermediates. Ultrashort pulse spectroscopy using transient fluorescence, absorption and Raman scattering methods are playing an important role in identifying these elusive short-lived species.

8.5.1 Role of Charge Transfer in Chemical Reactions

A primary step in many chemical reactions is the transfer of an electron between the reacting species. The dynamics of this step, often the first one in a chemical reaction, depends not only on the nature of the reactants but on the solvent environment. The exothermicity of the reaction, the geometric requirements for transfer, the solvent dependence of the energies of the various electronic states and the solvent reorganization energy, are some of the factors that have to be considered in the charge transfer dynamics. A number of picosecond laser studies have been carried out on the electron transfer dynamics between freely diffusing reactants [8.116–119], between methylene chains [8.116–123] or directly connected [8.124] reactants, between rigidly separated reactants using a 30 ps electron pulse source [8.126], and between metal centers of organometallic compounds [8.127–129]. The rates can be extremely rapid (ps) depending upon distance and solvent.

a) Hydrogen Atom Transfer by Electron Plus Proton Transfer

The electron transfer reaction can play an important role in more complicated reactions as well. For example, one important class of reactions, sometimes referred to as photoreduction, involves the transfer of a hydrogen atom from a ground state molecule to a photoexcited one. Although the overall reaction involves an $H^{\cdot}$ transfer it was proposed [8.130–134] that the actual reaction takes place in two steps, rather than a single reaction $H^{\cdot}$ transfer step (Scheme 8.2). The first step would be electron transfer, followed then by a proton transfer, the net being a hydrogen atom transfer.

This electron plus proton transfer pathway versus the direct $H^{\cdot}$ transfer pathway is favored in cases where one of the reactants, having a low ionization potential, acts as an efficient electron donor, e.g. amines. The electron affinity of the acceptor A is increased by its excitation to A^* and further drives the charge transfer process. To establish that a reaction has followed this route one can use transient absorption to track the appearance of the radical ion or charge transfer complex absorption spectra as the first chemical step following excitation. The radical ion spectra should then evolve into the free radical spectra as the proton transfer takes place.

A good example of photoreduction is the reaction of benzophenone (Bp) in its lowest triplet state with an amine such as N,N-dimethylaniline shown in Scheme 8.3 [8.131,133,135,136]. In non–hydrogen-bonding solvents, the electron transfer step was proposed on the basis of picosecond experiments to take place between Bp(T_1) and DMA(S_0) molecules that are separated by solvent molecules. The solvent-separated ion pair produced by the electron transfer is then thought to diffuse together to form the contact ion pair. This view is based on the observed time dependent spectral shift which is similar to that known for ion pairs of aromatic anions and alkali metals. The origin of the large blue shift ($1000\,cm^{-1}$) is the interaction between the $Bp^{-\cdot}$ anion and the amine $DMA^{+\cdot}$ cation in the contact ion pair. Further support for this interpretation is provided

$$A^* + BH \xrightarrow{e^-} A^- \cdots (BH)^+ \xrightarrow{H^+} AH^{\cdot} + B^{\cdot}$$

CT COMPLEX RADICALS **Scheme 8.2**

Scheme 8.3

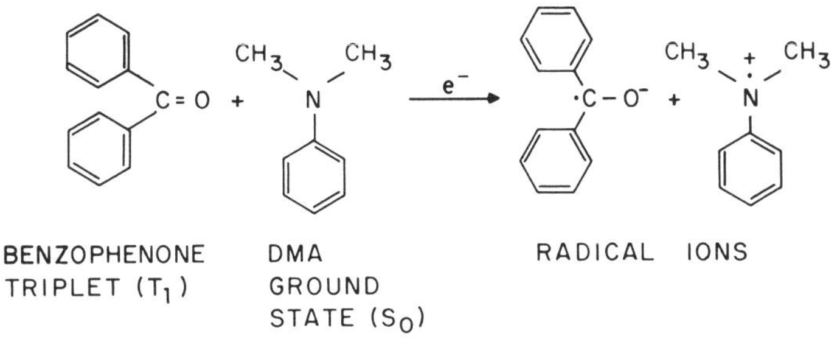

BENZOPHENONE TRIPLET (T_1) DMA GROUND STATE (S_0) RADICAL IONS

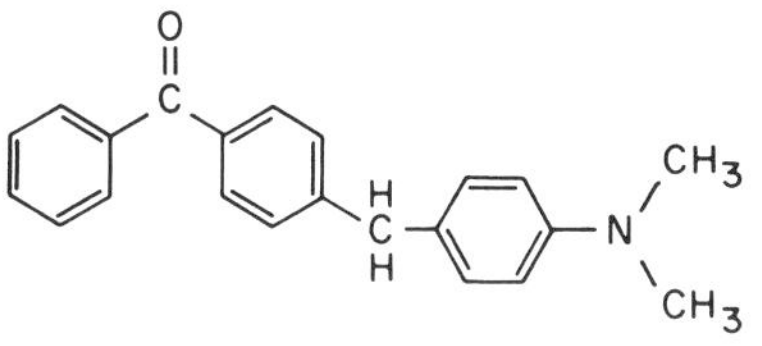

Scheme 8.4

by the spectrum observed following electron transfer in the hooked-together molecule [8.137,138], Scheme 8.4.

Following photoexcitation of the benzophenone part of this molecule in the polar solvent acetonitrile, it is found that the benzophenone radical ion $Bp^{-\cdot}$ appears in less than 20 ps. Its spectrum is the same as that of unconnected $Bp^{-\cdot}$ at early times. Whereas the free $Bp^{-\cdot}$ spectrum then shifts due to the formation of the contact ion pair, it is found that the spectrum of the hooked-together $Bp^{-\cdot}$ does not shift, presumably because the methylene chain prevents the collapse of the radical ions to the contact ion pair separation and geometry. It is furthermore found that the proton transfer step which produces the neutral free radical does not form in the hooked-together moiety. For the free Bp and DMA in acetonitrile the neutral free radical is observed to form only after the contact ion pair is formed. Thus, the proton transfer step might occur only over short distances, whereas the electron transfer can take place over distances of up to ~ 15 Å. It is also possible that the rate of H^+ transfer is slow compared with the diffusion of the solvent-separated ion pair to the proper distance and orientation for contact ion pair formation. There is support for the latter idea mainly in that proton transfer can occur over reasonably long distances but at a much slower rate than electron transfer [8.138]. Placing the hooked-together molecule in the nonpolar solvent benzene yields the absorption of the benzophenone free radical thus showing that the proton transfer step must have taken place. The possibility of a smaller barrier to produce the neutral free radical in the less polar solvent could be responsible for the proton transfer occurring in benzene and not in acetonitrile. That geometry is an important factor in the proton transfer step is based on the result that the benzophenone free radical is formed in acetonitrile when Bp and DMA are free but not hooked together. When free they can orient appropriately at the optimum distance for contact ion pair formation, and then proton transfer. Interesting results [8.138,139] on the effects of orientation and distance have also been obtained by separating the donor and acceptor molecules by two or three methylene (CH_2) groups. In addition, the variation of the donor and acceptor molecules, e.g. alkyl as well as aromatic amines and acceptor molecules, yield important insights [8.140–142] into the effects of ionization potential, steric hindrance, and salt effects on the nature of electron plus proton transfer.

This process of transferring a hydrogen atom via an electron and then a proton can be extended even further. There is picosecond spectroscopic evidence [8.143] that the transfer of a hydride, H^-, takes place by a sequence of electron then proton transfer followed by a second electron transfer giving the net hydride transfer reaction. The rules governing the transfer of several successive particles rather than the net particle, e.g. an $H^{\cdot}$ atom or hydride ion, is one of importance and one of increasing research activity.

b) Electron Donor – Acceptor Complexes

We have so far considered interactions between an excited and a ground state molecule leading to chemical reaction via electron and proton transfer. The donor and acceptor molecules were not complexed together, i.e. did not interact significantly prior to excitation of one of them. There is, however, a class of molecules that form electron donor-acceptor complexes in their ground electronic states, and which are important as intermediates in thermal and photochemical reactions [8.144–147]. One example [8.147] is the complex of indene (In), with an acceptor such as tetracyanoethylene (TCNE), Scheme 8.5, where the binding is provided by partial transfer of an electron from D to A, i.e. mixing of a charge transfer state into the no bonding state DA.

Excitation [8.147] into the charge transfer band produces the absorption spectra of the radical ions $In^{+\cdot}$ and $TCNE^{-\cdot}$. The radical ion pair decays within 60 ps. However, the time to return to the initial ground state complex was found to be 500 ps. This points to an intermediate state or structure (see Scheme 8.6) as a possibility since the system returns to its original ground state concentration. Whether it is an intermediate or dissociation of the radical ion pair into the neutral indene plus tetracyanoethylene which then recombine to form the ground state complex is unknown and under investigation. From these studies

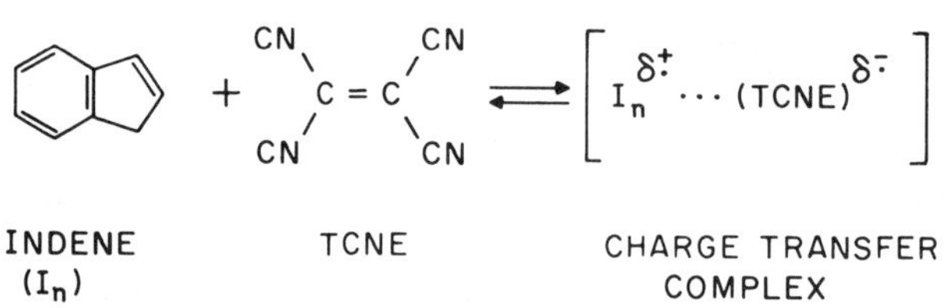

$$\text{INDENE } (I_n) + (CN)_2C{=}C(CN)_2 \text{ (TCNE)} \rightleftharpoons \left[I_n^{\delta+} \cdots (TCNE)^{\delta-} \right] \text{ CHARGE TRANSFER COMPLEX}$$

Scheme 8.5

Scheme 8.6

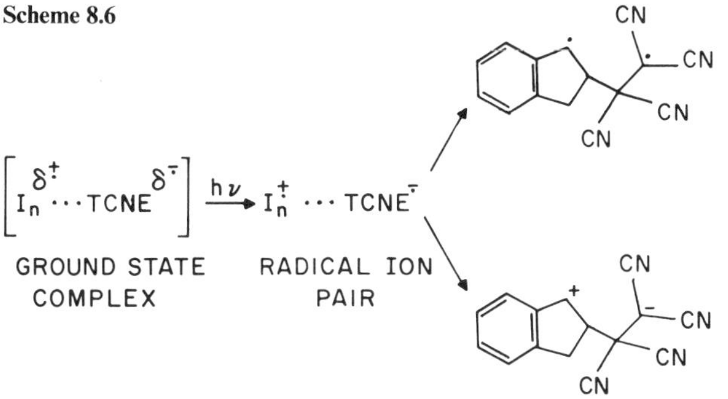

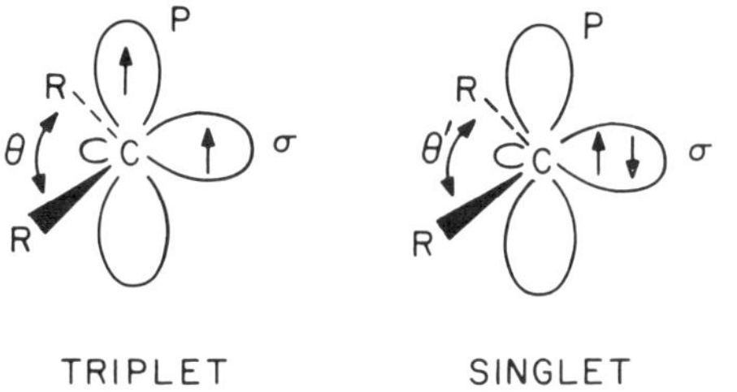

$$\left[\overset{\delta^{+}}{I_n}\cdots\overset{\delta^{-}}{TCNE}\right] \xrightarrow{h\nu} I_n^{+\cdot} \cdots TCNE^{-\cdot}$$

GROUND STATE COMPLEX — RADICAL ION PAIR

TRIPLET — SINGLET

Fig. 8.7. Triplet ($\sigma' \rho'$) and singlet ($\sigma^2 \rho^0$) electronic spin states of carbene

one gains insight into why excitation of the indene-TCNE complex produces no irreversible chemistry whereas other aromatic-TCNE complexes do undergo photochemical change.

8.5.2 Carbenes

A class of intermediates which has been the focus of an enormous amount of research in chemistry has been divalent carbon fragments, usually referred to as carbenes [8.148–154]. The implication of carbenes in many different reactions, the novelty of their reactions, as well as their interesting spectroscopy and widespread utility in microelectronics and semiconductor fabrication has maintained interest in these reactive species. Their unique chemical and physical properties derive from the presence of an unshared pair of electrons which result in two low-lying electronic states, a singlet and triplet (see Fig. 8.7).

For most carbenes the ground state is a triplet [8.149,152,155] with the singlet state often being thermally accessible. Since the chemistry [8.156–158] of the singlet and triplet differ, the dynamics of spin relaxation [8.159,160] is of importance in establishing the populations of these states, which in turn influence the reactions which take place. Picosecond laser studies [8.153,161–163] of a variety of carbenes where the attached groups are aromatic have yielded new insights into their chemical reactions.

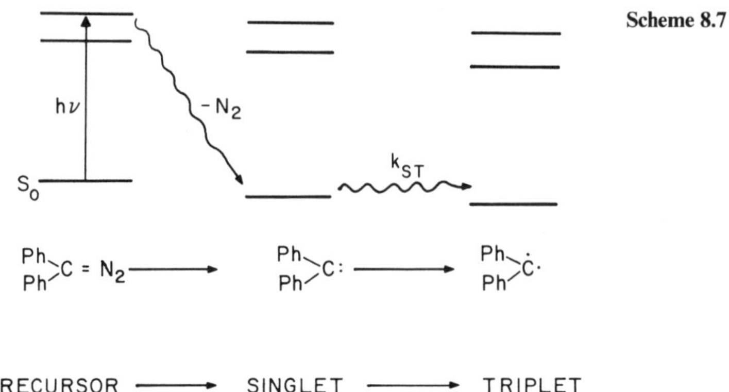

Scheme 8.7

a) Spin Relaxation in Carbenes – Solvent and Structure Effects

Common precursors to the carbene intermediate are diazo compounds. Photoexcitation with ultraviolet light releases nitrogen and the carbene fragment (Scheme 8.7). A typical example is the carbene fragment generated in its singlet state in less than 5 ps from the photoexcited precursor diphenyldiazomethane [8.153,165].

Intersystem crossing from the singlet to the ground triplet has been found to vary widely for a given carbene in different solvents [8.166–168], and to vary widely from one carbene to another [8.168]. For a given carbene the origin of the large solvent effect on spin relaxation has been traced [8.167,168] to the dependence of the singlet-triplet energy gap on the polarity of the solvent.

The difference in the electronic structures of the singlet and triplet states results in the polar singlet being more stabilized in a polar solvent than the triplet. This different stabilization of the two states leads to a decrease in the singlet-triplet energy gap as the solvent polarity increases. Using the values of k_{ST} and k_{TS} one can calculate the equilibrium constant, K_{eq}, between the singlet and triplet states of the carbene from the relation:

$$K_{eq} = \frac{k_{ST}}{k_{TS}} \ .$$

The free energy difference between these two states follows directly from K_{eq}:

$$\Delta G = -RT \ln K_{eq} \ .$$

From the value obtained for ΔG the energy gap between the singlet and triplet states can be estimated from the thermodynamic relation $\Delta G = \Delta H - T\Delta S$. Since the entropy difference, ΔS, is expected to be primarily due to the multiplicity difference between the singlet and triplet states, and the enthalpy difference, ΔH, and the energy gap, ΔE, are equal to a good approximation in liquids, we obtain,

$$\Delta G = \Delta H - T\Delta S = \Delta E - RT \ln 3 \ .$$

Therefore, values of the singlet-triplet equilibrium constant, K_{eq}, obtained from picosecond measurements of singlet to triplet, and triplet to singlet rate constants yielded estimates of energy gaps in these various solvents [8.163,168]. An interesting result was that the singlet to triplet radiationless transition was slower for smaller energy gaps. The more usual case in radiationless transitions is for the rate to become faster for smaller energy gaps due to the larger Franck-Condon factors. It was proposed [8.167–169] that the inverse gap effect observed in the various aromatic carbenes is a consequence of a small energy gap (<2500 cm^{-1}). For a small energy gap there is a sparse manifold of triplet vibronic states to which the singlet is coupled. As the gap increases there are more triplet vibronic states in the vicinity of the singlet, thereby enhancing the coupling and the intersystem crossing. This explanation accounts for the observations in the carbenes shown in Scheme 8.8.

The above ΔE_{ST} values are for the carbenes in acetonitrile solvent. The inverse dependence of singlet to triplet relaxation rate with energy gap was found to explain not only the solvent dependence for a given carbene but also the trend in rates from one carbene to another [8.166,168,169]. Comparison of the rate of $S \rightarrow T$ of different carbenes in the same solvent showed that the rate was slower for the smaller gap [8.168,169]. To complete the picture of the dynamics of aromatic carbenes, studies [8.170] were carried out on an aromatic carbene already discussed, namely dimesitylcarbene (DMC) shown in Scheme 8.9. The picosecond laser studies showed that the energy gap for DMC was sufficiently large, $>2500\,cm^{-1}$, to yield a normal energy gap rate law, i.e. rate increases as

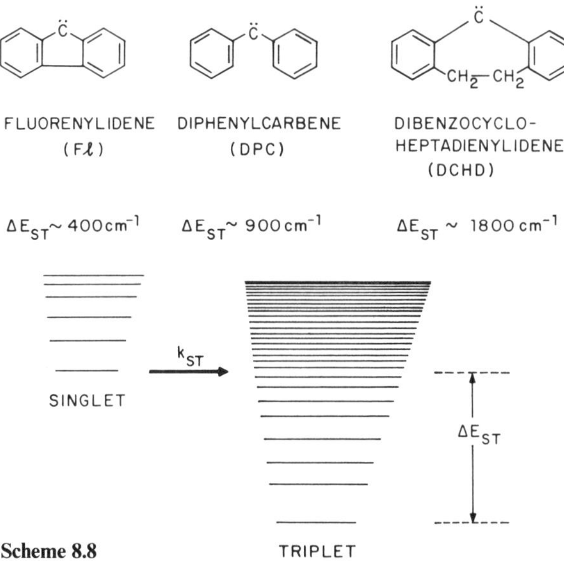

Scheme 8.8

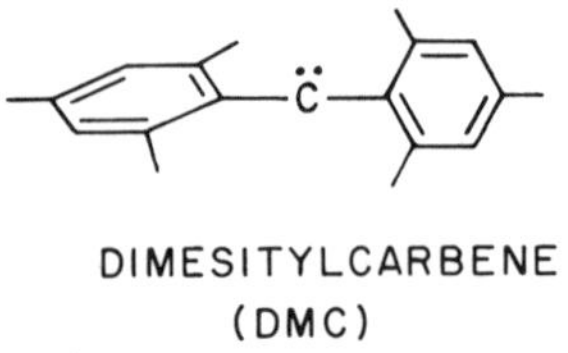

Scheme 8.9

DIMESITYLCARBENE
(DMC)

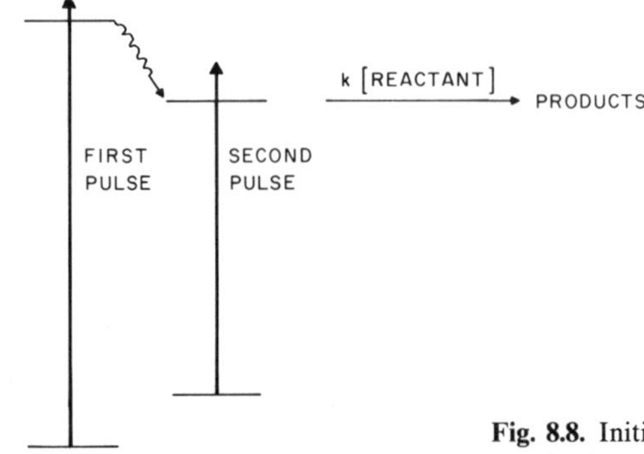

Fig. 8.8. Initiation of excited state photochemistry of a reactive intermediate via a two pulse laser sequence

the gap decreases. Thus for dimesitylcarbene in the polar solvents where the gap was smaller than in the nonpolar solvents, the intersystem crossing rate was indeed faster, as expected for a normal gap dependence. The interpretation of the kinetics of spin relaxation in these various aromatic carbenes in terms of the measured variation of the singlet-triplet energy gap with solvent and carbene structure explains the chemical reactivities of these carbenes as well. For example, the dimesitylcarbene is found to have a large energy gap which explains the observed variation in kinetics with solvent, and also why the reactions are predominantly from the ground triplet state.

b) Photochemistry of Chemical Intermediates

The studies of chemical intermediates have hitherto been restricted to reactions in their ground or thermally accessible electronic states due to their high reactivity. They generally do not live long enough in a reactive environment to permit populating enough of them in excited electronic states for study. With short-pulsed lasers [8.171–173] it becomes possible to quickly generate the intermediate and then, with a second pulse, to excite some fraction of them to an upper electronic state where their photochemistry can be studied, as shown in Fig. 8.8.

As can be anticipated this opens many new possibilities, since the optical and chemical properties can be very different from those of the ground states. In the aromatic carbene diphenylcarbene, it has been found [8.171,172] that the excited triplet molecule reacts with alcohols, which is unexpected for a triplet state but rather like a singlet state. An explanation [8.172] of this observation in terms of a molecular orbital scheme succeeded in accounting for the reactions of the ground triplet, singlet and excited triplet with alcohol. It has also been observed [8.173] that the excited triplet undergoes a charge transfer reaction with amines. This is a reaction which was previously not known for carbenes. From these few examples we see that the photochemistry of short-lived chemical intermediates opens up new possibilities which will further enhance research in this young field.

8.5.3 Nitrenes

A class of chemical intermediates that is analogous to the carbenes is the nitrenes [8.174–178] (Scheme 8.10). Rather than being centered on a carbon atom the unshared electrons are centered on a nitrogen atom, where the dots represent nonbonding electrons. Just as the carbene electronic arrangement generally leads to a ground triplet with a neighboring singlet, so does the nitrene. The reactions in carbene chemistry are found to have their parallel in nitrene chemistry [8.174–178]. However, up to now the research on nitrenes has been far less extensive than that on carbenes. In fact, there is considerable controversy as to whether a nitrene intermediate, rather than some alternative reaction path, is present in a number of reactions. Short-pulse spectroscopic techniques could make a substantial contribution to this field. On reaction which is known to form nitrenes and which provides a starting point for their study is the photodissociation of azides. In particular the picosecond laser induced dissociation of *p*-(dimethylamino) phenylazide has been observed [8.179] (Scheme 8.11). The initial photofragmentation step produces N_2 and excited singlet nitrene in less than 10 ps. The subsequent steps, also monitored by transient absorption, lead to a different singlet nitrene state, or perhaps a structurally rearranged nitrene (tautomer) before finally reaching the ground triplet nitrene in about 120 ps. Further work to identify the precursor to the triplet nitrene is being pursued. With the time resolved transient spectroscopic information on this nitrene it

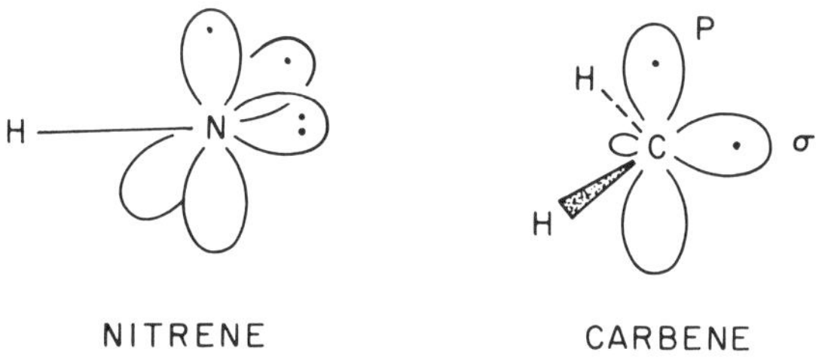

NITRENE CARBENE

Scheme 8.10

NITRENE

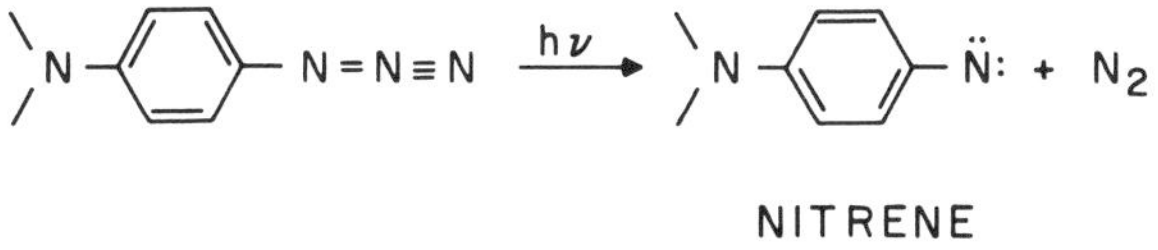

Scheme 8.11

should be possible to determine whether nitrene intermediates are in fact present in a number of important chemical reactions.

8.5.4 Photofragmentation of Singlet Oxygen Precursors

Oxygen in its lowest singlet state, $^1\Delta_g$, at about 1 eV, is significantly involved in chemical, physical, biological, atmospheric, and medical phenomena. Its spectroscopy, interesting chemistry and role in materials degradation has resulted in the development of an extremely active and productive area of research [8.180, 181]. A convenient photochemical source $^1O_2$ are compounds known as endoperoxides [8.180–185]. These compounds have a peroxide bridge —O—O— connecting two of the carbons in the molecule. As an example, (Scheme 8.12), the photoexcitation of the endoperoxide of the following anthracene derivative at an appropriate wavelength yields $^1O_2$.

An important aspect of this reaction is its description as an adiabatic photodissociation [8.186, 187], specified as having both the reactant and product on excited electronic surfaces. This identification was based [8.183] on the detection of $^1O_2$. Direct support for an adiabatic pathway was subsequently obtained from picosecond laser experiments [8.188, 189] where it was discovered that the aromatic fragment was produced in an excited electronic state within 5 ps after the endoperoxide photoexcitation. This reaction pathway is unusual in producing both fragments in excited electronic states. Based on the energetics of this pathway, it was determined that the $^1O_2$ was not produced in the higher $^1\Sigma_g^+$ state, which can decay to the lower $^1\Delta_g$, but rather in the $^1\Delta_g$ directly. Despite the novelty of this doubly excited fragment pathway, it was shown that it is a minor pathway [8.189] (less than a few percent) compared with the dominant pathway which produced the ground state anthracene derivative and $^1O_2$. However this dominant channel for generating the $^1O_2$ had an interesting feature [8.189] in that the final ground state aromatic fragment 1,4-dimethyl-9,10-diphenylanthracene was not produced directly from the photoexcited precursor.

From the appearance kinetics of the ground state aromatic fragment and the <5 ps appearance time of the excited aromatic fragment various routes were proposed [8.189] for the overall photodissociation. One route leads to the excited (S_1) fragment in less than 5 ps and the others to the delayed appearance of the ground state aromatic via two possible intermediates.

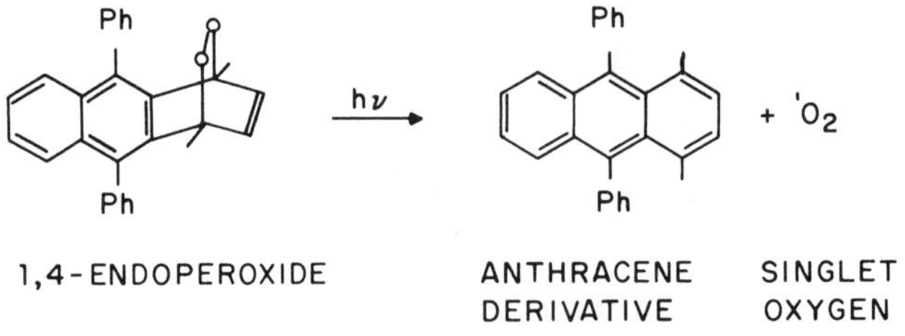

Ph Ph

$h\nu$ → + $^1O_2$

Ph Ph

1,4-ENDOPEROXIDE ANTHRACENE DERIVATIVE SINGLET OXYGEN

Scheme 8.12

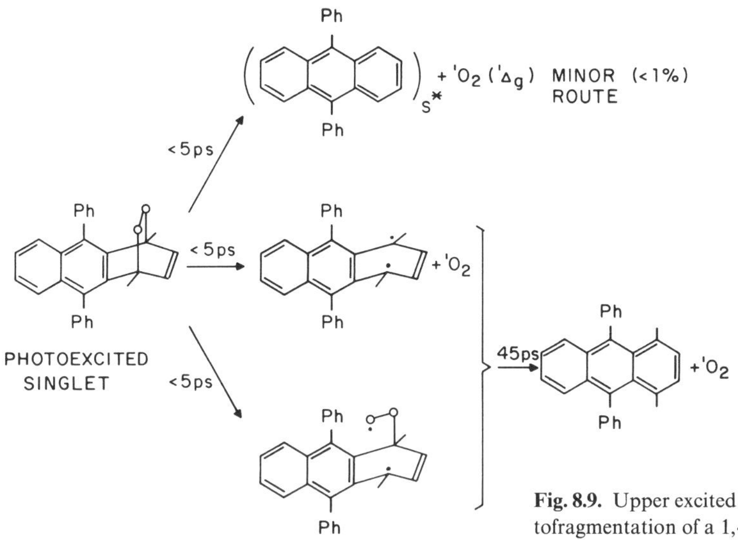

Fig. 8.9. Upper excited π, π^* state photofragmentation of a 1,4-endoperoxide

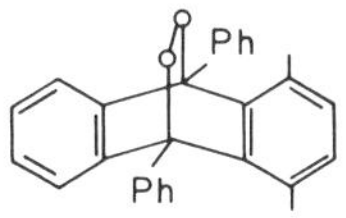

9,10 ENDOPEROXIDE **Scheme 8.13**

As shown in Fig. 8.9 one route involves breaking one of the carbon-oxygen bonds producing a biradical, perhaps excited, which subsequently undergoes a rupture of the second carbon-oxygen bond in 45 ps, yielding the fully conjugated aromatic and $^1O_2$. The other route produces the aromatic fragment in an isomeric form, i.e. some structure akin to Dewar benzene, which then rearranges to the ground state in 45 ps. Work on the related endoperoxide (Scheme 8.13) for which the peroxide bridge spans the central ring rather than an outer ring yielded similar results [8.190].

The rise time of the excited aromatic fragment was again less than 5 ps but the ground state aromatic appeared in 75 ps rather than 45 ps. This difference could be due to the different stabilities of the biradicals or of the isomers. Full resolution of these issues requires further investigation.

8.5.5 Biradicals

On cleaving a bond in a molecule, for example a carbon-carbon bond, a radical pair is produced. The subsequent chemistry is determined by the possibility of recombination which yields the original molecule versus other free radical reaction pathways. If the C—C bond is part of a ring in the original molecule then

Scheme 8.14

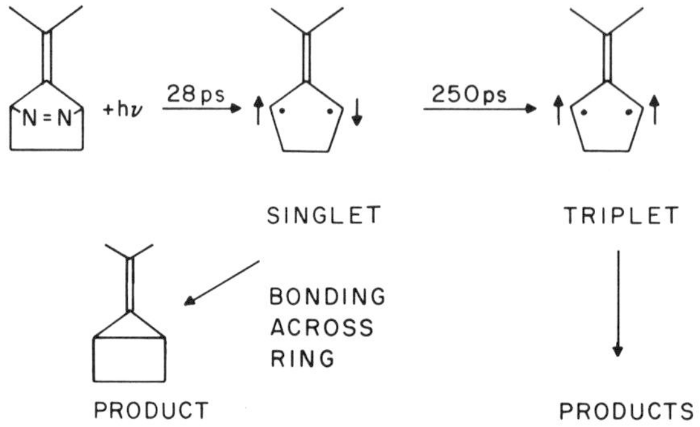

Scheme 8.15

the two radicals remain bound together and the fragment is called a biradical [8.174,191–193]. A simple example is given in Scheme 8.14. Two radical centers can interact easily in this hooked-together fragment to revert to the original molecule or to produce a different structure, i.e. an isomer. This biradical pathway is an important mechanism for molecular isomerization [8.174,191–193].

Using picosecond lasers the direct observation of a 1,3 biradical, i.e. one in which the radical centers are one center apart at positions one and three, has been achieved [8.194,195]. The biradical was identified by its laser induced fluorescence following photodissociation of the parent compound (Scheme 8.15). An interesting result of the work is the suggestion that both carbon-nitrogen bonds are broken simultaneously in the photodissociation process unlike in the ground state thermal reaction where the carbon-nitrogen bonds are broken sequentially.

The biradical, a versatile and frequently postulated intermediate, can be generated not only by bond cleavage as we have just seen, but also in the course of bond formation [8.196]. An example of an important class of photoreactions shown spectroscopically to involve a 1,4-biradical intermediate using picosecond laser methods [8.197–199] is given in Scheme 8.16.

To form a bond, and thus effect ring closure, the initially formed biradical which is in a triplet state must relax to the singlet spin state. In competition with ring closure is the breaking apart of the biradical to produce two charged fragments (Scheme 8.17).

From the picosecond kinetic studies it has been shown [8.199] that these charged fragments appear as the biradical decays. This result is significant in that it establishes that the biradical produces the charged fragments rather than

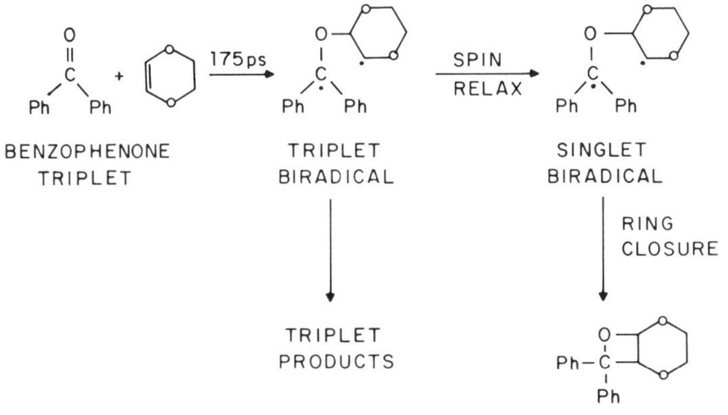

Scheme 8.16

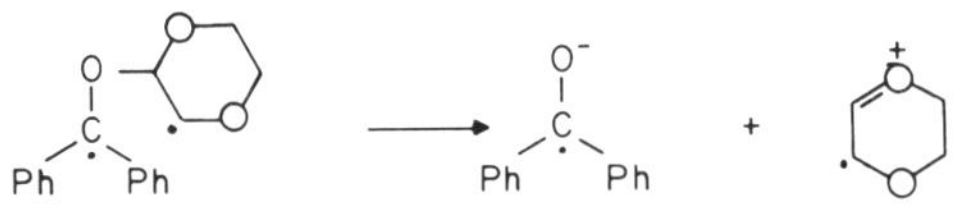

175ps

SPIN RELAX

BENZOPHENONE TRIPLET

TRIPLET BIRADICAL

SINGLET BIRADICAL

RING CLOSURE

TRIPLET PRODUCTS

Scheme 8.17

the other way around as had been previously proposed for these types of photoreactions.

8.6 Excited-State Proton Transfer

The transfer of a proton from one atom to another, either within a molecule, or between molecules constitutes an elementary chemical reaction. The acid-base properties of a molecule, i.e. its ability to donate or accept a proton, are sensitive to the electronic structure of the molecule. Although not surprising it is impressive nonetheless that the acid equilibrium constant can change by 5 to 10 orders of magnitude in going from the ground state to the lowest excited singlet state [8.200–203]. This strong connection between the acid-base properties and the electron distribution is also observed in dynamics where proton transfer can open up important channels for energy relaxation, structural change and intermolecular reactions [8.200–206].

The nature of the solvent is of major importance in acid-base reactions. The solvent can act as a reactant in the proton transfer or can interfere with intramolecular proton transfer due to competitive hydrogen-bonding interactions with the excited molecule. The enormous range of interaction strengths and variety of participating molecules extending from small polyatomics to organic dyes, to proteins and other biopolymers, make this an active area of research. Ultrashort pulse laser spectroscopy has become one of the chief ways for studying the rapid kinetics of proton transfer reactions.

8.6.1 The Naphthols – Intermolecular Proton Transfer

One of the simpler examples of intermolecular proton transfer is that of the hydroxy-substituted naphthalenes, the naphthols [8.201–203,207–212] shown in Scheme 8.18.

The spectroscopy and kinetics of the proton transfer reactions in the ground and excited states have been interpreted in terms of what is known as the Förster cycle. Using an OH-(hydroxy)-containing molecule as an example we can represent the cycle as shown in Fig. 8.10.

This cycle includes both the ground and excited state protonated and deprotonated forms. The formation and decay rates of the excited species ROH^* and RO^{-*} can be measured by their emissions. By adjusting the hydrogen ion concentration of the solution one can readily arrange to have the ground state molecules, prior to excitation, predominantly in the protonated form, ROH, or in the deprotonated form, RO^-. For 1-naphthol, a pH less than 7 yields a solution greater than 99.4% in ROH, whereas a pH greater than 12 yields a solution greater than 99.8% in RO^-. Carrying out the experiment in 1-naphthol at pH = 7, the deprotonation time of ROH^* was found to be 35 ps [8.210]. This value was obtained from the decay of ROH^* monitored at its emission wavelength (370 nm) and also from the rise of RO^{-*} monitored at its emission wavelength (540 nm). The identical values obtained are consistent with the Förster cycle description of the proton transfer reactions.

A striking result was the observation that the deprotonation time for the 2-naphthol [8.209] excited species ROH^* is about 300 times slower than for 1-naphthol [8.210]. A reasonable explanation of the sensitivity of deprotonation to the position of the OH group is the different charge transfer character associated with these positions [8.210]. For both naphthols the excited singlet state involves the transfer of electron density from the OH group into the naphthalene moiety, resulting in the observed increased acidity of the excited

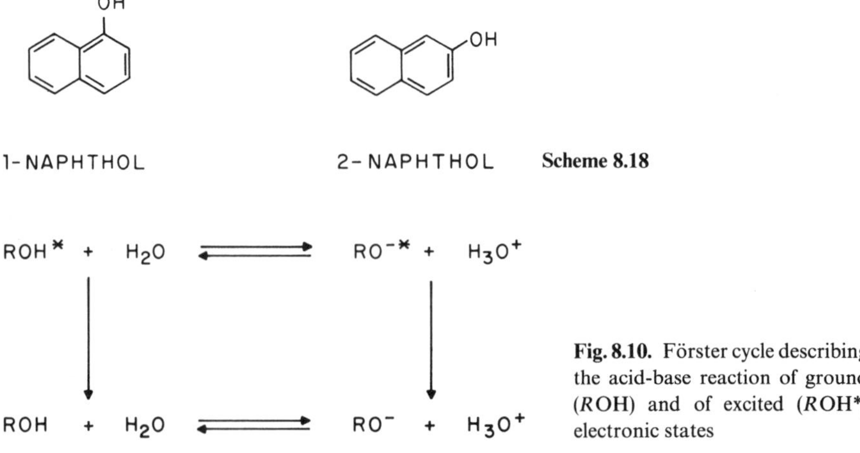

Scheme 8.18

Fig. 8.10. Förster cycle describing the acid-base reaction of ground (ROH) and of excited (ROH^*) electronic states

singlet naphthols relative to their ground states. The greater decrease in the electron density at the oxygen for the 1-naphthol would increase its acidity more. This is manifested by its larger excited state acid equilibrium constant and is consistent with its faster deprotonation kinetics.

An important question with regard to the proton transfer process is the nature of the accepting water molecules. Although the reaction, as shown in the Förster cycle, is written as if a single water molecule is involved, this is not intended as a statement of fact. By studying the kinetics of 2-naphthol in water/methanol mixtures at various temperatures, information on the size of the accepting water cluster has been obtained [8.211]. By using a Markov random-walk method for describing the time development of possible cluster configurations and comparing model calculations with measured kinetics a critical cluster size of four water molecules was obtained [8.211]. Earlier studies [8.213] on various aromatic alcohols and amines in water/alcohol mixtures had also proposed an accepting water cluster model, though the numbers differed, extending to clusters as large as ten water molecules. The qualitative results of both studies are nonetheless in agreement. It is of interest that earlier temperature-jump studies [8.214] of ground state acid-base reactions in water also led to the proposal of four water molecules for the stable hydration of a proton.

8.6.2 Intramolecular Proton Transfer

The apparently simple structural change associated with the shift of a proton (or hydrogen atom) from one position to another in a photoexcited molecule can provide an efficient channel for energy degradation [8.201–203,206]. This intramolecular chemical reaction can protect the photoexcited molecule from irreversible chemical reactions with neighboring molecules by dissipating the energy before intermolecular reactions can occur. There are many examples [8.201–203,206,215–233] of these photostabilizing molecules which play an important role in photochromic materials and in the photochemical stabilization of polymers. Although the following discussion is limited to picosecond experiments, it should be noted that important earlier studies of the spectroscopy of these molecules provided the foundation for the picosecond work.

a) Transfer in Non-Hydrogen-Bonding Solvents

The excited state intramolecular proton shift generally involves adjacent atoms that are hydrogen bonded to each other. The presence of a ground state intramolecular hydrogen-bonding interaction is known in many cases from infrared and NMR spectroscopy. Among the various molecules undergoing a rapid relaxation associated with an excited state proton transfer are those shown in Fig. 8.11.

Photoexcitation of these molecules changes the electronic charge distribution at the atoms connected by the hydrogen bond. For OHBP (*ortho*-hydroxybenzo-

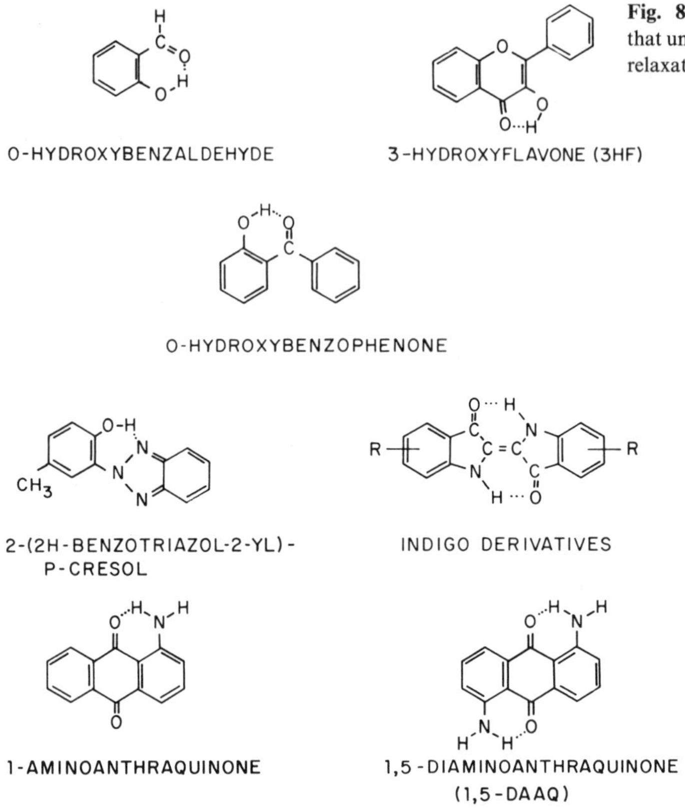

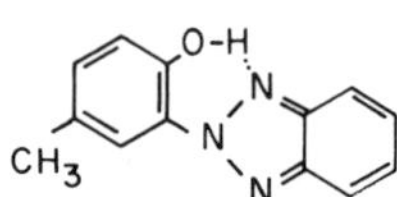

O-HYDROXYBENZALDEHYDE

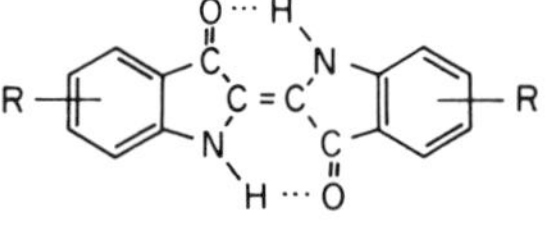

3-HYDROXYFLAVONE (3HF)

O-HYDROXYBENZOPHENONE

2-(2H-BENZOTRIAZOL-2-YL)-P-CRESOL

INDIGO DERIVATIVES

1-AMINOANTHRAQUINONE

1,5-DIAMINOANTHRAQUINONE (1,5-DAAQ)

Fig. 8.11. Some molecules that undergo rapid electronic relaxation

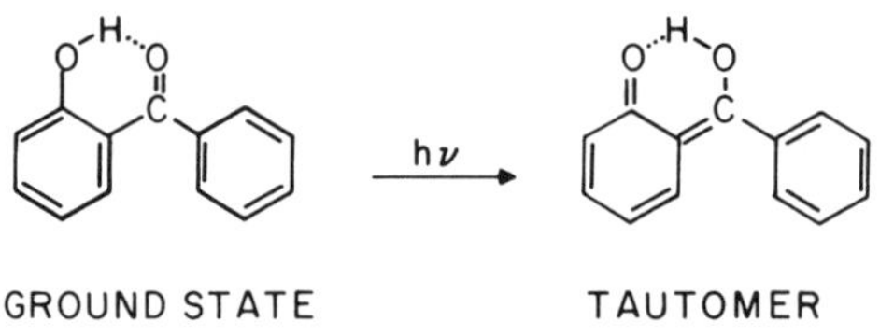

GROUND STATE $\xrightarrow{h\nu}$ TAUTOMER

Scheme 8.19

phenone) the photoexcited singlet is a π, π^* charge transfer state [8.206,235,236]. For this state the electronic charge distribution decreases at the hydroxy oxygen, thereby increasing its proton transfer ability (acidity). Photoexcitation also increases the electron density at the carbonyl oxygen, thereby enhancing its proton accepting ability (basicity). Both of these effects act to drive the proton on the hydroxy (OH) group to the carbonyl (C=O) group. This shift in the position of the proton quenches the photoexcited singlet and generates what can be represented, for complete transfer, as a tautomeric form [8.206,235–237] (Scheme 8.19). In alkanes this rate has been found to be less than 4 ps in one study [8.217], 6 ps and less than 10 ps in two other studies [8.221,238]. It is important to note that trace amounts of hydrogen-bonding impurities in the alkane solvents can greatly modify the observed kinetics and spectroscopy. This was shown [8.239] to be the case for 3-HF (3-hydroxyflavone). There is only a single rapid decay in pure alkanes whereas a minute quantity of water, alcohol or other impurities produced a slow decay component as well.

Although the excited singlet in OHBP decays in several picoseconds the ground state repopulation time was found [8.217] to be 35 ps. This observation indicates that intermediate structures, presumably the tautomeric form, and possibly triplet states are involved in the passage from the excited singlet to the initial ground state protonated form. To determine whether triplet states are involved in the relaxation process, a high concentration of 1-methylnaphthalene, a known triplet quencher, was placed in the hexane solution. At the concentration of the quencher used all of the OHBP triplets would be quenched within 20 ps. Knowing the triplet-triplet absorption of 1-methylnaphthalene it was determined [8.217] that less than 2% of OHBP decays into the triplet state.

A contrasting example to the one given above is found in the recent work [8.242] on indigo (see Fig. 8.11 for the structure). Using a novel application of time resolved infrared spectroscopy the time evolution of the intramolecular N—H and C=O stretching modes in electronically excited indigo molecules were monitored [8.242]. The excited singlet state was characterized as having enhanced hydrogen bonding relative to the ground state. Interestingly, the tautomer (proton transfer) state was not observed. If the tautomer were formed, and if it were of sufficient lifetime to be monitored, then one would expect to see a loss of the C=O band with a simultaneous growth of an O—H band. Neither change was observed, and thus thc qucstion ariscs of whether proton transfer, as a mechanism for excited state relaxation, occurs in the indigo system. To be consistent with the proton transfer mechanism, the lifetime of the tautomer state would have to be significantly less than the 40 ps lifetime found for the photoexcited singlet [8.242]. On the other hand, a mechanism equally consistent with the spectroscopic and kinetic data is one which does not require proton transfer. This latter mechanism, which does not involve an intermediate tautomer state, could occur if there are decay channels specific to the intramolecular hydrogen bond, i.e. the vibrations of the N—H group [8.242].

b) Transfer in Hydrogen-Bonding Solvents

The hydrogen bonding interactions between the solute and solvent molecules perturb the intramolecular hydrogen bond in the solute [8.206,217,221–225,228, 230,231,237–241]. A variety of structures can be envisioned which can alter the spectroscopy and dynamics by causing shifts in energy levels and changes in proton transfer barrier heights. Some of the possible structures include those having an intermolecular but no intramolecular hydrogen bond, those having an intramolecular bond solvated by the solvent OH group, a solvent coupling the solute C=O and OH groups, and structures intermediate to these. Using

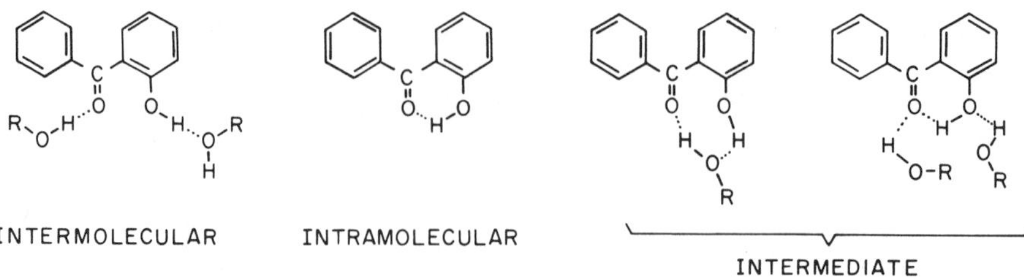

Fig. 8.12. Possible hydrogen bonding structures involving *ortho*-hydroxybenzopheneone (OHBP) in solution

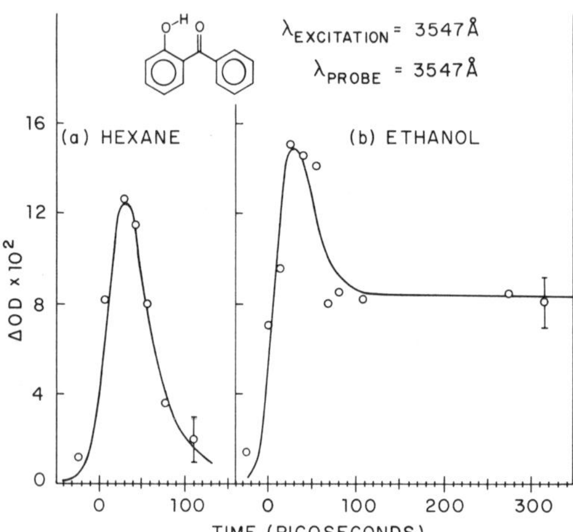

Fig. 8.13. Repopulation kinetics of *ortho*-hydroxybenzophenone

OHBP as an example [8.217,221], we note possible hydrogen bonding structures shown in Fig. 8.12.

To monitor the dynamics of excited-state proton transfer a variety of methods have been used to measure the singlet lifetime of OHBP in ethanol. One was based on fluorescence decay measured with a streak camera [8.217], another was a consecutive two photon fluorescence excitation method [8.238] ($S_0 \rightarrow S_1 \rightarrow S_2$) which yields the S_1 lifetime by measuring the low quantum yield fluorescence from S_2, and lastly a qualitative estimate from a transient absorption method [8.221]. These different methods yielded the values of 7, 30 and $\geqslant 10$ ps respectively for the singlet lifetime.

From measurements [8.217] of the repopulation kinetics of OHBP, it was found that there were two clear components (Fig. 8.13). The fast one gave a value of about 30 ps and the slow one 1.5 ns. The long-lived 1.5 ns transient was assigned to the triplet state. Was the triplet being populated by the decay of the singlet OHBP? Apparently not, since the rise time of the triplet was less than 15 ps which is significantly less than the ground state repopulation time of 35 ps.

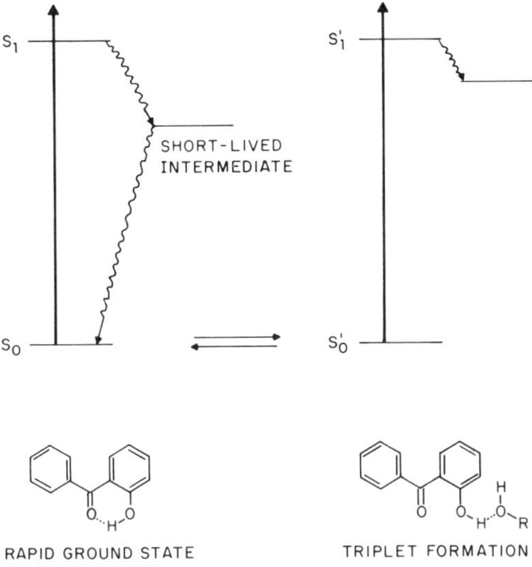

Fig. 8.14. Scheme describing two types of hydrogen bonding structures involving OHBP in alcohol, and their effect on excited state decay. Case 1 (*left*): The intramolecular hydrogen bonding species decays rapidly to the ground state via a short lived intermediate. Case 2 (*right*): The intermolecular hydrogen bonding species relaxes via an excited triplet state, and thus decays slowly to the ground state

One view [8.217,237] consistent with these results would be the presence of two distinct ground state species in ethanol which differ only in their hydrogen bonding interactions with the solvent, as shown in Fig. 8.14. One could be an intramolecularly hydrogen bonded species which on excitation to its excited singlet state rapidly decays to the ground state or to some intermediate structure via the radiationless channel provided by the intramolecular hydrogen bond. For these molecules no triplets are produced. The other ground state population could (at the time of excitation) be those molecules which have intermolecular hydrogen bonds to the solvent but do not have intramolecular hydrogen bonds. These intermolecularly hydrogen bonded molecules on excitation to their singlet states cross over to their triplet states with high efficiency as in the case of unsubstituted benzophenone.

The effects of different conformers in solution arising from the difference in hydrogen bonding have been proposed to be important factors in describing the kinetics of other solutes such as 3-HF [8.241], orthohydroxybenzaldehyde [8.228], anthraquinone derivatives [8.224] and various indigo dye derivatives [8.223]. It has been established in these various studies that the intramolecular and intermolecular hydrogen bond can both provide efficient pathways for relaxation.

Although proton transfer processes were described earlier as "simple" chemical reactions the results of these various picosecond laser studies on a variety of molecules have shown that they can be quite involved. A further example of this is the effect of deuterium isotope substitution on the dynamics. It turns out that the magnitude of the effect [8.243] has been observed to vary from a factor of the order of ten to no effect at all. This latter case which has been observed for 2-(2H-benzotriazol-2-yl)-*p*-cresol [8.225] and *ortho*-hydroxybenzophenone [8.244] indicates that the proton transfer is not the rate limiting step. One possibility is that prior structural arrangements are required to effect the proton transfer reaction.

c) Double Proton Transfer

Another mechanism by which a proton transfer within a molecule can be carried out involves the use of a solvent molecule, or another solute molecule to act as a transfer agent. The phenomenon, first discovered in 7-azaindole [8.245], involves the simultaneous motion of two protons (See Fig. 8.15). In the alkane solvent, 3-methylpentane, 7-azaindole at moderate concentrations ($\leqslant 10^{-2}$ M), forms primarily ground state hydrogen-bonded dimers. On photoexcitation in the ultraviolet a double proton transfer occurs which shifts the hydrogens from the five membered ring to the six membered ring. A new emission [8.245,246] due to the tautomer appears in the visible and can be used to follow the kinetics [8.247–251] of the reaction.

In solution at room temperature the double proton jump occurs in less than 5 ps [8.247]. At 77 K in a 3-methylpentane glass the thermal barrier crossing is slowed down to about 1 ns [8.247–248]. There remains a fast component, faster than 5 ps, due to decay to the tautomer directly from the initially prepared excited state which is significantly above the 500–700 cm^{-1} barrier. The result was supported by the observed dependence of the ratio of tautomer to dimer fluorescence on excitation wavelength. An interesting aspect of the 7-azaindole dimer

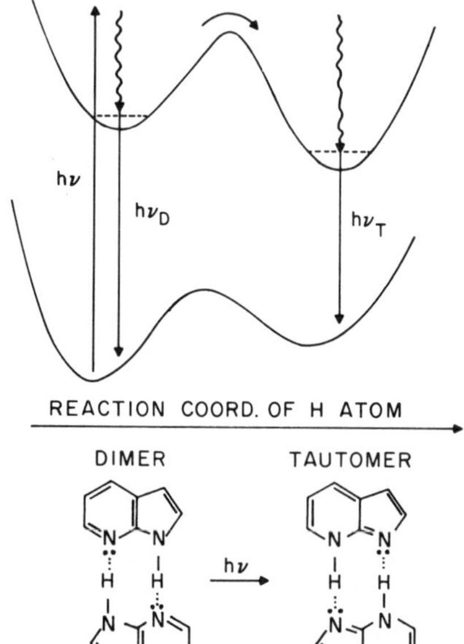

Fig. 8.15. Photophysics of the double proton transfer photo-reaction of the 7-azaindole dimer

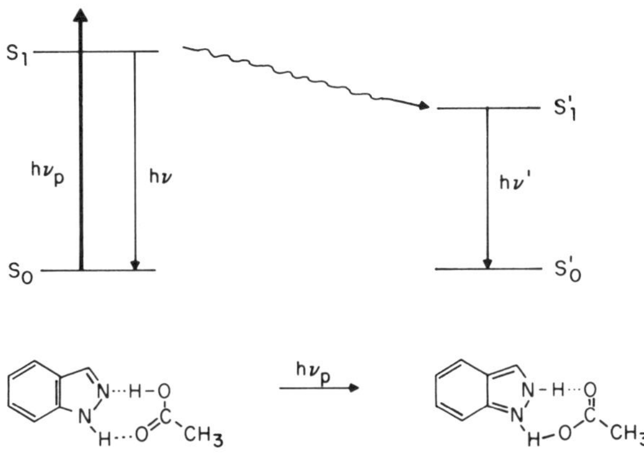

Fig. 8.16. Kinetic scheme describing the photo-induced double proton transfer reaction between indazole and acetic acid

is its similarity to the hydrogen-bonded base pairs of DNA. This similarity led to its being considered as a possible model for mutagenic changes brought about by the structure changing double proton transfer phenomenon.

Another example of the solvent mediating the transfer of a hydrogen is that of indazole in acetic acid [8.252] (Fig. 8.16). In the ground state the stable form of indazole has the hydrogen on the nitrogen in the 1 position. In acetic acid, unlike other solvents, a new emission appears that is assigned to an excited indazole with H in the 2 position. The picosecond kinetics and spectroscopy support the following scheme in which the acetic acid acts as a switching device to shift the hydrogen from the 1 to the 2 position of indazole. From the rise of the tautomer (2H-indazole) emission and decay of initially excited 1H-indazole, a double proton transfer time of 370 ps is obtained. The time scale of the reverse double proton transfer from the ground state tautomer to the ground state 1H-indazole has not been reported at this time.

8.7 Reactions in Bulbs and Beams

Although the focus of this chapter has been on reaction in liquids, it is useful to note that reactions in the gas phase and cooled molecular beams as seen in the stilbene studies can provide valuable information on the effects of the solvent on reactions. Multiphoton photofragmentation using picosecond lasers of different colors [8.253,254], and recently femtosecond lasers [8.255], as well as detection by picosecond CARS [8.256,257] and emission [8.258] produce new information not only on intramolecular relaxation in molecules but on the mechanisms for reactions. The technique of picosecond mass spectrometry is becoming increasingly important in probing chemical reactions. By carrying out the reactions in cooled molecular beams and probing by multiphoton ionization, i.e. picosecond mass spectrometry, one can investigate both time and energy resolution in reactants and products as a valuable complement to laser induced fluorescence and transient absorption or scattering probes. This has recently been carried out in a series of picosecond wavelength-dependent photodissociation studies [8.259,260] of the van der Waals phenol: benzene molecule and alkyl iodides.

8.8 Concluding Remarks

The reader has probably (hopefully) concluded at this point in the chapter that ultrafast laser spectroscopy is being applied to a large number of different chemical studies. The impact of technical advances in producing short pulses (from picoseconds to femtoseconds), tunability extending from the infrared to the x-ray region, high repetition rate systems, and greater ease and reliability of these short-pulse laser systems is just beginning to be felt. New and different questions on the nature of chemical intermediates can be, and are being considered, using these various techniques. It is certainly clear that in this chapter we have been able to sample only a narrow range of the forefront work going on in this area of chemistry; apologies to those whose fine work was not included due to oversight or lack of space.

Acknowledgements. The author wishes to thank Dr. E.V. Sitzmann for his much appreciated help in preparing this chapter, and the National Science Foundation, the Air Force Office of Scientific Research, and the Joint Services Electronic Program 29-85-K-0049 for their support.

References

8.1 S.A. Adelman: In *Advances in Chemical Physics*, ed. by I. Prigogine, S. Rice (Wiley, New York 1983) **LIII**, p. 61

8.2 D.H. Auston, K.B. Eisenthal (eds.): *Ultrafast Phenomena IV*, Springer Ser. Chem. Phys., Vol. 38 (Springer, Berlin, Heidelberg 1984)

8.3 K.B. Eisenthal (ed.): *Applications of Picosecond Spectroscopy to Chemistry* (Reidel, Dordrecht 1984)

8.4 J. Saltiel, J.L. Charlton: *Rearrangements in Ground and Excited States*, ed. by P. DeMeyo (Academic, New York 1980) Vol. III, p. 25

8.5 R.M. Hochstrasser: Pure Appl. Chem. **52**, 2683 (1980)

8.6 G. Orlandi, W. Siebrand: Chem. Phys. Lett. **30**, 352 (1975)

8.7 P. Tavan, K. Schulten: Chem. Phys. Lett. **56**, 200 (1978)

8.8 J.B. Birks: Chem. Phys. Lett. **54**, 430 (1978)

8.9 J.B. Birks, B.N.R. Tripathi, M.D. Lumb: Chem. Phys. **33**, 185 (1978)

8.10 R.M. Weiss, A. Warshel: J. Am. Chem. Soc. **101**, 6131 (1979)

8.11 M. Sumitani, N. Nakashima, K. Yoshihara, S. Nagakura: Chem. Phys. Lett. **51**, 183 (1977)
8.12 F. Heisel, J.A. Miehe, B. Sipp: Chem. Phys. Lett. **61**, 115 (1979)
8.13 F.E. Doany, B.I. Greene, Y. Liang, D.K. Negus, R.M. Hochstrasser: *Picosecond Phenomena II*, ed. by R.M. Hochstrasser, W. Kaiser, C.V. Shank, Springer Ser. Chem. Phys., Vol. 14 (Springer, Berlin, Heidelberg 1980) p. 254
8.14 J.R. Taylor, M.C. Adams, W. Sibbett: J. Photochem. **12**, 127 (1980)
8.15 J.A. Syage, W.R. Lambert, P.M. Felker, A.H. Zewail, R.M. Hochstrasser: Chem. Phys. Lett. **88**, 266 (1982)
8.16 T.J. Majors, U. Even, J. Jortner: J. Chem. Phys. **81**, 2330 (1984)
8.17 J.A. Syage, P.M. Felker, A.H. Zewail: J. Chem. Phys. **81**, 4706 (1984)
8.18 J. Troe: Chem. Phys. Lett. **114**, 241 (1985)
8.19 O. Teschke, E.P. Ippen, G. Holtom: Chem. Phys. Lett. **52**, 233 (1977)
8.20 B.I. Greene, R.M. Hochstrasser, R. Weisman: J. Chem. Phys. **71**, 544 (1979)
8.21 K. Yoshihara, A. Namiki, M. Sumitani, N. Nakashima: J. Chem. Phys. **71**, 2892 (1979)
8.22 F.E. Doany, B.I. Greene, R.M. Hochstrasser: Chem. Phys. Lett. **75**, 206 (1980)
8.23 M. Sumitani, K. Yoshihara: Bull. Chem. Soc. Japan **55**, 85 (1982)
8.24 V. Sundstrom, T. Gillbro: Chem. Phys. Lett. **109**, 538 (1984)
8.25 F.E. Doany, E.J. Heilweil, R. Moore, R.M. Hochstrasser: J. Chem. Phys. **80**, 201 (1984)
8.26 B.I. Greene, R.C. Farrow: J. Chem. Phys. **78**, 3336 (1983)
8.27 B.I. Greene, T.W. Scott: Chem. Phys. Lett. **106**, 399 (1984)
8.28 N.F. Scherer, J.W. Perry, F.E. Doany, A.H. Zewail: J. Phys. Chem. **89**, 894 (1985)
8.29 T.L. Gustafson, D.M. Roberts, D.A. Chernoff: J. Chem. Phys. **79**, 1559 (1983)
8.30 M. Sumitani, N. Nakashima, K. Yoshihara: Chem. Phys. Lett. **68**, 255 (1979)
8.31 M. Sumitani, K. Yoshihara: J. Chem. Phys. **76**, 738 (1982)
8.32 H.A. Kramers: Physica **7**, 284 (1940)
8.33 S. Chandrasekhar: Rev. Mod. Phys. **15**, 1 (1943)
8.34 J.S. McCaskill, R.G. Gilbert: Chem. Phys. **44**, 389 (1979)
8.35 S.P. Velsko, G.R. Fleming: J. Chem. Phys. **76**, 3553 (1982)
8.36 S.P. Velsko, D.H. Waldeck, G.R. Fleming: J. Chem. Phys. **78**, 249 (1983)
8.37 K.M. Keery, G.R. Fleming: Chem. Phys. Lett. **93**, 322 (1982)
8.38 G. Rothenberger, D.K. Negus, R.M. Hochstrasser: J. Chem. Phys. **79**, 5360 (1983)
8.39 V. Sundstrom, T. Gillbro: Chem. Phys. Lett. **109**, 538 (1984)
8.40 S. Russo, P.J. Thistlethwaite: Chem. Phys. Lett. **106**, 91 (1984)
8.41 M. Lee, G.R. Holtom, R.M. Hochstrasser: Chem. Phys. Lett. **118**, 359 (1985)
8.42 S.H. Courtney, G.R. Fleming: J. Chem. Phys. **83**, 215 (1985)
8.43a G. Maneke, J. Schroeder, J. Troe, F. Voss: Ber. Bunsen-Ges. Phys. Chem. **89** 896 (1985)
8.43b J. Schroeder, J. Troe: Chem. Phys. Lett. **116**, 453 (1985)
8.44 R.F. Grote, J.T. Hynes: J. Chem. Phys. **73**, 2715 (1980)
8.45 B. Bagchi, D.W. Oxtoby: J. Chem. Phys. **78**, 2735 (1983)
8.46 R.W. Zwanzig: Phys. Fluid **2**, 12 (1959)
8.47 P. Hanggi, F. Mojtabac: Phys. Rev. A **26**, 1168 (1982)
8.48 B. Carmeli, A. Nitzan: J. Chem. Phys. **80**, 3586 (1984)
8.49 R. Landauer, J.A. Swanson: Phys. Rev. **121**, 1668 (1961)
8.50 J.S. Langer: Ann. Phys. (NY) **54**, 258 (1969)
8.51 R.F. Grote, J.T. Hynes: J. Chem. Phys. **74**, 4465 (1981)
8.52 G. van der Zwan, J.T. Hynes: J. Chem. Phys. **77**, 1295 (1982)
8.53 M. Borkovec, B.J. Berne: J. Chem. Phys. **82**, 794 (1985)
8.54 P.L. Bhatnagar, E.P. Cross, M. Krook: Phys. Rev. **94**, 511, (1954)
8.55 J.L. Skinner, P.G. Wolynes: J. Chem. Phys. **69**, 2143, (1978); **72**, 4913, (1980)
8.56 B.J. Berne, J.L. Skinner, P.G. Wolynes: J. Chem. Phys. **73**, 4314 (1980)
8.57 D.K. Garrity, J.L. Skinner: Chem. Phys. Lett. **95**, 46 (1983)
8.58 R.O. Rosenberg, B.J. Berne, D. Chandler: Chem. Phys. Lett. **75**, 162 (1980)

8.59 J.A. Montgomery, Jr., D. Chandler, B.J. Berne: J. Chem. Phys. **70**, 4056 (1979)
8.60 C.V. Shank, E.P. Ippen, O. Teschke, K.B. Eisenthal: J. Chem. Phys. **67**, 5547 (1978)
8.61a D.P. Millar, K.B. Eisenthal: J. Chem. Phys. **83**, 5076 (1985)
8.61b D.L. Hasha, T. Eguchi, J. Jonas: J. Chem. Phys. **75**, 1571 (1981)
8.62 L.R. Khundkar, R.A. Marcus, A.H. Zewail: J. Phys. Chem. **87**, 2473 (1983)
8.63 T.J. Majors, U. Even, J. Jortner: J. Chem. Phys. **81**, 2330 (1984)
8.64 J. Syage, P.M. Felker, A.H. Zewail: J. Chem. Phys. **81**, 4706 (1984)
8.65 J. Troe: Chem. Phys. Lett. **114**, 241 (1985)
8.66 J.F. Shepanski, B.W. Keelan, A.H. Zewail: Chem. Phys. Lett. **103**, 9 (1983)
8.67 S.H. Courtney, G.R. Fleming, L.R. Khundkar, A.H. Zewail: J. Chem. Phys. **80**, 4559 (1984)
8.68 J. Troe, A. Amirav, J. Jortner: Chem. Phys. Lett. **115**, 245 (1985)
8.69 G.W. Robinson, W.A. Jalenak, D. Statman: Chem. Phys. Lett. **110**, 135 (1984)
8.70 E. Lippert, W. Luder, H. Boos: *Advances in Molecular Spectroscopy*, ed. by A. Mangini (Pergamon, Oxford 1962) p. 443
8.71 K. Rotkiewicz, K.H. Grellmann, Z.R. Grabowski: Chem. Phys. Lett. **19**, 315 (1973)
8.72 N. Nakashima, N. Mataga: Bull. Chem. Soc. Japan **46**, 3016 (1973)
8.73 W.S. Struve, P.M. Rentzepis, J. Jortner: J. Chem. Phys. **59**, 5014 (1973)
8.74 Z.R. Grabowski, K. Rotkiewicz, W. Rubaszewska, E. Kirkor-Kaminska: Acta Physica Polonica A **54**, 767 (1978)
8.75 Z.R. Grabowski, K. Rotkiewicz, A. Siemiarczuk, D. Cowley, W. Baumann: Nouveau Journal de Chimie **3**, 443 (1979)
8.76 W. Rettig. V. Bonacic-Koutecky: Chem. Phys. Lett. **62**, 115 (1979)
8.77 W. Rettig, G. Wermuth, E. Lippert: Ber. Bunsenges. Phys. Chem. **83**, 692 (1979)
8.78 Y. Wang, M. McAuliffe, F. Novak, K.B. Eisenthal: J. Phys. Chem. **85**, 3736 (1981)
8.79 D. Hupert, S. Rand, P. Rentzepis, P. Barbara, W. Struve, Z.R. Grabowski: J. Chem. Phys. **75**, 5714 (1981)
8.80 Y. Wang, K.B. Eisenthal: J. Chem. Phys. **77**, 6076 (1982)
8.81 F. Heisel, J. Miehe: Chem. Phys. Lett. **100**, 6076 (1982)
8.82 R.J. Visser, C. Varma, J. Konijneberg, P. Weisenborn: J. Mol. Structure **114**, 105 (1984)
8.83 J. Hicks, M. Vandersall, Z. Babarogic, K.B. Eisenthal: Chem. Phys. Lett. **116**, 18 (1985)
8.84 J. Hicks, M.T. Vandersall, E.V. Sitzmann, K.B. Eisenthal: Chem. Phys. Lett. **135**, 413 (1987)
8.85 E. Kosower: J. Am. Chem. Soc. **107**, 1114 (1985)
8.86 B.I. Greene: Chem. Phys. Lett. **79**, 51 (1981)
8.87 A. Warshel: J. Chem. Phys. **62**, 214 (1975)
8.88 T.L. Gustafson, D.M. Roberts, D.A. Chernoff: J. Chem. Phys. **79**, 1559 (1983)
8.89 J. Franck, E. Rabinowitch: Trans. Farad. Soc. **30**, 120 (1934)
8.90 E. Rabinowitch, W.C. Wood: Trans. Farad. Soc. **32**, 1381 (1936)
8.91 E. Rabinowitch: Trans. Farad. Soc. **33**, 1225 (1937)
8.92 R.M. Noyes: Progr. Reaction Kinetics **1**, 129 (1961)
8.93 K. Luther, J. Schroeder, J. Troe, U. Unterberg: J. Phys. Chem. **84**, 3072 (1980)
8.94 B. Otto, J. Schroeder, J. Troe: J. Chem. Phys. **81**, 202 (1984)
8.95 T.J. Chuang, G.W. Hoffman, K.B. Eisenthal: Chem. Phys. Lett. **25**, 201 (1974)
8.96 W.S. Struve: Chem. Phys. Lett. **51**, 603 (1977)
8.97 C.A. Langhoff, K. Gnadig, K.B. Eisenthal: Chem. Phys. **46**, 117 (1980)
8.98 S.D. Woodruff, W.S. Struve: Chem. Phys. Lett. **69**, 50 (1980)
8.99 D.F. Kelley, P.M. Rentzepis: Chem. Phys. Lett. **85**, 85 (1982)
8.100 C.A. Langhoff, B. Moore, M. DeMeuse: J. Am. Chem. Soc. **104**, 3576 (1982)
8.101 M. Berg, A.L. Harris, J.K. Brown, C.B. Harris: In *Ultrafast Phenomena IV*, ed. by D.H. Auston, K.B. Eisenthal, Springer Ser. Chem. Phys., Vol. 38 (Springer, Berlin, Heidelberg 1984) p. 300.
8.102 M. Berg, A.L. Harris, C.B. Harris: Phys. Rev. Lett., **54**, 951 (1985)
8.103 J. Tellinghuisen: J. Chem. Phys. **82**, 4012 (1985)
8.104 J.E. Selwyn, J.I. Steinfeld: Chem. Phys. Lett. **4**, 217 (1969)

8.105 D.J. Nesbitt, J.T. Hynes: J. Chem. Phys. **77**, 2130 (1982)
8.106 P. Bado, P.H. Berens, J.P. Bergsma, S.B. Wilson, K.R. Wilson: In *Picosecond Phenomena III*, ed. by K.B. Eisenthal, R.M. Hochstrasser, W. Kaiser, A. Laubereu, Springer Ser. Chem. Phys., Vol. 23 (Springer, Berlin, Heidelberg 1982) p. 260
8.107 C.L. Brooks III, M.W. Balk, S.A. Adelman: J. Chem. Phys. **79**, 784 (1983)
8.108 J.P. Bergsma, P.H. Berens, K.R. Wilson, D.R. Fredkur, E.J. Heller: J. Phys. Chem. **88**, 612 (1984)
8.109 P. Bado, C. Dupuy, J.P. Bergsma, K.R. Wilson: In *Ultrafast Phenomena IV*, ed. by D.H. Auston, K.B. Eisenthal, Springer Ser. Chem. Phys., Vol. 38 (Springer, Berlin, Heidelberg 1984) p. 296
8.110 D.F. Kelley, N.A. Abul-Haj, D.J. Jang: J. Chem. Phys. **80**, 4105 (1984)
8.111 D.F. Kelley, N.A. Abul-Haj: In *Ultrafast Phenomena IV*, ed. by D.H. Auston, K.B. Eisenthal, Springer Ser. Chem. Phys., Vol. 38 (Springer, Berlin, Heidelberg 1984) p. 292
8.112 P. Bado, C. Dupuy, D. Magde, K.R. Wilson, M.M. Malley: J. Chem. Phys. **80**, 5531 (1984)
8.113 N.A. Abul-Haj, D.F. Kelley: J. Chem. Phys. **84**, 1335 (1986)
8.114 A.L. Harris, M. Berg, C.B. Harris: J. Chem. Phys. **84**, 788 (1986)
8.115 J.M. Dawes, M.G. Skeats: Private communication
8.116 T.J. Chuang, K.B. Eisenthal: J. Chem. Phys. **62**, 2213 (1975)
8.117 N. Mataga, M. Ottolenghi: In *Molecular Association*, ed. by R. Foster (Academic, New York 1979), Vol. 2, p. 1 and references therein
8.118 D. Doizi, J. Mialocq: In *Ultrafast Phenomena IV*, ed. by D.H. Auston, K.B. Eisenthal, Springer Ser. Chem. Phys., Vol. 38 (Springer, Berlin, Heidelberg 1984) p. 377
8.119 T. Kakitani, N. Mataga: J. Phys. Chem. **89**, 8 (1985) and references therein
8.120 T.J. Chuang, R.J. Cox, K.B. Eisenthal: J. Am. Chem. Soc. **96**, 6828 (1974)
8.121 M. Migita, T. Okada, N. Mataga, Y. Sakata, S. Misumi, N. Nakashima, K. Yoshihara: Bull. Chem. Soc. Japan **54**, 3304 (1980) and references therein
8.122 M.K. Crawford, Y. Wang, K.B. Eisenthal: Chem. Phys. Lett. **79**, 529 (1981)
8.123 T. Okada, I. Karaki, E. Matsuzawa, N. Mataga, Y. Sakata, S. Misumi: J. Phys. Chem. **85**, 3957 (1981)
8.124 Y. Wang, M.C. Crawford, K.B. Eisenthal: J. Am. Chem. Soc. **104**, 5874 (1982) and references therein
8.125 T. Okada, M. Kawai, T. Ikemachi, N. Mataga, Y. Sakata, S. Misumi, S. Shionoya: J. Phys. Chem. **88**, 1976 (1984)
8.126 J.R. Miller, L.T. Calcaterra, G.L. Closs: J. Am. Chem. Soc. **106**, 3047 (1984) and references therein
8.127 T.L. Netzel, P. Kroger, C.K. Chang, I. Fujita, J. Fajer: Chem. Phys. Lett. **67**, 223 (1979)
8.128 C. Creutz, P. Kroger, T. Matsubara, T.L. Netzel, N. Sutin: J. Am. Chem. Soc. **101**, 5442 (1979)
8.129 B.T. Reagor, D.F. Kelley, D.H. Huchital, P.M. Rentzepis: J. Am. Chem. Soc. **104**, 7400 (1982)
8.130 S.G. Cohen, J.I. Cohen: J. Am. Chem. Soc. **89**, 164 (1967)
8.131 R.F. Davidson, P.F. Lambeth: Chem. Commun. 511 (1968)
8.132 P.L. Wagner, A. Kemppainen: J. Am. Chem. Soc. **91**, 3085 (1969)
8.133 S.G. Cohen, A. Parola, G.H. Parsons, Jr.: Chem. Rev. **73**, 141 (1973)
8.134 P.J. Wagner: Topics Curr. Chem. **66**, 1 (1976)
8.135 S. Arimitsu, H. Masuhara: Chem. Phys. Lett. **22**, 543 (1973)
8.136 J.D. Simon, K.S. Peters: J. Am. Chem. Soc. **103**, 6403 (1981)
8.137 J.D. Simon, K.S. Peters: J. Am. Chem. Soc. **104**, 6142 (1982)
8.138 H. Masuhara, Y. Maeda, N. Mataga, K. Tomita, H. Tatemitsu, Y. Sakata, S. Misumi: Chem. Phys. Lett. **69**, 182 (1980)
8.139 H. Masuhara, N. Tamai, N. Mataga, F.C. De Schryver, J. Vandendriessche: J. Am. Chem. Soc. **105**, 7256 (1983)
8.140 J.D. Simon, K.S. Peters: Acc. Chem. Res. **17**, 277 (1984) and references therein
8.141 H. Ohtani, T. Kobayashi, K. Suzuki, S. Nagakura: J. Chem. Soc. Japan **10**, 1479 (1984)
8.142 M. Martin, H. Miyasaka, A. Karen, N. Mataga: J. Phys. Chem. **89**, 182 (1985)

8.143 K.S. Peters, E. Pang, J. Rudzki: J. Am. Chem. Soc. **104**, 5535 (1982)
8.144 R. Foster (ed.): *Organic Charge Transfer Complexes* (Academic, New York 1969).
8.145 J.K. Kochi: *Organometallic Mechanisms and Catalysis*, Part 3 (Academic, New York 1978)
8.146 F.D. Lewis: Acc. Chem. Res. **12**, 152 (1979)
8.147 E.F. Hilinski, J.M. Masnovi, J.K. Kochi, P.M. Rentzepis: J. Am. Chem. Soc. **106**, 8071 (1984) and references therein
8.148 W. Kirmse: *Carbene Chemistry*, 2nd ed. (Academic, New York 1971)
8.149 A.M. Trozzolo: Acc. Chem. Res. **1**, 329 (1968)
8.150 R.A. Moss, M. Jones, Jr. (eds.): *Carbenes*, Vol. 1 (Wiley, New York 1973)
8.151 R.A. Moss, M. Jones, Jr. (eds.): *Carbenes*, Vol. 2 (Wiley, New York 1973)
8.152 H.D. Roth: Acc. Chem. Res. **10**, 85 (1977)
8.153 K.B. Eisenthal, R.A. Moss, N.J. Turro: Science **225**, 1439 (1984)
8.154 M. Platz (ed.): *Recent Aspects of Carbene Chemistry*, Tetrahedron **41**, No. 8 (1985)
8.155 A.M. Trozzolo, E. Wasserman: In [Ref. 8.151, Chap. 5, p. 185]
8.156 P.S. Skell, R.C. Woodworth: J. Am. Chem. Soc. **78**, 4496 (1956)
8.157 P.S. Skell, A.Y. Garner: J. Am. Chem. Soc. **78**, 5430 (1956)
8.158 P.P. Gaspar, G.S. Hammond: In [Ref. 8.151, Chap. 6, p. 207]
8.159 G.L. Closs: In [Ref. 8.151, Chap. 4, p. 159]
8.160 G.L. Closs, B.E. Rabinow, J. Am. Chem. Soc. **98**, 8190 (1976)
8.161 S.C. Lapur, B.E. Brauer, G.B. Schuster: J. Am. Chem. Soc. **106**, 2092 (1984)
8.162 J.J. Zupancic, P.B. Grasse, S.C. Lapur, G.B. Schuster: In *Recent Aspects of Carbene Chemistry*, ed. by M. Platz, Tetrahedron **41**, 1471 (1985) and references therein
8.163 K.B. Eisenthal, N.J. Turro, E.V. Sitzmann, I.R. Gould, G. Hefferon, J. Langan, Y. Cha: In *Recent Aspects of Carbene Chemistry*, ed. by M. Platz, Tetrahedron **41**, 1543 (1985) and references therin
8.164 E.V. Sitzmann, K.B. Eisenthal: In *Applications of Picosecond Spectroscopy to Chemistry*, ed. by K.B. Eisenthal (Reidel, Dordrecht 1984) p. 41 and references therein
8.165 C. Dupuy, G.M. Korenowski, M. McAuliffe, W.M. Hetherington III, K.B. Eisenthal: Chem. Phys. Lett. **77**, 272 (1981)
8.166 P.B. Grasse, B.E. Brauer, J.J. Zupancic, K.J. Kaufmann, G.B. Schuster: J. Am. Chem. Soc. **105**, 6833 (1983)
8.167 E.V. Sitzmann, J.G. Langan, K.B. Eisenthal: J. Am. Chem. Soc. **106**, 1868 (1984)
8.168 J.G. Langan, E.V. Sitzmann, K.B. Eisenthal: Chem. Phys. Lett. **110**, 521 (1984)
8.169 E.V. Sitzmann, J.G. Langan, Z.Z. Ho, K.B. Eisenthal: In [Ref. 8.2, p. 330]
8.170 E.V. Sitzmann, J.G. Langan, K.B. Eisenthal: To be published
8.171 Y. Wang, E.V. Sitzmann, F. Novak, C. Dupuy, K.B. Eisenthal: J. Am. Chem. Soc. **104**, 3238 (1982)
8.172 E.V. Sitzmann, Y. Wang, K.B. Eisenthal: J. Phys. Chem. **87**, 2283 (1983)
8.173 E.V. Sitzmann, J.G. Langan, K. Eisenthal: Chem. Phys. Lett. **102**, 446 (1983)
8.174 N.J. Turro: *Modern Molecular Photochemistry* (Benjamin/Cummings, California 1978)
8.175 W. Lwowski: In *Reactive Intermediates*, ed. by M. Jones, Jr. R.A. Moss (Wiley, New York 1978, 1981) Vols. 1, 2
8.176 E.F.V. Scriven: In *Reactive Intermediates*, ed. by R.A. Abramovitch (Plenum, New York 1982) Vol. 2
8.177 A. Padura, P.H.J. Carlson: In [Ref. 8.176]
8.178 C. Wentrup: In [Ref. 8.176, Vol. 1]
8.179 T. Kobayashi, H. Ohtani, K. Suzuki, T. Yamsaka: J. Phys. Chem. **89**, 776 (1985)
8.180 C.S. Foote: Acc. Chem. Res. **1**, 104 (1968)
8.181 H.H. Wasserman, R.W. Murray (eds.): *Singlet Oxygen* (Academic, New York 1979)
8.182 J. Rigaudy, C. Breliere, P. Scribe: Tetra. Lett. 687 (1978)
8.183 W. Drews, R. Schmidt, H.D. Brauer: Chem. Phys. Lett. **70**, 84 (1980)
8.184 R. Schmidt, K. Schaffner, W. Trost, H.D. Brauer: J. Phys. Chem. **88**, 956 (1984) and references therein

8.185 N.J. Turro, M.F. Chow, J. Rigandy: J. Am. Chem. Soc. **103**, 7218 (1981)
8.186 D.R. Kearns: Chem. Rev. **71**, 395 (1971) and references therein
8.187 N.J. Turro, J. McVey, R. Ramamurthy, P. Lechtken: Angew. Chemie **91**, 597 (1979)
8.188 S.Y. Hou, C.G. Dupuy, M.J. McAuliffe, D.A. Hrovat, K.B. Eisenthal: J. Am. Chem. Soc. **103**, 6982 (1981)
8.189 E.V. Sitzmann, C.G. Dupuy, Y. Wang, K.B. Eisenthal: In [Ref. 8.2, p. 168]
8.190 C.G. Dupuy, D.A. Hrovat, J.G. Langan, E.V. Sitzmann, T.A. Jenny, K.B. Eisenthal, N.J. Turro: J. Phys. Chem. **90**, 5168 (1986)
8.191 C. Wentrup: *Reactive Molecules* (Wiley, New York 1984)
8.192 J.A. Berson: Acc. Chem. Res. **11**, 446 (1987)
8.193 W.T. Borden (ed.): *Diradicals* (Wiley, New York 1982)
8.194 D.F. Kelley, P.M. Rentzepis, M.R. Mazur, J.A. Berson: J. Am. Chem. Soc. **104**, 3764 (1982)
8.195 D.F. Kelley, P.M. Rentzepis: J. Am. Chem. Soc. **105**, 1820 (1983)
8.196 T. Okada, K. Kida, N. Mataga: Chem. Phys. Lett. **88**, 157 (1982)
8.197 S.C. Freilich, K.S. Peters: J. Am. Chem. Soc. **103**, 6255 (1981)
8.198 R.A. Caldwell, T. Majima, C. Pac: J. Am. Chem. Soc. **104**, 629 (1982)
8.199 S.C. Freilich, K.S. Peters: J. Am. Chem. Soc. **107**, 3819 (1985)
8.200 Th. Förster: Z. Electrochem. **54**, 43 (1950)
8.201 A. Weller: Prog. React. Kinet. **1**, 188 (1961)
8.202 E. Vander Donckt: Prog. React. Kinet. **5**, 273 (1970)
8.203 N. Mataga, T. Kubota: *Molecular Interactions and Electronic Spectra* (Dekker, New York 1970)
8.204 N.C. Yang, C. Rwas: J. Am. Chem. Soc. **83**, 2213 (1961)
8.205 J. Faure, J. Joussot-Dubien: J. Chim. Phys. Phys. Chim. Biol. **63**, 621 (1966)
8.206 W. Klopfer: Advan. Photochem. **10**, 311 (1977)
8.207 S.G. Schulman: *Fluorescence and Phosphorescence Spectroscopy* (Pergamon, Oxford 1977)
8.208 D. Huppert, E. Kolodney: Chem. Phys. **63**, 401 (1981)
8.209 C.M. Harris, B.K. Selinger: J. Phys. Chem. **84**, 891 (1980)
8.210 S.P. Webb, S.W. Yeh, L.A. Philips, M.A. Tolbert, J.H. Clark: J. Am. Chem. Soc. **106**, 7286 (1984)
8.211 J. Lee, R.D. Griffin, G.W. Robinson: J. Chem. Phys. **82**, 4920 (1985)
8.212 A.J. Campillo, J.H. Clark, S.L. Shapiro, K.R. Winn: In *Picosecond Phenomena*, ed. by C.V. Shank, E.P. Ippen, S.L. Shapiro, Springer Ser. Chem. Phys., Vol. 4 (Springer, Berlin, Heidelberg 1978) p. 319
8.213 D. Hupert, E. Kolodney, M. Gutman, E. Nachliel: J. Am. Chem. Soc. **104**, 6949 (1982)
8.214 M. Eigen, L. De Maeyer: Proc. Roy. Soc. A **247**, 505 (1958)
8.215 R.S. Becker, W.F. Richey: J. Am. Chem. Soc. **89**, 1298 (1967)
8.216 K.K. Smith, K.J. Kaufmann: J. Phys. Chem. **82**, 2286 (1978)
8.217 S.Y. Hou, W.M. Hetherington III, G.M. Korenowski, K.B. Eisenthal: Chem. Phys. Lett. **68**, 282 (1979)
8.218 T. Kobayashi, E.O. Dengenkolb, P.M. Rentzepis: J. Phys. Chem. **83**, 2431 (1979)
8.219 P.J. Thistlethwaite, G.J. Woolfe: Chem. Phys. Lett. **63**, 401 (1979)
8.220 P.F. Barbara, P.M. Rentzepis, L.E. Brus: J. Am. Chem. Soc. **102**, 2786 (1980)
8.221 C. Merritt, G.W. Scott, A. Gupta, A. Yavrouian: Chem. Phys. Lett. **69**, 169 (1980)
8.222 G.J. Woolfe, P.J. Thistlethwaite: J. Am. Chem. Soc. **103**, 3849 (1981)
8.223 S. Schneider, E. Lill, P. Hefferle, F. Dorr: Nuovo Cimento **63B**, 411 (1981)
8.224 H. Inoue, M. Hida, N. Nakashima, K. Yoshihara: J. Phys. Chem. **86**, 3184 (1982)
8.225 A.L. Huston, G.W. Scott, A. Gupta: J. Chem. Phys. **76**, 2978 (1982)
8.226 A.U. Khan, M. Kasha: Proc. Natl. Acad. Sci. USA **80**, 1767 (1983)
8.227 A.J.G. Strandjord, S.H. Courtney, D.M. Friedrich, P.F. Barbara: J. Phys. Chem. **87**, 1125 (1983)
8.228 S. Nagaoka, N. Hirota, M. Sumitani, K. Yoshihara: J. Am. Chem. Soc. **105**, 4220 (1983)
8.229 D. McMorrow, M. Kasha: Proc. Natl. Acad. Sci. USA **81**, 3375 (1984)
8.230 D.J. Jang, D.F. Kelley: J. Phys. Chem. **89**, 209 (1985)

8.231 A.J.G. Strandjord, D.E. Smith, P.F. Barbara: J. Phys. Chem. **89**, 2362 (1985) and references therein
8.232 S.R. Flour, P.F. Barbara: J. Phys. Chem. **89**, 4489 (1985)
8.233 D.B. O'Connor, G.W. Scott, D.R. Coulter, A. Gupta, S.P. Webb, S.W. Yeh, J.H. Clark: To be published
8.234 J.R. Merrill: J. Phys. Chem. **65**, 2023 (1961)
8.235 A. Beckett, G. Porter: Trans. Farad. Soc. **59**, 2038, 2051 (1963)
8.236 G. Porter, P. Suppau: Trans. Farad. Soc. **61**, 1664 (1965)
8.237 A.A. Lamola, L.J. Sharp: J. Phys. Chem. **70**, 2634 (1966)
8.238 K.J. Choi, L.A. Hallidy, M.R. Topp: In *Picosecond Phenomena II*, ed. by R.M. Hochstrasser, W. Kaiser, C.V. Shank, Springer Ser. Chem. Phys., Vol. 14 (Springer, Berlin, Heidelberg 1980) p. 131
8.239 D. McMorrow, M. Kasha: J. Phys. Chem. **88**, 2235 (1984)
8.240 D. Ford, P.J. Thistlethwaite, G.J. Woolfe: Chem. Phys. Lett. **69**, 246 (1980)
8.241 A.J.G. Strandjord, P.F. Barbara: J. Phys. Chem. **89**, 2355 (1985)
8.242 T. Elsaesser, W. Kaiser, W. Luttke: J. Chem. Phys. **90**, 2901 (1986)
8.243 K. Wiberg: Chem. Rev. **55**, 713 (1955)
8.244 C. Dupuy: Doctoral Thesis, Columbia University, 1982
8.245 C.A. Taylor, M.A. El-Bayoumi, M. Kasha: Proc. Natl Acad. Sci. USA **63**, 253 (1969)
8.246 K.C. Ingham, M. Abu-Eigheit, M.A. El-Bayoumi: J. Am. Chem. Soc. **93**, 5023 (1971)
8.247 W.M. Hetherington III, R.H. Micheels, K.B. Eisenthal: Chem. Phys. Lett. **66**, 230 (1979)
8.248 H. Bulska, A. Grabowska, B. Pakuļa, J. Sepiol, J. Waluk, U.P. Weld: J. of Lumin. **29**, 65 (1984)
8.249 K. Fuke, H. Yoshiuchi, K. Kaya: J. Phys. Chem. **88**, 5840 (1984)
8.250 K. Tokumura, Y. Watanabe, M. Itoh: Chem. Phys. Lett. **111**, 379 (1984)
8.251 J. Waluk, A. Grabowska, B. Pakula, J. Sepiol: J. Phys. Chem. **88**, 1160 (1984)
8.252 M. Noda, N. Hirota, M. Sumitani, K. Yoshihara: J. Phys. Chem. **89**, 399 (1985)
8.253 D.A. Gobeli, J.R. Morgan, R.J. St. Pierre, M.A. El-Sayed: J. Phys. Chem. **88**, 178 (1984)
8.254 D.A. Gobeli, J.D. Simon, M.A. El-Sayed: J. Phys. Chem. **88**, 3949 (1984)
8.255 J.M. Wiesenfeld, B.I. Greene: Phys. Rev. Lett. **51**, 1745 (1983)
8.256 W.M. Hetherington III, G.M. Korenowski, K.B. Eisenthal: Chem. Phys. Lett. **77**, 279 (1981)
8.257 L.S. Goldberg: In *Picosecond Phenomena III*, ed. by K.B. Eisenthal, R.M. Hochstrasser, W. Kaiser, A. Laubereau Springer Ser. Chem. Phys., Vol. 23 (Springer, Berlin, Heidelberg 1982) p. 94
8.258 B.B. Craig, W.L. Faust, L.S. Goldberg, P.E. Schoen, R.G. Weiss: In *Picosecond Phenomena II*, ed. by R.M. Hochstrasser, W. Kaiser, C.V. Shank, Springer Ser. Chem. Phys., Vol. 14 (Springer, Berlin, Heidelberg 1980) p. 253
8.259 J.L. Knee, L.R. Khundkar, A.H. Zewail: J. Chem. Phys. **82**, 4715 (1985)
8.260 J.L. Knee, L.R. Khundkar, A.H. Zewail: J. Chem. Phys. **83**, 1996 (1985)

9. Biological Processes Studied by Ultrafast Laser Techniques

Robin M. Hochstrasser and Carey K. Johnson

With 23 Figures

In writing this review of fast kinetic and spectroscopic studies in biology we have attempted to put the new pulsed laser results into a broad biological context. In this way it is hoped that those not so familiar with the biology will be familiarized with at least a few of the basic questions that modern laser technology might address.

The subject matter is confined on the whole to subnanosecond events which are studied by picosecond and subpicosecond laser methods. This choice has resulted in the omission of a significant amount of vitally important optical and laser work that laid the framework upon which the ultrafast spectroscopic experiments could be designed.

The choice of topics reflects our own interests and the article does not represent a comprehensive treatment of the whole subject of ultrafast studies of systems relevant to biological questions. We have dealt with the three main biological topics: heme protein dynamics, photosynthesis, and the operation of rhodopsin and bacteriorhodopsin. There is a brief section on electron transfer to provide some of the physics essential to the main topics.

9.1 Heme Proteins

Hemoglobin (Hb) has the remarkable ability to assume the dual role of collecting oxygen in the lungs and delivering it efficiently to the distant tissues. To accomplish this, in addition to having a variety of special properties, the molecule must bind oxygen efficiently at ambient pressures and transport it efficiently to the lower-pressure tissue regions. It is the famous sigmoid shaped relation between the degree of saturation and the oxygen pressure, and its sensitivity to pH which make such behavior possible. The protein binds or delivers oxygen at a rate that depends on how much is already bound: This has been termed the cooperative binding of oxygen.

The x-ray diffraction results for Hb and the fully oxygenated form, HbO_2 (though it has four oxygen molecules attached) show that the two structures have significant differences. This fact suggested the so-called two-state model of hemoglobin [9.1] in which the kinetics of oxygen binding is modeled by assuming the existence of just two basic protein structures, R and T, each of which can bind oxygen and whose various partially oxygenated forms can interconvert. One of these, the T form, is considered to have lower affinity for O_2 than the other. The structural differences between the equilibrium Hb and HbO_2 systems, which are generally regarded to have the T and R forms, involve mainly a change in the way the various components of the protein are bound. However, as will be seen later, there are both global and more subtle structural differences between the deoxy and oxy forms and there exist important questions concerning the dynamical relationships between those changes [9.2–5].

Hemoglobin [9.6] is a 65 000 molecular weight protein consisting of four subunits. Each subunit forms a nest for one heme (an iron porphyrin) and each iron can bind one oxygen. The Fe(II) is five coordinate in the paramagnetic deoxy form having four porphyrin pyrrole nitrogen ligands and one histidine ligand called the proximal histidine. The four subunits are bound together into the quaternary structure of hemoglobin which is different for T and R as described above. There are in fact only two types of subunits, or chains, so that each Hb molecule consists of two α-chains and two β-chains. Structural features regarding the subunit configuration are termed tertiary, so that α and β chains have different tertiary structure. The terms primary and secondary structure are used to describe the amino acid sequences and the configurations of these side chains respectively.

There are other heme proteins for which O_2 binding is functional. One example is myoglobin (Mb) which is used to store oxygen rather than to transport it. The tertiary structure of Mb is quite similar to that of isolated α-chains. Since Mb is a single subunit, its oxygen binding can serve as an important comparison with which to attain a better understanding of Hb. The oxygen binding takes place in a relatively hydrophobic region of the protein known as the heme pocket, which has similar constitution for Hb and Mb.

The literature on Hb and Mb is enormous [9.6–8]. The current understanding is the result of the work of many groups working over many years. Nevertheless there remain certain questions for which detailed answers do not exist, the most evident being the nature, on an atomic scale, of cooperativity. We must ask what electronic factors are critical to the behavior of Hb as well as what effects the atomic motions have on its properties. The quantum mechanical description of the reaction between oxygen and the heme has not yet been settled. We do not yet know the detailed pathways of the kinetics of the quaternary structure change linking T and R informs. To obtain the full answer to these questions information will be required about the molecule on a wide range of time scales including the picosecond and subpicosecond regimes. It is well known that the time scale for the operation and behavior of Hb cannot be pinned down to any narrow region. In fact, as shown in Fig. 9.1, the various responses of Hb span about eighteen orders of magnitude of time ranging from the femtosecond photophysics and chemistry to the 10 day useful lifetime for Hb molecules in humans.

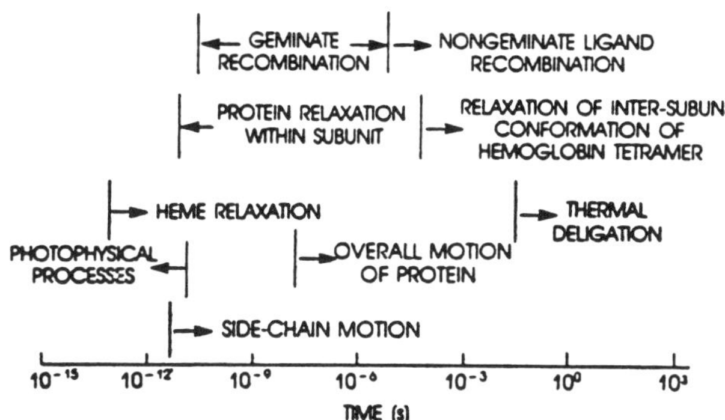

Fig. 9.1. Timescales for various properties which are vital for the function of hemoglobin. [We are indebted to W.A. Eaton for the idea of this diagram]

One might now ask whether ultrashort pulsed laser experiments can contribute anything to understanding such a complex system. There are two good reasons to believe that the answer is positive. One involves the heme properties and another the protein structure fluctuations. When diamagnetic six coordinated hemes such as the CO derivatives of myoglobin or hemoglobin are irradiated with visible or ultraviolet light, the diatomic molecule is released leading to a five coordinate (paramagnetic) Fe(II). The rate of photodissociation appears to be at least in the hundreds of femtosecond time scale or faster [9.10, 11] so that it is possible, by using light pulses having various lengths, to prepare various structures of the deoxyhemeprotein. Transient spectroscopic methods can then be used to follow the further evolution of these structures. When quite short light pulses are used in the photolysis, the expectation is that the tertiary and quaternary structure of the protein just after the pulse will be very similar to the six coordinate structure. One would therefore be afforded the opportunity of studying directly with short pulse methods the decay of this metastable form into the equilibrium deoxy molecule. This is the photoinitiated R $\rightarrow$ T transition. In addition to such heme initiated processes the secondary structure is undergoing fluctuations on the time scale of hindered molecular rotations. These motions also can be studied by various fast spectroscopic techniques and assessed as to their importance in the behavior of heme proteins. This is specially important since it has been pointed out [9.11, 12] that these diatomic molecules would be unable to pass efficiently between the heme and the solvent surrounding the protein if the protein were a rigid structure fixed at the crystallographic configuration.

9.1.1 Optical Spectroscopy of Heme Proteins

The visible and near-ultraviolet spectra of heme proteins correspond to the Q-band and Soret band whose peak wavelengths are distinctive of the electronic structure and coordination of the heme. Figure 9.2 shows the spectra of Hb,

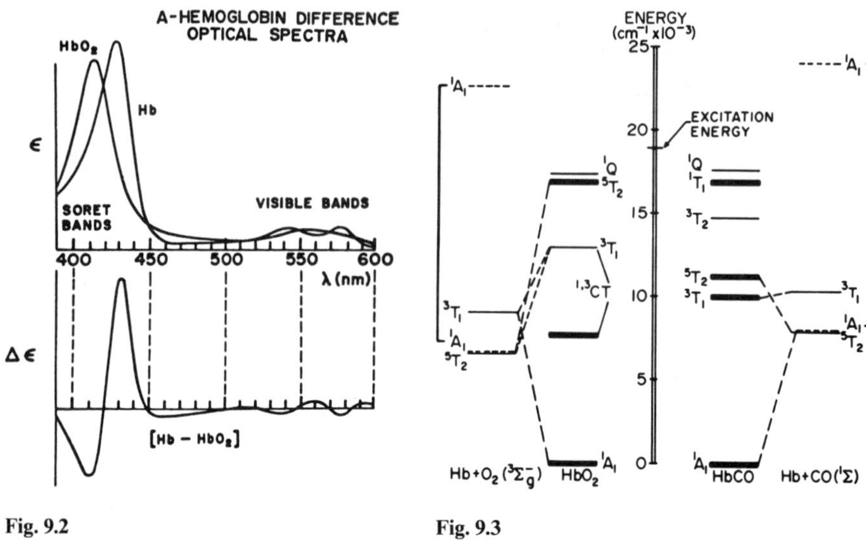

Fig. 9.2

Fig. 9.3

Fig. 9.2. Make up of optical difference spectra of Hb:HbO_2 in the Soret and visible band region

Fig. 9.3. Some of the electronic states of Hb, HbCO and HbO_2 relevant to the photoprocesses discussed in the text. A detailed discussion of the states can be found in [9.13,14]. The correlation lines are drawn to conserve spin and orbital symmetry

HbO_2 and their difference spectrum. This equilibrium difference spectrum is what would be observed in a transient experiment if the HbO_2 originally present were to be photolyzed to Hb having the equilibrium structure.

Picosecond spectra of heme proteins were initially reported by *Greene* et al. [9.13] for HbCO and HbO_2 in one of the first applications of a fully multiplexed picosecond double-beam spectrometer. These experiments at ca. 6 ps resolution showed that the difference spectra were similar but distinctly altered from those at equilibrium. The significant broadening of the Soret band could have resulted from a distribution of metastable Hb structures created by absorption of many photons or from the presence of time-unresolved short-lived intermediates. A first attempt was made to establish a relationship between the photochemical pathway for dissociation and the electronic structure and relaxation dynamics in the heme. It was proposed that the dissociation of CO occurs after rapid internal conversion to the $^3T_1$ ligand field state of HbCO whereas O_2 dissociation requires excitation of a triplet charge-transfer state of HbO_2. Figure 9.3 shows some of heme protein states relevant to these experiments. The picosecond spectra were improved in later studies by *Chernoff* et al. [9.14,15] in which it was also discovered that the oxygen from dissociated HbO_2 recombined with the Hb on the time scale of a few hundred picoseconds. Some of these experiments are shown in Fig. 9.4. This recombination is sufficiently fast that it was proposed to be geminate, involving the pairing of O_2 with the Fe from which it had been

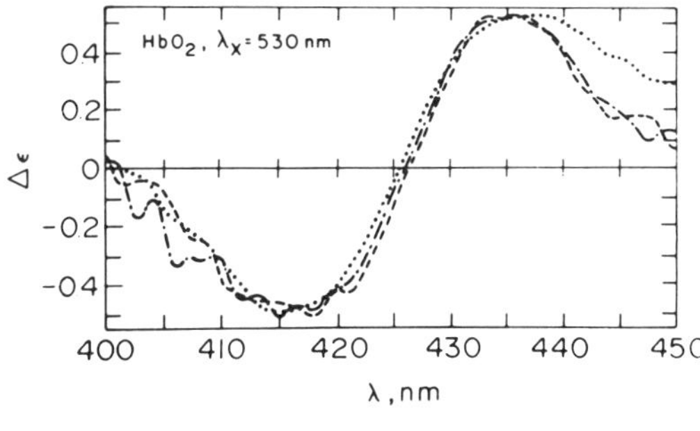

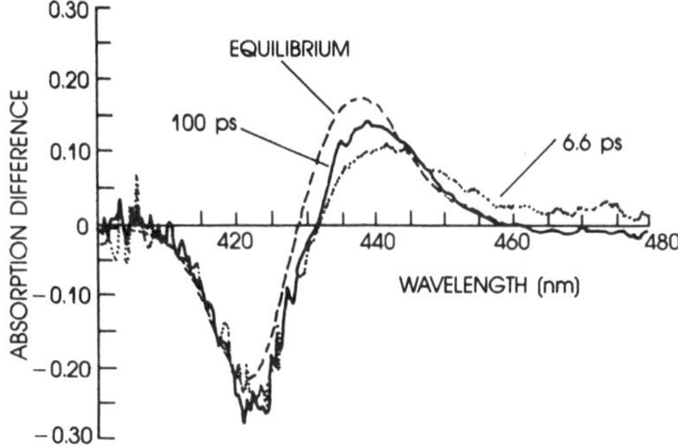

Fig. 9.4. Typical spectra of photolyzed HbO_2 (*upper*; after [9.14]) and MbCO (*lower*; after [9.27]). For HbO_2 spectra are shown at 6 ps (····), 93 ps (–·–·–) and 1.2 ns (– – –)

dissociated. In this work the ligand field states $^3T_1$ and $^5T_2$ were considered most likely precursor states for the photodissociation. We will return later to questions relating to the photophysics.

It was apparent from all the transient absorption and kinetics studies using 6–10 ps pulses, that the actual dissociation rate was too fast to measure using such time resolution [9.13–17,32]. By means of experiments with a subpicosecond laser *Shank* et al. [9.9] concluded that a photo-process takes place in less than 1 ps. More recently *Martin* et al. [9.10] used 250 fs pulses to measure the rate at which HbCO and MbCO were bleached by 307 nm radiation. They found that the bleaching evolution was pulse-width limited but that the appearence of transient absorption at a wavelength corresponding to the peak absorption of the deoxy species took 350 fs. Their result is shown in Fig. 9.5. *Martin* et al. [9.18] have also reported comparable results for HbO_2 *with* 100 fs pulses at 580 nm for excitation. These studies show that during the first few picoseconds there are photophysical processes occurring in the heme (see below) but the system appears to settle down a few picoseconds after photolysis. At this point the spectrum is quite similar to that of the equilibrium deoxy form. In the case of Hb, a spectral shift of a few nm is seen even after 2 ns pulse excitation [9.19], but this is not the case with Mb where the ns difference spectrum is identical with that at equilibrium [9.20].

While there has been a good deal of work on the transient optical spectra of the photolysis products of these proteins in the nanosecond, picosecond and

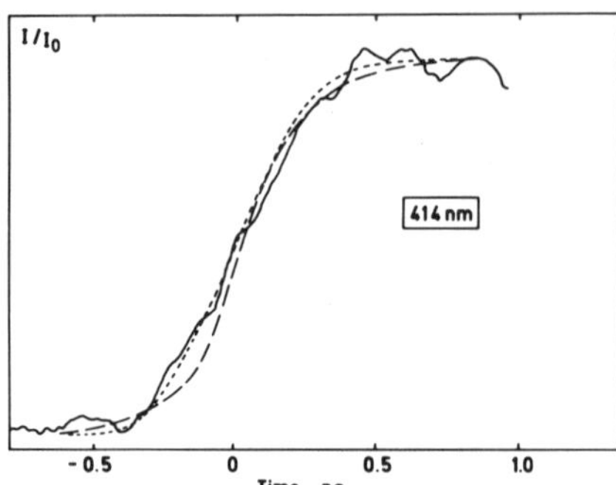

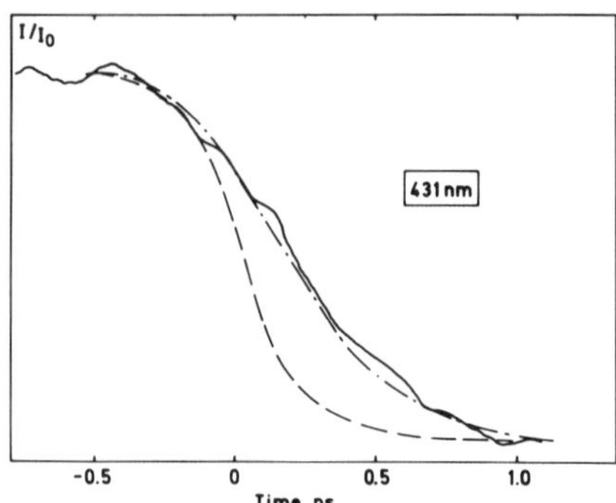

Fig. 9.5. Normalized kinetic results obtained after photodissociation of HbO_2 solution at 309 nm with 100 fs pulses. *Left*: heavy lines induced transmission (I/I_0 at 414 nm showing a rise time limited by the pulse duration as indicated by the dotted line (····) which represents the integral of the cross-correlation as determined in a separate experiment. The (– – –) line is the best fit of the experimental trace (heavy curve) assuming an exponential pulse of duration 105 fs and an instantaneous bleaching. *Right*: induced absorption at the maximum positive change in the stable difference spectrum (431 nm) corresponding to the appearance of deoxy Hb. The heavy line is experimental. The (– – –) and (–·–·–) curves correspond respectively to the computed instaneous response from the upper panel and to the best fit with a time constant $\tau_{Fe} = 250$ fs. This time constant cannot be due to the group velocity dispersion which slows up as a pure delay and is taken into account in the zero position. The abscissa is 2 ps full scale. Identical results were obtained using 580 nm excitation pulses. (After [9.10])

femtosecond regimes, as indicated above, there are few experiments that carefully monitor the evolution of these spectra between the picosecond and nanosecond timescales. The spectra reported for these two timescales are different [9.13–15, 19, 20, 32] which, if correct, would imply that structural changes occur on the timescale 8 ps to 5 ns. However, this is not case as is discussed below on the basis of new results which span most of this region.

Recent results by *Janes* et al. (9.267) have shown that the transient difference spectra of MbCO and HbCO transients have fully evolved to those observed previously with nanosecond time resolution within 30 ps following photolysis with 532 nm or 355 nm pulses of 30 ps duration. In addition, the spectra of Mb with overlapped 30 ps pump and probe pulses were found to be identical with those obtained for steady-state methods. This suggests that for Mb, significant changes in heme structure such as might result from direct coupling to the protein via the proximal histidine linkage and van der Waals contacts are completed within the timescale of 30 ps. For HbCO photolysis, the results indicate no change is occurring in heme structure from about 30 ps to 6.5 ns. This finding, together with the results of nanosecond kinetic studies [9.19, 20], suggests that protein conformational changes within the R quaternary structure that influence the rate of ligand rebinding take place in less than 30 ps.

9.1.2 Structural Information

In order to investigate further the possible relationships between picosecond laser experiments and questions relevant to the mechanism of cooperativity, it is necessary to consider in more detail the protein structure itself. There is little direct experimental information about the dynamical aspects of such structures in solution. One therefore relies on the information from x-ray diffraction of crystals as a starting point for discussions of structural dynamics. Since these crystals contain solvent which allows reactants to diffuse in and out of the heme pockets, chemical evidence has accumulated that the ligand kinetics and hence the structure in the crystal could be quite similar to average structures in solution.

The twofold symmetry of Hb results in there being two α and two β subunits, labeled α_1, α_2 and β_1, β_2. Ligand formation results in a variety of small changes in the subunit (tertiary) structure but there is also a significant alternation in the packing between the symmetry-related dimers $\alpha_1\beta_1$ and $\alpha_2\beta_2$. It is this quaternary structure change that is believed to be associated with cooperativity. Some of the notable changes of tertiary structure that occur on ligand formation are as follows: The hemes move further into the hydrophobic pockets; the proximal histidine, which is tilted to the heme plane becomes more symmetrically positioned; the F-helix, to which this histidine is bonded, moves relative to the heme in each subunit. The quaternary changes involve the nature of $\alpha_1\beta_2$ (or equivalently $\alpha_2\beta_1$) contacts. In particular, there are more salt bridges binding the $\alpha_1\beta_2$ subunits in the deoxy form (T-structure) [9.21]. This structure has a lower affinity for O_2 and is thermodynamically more stable than the R structure. *Gelin* and *Karplus* [9.22] have suggested that the change in the distances between the porphyrin nitrogens and the proximal histidine carbons, leading to the occurrence of a more tilted histidine-heme configuration, as mentioned above, is the cause of the low affinity of the T state of Hb. It follows that the movement of the proximal histidine so that it stands upright, would be one of the necessary dynamical steps in the optically initiated $T \rightarrow R$ transition. Since this ligand is so intimately coupled to the heme, one would expect that heme spectra will be sensitive to the "uprightness" of the histidine. The F-helix can be seen in Fig. 9.6 coupled to the iron through the proximal histidine (F7). In the deoxy form with the iron significantly out of the porphyrin plane, the F-helix is slightly further from the heme compared with the liganded structure [9.21]. Obviously there are many questions of varying complexity regarding the dynamics of this structural change. An important point is that the F-helix moves only about 0.15 A relative to the heme in myoglobin. It has therefore become customary, especially in new experiments, to compare results obtained for myoglobin and hemoglobin.

Following photolysis of MbCO, a species closely resembling but reportedly not identical with equilibrium deoxymyoglobin is formed. The picosecond transient-absorption experiments are sensitive to changes in heme structure, but the observations are not readily translated into specific alterations in structure. The

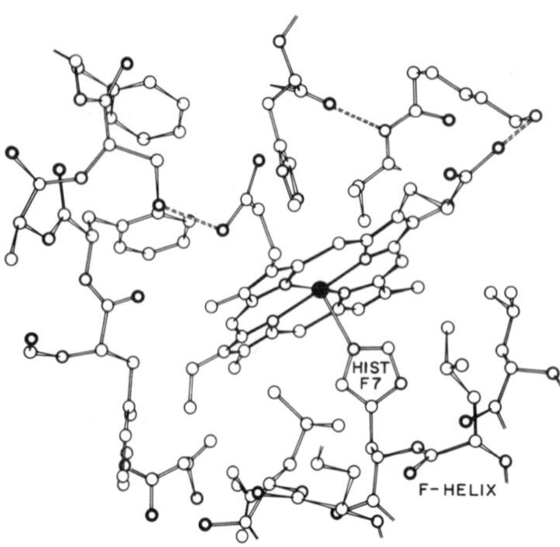

Fig. 9.6. A view of the Mb structure showing the heme (*solid lines*), the proximal histidine (F7) and the F-helix region

results of the subpicosecond kinetic study suggest that a deoxy-like species is formed with a time constant of 350 fs. Like all single wavelength kinetic studies, the observation of a 350 fs transient has other possible interpretations. One alternative is that an excited state of deoxy is formed very quickly, with the ligand being ejected in less than 150 fs. The observed transient would then represent the electronic relaxation of the excited state of deoxy. Another possibility is that the experiments also detect hot molecules [9.23]. A detailed study of the optical spectra at 100 fs should clear up these questions. The attractive suggestion [9.10] that the ligand moves out of the plane in 350 fs is one that we shall adopt in this discussion.

9.1.3 Transient Raman Spectroscopy

Information about the R → T transition of hemoglobin can also be obtained through studies of resonant Raman scattering from the photolysis products. The vibrational transitions involved in resonant scattering of visible and near ultraviolet light correspond to porphyrin skeletal motions and to ligand-porphyrin motions which influence the π-electron structure. As a result of the work of many investigators on vibrational spectra of model compounds, it has become feasible to associate an observed vibrational transition with a particular nuclear motion in the heme. For example, the mode near $1540\,cm^{-1}$ in hemes is certainly one which involves the pyrrole nitrogens, and it is known to be sensitive to the doming of the heme in model compounds of known structure [9.4]. Accordingly, its frequency can be expected to signal deviations of the Fe from its equilibrium configuration in diamagnetic metastable forms of Hb. A motion of the proximal histidine against the Fe (primarily) is certainly connected with the frequency at $220\,cm^{-1}$ and it can be expected that this frequency be sensitive to the configuration forced on the proximal histidine which is expected to change at some point after photolysis. Other vibrational transitions are particularly sensitive to the spin state of the molecule. It is not surprising therefore that a significant research effort was made in the past few years to obtain picosecond transient Raman spectra and to understand the transition frequencies in structural terms by appealing to the correlations found in model compounds. While this approach is reasonable, it should not be considered to yield proof of structure at this stage. Vibrational transitions can be shifted for reasons other than those identified in model compounds. However, the technique is eminently justifiable and may indeed lead to more definite identification of structural changes once the transition frequencies for the intermediates become known for many isotopically substituted hemes, and when these sets of frequencies are used to find or optimize force fields for the heme and its associated ligands.

Raman scattering can be accomplished with picosecond time resolution by a variety of methods. Each relies on the fact that in condensed media the pure dephasing of optical transitions occurs on the femtosecond time scale so that the spontaneous Raman emission persists for an extremely short time even for resonant excitation. It follows that the medium itself acts as a timing gate for such emission so that it is possible to create transients and study their Raman spectra with a single laser pulse. In fact this was the method used in the first picosecond Raman experiments on heme proteins [9.25].

The study of the time evolution of Raman spectra on the picosecond time scale is somewhat more difficult but has been accomplished: Pump-probe Raman experiments have been reported with both high repetition rate synch-pumped cavity dumped lasers [9.26] as well as with 20 Hz pulsed YAG lasers [9.27]. The principal difficulty with such experiments, which limits their applicability, is the occurrence of nonlinear processes and spontaneous emission having lifetimes longer than the dephasing time. Thus picosecond optical gating may have to be part of a more versatile time resolved Raman instrument. In order to measure resonance Raman spectra, a probing average power of ~ 1 mW is required. On the other hand peak power densities in excess of $\sim 30\,\mathrm{GW/cm^2}$ should be avoided. This implies that the repetition rate must be increased approximately in proportion to the inverse of the pulse duration. Thus with 30 ps pulses the repetition rate can be reduced to ~ 20 Hz if necessary but with 100 fs pulses the repetition rate would have to be at least 600 Hz in order to achieve the same average power without exceeding the photodamage limit.

The first picosecond Raman spectra of heme proteins were due to *Terner* et al. [9.25] who obtained the resonantly enhanced spectrum of HbCO with 30 ps pulses each of which formed both the photolysis and Raman probe source. A Hb-like species was formed within the 30 ps pulse. The core marker bands near $1540\,\mathrm{cm^{-1}}$ were probed in these experiments and the observed reduction in frequency compared with equilibrium Hb was used as evidence that the heme core size was expanded. Later work [9.28] showed that this metastable form persisted for at least 20 ns but not as long as 300 ns.

The situation for the core marker bands is different for MbCO photolysis products. *Johnson* and co-workers [9.29,30] showed that shifts comparable to those in HbCO were present at 30 ps, but with 7 ns pulses the Mb equilibrium spectra were already present.

In recent work *Johnson* and co-workers [9.30] have reported the first picosecond time resolved (pump-probe) Raman spectra of hemeproteins. This work has produced preliminary evidence that the core marker band of Mb from MbCO corresponds to a metastable state between 30 and 100 ps which by 400 ps has reverted to become indistinguishable from the Raman transition of equilibrium Mb.

The other Raman mode which has received detailed study is the Fe-histidine stretch at $\sim 220\,\mathrm{cm}^{-1}$. The Fe-histidine Raman frequency of the photolysis product of MbCO assumes the value for equilibrium Mb by 30 ps according to *Findsen* et al. [9.33]. On the other hand for HbCO photolysis, a shifted Raman transition is observed by 25 ps which persists into the nanosecond time regime at which point tertiary structural alterations begin to dominate the relaxation of the metastable form.

While the core marker bands correlate well with the iron displacement in a large group of model compounds, a structural feature that correlates with the Fe-histidine stretch has not been determined by such an analysis. *Findsen* et al. [9.33] have suggested a sensitivity of this motion to the tilting of the proximal histidine. Such a relationship would have important consequences, since time resolved experiments on the Fe-histidine stretch would then permit direct tests of the *Gelin* and *Karplus* [9.22] idea that the low affinity of the T-state is significantly influenced by the tilt of the histidine. However, the metastable (30 ps) form of Mb, from photolysis of MbCO, shows an Fe-histidine transition quite similar to that of the equilibrium Mb whereas there is a significant (5°) histidine tilt expected on deligation [9.21]. Thus it may be necessary to look for intersubunit constraints to account for the changes in the Fe-histidine stretch frequency of Hb. The relaxation of the core marker transitions of photogenerated Mb to their equilibrium values occurs on the subnanosecond time scale whereas for Hb it takes 100s of ns. This suggests that the Fe in Mb is not fully out the heme plane until 100–400 ps, whereas in photogenerated Hb the restriction on the core size expansion is maintained for a much longer period.

The results of resonance Raman experiments on the core marker bands of MbCO discussed above are obviously difficult to reconcile with the transient optical studies described earlier. The Raman and optical absorption are directly related so the observation of a vibrational frequency shift implies that some change in the optical spectrum should occur, but none was observed in the range 30 ps to 6.5 ns. Two possible explanations were suggested for this discrepancy [9.267]: One considers local heating of the heme by the optical source, which will be discussed in more detail below, and the other is that the structural change that shifts the core marker band does not affect the Soret absorption spectrum whose transition strength is largely dominated by other types of porphyrin vibrations. Careful studies of any absorption spectral changes in the visible bands are therefore needed in the time regime 30 ps.

Given the recent advances in picosecond Raman methodology that have occurred in the past year or so, there is reason to suppose that the inferences drawn from this form of spectroscopy will ultimately provide a coarse grain structural map of the dynamics occurring near the heme.

9.1.4 Picosecond Transient Infrared Spectroscopy

The infrared spectra of carboxyheme proteins in the region of the CO stretching frequency are known to provide plausible indications of heme-ligand structures [9.268]. A new method of determining infrared spectra with picosecond time resolution was recently reported [9.269] and used [9.270] to obtain iron-carbonyl geometries in MbCO and HbCO. The angle between the carbonyl axis and the normal to the heme plane was found to be 18° for the 1951 cm^{-1} band of HbCO and 20° and 35° respectively for the 1944 cm^{-1} and 1933 cm^{-1} bands of MbCO. These results suggest structures, consistent with recent x-ray diffraction [9.271], in which the Fe—C bond is tilted to the heme normal and the Fe—C—O angle is significantly different from zero. Furthermore, it was also found that in protohemes [9.269, 270] the CO is significantly inclined to the heme normal and that the amount of inclination depends on the different solvent structures obtained as the viscosity is varied.

9.1.5 Protein Dynamics

The understanding of cooperativity dynamics at an atomic level will require experiments that focus on each portion of the protein and not just on the heme group and its ligands. The amino acids tryptophan, tyrosine and phenylalanine offer the most obvious opportunities for ultrashort time scale experiments because they have absorption bands that are accessible by current pulsed lasers [9.34–36].

Tryptophan is in many respects the most attractive candidate, both because of its relatively low energy and high extinction coefficient absorption spectrum and because myoglobin has only two, and hemoglobin only six, tryptophans with some of them strategically located at interesting structural points. However there are possible disadvantages of using tryptophan as a probe since it presents some unsolved problems with regard to its own photophysical properties [9.37–40]. It is of interest to study both the overall motion of tryptophans and the electronic changes which might occur as their location and environment is altered. The variety of techniques that can be utilized for such studies include fluorescence spectroscopy (spectra, lifetimes and anisotropy), nonlinear spectroscopies (polarization methods and transient gratings) and conventional transient absorption spectroscopy. Some of these methods signal only the properties of the electronically excited states while others, such as polarization methods and transient absorption, can yield ground state behavior. However most work to date has involved fluorescence.

9.1.6 Fluorescence of Tryptophans in Mb and Hb

Pulsed laser techniques are needed to study tryptophan fluorescence in heme proteins as a result of very efficient energy transfer to the heme. The low fluorescence quantum yields for tryptophan in heme proteins suggested that the corresponding lifetimes are in the picosecond range [9.41]. Thus, to obtain direct information on the lifetimes and motional characteristics of such tryptophans it was necessary to devise experiments by which the fluorescence and its polarization could be measured with picosecond accuracy.

Sperm whale (SW) Mb is known to contain the two tryptophan residues, Trp-7 and Trp-14. Thus, it is expected that the emission will be dominated by two decays. The heme is considerably further from Trp-7 than from Trp-14 (20 Å compared with 15 Å) so that the effects of energy transfer on these excited states are expected to be different. Tuna Mb has an x-ray structure very similar to SW Mb but has only one tryptophan, Trp-14, and could provide a simpler system with which to understand the fluorescence properties.

Qualitatively the experimental results on various myoglobins (see Fig. 9.7) are in accord with the predictions based on the structure. Both Mb and MbCO show fluorescence decays which are nonexponential but, when fit to a double exponential, yield two components having equal weight and lifetimes of 20 ps and 120 ps, corresponding respectively to the Trp-14 and Trp-7 energy transfer decay constants [9.42].

In principle, the fluorescence decay of *each* Trp in a heme protein is not expected to be exponential because the relative motion of the Trp and heme causes the (rate limiting) energy transfer rate coefficient to be time dependent. The expected character of the decay profile I_f is not obvious, since if k_f is the fluorescence decay constant, k_r the radiative rate constant and $k_{ET}(\tau)$ the energy transfer rate coefficient at time τ, I_f has the form:

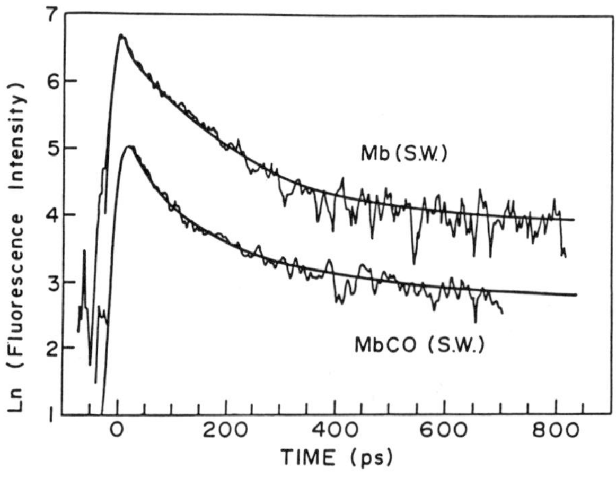

Fig. 9.7. The fluorescence spectra of Mb and MbCO. (Note the logarithmic scale.) The time resolution is ca. 15 ps in these experiments. Most of the fluorescence (ca. 95%) occurs on the ps timescale. (After [9.42,45])

$$I_f = k_r \left\langle \exp\left[- \int_{-\infty}^{t} (k_f + k_{ET}(\tau)) d\tau \right] \right\rangle . \tag{9.1}$$

In this expression the angle brackets imply an average over the ensemble of protein molecules each of which exhibits a particular k_{ET} at time τ. The value of $k_{ET}(\tau)$ can be calculated from Forster theory, which implies that:

$$k_{ET}(\tau) = \text{const} \cdot \frac{K(\tau)}{R^6(\tau)} \tag{9.2}$$

where $K(\tau)$ and $R(\tau)$ are the orientational factor [9.43] and Trp-heme separation, respectively.

Because Mb and Hb contain more than one Trp it is difficult to identify with confidence the presence of nonexponential decays from the individual Trps. However, the Mb from yellow fin tuna [9.44] contains only Trp-14 and can therefore be used to study motional effects on the fluorescence decay. The subnanosecond decay profile, obtained with a streak camera shows a distinct nonexponential decay which is approximately describable as a double exponential, 90% of $\tau = 31$ ps and 8% of $\tau = 132$ ps [9.38]. It should be possible to bring results of this nature into sharper relationship with theory by means of molecular dynamics simulations and recent efforts along these lines will be described later.

Considerably more work is needed to understand in detail the fluorescence of Trp in Mb and Hb. The decay characteristics are excitation wavelength dependent [9.45] suggesting that competitive photophysical processes also require elucidation before such probes can be used effectively. Selective excitation of one of the tryptophans might conceivably occur if different spectra result from the different environments, but it is also possible that the two tryptophans undergo photophysical processes with different efficiencies and wavelength dependences. Here is an area where subpicosecond fluorescence methodology could begin to make a significant impact.

The fluorescence anisotropy is a direct measure of orientational relaxation is as much as it measures:

$$\tfrac{2}{5}\langle P_2(\hat{\mu}_a(0) \cdot \hat{\mu}_e(t)\rangle , \tag{9.3}$$

where $\hat{\mu}_a$ and $\hat{\mu}_e$ are the absorption and emission unit transition dipoles, P_2 is the second Legendre polynomial and $\langle \ldots \rangle$ again indicates an ensemble average. For the case of Trps in Mb and Hb the energy transfer rate is time dependent, in principle, and the anisotropy is more correctly written as:

$$r(t) = \frac{2}{5} \frac{\left\langle \exp\left[- \int_{-\infty}^{t} k(\tau) d\tau \right] P_2(\hat{\mu}_a(0) \cdot \hat{\mu}_e(t)) \right\rangle}{I(t)} , \tag{9.4}$$

where $k(\tau) = k_f + k_{ET}(\tau)$ and $I(t)$ is the total emission rate at time t. Obviously the possible angular motions of fluorescent molecules span a wide range of frequencies. In the diffusive regime, such as would prevail when many solvent collisions occur for each increment of angular motion of the solute, it is possible to obtain explicit predictions for these correlation functions in terms of the anisotropic rotational diffusional coefficients [9.46]. The anisotropy is obtained experimentally as:

$$r(t) = \frac{I_{\parallel}(t) - I_{\perp}(t)}{I(t)} , \tag{9.5}$$

where $I_{\parallel}(t)$ and $I_{\perp}(t)$ are the fluorescence signals at time t after excitation for emitted polarization either parallel or perpendicular to the pump polarization.

The function $r(t)$ was recently obtained for the myoglobin tryptophan emission region with a time resolution of 16 ps [9.42]. The results showed a decay which was fit approximately by a single exponential decay constant of 830 ps. The results, shown in Fig. 9.8, do not show any discontinuous change in the anisotropy near $t = 0$, so that the experimentally available time resolution (15 ps) is probably much longer than the time required to achieve the diffusional regime. The observed $r(0)$ values, obtained by extrapolation of diffusion controlled data for all myoglobins studied until now, lie in the range 0.13–0.17. Such low values arise because of the differences in the directions of the absorption and emission dipoles [9.47].

The experimental anisotropy yields insufficient information to separate the contributions of Trp-7 and Trp-14, but since the lifetimes are known, there are

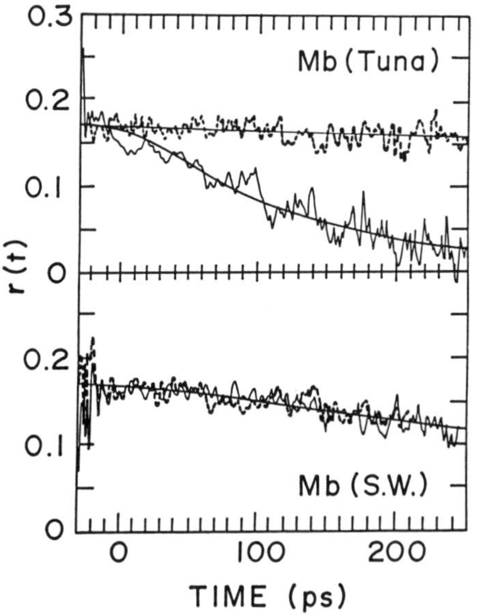

Fig. 9.8. Fluorescence anisotropy at 15 ps time resolution for Mb. *Lower*: Sperm whale Mb; solid line is met-Mb, dashed line is apo-met Mb. *Upper*: yellow fin Tuna Mb; solid line is met-Mb, dashed line is apo-met Mb. (After [9.42,45])

certain definite conclusions that can be drawn. First, the anisotropy at long times is dominated by the longer lived component which was assigned to Trp-7 on the basis of the energy transfer rate calculation. It can be concluded that Trp-7 has a rotational relaxation time, τ_R, of about 800 ps. Because of the form of $r_{tot}(t)$ for the system of two Trps:

$$r_{tot}(t) = \frac{r_7(t)I_7(t) + r_{14}(t)I_{14}(t)}{I(t)}\,, \tag{9.6}$$

where, for example, $r_7(t)$ has the form of (9.4) for Trp-7 alone, the information available about $r_{14}(t)$ is limited by the signal-to-noise of the experiment. If $r_{14}(t)$ decayed exponentially with a time constant less than 25 ps then $r_{tot}(t)$ would be non-monotonic, showing at 10–20 ps a 15% drop below the indicated exponential decay line. After a few lifetimes of Trp-14 the signal would become dominated by Trp-7. The data are probably good enough to exclude this but if the rotational relaxation time of Trp-14 exceeded $\sim$35 ps, the deviation from monotonic behavior would be so slight (less than 5%) as to be undetectable. Thus these studies allowed the conclusion: $\tau_R{}^{(7)} \approx 800$ ps; $\tau_R{}^{(14)} \geqslant 35$ ps.

In recent work [9.45] we showed that different results for the fluorescence anisotropy are obtained when a wavelength of 287 nm (rather than 265 nm) was used to excite the sample. In that case it appears as if the two tryptophans are not equally excited so that the effects on the anisotropy of a rapid rotational relaxation ($\sim$50 ps) from a tryptophan was readily observed.

The fluorescence of tuna Mb, which contains only Trp-14, was also obtained by 265 nm excitation and found to be dominated (91%) by a 31 ps component, supporting the conjecture that Trp-14 in the Mb containing both Trps is the short-lived component. In the tuna Mb the anisotropy decayed with a time constant of 110 ps, consistent with the analysis of $r_{tot}(t)$ given in the previous paragraph. Of course the structure of tuna Mb is not known, so these conclusions must be considered tentative.

Clearly much experimental work can still be done with picosecond pulses to evaluate motional characteristics of hemeproteins. In the case of heme proteins the energy transfer introduces a time dependent rate constant which allows contributions of different Trps to be separated and which, in principle, can be used to learn more about the potential surface describing the motions of the individual amino acid probes [9.48]. For example, the anisotropy decay in tuna Mb is much longer when the heme is removed such that energy transfer can no longer occur [9.42]. This has been attributed to the fact that certain conformations, while they may contribute significantly to the equilibrium ensemble average might not influence the anisotropy because they are caused to decay by energy transfer, whereas others which might not contribute much to the equilibrium ensemble could have favorable geometric factors to minimize energy transfer.

9.1.7 Molecular Dynamics Simulations

In a review of ultrafast processes in heme proteins it is important to consider the predictions from numerical experiments. Currently, molecular dynamics simulations offer the best opportunity to compare time resolved experiments with theory in cases where the evolution might be rate limited by alterations in the structure of the protein. In such simulations an empirical potential function for the whole protein is used to generate the forces needed to integrate Newton's equations for the set of nuclei. The calculations can incorporate directional hydrogen-bond potentials, dihedral angle distortions, electrostatic effects, and van der Waals interactions. Numerous reviews have been written on applications of these methods to proteins [9.49,50].

A first step in simulating structural changes induced by light in hemoglobin by molecular dynamics calculations was taken recently [9.51]. A structural response to the photodissociation of CO from HbCO was simulated by suddenly changing the Fe—CO potential function from a bonding to a dissociative one whilst running normal trajectories. The suddenly-formed local potential would force the Fe to an equilibrium position displaced by 0.55 Å from the heme plane in the absence of the protein. The simulations required input from quantum mechanical calculations of the iron out-of-plane motion. The trajectory was run for a whole subunit of HbCO (1271 atoms), and the equations of motion were integrated in steps of 2 fs. It was found that the iron displaces from the porphyrin plane with a time constant of 70–210 fs. These results reproduced in Fig. 9.9 support the interpretation of optical absorption studies mentioned earlier [9.10], that the iron could be displaced within 350 fs after the photolysis pulse.

The fluorescence of tryptophans in Mb and their concomitant energy transfer to the heme are dependent on the details of protein motion and therefore represent tests of molecular dynamics simulations. The fluorescence anisotropy in the absence of energy transfer is readily calculated from the trajectory of the transition dipole orientation obtained directly from a simulation [9.52]. For a heme protein the trajectory can be utilized to calculate $r(t)$ from (9.4).

Recently, a simulation was run for myoglobin with attention focused on the two Trps [9.53]. The calculation of the fluorescence evolution showed a non-exponential behavior which simplifies to the sum of two approximate exponential decays, one for each Trp. The calculation shows that Trp-14 decays faster than Trp-7, supporting the interpretation given to the experiments. Both Trps are subject to fluctuations in their distance and angular relation to the heme resulting in significant stochastic variations in the rate of fluorescence emission. The decay curve for Trp-14 showed slight deviations from exponential indicating that the average energy transfer rate, $\langle k_{ET}(t) \rangle$, is still somewhat smaller than the frequency of conformational fluctuations. Obviously, if the energy transfer time were very long compared with these fluctuations in structure, the decay constant would correspond to that from an average conformation and it would be exponential.

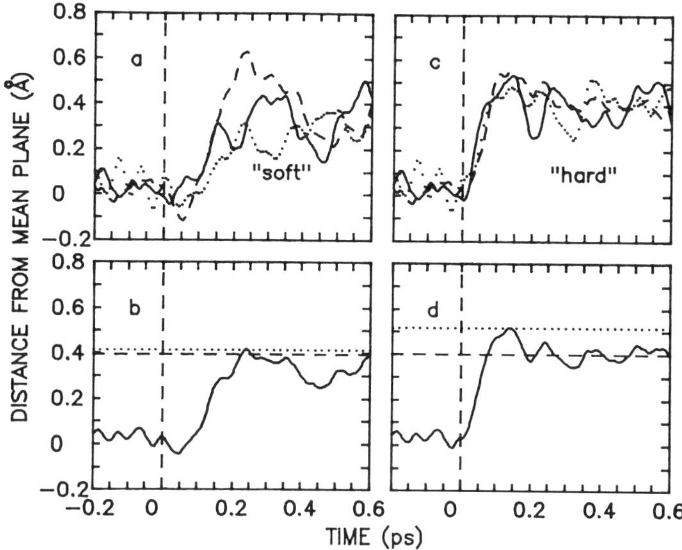

Fig. 9.9a–d. Photolysis of Hb_α subunit. The distance of the iron from the mean plane of the 4 pyrrole nitrogens is plotted versus time at 2 fs intervals. The vertical dashed line indicates the time at which the heme potential function is changed to simulate the initiation of photodissociation of the CO ligand. (**a**) The solid, dashed, and dotted curves correspond to photolysis at 30, 40, and 50 ps into the trajectory of the CO complex. (**b**) The solid curve is the average displacement at each time step following photolysis calculated from the three curves in (**c**). The horizontal dotted line is the average displacement calculated from the last 40 ps of a 70 ps trajectory of the α subunit of human deoxy-Hb A. The horizontal dashed line is the average displacement calculated for 9 ps, beginning at 1 ps after the potential function was changed to simulate photolysis. (**c** and **d**) Same as (**a**) and (**b**) except that a harder potential was used. (After [9.51])

The calculation of the fluorescence anisotropy has yielded some interesting challenges for picosecond spectroscopy. The simulation in Fig. 9.10 shows a nanosecond time scale for the decay of anisotropy except for the first picosecond or so, during which time there is a rapid drop to ~75% of the initial value. In these calculations no account is taken of electronic or vibrational relaxation effects so that the initial anisotropy is arbitrarily set at 2/5 corresponding to the case where the classical absorption and emission dipoles are parallel in the molecular frame. This rapid decay, which has not yet been observed in the laboratory, corresponds to the region of quasi-free rotation. The total anisotropy at longer times is similar to the experimental observations for sperm-whole met-Mb (shown in Fig. 9.8) whose coordinates were used to initiate the trajectory. The rotational relaxation (~20 ns) due to the overall motion of the whole protein is not included in the calculation. However the experimental indications that $r(t)$ for tryptophan-14 decays significantly more quickly than the total [9.42,45,48] are not predicted by the calculation.

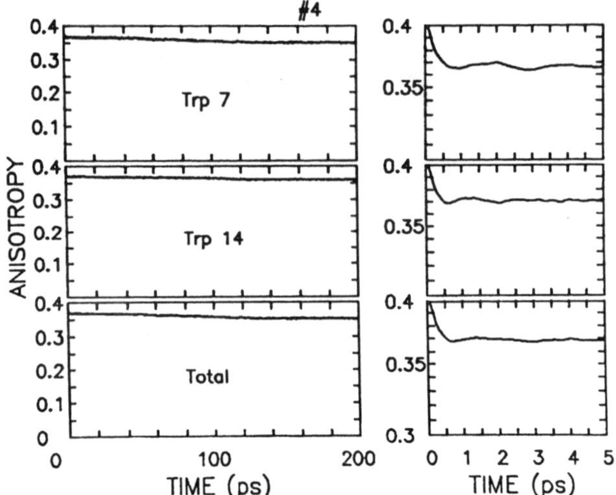

Fig. 9.10. Theoretical calculations of the fluorescence anisotropy for Trp 7 and Trp 14 in Mb obtained from molecular dynamics trajectories to evaluate Eq. (9.4). (After [9.53])

9.1.8 Geminate Recombination

After photolysis the dissociated ligand may recombine with the iron to which it was originally bonded; this is the geminate recombination. Since the concentration of ligands in the solution including the protein is known, the diffusion controlled rate of ligand binding can be estimated. Furthermore, the binding rate can be tested as to its dependence on the concentration of free ligand and the geminate part should be invariant. The time scales involved are such that recovery in unpressurized solutions on the timescale of a few nanoseconds or less can be confidently ascribed to geminate recombination.

At reduced temperatures the geminate rebinding can be slowed down and studied without the need for picosecond time resolution [9.54]. In the present review we consider only the dynamics under ambient conditions and leave aside the interesting question of how to connect these results with those obtained from frozen proteins.

The recombination of ligands with photolyzed heme proteins is influenced by both tertiary and quaternary structure changes. According to *Hofrichter* et al. [9.19] the quaternary structural change and the geminate recombination occurs on a time scale of 40–50 ns during which time all the released ligands have either left the heme pocket or have recombined. In the case of HbCO there does not appear to be any more than 10–15% of a subnanosecond recombination process occurring [9.13,14]. A similar situation prevails with MbCO [9.13,55]. A tertiary structural change is apparently occurring on the timescale of tens of nanoseconds. The change could be a result of the ligand leaving the heme pocket [9.19]. It seems reasonable to conjecture that many picoseconds after photolysis, the CO is somehow constrained to remain in the heme pocket where it is weakly bound to one or more sites [9.56,57]. It undergoes a slow reaction with the heme ($\sim$170 ns) which consumes about 30% of the ligand, the other 70% leaving the heme pocket with a time constant of $\sim$100 ns, accompanied by a tertiary structural change.

The situation with O_2 and NO ligands is quite different since they were discovered to show efficient subnanosecond geminate recombinations [9.14,58, 59]. For example, in the case of HbO_2 about 50–70% of photolyzed O_2 appears to recombine in a few hundred picoseconds. Why are the geminate reactions of O_2 and NO so much faster or efficient than that of CO? One obvious possibility is that the activation barriers for the ligand reaction at the iron are lowered for O_2 and NO ligands. *Cornelius* et al. [9.59] have suggested that one factor in lowering this barrier could be the reduction of the effective spin restrictions on the reaction that might prevail for CO($^1\Sigma$) combining with five ligand heme (^{5}A) to yield six ligand heme (^{1}A). Another contribution to the different rates of geminate rebinding of O_2 compared with CO in Mb and Hb was recently deduced from the infrared spectroscopy results discussed in Sect. 9.1.4. The occurrence of subnanosecond recombination of O_2 to Hb [9.14, 15] and Mb [9.272] suggests that O_2 may be presented a significantly lower steric barrier to recombination than CO. It was suggested that O_2 may bind when it approaches the iron on trajectories for which the angle between the O_2 axis and the heme normal is large, whereas the uprightness of the FeCO bond may require that the protein expand to allow the CO to bond. On the other hand, CO recombines to photolyzed protohemes in ca. 60 ps when the solvent has sufficiently high viscosity [9.273]. Much more work is required in this area. It will be possible to bring very accurate spectroscopic methods to bear on these reactions to observe directly the states of the diatomic molecule as well as those of the heme. A recent article by *Frauenfelder* and *Wolynes* [9.60] has summarized a selection of the issues pertinent to the geminate binding.

9.1.9 Photophysics of Hemes and Heme Proteins

A significant literature is beginning to take shape concerned with the photochemical and photophysical properties of metalloporphyrins. However the iron porphyrins still present many unsolved problems, largely because the photophysical processes are so fast. The Fe (II) and Fe (III) porphyrins have extremely small fluorescence yields and therefore undergo subpicosecond relaxation processes. These relaxation processes are believed to result from the interactions of states responsible for the strong visible and Soret bands with the $\Pi\Pi^*$; these are charge transfer and ligand field states that are suspected or known to lie at lower energies [9.61]. Very little information is available about the actual bottlenecks in the relaxation pathways. It is hoped that this situation can be remedied through femtosecond laser experiments.

Some spectral information was obtained for Fe(III)TPP Cl with 6 ps pulses [9.62]. A spectrum in the Soret region having a lifetime of 50 ± 20 ps was attributed to a triplet state. However most of the excitation followed a path back to the ground state on the subpicosecond time scale. More recently, Greene [9.63] has used 150 fs pulses to explore the same system. A transient spectrum corresponding to this triplet was observed to have a lifetime of 25 ± 10 ps. But this work also identified the subpicosecond transient with a lifetime of 750 fs which was suggested to be a singlet charge-transfer state. These transients were formed much faster than the 150 fs time resolution regardless of whether excitation was in the visible or the uv bands. This emphasizes the fs nature of relaxation processes in such systems. However, it must be realized that such changes occur on time scales faster than collisions can remove the vibrational energy excess, so the interpretations in terms of a few states need careful consideration (see below).

In order to obtain a complete picture from experiments on the molecular dynamics following photolysis, it is necessary to understand the photophysical processes which can occur on the same time scales. The first evidence for photophysics in the heme proteins was obtained by *Cornelius* and coworkers [9.31,32] who discovered that the spectrum of Mb from photolyzed MbO_2 did not appear immediately but instead, a species absorbing at 455 nm was observed. In this same study it was also reported that Mb itself displayed a short-lived transient.

The complexities of the early photophysics is further evidenced by the femtosecond work of *Martin* et al. [9.18]. In all the compounds they have studied they find that the Soret bleaching kinetics tracks the excitation pulse. Does the occurrence of bleaching imply that photodissociation has occurred? Obviously the answer is no: The bleaching of all absorption bands is instantaneous if the ground state molecules are directly transferred to another state by the excitation pulse. In addition, all species studied showed a deoxy-like species appearing in 300 fs. It seems unlikely that MbX_2 and HbX_2 derivatives ($X_2 = CO, NO, O_2$) would show the same excited state lifetime, so this result hints that the dissociation is much faster than 300 fs, perhaps as fast as the bleaching. However, in our view this is still an open question. *Martin* et al. [9.18] favor the interpretation that dissociation occurs in a time less than the shortest pulsewidth used of 50 fs. These authors also observe and associate a 1.5 ps lifetime to a 455 nm transient similar to that reported by *Cornelius* et al. [9.31] and assigned as an excited state of the five liganded species. Rather than generate the deoxy-like spectrum, this species decays directly back to the HbX_2 ground state. In addition, a transient precurser to the deoxy-like spectrum is observed at 480 nm which exhibits a lifetime of 350 ps. These transient intermediates must be assigned to either excited states of five or six liganded heme or to structural isomers of the ground states of these species, or to vibrationally hot states of any of these intermediates (see below).

A number of possible schemes are consistent with all of the available kinetic data. The one which was suggested by *Martin* et al., has the initially excited HbX_2 undergoing less than 150 fs dissociation into two metastable states of Hb (Hb_I and Hb_{II}). The Hb_{II} which absorbs at 480 nm (not at 455 nm) reforms HbX_2 in 2.5 ps, while the Hb_I, absorbing at 455 nm, is a precursor to the metastable

$Hb^{\dagger}$ which persists beyond the time period of the photophysics. $Hb^{\dagger}$ ultimately yields Hb as a result of tertiary and quaternary structure effects. Alternatively the species Hb_{II} might be an excited state of HbX_2 which does not dissociate and has a lifetime of 2.5 ps. The possibility should also be considered that both the observed absorption transients (455 nm and 480 nm) are actually excited states of HbX_2 and that the photodissociation actually takes 350 fs.

The absorption of light by heme proteins clearly initiates a complex series of photophysical and photochemical events which will require considerable experimental and theoretical effort to unravel. These transients absorb light in similar spectral regions to the starting materials so that secondary photochemical processes are expected to occur with high efficiency when heme proteins are irradiated with light pulses whose widths are longer than a few picoseconds. For example, the work of *Cornelius* et al. [9.31] with 6 ps pulses indicated that photodissociation could occur by absorption of a second photon by an excited state of MbO_2; this would not necessarily occur with 50 fs pulses. It follows that the dynamical and spectral properties of liganded heme proteins can be expected to be dependent on the nature of the pulses used in their photolysis.

9.1.10 Laser Induced Heating

The absorption of photons by a heme must result in an increase in temperature. If there were insufficient time for the chromophore to transfer its acquired energy and if the internal conversion and intramolecular vibrational energy redistribution were very fast compared with the thermal diffusion rate, the absolute temperature T of the molecule would be given by:

$$\sum_{j=1}^{3N-6} \frac{\varepsilon_j}{e^{\varepsilon_j/kT} - 1} = \langle E \rangle = nh\nu + \langle E_v \rangle , \tag{9.7}$$

where n is the number of absorbed photons of frequency ν, ε_j the energy of the jth vibrational state and $\langle E_v \rangle$ the average energy per molecule before excitation. The sum is over all the modes that are involved up to the time of the measurement. The temperature for $n = 1$ and $\lambda = 530$ nm is calculated to be 520 K for a typical porphyrin initially at 300 K. The possible static effects of significant temperature increases on Raman spectra were considered by *Asher* and *Murtaugh* [9.64], but there would be expected to be large effects on the optical spectra as well, and important dynamical effects due to cooling. The importance of such processes on picosecond spectroscopy is greatly dependent on the rate at which the number of modes containing the energy is increased by incorporating the surrounding medium. Since the hemes in Mb and Hb have some 90 van der Waals contacts, it can be expected that their behavior be comparable with molecules in solution or glasses.

There have been a number of reports of transitory hot molecules in solutions. Hot stilbene molecules were proposed to be the cause of the spectral narrowing of the S_1 absorption which did not settle down until ~30 ps after excitation with 6 ps pulses [9.65,66]. More recently *Gottfried* et al. [9.67] and *Kaiser* and *Sellmeier* have shown that anthracene molecules in solution, when pumped with ir pulses, take ~20 ps to cool. In experiments with azulene it was shown that cooling of the molecules continues for many tens of ps [9.68]. These results suggest that one should take very seriously the effects of molecular heating when interpreting ultrafast experiments of any type. It is easy to see that as a result of heating, difference spectra in the Soret region would appear in the long wavelength edge near to where the transients reported by *Cornelius* et al. [9.31,32,62] and *Martin* et al. [9.18] are located. It is also possible that such local heating can alter the early stages of the geminate reactions between heme and diatomics. This is another area where molecular dynamics simulations can have predictive value.

Recently, *Henry* et al. [9.69] have carried out molecular dynamics simulations on Mb and Cyt-c which predict readily measurable cooling rates following deposition of the energy from 530 nm or 353 nm photons into the ground state vibrations of the Π-electron skeleton. The calculated decays show significant cooling on time scales of 2–20 ps for all trajectories. Figure 9.11 shows the predicted temperature variations as a function of time for Mb and Cyt-c. The results predict that the effects of heating remain significant even for experiments with 30 ps pulses.

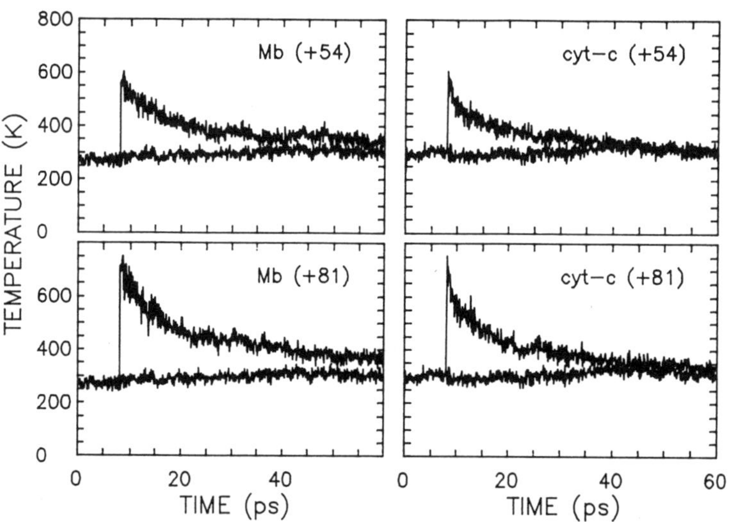

Fig. 9.11. Average kinetic temperatures of the 24 porphin skeleton atoms as a function of time. The plotted quantity is calculated from the total kinetic energy using $3Nk_gT/2 = \Sigma m_i v_i^2/2$ ($N = 24$). The upper plot in each panel shows the average of the results from 8 simulations and the lower plot shows the average of the results from corresponding simulations in which no energy was injected. Results are shown for injection of 54 and 81 kcal mole^{-1} corresponding to 530 and 353 nm photons. (After [9.69])

The remarks of this section are applicable to picosecond and subpicosecond spectroscopy of all the systems treated in this article and they emphasize the need for high quality spectroscopy as well as kinetic studies.

The calculations indicate that the effects of molecular heating by the laser pulse excitation should influence spectroscopic results from picosecond or subpicosecond experiments. The two forms of spectroscopy that are most often used to study heme proteins on this time scale are resonance Raman scattering and optical absorption. Raman transition frequencies change with temperature because of anharmonic coupling of modes having thermal populations. Usually the frequency shift has an Arrhenius temperature behavior. For the $1600\,cm^{-1}$ transition of a nickel porphyrin the activation energy is $\sim 500\,cm^{-1}$ and the pre-exponential factor is $76\,cm^{-1}$ [9.64]. This vibration corresponds to one of the so-called core marker bands which have been extensively studied in hemeproteins [9.24]. These data indicate by extrapolation that at 450 K the band would be shifted by $8\,cm^{-1}$ from the transition at 300 K. Presumably this full effect should be sensed by a subpicosecond pulse if the experimental method has sufficient signal-to-noise to permit the identification of peak positions of uncertainty-broadened transitions. It should be possible to observe significant shifts of the Raman transitions as a function of time. A number of Raman experiments have been carried out with 30 ps pulses for which the maximum observable Raman shift as a result of heating was estimated to be $\sim 2\,cm^{-1}$ [9.29]. In this case the calculations predict that the shift should be present during the pump pulse period. In recent work [9.274] *Petrich* et al. have obtained subpicosecond Raman spectra of photolyzed HbCO which show substantial shifts on the timescale 0.2 ps to 10 ps. These authors attribute the shifts to cooling of the heme.

Temperature effects on absorption spectra can also evaluated. When the pumping laser pulse produces electronically excited states which have characteristic spectra outside of the range of ground state absorptions, then a careful study of the spectral shape should provide information about the cooling of the molecule. On heating, absorption bands will typically broaden while retaining the same integrated intensity, shift slightly, or develop hot band structure if the Franck-Condon parameters are suitable. The experimental situation with heme proteins and porphyrins is complex experimentally because the stable and transient species all absorb in the same region and it is difficult to distinguish photophysical from photochemical processes on the basis of transient optical-difference spectroscopy. Most of the accurate picosecond time-resolved spectra of hemes have involved difference spectra in the region of the Soret band near 420 nm. Photolysis always yields a diminution of the peak intensity and new absorption on the wings of the band (see Fig. 9.2). Therefore heating would directly influence the observed heme protein transient absorption spectra. Recent studies on absorption transients of iron porphyrins [9.62,63] show spectra that contain features like those expected for heating, namely an apparent bleaching of the Soret band and the appearance of positive $\Delta(OD)$ on the wings of the absorption band.

9.1.11 Summary

The R $\rightarrow$ T transition in hemoglobin can be triggered by optical pulses that dissociate the ligand (O_2, CO) in $\sim$350 fs. Optical studies of heme properties have not detected further changes between tens of ps and a few ns. It is thus believed that the quaternary structure alterations occur later and that the protein structural changes within the R structure that influence the rate of ligand rebinding are completed by tens of picoseconds.

Fluorescence methods are being employed to study side-chain motion and energy transfer in proteins and to test molecular dynamics simulations. The simulations show rapid (<1 ps) decays in fluorescence anisotropy as a result of the interrupted free rotation of the chromophore. While such phenomena have not yet been observed in proteins, recent work on aniline in solutions [9.70, 71] has detected inertial rotation by fluorescence anisotropy. Subpicosecond fluorescence methods are clearly needed for the study of protein side-chain dynamics. The connections between molecular motion, such as measured by amino acid residue fluorescence, and the operation or allostery of hemoglobin remains to be established.

The photophysics of hemes in heme proteins contains a wealth of ultrafast phenomena which require a deeper understanding in terms of the electronic and nuclear structure. Present indications are that the photophysics is completed within a few ps following dissociation of the diatomic molecule. However, distinctions between photophysics and geminate chemical processes are difficult to establish [9.59] and require further study. The nature of the electronically based potential barriers to ligand formation and the identification of times cales where spin conservation is important remain to be elucidated experimentally and theoretically.

9.2 Photosynthesis

Photosynthesis is the conversion of light quanta into chemical energy by the oxidation of a hydrogen donor, and the reduction of CO_2. For purple and green bacteria the H donor is H_2S, or organic compounds such as succinate, while for green plants it is H_2O. To convert light into chemical energy, photosynthetic organisms must first absorb light and then rapidly convert the excitation into chemical potential before its energy is lost by radiative or nonradiative decay. Light is absorbed in these organisms by molecules called antenna pigments, predominantly chlorophylls supplemented by accessory photosynthetic pigments such as carotenoids and phycobiliproteins. Energy transfer in an array of antenna pigments brings the excitation to a reaction center where it is trapped. Here a series of electron transfers occurs, leading, in green plants, to the reduction of the nicotinamide adenine dinucleotide phosphate cation ($NADP^+$), the conversion of adenosine diphosphate (ADP) to adenosine triphosphate (ATP), called photophosphorylation, and the cleavage of water. These reactions, summarized as

$$2H_2O + 2NADP^+ \rightarrow 2NADPH + 2H^+ + O_2$$

are the so-called light reactions. The utilization of ATP and NADPH to reduce CO_2 proceeds with or without light (although regulated by light) by the photosynthetic dark reactions. Photosynthesis in purple or green bacteria proceeds by a less complex mechanism to reduce nicotine adenine dinucleotide (NAD^+) and phosphorylate ADP in green bacteria. In purple bacteria where the primary electron acceptor does not reduce NAD^+, photosynthesis drives the phosphorylation of ADP to ATP, which is ultimately used to reduce NAD^+ [9.95].

Photosynthesis has been the subject of intense research efforts throughout this century [9.96]. However, the details of the relationship between structure and electron-transfer mechanisms in bacterial reaction centers have only recently become understood in significant detail, while in green-plant reaction centers, many crucial links in the primary photochemistry remain to be established. The understanding of time scales involved in the photochemistry has depended on the development of ultrafast spectroscopic techniques supplemented by ESR and ENDOR. The ultrashort processes in photosynthesis are the energy transfer and trapping steps, and the initial electron-transfer reactions, which constitute the primary photochemistry of reaction centers. The approach to studying ultrafast reaction-center photochemistry has been to study preparations of reaction centers where interference from antenna pigments is reduced or eliminated. With the application of subpicosecond laser pulses, it has become possible to access the intial electron-transfer event. The ultrafast laser techniques applied to photosynthetic systems have primarily been fluorescence decay measurements as probes of antenna energy transfer and trapping, and transient absorption and linear dichroism measurements as probes of reaction centers. To achieve a full understanding of the ultrashort events in photosynthesis will require the correlation of information from a variety of ultrashort laser spectroscopies with detailed structural studies of reaction centers.

Since pure reaction-center preparations stripped of antenna pigments have been successfully prepared from bacterial photosystems but not from photosystems of green plants, the primary photochemistry of bacterial reaction centers is understood in greater detail. Our approach will be to discuss first the primary photochemistry in bacterial reaction centers, secondly antenna energy transfer, and thirdly, the primary photochemistry of green-plant photosystems.

9.2.1 Electron Transfer

Electrons are transferred in biological systems as a means of storing chemical energy. In photosynthetic reaction centers, electronic excitation is converted to chemical oxidation-reduction potential through a series of electron transfers. In mitochondrial membranes, chemical redox potential is released in an electron-transfer chain aimed at bringing about oxidative phosphorylation.

In photosynthetic reaction centers, electron transfer reactions have been observed which have vastly different reaction rates. A fundamental understanding of electron transfer requires knowledge of microscopic biological structure, potential energy surfaces, nuclear configuration, and electronic-vibrational coupling all of which participate in controlling the rates of electron-transfer reactions.

The theory of electron transfer rates in chemical and biochemical systems has been the subject of intense study since the 1950s. A classical rate theory of electron transfer was developed by *R.A. Marcus* [9.72,73]. The treatment assumes classical motion of nuclei along two intersecting potential energy surfaces. The probability of crossing from the potential surface of the electron donor to that of the acceptor when the interaction between the two surfaces is weak is given by the Landau-Zener formula. A rate expression analogous to that of absolute rate theory results:

$$k = \kappa Z \exp\left(-\frac{\Delta G^*}{kT}\right), \tag{9.8}$$

where κ contains the square of the electron transfer matrix element, Z is a vibrational frequency, $\sim 10^{13}\,\mathrm{s}^{-1}$ for a first order reaction, and ΔG^* is the free energy of activation. The Marcus expression for ΔG^*,

$$\Delta G^* = \frac{(\Delta G^0 + \lambda)^2}{4} \tag{9.9}$$

involves the free energy of reaction ΔG^0, and λ, which is the energy to reorganize the system from the equilibrium position of reactants to the equilibrium position of products without electron transfer, consisting of a contribution from internal vibrations of donor and acceptor and a contribution from low frequency solvent interactions. This form of ΔG leads to the celebrated Marcus inverted region, where the electron-transfer rate decreases with increasing driving force. When κ is large ($\kappa \approx 1$), the system crosses from one zero-order potential energy surface to another with unit probability, and the reaction is called adiabatic (only the ground potential energy surface of the interacting system is traversed). For small κ, the system may traverse the interaction region many times on the donor's zero-order surface before electron transfer occurs, and the reaction is called "nonadiabatic".

Is classical electron-transfer rate theory adequate to describe biological electron-transfer processes? Biological electron-transfer reactions bring new aspects to the electron-transfer problem as originally considered by Marcus. Firstly, many important biological electron transfers are thought to be non-adiabatic ($\kappa \ll 1$), due to the large distances (> 5–10 Å) over which the electron must be transferred, whereas Marcus' theory, as originally conceived, was an adiabatic theory. Secondly, while Marcus' theory regards molecular vibrations classically, it was necessary in describing biological electron transfer to include configurational changes in high-frequency vibrational modes [9.74, 75]. Thirdly, the primary photo-induced electron transfer in reaction centers has been observed [9.76,77] to proceed at a rate approaching $10^{12}\,s^{-1}$, close to the theoretical maximum of $\sim 10^{13}\,s^{-1}$ [9.78]. This rate is comparable to the rate of vibrational relaxation following excitation. In this case, the competition between vibrational relaxation and electron transfer requires attention [9.79,80].

Classical electron transfer rate theory was challenged by the discovery of *DeVault* and *Chance* [9.81,82] that the rate constant for cytochrome oxidation in the photosynthetic bacterium *Chromatium vinosum*, about $10^{-6}\,s^{-1}$ at room temperature, becomes nearly temperature independent below 100 K. *Hopfield* [9.74,83] and *Jortner* et al. [9.75,79,84–86] developed theories with quantum treatments of vibrations. Although they treated coupling with high frequency vibrations somewhat differently, both incorporate a transition from tunneling [9.87] at low temperature to activated transfer at high temperatures. Their rate expressions, which have the form

$$W = \frac{2\pi}{\hbar}|V|^2 F \ , \qquad (9.10)$$

where V is the electronic coupling matrix element and F is a real number incorporating vibrational wave-function overlap (Franck-Condon factor), were used to fit the cytochrome oxidation data of DeVault and Chance.

With mode-locked lasers, electron-transfer reactions faster by several orders of magnitude than the DeVault-Chance reaction could be studied. The electron-transfer rate from bacteriopheophytin to quinone in reaction centers, $\sim 5 \times 10^9\,s^{-1}$, was reported to be nearly independent of temperature between 4 and 300 K [9.88] or to decrease slightly with increasing temperature in this range [9.89]. *Redi* and *Hopfield* [9.90] and *Buhks* et al. [9.75,80,84] accounted for this temperature independence by adjusting the vibrational coupling energy to cancel the electronic energy gap. The result is an activationless process that can be described without recourse to nuclear tunneling. *Warshel* [9.91] has applied these theories to bacterial photosynthesis by calculating potential energy surfaces for the primary electron donor, a bacteriochlorophyll dimer, and the bacteriopheophytin intermediate electron acceptor. Two coordinates were considered: one the separation between the two components of the dimer, and the other a vibrational coordinate of the acceptor. Coordinate shifts along both were considered important for fast forward electron transfer and slower reverse electron transfer. More elaborate theories have been developed which incorporate several molecular vibrational modes. The inclusion of low-frequency modes as well as higher frequency intramolecular modes seems to account for the observed temperature dependence [9.92]. In another modification vibrational modes are allowed to change frequency and to undergo equilibrium coordinate shifts during the electron transfer reaction [9.93].

The rates of the primary electron-transfer steps in *R. viridis* reaction centers have recently been measured with subpicosecond resolution [9.112]. Picosecond electron transfer was observed. These are the fastest electron-transfer rates measured to date. Clearly the initial electron transfer competes effectively with vibrational relaxation. *Jortner* and co-workers [9.79,80,94] have outlined a theory allowing activationless electron transfer from an initial state not thermally equilibrated. Whether this theory can account for the temperature dependence and excitation-energy dependence of picosecond and subpicosecond electron transfer in reaction centers remains an open question.

9.2.2 Bacterial Reaction Centers

The most thoroughly investigated of bacterial photosynthetic systems have been these of the purple non-sulfur bacterium *Rhodopseudomonas* and *Rhodobacter*. Methods of preparing pure reaction-center extracts unencumbered by antenna pigments have been developed. These reaction centers consist of three polypeptide chains of molecular weight 20 000–30 000, designated H, L, and M, of which the H subunit is not essential to the photochemistry [9.95–97]. The LM particle contains four bacteriochlorophyl (BChl) molecules, two bacteriopheophytins (BPh), which are chlorophylls without the central Mg atom, two quinones (Q) (ubiquinone, in *R. sphaeroides*), and an iron atom. In experiments on these preparations, reaction centers are excited directly by ultrashort light pulses, rather than by energy transfer from chlorophyll pigments. The validity of this approach depends on the degree to which photoexcitation of reaction-center components other than the primary electron donor interferes with the primary photochemistry.

Structure of Bacterial Reaction Centers. Since crystallization of the reaction centers proved difficult, the orientation of the components in the reaction center was investigated first by spectroscopic techniques [9.98]. Evidence from several sources, including ESR and ENDOR, led to the belief that two BChl molecules are paired together to form a dimer called the special pair. The special pair was identified as P870, the primary electron donor detected by bleaching at 870 nm upon illumination of purple bacteria. Our understanding of the structure of reaction centers was sharpened recently by the successful crystallization and x-ray structure analysis at 3 Å resolution of the reaction center of *Rhodopseudomonas viridis* [9.99,100]. This structure, shown in Fig. 9.12 opens the way for the first time to relate in detail ultrafast kinetics in photosynthetic electron transfer to the spatial arrangement within the reaction center. The structure shows at the center the special pair of BChl molecules. About 13 Å away (Mg–Mg distance) are the monomeric BChl molecules, in contact with the BPh molecules (11 Å Mg–Mg distance). A quinone (MQ) is located near one of the BPhs and near a non-heme Fe atom. Another quinone near the other BPh may have been lost during crystal preparation. At this level of resolution, the structure appears

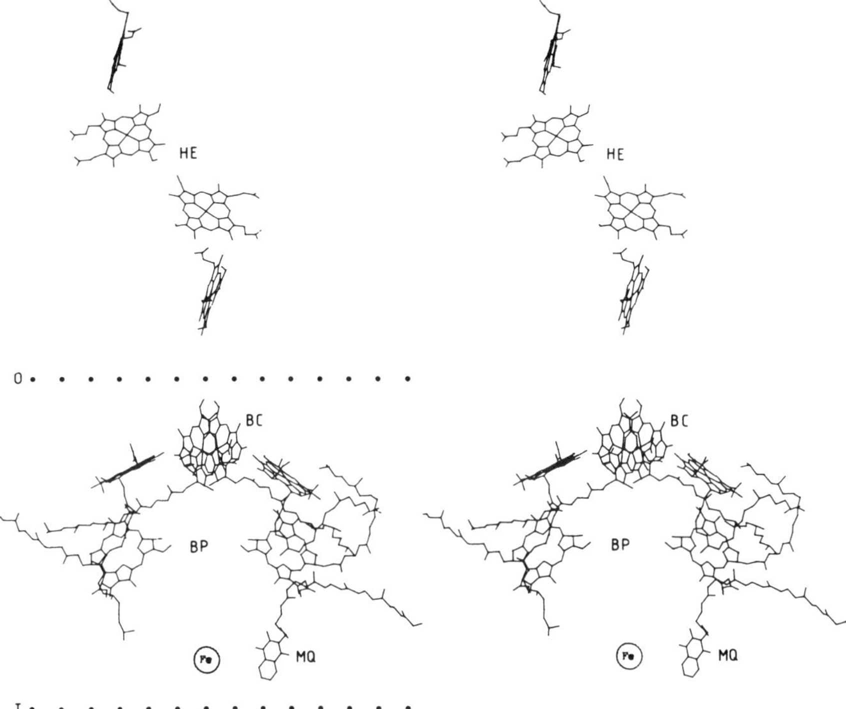

Fig. 9.12. Stereo representation of the spatial arrangement of prosthetic groups in the bacterial reaction center of *Rhodospcudomonas viridis*. Shown arc four hcmc groups (IIE), thc bactcriochlorophyll special pair (BC) and two monomeric bacteriochlorophylls, two bacteriopheophytins (BP, menaquinone (MQ), and the non-heme Fe. The lines O and I mark the supposed outer and inner membrane boundaries. (By permission of H. Michel and J. Deisenhofer.)

to show two symmetric electron-transfer pathways. However, in low temperature absorption spectra, where two Bph Q_x and Q_y absorption bands are discernable, bleaching is observed in only one band [9.101,102]. The electron-transfer pathway thus seems to follow one side of the structure shown in Fig. 9.12. The heme groups are part of the cytochrome subunit of the reaction center and presumably function in the cytochrome reduction of the special pair. More recently, structures were also reported for *R. sphaeroides* [9.103, 275, 276].

The primary photochemistry of the purple bacterium reaction center is represented schematically in Fig. 9.13. Our understanding of the kinetics of the early electron-transfer steps in Fig. 9.13 comes principally from laser flash photolysis. In Table 9.1 the absorption maxima of the components of *R. viridis* and *R. sphaeroides* are tabulated.

Primary Electron Transfer. In order to trace the electron-transfer mechanism back to the primary donor, it was necessary to use flash-photolysis techniques.

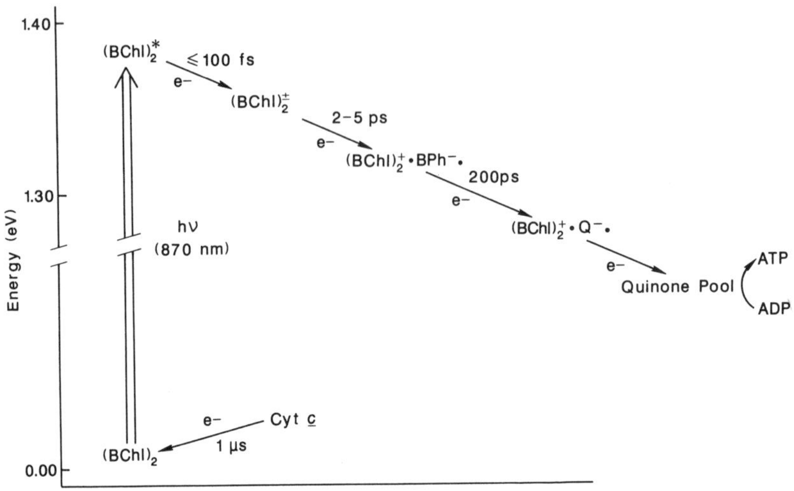

Fig. 9.13. Energy-level scheme and lifetimes of intermediates in the electron-transfer sequence in bacterial photosynthesis

Table 9.1. Absorption maxima for reaction-center components

	R. sphaeroides		*R. viridis*	
	Q_x	Q_y	Q_x	Q_y
$(BChl)_2$	610 nm	870 nm	620 nm	960 nm
BChl	600 nm	800 nm	610 nm	830 nm
BPh	545 nm	760 nm	545 nm	790 nm

This could be done with nanosecond lasers by artificially reducing the quinone electron acceptor [9.104], or it could be done with ultrashort laser pulses [9.105–108]. These experiments showed that the primary donor was bleached within the time-resolution of the laser pulses, but allowed the electron-transfer rate to quinone to be measured [9.109,110]. As shorter light pulses have become available, it has become possible to identify earlier events in the primary photochemistry.

Recently experiments with 100 to 150 fs laser pulses were reported on reaction centers of *R. viridis* and *R. sphaeroides* [9.76,111–113,277,278]. In studies on *R. viridis* [9.76] with a chemically reduced quinone, 150 fs pulses from a colliding-pulse dye laser excited the Q_x absorption band of $(BChl)_2$ and the overlapping Q_x band of the BChl monomers at low excitation levels. Pulse-limited bleaching and recovery of absorption were attributed to excitation of $(BChl)_2$ followed by charge separation within the special pair. An increase in absorption with a 5 ps risetime was interpreted as electron transfer from the special pair to a bacteriopheophytin [9.76]. The 5 ps risetime of BPh^- absorption was also observed in transient-absorption experiments at other wavelengths with 4 ps pulses [9.76]. Similarly, with direct excitation of the special pair at 950 nm in pre-reduced reaction centers, bacteriopheophytin reduction was found to occur simultaneously with changes in the BChl *b* monomer absorption around 830 nm [9.277]. Both were observed with 6-ps kinetics (at 450-fs time resolution), suggesting that only a single electron-transfer step occurs to reduce BPh.

Working with *R. viridis* reaction centers in their natural oxidation state, researchers in two groups found 2.8-ps disappearance of BPh and appearance of BPh^- [9.111,278]. The special-pair excited state was observed to decay concurrently in 2.8 ps [9.279]. Again, evidence was found for only one electron-transfer step leading to BPh^-. The fastest event observed in these experiments was bleaching of the special-pair absorption. Regardless of whether the reaction centers were excited in the special-pair, monomeric BChl, or BPh Q_y absorption bands, bleaching of the special pair was observed within the 100-fs resolution of the experiment [9.279].

Experiments on *R. sphaeroides* [9.111–114,77] also resolved the initial electron transfer from the special pair. With 150 fs excitation pulses at 865 nm to excite the Q_y band of the *R. sphaeroides* $(BChl)_2$ special pair, electron transfer to BPh was observed in 2.8 ps [9.111,112]. A simultaneous 2.8 ps decay of 970–950 nm stimulated emission and 2.8 ps absorption rise at 1240 nm, attributed to the decay of the special-pair excited state and formation of the $(BChl)_2^{+\cdot}$ radical cation, suggested that electron transfer occurs directly from the special pair to BPh. Similarly a 4 ps decay of stimulated emission around 920 nm was observed after excitation of *R. sphaeroides* with 0.8 ps, 610 nm pulses [9.77]. Within experimental error, this lifetime matches that for the formation of BPh^- and $(BChl)_2^+$ formation [9.114,77] and $(BChl)_2$ fluorescence decay [9.115]. No evidence was found for electron transfer to BChl prior to the formation of BPh^- [9.111–113,77].

Since the absorption spectra of *R. sphaeroides* and *R. viridis* differ in several details, one may ask whether there are differences in the early electron-transfer kinetics of these species. Experiments to compare the optical absorption changes during the first few picoseconds following excitation have uncovered no discernable differences in the kinetics of electron transfer from $(BChl)_2$ to BPh [9.111–113,278]. The decay of excitation from the special pair, and the appearance of BPh^- absorption occur with a 2.5–3.0 ps time constant in both species with the quinone in its oxidized state, and in 4.4–6.0 ps with a chemically reduced quinone [9.76,111–114,277,278].

These results can be compared with the crystallographic structure of the *R. viridis* reaction center (Fig. 9.12). The structure suggests an electron transfer pathway:

$$(BChl)_2 \rightarrow BChl \rightarrow BPh \rightarrow Q \quad .$$

Is an electron transferred from the special pair to the BChl monomer in a separate step preceding reduction of BPh? No convincing evidence for a state $(BChl)_2^+$

$BChl^-$ has been found [9.112,116,117]. Apparent bleaching around 800 nm in earlier reports [9.119] may have resulted from intense excitation, or from a blue shift of BChl absorption upon electron transfer to BPh, and not from bleaching of BChl. Observations near the isosbestic point of this blue shift [9.112,116,117] do not show the bleaching of BChl absorption that would be expected upon electron transfer to BChl.

The results of femtosecond experiments together with the bacterial reaction center structural determination raise the challenge of relating the mechanism of the primary electron transfer to the known structures. This challenge is currently stimulating a fascinating debate [9.279–282]. Is the electron transferred from the special pair to BPh by super-exchange via virtual states of the BChl monomer [9.280,281]? Or, is it transferred sequentially by a slow step (2.8 ps) to BChl, followed by electron transfer to BPh that is fast enough (say ~0.5 ps) so that the intermediate $BChl^-$ could not be detected [9.279]? *Scherer* and *Fischer* [9.282] proposed a third possibility: that the state formed in 2.8 ps is a charge-transfer state of monomeric BChl and BPh, $BChl^+BPh^-$, which is excitonically coupled to the special pair excited state. This state must decay to the radical pair $(BChl)_2^+BPh^-$ in less than 1 ps. The issue is still unresolved.

What is the role of the BChl monomer if it is not an intermediate in the electron-transfer pathway? While calculations intimated that the state $(BChl)_2^+$ $BChl^-$ lies above the $(BChl)_2^\pm$ charge-transfer state, [9.118] it has been suggested that the role of the monomer is to lower the barrier for electron transfer to BPh [9.77]. Furthermore, the nature of the state responsible for the 2.8 ps kinetics is not known. It may be an excited state of $(BChl)_2$, or a charge-transfer state $(BChl)_2^\pm$, or perhaps involves some BChl monomer character [9.119,120]. Delocalization of excited states of the special pair by mixing with monomer states [9.118] could serve to facilitate electron transfer to BPh.

It is interesting to compare the time-resolved studies of the primary photochemistry with recent frequency-domain experiments on *R. sphaeroides* reaction centers by hole-burning [9.121,122] and accumulated photon-echo experiments [9.121] at low temperature. Hole widths of around 400 cm^{-1} in the Q_y band of the $(BChl)_2$ dimer were recorded by both groups, who concluded that this width represents the homogeneous width of transition. If there is no contribution of pure dephasing at these temperatures (<2 K), this width corresponds to a 25 fs lifetime. This was interpreted as the lifetime for charge separation within the $(BChl)_2$ dimer. However a recent theory and analysis of these hole-burning results by *Small* and co-workers [9.283–285] showed that the primary electron donor states exhibit strong electron-phonon coupling characteristic of charge-transfer states. The large hole width is explained by electron-phonon coupling, so that an ultrafast charge separation need not be invoked. Similarly large hole widths were observed for the primary donor of photosystem I, with a narrow zero-phonon hole, consistent with the strong electron-phonon coupling model [9.283].

Electron Transfer to Quinone. The electron transfer *from* BPh to quinone (ubiquinone in *R. sphaeroides*, menaquinone in *R. viridis*) has been the subject of numerous studies [9.88,89,100,101,123–125]. Transient absorption attributed to BPh was found to decay in 200 ps at room temperature for reaction centers of *R. sphaeroides* and *R. viridis*, and in shorter times at lower temperatures.

Recently *Holten* and coworkers have published a series of papers on the kinetics of the $(BChl)_2^+ BPh^- \rightarrow (BChl)_2^+ Q^-$ electron transfer [9.102,126–132). Figure 9.14 illustrates the transient absorption spectrum after 600 nm excitation of *R. sphaeroides* before (30 ps) and after (1.6 ns) electron transfer from BPh to Q [9.130]. The kinetics of this electron transfer were measured by the decay of bleaching at 545 nm due to the BPh Q_x band and the decay of absorption at 695 nm from the BPh^- absorption. The temperature dependence of the electron-transfer kinetics is shown in Fig. 9.15. The lifetime decreases from 230 ps at 295 K to about 100 ps at 100 K and stays constant at lower temperatures [9.130]. This

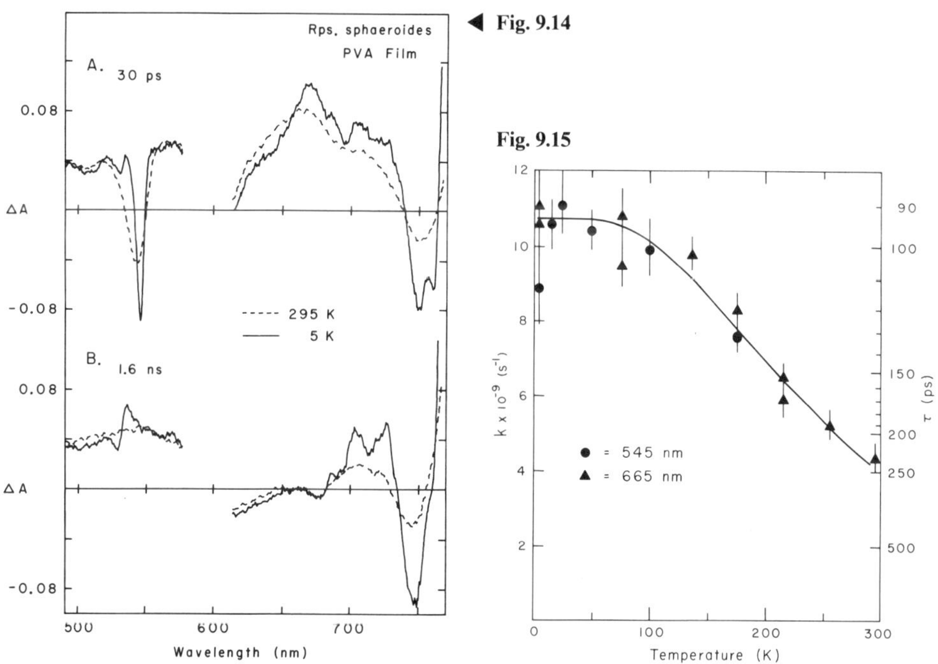

Fig. 9.14. Absorption change in reaction center of *R. sphaeroides* 30 ps (A) and 1.6 ns (B) after excitation at 600 nm. (From *Kirmaier* et al. [9.102])

Fig. 9.15. Temperature dependence of the rate of electron transfer from bacteriopheophytin to quinone in *R. sphaeroides*, measured by the decay of bleaching at 545 nm and the decay of transient absorption at 665 nm. The data were fit to a rate expression of *T. Kakitani* and *H. Kakitani*, [9.93] (*solid curve*). (From Kirmaier et al. [9.102])

behavior was compared with activationless electron-transfer theories [9.79,85, 86,92,93]. Given the observed temperature dependence these theories imply that $300\,\text{cm}^{-1}$ vibrational modes are responsible for structural changes of donor and acceptor, or that these modes change frequency as a result of structural changes along coordinates of higher-energy vibrations. On the other hand, a recent study of the coupling of protein phonon modes to the electron transfer led to the conclusion that the primary contribution to the nuclear reorganization energy is from $\sim 100\,\text{cm}^{-1}$ phonons [9.286]. The electron transfer to Q was also measured for *R. viridis* and was found to occur with a lifetime of 175 ps at 285, 76, and 5 K [9.132]. The structural or chemical basis for the difference between this temperature dependence and that of *R. sphaeroides* remains to be established.

Experiments are only beginning to broach the question of the effect of the protein on the dynamics of electron transfer, and of the motion or structural changes of the reaction-center pigments during the electron-transfer sequence. As the relationship between structure and dynamics is clarified, these questions will provide an intriguing subject of study.

9.2.3 Antenna Energy Transfer

The task of collecting light energy from the sun is performed by the antenna pigments. In green plants and algae, these are found in the thylakoid membranes of chloroplasts. In photosynthetic bacteria, they are found in intracytoplasmic membranes or in special vesicles in the cell [9.95]. The principal antenna pigments are chlorophyll *a* and *b* in plants, chlorophyll *c* in some algae, and bacteriochlorophyll *a*, *b* or *c* in bacteria. Other pigments, called accessory pigments, are carotenoids and phycobiliproteins. Following light absorption by a pigment molecule, the electronic excitation is transferred until it is trapped by a reaction center. This energy transfer has been described in the language of molecular excitons [9.133,134] as incoherent hopping by Forster energy transfer [9.135,136].

Energy Transfer in Chlorophyll Arrays. Although the light harvesting arrays transfer excitation energy efficiently, they also fluoresce, thereby providing a natural probe of the pathways leading to the excitation of the reaction centers [9.137]. Mode-locked lasers have been used to excite fluorescence, which was detected first by gating [9.138–140] and later by streak cameras [9.141–144]. These early studies necessitated high excitation intensities, which led to anomalously short lifetimes due to the mutual annihilation of excited states [9.145–147].

Significant advances became possible with the use of time-correlated single-photon counting and synchronously pumped lasers [9.148]. The approach of this work has been to fit the recorded fluorescence decay times to sums of exponential decay components. The challenge posed by these results is to assign the decays to components of the light-harvesting or antenna arrays. Most of these studies have explored fluorescence from green plants or algae, where two photosystems, photosystem I (PS I) and photosystem II (PS II), are operative. Since each has associated antenna pigments, the fluorescence may contain contributions from several components of a complex system of pigments.

The overall fluorescence yield is sensitive to the state of the reaction centers. The limiting low fluorescence quantum yield occurs at low excitation levels with open reaction centers. At high light intensity, some reaction centers are unable to be involved in energy transfer because they are undergoing photochemistry. These reaction centers are closed. Under these conditions, the photochemical quantum yield is at a minimum, and the fluorescence quantum yield is at its maximum value. A similar increase in fluorescence quantum yield can be induced by chemically blocking electron transfer in reaction centers of PS II.

Applications of time-correlated single-photon counting methods to these systems showed that three exponential decays are required to fit the fluorescence decays adequately [9.149–154]. At low incident intensity, decay times of about 100, 400 ps, and 1–2 ns were typically measured, with a relatively small contribution from the slowest component. In the highly fluorescent state, induced by chemical inhibition, by high excitation intensity, or by potentiometric titration [9.154], the quantum yield of the slow component is enhanced several fold, accompanied by an approximately two-fold increase in lifetime. The slow component is thus responsible for the increased fluorescence yield. The other two decay components are much less affected by closing PS II reaction centers. Efforts were made to explain this behavior and to assign the fluorescence decay components in terms of models of the light-harvesting and antenna arrays of green plants. It was proposed that the fast decay component was from Chl *a* antenna molecules associated with PS I [9.151], PS II [9.153,155] or both [9.149,150, 156]. It was suggested that the middle component derives from light-harvesting Chl *a*/*b* protein complexes loosely associated with PS II [9.149,150,153], or from isolated PS II units [9.155,156].

Our understanding of the origins of the flourescence-decay components has been enhanced by several recent experiments [9.157–161]. In the experiments by *Fleming* and collaborators [9.159], fluorescence decays were measured for mutant strains of the green alga *Chlamydomonas reinhardii* which lacked either PS I, PS II, or both. The 680 nm fluorescence decays from wild type *C. reinhardii*, and from strains lacking one or both photosystems were fit by triple exponential decays. The fluorescence decay of wild algae could be simulated as a sum of the fluorescence decays of the three mutant strains with approximately equal weights from PS I and PS II, and a small contribution from the mutant strain lacking both photosystems. It was concluded that the decay time of about 100 ps in these and similar experiments, is an average of a shorter decay time (53 ps) and a slightly longer decay time (152 ps), due to Chl *a* molecules closely associated with PS I and with PS II respectively. Fast decay originating in both photosystems was also found in the fluorescent emission from the green alga *Chlorella vulgaris* [9.160]. The fast decay component consisted of emission around 685 nm attributed to Chl associated with PS II decaying in about 180 ps, and emission around 700 nm attributed to Chl associated with PS I decaying in about 80 ps. A similar 700 nm emission contribution to the fast decay was reported for *Chlorella pyrenoidosa*, whereas the longer decay components were found to emit around 685 nm, characteristic of PS II [9.161]. The excitation spectrum of the 180 ps component suggested that Chl *a* antenna are coupled to the Chl *a/b* light-harvesting complex [9.160]. The middle decay time of about 400 ps was assigned to larger Chl array systems less well coupled to the photosystem [9.159],

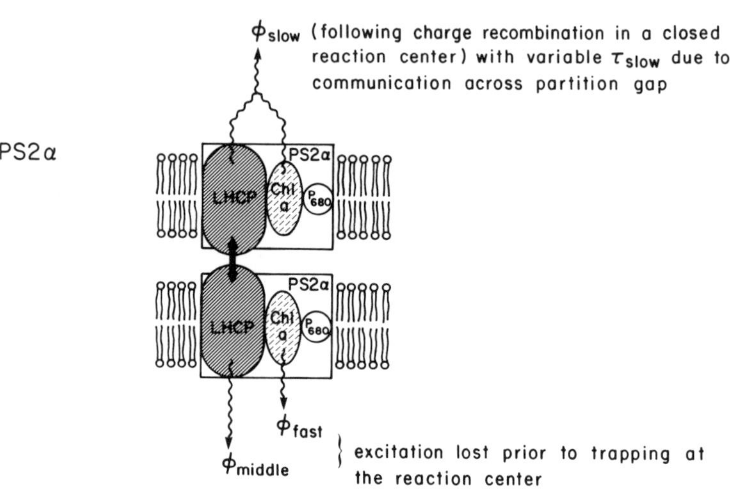

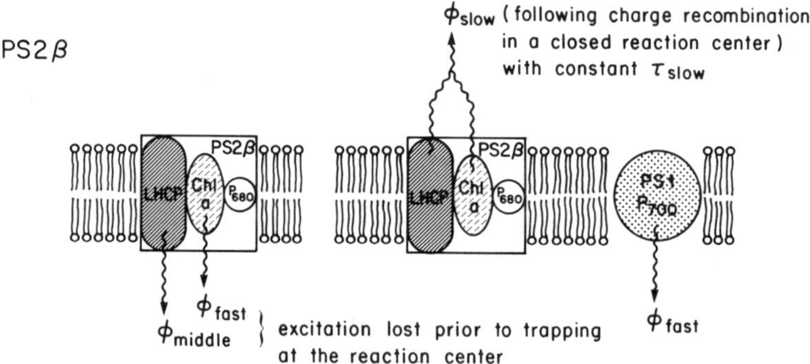

Fig. 9.16. Model of the antenna systems in thylakoid membranes. Two types of photosystem II reaction centers are proposed: PS2α centers in stacked thylakoids and PS2β centers in unstacked thylakoids. Excitation is trapped by P680 in photosystem II, and P700 in photosystem I. The fast fluorescence decay component, ϕ_{fast}, is attributed to chlorophyll *a* (Chl a) antennae of photosystem II and to light-harvesting pigments of photosystem I (PS1). The middle decay component, ϕ_{middle} is attributed to chlorophyll *a/b* light-harvesting complexes (LHCP) of photosystem II. The slow component, ϕ_{slow}, follows from charge recombination in photosystem II reaction centers. (From *Karukstis* and *Sauer* [9.158])

or to PS II associated with a different antenna system than that responsible for the fast decay [9.160]. *Fleming* and co-workers also calculated the single-step energy-transfer time and the average number of visits made by an excitation to the photosystem before trapping [9.159]. They estimated single-step times of 100–700 fs for PS I, and an average of 2–4 visits to the reaction center before trapping of the excitation.

Sauer and co-workers have investigated the origin of the variable fluorescence quantum yield by probing the dependence of the fluorescence decay kinetics on a number of chemical and physical factors [9.152,154,157,158,162]. Since the variable fluorescence of the long-decay component is sensitive to the state of the reaction center of PS II, they attribute this fluorescence to closed reaction centers. In this picture, excitation which is trapped at a closed PS II reaction center leads first to charge separation, but, since the electron-transfer pathway is closed, charge recombination results, repopulating an excited state which may transfer its energy back to the antenna system where fluorescence may occur. Two forms of PS II are hypothesized, PS II_α centers, which can transfer excitation to other reaction centers through light-harvesting arrays, and PS II_β centers, which are isolated from other reaction centers [9.154,158]. This model is illustrated in Fig. 9.16. The fast fluorescence decay is attributed to both PS I and PS II, in agreement with recent results [9.159,160]. The middle decay time in this model is associated with light-harvesting chlorophyll *a*/*b* arrays. The slow component is supposed to result from charge recombination.

Energy Transfer in Phycobilisomes. Cyanobacteria (blue-green algae) and red algae utilize antenna pigments called phycobilins packed into complexes called phycobilisomes, which are attached to the photosynthetic membranes. Phycobilisomes contain several hundred bilin chromophores, linear tetrapyrroles attached to the protein. These proteins are organized into disks which are themselves stacked into rods, with disks containing shorter-wavelength pigments on one end, and longer-wavelength pigments at the other end next to a central core [9.163]. Thus the shorter-wavelength absorbers, phycoerythrins (PE), 570 nm, are on the outside, phycocyanins (PC), 630 nm, within them, allophycocyanins (APC), 650 nm, in the core, followed by Chl *a* inside the photosynthetic membrane. This organized structure efficiently transfers energy into the core. Since the structures are well understood, and the fluorescence of each component is distinct, detailed studies of energy transfer in phycobilisomes have been undertaken. With ultrashort pulses and streak-camera detection it was possible to observe the rise and decay of fluorescence from each of these components in the red alga *Porphyridium cruentum* following excitation at 530 nm of PE [9.164, 165]. Recently obtained fluorescence spectra of *P. cruentum* are shown in Fig. 9.17 as a function of time following excitation of PE [9.166]. The progression of the fluorescence from PE to PC, then to APC, and finally to Chl *a* is clearly distinguishable in the 500 ps time span. Decay times for the first two steps were measured to be 60 ps and 40 ps respectively [9.167]. The rise-time of PC fluores-

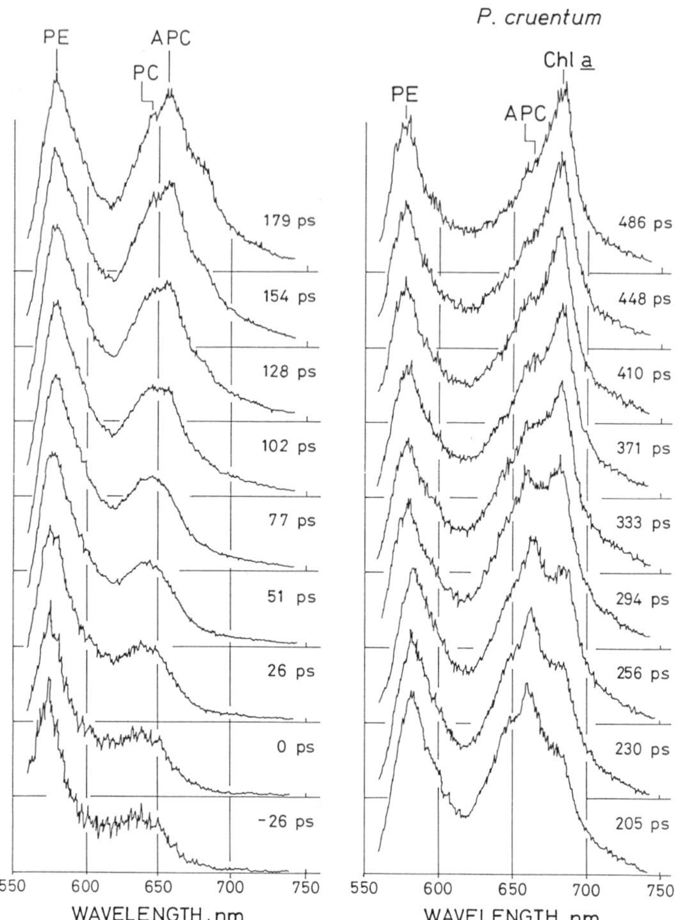

Fig. 9.17. Fluorescence spectra of phycobilisomes and chlorophyll *a* in *P. cruentum* after excitation at 540 nm, time-resolved by time-correlated single photon counting. (From *Yamazaki* et al. [9.166])

cence approximately matched the decay of PE fluorescence, and the rise-time of APC fluorescence approximately fit the decay of PC. Fast decay of absorption anisotropy, in 12 ps, was attributed to energy transfer within PE protein units.

In the cyanobacterium *Anacystis nidulans*, also known as *Synechococcus*, energy transfer from PE, excited at 570 nm, to the terminal acceptor PC occurs in 56 ps, as measured with a streak camera [9.168]. Since energy transfer within isolated phycobiliproteins was measured to be faster than the 8 ps time resolution, disk-to-disk energy transfer was thought to be the rate limiting step. The rate of interdisk energy transfer was deduced by comparing fluorescence or absorption lifetimes in wild-type *Synechococcus* with decays in mutant strains lacking PE or PC disks [9.168–170]. Differences of from 30–50 ps in the transients were observed. Similar times were observed for fluorescence decay of PC and rise of APC fluorescence. This time was thought to be the rate-limiting energy transfer from PC to the APC core [9.170,171]. An approximately 10 ps absorption recovery and a 10 ps absorption anisotropy decay component were thought to be due to energy transfer within phycobiliprotein units [9.170], or to energy transfer from APC to a terminal acceptor [9.163].

Some attention has been given to the functional form of fluorescence decay in phycobilisomes. Decays have been fit to multiple exponential functions and to $\exp(-At^{1/2})$. *Yamazaki* et al. [9.166] analyzed the time-resolved fluorescence spectra from *P. cruentum* and *A. nidulans* into components due to each phycobiliprotein. They found that an $\exp(-At^{1/2})$ decay, which is anticipated for Forster energy transfer under certain conditions, fit the decay kinetics well.

9.2.4 Reaction Centers in Green Plants

In green plants reaction centers of two interconnected photosystems, PS I and PS II, work together in an electron-transfer chain. Although particles containing one or the other photosystem have been prepared, it has not proved feasible to remove all the antenna pigments from these particles. As a result, studies of isolated green-plant reaction centers comparable to those of bacterial reaction centers have not been possible.

Chemical reduction or background illumination of particles of chloroplasts enriched in PS I or PS II, allow the characterization of the early electron donor and acceptors in PS I and II [9.109,172]. For example, the difference spectrum for oxidation of the primary donor in PS I has been obtained [9.109]. The primary donor P700, absorbing around 700 nm, is now regarded as a Chl *a* monomer [9.173]. Evidence has also been found for a Chl intermediate electron acceptor [9.174]. Picosecond studies of PS I enriched particles found formation of $P700^+$ within 15–30 ps of excitation [9.175–179]. This was thought to be the time for excitation trapping at the reaction center and initial charge separation. The initial electron acceptor decayed in about 100–200 ps [9.179,180]. The difference spectrum for this decay appeared to be that of a Chl *a* anion, although shifted 30 nm to the red [9.177]. This result appears to confirm the identity of the intermediate electron acceptor as Chl *a* (rather than a pheophytin, as in bacterial reaction centers), which transfers an electron in about 200 ps to an acceptor thought to be an Fe-S center [9.109].

The ultrashort processes of PS II are even less well understood than those of PS I. The electron-transfer pathway in PS II seems to bear closer resemblance to that in bacterial reaction centers than does the electron-transfer system in PS I [9.173]. The intermediate electron acceptor, for example, is thought to transfer an electron on to a quinone, in this case plastoquinone. The ultrashort photochemistry in PS II has not yet been probed directly, although the reduction of pheophytin has been shown by nanosecond techniques to occur in less than 2 ns [9.181].

9.3 Rhodopsin and Bacteriorhodopsin

9.3.1 Introduction and Background

The highly specialized biological processes responsible for vision are based on the photochemistry of a class of proteins known as rhodopsins. Ultrashort light pulses have helped generate a detailed knowledge of the natural function of this class of biological molecules which react sensitively to light. The light-sensitive membrane protein found in the cells responsible for vision contains in all known cases the molecule retinal or 3-dehydroretinal attached to the protein by a Schiff-base linkage. In vertebrates the visual pigment rhodopsin is located in disk membranes in the outer segment of rod and cone cells. *George Wald* [9.182] first demonstrated the importance of retinal in visual transduction and showed that retinal isomerizes from the 11-*cis* to the all-*trans* configuration during the bleaching sequence initiated by optical excitation of the pigment.

Another light-sensitive protein was discovered in the purple membrane protein of the bacterium *Halobacterium halobium* [9.183]. These purple membranes, which appear when the bacterium is deprived of oxygen, allow it to synthesize ATP in the absence of oxygen. *Stoeckenius* and co-workers [9.183] showed that the purple membranes contain retinal and constitute a protein remarkably similar to rhodopsin, called bacteriorhodopsin (bR). As for rhodopsin, the photochemistry of bacteriorhodopsin results in isomerization. Light-adapted bacteriorhodopsin was shown by extraction of the chromophore from the protein, to contain all-*trans* retinal while extraction of retinal from the dark-adapted bacteriorhodopsin yields a mixture of 13-*cis* and all-*trans* retinal [9.184]. *H. halobium* contains two other proteins with retinal chromophores [9.185]. Halorhodopsin serves as a chloride pump [9.186], and slow cycling rhodopsin initiates phototactic response [9.187].

9.3.2 Rhodopsin and Bacteriorhodopsin Photochemistry

Rhodopsin Bleaching Sequence. The sequence of events which follows light absorption by rhodopsin leads to the release of retinal from the protein. This is the bleaching sequence shown in Fig. 9.18. Intermediates in the bleaching sequence were identified first by their low-temperature absorption spectra [9.188,189]. As the temperature is raised after photoexcitation of rhodopsin at low temperatures, a series of intermediates can be identified by their shifted absorption spectra, beginning with the red-shifted species bathorhodopsin. Each intermediate is stable below a transition temperature. In vertebrates the sequence terminates with retinal in the all-*trans* configuration ejected from the protein opsin. *In vivo* all-*trans* retinal is isomerized enzymatically back to the 11-*cis* configuration, which binds to opsin to regenerate rhodopsin. The configuration of retinal in rhodopsin was shown to be 11-*cis* by the reconstitution of rhodopsin from opsin and 11-*cis* retinal [9.190]. The isomer 9-*cis* retinal also binds to opsin, forming

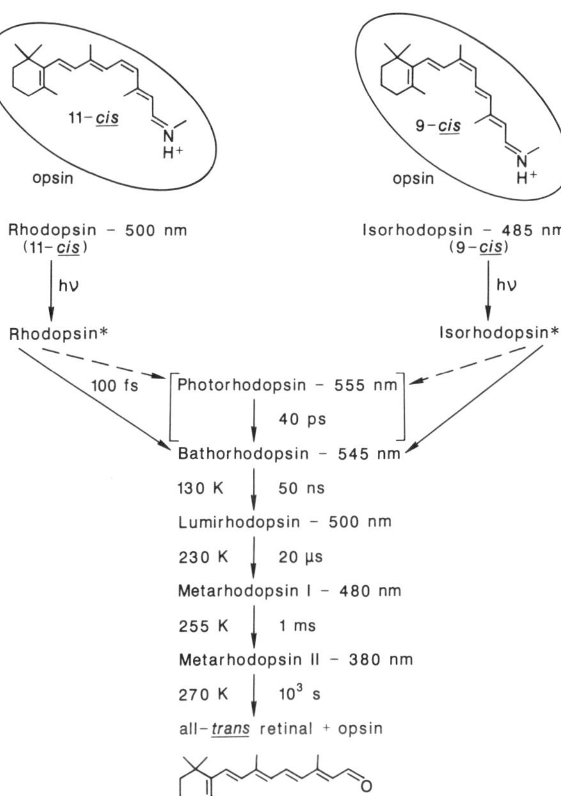

Fig. 9.18. The photobleaching sequence in rhodopsin. The transition temperature is the temperature below which an intermediate is stable. The possible formation of a precursor to bathorhodopsin, photorhodopsin [9.235], is represented by dashed lines

the artificial pigment isorhodopsin [9.182]. Since both isorhodopsin and rhodopsin are converted photochemically to bathorhodopsin [9.185,191], it has been long believed that the conformation of retinal in bathorhodopsin is all-*trans* [9.192].

While the sequence of visual intermediates was first identified spectroscopically as a sample of rhodopsin which had been exposed to light was warmed, flash photolysis was required for the determination of lifetimes. The lifetimes of the intermediates were investigated beginning over a decade ago by laser flash photolysis [9.193–196]. Lifetimes of the intermediates in the visual sequence are shown in Fig. 9.18. Further discussion of the visual bleaching sequence is available in a number of review articles [9.197–200].

Bacteriorhodopsin Photocycle. The intermediates in the bacteriorhodopsin photocycle, like those in the rhodopsin sequence, have been studied by low-temperature spectrophotometry. In many respects the sequence of intermediates in bacteriorhodopsin has proved to be parallel to the sequence in rhodopsin. A red-shifted photoproduct K is formed at low temperatures, and decays to other, spectrally

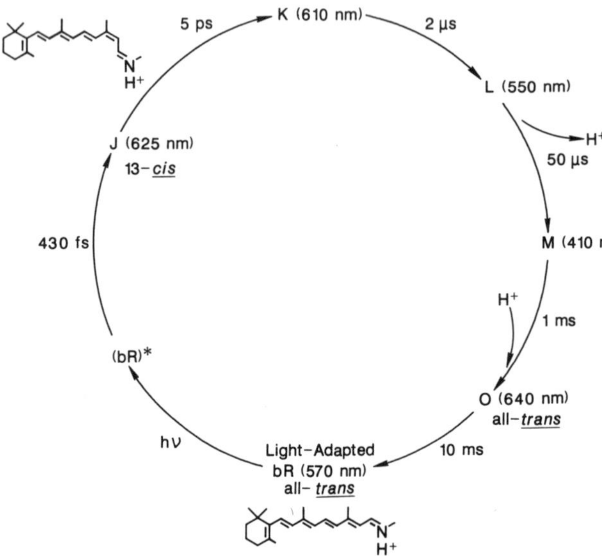

Fig. 9.19. The photocycle of light-adapted bacteriorhodopsin (bR). Recent data are incorporated for the formation times of the intermediates *J* and *K* [9.256,257]

distinct, species as the sample is warmed (see Fig. 9.19). Unlike the rhodopsin sequence, however, bacteriorhodopsin photochemistry is cyclic, and does not result in the expulsion of the chromophore from the protein. Flash photolysis and transient spectroscopy were used to detect intermediates in the bacteriorhodopsin photocycle [9.201–204]. A detailed discussion of the bacteriorhodopsin photocycle is given in *Ottolenghi's* review [9.197].

Isomerization from the all-*trans* configuration in light-adapted bacteriorhodopsin (bR) to 13-*cis* is known to occur sometime during the bacteriorhodopsin photocycle since chromophore extraction from the intermediate *M* yields 13-*cis* retinal [9.205]. The identification of isomerization with the primary photochemical event was proposed first by analogy with rhodopsin [9.192], although no isorhodopsin analogue exists for bacteriorhodopsin from which to argue for isomerization based on a common batho intermediate.

Resonance Raman Spectroscopy. Resonance Raman spectroscopy has played a major role in identifying intermediates in the rhodopsin sequence and in the bacteriorhodopsin cycle, and has produced the most detailed structural information yet available for rhodopsin and the visual intermediates, providing strong evidence for a twisted all-*trans* conformation of retinal in bathorhodopsin [9.206], and for the 11-*cis* and 9-*cis* conformations in rhodopsin and isorhodopsin respectively [9.207,208]. This work, as well as resonance Raman evidence that the Schiff base of retinal is protonated in both rhodopsin and bathorhodopsin [9.209,210], has been invoked to restrict models of primary processes of vision to those incorporating isomerization of a protonated retinal Schiff base, and strongly suggests that the primary photophysical process does not include transfer of a proton to or from the retinal Schiff base (although this interpretation is not universally accepted [9.211]).

The resonance Raman spectrum of the bacteriorhodopsin intermediate *K* has been obtained at room temperature [9.212] and at 77 K [9.213,214]. These studies indicate that *K* contains 13-*cis* protonated retinal Schiff base. Thus, as for rhodopsin, the resonance Raman evidence indicates convincingly that isomerization plays a role in the primary photochemistry. Proton transfer from the Schiff base in bacteriorhodopsin occurs later, with the formation of the intermediate *M*.

Methods of obtaining time-resolved Raman spectra on microsecond and millisecond time scales were developed to obtain Raman spectra of bacteriorhodopsin intermediates as a function of time after photoexcitation. *El-Sayed* and his group observed the deprotonation of the Schiff base and the appearance of *M* over a microsecond time scale [9.215], and the decay of the unprotonated Schiff base associated with *M* on a millisecond time scale [9.216] by chopping a cw excitation beam. *Marcus* and *Lewis* [9.217], by varying the flow rate of sample through the excitation beam, observed the kinetics of Schiff base deprotonation and the appearance of bands associated with the intermediate *M*. Photolytic interconversion of intermediates in the bR photocycle and the photolytic generation of new species from photocycle intermediates was detected by time-resolved Raman spectroscopy by *Grieger* and *Atkinson* [9.218]. The extension of time-resolved Raman techniques to the picosecond time scale is an important recent advance in the study of the photochemistry and subsequent dynamics of bacteriorhodopsin and will be discussed in the next section, along with other picosecond methods.

Models of Early Steps of Visual Transduction and Proton Pumping. Models of the primary photochemistry of rhodopsin [9.192,219–221] and bacteriorhodopsin [9.220,222,287,288] in which photo-isomerization of retinal occurs rapidly, leading to separation of charge between the protonated Schiff base and a counter ion in the protein, account qualitatively for the observed red shift of the first intermediate and are consistent with the resonance Raman evidence and the thermochemistry of the visual process. *Warshel*, in a series of papers dealing with both bacteriorhodopsin and rhodopsin [9.219,221,222], examined the effects of electrostatic interactions on the ground- and excited-state potential energy surfaces of retinal. These interactions were proposed to alter the potentials so as to stabilize the retinal chromophore in a twisted geometry, as shown in Fig. 9.20, with resulting separation of charge. The reaction pathway was envisioned arriving at charge separation during the primary photochemical process by twisting of the retinal double-bond coordinate either leading to isomerization or to a somewhat twisted form of the original retinal conformation. *Honig* et al. [9.220]

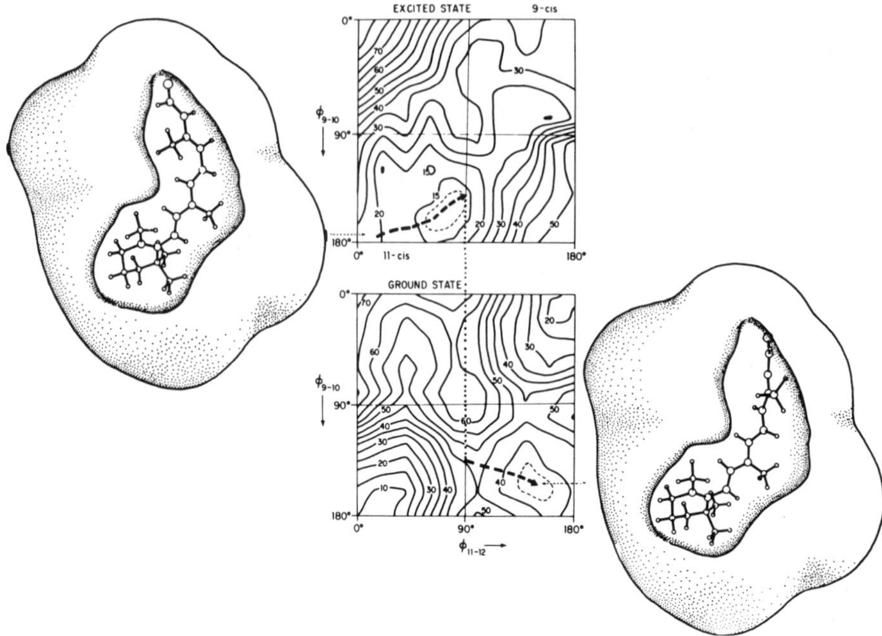

Fig. 9.20. Excited-state and ground-state potential surfaces calculated for rhodopsin. A pathway is shown (*dashed and dotted lines*) for the formation of bathorhodopsin in a twisted configuration by concerted motion about two double bonds. (From *Warshel* and *Barboy*, [9.223])

argued for isomerization leading to charge separation, with subsequent proton transfer. In this picture, isomerization breaks an ionic bond between the protonated Schiff base and a counter ion in the protein and stores energy by charge separation. It was suggested that a subsequent proton transfer, perhaps near the counter ion, generates the intermediate bathorhodopsin. *Warshel* and *Barboy*'s treatment [9.221] of the primary step of vision, simulating protein constraints and electrostatic interactions during isomerization, yielded ground and excited potential surfaces along two double-bond rotational coordinates (see Fig. 9.20). These surfaces account for energy storage during the primary step and permit calculation of resonance Raman intensities of the intermediate bathorhodopsin.

For bacteriorhodopsin, *Schulten* and *Tavan* [9.287] proposed that proton pumping in *H. halobium* involves concerted photoisomerization of light-adapted bacteriorhodopsin about the 13–14 double bond and 14–15 single bond. Their model is based on semi-empirical calculations and has been refined in subsequent work [9.288–290]. In this model, protonation or deprotonation of the Schiff base for the various intermediates, and the position of counterions conspire to direct the dark and light-adapted photocycles along specific C—C bond torsions. As of this writing, the question of the configuration, s-*cis* or s-*trans*, about the 14–15 single bond in the intermediate *K*, is still in dispute [9.291,292].

9.3.3 Picosecond Experiments

a) Visual Pigments

Optical Absorption Studies. Until picosecond lasers were used to study rhodopsin photochemistry, the time scale of the events involved was only known to be less than a microsecond. The first application of picosecond lasers in biology [9.223], was the observation of transient absorption at 560 nm within a few picoseconds of light absorption by rhodopsin, showing that the formation of a photoproduct of rhodopsin is very fast. Since then our understanding of the ultrafast photochemistry and its relation to biologically important events has been refined by a number of experiments with ultrashort pulses. The picosecond work on rhodopsin up to about 1980 is described by *Peters* and *Leontis* [9.224]. A recent article by *Yoshizawa* and *Shichida* summarizes picosecond experiments through 1984 [9.225].

Early picosecond studies [9.223,226–228] detected a red-absorbing intermediate, identified as bathorhodopsin, appearing faster than the 6 ps or so resolution of these experiments. If the primary photochemical event is *cis-trans* isomerization, these experiments showed that this isomerization occurs with extraordinary rapidity. A molecular dynamics simulation based on the so-called "bicycle-pedal" model for isomerization involving several bond coordinates, however, calculated an isomerization time of 0.2 ps [9.229]. This calculation and other simulations demonstrated that it is conceivable for isomerization to occur in less than 1 ps (see Fig. 9.21). Another simulation [9.230] resulted in a calculated isomerization time of 2 ps along a quantum-chemically calculated excited-state potential surface for the C_{11}–C_{12} coordinate only. Objections that 1 ps or less is too short a time for isomerization have thus been laid aside.

The observation of a picosecond-time-scale intermediate whose absorption is consistent with the low-temperature spectrum of bathorhodopsin led to efforts to resolve the formation of this intermediate. *Peters* et al. [9.228] cooled the sample to 4 K and observed that the rise time of a batho intermediate slowed to 36 ps. Their temperature-dependent data on the kinetics of the formation of this intermediate showed non-Arrhenius behavior at low temperatures. To test the idea that proton tunneling might be responsible for such behavior, they examined the kinetic behavior of rhodopsin washed and suspended in D_2O. The formation time of the deuterated batho intermediate was reported to be significantly slower (260 ps at 4 K), also with a non-Arrhenius temperature dependence. This experiment led to a re-examination of the models of the primary photophysical and photochemical events in vision and seemed to some authors to contradict earlier photochemical and resonance Raman evidence for isomerization. Other authors preferred to assign the primary event of vision to isomerization, and to explain the isotope effect by later events in the protein. *Honig* et al. [9.220] and *Warshel* and *Barboy* [9.219,221] thought that isomerization occurs extremely quickly, even at 4 K, on a barrierless excited-state potential surface, producing a ground-state intermediate that decays to bathorhodopsin by proton transfer in the

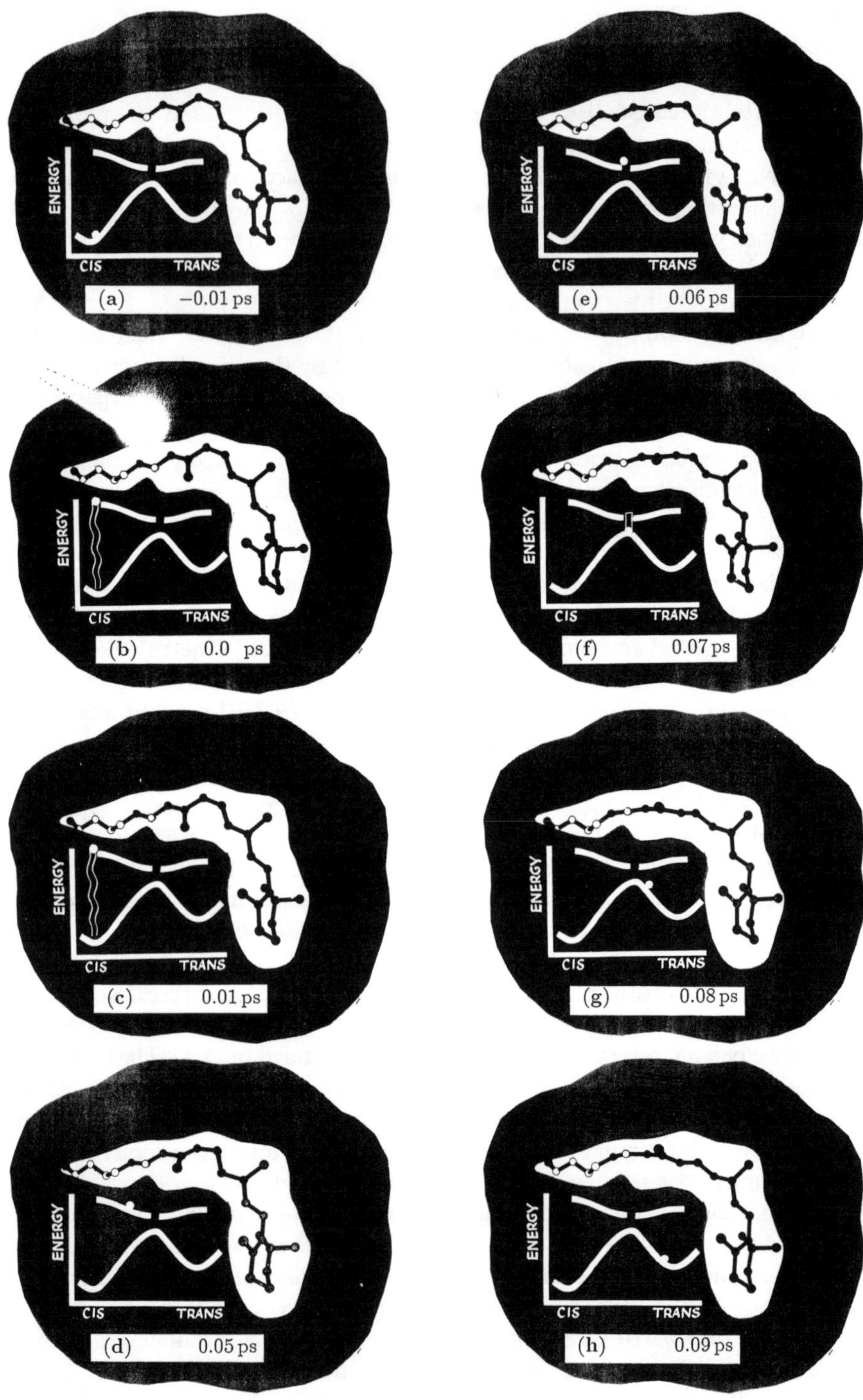

Fig. 9.21. Caption see opposite page

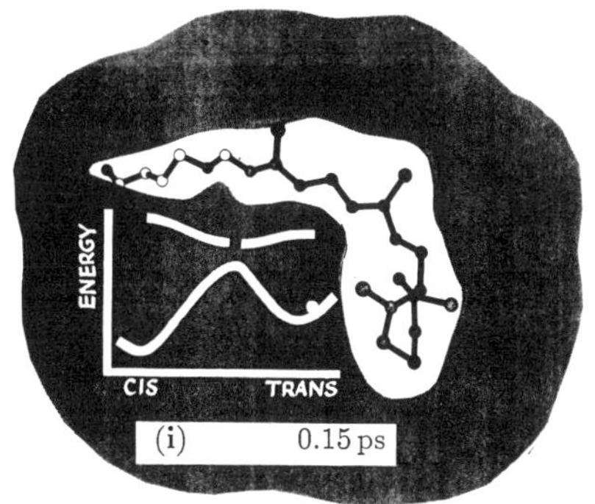

Fig. 9.21a–i. Molecular dynamics simulation of photoisomerization of retinal in rhodopsin. The structure of the chromophore and trajectory along the potential surfaces are illustrated from 0.01 ps before absorption of a photon to 0.15 ps after absorption. (We are grateful to A. Warshel for supplying this figure)

process observed by Peters and co-workers. This idea is consistent with the evidence of low-temperature fluorescence kinetics [9.231] that the precursor to bathorhodopsin is a ground-state species. The low-temperature isotope effect in bathorhodopsin, however, should be confirmed with low-intensity excitation pulses. In view of the suggestion (see below) that an isotope effect is induced in bacteriorhodopsin by high-intensity excitation, further work is needed to clarify the role of proton transfer in the formation of bathorhodopsin.

Since the longest-standing argument for primary photoisomerization is based on the common batho intermediate, apparently formed from both rhodopsin and isorhodopsin (9-*cis* retinal rhodopsin), several recent experiments have compared the transient absorption spectra of bathorhodopsin formed from rhodopsin and from isorhodopsin. The observation that bathorhodopsin appears within 3 ps of excitation of both rhodopsin and isorhodopsin [9.232] strengthened the argument for the primary photoisomerization mechanism based on a common batho intermediate. One analysis, which compared the transient spectra 85 ps after excitation [9.233], indicates a possible 10 nm shift between batho intermediates. Other workers, however, concluded that their spectra of bathorhodopsin from rhodopsin and isorhodopsin were identical [9.234]. They suggested that the difference between these results could be due to differing methods of preparing isorhodopsin.

Since the protein pocket must be slightly different in isorhodopsin and in rhodopsin, their transient absorptions may be expected to differ at very early times. Protein relaxation may then lead to a common intermediate for rhodopsin and isorhodopsin. The rate of protein relaxation around the isomerized retinal molecule is not known at present, but may be reflected by the reported decay of a red-shifted precursor to bathorhodopsin [9.235]. This precursor, called photorhodopsin (see Fig. 9.18), was reported to decay to bathorhodopsin 40 ps after excitation of bovine rhodopsin and between 250 and 500 ps after excitation of squid rhodopsin [9.235]. Confirmation of the existence of a ground-state precursor to bathorhodopsin would contribute to refining models of the primary events in vision.

Fluorescence and Raman Evidence. The picosecond fluorescence studies of rhodopsin reported by *Doukas* et al. [9.231,236,237], are consistent with isomerization occurring at a rate of 10 ps^{-1}. A similar fluorescence decay measurement on rhodopsin prepared with a synthetic retinal incorporating a bridge preventing isomerization, however, yielded very different results [9.238], without the very fast fluorescence decay occurring in natural rhodopsin. These results support photoisomerization as the primary photochemical event, though its rate constant can only be estimated from them.

One picosecond Raman study of rhodopsin has been reported, also pointing to rapid photoisomerization. Using a 30 ps pulse to excite and probe the sample, *Hayward* et al. [9.239] obtained resonance Raman spectra that resembled the bands reported for bathorhodopsin with cw lasers. The pulse intensity necessary for this experiment, however, exceeds by at least an order of magnitude the limits for the linear excitation regime. Nevertheless, this study also implicates rapid photoisomerization.

Multiphoton Processes. The danger of generating intermediates whose properties are not biologically relevant is illustrated by work on squid rhodopsin. Measurements of absorbance changes as a function of laser power showed that saturation of the absorption occurs above about 0.03 GW/cm^2, or 0.1 to 0.2 photons per molecule [9.240]. Although some earlier studies [9.226,241–243] reported that transient absorption was linear in laser intensity, all preceding experiments used intensities that exceeded these limits, often by one to two orders of magnitude. The apparently contradictory results, especially concerning the possibility of a blue-absorbing precursor to bathorhodopsin called hypsorhodopsin [9.232,241–244], left the nature of the primary photophysics and photochemistry of vision in doubt. *Yoshizawa* and *Shichida* [9.225] pointed out that the high laser excitation intensity might induce non-physiological photochemistry such as multi-photon processes. The study of *Matuoka* et al. [9.240] detected a blue-shifted intermediate in squid rhodopsin only when higher laser intensities were used. They concluded that hypsorhodopsin is not an early intermediate in the visual sequence, but rather an artifact of laser intensity, to which squid seems more susceptible than bovine rhodopsin.

b) Purple Membrane Pigments

Optical Absorption. The issues relating to the ultrafast dynamics of bacteriorhodopsin are analogous to those regarding rhodopsin, and concern the nature of the primary photophysics and photochemistry, the nature and lifetimes of early intermediates, the rate of isomerization, the possibility of proton transfer, and the existence of a possible isotope effect. The earliest picosecond work showed that the primary photochemistry occurred within a few picoseconds [9.245]. Transient absorption showed a 6 ps pulse-limited absorption rise time at 635 nm at room temperature [9.246] and pulse-limited bleaching at 570 nm at 68 K followed by a 20 ps recovery [9.247]. The transient absorption spectrum 13 ps after excitation [9.246] showed the existence of a red-shifted absorber nearly matching the microsecond absorption spectrum of the intermediate K [9.201]. The rise time of transient absorption was measured by subpicosecond techniques at 615 nm to be 1.0 ps $\pm$ 0.5 ps [9.248]. Thus the batho intermediate of bacteriorhodopsin appears rapidly, like that of rhodopsin. An intermediate preceding K, analogous to the precursor of bathorhodopsin, was also detected [9.249–251]. Upon excitation of bacteriorhodopsin, a transient absorption was observed within the 6 ps pulse at 660 and 700 nm. This transient appeared to decay in about 11 ps, simultaneous with an 11 ps rise time of transient absorption around 600 nm [9.249]. These results suggested a slightly red-shifted precursor J to the batho intermediate K. The appearance of transient absorption at 630 nm was measured with 4–8 ps pulses to be less than 2 ps [9.252]. The relaxation of transient bleaching at 576 nm was also reported to be <2 ps. Deuteration did not significantly change this result, and, in particular the 18 ps transient absorption rise time previously observed around 570 nm [9.249] in deuterated bR, was not reproduced. The reverse transformation from K to bacteriorhodopsin was measured at low temperature and was reported to occur within the 30 ps excitation pulse duration for both protonated [9.253] and deuterated [9.254] samples. The results were interpreted to imply that the reverse transformation does not proceed through the J intermediate.

The formation time of K is reportedly slowed to 36 ps at 4 K, with a non-Arrhenius temperature dependence [9.249]. Deuteration was reported to slow the formation time of K to 18 ps at room temperature and 88 ps at 4 K. These results led the authors to propose, as they did for bathorhodopsin [9.228], that proton transfer is responsible for the formation of K. While transfer of Schiff-base proton seems inconsistent with resonance Raman evidence [9.216–218], other proposals, analogous to those for rhodopsin, accounted for the reported isotope effect by suggesting that transfer of another proton near the chromophore is associated with the $J \rightarrow K$ transition following photoisomerization [9.192,255]. In light of the work of *Kaiser* and coworkers [9.256,257], however, the conclusion that proton transfer must be associated with the formation of K is incorrect.

Studies of bacteriorhodopsin, like those of rhodopsin, may be subject to multiphoton processes if high intensity pulses are used. Many of the early picosecond experiments used very high laser excitation intensities which can generate non-biologically relevant photochemical processes for example by multiphoton excitation. *Shichida* et al. [9.250] found that the transient absorption at 630 nm becomes saturated by pulse excitation intensities at 532 nm greater than 0.02 GW/cm^2, much lower than most of the previous picosecond studies employed.

With improved time resolution it has been possible to resolve the initial events at room temperature. Subpicosecond processes have been observed, and these are discussed below in the section on subpicosecond spectroscopy. The formation of the K state has also been studied. In one of these studies, *Kaiser*

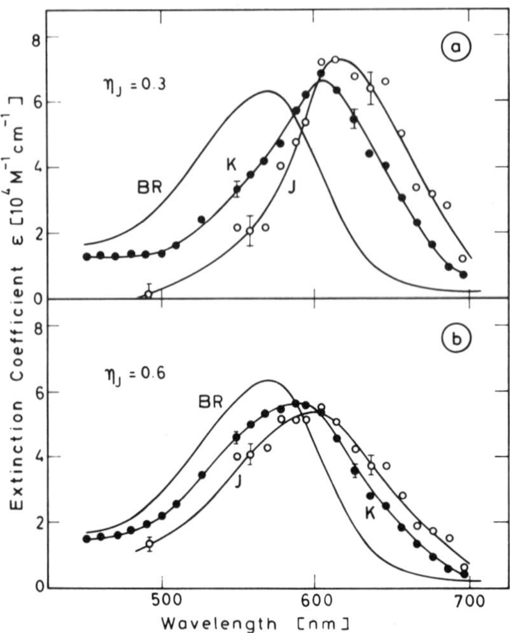

Fig. 9.22. Absorption spectra of bacteriorhodopsin (BR) and the intermediates J and K, calculated from transient absorption spectra assuming a quantum yield for the formation of J, $\eta_J = 0.3$ (**a**), and $\eta_J = 0.6$ (**b**). (From *Polland* et al. [9.257])

and co-workers [9.256,257,293] recently measured the rate of the $J \to K$ transformation. Figure 9.22 shows the calculated absorption spectra of the intermediates J and K. Independent of deuteration, the intermediate K was observed to appear with a 5 ps rise time as J decays. *Sharkov* et al. reported a value of 3 ± 0.5 ps [9.258] and a decay of 3.2 ± 0.2 ps was reported by *Petrich* et al. [9.294]. The report that the rate of formation of K is unaffected by deuteration [9.257], leaves the nature of the $J \to K$ transition an open question. This step may be due to protein relaxation around the twisted 13-*cis* conformation of the chromophore, or merely vibrational relaxation of the chromophore [9.256,257].

Picosecond Raman Spectroscopy. Picosecond Raman techniques were introduced to the study of bacteriorhodopsin by *El-Sayed* and co-workers [9.259–261]. Pulses from a synchronously mode-locked laser were used to excite the bacteriorhodopsin sample and generate Raman scattering. Subtraction of the Raman spectrum of bacteriorhodopsin yielded the signal from the intermediate generated within the 40 ps pulse. The Raman spectrum of the intermediate is characteristic of 13-*cis* retinal, but differs somewhat from the Raman spectrum of K at low temperatures [9.213], and appears to evolve between 40 ps and 15 ns. Pump-probe time-resolved Raman spectroscopy with 5 ps resolution was recently applied to the study of bacteriorhodopsin dynamics by *Atkinson* and co-workers [9.262], who observed conformational changes during the first 40 ps after photoexcitation. One Raman band (1194 cm^{-1}) was observed to grow in intensity in 40 ps, while other bands, the hydrogen out-of-plane modes (800–

1100 cm^{-1}) were found to appear within the 5 ps time resolution. The former band was interpreted to signal the appearance of the 13-*cis* isomer, while the fast rise time of the hydrogen out-of-plane bands was attributed to formation of a twisted intermediate state. These results are consistent with the formation of a twisted ground-state species in <5 ps and subsequent relaxation to a conformation more nearly resembling 13-*cis* retinal in 40 ps.

Subpicosecond Spectroscopy. A recent series of studies of the primary events in bacteriorhodopsin photochemistry was completed by workers in *Kaiser*'s group [9.256,257,263,293]. Using a colliding-pulse mode-locked laser with 160 fs resolution, these workers measured the transient absorption in bacteriorhodopsin at 620 nm. Figure 9.23 shows their results for both protonated and deuterated bacteriorhodopsin. The rise time of the absorption was 430 ± 50 fs in both cases, in agreement with the rise time of 0.7 ± 0.3 ps measured at several red wavelengths with 0.3 pulses from a Nd:glass laser. This step is regarded as the formation time of the intermediate *J*. *Sharkov* et al. [9.258] reported the formation time of *J* to be 0.7 ± 0.3 ps measured with 0.6 ps pulses from an amplified CPM laser. These results are in agreement with earlier estimates of the fluorescence lifetime of bacteriorhodopsin at room temperature [9.264,265].

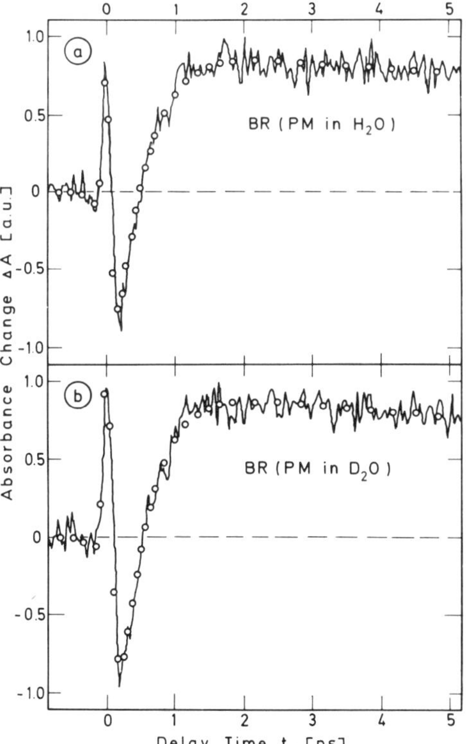

Fig. 9.23. Absorption change at 620 nm in purple membranes of bacteriorhodopsin suspended in (**a**) H_2O, and (**b**) D_2O, with 620 nm, 160 fs excitation. The open circles are a fit by rate equations with time constants of 430 fs and 5 ps. (From *Nuss* et al. [9.256])

The excited-state dynamics in bacteriorhodopsin was the subject of very recently reported experiments by *Kaiser* and co-workers [9.293]. The decay of the initially populated Franck-Condon states on the excited potential surface was detected with a lifetime of 200 ± 70 fs by gain observed with 695 to 735 nm probing pulses following 620-nm excitation with 80 fs pulses. With longer wavelength (850 to 930 nm) probing pulses, a 500-fs decay of the gain was observed, corresponding to the previously reported formation time of the *J* intermediate. The 200-fs decay is interpreted as motion along the reaction coordinate from the initially prepared excited state to the twisted configuration corresponding to the minimum of the S_1 excited state. Absorption due to *J* appeared in 500 fs, in agreement with earlier results, and no faster component could be detected in absorption, leading the authors to conclude that the 200-fs decay must be an excited-state process.

The finding that deuteration has no perceptible effect on the rate of formation of *J* [9.256,257], supports the view that just as proton transfer to or from the Schiff base is not responsible for the $J \to K$ transformation, neither is it involved in the formation of *J*. The transient absorption and fluorescence decay of bacteriorhodopsin containing a synthetic retinal with a bridge preventing isomerization was also investigated [9.263]. An excited-state lifetime of 10 ps was measured. Since inhibition of isomerization extends the lifetime of the excited state from about 0.4 ps to 10 ps, the formation of the first intermediate *J* was attributed to isomerization of the retinal chromophore. The extent of conformation change of the chromophore in the formation of *J* is not yet clear. The chromophore may first attain a twisted 13-*cis*–like structure, *J*, which subsequently relaxes and forms the structure observed in the cw Raman spectra of the intermediate *K* [9.212–214].

One may ask whether bacteriorhodopsin in its dark-adapted state undergoes a similar photochemistry. In fact, the difference between the photochemistry of bacteriorhodopsin in its light-adapted and dark-adapted states seems to be a key to understanding the mechanism of proton pumping, since dark-adapted bacteriorhodopsin is ineffectual in proton pumping in contrast to light-adapted bacteriorhodopsin. Interestingly, in work by *Petrich* and co-workers [9.294] the subpicosecond and early picosecond kinetics of these two forms of bacteriorhodopsin appear to be virtually identical. In both cases, absorption at 460 nm appears within 50 fs and decays in 500 ± 40 fs, after excitation at 612 nm with 120 fs pulses [9.294]. This corresponds well with other measurements of the formation time of *J* [9.256–258,293]. Decay of *J* to *K* was observed in 3.2 ± 0.2 ps for both light- and dark-adapted bacteriorhodopsin. Dark-adapted bacteriorhodopsin contains about 50% all-*trans* and 50% 13-*cis* retinal Schiff base [9.184]. The structure of the latter is considered to be 13-*cis*, 15-*cis* [9.291]. The early kinetics of 13-*cis*, 15-*cis* bacteriorhodopsin were thus estimated to be identical to the kinetics of all-*trans* (light-adapted) bacteriorhodopsin to within a factor of 1.5. *Petrich* et al. suggest [9.294] that the formation of the *K* intermediate of dark-adapted baceriorhodopsin may consist of isomerization about the C15-N double bond, or some other structural change leading to a charge separation similar to that in light-adapted bacteriorhodopsin, and thus to similar spectral shifts. In this view, it is not the structure of the K intermediate but rather the protonation state of the protein that is responsible for the inability of dark-adapted bacteriorhodopsin to pump protons [9.295].

Halorhodopsin. The bacterium *Halobacterium halobium* contains in addition to the proton-pumping protein bacteriorhodopsin, a chloride pump called halorhodopsin. The rise time of its batho intermediate was measured by transient absorption and fluorescence decay [9.266]. An excited state was found with a lifetime of 5 ± 2 ps, decaying to a red-shifted intermediate whose absorption does not shift for at least 2 ns. Interestingly, there is no evidence for a short-lived precursor to this batho intermediate, in contrast to both rhodopsin and bacteriorhodopsin.

9.3.4 Summary

The resonance Raman spectra, the identity or near identity of the batho intermediates of rhodopsin and isorhodopsin, and the different behavior of rhodopsin and bacteriorhodopsin with sterically fixed retinals, have led to the view that isomerization is the primary photochemical event in both rhodopsin and bacteriorhodopsin. The rate of isomerization was recently measured directly in bacteriorhodopsin [9.256,293,294]. These results also show that the first intermediate decays in 3 ps with no deuteration effect. Models of the primary events in bacteriorhodopsin will need to account for this recent evidence. Although several reports suggest that a similar transformation from a bathorhodopsin precursor (photorhodopsin) to bathorhodopsin occurs in rhodopsin, it is not known whether an isotope effect exists for this step in rhodopsin. Studies at room temperature with low intensity excitation pulses are required. Furthermore, it is not known how the protein constrains the chromophore immediately after isomerization, and how the chromophore and protein interact after that.

Over a decade of work on the picosecond photophysics and photochemistry of rhodopsin and bacteriorhodopsin has generated a more thorough understanding of the nature of the events which make possible vision and proton pumping. Studies at picosecond and subpicosecond time resolution of both processes will clarify further the ultrafast behavior of the chromophore and its interaction with its protein environment.

Acknowledgements. This research was supported by grants from NSF and NIH.

References

9.1 J. Monod, J. Wyman, J.P. Changeux: J. Mol Biol. **12**, 88 (1965)
9.2 M.F. Perutz: Nature (London) **228**, 726 (1970)
9.3 R.G. Shulman, J.J. Hopfield, S. Ogawa: Quart. Rev. Biophys. **8**, 325 (1975)

9.4 S.J. Edelstein: Ann. Rev. Biochem. **44**, 209 (1975)
9.5 A. Szabo, M. Karplus: J. Mol. Biol. **72**, 163 (1972)
9.6 E. Antonini, M. Brunori: *Hemoglobin and Myoglobin and their Reactions with Ligands* (North-Holland, Amsterdam 1971)
9.7 Chien Ho (ed.): *Hemoglobin and Oxygen Binding* (North-Holland, Amsterdam, 1982)
9.8 H.F. Bunn, B.B. Forget: *Hemoglobin: Molecular, Genetic, and Clinical Aspects* (Saunders Phila delphia 1986)
9.9 C.V. Shank, E.P. Ippen, R. Bersohn: Science **193**, 50 (1977)
9.10 J.L. Martin, A. Migus, C. Poyart, Y. Lecarpentier, R. Astier, A. Antonetti: Proc. Natl. Acad. Sci. USA **80**, 173 (1983)
9.11 M.F. Perutz, F.S. Mathews: J. Mol. Biol. **121**, 199 (1976)
9.12 D.A. Case, M. Karplus: J. Mol. Biol. **132**, 343 (1979)
9.13 B.I. Greene, R.B. Weisman, R.M. Hochstrasser, W.A. Eaton: Proc. Natl. Acad. Sci USA **75**, 5255 (1978)
9.14 D.A. Chernoff, R.M. Hochstrasser, A.W. Steele: Proc. Natl. Acad. Sci. USA **77**, 5606 (1980)
9.15 D.A. Chernoff, R.M. Hochstrasser, A.W. Steele: In *Hemoglobin and Oxygen Binding*, ed. by Chien Ho (North-Holland, Amsterdam 1982) p. 345
9.16 A.H. Reynolds, S.D. Rand, P.M. Rentzepis: Proc. Natl. Acad. Sci. USA **78**, 2292 (1981)
9.17 J.A. Hutchison, L.J. Noe: J. Quant. Electr. QE **20**, 1353 (1984)
9.18 J.L. Martin, A. Migus, C. Poyart, Y. LeCarpentier, A. Astier, A. Antonetti: In *Ultrafast Phenomena IV*, ed. by D.H. Auston, K.B. Eisenthal, Springer Ser. Chem. Phys., Vol. 38 (Springer, Berlin, Heidelberg 1984) p. 447
9.19 J. Hofrichter, J.H. Sommer, E.R. Henry, W.A. Eaton: Proc. Natl. Acad. Sci. USA **80**, 2235 (1983)
9.20 E.R. Henry, J.H. Sommer, J. Hofrichter, W.A. Eaton: J. Mol. Biol. **166**, 443 (1983)
9.21 J. Baldwin, C. Chothia: J. Mol. Biol. **129**, 175 (1979)
9.22 B.R. Gelin, M. Karplus: Proc. Natl. Acad. Sci. USA **74**, 801 (1977)
9.23 E. Henry, W.A. Eaton, R.M. Hochstrasser: *Molecular Dynamics Simulations of Cooling in Laser Excited Heme Proteins*, Proc. Natl. Acad. Sci. USA (in press)
9.24 S. Choi, T.G. Spiro, K.C. Langry, K.M. Smith, D.L. Budd, G.N. Lamar: J. Am. Chem. Soc. **104**, 4345 (1982);
T.G. Spiro: In *Iron Porphyrins*, ed. by Lever and Gray (Addison-Wesley, Reading MA 1983) p. 89
9.25 J. Terner, T.G. Spiro, M. Nagumo, M.F. Nicol, M.A. El-Sayed: J. Am. Chem. Soc. **102**, 3238 (1980)
9.26 T.L. Gustaffson, D.M. Roberts, D.A. Chernoff: J. Chem. Phys. **79**, 1559 (1983)
9.27 C.K. Johnson, G.A. Dalickas, S.A. Payne, R.M. Hochstrasser: Pure & Appl. Chem. **57**, 195 (1985)
9.28 J. Terner, J.D. Stong, T.G. Spiro, M. Nagumo, M.F. Nicol, M.A. El-Sayed: Proc. Natl. Acad. Sci. USA **78**, 1313 (1981)
9.29 S. Dasgupta, T.G. Spiro, C.K. Johnson, G.A. Dalickas, R.M. Hochstrasser: Biochemistry, **24**, 5295 (1985)
9.30 S. Dasgupta, T.G. Spiro, C.K. Johnson, G.A. Dalickas, R.M. Hochstrasser: To be published
9.31 P. Cornelius, A.W. Steele, D.A. Chernoff, R.M. Hochstrasser: Proc. Natl. Acad. Sci. USA **78**, 7526 (1981)
9.32 P.A. Cornelius, R.M. Hochstrasser: In *Picosecond Phenomena III*, ed. by K.B. Eisenthal, R.M. Hochstrasser, W. Kaiser, A. Laubereau, Springer Ser. Chem. Phys., Vol. 23 (Springer, Berlin, Heidelberg 1982) p. 288
9.33 E.W. Findsen, T.W. Scott, M.R. Chance, J.M. Friedman, M.R. Ondrias: J. Amer. Chem. Soc. **107**, 3355 (1985)
9.34 I. Munro, I. Pecht, L. Stryer: Proc. Nat. Acad. Sci. USA **76**, 56 (1976)
9.35 J. Ross, K. Rousslang, L. Brand: Biochemistry **20**, 4361 (1981)
9.36 J.M. Beecham, L. Brand: Ann. Rev. Biochem. **54**, 43 (1985)
9.37 D. Rayner, A. Szabo: Can. J. Chem. **56**, 743 (1978)

9.38 A. Szabo, D. Rayner: J. Am. Chem. Soc. **102**, 554 (1980)
9.39 J. Petrich, M. Chang, D. McDonald, G. Fleming: J. Am. Chem. Soc. **105**, 3819 (1983)
9.40 J. Petrich, M. Chang, D. McDonald, G. Fleming: J. Am. Chem. Soc. **105**, 3824 (1983)
9.41 G. Weber, F.J. Teale: Disc. Farad Soc. **27**, 134 (1959)
9.42 D.K. Negus, R.M. Hochstrasser: Proc. Nat. Acad. Sci. USA **81**, 4399 (1984)
9.43 V.M. Agranowitch, M.D. Galanin: *Electronic Excitation Energy Transfer in Condensed Matter* (Elsevier, Amsterdam 1984)
9.44 G.J. Foshmire, W.D. Brown: Comp. Biochem. Physiol., B **55**, 293 (1976)
9.45 D.K. Negus: Ph.D. Dissertation, Univ. of Penn. 1985
9.46 A. Szabo: J. Chem. Phys. **81**, 150 (1984)
9.47 B. Valeur, G. Weber: Photochem and Photobiol. **25**, 441 (1977)
9.48 D.K. Negus, R.M. Hochstrasser: J. Lumin. **31**, 3 (1984)
9.49 M. Karplus, J.A. McCammon: CRC Crit. Rev. Biochem. **9**, 293 (1981)
9.50 M. Karplus, J.A. McCammon: Am. Rev. Biochem. **53**, 263 (1983)
9.51 E.R. Henry, M. Levitt, W.A. Eaton: Proc. Natl. Acad. Sci. USA **82**, 2034 (1985)
9.52 T. Ichiya, M. Karplus: Biochemistry **22**, 2884 (1983)
9.53 E. Henry, R.M. Hochstrasser: (Abstract: Biophys. Soc. Meeting 1986) in preparation
9.54 N. Alberding, S.S. Chan, L. Eisenstein, H. Frauenfelder, D. Good, J.C. Gunsdus, T.M. Nordlund, M.F. Perutz, A.H. Reynolds, L.B. Sorensen: Biochem. **17**, 43 (1978);
D.D. Dlott, H. Frauenfelder, P. Langer, H. Roder, E.E. Dilono: Proc. Natl. Acad. Sci. USA **80**, 623 (1983)
9.55 L.J. Noe, W.G. Eisert, P.M. Rentzepis: PNAS **75**, 573 (1978)
9.56 L. Lindquist, S. El Moshni, F. Tfibel, B. Alpert, C.J. Andre: Chem. Phys. Lett. **79**, 525 (1982)
9.57 R. Caterall, D.A. Duddell, R.J. Morris, J.T. Richards: Biochem. Biophys. Acta. **705**, 256 (1982)
9.58 J.M. Friedman, T.N. Scott, G.J. Fisanick, S.R. Simon, E.W. Findsen, M.R. Ondrias, V.W. McDonald: Science **229**, 187 (1985)
9.59 P. Cornelius, A. Steele, R.M. Hochstrasser: J. Mol. Biol. **163**, 119 (1983)
9.60 H. Frauenfelder, P.G. Wolynes: Science **229**, 337–345 (1986)
9.61 W.A. Eaton, L.K. Hanson, P.J. Stephens, J.C. Sutherland, J.B.R. Dunn: J. Am. Chem. Soc. **100**, 4991 (1978);
W.A. Eaton, J. Hofrichter: Meth. Enzym. **76**, 175 (1981)
9.62 P.A. Cornelius, A.W. Steele, D.A. Chernoff, R.M. Hochstrasser: Chem. Phys. Lett. **82**, 9 (1981)
9.63 B.I. Greene: Chem. Phys. Lett. **117**, 191 (1985)
9.64 S.A. Asher, J. Murtaugh: J. Am. Chem. Soc. **105**, 7244 (1983)
9.65 B.I. Greene, R.B. Weisman, R.M. Hochstrasser: Chem. Phys. Lett. **62**, 427 (1979)
9.66 F.E. Doany, B.I. Greene, R.M. Hochstrasser: Chem. Phys. Lett. **75**, 206 (1980)
9.67 N.H. Gottfried, A. Sellmeier and W. Kaiser, Chem. Phys. Lett. **111**, 326 (1984)
9.68 W. Kaiser, A. Sellmeier: In *Time Resolved Vibrational Spectroscopy*, ed. by A. Laubereau, M. Stockburger Springer Proc. Phys., Vol. 4 (Springer, Berlin, Heidelberg 1985)
9.69 E. Henry, W.A. Eaton, R.M. Hochstrasser: Proc. Natl. Acad. Sci. USA **83**, 8982–8986 (1986)
9.70 R.M. Hochstrasser: Hyperfine Interactions **38**, 635–650 (1987)
9.71 A.B. Myers, P.L. Holt, M.A. Pereira and R.M. Hochstrasser: J. Chem. Phys. **86**, 5146–5153 (1987)
9.72 R.A. Marcus: J. Chem. Phys. **24**, 966–978 (1956); Annu. Rev. Phys. Chem. **15**, 155–196 (1964); J. Chem. Phys. **43**, 679–701 (1965)
9.73 R.A. Marcus: Biochim. Biophys. Acta **811**, 265–322 (1985)
9.74 J.J. Hopfield: Proc. Natl. Acad. Sci. USA **71**, 3640–3644 (1974)
9.75 J. Jortner: J. Chem. Phys. **64**, 4860–4867 (1976)
9.76 W. Zinth, M.C. Nuss. M.A. Franz, W. Kaiser, H. Michel: In *Antennas and Reaction Centers of Photosynthetic Bacteria*, ed. by M.E. Michel-Beyerle, Springer Ser. Chem. Phys., Vol. 42 (Springer, Berlin, Heidelberg 1985) pp. 286–291
9.77 N.W. Woodbury, M. Becker, D. Middendorf, W.W. Parson: Biochemistry **24**, 7516–7521 (1985)

9.78 R.A. Marcus: In *Tunneling in Biological Systems*, ed. by B. Chance, J.R. Schrieffer, N. Sutin (Academic, New York 1979) pp. 109–127
9.79 J. Jortner: J. Am. Chem. Soc. **102**, 6676–6686 (1980)
9.80 E. Buhks, J. Jortner: FEBS Lett. **109**, 117–120 (1980)
9.81 D. DeVault, B. Chance: Biophys. J. **6**, 825–847 (1966)
9.82 D. DeVault, J.H. Parkes, B. Chance: Nature **215**, 642–644 (1967)
9.83 J.J. Hopfield: Biophys. J. **18**, 311–321 (1977)
9.84 E. Buhks, M. Bixon, J. Jortner: Chem. Phys. **55**, 41–48 (1981)
9.85 J. Jortner: Biochim. Biophys. Acta **594**, 193–230 (1980)
9.86 M. Bixon, J. Jortner: Faraday Discuss. Chem. Soc. **74**, 17–29 (1982)
9.87 D. DeVault: *Quantum-Mechanical Tunnelling in Biological Systems*, 2nd ed. (Cambridge Univ. Press, Cambridge 1984)
9.88 K. Peters, P. Avouris, P.M. Rentzepis: Biophys. J. **23**, 207–217 (1978)
9.89 C.C. Schenck, W.W. Parson, D. Holten, M.W. Windsor, A. Sarai: Biophys. J. **36**, 479–489 (1981)
9.90 M. Redi, J.J. Hopfield: J. Chem. Phys. **72**, 6651–6660 (1980)
9.91 A. Warshel: Proc. Natl. Acad. Sci. USA **77**, 3105–3109 (1980)
9.92 A. Sarai: Chem. Phys. Lett. **63**, 360–366 (1979); Biochim. Biophys. Acta **589**, 71–83 (1980)
9.93 T. Kakitani, H. Kakitani: Biochim. Biophys. Acta **635**, 498–514 (1981)
9.94 M. Bixon, J. Jortner: Faraday Discuss. Chem. Soc. **74**, 17–29 (1982)
9.95 For an introduction to photosynthesis, see R.K. Clayton: *Photosynthesis, Physical Mechanisms and Chemical Patterns* (Cambridge Univ. Press, Cambridge 1980)
9.96a See reviews of recent work in R. Govindjee (ed.): *Photosynthesis*, Vols. 1, 2 (Academic, New York 1982);
9.96b F.K. Fong (ed.): *Light Reaction Path in Photosynthesis* (Springer, Berlin, Heidelberg 1982)
9.97 S. Kaplan, C.J. Arntgen: in Govindjee, [9.96a], Vol. 1, pp. 65–151
9.98 J. Breton and A. Vermeglio: in Govindjee, [9.96a], Vol. 1, pp. 153–194
9.99 H. Michel: J. Mol. Biol. **158**, 567–572 (1982)
9.100 J. Deisenhofer, O. Epp, K. Miki, R. Huber, H. Michel: J. Mol. Biol. **180**, 385–398 (1984)
9.101 R.K. Clayton, T. Yamamoto: Photochem. Photobiol. **24**, 67–70 (1976)
9.102 C. Kirmaier, D. Holten, W.W. Parson: Biochim. Biophys. Acta **810**, 33–48 (1985)
9.103 J.P. Allen, G. Feher, T.O. Yeates, D.C. Rees, D.S. Eisenberg, J. Deisenhofer, H. Michel, R. Huber: Biophys. J. **49**, 583a (1986)
9.104 W.W. Parson R.K. Clayton, R.J. Cogdill: Biochim. Biophys. Acta **387**, 268–278 (1975); R.J. Cogdell, T.G. Monger, W.W. Parson: Biochim. Biophys. Acta **408**, 189–199 (1975)
9.105 P.L. Dutton, K.J. Kaufmann, B. Chance P.M. Rentzepis: FEBS Lett. **60**, 275–280 (1975)
9.106 K.J. Kaufmann, P.L. Dutton, T.L. Netzel, J.S. Leigh, P.M. Rentzepis: Science **188**, 1301–1304 (1975)
9.107 K.J. Kaufmann, K.M. Petty, P.L. Dutton, P.M. Rentzepis: Biochem. Biophys. Res. Commun. **70**, 839–845 (1976)
9.108 M.G. Rockley, M.W. Windsor, R.J. Cogdell, W.W. Parson: Proc. Natl. Acad. Sci. USA **72**, 2251–2255 (1975)
9.109 W.W. Parson, B. Ke; in Govindjee, [9.96a], Vol. 1, pp. 331–385
9.110 T.L. Netzel: In *Biological Events Probed by Ultrafast Laser Spectroscopy*, ed. by R.R. Alfano, Academic, New York 1982)
9.111 W. Zinth, J. Dobler and W. Kaiser: In *Ultrafast Phenomena V*, ed. by G.R. Fleming and A.E. Siegman, Springer Ser. Chem. Phys., Vol. 46 (Springer, Berlin, Heidelberg 1986) pp. 379–383.
9.112 J.-L. Martin, J. Breton, A.J. Hoff, A. Migus, A. Antonetti: Proc. Natl. Acad. Sci. USA **83**, 957–961 (1986)
9.113 J. Breton, J.-L. Martin, A. Migus, A. Antonetti, A. Orszag: In *Ultrafast Phenomena V*, ed. by G.R. Fleming and A.E. Siegman, Springer Ser. Chem. Phys., Vol. 46 (Springer, Berlin, Heidelberg 1986) pp. 393–397

9.114 D. Holten, C. Hoganson, M.W. Windsor, C.C. Schenck, W.W. Parson, A. Migus, R.L. Fork, C.V. Shank: Biochim. Biophys. Acta **592**, 461–477 (1980)
9.115 V.Z. Paschenko, B.N. Korvatovskii, A.A. Kononenko, S.K. Chamorosky, A.B. Rubin: FEBS Lett. **191**, 245–248 (1985)
9.116 C. Kirmaier, D. Holten, W.W. Parson: FEBS Lett. **185**, 76–82 (1985)
9.117 A.Y. Borisov, R.V. Danielius, S.D. Kudzmauskas, A.S. Piskarskas, A.P. Razjivin, V.A. Sirutkaitis, L.L. Valkunas: Photobiochem. Photobiophys. **6**, 33–38 (1983)
9.118 W.W. Parson, A. Scherz, A. Warshel: In *Antennas and Reaction Centers of Photosynthetic Bacteria*, ed. by M.E. Michel-Beyerle, Springer Ser. Chem. Phys., Vol. 42 (Springer, Berlin, Heidelberg 1985) pp. 122–130
9.119 V.A. Shuvalov, A.V. Klevanik: FEBS Lett. **160**, 51–55 (1983)
9.120 W.W. Parson: Annu. Rev. Biophys. Bioeng. **11**, 57–80 (1982)
9.121 S.R. Meech, A.J. Hoff, D.A. Wiersma: Chem. Phys. Lett. **121**, 287–292 (1985)
9.122 S.G. Boxer, D.J. Lockhart, T.R. Middendorf: Chem. Phys. Lett. **123**, 476–482 (1986)
9.123 D. Holten, M.W. Windsor, W.W. Parson, J.P. Thornber: Biochim. Biophys. Acta **501**, 112–126 (1978)
9.124 M.J. Pellin, C.A. Wraight, K.J. Kaufmann: Biophys. J. **24**, 361–367 (1978)
9.125 C.C. Schenck, W.W. Parson, D. Holten, M.W. Windsor: Biochim. Biophys. Acta **635**, 383–392 (1981)
9.126 C. Kirmaier, D. Holten, R. Feick, R.E. Blankenship: FEBS Lett. **158**, 73–78 (1983)
9.127 C. Kirmaier, D. Holten, W.W. Parson: Biochim. Biophys. Acta **725**, 190–202 (1983)
9.128 C. Kirmaier, D. Holten, L.J. Mancino, R.E. Blankenship: Biochim. Biophys. Acta **765**, 138–146 (1984)
9.129 C.E.D. Chidsey, C. Kirmaier, D. Holten, S.G. Boxer: Biochim. Biophys. Acta **766**, 424–437 (1984)
9.130 C. Kirmaier, D. Holten, W.W. Parson: Biochim. Biophys. Acta **810**, 49–61 (1985)
9.131 P. Maroti, C. Kirmaier, C. Wraight, D. Holten, R.M. Pearlstein: Biochim. Biophys. Acta **810**, 132–139 (1985)
9.132 C. Kirmaier, D. Holten, W.W. Parson: Biochim. Biophys. Acta, in press.
9.133 R.S. Knox: In *Bioenergetics of Photosynthesis*, ed. by Govindjee (Academic, New York 1975) pp. 183–221
9.134 R.S. Knox: Top. Photosyn. **2**, 55–97 (1977)
9.135 T. Forster: In *Modern Quantum Chemistry*, Pt. 3, ed. by O. Sinanoglu (Academic, New York 1965) pp. 93–137
9.136 R.M. Pearlstein: In *Photosynthesis*, Vol. 1, ed. by Govindjee (Academic, New York 1982) pp. 293–330
9.137 K.K. Karukstis, K. Sauer: J. Cell. Biochem. **23**, 131–158 (1983)
9.138 H. Merkelo, S.R. Hartmann, T. Mar, G.S. Singhal, Govindjee: Science **164**, 301–302 (1969)
9.139 M. Seibert, R.R. Alfano, S.L. Shapiro: Biochim. Biophys. Acta **292**, 493–495 (1973)
9.140 M. Seibert, R.R. Alfano: Biophys. J. **14**, 269–283 (1974)
9.141 V.Z. Paschenko, S.P. Protasov, A.B. Rubin, K.N. Timofeev, L.M. Zamazova, L.B. Rubin: Biochim. Biophys. Acta **408**, 143–153 (1975)
9.142 G.S. Beddard, G. Porter, C.J. Tredwell, J. Barker: Nature **258**, 166–168 (1975)
9.143 V.H. Kollman, S.L. Shapiro, A.J. Campillo: Biochem. Biophys. Res. Commun. **63**, 917–922 (1975)
9.144 F. Pellegrino, R.R. Alfano: in Alfano [9.110], pp. 27–53
9.145 D. Mauzerall: Biophys. J. **16**, 87–91 (1976)
9.146 A.J. Campillo, S.L. Shapiro, V.H. Kollman, K.R. Winn, R.C. Hyer: Biophys. J. **16**, 93–97 (1976)
9.147 J. Breton, N.E. Geacintov: Biochim. Biophys. Acta **594**, 1–32 (1980)
9.148 G.S. Beddard, G.R. Fleming, G. Porter, G.F.W. Searle, J.A. Synowiec: Biochem. Biophys. Acta **545**, 165–174 (1979)
9.149 W. Haehnel, J.N. Nairn, P. Reisberg, K. Sauer: Biochim. Biophys. Acta **680**, 161–173 (1982)

9.150 R.J. Gulotty, G.R. Fleming, R.S. Alberte: Biochim. Biophys. Acta **682**, 322–331 (1982)
9.151 D. Magde, S.J. Berens, W.L. Butter: Proc. SPIE-Int. Soc. Opt. Eng. **322**, 80–86 (1982)
9.152 J.A. Nairn, W. Haehnel, P. Reisberg, K. Sauer: Biochim. Biophys. Acta **682**, 420–429 (1982)
9.153 W. Haehnel, A.R. Holzwarth, J. Wendler: Photochem. Photobiol. **37**, 435–443 (1983)
9.154 K.K. Karukstis, K. Sauer: Biochim. Biophys. Acta **722**, 364–371 (1983)
9.155 W.L. Butler, D. Magde, S.J. Berens: Proc. Natl. Acad. Sci. USA **80**, 7510–7514 (1983)
9.156 S.J. Berens, J. Scheele, W.L. Butler, D. Magde: Photochem. Photobiol. **42**, 51–57 (1985)
9.157 K.K. Karukstis, K. Sauer: Biochim. Biophys. Acta **776**, 148–155 (1984)
9.158 K.K. Karukstis, K. Sauer: Biochim. Biophys. Acta **806**, 374–388 (1985)
9.159 R.J. Gulotty, L. Mets, R.S. Alberte, G.R. Fleming: Photochem. Photobiol. **41**, 487–496 (1985)
9.160 A.R. Holzwarth, J. Wendler, W. Haehnel: Biochim. Biophys. Acta **807**, 155–167 (1985)
9.161 I. Yamazaki, M. Mimuro, N. Tamai, T. Yamazaki, Y. Fujita: FEBS Lett. **179**, 65–68 (1985)
9.162 P. Reisberg, J.A. Nairn, K. Sauer: Photochem. Photobiol. **36**, 657–661 (1982)
9.163 A.N. Glazer: Annu. Rev. Biophys. Biophys. Chem. **14**, 47–77 (1985)
9.164 G. Porter, C.J. Tredwell, G.F.W. Searle, J. Barker: Biochim. Biophys. Acta **501**, 232–245 (1978)
9.165 G.F.W. Searle, J. Barker, G. Porter, C.J. Tredwell: Biochim. Biophys. Acta **501**, 246–256 (1978)
9.166 I. Yamazaki, M. Mimuro, T. Murao, T. Yamazaki, K. Yohihara, Y. Fujita: Photochem. Photobiol. **39**, 233–240 (1984)
9.167 J. Wendler, A.R. Holzwarth, W. Wehrmeyer: Biochim. Biophys. Acta **765**, 58–67 (1984)
9.168 A.N. Glazer, S.W. Yeh, S.P. Webb, J.H. Clark: Science **227**, 419–423 (1985)
9.169 T. Gillbro, A. Sandstrom, V. Sundstrom, J. Wendler, A.R. Holzwarth: Biochim. Biophys. Acta **808**, 52–65 (1985)
9.170 T. Gillbro, A. Sandstrom, V. Sundstrom, A.R. Holzwarth: FEBS Lett. **162**, 64–68 (1983)
9.171 G.W. Suter, P. Mazzola, J. Wendler, A.R. Holzwarth: Biochim. Biophys. Acta **766**, 269–276 (1984)
9.172 B. Ke, E. Dolan, V.A. Shurvalov, V.V. Klimov: In Alfano [9.110], pp. 55–77
9.173 C.A. Wraight: In Govindjee, [9.96a], pp. 17–61
9.174 K. Sauer, P. Mathis, S. Acker, J.A. Van Best: Biochim. Biophys. Acta **503**, 120–134 (1978)
9.175 M.D. Il'ina, V.V. Krasauskas, R.J. Rotomskis, A.Y. Borisov: Biochim. Biophys. Acta **767**, 501–506 (1984)
9.176 K. Kamogawa, A. Namiki, N. Nakashima, K. Yoshihara, I. Ikegami: Photochem. Photobiol. **34**, 511–516 (1981)
9.177 V.A. Shuvalov, A.V. Klevanik, A.V. Sharkov, P.G. Kryukov, B. Ke: FEBS Lett. **107**, 313–316 (1979)
9.178 J.M. Fenton, M.J. Pellin, Govindjee, K.J. Kaufmann: FEBS Lett. **100**, 1–4 (1979)
9.179 L.B. Giorgi, B.L. Gore, D.R. Klug, G. Porter, J. Barber: Biochem. Soc. Trans. **14**, 47–48 (1986)
9.180 V.A. Shuvalov, B. Ke, E. Dolan: FEBS Lett. **100**, 5–8 (1979)
9.181 V.A. Shuvalov, V.V. Klimov, E. Dolan, W.W. Parson, B. Ke: FEBS Lett. **118**, 279–282 (1980)
9.182 G. Wald: Nature **219**, 800–807 (1968); Science **162**, 230–239 (1968)
9.183 D. Oesterhelt, W. Stoeckenius: Nature New Biol. **233**, 149–152 (1971)
9.184 D. Oesterhelt, M. Meentzen, L. Schumann: Eur. J. Biochem. **40**, 453–463 (1973)
9.185 J.L. Spudich, R.A. Bogomolni: Biophys. J. **43**, 243–246 (1983)
9.186 B. Schobert, J.K. Lanyi: J. Biol. Chem. **257**, 10306–10313 (1982)
9.187 R.A. Bogomolni, J.L. Spudich: Proc. Natl. Acad. Sci. USA **79**, 6250–6254 (1982)
9.188 T. Yoshizawa, G. Wald: Nature **197**, 1279 (1963)
9.189 T. Yoshizawa, Y. Shichida: Methods Enzym. **81**, 333–354 (1982)
9.190 G. Wald, P.K. Brown: Proc. Natl. Acad. Sci. USA **36**, 84–92 (1950)
9.191 R. Hubbard, A. Kropf: Proc. Natl. Acad. Sci. USA **44**, 130–139 (1958)
9.192 T. Rosenfeld, B. Honig, M. Ottolenghi, J. Hurley, T.G. Ebrey: Pure Appl. Chem. **49**, 341–351 (1977)
9.193 R.A. Cone: Nature (New Biol.) **236**, 39–43 (1972)
9.194 T. Rosenfeld, A. Alchalel, M. Ottolenghi: Nature **240**, 482–483 (1972)

9.195 R. Bensasson, E.J. Land, T.G. Truscott: Nature **258**, 768–770 (1975)
9.196 C.R. Goldschmidt, M. Ottolenghi, T. Rosenfeld: Nature **263**, 169–171 (1976)
9.197 M. Ottolenghi: Adv. Photochem. **12**, 97–200 (1980)
9.198 R.R. Birge: Annu. Rev. Biophys. Bioeng. **10**, 315–4354 (1981)
9.199 B. Honig: Annu. Rev. Phys. Chem. **29**, 31–57 (1978)
9.200 V. Balogh-Nair, K. Nakanishi: In *Stereochemistry*, ed. by Ch. Tamm (Elsevier, Amsterdam 1982) pp. 283–334
9.201 R.H. Lozier, R.A. Bogomolni, W. Stoeckenius: Biophys. J. **45**, 955–963 (1975)
9.202 M.C. Kung, D. Devault, B. Hess, D. Oesterhelt: Biophys. J. **45**, 907–911 (1975)
9.203 N. Dencher, M. Willms: Biophys. Struct. Mech. **1**, 259–271 (1975)
9.204 B. Chance, M. Porte, B. Hess, D. Oesterhelt: Biophys. J. **45**, 913–917 (1975)
9.205 M.J. Pettei, A.P. Yudd, K. Nakanishi, R. Henselman, W. Stoeckenius: Biochem. **16**, 1955–1959 (1977)
9.206 G. Eyring, B. Curry, A. Broek, J. Lugtenburg, R. Mathies: Biochem. **21**, 384–393 (1982)
9.207 R. Mathies, A.R. Oseroff, L. Stryer: Proc. Natl. Acad. Sci. USA **73**, 1–5 (1976)
9.208 R.H. Callender, A. Doukas, R. Crouch, K. Nakanishi: Biochem. **15**, 1621–1629 (1976)
9.209 B. Aton, A.G. Doukas, D. Narva, R.H. Callender, U. Dinur, B. Honig: Biophys. J. **29**, 79–94 (1980)
9.210 G. Eyring, R. Mathies: Proc. Natl. Acad. Sci. USA **76**, 33–37 (1979)
9.211 C. Sandorfy: Int. J. Quant. Chem. **24**, 907–915 (1984)
9.212 J. Terner, C.-L. Hsieh, A.R. Burns, M.A. El-Sayed: Proc. Natl. Acad. Sci. USA **76**, 3046–3050 (1979)
9.213 M. Braiman, R. Mathies: Proc. Natl. Acad. Sci. USA **79**, 403–407 (1982)
9.214 J. Pande, R.H. Callender, T.G. Ebrey: Proc. Natl. Acad. Sci. USA **78**, 7379–7382 (1981)
9.215 A. Campion, J. Terner, M.A. El-Sayed: Nature **265**, 659–661 (1977); Biophys. J. **20**, 269–275 (1977)
9.216 J. Terner, A. Campion, M.A. El-Sayed: Proc. Natl. Acad. Sci USA **74**, 5212–5216 (1977)
9.217 M.A. Marcus, A. Lewis: Science **195**, 1328–1330 (1977)
9.218 I. Grieger, G.H. Atkinson: Biochem. **24**, 5660–5665 (1985)
9.219 A. Warshel: Proc. Natl. Acad. Sci. USA **75**, 2558–2562 (1978); R.M. Weiss, A. Warshel: J. Am. Chem. Soc. **101**, 6131–6133 (1979)
9.220 B. Honig, T.G. Ebrey, R.H. Callender, U. Dinur, M. Ottolenghi: Proc. Natl. Acad. Sci. USA **76**, 2503–2507 (1979)
9.221 A. Warshel, N. Barboy: J. Am. Chem. Soc. **104**, 1469–1476 (1982)
9.222 A. Warshel: Photochem. Photobiol. **30**, 285–290 (1979)
9.223 G.E. Busch, M.L. Applebury, A.A. Lamola, P.M. Rentzepis: Proc. Natl. Acad. Sci. USA **69**, 2802–2806 (1972)
9.224 K. Peters, N. Leontis: In Alfano [9.110], pp. 259–269
9.225 T. Yoshizawa, Y. Shichida: In *Biomolecules*, ed. by C. Nagata (Elsevier, Amsterdam 1985)
9.226 V. Sundstrom, P.M. Rentzepis, K. Peters, M.L. Applebury: Nature **267**, 645–646 (1977)
9.227 B.H. Green, T.G. Monger, R.R. Alfano, B. Aton, R.H. Callender: Nature **269**, 179–180 (1977)
9.228 K. Peters, M.L. Applebury, P.M. Rentzepis: Proc. Natl. Acad. Sci. USA **74**, 3119–3123 (1977)
9.229 A. Warshel: Nature **260**, 679–683 (1976)
9.230 R.R. Birge, L.M. Hubbard: J. Am. Chem. Soc. **102**, 2195–2205 (1980)
9.231 A.G. Doukas, J.R. Junnarkar, R.R. Alfano, R.H. Callender, V. Balogh-Nair: Biophys. J. **47**, 795–798 (1985)
9.232 T.G. Monger, R.R. Alfano, R.H. Callender: Biophys. J. **27**, 105–115 (1979)
9.233 J.D. Spalink, A.H. Reynolds, P.M. Rentzepis, W. Sperling, M.L. Applebury: Proc. Natl. Acad. Sci. USA **80**, 1887–1891 (1983)
9.234 J.E. Rudzki, K.S. Peters: Biochemistry **23**, 3843–3848 (1984)
9.235 Y. Shichida, S. Matuoka, T. Yoshizawa: Photobiochem. Photobiophys. **7**, 221–228 (1984)
9.236 A.G. Doukas, P.Y. Lu, R.R. Alfano: Biophys. J. **35**, 547–550 (1981)

9.237 A.G. Doukas, M.R. Junnarkar, R.R. Alfano, R.H. Callender, T. Kakitani, B. Honig: Proc. Natl. Acad. Sci. USA **81**, 4790–4794 (1984)
9.238 J. Buchert, V. Stefancic, A.G. Doukas, R.R. Alfano, R.H. Callender, J. Pande, H. Akita, V. Balogh-Nair, K. Nakanishi: Biophys. J. **43**, 279–283 (1983)
9.239 G. Hayward, W. Carlsen, A. Siegman, L. Stryer: Science **211**, 942–944 (1980)
9.240 S. Matuoka, Y. Shichida, T. Yoshizawa: Biochim. Biophys. Acta **765**, 38–42 (1984)
9.241 Y. Shichida, T. Kobayashi, H. Ohtani, T. Yoshizawa, S. Nagakura: Photochem. Photobiol. **27**, 335–341 (1978)
9.242 T. Kobayashi: FEBS Lett. **106**, 313–316 (1979)
9.243 T. Kobayashi: Photochem. Photobiol. **32**, 207–215 (1980)
9.244 Y. Shichida, T. Yoshizawa, T. Kobayashi, H. Ohtanik, S. Nagakura: FEBS Lett. **80**, 216–241 (1977)
9.245 T.G. Ebrey: in Alfano [9.110], pp. 271–280
9.246 K.J. Kaufmann, P.M. Rentzepis, W. Stoeckenius, A. Lewis: Biochem. Biophys. Res. Commun. **68**, 1109–1115 (1976)
9.247 K.J. Kaufmann, V. Sundstrom, T. Yamane, P.M. Rentzepis: Biophys. J. **22**, 121–123 (1976)
9.248 E.P. Ippen, C.V. Shank, A. Lewis, M.A. Marcus: Science **200**, 1279–1281 (1978)
9.249 M.L. Applebury, K.S. Peters, P.M. Rentzepis: Biophys. J. **23**, 375–382 (1978)
9.250 Y. Shichida, S. Matuoka, Y. Hidaka, T. Yoshizawa: Biochim. Biophys. Acta **723**, 240–246 (1983)
9.251 T. Kobayashi, H. Ohtani, J. Iwai, A. Ikegami, H. Uchiki: FEBS Lett. **162**, 197–200 (1983)
9.252 T. Gillbro, V. Sundstrom: Photochem. Photobiol. **37**, 445–455 (1983)
9.253 T. Iwasa, Y. Suzuki, T. Nakayama, F. Tokunaga, M. Hirai: J. Phys. Soc. Jpn. **53**, 2851–2856 (1984)
9.254 P.G. Kryukov, Y.A. Lazarev, Y.A. Matveetz, E.L. Terpugov, L.N. Chekulaeva, A.V. Sharkov: Stud. Biophys. **83**, 101–108 (1981)
9.255 U. Dinur, B. Honig, M. Ottolenghi: Photochem. Photobiol. **33**, 523–527 (1981)
9.256 M.C. Nuss, W. Zinth, W. Kaiser, E. Kolling, D. Oesterhelt: Chem. Phys. Lett. **117**, 1–7 (1985)
9.257 H.-J. Polland, M.A. Franz, W. Zinth, W. Kaiser, B. Kolling, D. Oesterhelt: Biophys. J. (1985)
9.258 A.V. Sharkov, A.V. Pakulev, S.V. Chekalin, U.A. Matveetz: Biochim. Biophys. Acta **808**, 94–102 (1985)
9.259 C.-L. Hsieh, M. Nagumo, M. Nicol, M.A. El-Sayed: J. Phys. Chem. **85**, 2714–2717 (1981)
9.260 C.-L. Hsieh, M.A. El-Sayed, M. Nicol, M. Nagumo, J.-H. Lee: Photochem. Photobiol. **38**, 83–94 (1983)
9.261 M.A. El-Sayed: Pure Appl. Chem. **57**, 187–193 (1985)
9.262 G.H. Atkinson, T.L. Brack, D. Blanchard, G. Rumbles, L. Siemankowski: In *Ultrafast Phenomena V*, ed. by G.R. Fleming and A.E. Siegman, Springer Series Chem. Phys., Vol. 46 (Springer, Berlin, Heidelberg 1986) pp. 409–412
9.263 H.-J. Polland, M.A. Franz, W. Zinth, W. Kaiser, E. Kolling, D. Oesterhelt: Biochim. Biophys. Acta **767**, 635–639 (1984)
9.264 R.R. Alfano, W. Yu, R. Govindjee, B. Becher, T.G. Ebrey: Biophys. J. **16**, 541–545 (1976)
9.265 A.V. Sharkov, Y.A. Matveetz, S.V. Chekalin, A.V. Konyashchenko, O.M. Brekhov, B.Y. Rootskoy: Photochem. Photobiol. **38**, 108–111 (1983)
9.266 H.-J. Polland, M.A. Franz, W. Zinth, W. Kaiser, P. Hegemann, D. Oesterhelt: Biophys. J. **47**, 55–59 (1985)
9.267 S.M. Janes, G.A. Dalickas, W.A. Eaton, R.M. Hochstrasser: "Picosecond Transient Absorption Study of Photodissociated Carboxy Hemoglobin and Myoglobin". Biophys. J. (in press)
9.268 M.W. Makinen, R.A. Houtchens, W.S. Caughey: Proc. Natl. Acad. Sci. USA **76**, 6042–6046 (1979)
9.269 J.N. Moore, P.A. Hansen, R.M. Hochstrasser: Chem. Phys. Lett. **138**, 110–114 (1987)
9.270 J.N. Moore, P.A. Hansen, R.M. Hochstrasser: "Iron-Carbenyl Bond Geometry of MbCO and HbCO in Solution Determined by Picosecond Time Resolved Infrared Spectroscopy". Proc. Natl. Acad. Sci. USA (in press)

9.271 J. Kuriyan, S. Wilz, M. Karplus, G.A. Petsko: J. Mol. Biol. **192**, 133–154 (1986)
9.272 J.L. Martin, A. Migus, C. Poyart, Y. Le Carpentier, A. Antonetti A. Orzag: Biophys. Biochem. Res. Comm. **107**, 803–810 (1982)
9.273 P.A. Hansen, J.N. Moore, R.M. Hochstrasser: "Picosecond Transient Optical-Infrared Spectroscopy of Protoheme Photodissociation". Submitted to Chem. Phys.
9.274 J.W. Petrich, J.L. Martin, D. Houde, C. Poyart, A. Orzag: "Time Resolved Raman Spectroscopy with Subpicosecond Resolution; Vibrational Cooling and Delocalization of Strain Energy in Photodissociation HbCO". Biochem. (in press)
9.275 C.H. Chang, M. Schiffer, D. Tiede, U. Smith, J. Norris: J. Mol. Biol. **186**, 201–203 (1985)
9.276 J.P. Allen, G. Feher, T.O. Yeates, D.C. Rees, J. Diesenhafer, H. Michel, R. Huber: Proc. Natl. Acad. Sci. USA **83**, 8589–8593 (1986);
J.P. Allen, G. Feher, T.O. Yeates, H. Komiya, D.C. Rees: Proc. Natl. Acad. Sci. USA **84**, 5730–5734 (1987)
9.277 M.R. Wasielewski, D.M. Tiede: FEBS Lett. **204**, 368–372 (1986)
9.278 J. Breton, J.-L. Martin, A.J. Hoff, A. Migus, A. Antonetti: Proc. Natl. Acad. Sci. USA **83**, 5121–5125 (1986)
9.279 R. A. Marcus: Chem. Phys. Lett. **133**, 471–477 (1987)
9.280 A. Ogrodnik, N. Remy-Richter, M.E. Michel-Beyerle, R. Feick: Chem. Phys. Lett. **135**, 576–580 (1987)
9.281 M. Bixon, J. Jortner, M.E. Michel-Beyerle, A. Ogrodnik, W. Lersch: Chem. Phys. Lett. **140**, 626–630 (1987)
9.282 P.O.J. Scherer, S.F. Fischer: Chem. Phys. Lett. **141**, 179–185 (1987)
9.283 J.K. Gillie, B.L. Fearez, J.M. Hayer, G.J. Small, J.H. Golbeck: Chem. Phys. Lett. **134**, 316–322 (1987)
9.284 J. M. Hayes, G.J. Small: J. Phys. Chem. **90**, 4928–4931 (1986)
9.285 J.M. Hayes, J.K. Gillie, D. Tang, G.J. Small: Biochim. Biophys. Act., to be published
9.286 M. Bixon, J. Jortner: J. Phys. Chem. **90**, 3795–3800 (1986)
9.287 K. Schulten, P. Tavan: Nature **272**, 8586 (1978)
9.288 K. Schulten, Z. Schulten, P. Tavan: In *Information and Energy Transduction in Biological Membranes*, ed. by L. Bolis, E.J.M. Melmreich, and H. Passaw (Allen R. Liss, Inc., New York 1984)
9.289 P. Tavan, K. Schulten, D. Oesterhelt: Biophys. J. **47**, 349–355 (1985); Biophys. J. **47**, 415–430 (1985)
9.290 P. Tavan, K. Schulten: Biophys. J. **50**, 81–89 (1986)
9.291 S.O. Smith, J. Lugtenburg, R.A. Mathies: J. Membrane Biol. **85**, 95–109 (1985)
9.292 K. Gerwert, F. Siebert: EMBO J. **5**, 805–811 (1986)
9.293 J. Dobler, W. Zinth, W. Kaiser, D. Oesterhelt: Chem. Phys. Lett. **144**, 215–220 (1988)
9.294 J. W. Petrich, J. Breton, J.L. Martin, A. Antonetti: Chem. Phys. Lett. **137**, 369–375 (1987)
9.295 H.W. Trissl, W. Gartner: Biochem. **26**, 751–758 (1987)

Addendum A: Generation of Ultrashort Optical Pulses

Charles V. Shank

Since the writing of original chapter (Chap. 2) some significant advances have taken place in femtosecond optical-pulse generation. Progress has been made on developing new mode-locking techniques and extending the wavelength ranges over which optical pulses have been obtained.

Numerous groups have used external cavities to provide feedback into a cavity to shape the pulse in the mode-locking process. The soliton mode-locking technique has been extended to the wavelength region where solitons cannot form. This new technique is called additive pulse mode-locking and has been described both theoretically [A.1] and experimentally [A.2–5].

Passive mode-locking has taken on a new twist with self mode-locking in the Ti: Sapphire laser [A.6]. Synchronous mode-locking of this laser has led to tunable mode-locked pulses of a duration well under 100 femotseconds [A.7–10]. Traveling-wave amplification of spontaneous emission [A.11] has led to optical pulses as short as 75 fs. Pulse amplification, continuum generation and pulse compression have led to optical pulses in the blue-green region of the spectrum with a duration of less than 8 fs [A.12].

Passive mode-locking with the help of bulk semiconductors and quantum wells was applied to generate femtosecond pulses in color-center lasers [A.13, 14]. Subpicosecond pulses at new wave-lengths were generated by nonlinear intra-cavity frequency conversion of dye lasers. Efficient second-harmonic generation gives pulses around 310 nm with a repetition rate of 100 MHz and low-intensity fluctuations [A.15–17]. Pulse trains in the near-infrared were produced by parametric frequency mixing [A.18] (Chap. 3).

Finally colliding-pulse mode-locking techniques have been applied to form a monolithic quantum-well semiconductor laser [A.19].

References

A.1 E. P. Ippen, H.A. Haus, L.Y. Liu: J. Opt. Soc. Am. *6*, 1736 (1989)

A.2 J. Goodberlet, J. Wang, J.G. Fujimoto, P.A. Schultz: Opt. Lett. *14*, 1125 (1989)

A.3 P.M.W. French, J.A.R. Williams, J.R. Taylor: Opt. Lett. *14*, 686 (1989)

A.4 X. Zhu, P.N. Kean, W. Sibbett: Opt. Lett. *14*, 1192 (1989)

A.5. X. Zhu, W. Sibbett: J. Opt. Soc. Am. *B7*, 2187 (1990)

A.6 D.E. Spence, P.N. Kean, W. Sibbett: Opt. Lett. *16*, 42 (1991)

A.7 Ch. Spielmann, F. Krausz, T. Brabec, E. Wintner, A.J. Schmidt: Opt. Lett. *16*, 1180 (1991)
A.8 N. Sarukura, Y. Ishida, H. Nakano: Opt. Lett. *16*, 153 (1991)
A.9 J. Squier, F. Salin, S. Coe, P. Bado, G. Mourou: Opt. Lett. *16*, 85 (1991)
A.10 F. Krausz, T. Brabec, C. Spielmann: Opt. Lett. *16*, 235 (1991)
A.11 J. Hebling, J. Kuhl: Opt. Lett. *14*, 278 (1989)
A.12 R. W. Schoenlein, J.-Y. Gigot, M.T. Portella, C.V. Shank: Appl. Phys. Lett, *58*, 801 (1991)
A.13 M.N. Islam, E.R. Sunderman, I. Bar-Joseph, N. Sauer, T.Y. Chang: Appl. Phys. Lett. *54*, 1203 (1989)
A.14 M.N. Islam, E.R. Sunderman, C.E. Soccolich, I. Bar-Joseph, N. Sauer, T.Y. Chang, B.I. Miller: IEEE J. QE-*25*, 2454 (1989)
A.15 G. Focht, M.C. Downer: IEEE J. QE-*24*, 431 (1988)
A.16 F. Laermer, J. Dobler, T. Elsaesser: Opt. Commun. *67*, 58 (1988)
A.17 D.C. Edelstein, E.S. Wachman, L.K. Cheng, W.R. Bosenberg, C.L. Tang: Appl. Phys. Lett. *52*, 2211 (1988)
A.18 D.C. Edelstein, E.S. Wachman, C.L. Tang: Appl. Phys. Lett. *54*, 1728 (1989)
A.19 M.C. Wu, Y.K. Chen, T. Tanbun-Ek, R.A. Logan, M.A. Chin, G. Raybon: Apply. Phys. Lett. *57*, 759 (1990)

Addendum B: Optical Nonlinearities with Ultrashort Pulses

Alfred Laubereau

Since the original Chapter 3 was written a variety of research activities was carried out on nonlinear optical phenomena with ultrashort laser pulses. A very large number of publications has appeared some of them will be summarized in the following.

B.1 Three-Wave Interactions

B.1.1 Second-Harmonic and Sum-Frequency Generation

The demand for frequency-converted fs pulses has renewed the interest in second-harmonic generation (SHG) of ultrashort laser pulses. The inherent limitations of the process, e.g. spectral bandwidth of the phase-matching condition, group velocity, walk-off, and competition with other nonlinear effects are well-understood from early [B.1] and more recent work [B.2]. New materials have been analysed for their potential to frequency-double fs pulses, e.g. beta-bariumborate (β-BaB_2O_4, BBO) [B.3] and lithiumtriborate (LiB_3O_5, LBO) [B.4]. A series of experiments has shown that intracavity frequency doubling by very thin crystals allows to fulfill the requirements for broad-band phase-matching and a large conversion efficiency. The crystal is placed into the resonator of a colliding-pulse mode-locked dye laser and the output coupling factors are adjusted to obtain optimum SHG output. For KDP, pulses of 170 fs were reported for a 1 mm crystal with conversion efficiencies below 1 % in [B.5] while *Laermer* et al [B.6] obtained shorter and more intense uv pulses of 120 fs with an average power of 1 mW at 314 nm for a thinner crystal of 100 μm (repetition rate 80 MHz; nominal efficiency 25 %). The superior properties of BBO for intracavity SHG were shown by *Edelstein* et al [B.7] who reported 43 fs uv pulses with an average power of several mW for a 55-μm-thick crystal; the same researchers obtained somewhat longer pulses near 100 fs with 20 mW in the uv for a BBO specimen of 150-μm length.

Highly promising are the theoretical results of several investigators [B.8–10] who proposed new schemes for SHG where thicker crystals can be used also in the fs region. The basic clue is the angular dispersion of SHG, i.e. the different phase-matching angles for different frequency components are exploited. The necessary angle distribution of the Fourier components of the input pulse can be generated by the help of an optical grating or prism, and has to match the properties of the nonlinear crystal. This method is equivalent to a cancelation of the group-velocity mismatch. Calculations for BBO with type-I phase-matching predict that specimens of a few mm length may be used without distortion of the SH pulse duration.

In the picosecond time domain new materials [B.11, 12] and applications [B.13–17] were reported recently. For example, passive mode-locking via the nonlinear loss imposed by SHG on the fundamental frequency component was discussed [B.13, 14]. Generation of higher harmonics was studied in [B.18–20]. The switching potential of SHG [B.15–17] is closely related to the time-gating applications of parametric sum-frequency generation. The latter was expanded in recent years in context with coherent-pulse-propagation studies [B.21–23], femtosecond luminescence spectroscopy [B.24–26] and time-resolved infrared spectroscopy [B.27, 28].

B.1.2 Difference-Frequency Generation and Stimulated Parametric Amplification

The increasing interest in time-resolved spectroscopy of semiconductors, vibrational spectroscopy of molecular systems and in optical-fiber communication has strongly promoted techniques to generate infrared laser pulses. A versatile method in this context is parametric down-conversion starting from available laser sources in the visible and near infrared. Considerable progress was achieved in recent years with the help of new materials. The superior performance of potassium titanylphosphate ($KTiOPO_4$, KTP) for difference-frequency generation (DFG) was demonstrated by *Vanherzeele* [B.29] for pulses of 2–4 ps duration. Mixing of the tunable output of 150–250 mW of a synchronously (sync)-pumped dye laser (0.56–1.05 μm) with longer green pulses derived from a cw mode-locked Nd:YLF laser yielded at an average power of 7 mW in the wavelength range of 1.2–4.5 μm with a 100-MHz repetition rate. These numbers correspond to a photon-conversion efficiency of several percent.

Efficient down-conversion of femtosecond pulses in KTP (3.4 mm long) and BBO (5 mm) was reported in [B.30]. A cavity dumped sync-pumped dye laser was applied with 65 mW average power and a repetition rate of 3.8 MHz together with green picosecond pump pulses. Tunable infrared pulses of 94 fs were generated in the 1.4–1.6 μm region with 3 mW average power. The pulse shortening of $\simeq 30\,\%$ of the mixing process and a quantum yield larger than 1 of the low-frequency input component (Nd:YAG) indicate stimulated parametric amplification; i.e., the pump intensity level was already above the low-intensity regime of DFG. For the conventional nonlinear material $LiNbO_3$ (3 mm long) the ir generation was weaker by two orders of magnitude under similar experimental conditions [B.31].

A somewhat different approach was pursued in [B.32] where continuum generation provided a broad-band mixing partner for sub-ps dye-laser pulses. The same scheme was previously investigated in [B.33–36]. Down-conversion in a $LiIO_3$ crystal (0.5 mm long) served for spectral selection due to the available phase-matching bandwidth and generation of tunable MID-ir pulses of 350 fs duration and 10^{-10} J energy (efficiency 10^{-7}). Although Fedju and Rothberg termed their process "seeded parametric amplification" the small quantum yield suggests DFG in the low-intensity case as the relevant mechanism. Down-conversion with considerably higher quantum efficiency of 10 % was reported by *Elsaesser* and *Nuss* [B.37]. They mixed the amplified pulses of a colliding-pulse mode-locked dye laser (620 nm) with the output of a femtosecond travelling-wave dye laser at 690 nm in a 3 mm-long $LiIO_3$ crystal. Stable infrared pulses around 5 μm were obtained with less than 400-fs duration, 10-nJ energy and a 8-kHz repetition rate. In this context the question of broad-band down-conversion may be raised for generating infrared continua. An experimental realization in the picosecond time domain was reported recently [B.38]. Starting with a pulsed Nd:YAG-laser and dye-laser pulses of 175 cm^{-1} bandwidth (25 ps, 150 μJ, 10 Hz), DFG in a 3-cm long $LiIO_3$ specimen provided 100 cm^{-1} broad-band pulses around 4.8 μm with 4.2 % quantum efficiency of the dye-laser photons. The generation of intense picosecond pulses over a large tuning range was demonstrated by *Sukowski* and *Seilmeier* [B.39] using high-gain parametric amplification in BBO. The third harmonic of a pulsed Nd:YAG laser (353 nm, 3.2 mJ, 18 ps) and amplified spontaneous emission of a dye cell with tunable spectral selection served as input radiation. Parametric amplification was performed in a system of three BBO crystals, each of 8 mm length. The gain by a factor of 10^7 allowed the generation of 24 ps pulses with 10 cm^{-1} spectral width and large pulse energy ≤ 65 μJ corresponding to a quantum yield of the pump source of 11 %. The frequency range (3,500 to 25,000 cm^{-1}) of the tunable pico second pulses covers the full visible range ("signal") and a large fraction of the infrared spectrum ("idler"); it exceeds previous results with other materials. A review of BBO properties and applications can be found in [B.40]. High-gain parametric amplification in KTP was discussed in [B.41].

B.1.3 Optical Parametric Generators and Oscillators

The progress in the field of nonlinear materials and short-pulse laser sources was accompanied by corresponding improvements of parametric-generator devices, i.e. high-gain, travelling-wave parametric amplification starting from quantum noise. Results for BBO and KTP towards the generation of picosecond pulses in the visible and infrared (0.4–4 μm) were reported in [B.42–44], while conventional $LiNbO_3$ was investigated in [B.45]. Further down-conversion of short pulses (150 ps) of the mode-locked erbium laser (2.79 or 2.94 μm) was investigated by *Vodopyanov* et al [B.46, 47]. Parametric-generator devices using zinc germanium diphosphide ($ZnGeP_2$) or gallium selenide (GaSe) benefit from long samples (a few cm) and the large figure of merit of these materials, which exceeds that of common nonlinear crystals by more than an order of magnitude. As a result tunable, intense pulses of 0.3 mJ were generated in the mid-infrared, 3.5–18 μm, with quantum efficiencies up to 17 %.

The potential of the optical parametric oscillator (OPO) for the generation of intense, tunable radiation at a moderate pump level has renewed the interest in this nonlinear device. For ps and fs applications synchronous pumping by a train of excitation pulses is required. Singly-resonant cavities with a slightly off-axis phase-matching geometry are used, in general, for practical reasons. The favourable intensity level of short pump pulses (note the higher threshold of crystal damage for shorter pulses) allows to overcome the higher threshold as compared with doubly-resonant configurations; the latter also suffer from stability constraints [B.48].

With the help of a pulsed solid-state laser with passive mode-locking and 5 Hz repetition rate, *Piskarskas* and co-workers [B.49] studied the performance of several well-established nonlinear materials in pico second OPO's, e.g. $LiNbO_3$, CsH_2AsO_4 and $Ba_2NaNb_5O_{15}$ ("BANANA"). A tuning range of 0.65–3.0 μm was achieved with conversion efficiencies of 5–12 % for pump sequences of 10–20 green pulses with 2.5 mJ total energy. The OPO pulse duration was 10 ps; i.e., no shortening of the parametric output occured with respect to the pump source. A somewhat larger efficiency up to 30 % was reported in [B.50] for an OPO with a 7.2-mm long BBO in the tuning range of 0.68 to 2.4 μm. A constant output duration of 75 ps was reported that originates from the longer pump pulses in this study.

Similar results, as quoted above for parametric amplification [B.39], were reported for an OPO equipped with a 12 mm BBO specimen [B.51]. Pumping was achieved by trains of approximately ten frequency-tripled pulses (360 nm, 10 mJ, <25 ps) of a Nd:YAP laser. The 20 ps OPO output could be tuned from 0.406 to 3.17 μm with 8.5 % efficiency and a spectral bandwidth of only 0.24 nm applying an optical grating as a wavelength selector in the OPO cavity. Considerably higher repetition rates in the kHz range have been accomplished by several investigators with cw pump-laser sources operated in the Q-switch mode to meet the power demands of the OPO devices [B.48, 52–54]. Tuning ranges of several thousand cm^{-1} in the VIS and NIR were reported for BBO and BANANA with pulse durations of 15–45 ps and conversion efficiencies up to 20 % at degeneracy. For shorter pump pulses delivered by a fiber compressor, 3.5 ps pulses were generated in a BBO-OPO in the range of 0.95–1.2 μm with 3 kW peak power and 3 kHz repetition rate (efficiency up to 13 %) [B.52]. Very recently a KTP-OPO with rapid (computer-controlled) angle tuning from 1.2 to 1.8 μm was accomplished producing pulses of 2 ps, 6 cm^{-1} and 50 nJ with a 70-Hz repetition rate [B.55].

An important step towards broadly-tunable femtosecond pulses was the development of the femtosecond OPO [B.56–60]. *Edelstein* et al [B.56, 57] incorporated a 1.4-mm Long KTP crystal with focussing mirror optics of an OPO in the cavity of a femtosecond CPM dye laser to obtain mW average output power in the signal (0.82–0.92 μm) or idler component (1.9–2.54 μm). A pulse duration of 220 fs was measured at 0.85 μm while the CPM pump pulses were stretched to 173 fs since the group-velocity dispersion of the KTP crystal was not fully compensated by the prism compressor of the dye laser. The same researchers obtained with additional dispersion compensation in the OPO resonator pulses of 105 fs at 0.84 μm [B.57]. Active length stabilisation of the OPO cavity within ± 10 nm was necessary to maintain a stable parametric output of typically 2 mW (15 % energy conversion).*) A different femtosecond OPO device with pulsed operation was demonstrated by *Laenen* et al [B.59]. They worked with a 5.8 mm BBO crystal with long trains of subpicosecond pulses (0.8 ps) that were derived from a feedback-controlled Nd:glass laser with 8–15 Hz repetition rate. A double pass of the pump pulses with partial compensation of the group-velocity dispersion was decisive for the generation of parametric output pulses of $\simeq 25$ nJ (energy conversion 3 %) and almost constant duration of 200 fs in the tuning range of 8,700 cm^{-1} (0.7 to 1.8 μm). Close to degeneracy at 1.076 μm further pulse shortening to 65 fs was observed and explained by chirp reversal of the idler pulses and self-compression via the normal dispersion of the BBO crystal and other cavity elements [B.60].

B.2 Four-Wave Interactions

B.2.1 Fully Degenerate Four-Wave Mixing

The search for new nonlinear materials has led many investigations to the technique of degenerate four-wave mixing (DFWM). For the measurement of the third-order nonlinear susceptibility x_3 the phase-conjugation geometry has found frequent application. A recent review was given by *Maloney* et al [B.61] for conjugated organic molecules. The nonlinear optical properties of semiconducting organic polymers and of nematic liquid crystals were described in [B.62 and 63]. A high efficiency of the nonlinear interaction was reported for the conventional media CS_2 and potassium vapour [B.64, 65]. Self-defocusing of 45 ps CO_2-laser pulses was observed in the fir in InSb [B.66]. Applications of DFWM (with a phase-conjugation geometry) for the synthesis and analysis of laser pulses were theoretically discussed [B.67, 68]. The potential of phase-conjugation devices for parallel optical computation on the picosecond time scale was demonstrated in [B.69, 70].

B.2.2 Four-Wave Mixing with Raman Resonances

Partially degenerate four-wave mixing (FWM) both in resonant and nonresonant situations is a powerful technique to study dynamical properties of physi-

*) Note added in proof: further improvements were reported very recently by Pelouch et al (Optics Lett. *17*, 1070 (1992)) and by Mak et al (Optics Lett. *17*, 1006 (1992)).

cal systems. In context with Raman (difference frequency) resonances the interaction is generally termed coherent Stokes Raman scattering (CSRS). A general theoretical treatment predicting new spectral features in field-correlated mixing experiments with incoherent light was developed by *Dugan* and *Albrecht* [B.71]. Some of the predictions representing a new type of Rabi-detuning oscillations were observed by *Dugan* and *Albrecht* in CSRS-signal transients of benzene on the sub-picosecond time scale using nanosecond dye-laser pulses and narrow-band detection of the coherent scattering with 0.5 cm^{-1} resolution [B.72]. Related spectroscopic techniques were used by *Kobayashi* et al [B.73] to study femtosecond relaxation processes with incoherent light in polydiacetylene, CS_2, and other systems. Polarization effects in CSRS on the fs time-scale were discussed by *Etchepare* et al [B.74]. They showed that the electronic, orientational ("Kerr-like") and oscillatory contributions to x_3 possess different polarization properties that can be exploited experimentally. As a special case of partially degenerate FWM, CSRS can be carried out with backscattering, i.e. in the phase-conjugation geometry, since the interaction is automatically phase-matched. The process has been called "Raman-induced phase-conjugation technique" and applied to studies of vibrational modes on the picosecond time-scale in CS_2, nitrobenzene and crystalline materials [B75, 76].

In the case of low-frequency Raman resonances and with excitation by intense fs pulses, partially degenerate FWM is often termed impulsive stimulated Raman scattering. A recent review of applications to molecular dynamics in liquids was given in [B.77] (see also Chap. 6). Of special interest are strongly damped and overdamped intermolecular modes, e.g. librations in CS_2 that are readily accessible with this technique. More recently multiple-pulse excitation was demonstrated to select librational contributions for the signal transients [B.78] as compared to the electronic nonlinearity. These more advanced excitation schemes with a pulse-sequence frequency of 2.4 THz ($\simeq 80$ cm^{-1}) provide a transition from impulsive stimulated Raman spectroscopy to coherent Raman scattering with a well-defined frequency difference of the two excitation pulses ("laser" and "Stokes"). A microscopic theory for the analysis of coherent scattering experiments with multiple-pulse excitation was worked out in [B.79].

The anti-Stokes counterpart of coherent Raman scattering, i.e. CARS, was studied on the fs time-scale by several groups. Terahertz beats due to the excitation of multiple vibrational transitions in several organic liquids were observed by *Zinth* et al [B.80]. The high beat frequencies reflect the improved temporal resolution and the enlarged spectral width of the excitation process for fs pulses, as compared with analoguous earlier observations on a picosecond time scale. Similar results were subsequently obtained in [B.81] and [B.82]. The influence of the coherence properties of the laser pulses on fs-CARS was investigated in [B.83]. *Holzapfel* et al. showed that temporal oscillations of the signal transient can originate from the phase modulation of the laser pulses. Three-colour CARS was demonstrated on the fs time-scale by *Fickenscher* et al [B.84, 85]. Here all frequency degeneracies are removed. The approach avoids some of the coherent coupling artifacts and excels by its high experimental accuracy and sensitivity. Proper choice of the polarization geometry also allows to suppress individual scattering mechanisms, e.g. the non-resonant (electronic) or resonant isotropic (pure vibrational) contributions [B.86]. Polarization effects were also discussed in [B.81] and [B.87]. Picosecond CARS activities were recently reviewed by *Koroteev* et al [B.88]; a detailed discussion is presented in Chap. 6.

Concluding this sub-section several interesting applications of FWM to novel mode-locking techniques should be mentioned. They greatly facilitate fs-pulse generation. In one series of investigations termed "additive pulse mode-locking" or "coupled-cavity mode-locking" the nonlinear phase shifts in an additional cavity or interferometer were used for the generation of picosecond and femtosecond laser pulses [B.89–92]. In a second group of studies the nonlinear beam deformation (weak self-focusing) was exploited in context with diffraction losses from apertures or slits for efficient pulse shortening called "Kerr-lens mode-locking" or "self mode-locking" [B.93–95]. The significance of self-phase modulation of the laser rod in combination with group-velocity dispersion was demonstrated for the Ti:Sapphire laser yielding pulses as short as 33 fs [B.96, 97]. Several of the important mode-locking applications involve nonlinear fiber pulse propagation and will be discussed below; for more details the reader is referred to Chap. 2 on pulse generation techniques.

B.3 Short Pulse Nonlinearities with Special Boundary Conditions

B.3.1 Interfaces

Second-harmonic (SH) and sum-frequency generation was shown in recent years to be a powerful spectroscopic tool for studies of the surface structure, absorption of molecules and chemical processes at interfaces. We recall that three-wave interactions are surface-specific in centrosymmetric media. In many of the investigations ultrashort pulses are involved [B.98] since higher intensities are applicable for sufficiently short times increasing the efficiency of the nonlinear detection scheme. Of particular interest are time-resolved second-harmonic studies performed recently with ultrashort time resolution [B.99–103]. For instance, the picosecond dynamics of an isomerisation reaction at a liquid-air interface (*Sitzmann* and *Eisenthal* [B.99]) and of the desorption of a dye monolayer at the surface of fused silica (*Meech* and *Yoshihara* [B.100]) were studied with large experimental sensitivity. The advantage of a total internal reflection technique at solid-liquid interfaces was demonstrated observing the isomerisation of a triphenylmethane dye absorbed on SiO_2 [B.101]. A femtosecond SH study of the GaAs surface was performed in [B.103]. The rapid decrease of the SH intensity during a femtosecond pulse was taken as evidence for fast atomic disorder induced by electronic excitation within a (relatively cold) lattice immediately before melting of the GaAs crystal. The significance of the laser fluence for the melting process of GaAs induced by femtosecond pulses was recognized by *Sokolowski-Tinten* et al [B.104]. They obtained evidence for classic thermal melting at a low laser fluence while an ultrafast nonthermal transformation is operative in the high-fluence regime above 0.34 J/cm^2.

Infrared-visible sum-frequency generation (SFG) was used by *Shen* and coworkers [B.105–108] in a series of investigations to study monolayers of absorbed molecules on vapour-liquid and solid-liquid interfaces. The CH-stretching region was probed in this surface-specific nonlinear vibrational spectroscopy. Different techniques were described to determine the phase of the nonlinear susceptibility from polarisation effects and hence the polar orientations of atomic groups of the absorbed molecules. At the methanol liquid-vapour interface the CH_3 groups were found to point away from the liquid with a broad orientational distribution [B.108].

The generalization of surface-SFG to time-resolved vibrational spectroscopy was performed for the first time by *Harris* and coworkers [B.109–112]. Using infrared excitation pulses of 3 ps in the CH-stretching region ($\simeq$3000 cm^{-1}) vibrational population lifetimes of absorbed molecules on Ag surfaces were measured. Values of 55 to 90 ps were reported for methylthiolate in the temperature range of 380 – 110 K [B.111]. An additional, fast component of the signal transient was related later to a coherence peak artifact [B.112]. These investigations benefit from the resonance enhancement of x_2 at the silver surface. A competing technique for time-resolved vibrational spectroscopy at surfaces was recently demonstrated measuring infrared absorption changes (*Heilweil* et al, [B.113–115]).

Third-order interactions at interfaces have been studied only occasionally [B.116, 117]. *Casstevens* et al [B.117] studied degenerate FWM under phase-conjugation conditions for evaporated films and Langmuir-Blodgett layers of dye molecules with sub-ps pulses. Resonance enhancement of x_3 and the significance of exciton-exciton annihilation were pointed out. A recent review of nonlinear surface interactions was presented in [B.118].

B.3.2 Quantum Confined Structures

The potential of multiple quantum-well devices for applications in electrooptics and optical data processing has stimulated numerous investigations; these topics are treated in other chapters of the present volume (e.g., in Chap. 4). The following discussion concentrates on the nonlinear-optical aspect.

The properties of low-temperature grown (300° C) GaAs quantum wells were studied on the fs time-scale by *Knox* et al [B.119]. Investigating the nonlinear-optical saturation response, a broadened excitonic resonance was found. Comparison with parallel-field transport properties substantiated that a subpicosecond non-radiative decay dominates the carrier recombination. *Klann* et al [B.120] measured the optical nonlinearities of PbSe on a picosecond time scale in the wavelength range from 4 to 8 μm. Excitation of the electron-hole plasma leads to a blue shift of the absorption edge and to large refractive-index changes up to 10^{-1} for a carrier density of 4×10^{17} cm^{-3}. *Klann* et al. pointed out that band filling by both electrons and holes contributes to the nonlinearity. *Stark* et al [B.121] investigated excitons under extreme magnetic confinement in GaAs/AlGaAs multiple quantum wells with fs time resolution. The applied perpendicular magnetic field confines the quasi-two-dimensional states into quasi-zero dimensions for large field strengths (up to 12 T). The observed nonlinear-optical response agrees qualitatively with calculations describing Coulomb-coupled Landau orbitals [B.122]. *Stark* et al. [B.121] presented evidence for the first time that at high magnetic fields the 1s electron-hole pairs behave like a gas of noninteracting particles.

Fast optical switching with 33 ps was demonstrated experimentally exploiting the optical bistability resulting from the large third order nonlinear susceptibility [B.123]. Theoretical predictions considering the nonlinearity due to the virtual charge polarization in a quantum well structure suggest even shorter switching times of 100 fs [B.124]. For a monostable GaAs/AlGaAs device sub-ps switching was found previously [B.125]. Measurements of non-degenerate FWM in AlGaAs ridge waveguides with picosecond time resolution yielded values of the tensor elements of $x^{(3)}$ [B.126].

Optical bistability with a 25-ps switching time and large hysteresis was found by *Yumoto* et al [B.127] in CdS_xSe_{1-x}-doped glasses. A large nonlinearity, $x^{(3)} \simeq 1.3 \times 10^{-9}$ esu was estimated and attributed to the band-filling effect. An optical device with guided modes in silicon films displayed switching within a few hundred picoseconds [B.128]. The properties of CdTe quantum dots in glass were studied by *Esch* et al [B.129]; excitation of the microcrystallites with a diameter <80 Å by pulses ranging from femtoseconds to picoseconds allowed to distinguish various nonlinear mechanisms. For recent reviews of short-pulse activities on quantum-well structures the reader is referred to [B.130] and [B.131].

B.4 Optical Fibers and Soliton Propagation

Nonlinear-propagation phenomena in optical fibers have intrigued numerous investigators during the past few years. Considerable progress was achieved towards the realization of long-range fiber communication systems [B.132]. The principal nonlinearities include self-phase modulation (SPM) [B.133], stimulated Raman scattering, modulational instability [B.134] and four-wave mixing. Applications cover temporal pulse compression, optical amplification and generation of tunable laser pulses. The role of higher-dispersive and nonlinear-propagation processes were theoretically treated in [B.135] for the evolution of fs pulses in single-mode fibers. Bistability in a fiber-optic ring resonator [B.136], nonlinear-optical switching for logic-gate applications [B.137], phase conjugation during second-harmonic generation [B.138] and applications of SFG in doped optical fibers for time-resolved fluorescence spectroscopy [B.139] were considered in several publications. A new theoretical approach for nonlinear fiber pulse propagation termed pseudospectral method was developed [B.140]. The spatial analogue to temporal soliton formation, i.e. self-trapping of soliton beams was recently observed in planar waveguides [B.141]. A comprehensive monograph on nonlinear fiber optics was published by *Agrawal* [B.142].

Important progress was recently achieved in the field of passive switching techniques with optical fibers applicable to optical data processing and passive mode-locking of lasers. *Blow* et al [B.143] and *Islam* et al [B.144] simultaneously demonstrated the ultrafast switching of solitons in an all-fiber nonlinear interferometer set-up. In particular, the Ti:sapphire laser [B.145–149] and Nd:glass laser [B.150, 151] have benefited largely from new fiber-based mode-locking concepts with improved stability and pulse shortening to values below 100 fs. For details the reader is referred to Chap. 2.

The influences of stimulated Raman scattering and other types of FWM effects on fiber pulse propagation were studied in numerous papers. One of the motivations is the search for possibilities to overcome linear losses in long-range fiber communication systems by nonlinear pumping schemes. In fact, considerable progress in this respect was achieved by *Mollenauer* and coworkers [B.132]. These investigators have demonstrated the feasibility of undistorted soliton transmission at 1.6 μm over more than 9000 km for bit rates of several Gbits/s. Exact loss compensation of the $\simeq 50$ ps soliton pulses in a single-mode fiber loop was adjusted by low-gain stimulated-Raman amplification using a colour-center pump laser at 1.47 μm.

A second motivation for the study of four-wave mixing effects is pulse shortening. An analytical solution for compressed soliton-like Stokes pulses was derived in [B.152] assuming negligible pump depletion. Tunable femtosecond pulses of 150–500 fs with peak powers in the kW regime were generated by *Gouveia-Neto* et al [B.153] in a single-mode optical fiber by means of a soliton temporal narrowing and Raman frequency conversion. Similar results were obtained in [B.154, 155]. *Grudinin* et al [B.156] reported stimulated Raman scattering excitation of 18 fs pulses in the 1.6 μm region with pumping by a Nd:YAG laser. Generation of SRS solitons of 70–100 fs in multimode fibers was observed by the same group [B.157]. The influence of group-velocity dispersion on Nd:YAG-laser pulses depleted by multiple-order stimulated Raman scattering in a single-mode fiber was studied by *Heritage* et al [B.158]. A stabilized chirp was noted, suitable for ultrastable pulse compression in a grating pair from 70 ps to 550 fs.

Agrawal inspected the effect of gain dispersion and stimulated Raman scattering of soliton amplification. Under high-gain conditions novel temporal and spectral features were predicted with a splitting of the input pulse in several subpulses of $\simeq 50$ fs [B.159]. These theoretical results are consistent with earlier observations of the break-up of femtosecond pulses in Er-doped fibers [B.160–162], analysed in [B.163]. The noted soliton collapse originates from the competition between SPM and fiber dispersion with the finite bandwidth of stimulated Raman amplification. Undistorted femtosecond soliton amplification is unstable and restricted to small amplification, e.g., short fibers [B.164].

Applying SRS as the amplification process in ring resonators, fiber Raman soliton lasers were constructed and extensively used to generate fs pulses around 1.3 μm. *Islam* et al [B.165] obtained 250 fs pulses with a two-fiber ring cavity pumped by 10 ps pulses of a color-center laser. *Zyssat* et al [B.166] accomplished a minimum pulse duration of 77 fs tunable over 1.37–1.49 μm using l-ps pump pulses of an infrared dye laser. The amplitude and timing stability of a fiber Raman-laser device was recently investigated by *Keller* et al [B.167].

References

B.1 W.H. Glenn: IEEE J. QE-*5*, 284 (1969)
B.2 D. Kühlke, U. Herpers: Opt. Commun. *69*, 75 (1988)
B.3 Y. Ishida, T. Yajima: Opt. Commun. *62*, 197 (1987)
B.4 W.S. Pelouch, T. Ukachi, E.S. Wachman, C.L. Tang: Appl. Phys. Lett. *57*, 111 (1990)
B.5 G. Focht, M.C. Downer: IEEE J. QE-*24*, 431 (1988)
B.6 F. Laermer, J. Dobler, T. Elsaesser: Opt. Commun. *67*, 58 (1988)
B.7 D.C. Edelstein, E.S. Wachman, L.K. Cheng, W.R. Bosenberg, C.L. Tang: Appl. Phys. Lett. *52*, 2211 (1988)
B.8 O.E. Martinez: IEEE J. QE-*25*, 2464 (1989)
B.9 T.R. Zhang, Heung Ro Choo, M.C. Downer: Appl. Opt. *29*, 3927 (1990)
B.10 G. Szabo, Z. Bor: Appl. Phys. B *50*, 51 (1990)
B.11 D. Josse, R. Hierle, I. Ledoux, J. Zyss: Appl. Phys. Lett. *53*, 2251 (1988)
B.12 R.R. Alfano, Q.Z. Wang, T. Jimbo, P.P. Ho, R.N. Bhargava, B.J. Fitzpatrick: Phys. Rev. A *35*, 459 (1987)
B.13 K.A. Stankov, J. Jethwa: Opt. Commun. *66*, 41 (1988)
B.14 J.R.M. Barr, D.W. Hughes: Appl. Phys. B *49*, 323 (1989)
B.15 K.M. Yoo, Qirong Xing, R.R. Alfano: Optics Lett. *16*, 1019 (1991)
B.16 Yao Li, Leming Wang, Perry Neos, Gang Zhang, X.C. Liang, R.R. Alfano: Optics Lett. *14*, 347 (1989)
B.17 J. Collet, T. Amand: Opt. Commun. *62*, 353 (1987)
B.18 P. Qiu, A. Penzkofer: Appl. Phys. B *45*, 225 (1988)
B.19 A. Penzkofer, F. Ossig, P. Qiu: Appl. Phys. B *47*, 71 (1988)
B.20 I.A. Begishev, R.A. Ganeev, A.A. Gulamov, E.A. Erofeev, Sh.R. Kamalov, T. Usmanov, A.D. Khadzhaev: Sov. J. Quant. Electron. *18*, 224 (1988)
B.21 K. Bratengeier, H.-G. Purucker, A. Laubereau: Chem. Phys. Lett. *70*, 393 (1989)
B.22 A. Seilmeier, M. Wörner, H.-J. Hübner, W. Kaiser: Appl. Phys. Lett. *53*, 2468 (1988)
B.23 M. Woerner, A. Seilmeier, W. Kaiser: Opt. Lett. *14*, 636 (1989)
B.24 J. Shah: IEEE J. QE-*24*, 276 (1988)
B.25 M. Marconelli, G.R. Fleming: J. Chem. Phys. *89*, 875 (1988)
B.26 A. Mokhtari, J. Chesnoy, A. Laubereau: Chem. Phys. Lett. *155*, 593 (1989)
B.27 J.N. Moore, P.A. Hansen, R.M. Hochstrasser: Chem. Phys. Lett. *138*, 110 (1987)

B.28 J.C. Owrutsky, Y.R. Kim, M. Li, M.J. Sarisky, R.M. Hochstrasser: Chem. Phys. Lett. *184*, 368 (1991)
B.29 H. Vanherzeele: Opt. Lett. *14*, 728 (1989)
B.30 K. Kurokawa, M. Nakazawa: Appl. Phys. Lett. *55*, 7 (1989)
B.31 K. Kurokawa, M. Nakazawa, T.A. Caughey: Opt. Commun. *75*, 413 (1990)
B.32 T.M. Jedju, L. Rothberg: Appl. Opt. *27*, 615 (1988)
B.33 I. Ledoux, J. Badan, J. Zyss, A. Migus, D. Hulin, J. Etchepare, G. Grillon, A. Antonetti: J. Opt. Soc. Am. B *4*, 987 (1987)
B.34 J. Zyss, I. Ledoux, J. Badan, J.L. Oudar, J. Etchepare, D. Hulin, A. Migus, A. Antonetti: Rev. Phys. Appl. *22*, 1229 (1987)
B.35 J.H. Glowina, J. Misevich, P.P. Sorokin: Opt. Lett. *12*, 19 (1987)
B.36 W.H. Knox: J. Opt. Soc. Am. B *4*, 1771 (1987)
B.37 T. Elsaesser, M.C. Nuss: Opt. Lett. *16*, 411 (1991)
B.38 E.J. Heilweil: Opt. Lett. *14*, 551 (1989)
B.39 U. Sukowski, A. Seilmeier: Appl. Phys. B *50*, 541 (1990)
B.40 D.N. Nikogosyan: Appl. Phys. A *52*, 359 (1991)
B.41 H. Vanherzeele: SPIE *1104*, 44 (1989)
B.42 J.Y. Huang, J.Y. Zhang, Y.R. Shen: Appl. Phys. Lett. *57*, 1961 (1990)
B.43 W. Joosen, H.J. Bakker, L.D. Noordam, H.G. Muller, H.B. van Linden van den Heuvell: J. Opt. Soc. Am. B *8*, 2087 (1991)
B.44 H. Vanherzeele: Appl. Opt. *29*, 2246 (1990)
B.45 M. Ignatavichyus, G. Orshevskii, V. Potsyunas, E. Skardzhyus: Sov. J. Quant. Electron. *19*, 318 (1989)
B.46 K.L. Vodopyanov, V.G. Voevodin, A.I. Gribenyukov, L.A. Kulevskii: Sov. J. Quant. Electron. *17*, 1159 (1987)
B.47 K.L. Vodopyanov, L.A. Kulevskii, V.G. Voevodin, A.I. Gribenyukov, K.R. Allakverdiev, T.A. Kerimov: Opt. Commun. *83*, 322 (1991)
B.48 G.T. Maker, A.I. Ferguson: Appl. Phys. Lett. *56*, 1614 (1990)
B.49 G. Ionushauskas, A. Piskarskas, V. Sirutkaitis, A. Yuozapavichyus: Sov. J. Quant. Electron. *17*, 1303 (1987)
B.50 L.J. Bromley, A. Guy, D.C. Hanna: Opt. Commun. *67*, 316 (1988)
B.51 S. Burdulis, R. Grigonis, A. Piskarskas, G. Sinkevicius, V. Sirutkaitis, A. Fix, J. Nolting, R. Wallenstein: Opt. Commun. *74*, 398 (1990)
B.52 A.S. Piskarskas, V.J. Smilgevicius, A.P. Umbrasas, J.P. Juodisius: Opt. Lett. *14*, 557 (1989)
B.53 A. Piskarskas, V. Smilgevicius, A. Umbrasas: Opt. Commun. *73*, 322 (1989)
B.54 A. Piskarskas, V. Smilgevicius, A. Umbrasas, A. Fix, R. Wallenstein: Opt. Commun. *77*, 335 (1990)
B.55 K. Wolfrum, R. Laenen, A. Laubereau: Opt. Commun. in press
B.56 D.C. Edelstein, E.S. Wachman, C.L. Tang: Appl. Phys. Lett. *54*, 1728 (1989)
B.57 E.S. Wachman, D.C. Edelstein, C.L. Tang: Opt. Lett. *15*, 136 (1990)
B.58 G. Mak, Q. Fu and H.M. van Driel, Appl. Phys. Leth. *60*, 542 (1992)
R. Laenen, H. Graener, A. Laubereau: Opt. Commun. *77*, 226 (1990)
B.59 R. Laenen, H. Graener, A. Laubereau: Opt. Lett. *15*, 971 (1990)
B.60 R. Laenen, H. Graener, A. Laubereau: J. Opt. Soc. Am. B *8*, 1085 (1990)
B.61 C. Maloney, H. Byrne, W.M. Dennis, W. Blau: Chem. Phys. *121*, 21 (1988)
B.62 P.-A. Chollet, F. Kajzar, J. Messier, J.-M. Nunzi, D. Grec: Rev. Phys. Appl. *22*, 1221 (1987)
B.63 J.P. Jiang, P.M. Rentzepis: J. Appl. Phys. *70*, 5125 (1991)
B.64 B. Van Wonterghem, T.E. Dutton, S.M. Saltiel, P.M. Rentzepis: J. Appl. Phys. *64*, 4329 (1988)
B.65 T. Mikropoulos, Z. Pan: Appl. Phys. B *50*, 19 (1990)
B.66 Mansoor Sheik-Bahae: IEEE J. QE-*23*, 1974 (1987)
B.67 C. Joubert, M.L. Roblin, R. Grousson: Appl. Opt. *28*, 4604 (1989)
B.68 G. Manneberg: J. Opt. Soc. Am. B *4*, 313 (1987)
B.69 Yao Li, G. Eichmann, R. Dorsinville, R.R. Alfano: Opt. Lett. *13*, 178 (1988)

B.70 Yao Li, M. Turner, P. Neos, R. Dorsinville, R.R. Alfano: Opt. Lett. *14*, 773 (1989)
B.71 M.A. Dugan, A.C. Albrecht: Phys. Rev. A *43*, 3877 (1991)
B.72 M.A. Dugan, A.C. Albrecht: Phys. Rev. A *43*, 3922 (1991)
B.73 T. Kobayashi, T. Hattori, A. Terasaki, K. Kurokawa: Rev. Phys. Appl. *22*, 1773 (1987)
B.74 J. Etchepare, G. Grillon, J. Arabat: Appl. Phys. B *49*, 425 (1989)
B.75 R. Dorsinville, P. Delfyett, R.R. Alfano: Appl. Opt. *26*, 3655 (1987)
B.76 P.J. Delfyett, R. Dorsinville, R.R. Alfano: Phys. Rev. B *40*, 1885 (1989)
B.77 S. Ruhman, A.G. Joly, B. Kohler, L.R. Williams, K.A. Nelson: Rev. Phys. Appl. *22*, 1717 (1987)
B.78 A.M. Weiner, D.E. Leaird, G.P. Wiederrecht, K.A. Nelson: J. Opt. Soc. Am. B *8*, 1264 (1991)
B.79 Y.J. Yan, S. Mukamel: J. Chem. Phys. *94*, 997 (1991)
B.80 R. Leonhardt, W. Holzapfel, W. Zinth, W. Kaiser: Rev. Phys. Appl. *22*, 1735 (1987)
B.81 T. Joo, M.A. Dugan, A.C. Albrecht: Chem. Phys. Lett. *177*, 4 (1991)
B.82 H. Okamoto, K. Yoshihara: Chem. Phys. Lett. *177*, 568 (1991)
B.83 W. Holzapfel, R. Leonhardt, and W. Zinth: Appl. Phys. B *47*, 307 (1988)
B.84 M. Fickenscher, A. Laubereau: J. Raman Spectrosc. *21*, 857 (1990)
B.85 P. Aechtner, M. Fickenscher, A. Laubereau: Ber. Bunsenges. Phys. Chem. *94*, 399 (1990)
B.86 W. Li, H.-G. Purucker, A. Laubereau: Opt. Commun. *94*, 300 (1992)
B.87 R. Torre, R. Righini, P. Foggi, L. Angeloni: Appl. Phys. B *52*, 132 (1991)
B.88 V.F. Kamalov, N.I. Koroteev, B.N. Toleutaev: In *Time-Resolved Spectroscopy*, ed. by R.J.H. Clark and R.E. Hester, (Wiley & New York 1989) For more recent results see G. Marowsky, V.V. Smirnov (eds.): *Coherent Raman Spectroscopy: Recent Advances*. Springer Proc. Phys. *63* (Springer, Berlin, Heidelberg 1992)
B.89 F. Onelette, M. Piche: Opt. Commun. *60*, 99 (1986)
B.90 K.J. Blow, P. Wood: J. Opt. Soc. Am. B *5*, 629 (1988)
B.91 E.P. Ippen, H.A. Haus, L.Y. Liu: J. Opt. Soc. Am. B *6*, 1735 (1989)
B.92 P.M. French, S.M. Kelley, J.R. Taylor: Opt. Lett. *15*, 378 (1990)
B.93 D.E. Spence, P.N. Kean, W. Sibbett: Opt. Lett. *16*, 42 (1991)
B.94 D.K. Negus, L. Spinelli, N. Goldblatt, G. Fenguet: Proc. Adv. Solid State Lasers (1991), paper PDP4
B.95 U. Keller, G.W. t'Hooft, W.H. Knox, J.E. Cunningham: Opt. Lett. *16*, 1022 (1991)
B.96 Ch. Spielmann, F. Krausz, T. Brabec, E. Wintner, A.J. Schmidt: Opt. Lett. *16*, 1180 (1991)
B.97 F. Krausz, Ch. Spielmann, T. Brabec, E. Wintner, A.J. Schmidt: Opt. Lett. *17*, 204 (1992)
B.98 S. Janz, K. Pedersen, H.M. van Driel: Phys. Rev. B *44*, 3943 (1991)
B.99 E.V. Sitzmann, K.B. Eisenthal: J. Phys. Chem. *92*, 4579 (1988); J. Chem. Phys. *90*, 2831 (1989)
B.100 S.R. Meech, K. Yoshihara: Chem. Phys. Lett. *154*, 30 (1989); J. Phys. Chem. *94*, 4913 (1990)
B.101 S.R. Meech, K. Yoshihara: Chem. Phys. Lett. *174*, 423 (1990)
B.102 M.C. Goh, K.B. Eisenthal: Chem. Phys. Lett. *157*, 101 (1989)
B.103 S.V. Govorkov, I.L. Shumay, W. Rudolph, T. Schröder: Opt. Lett. *16*, 1013 (1991)
B.104 K. Sokolowski-Tinten, H. Schulz, J. Bialkowski, D. von der Linde: Appl. Phys. A *53*, 227 (1991)
B.105 P. Guyot-Sionnest, J.H. Hunt, Y.R. Shen: Phys. Rev. Lett. *59*, 1597 (1987)
B.106 P. Guyot-Sionnest, R. Superfine, J.H. Hunt, Y.R. Shen: Chem. Phys. Lett. *144*, 1 (1988)
B.107 R. Superfine, J.Y. Huang, Y.R. Shen: Opt. Lett. *15*, 1276 (1990)
B.108 R. Superfine, J.Y. Huang, Y.R. Shen: Phys. Rev. Lett. *66*, 1066 (1991)
B.109 A.L. Harris, N.J. Levinos: J. Chem. Phys. *90*, 3878 (1989)
B.110 A.L. Harris, L. Rothberg, L.H. Dubois, N.J. Levinos, L. Dhar: Phys. Rev. Lett. *64*, 2086 (1990)
B.111 A.L. Harris, L. Rothberg, L. Dhar, N.J. Levinos, L.H. Dubois: J. Chem. Phys. *94*, 2438 (1991)
B.112 A.L. Harris, L. Rothberg: J. Chem. Phys. *94*, 2449 (1991)
B.113 E.J. Heilweil, R.R. Cavanagh, J.C. Stephenson: J. Chem. Phys. *89*, 5342 (1988)
B.114 J.D. Beckerle, M.P. Casassa, R.R. Cavanagh, E.J. Heilweil, J.C. Stephenson: J. Chem. Phys. *90*, 4619 (1989)

B.115 J.D. Beckerle, M.P. Casassa, R.R. Cavanagh, E.J. Heilweil, J.C. Stephenson: Phys. Rev. Lett. *64*, 2090 (1990)
B.116 H.W. Li, G. Martinelli, Cl. Froehly, A. Lacourt: Opt. Commun. *75*, 321 (1990)
B.117 M.K. Casstevens, M. Samoc, J. Pfleger, P.N. Prasad: J. Chem. Phys. *92*, 2019 (1990)
B.118 Y.R. Shen: Nature *337*, 519 (1989)
B.119 W.H. Knox, G.E. Doran, M. Asom, G. Livescu, R. Leibenguth, S.N.G. Chu: Appl. Phys. Lett. *59*, 1491 (1991)
B.120 R. Klann, R. Buhleier, Th. Elsaesser, A. Lambrecht: Appl. Phys. Lett. *59*, 885 (1991)
B.121 J.B. Stark, W.H. Knox, D.S. Chemla, W. Schäfer, S. Schmitt-Rink, C. Stafford: Phys. Rev. Lett. *65*, 3033 (1990)
B.122 C. Stafford, S. Schmitt-Rink, W. Schäfer: Phys. Rev. B *41*, 10000 (1990)
B.123 G.D. Boyd, A.M. Fox, D.A.B. Miller, L.M.F. Chirovsky, L.A. D'Asaro, J.M. Kuo, R.F. Kopf, A.L. Lentine: Appl. Phys. Lett. *57*, 1843 (1990)
B.124 M. Yamanishi, M. Kurosaki: IEEE J. QE-*24*, 325 (1988)
B.125 D. Hulin, A. Antonetti, M. Joffre, A. Migus, A. Mysyrowicz, N. Peyghambarian, H.M. Gibbs: Rev. Phys. Appl. *22*, 1269 (1987)
B.126 H.Q. Le, D.E. Bossi, K.B. Nichols, W.D. Goodhue: Appl. Phys. Lett. *56*, 1008 (1990)
B.127 J. Yumoto, S. Fukushima, K. Kubodera: Opt. Lett. *12*, 832 (1987)
B.128 H. Chelli, A. Koster, N. Paraire, F. Pardo, H. Sauer, M. Carton, S. Laval: Rev. Phys. Appl. *22*, 1273 (1987)
B.129 V. Esch, B. Fluegel, G. Khitrova, H.M. Gibbs, Xu Jiajin, K. Kang, S.W. Koch, L.C. Liu, S.H. Risbud, N. Peyghambarian: Phys. Rev. B *42*, 7450 (1990)
B.130 W.H. Knox, J.E. Henry, K.W. Goossen, K.D. Li, B. Tell, D.A.B. Miller, D.S. Chemla, A.C. Gossard, J. English, S. Schmitt-Rink: IEEE J. QE-*25*, 2586 (1989), D.S. Chemla, S. Schmitt-Rink, W.H. Knox, D.A.B. Miller, K.W. Goossen, G. Hasnain: Phys. Stat. Sol. *159*, 11 (1990)
B.131 H.M. Gibbs, G. Khitrova, N. Peygham barian (eds.): *Nonlinear Photonics*, Springer Ser. Electron. Photonics, Vol. 30 (Springer, Berlin, Heidelberg 1991)
B.132 L.F. Mollenauer, K. Smith: Opt. Lett. *13*, 675 (1988), L.F. Mollenauer, M.J. Neubelt, S.G. Evangelides, J.P. Gordon, J.R. Simpson, L.G. Cohen: Opt. Lett. *15*, 1203 (1990); Electron. Lett. *27*, 178 (1991)
B.133 P.N., K. Smith, W. Sibbett: IEE Proc. *134*, 163 (1987)
B.134 K. Tai, A. Hasegawa, A. Tomita: Phys. Rev. Lett. *56*, 135 (1986)
B.135 E. Bourkoff, W. Zhao, R.I. Joseph, D.N. Christodoulides: Opt. Lett. *12*, 272 (1987)
B.136 R.M. Shelby, M.D. Levenson, S.H. Perlmutter: J. Opt. Soc. Am. B *5*, 347 (1987)
B.137 R. Normandin: Can. J. Phys. *65*, 913 (1987)
B.138 B.Ya. Zel'dovich, Yu.E. Kapitskii: JETP Lett. *51*, 441 (1990)
B.139 E.S. Moraes, M.M. Opalinska, A.S.L. Gomes, C.B. de Araujo: Opt. Commun. *84*, 279 (1991)
B.140 P.L. François: J. Opt. Soc. Am. B *8*, 276 (1991)
B.141 S. Maneuf, F. Reynaud: Opt. Commun. *66*, 325 (1988)
B.142 G.P. Agrawal: *Nonlinear Fiber Optics* (Academic, San Diego 1989)
B.143 K.J. Blow, N.J. Doran, B.K. Nayar: Opt. Lett. *14*, 754 (1989)
B.144 M.N. Islam, E.R. Sunderman, R.H. Stolen, W. Pleibel, J.R. Simpson: Opt. Lett. *14*, 811 (1989)
B.145 J. Mark, L.Y. Liu, K.L. Hall, H.A. Haus, E.P. Ippen: Opt. Lett. *14*, 48 (1989)
B.146 J. Goodberlet, J. Wang, J.G. Fujimoto, P.A. Schulz: Opt. Lett. *14*, 1125 (1989)
B.147 J. Goodberlet, J. Jacobson, J.G. Fujimoto P.A. Schulz, T.Y. Fan: Opt. Lett. *15*, 504 (1990)
B.148 J.M. Liu, J.K. Chee: Opt. Lett. *15*, 685 (1990)
B.149 L. Yan, J.D. Ling, P.-T. Ho, C.H. Lee: Opt. Lett. *11*, 502 (1986)
B.150 F. Krausz, Ch. Spielmann, T. Brabec, E. Wintner, A.J. Schmidt: Opt. Lett. *15*, 1082 (1990)
B.151 Ch. Spielmann, F. Krausz, T. Brabec, E. Wintner, A.J. Schmidt: Appl. Phys. Lett. *58*, 2470 (1991)
B.152 A. Höök, D. Anderson, M. Lisak: Opt. Lett. *13*, 1114 (1988)
B.153 A.S. Gouveia-Neto, A.S.B. Sombra, J.R. Taylor: Opt. Commun. *68*, 139 (1988)

B.154 J.D. Kafka, T. Baer: Opt. Lett. *12*, 181 (1987)
B.155 J.K. Chee, J.M. Liu, M.N. Kong: Optics Lett. *14*, 1074 (1989)
B.156 A.B. Grudinin, E.M. Dianov, D.V. Korobkin, A.M. Prokhorov, V.N. Serkin, D.V. Khaidarov: JETP Lett. *45*, 260 (1987)
B.157 A.B. Grudinin, E.M. Dianov, D.V. Korbkin, A.M. Prokhorov, D.V. Khaidarov: JETP Lett. *47*, 356 (1988)
B.158 J.P. Heritage, A.M. Weiner, R.J. Hawkins: Opt. Commun. *67*, 367 (1988)
B.159 G.P. Agrawal: Opt. Lett. *16*, 226 (1991)
B.160 Yu. Khrushchev, A.B. Grudinin, E.M. Dianov, D.V. Korobkin, V.A. Semenov, A.M. Prokhorov: Electron. Lett. *26*, 456 (1990)
B.161 M. Nakazawa, K. Kurokawa, H. Kubota, K. Suzuki, Y. Kimura, Appl. Phys. Lett. *57*, 653 (1990)
B.162 X. Zhou, W. Sibbett: IEEE J. QE-*27*, 101 (1991)
B.163 I.R. Gabitov, M. Romagnoli, S. Wabnitz: Appl. Phys. Lett. *59*, 1811 (1991)
B.164 K.J. Blow, N.J. Doran, D. Wood: J. Opt. Soc. Am. B *5*, 1301 (1988)
B.165 M.N. Islam, L.F. Mollenauer, R.H. Stolen: In *Ultrafast Phenomena* V, ed. by G.R. Fleming and A.E. Siegman, Springer Ser. Chem. Phys., vol. 46 (Springer, Berlin, Heidelberg 1986)
B.166 B. Zysset, P. Beaud, W. Hodel: Appl. Phys. Lett. *50*, 1027 (1987)
B.167 U. Keller, K.D. Li, M. Rodwell, D.M. Bloom: IEEE J. QE-*25*, 280 (1989)

Addendum C: Ultrashort Interactions in Solids

Dietrich von der Linde

During the five years after the writing of the original chapter 4 a stunning growth of the field has occurred, and an enormous amount of literature on the subject has been published. The attempt of an update can hardly be more than an enumerative, but non-exhaustive list of references with just a few comments added here and there.

Reference is made to several pertinent conference proceedings [C.1–5], special journal issues [C.6–8], and review articles [C.9–12] which may also be helpful in providing a picture of recent developments.

The ability to fabricate semiconductor heterostructures with almost atomic precision has added important new dimensions to semiconductor physics and technology. By suitably choosing the composition and the thickness of semiconductor layer structures, superlattices and quantum-well structures can be designed which possess new "artificial" electronic energy levels, not present in the homogeneous bulk materials. During the last few years semiconductor quantum-well structures and superlattices have attracted an increasing interest, and studies of various aspects of the dynamics and relaxation of electronic excitations in such systems have played an important role. Examples include *intersubband carrier relaxation* [C.13–15], *trapping of carriers from barriers into wells* [C.16], and, in particular, *tunneling processes.*

Tunneling Processes: Tunneling of electronic carriers out of quantum wells and between quantum wells has been studied by many workers [C.17–36]. Much attention has been payed to the properties of *asymmetric double quantum wells.* By applying an external electric field the electronic levels of wells can be shifted with respect to each other and brought into resonance. For example, *Oberli* et al. [C.24] have used picosecond time-resolved luminescence to study the tunneling time as a function of the bias field. They observed a dramatic decrease of the tunneling time when the applied field corresponded to a resonant condition. The observed tunneling times are generally too long to be attributable to coherent tunneling processes, indicating that electronic relaxation processes have a significant effect on tunneling in semiconductor quantum wells. However, very recently *Leo* et al. [C.36] were able to observe coherent oscillations of an electronic wave packet between the two wells of an asymmetric double well structure.

Intervalley Scattering: Ultrafast intervalley scattering processes in semiconductors, mostly bulk GaAs, have been studied by a variety of optical techniques [C.37–53] including non-time resolving techniques [C.43,46,47]. *Shah* et al. [C.40] used subpicosecond time-resolved luminescence spectroscopy to measure the scattering rates of electrons between the Γ and the L valleys of GaAs. They found a Γ-L transfer time of about 100 fs, and a much slower time constant of ≈ 2 ps for the return from the L to the Γ valley. *Bigot* et al. [C.51] studied the energy dependence of the Γ-X intervalley scattering in GaAs. They observed scattering times between 20 and 50 fs with energy resonances (i.e., maxima of the scattering time) near the X-valley minimum. An interestering special type of intervalley scattering occuring in type-II quantum wells and superlattices has been studied by *Göbel* and coworkers [C.48,52]. In type-II structure transitions between the valleys are associated with spatial charge transfer across the interfaces of the layers of the quantum well or the superlattice. Another particularly interesting process is *surface intervalley scattering* which has been observed by *Haight* and *Silberman* [C.50] who used picosecond time-resolved photoemission of electrons. Very recently *Ferry* et al. [C.53] have discussed the role of collisional retardation on intervalley scattering. This effect comes into play when the transition times become comparable with the reciprocal phonon frequencies.

Carrier Relaxation: Rather extensive literature has accumulated on various other aspects of carrier relaxation, e.g. *carrier thermalization*, *energy relaxation*, *carrier* cooling, etc. [C.54–84]. Increasing attention has been payed to the *effect of spatial superstructures* on the relaxation mechanisms; carrier relaxation in bulk materials and quantum wells, quantum wires, etc. [C.85–92] was studied and compared. The role of *hot phonons* and *electronic screening* on carrier cooling in polar materials has also been analyzed [C.93–100]. Considerable amount of work is concerned with carrier relaxation in *disordered and amorphous semiconductors* [C.101–105].

Coherent Effects: A major direction in ultrafast optical spectroscopy of solids during the last few years has been the use of time-resolved four-wave mixing techniques to study transient coherent optical effects in semiconductors, that is to say, phenomena associated with the short-lived, macroscopic dielectric polarization generated during the interaction of an optical pulse with the electronic states of the material. The decay of this induced polarization represents the very first step in the hierarchy of relaxation processes leading to thermal equilibrium.

Time-resolved four-wave mixing has been used to measure the *dephasing time* (polarization decay time) for different electronic excitation [C.106–122]. An outstanding example is the measurement of the extremely fast dephasing times of the polarization associated with *free carrier excitations* [C.111]. Decay times as short as 3.5 fs have been measured, and this work undoubtedly represents a milestone in ultrafast optical spectroscopy. Other interesting developments include the observation of *photon echos* from excitonic excitations in semiconductors [C.116], the discovery of the importance of coherent polarization interaction effects (local-field effects) in time-resolved four-wave mixing in dense media [C.120], and *quantum beat phenomena* associated with interference of the coherent polarization of quantum states of different energy [C.123–129].

Coherent Oscillations, Stark Effect: When the time-resolution in a pump-probe experiment is shorter than the relaxation time of the macroscopic optical polarization (dephasing-time) phenomena are observed, which appear to be puzzeling at first sight. For example, the absorption spectrum seen by a continuum probe pulse is changed before the pump-pulse proper has arrived [C.129]. These phenomena have been analyzed in great detail and were shown to be a result of *coherent interaction of the pump and probe pulses* with the electronic states of the system [C.129–133].

In this context observations of the *optical Stark effect* of exciton [C.134–139] and continuum states [C.138] should also be mentioned. The optical Stark effect of excitons may be of interest from the point of view of device applications because it permits switching of the optical properties near excitonic resonances with subpicosecond response time.

Phonons: Several papers have dealt with the study of *non-equilibrium populations* of *phonons* and with measurements of *phonon relaxation* [C.140–145]. A particularly interesting piece of work is the paper of *Cho* et al. [C.145] who impulsively excited LO phonons at the surface of GaAs and studied the phonon lifetime (dephasing time) by directly monitoring the coherent phonon amplitude in the time domain.

Metals: Measurements of the ultrafast dynamics of electrons in metals have been reported in a number of papers [C.146–161]. Particular attention has been paid to the phenomenon of *anomalous heating*, i.e. the generation of a non-equilibrium distribution of electrons characterized by an electron temperature higher than the lattice temperature. *Electron cooling* [C.149,150] and *heat transport* [C.151] of photoexited non-equilibrium electrons has been studied. Relaxation of non-equilibrium electrons provides important information about the *electron-phonon coupling* in metals [C.160]. An interesting technique has been introduced by *Lagendijk* and coworkers who used the excitation of *surface plasmons* by attenuated total reflection to study electronic relaxation in thin metal films [C.159].

Phase Transformations: Several papers have appeared which deal with solid-liquid phase transformations induced by ultrashort laser pulses [C.162–173]. Still more evidence has been obtained indicating that for laser pulses of 20-ps duration or longer the phase transformation can be described as a thermal-melting process [C.164,170]. On the other hand, there are now quite a number of papers showing that with femtosecond laser pulses transitions from the crystalline state to a disordered, metallic-liquid state can be induced in 100 fs or less in many materials [C.163,165–169]. These results cannot be explained with a thermal model and it has been suggested that any electronic mechanism different from the mechanisms of thermal melting is responsible for these subpicosecond phase transformation [C.171–173]. Very recently *Sokolowski-Tinten* et al. [C.171] have shown that with femtosecond excitation both the (slow) normal melting process and an ultrafast phase transformation can be observed, depending on the energy of the laser pulses.

References

C.1 *Ultrafast Phenomena* V, ed. by G.R. Fleming, A.E. Siegman, Springer Ser. Chem. Phys., Vol. 46 (Springer, Berlin, Heidelberg 1986)

C.2 *Ultrafast Phenomena* VI, ed. by T. Yajima, K. Yoshihara, C.B. Harris, S. Shionoya, Springer Ser. Chem. Phys., Vol. 48 (Springer, Berlin, Heidelberg 1988)

C.3 *Ultrafast Phenomena* VII, ed. by C.B. Harris, E.P. Ippen, G.A. Mourou, A.H. Zewail, Springer Ser. Chem. Phys., Vol. 53 (Springer, Berlin, Heidelberg 1990)

C.4 *Ultrafast Phenomena* VIII

C.5 Proc. 5th Int'l. Conf. on Hot Carriers in Semicond. Sol. State Electr. **31** (1988) Proc. 6th Int'l. Conf. on Hot Carriers in Semicond. Sol. State Electr. **32** (1989)

C.6 IEEE J. QE. **24** (1988)

C.7 J. Mod. Opt. **35** (1988)

C.8 IEEE J. QE. **25** (1989)

C.9 S.A. Lyon: J. Luminesc. **35**, 121 (1986)

C.10 J. Shah: Superlatt. & Microstr. **6**, 293 (1989)

C.11 J. Shah: Solid State Electr. **32**, 1051, (1989)

C.12 E.D. Göbel: *Festkörperprobleme/Advances in Solid State Physics* **30**, 269 (Vieweg, Brannschweig 1990)

C.13 D.Y. Oberli, D.R. Wake, M.V. Klein, J. Klem, T. Henderson, H. Morkoc: Phys. Rev. Lett. **59**, 696 (1987)

C.14 A. Seilmeier, H.J. Hübner, G. Abstreiter, G. Weimann, W. Schlapp: Phys. Rev. Lett. **59**, 1345 (1987)

C.15 M.C. Tatham, J.F. Ryan, C.T. Foxon: Phys. Rev. Lett. **63**, 1637 (1989)

C.16 H.J. Polland, K. Leo, K. Rother, K. Ploog, J. Feldmann, G. Peter, E.O. Göbel, K. Fujiwara, T. Nakayama, Y. Ohta: Phys. Rev. B **38**, 7635 (1988)

C.17 H.J. Polland, K. Köhler, L. Schultheis, J. Kuhl, JE. O Göbel, C.W. Tu: Superlatt. & Microstr. **2**, 309 (1986)

C.18 M. Tsuchija, T. Matsusue, H. Sakaki: Phys. Rev. Lett. **59**, 2356 (1987)

C.19 H. Guo, K. Diff, G. Neofotistos, J.D. Gunton: Appl. Phys. Lett. **53**, 131 (1988)

C.20 G. Livescu, A.M. Fox, D.B.A. Miller, T. Sizer, W.H. Knox, A.C. Gossard, J.H. English: Phys. Rev. Lett. **63**, 438 (1989)

C.21 T.B. Norris, X.J. Song, W.J. Schaff, L.F. Eastman, G. Wicks, G.A. Mourou: Appl. Phys. Lett. **54**, 60 (1989)

C.22 M.K. Jackson, M.B. Johnson, D.H. Chow, T.C. McGill, C.W. Nieh: Appl. Phys. Lett. **54**, 552 (1989)

C.23 D.Y. Oberli, J. Shah, T.C. Damen, C.W. Tu, T.Y. Chang: Appl. Phys. Lett. **56**, 1239 (1989)

C.24 D.Y. Oberli, J. Shah, T.C. Damen, C.W. Tu, T.Y. Chang, D.A.B. Miller, J.E. Henry, R.F. Kopf, N. Sauer, A.E. Di Giovanni: Phys. Rev. B **40**, 3028 (1989)
C.25 M.G.W. Alexander, W.W. Rühle, R. Sauer, W.T. Tsang: Appl. Phys. Lett. **55**, 885 (1989)
C.26 T.B. Norris, N. Vodjani, B. Vinter, C. Weisbuch, G. Mourou: Phys. Rev. B **40**, 1392 (1989)
C.27 H. Liu, R. Ferreira, G. Bastard, C. Delalande, J.F. Palmier, B. Etienne: Appl. Phys. Lett. **54**, 2082 (1989)
C.28 M.G.W. Alexander, M. Nido, K. Reimann, W.W. Rühle: Appl. Phys. Lett. **55**, 2517 (1989)
C.29 T. Fukuzawa, E.E. Mendez, J.M. Hong, Phys. Rev. Lett. **64**, 3066 (1990)
C.30 M. Nido, M.G.W. Alexander, W.W. Rühle, T. Schweizer, K. Köhler: Appl. Phys. Lett. **56**, 335 (1990)
C.31 B. Deveaud, F. Clerot, A. Chomette, A. Regreny, R. Ferreira, G. Bastard, B. Sermage: Europhys. Lett. **11**, 367 (1990)
C.32 K. Leo, J. Shah, J.P. Gordon, T.C. Damen, D.A.B. Miller, C.W. Tu, J.E. Cunningham: Phys. Rev. B **42**, 7065 (1990)
C.33 K. Leo, J. Shah, E.O. Göbel, T.C. Damen, K. Köhler, P. Ganser: Appl. Phys. Lett., **56**, 2031 (1990)
C.34 M.G.W. Alexander, M. Nido, W.W. Rühle, K. Köhler: Phys. Rev. B **41**, 12295 (1990)
C.35 C. Tanguy, B. Deveaud, A. Regreny, D. Hulin, A. Antonetti: Appl. Phys. Lett **58**, 1283 (1991)
C.36 K. Leo, J. Shah, E.O. Göbel, T.C. Damen, S. Schmitt-Rink, W. Schäfer, K. Köhler: Phys Rev. Lett. **66**, 201 (1991)
C.37 W.Z. Lin, J.G. Fujimoto, E.P. Ippen, R.A. Logan: In [Ref. C.1, p. 193]
C.38 W.-Z. Lin, J.G. Fujimoto, E. Ippen: Appl. Phys. Lett. **50**, 124 (1987)
C.39 R.W. Schönlein, W.Z. Lin, E.P. Ippen, J.G. Fujimoto: Appl. Phys. Lett. **51**, 1442 (1987)
C.40 J. Shah, B. Deveaud, T.C. Damen, W.T. Tsang, A.C. Gossard, P. Lugli: Phys. Rev. Lett. **59**, 2222 (1987)
C.41 W.Z. Lin, R.W. Schoenlein, J.G. Fujimoto, E.P. Ippen: IEEE J. QE. **24**, 267 (1988)
C.42 P.C. Becker, H.L. Fragnito, C.H. Brito Cruz, J. Shah, R.L. Fork, J.E. Cunningham, J.E. Henry, C.V. Shank: Appl. Phys. Lett. **53**, 2089 (1988)
C.43 N.D. Mirlin, I. Ya. Karlik, V.F. Sapega: Solid State Commun. **65**, 171 (1988)
C.44 X.Q. Zhou, G.C. Cho, U. Lemmer, W. Kütt, K. Wolter, H. Kurz: Solid State Electr. **32**, 1591 (1989)
C.45 H. Kalt, W.W. Rühle, K. Reimann: Solid State Electr. **32**, 1819 (1989)
C.46 R. Ulbrich, J.A. Kash, J.C. Tsang: Phys. Rev. Lett. **62**, 949 (1989)
C.47 S. Zollner, J. Kircher, M. Cardona, S. Gopalan: Solid State Electr. **32**, 1585 (1989)
C.48 J. Feldmann, R. Sattmann, E.O. Göbel, J. Kuhl, J. Hebling, K. Ploog, R. Muralidharan, P. Dawson, C.T. Foxon: Phys. Rev. Lett. **62**, 1892 (1989)
C.49 J. Feldmann, R. Sattmann, E.O. Göbel, J. Kuhl, J. Hebling, K. Ploog, R. Muralidharan, P. Dawson, C.T. Foxon: Solid State Electr. **32**, 1713 (1989)
C.50 R. Haight, J.A. Silberman: Phys. Rev. Lett. **62**, 815 (1989)
C.51 J.Y. Bigot, M.T. Portella, R.W. Schoenlein, J.E. Cunningham, C.V. Shank: Phys. Rev. Lett. **65**, 3429 (1990)
C.52 J. Feldmann, J. Nunnenkamp, G. Peter, E.O. Göbel, J. Kuhl, K. Ploog, P. Dawson, C.T. Foxon: Phys. Rev. B **42**, 5809 (1990)
C.53 D.K. Ferry, A.M. Kriman, H. Hidas, S. Yamaguchi: Phys. Rev. Lett. **67**, 633 (1991)
C.54 R.A. Höpfel, J. Shah, A.C. Gossard: Phys. Rev. Lett. **56**, 765 (1986)
C.55 M.S. Stix, M.P. Kesler, E.P. Ippen: Appl. Phys. Lett **48**, 1722 (1986)
C.56 J. Collet, J.L. Oudar, T. Amand: Phys. Rev. B **34**, 5443 (1986)
C.57 M.J. Rosker, F.W. Wise, C.L. Tang: Appl. Phys. Lett. **49**, 1726 (1986)
C.58 R.A. Höpfel, J. Shah, D. Block, A.C. Gossard: Appl. Phys. Lett. **48**, 148 (1986)
C.59 M.C. Nuss, Z. Zinth, W. Kaiser: Appl. Phys. Lett **49**, 1717 (1986)
C.60 M.C. Nuss, D.H. Auston, F. Capasso: Phys. Rev. Lett. **58**, 2355 (1987)
C.61 W.Z. Lin, J.G. Fujimoto, E.P. Ippen, R.A. Logan: Appl. Phys. Lett. **51**, 161 (1987)

C.62 D.C. Edelstein, C.L. Tang, A.J. Nozik: Appl. Phys. Lett. **51**, 48 (1987)
C.63 F.W. Wise, I.A. Walmsley, C.L. Tang: Appl. Phys. Lett. **51**, 605 (1987)
C.64 R.W. Schoenlein, W.Z. Lin, E.P. Ippen, J.G. Fujimoto: Appl. Phys. Lett. **51**, 1442 (1987)
C.65 W.W. Rühle, H.J. Polland: Phys. Rev. B **36**, 1683 (1987)
C.66 J.F. Young, Kam Wan, A.J. Spring Thorpe, P. Mandeville: Phys. Rev. B **36**, 1316 (1987-I)
C.67 Wei-Zhu Lin, R.W. Schoenlein, J.G. Fujimoto, E.P. Ippen: IEEE J. QE. **24**, 267 (1988)
C.68 W.W. Rühle, H.J. Polland, E. Bauser, K. Ploog, C.W. Tu: Solid State Electr. **31**, 407 (1988)
C.69 R.A. Höpfel, J. Shah, P.A. Wolff, A.C. Gossard: Phys. Rev. B **37**, 6941 (1988-II)
C.70 R.J. Bäuerle, T. Elsaesser, W. Kaiser, H. Lobentanzer, W. Stolz, K. Ploog: Phys. Rev. B **38**, 4307 (1988-I)
C.71 W.H. Knox, D.S. Chemla, G. Livescu, J.E. Cunningham, J.E. Henry: Phys. Rev. Lett. **61**, 1290 (1988)
C.72 K.T. Tsen, H. Morkoc: Phys. Rev. B **38**, 5615 (1988-I)
C.73 S.M. Goodnick, P. Lugli: Phys. Rev. B **38**, 10135 (1988-I)
C.74 H. Roskos, B. Rieck A. Seilmeier, W. Kaiser: Appl. Phys. Lett. **53**, 2406 (1988)
C.75 V.S. Williams, G.R. Olbright, B.D. Fluegel, S.W. Koch, N. Peyghambarian: J. Mod. Opt. **35**, 1979 (1988)
C.76 B. Rieck, M. Goldstein, H. Roskos, A. Seilmeier, W. Kaiser, G.G. Baumann: Solid State Electr. **32**, 1405 (1989)
C.77 H. Roskos, B. Rieck, A. Seilmeier, W. Kaiser: Solid State Electr. **32**, 1437 (1989)
C.78 T. Elsässer, R.J. Bäuerle, W. Kaiser: Phys. Rev. B **40**, 2976 (1989)
C.79 P.M. Fauchet, D. Hulin: J. Opt. Soc. Am. B **6**, 1024 (1989)
C.80 J.H. Collet, W.W. Rühle, M. Pugnet, K. Leo, A. Million: Phys. Rev. B **40**, 12296 (1989-II)
C.81 Dai-sik Kim, P.Y. Yu: Appl. Phys. Lett. **56**, 2210 (1990)
C.82 C.J. Stanton, D.W. Bailey, K. Hess: Phys. Rev. Lett. **65**, 231 (1990)
C.83 R.P. Stanley, J. Hegarty, R. Fischer, J. Feldmann, E.O. Göbel, J. Cryst. Growth *101*, 683 (1990)
C.84 T. Elsässer, J. Shah, L. Rota, P. Lugli: Phys. Rev. Lett. **66**, 1757 (1991)
C.85 A.K. Sood, J. Menendez, M. Cardona, K. Ploog: Phys. Rev. Lett **54**, 2111 (1985)
C.86 A.K. Sood, J. Menendez, M. Cardona, K. Ploog: Phys. Rev. Lett **54**, 2115 (1985)
C.87 H. Lobentanzer, W.W. Rühle, H.J. Polland, W. Stolz, K. Ploog: Phys. Rev. B **36**, 2954 (1987)
C.88 K. Leo, W.W. Rühle, H.J. Queisser, K. Ploog: Phys. Rev. B **37**, 7121 (1988)
C.89 K. Leo, W.W. Rühle, H.J. Queisser, K. Ploog: Appl. Phys. A **45**, 35 (1988)
C.90 J.F. Ryan, M. Tatham: Sol. State Electr. **32**, 1429 (1989)
C.91 B.K. Ridley: Phys. Rev. B **39**, 5282 (1989)
C.92 J.K. Jain, S. Das Sarma: Phys. Rev. Lett. **62**, 2305 (1989)
C.93 W. Pötz: Phys. Rev. B **36**, 5016 (1987)
C.94 H. J. Polland, W.W. Rühle, K. Ploog, C.W. Tu: Phys. Rev. B **36**, 7722 (1987)
C.95 P. Lugli: Sol. State Electron. **31**, 667 (1988)
C.96 S. Dos Sarmas, J.K. Jain, R. Jalabert: Phys. Rev. B **37**, 6290 (1988-I)
C.97 R.P. Joshi, D.K. Ferry: Phys. Rev. B **39**, 1180 (1989)
C.98 W.W. Rühle, K. Leo, E. Bauser: Phys. Rev. B **40**, 1756 (1989)
C.99 J. Collet: Phys. Rev. B **39**, 7659 (1989)
C.100 Dai-sik Kim, P. Yu: Phys. Rev. Lett. **64**, 946 (1990)
C.101 S. Shevel, R. Fischer, E.O. Göbel, G. Noll, P. Thomas, C. Klingshirn: J. Luminesc. **37**, 45 (1987)
C.102 Z. Vardeny, J. Tauc: In *Disordered Semiconductors*, ed. by M.A. Kastner, G.A. Thomas (Plenum, New York 1987) p. 339
C.103 R. Fischer, E.O. Göbel, G. Noll, P. Thomas, A. Weller: In *Adv. Disordered Semiconductors*, Vol. 2, *Hopping and Related Phenomena*, ed. by H. Fritsche, M. Pollak (World Scientific, Singapore 1990) p. 403

C.104 G. Noll, E.O. Göbel: J. Non-Crystal Solids **97** & **98**, 141 (1987)
C.105 G. Noll, E.O. Göbel, U. Siegner: In [Ref. C.2, p. 240]
C.106 L. Schultheis, J. Kuhl, A. Honold, C.W. Tu: Phys. Rev. Lett. **57**, 1635 (1986)
C.107 L. Schultheis, J. Kuhl, A. Honold, C.W. Tu: Phys. Rev. Lett. **57**, 1797 (1986)
C.108 L. Schultheis, A. Honold, J. Kuhl, K. Köhler, C.W. Tu: Phys. Rev. B **34**, 9027 (1986)
C.109 J. Hegarty, M. Sturge: Surf. Sci. **196**, 555 (1988)
C.110 J. Hegarty, K. Tai, W.T. Tsang: Phys. Rev. B **38**, 7943 (1988)
C.111 P.C. Becker, H.L. Fragnito, C.H. Brito Cruz, R.L. Fork, J.E. Cunningham, J.E. Henry, C.V. Shank: Phys. Rev. Lett. **61**, 1647 (1988)
C.112 H.-E. Swoboda, F.A. Majumder, C. Weber, R. Renner, C. Klingshirn, G. Noll, E.O. Göbel, S. Permogorov, A. Reznitsky: Proc. Int'l. Conf. Phys. Semicond., ed. by W. Zawadski (Inst. of Phys., Warsaw 1988) p. 1323
C.113 J. Kuhl, A. Honold, L. Schultheis, C.W. Tu: *Festkörperprobleme/Advances in Solid State Physics 29*, 157 (Vieweg, Braunschweig 1989)
C.114 C. Dörnfeld, J.M. Hvam: IEEE J. QE. **25**, 904 (1989)
C.115 K. Leo, E.O. Göbel, T.C. Damen, J. Shah, S. Schmitt-Rink, W. Schäfer, J.F. Müller, K. Köhler, P. Ganser: Phys. Rev. B, *44*, 5726 (1991)
C.116 G. Noll, U. Siegner, S. Shevel, E.O. Göbel, Phys. Rev. Lett. **64**, 792 (1990)
C.117 M. Wegener, D.S. Chemla, S. Schmitt-Rink, W. Schäfer: Phys. Rev. A **42**, 5675 (1990)
C.118 G. Noll, U. Siegner, E.O. Göbel, H. Schwab, R. Ranner, C. Klingshirn: J. Cryst. Growth *101*, 731 (1991)
C.119 V.L. Gurevich, M.I. Muradov, D.A. Parshin: Europhys. Lett. in press
C.120 K. Leo, M. Wegener, J. Shah, D.S. Chemla, E.O. Göbel, T.C. Damen, S. Schmitt-Rink, W. Schäfer: Phys. Rev. Lett. **65**, 1340 (1990)
C.121 M.D. Webb, S.T. Cundiff, D.G. Steel: Phys. Rev. Lett. **66**, 934 (1991)
C.122 J.Y. Bigot, M.T. Portella, R.W. Schoenlein, J.E. Cunningham, C.V. Shank: Phys. Rev. Lett. **67**, 636 (1991)
C.123 E.O. Göbel, K. Leo, T.C. Damen, J. Shah, S. Schmitt-Rink, W. Schäfer, J.F. Müller, K. Köhler: Phys. Rev. Lett. **64**, 1801 (1990)
C.124 K. Leo, T.C. Damen, J. Shah, E.O. Göbel, K. Köhler: Appl. Phys. Lett. **57**, 19 (1990)
C.125 V. Langer, H. Stolz, W. von der Osten: Phys. Rev. Let. **64**, 854 (1990)
C.126 B.F. Feuerbacher, J. Kuhl, R. Eccleston, K. Ploog: Solid State Commun. **74**, 1279 (1990)
C.127 S. Bar-Ad, I. Bar-Joseph: Phys. Rev. Lett. **66**, 2491 (1991)
C.128 H. Stolz, V. Langer, E. Schreiber, S. Permogorov, W. von der Osten: Phys. Rev. Lett. **67**, 679 (1991)
C.129 B. Fluegel, N. Peyghambarian, G. Olbright, M. Lindberg, S.W. Koch: Phys. Rev. Lett. **59**, 2588 (1987)
C.130 C.H. Brito Cruz, J.P. Gordon, P.C. Becker, R.L. Fork, C.V. Shank: IEEE J. QE. **24**, 261 (1988)
C.131 M. Joffre, D. Hulin, A. Migus, A. Antonetti, C. Benoit à la Guillaume, N. Peyghambarian, M. Lindberg, S.W. Koch: Opt. Lett. **13**, 276 (1988)
C.132 J. P. Sokoloff, M. Joffre, B. Fluegel, D. Hulin, M. Lindberg, S.W. Koch, A. Migus, A. Antonetti, N. Peyghambarian: Phys. Rev. B **38**, 7615 (1988-I)
C.133 M. Lindberg and S.W. Koch: Phys. Rev. B **38**, 7607 (1988-I)
C.134 A. Mysyrowic, D. Hulin, A. Antonetti, A. Migus, W.T. Masselink, H.M. Morkoc: Phys. Rev. Lett. **56**, 2748 (1986)
C.135 A. von Lehmen, D.S. Chemla, J.E. Zucker, J.P. Heritage: Opt. Lett. **11**, 609 (1986)
C.136 M. Joffre, D. Hulin, A. Migus, A. Antonetti: J. Mod. Opt. **35**, 1951 (1988)
C.137 W.H. Knox, D.S. Chemla, D.A.B. Miller, J.B. Stark, S. Schmitt-Rink: Phys. Rev. Lett. **62**, 1189 (1989)
C.138 N. Peyghambarian, S.W. Koch, M. Lindberg, B. Fluegel, M. Joffre: Phys. Rev. Lett. **62**, 1185 (1989)

C.139 S.G. Lee, P.A. Harten, J.P. Sokoloff, R. Jim, B. Fluegel, K.E. Meissner, C.L. Chuang, J.N. Polky, G.A. Pubanz: Phys. Rev. B **43**, 1719 (1991-I)
C.140 W.E. Bron, J. Kuhl, B.K. Rhee: Phys. Rev. B **34**, 6961 (1986)
C.141 K.T. Tsen, H. Morkoc: Phys. Rev. B **34**, 4412 (1986)
C.142 J.A. Kash, S.S. Jha, J.C. Tsang: Phys. Rev. Lett. **58**, 1869 (1987)
C.143 J.A. Kash, J.C. Tsang: Sol. State Electr. **31**, 419 (1988)
C.144 T. Juhasz, W.E. Bron: Phys. Rev. Lett. **63**, 2385 (1989)
C.145 G.C. Cho, W. Kütt, H. Kurz: Phys. Rev. Lett. **65**, 764 (1990)
C.146 G.L. Eesley: Phys. Rev. B **33**, 2144 (1986)
C.147 C.A. Paddock, G.L. Eesley: Opt. Lett. **11**, 273 (1986)
C.148 C.A. Paddock, G.L. Eesley: J. Appl. Phys. **60**, 285 (1986)
C.149 H.E. Elsayed-Ali, T.B. Norris, M.A. Pessot, G.A. Mourou: Phys. Rev. Lett. **58**, 1212 (1987)
C.150 R.W. Schoenlein, W.Z. Lin, J.G. Fujimoto: Phys. Rev. Lett. **58**, 1680 (1987)
C.151 S.D. Brorson, J.G. Fujimoto, E.P. Ippen: Phys. Rev. Lett. **59**, 1962 (1987)
C.152 B.M. Clemens, G.L. Eesley, C.A. Paddock: Phys. Rev. B **37**, 1085 (1988)
C.153 M. van Exter, A. Lagendijk: Phys. Rev. Lett. **60**, 49 (1988)
C.154 A. Miklós and A. Lörincz: J. Appl. Phys. **63**, 2391 (1988)
C.155 R.W. Schoenlein, J.G. Fujimoto, G.L. Eesley, T.W. Capehart: Phys. Rev. Lett. **61**, 2596 (1988)
C.156 M.B. Agranat, S.I. Anisimov, B.I. Makshantsev: Appl. Phys. B **47**, 209 (1988)
C.157 P.B. Corkum, F. Brunel, N.K. Sherman, T. Srinivasan-Rao: Phys. Rev. Lett. **61**, 2886 (1988)
C.158 A. Miklós, A. Lörincz: Appl. Phys. B **48**, 261 (1989)
C.159 R.H.M. Groeneveld, R. Sprik, A. Lagendijk: Phys. Rev. Lett. **64**, 784 (1990)
C.160 S.D. Brorson, A. Kazeroonian, J.S. Moodera, D.W. Face, T.K. Cheng, E.P. Ippen, M.S. Dresselshaus, G. Dresselshaus: Phys. Rev. Lett. **64**, 2172 (1990)
C.161 H.E. Elsayed-Ali, T. Juhasz, G.O. Smitz, W.E. Bron: Phys. Rev. B **43**, 4488 (1991)
C.162 Y. Kanemitsu, Y. Ishida, I. Nakada, H. Kuroda: App. Phys. Lett. **48**, 209 (1986)
C.163 H.W.K. Tom, G.D. Aumiller, C.H. Brito-Cruz: Phys. Rev. Lett. **60**, 1438 (1988)
C.164 D. von der Linde, B. Danielzik, K. Sokolowski-Tinten, P. Harten: In *Ultrafast Phenomena VI*, ed. by T. Yajima, K. Yoshihara, C.B. Harris, S. Shionoya, Springer Ser. Chem. Phys., Vol. 48 (Springer, Berlin, Heidelberg 1988)
C.165 D.H. Reitze, X. Wang, H. Ahn, M.C. Downer: Phys. Rev. B **40**, 11, 986 (1989)
C.166 D.H. Reitze, H. Ahn, X. Wang, M.C. Downer: In *Ultrafast Phenomena VII*, ed. by C.B. Harris, E.P. Ippen, G.A. Mourou, A.H. Zewail, Springer Ser. Chem. Phys., Vol. 53 (Springer, Berlin, Heidelberg 1990) p. 113
C.167 T. Schröder, W. Rudolph, S.V. Govorkov, I.L. Shumay: Appl. Phys. A **51**, 49 (1990)
C.168 J.K. Wang, P. Saeta, Y. Siegal, E. Mazur, N. Bloembergen: In *Ultrafast Phenomena VII*, ed. by C.B. Harris, E.P. Ippen, G.A. Mourou, A.H. Zewail, Springer Ser. Chem. Phys., Vol. 53 (Springer, Berlin, Heidelberg 1990) p. 321
C.169 H. Schulz, J. Bialkowski, K. Sokolowski-Tinten, D. von der Linde: In *Ultrafast Phenomena VII*, ed. by C.B. Harris, E.P. Ippen, G.A. Mourou, A.H. Zewail, Springer Ser. Chem. Phys., Vol. 53 (Springer, Berlin, Heidelberg 1990) p. 365
C.170 D. von der Linde: In *Resonances—A Volume in Honor of the 70th Birthday of Nicolaas Bloembergen*, ed. M.D. Levenson, E. Mazur, P.S. Pershan, Y.R. Shen (World Scientific, Singapore, 1990)
C.171 K. Sokolowski-Tinten, H. Schulz, J. Bialkowski, D. von der Linde: Appl. Phys. A **53**, 227 (1991)
C.172 P. Saeta, J.-K. Wang, Y. Siegal, N. Bloembergen, E. Mazur: Phys. Rev. Lett. **67**, 1023 (1991)
C.173 S.V. Govorkov, I.L. Shumay, W. Rudolph, T. Schröder: Opt. Lett. **16**, 1013 (1991)

Addendum D: Ultrafast Optoelectronics

David H. Auston

Research in the area of ultrafast optoelectronics has continued since the writing of the original Chapter 5 to produce important advances in the generation and detection of ultrashort electromagnetic pulses. It has led to many new applications to measurements of the properties of materials, devices, and circuits. In this update, we will briefly describe some of the key developments in the following specific topics of ultrafast optoelectronics: (1) materials and generation techniques, (2) electro-optic and photoconductive sampling, and (3) the generation and detection of terahertz radiation and its applications to physics and spectroscopy. Although all of these topics have been the subject of much activity, the greatest amount of interest has been focussed on the new technology for generating and detecting terahertz radiation.

In addition to review papers [D.1–5], the proceedings of the conferences on "Picosecond Electronics and Optoelectronics" and "Ultrafast Phenomena" are useful accounts of current research activity in ultrafast optoelectronics. The IEEE Journal of Quantum Electronics has a special issue devoted to ultrafast optics every two years.

D.1 Materials and Generation Techniques

The investigation of new materials having greater speeds of response and sensitivity continues to be a high priority topic of research in ultrafast optoelectronics [D.6–14]. In the case of fast photoconducting materials, it has usually been necessary to make a substantial compromise in sensitivity to attain the necessary speed. Radiation-damaged silicon-on-sapphire continues to be a widely used material, primarily because of its fast response, its high dark resistivity, and as an insulating substrate. Measurements [D.6] of the speed of response of this material show it to have an effective lifetime of 600 fs. Similar measurements have been made in radiation-damaged gallium arsenide [D.7].

A new photoconducting material that has recently received considerable attention is arsenic-rich gallium arsenide that has been grown at low temperature by molecular beam epitaxy [D.8–12]. This material is non-stoichimetric and has a high concentration of defects which are thought to act as traps for the capture of free carriers. Its photocurrent response time has been estimated

[D.8–10] to be less than 400 fs and its mobility to be approximately 120 to 150 cm^2/Vs, which is 4 to 5 times larger than that in radiation-damaged silicon-on-sapphire. The expectations are that this material will find applications as both a source of sub-picosecond electrical pulses and a photoconducting sampling gate for detecting them. In one such application, this material has been used to make an interdigitated detector having a response time of 1.2 ps and an equivalent bandwidth of 375 GHz [D.11]. In another case, it has been used to generate high-voltage picosecond electrical pulses having an amplitude of 825 V and a rise-time of 1.4 ps [D.12].

In the area of electro-optic materials, the most recent development is the use of poled polymer thin films [D.15,16]. These materials have electro-optic coefficients that can be comparable to lithium tantalate and have the added property that their dielectric constant is relatively low. They can be readily fabricated in thin-film form on a wide range of insulating substrates, making them useful for applications to integrated optics as well as ultrafast optoelectronics. Speed measurements show them to have a response time that is as fast as 600 fs [D.15].

Lithium tantalate continues to be the preferred material for electrooptic sampling [D.20]. Its use for the generation of terahertz radiation by electro-optic Cherenkov radiation shows that it has an intrinsic speed of response of approximately 300 fs [D.26], being limited primarily by damping associated with far-infrared modes of the lattice vibrations. The electro-optic properties of gallium arsenide have made possible the direct probing of high-speed electrical waveforms in analogue and digital integrated circuits which are fabricated on gallium arsenide substrates [D.17].

D.2 Electro-optic and Photoconductive Sampling

As a tool for probing the properties of very-high-speed materials, devices and circuits, sampling techniques which use both photoconductors and electro-optic materials continue to be widely used [D.15–24]. The short aperture times, good signal-to-noise ratios, and non-intrusive nature of these techniques account for their many applications. In spite of this fact, the development of a complete commercial measurement system employing picosecond optoelectronic techniques has yet to be fully realized. The main reason for this is the cost and complexity of the laser systems usually employed for this purpose. Recent advances in the area of short pulse semiconductor lasers will probably help speed this development [D.19].

A novel approach to the use of picosecond lasers for probing circuits is the use of a photoemissive cathode for converting short optical pulses to picosecond electron beams. This has been used to make a system for e-beam probing of silicon and gallium arsenide integrated circuits [D.13].

A reflective electro-optic sampling technique has recently been developed for studies of the dynamical properties of materials [D.24,59]. This has been used to investigate the properties of optically injected electron-hole pairs in semiconductor space charge fields [D.24], and the coupling of the electro-optic response to lattice vibrations [D.59].

D.3 Generation and Detection of Terahertz Radiation

The area of ultrafast optoelectronics that has received the most attention in the last six years is the generation and detection of THz radiation and its application to measurements of materials [D.25–44]. Work on this topic began prior to 1985 and has since greatly expanded to include a wide range of new generation techniques and applications.

The basic concept is an extension of the use of fast photoconductors to generate electrical pulses which consists of placing a high-speed photoconductor in a fast transmission line and illuminating it with a short optical pulse. To generate THz radiation, the photocurrent signal is not coupled to a fast transmission line, but instead, radiates directly into free space. The fact that a high-speed transmission line is no longer necessary makes possible the generation of very-broad-band signals with extremely fast rise times. The devices that are used to generate THz radiation by this method are often referred to as photoconducting antennas.

Many different materials and device configurations have been used. Materials include radiation-damaged semiconductors [D.29], bulk semiconductors [D.31], semiconductor surface depletion layers [D.30], p-i-n diodes [D.32], strained-layer superlattices [D.47], and Schottky barriers [D.48], electro-optic materials [D.26,28], and surface-dipole distributions [D.14,55]. Geometric configurations have included Hertzian dipoles, resonant dipoles [D.27], tapered horns, log-periodic spirals [D.36] uniform apertures [D.29], and arrays [D.33].

Photoconducting antennas have also been employed to detect these signals. The use of fast photoconductive sampling gates mounted in small antennas has permitted time-resolved detection of the THz signals with excellent sensitivity.

Direct detection with bolometers has also been used to make inteferometric measurements of the signals [D.35]. These latter measurements have revealed signals that have effective pulse-widths as short as 160 fs and 3 db band-widths of 3 to 4 THz and useful signals as high as 6 to 7 THz.

The peak-power generation capability of the large-aperture devices is an important property. Scaling to large apertures, high bias fields, and high optical pulse energies, makes possible the generation of pulses in the range of many kV/cm peak amplitude. For the devices which use an external bias field applied to a large-aperture semiconductor-surface emitter, measurements have shown that the peak amplitude of the radiated electric pulse can be comparable to the magnitude of the applied bias field [D.34]. This is equivalent to a free-space radiating switch which releases the stored energy of the static bias field in the form of a freely propagating electrical pulse.

Another useful property of the large-aperture devices is their directionality. The THz beams are diffraction-limited and collimated and have beam-widths that can be only a few degrees for devices with cross-section measurements of 1 cm [D.31].

Arrays of photoconducting antennas have resulted in a novel steering property by which the direction of the radiated THz beam can be controlled electronically [D.33].

In cases where the bias field arises from internal fields in semiconductors such as space charge depletion-layers, Schottky barriers and strained-layer superlattices, the emitted THz radiation can be a valuable probe of the field. For example, it has been used to measure the surface-contact potential in Schottky barriers [D.43], and the polarity of the piezoelectric fields in strained-layer superlattices [D.47].

The basic mechanism for the generation of the THz signals appears to be primarily due to the rapid current transient produced by the optical injection of electron-holes pairs into a high-field region of a semiconductor. In addition to this photocurrent mechanism, a second contribution to the signal has been proposed arising from a nonlinear virtual charge polarization [D.44]. This latter mechanism is expected to have a much faster rise-time and may dominate the signal on very short time scales. Since it does not require the generation of real electron-hole pairs, it can also occur with below band-gap excitation. Recent measurements in gallium arsenide with tunable fs excitation do indeed show that a THz signal can be generated in the region below the band gap [D.49].

The technology of optoelectronic THz generation has proven to be a resourceful tool for measuring the electronic response of materials over a wide range of frequencies that are not readily accessible by other methods [D.45–60].

One of the most active areas of application has been the measurement of the properties of the new high-temperature superconductors [D37,38]. Using high-T_c materials to make high-speed transmission lines, it has been possible to measure the surface impedance of these materials over a wide range of frequencies and temperatures [D.38]. The substrates used for high-T_c materials have also been measured by THz spectroscopy [D.56]. Freely propagating beams of THz signals have also been used to measure the properties of silicon [D.53], gallium arsenide [D.53], and water vapor [D.57]. Also studied by this approach is the coherent resonances due to charge oscillations in coupled quantum wells [D.60].

References

D.1 D.H. Auston: Picosecond photoconductivity: High speed measurements of devices and materials, *Probing High Speed Electrical Signals*, ed. by R.B. Marcus (Academic Sam Diego 1989)

D.2 D.H. Auston: Probing semiconductors with femtosecond pulses. Physics Today *43*, 46–54, (February 1990)

D.3 G.A. Mourou, D.M. Bloom, C.-H. Lee (eds.): *Picosecond Electronics and Optoelectronics*, Springer Ser. Electrophys. Vol. 21 (Springer, Berlin, Heidelberg 1985)

D.4 F.J. Leonberger, C.H. Lee, F. Capasso, H. Morkoc (eds.): *Picosecond Electronics and Optoelectronics*, Springer Ser. Electron. Photon. Vol. 24 (Springer, Berlin, Heidelberg 1987)

D.5 G. Sollner, J. Shah (eds.): *Picosecond Electronics and Optoelectronics*, Proc. Optical Society of America Topical Meeting, Vol. 9 (1991)

D.6 F.E. Doany, D. Grischkowsky, C.-C. Chi: Appl. Phys. Lett. *50*, 460–462 (1987)

D.7 M. Lambsdorff. J. Kuhl: Appl. Phys. Lett. *58*, 1881 (1991)

D.8 S. Gupta, J. Pamulpati, J. Chwalek, P.K. Bhattacharya, G. Mourou: "Subpicosecond photoconductivity in III–V compound semiconductors using low temperature MBE growth techniques *Ultrafast Phenomena* VII, ed. by C.B. Harris, E.P. Ippen, G.A. Mourou, A.H. Zewail, Springer Ser. Chem. Phys., Vol. 53 (Springer, Berlin, Heidelberg 1990) pp. 297–299

D.9 J.M. Chwalek, J.F. Whitaker, G.A. Mourou: Low temperature epitaxially-grown GaAs as a high-speed photoconductor for terahertz spectroscopy, in: *Picosecond Electronics and Optoelectronics*, Technical Digest Series (Optical Society of America, Washington, DC 1991) pp. 15–19

D.10 S. Gupta, M.Y. Frankel, J.A. Valdmanis, J.F. Whitaker, G.A. Mourou: Appl. Phys. Lett. *59*, 3276 (1991)

D.11 Y. Chen, S. Williamson, T. Brock, F.W. Smith, A.R. Calawa: Appl. Phys. Lett. *59*, 1984 (1991)

D.12 T. Motet, J. Nees, S. Williamson, G. Mourou: Appl. Phys. Lett *59*, 1455–1457 (1991)

D.13 P.G. May, J.-M. Halbout, G.L.-T. Chiu: IEEE J. QE-*24*, 234 (1988)

D.14 D. Grischkowsky, I.N. Duling, J.C. Chen, C.-C. Chi: Phys. Rev. Lett. *59*, 1663 (1987)

D.15 P.M. Ferm, C. Knapp, C.-J. Wu, J.T. Yardley, B.B. Hu, X.-C. Zhang, D.H. Auston: Appl. Phys. Lett. *59*, 2651 (1991)

D.16 J.I. Thakara, D.M. Bloom, B.A. Auld: Appl. Phys. Lett. *59*, 1159 (1991)

D.17 K.J. Weingarten, M.J.W. Rodwell, D.M. Bloom: IEEE J. QE-*24*, 198 (1988)

D.18 D.R. Grischkowsky, M.B. Ketchen, C.-C. Chi, N. Duling, N.J. Halas, J.-M. Halbout, P.G. May: IEEE J. QE-*24*, 221 (1988)

D.19 M. Heutmaker, T.B. Cook, B. Bosacchi, J.M. Wiesenfeld, R.S. Tucker: IEEE J. QE-*24*, 226 (1988)

D.20 J.A. Valdmanis: Electron. Lett. *23*, 1308 (1987).

D.21 C.J. Madden, R.A. Marsland, M.J.W. Rodwell, D.M. Bloom: Appl. Phys. Lett. *54*, 1019 (1989)

D.22 M. Lambsdorf, M. Klingenstein, J. Kuhl, C. Moglestue, J. Rozezweig, A. Axmann, J. Schneider, A. Hutsmann, H. Leier, A. Forchel: Appl. Phys. Lett. *58*, 1410 (1991)

D.23 M. Klingenstein, J. Kuhl, J. Rozezweig, C. Moglestue, A. Axmann: Appl. Phys. Lett. *58*, 2503 (1991)

D.24 L. Min, R.J.D. Miller: Appl. Phys. Lett. *56*, 524 (1991)

D.25 A.C. Warren, N. Katzenellenbogen, D. Grischkowsky, J.M. Woodall: Appl. Phys. Lett. *58*, 1512 (1991)

D.26 D.H. Auston, M.C. Nuss: IEEE J. QE-*24*, 184–197 (1988)

D.27 P.R. Smith, D.H. Auston, M.C. Nuss: IEEE J. QE-*24*, 255–260 (1988)

D.28 B.B. Hu, X.-C. Zhang, D.H. Auston, P.R. Smith: Appl. Phys. Lett. *56*, 506 (1990)

D.29 B.B Hu, J.T. Darrow, X.-C. Zhang, D.H. Auston P.R. Smith: Appl. Phys. Lett. *56*, 886 (1990)

D.30 X.-C. Zhang, B.B. Hu, J.T. Darrow, D.H. Auston: Appl. Phys. Lett. *56*, 1011 (1990)

D.31 J.T. Darrow, B.B. Hu, X.-C. Zhang, D.H. Auston: Opt. Lett. *15*, 323 (1990)

D.32 L. Xu, X.-C. Zhang, D.H. Auston: Appl. Phys. Lett. *59*, 3357 (1991)

D.33 N. Froberg, M. Mack, B.B. Hu, X.-C. Zhang, D.H. Auston: Appl. Phys. Lett. *58*, 446 (1991)
D.34 J.T. Darrow, X.-C. Zhang, D.H. Auston: Appl. Phys. Lett. *58*, 25–27 (1991)
D.35 B.I. Greene, J.F. Federici, D.R. Dykaar, R.R. Jones, P.H. Bucksbaum: Appl. Phys. Lett. *59*, 893–895 (1991)
D.36 D.R. Dykaar, B.I. Greene, J.F. Fedel, A.F.J. Levi, L.N. Pfeiffer, R.F. Kopf: Appl. Phys. Lett. *59*, 262–264 (1991)
D.37 M.C. Nuss, K.W. Goossen, P.M. Mankiewich, M.L. O'Malley: Appl. Phys. Lett. *58* 2561 (1991)
D.38 M.C. Nuss, P.M. Mankiewich, M.L. O'Malley, E.H. Westerwick: Phys. Rev. Lett. *66*, 3305 (1991).
D.39 M.C. Nuss, K.W. Goossen, J.P. Gordon, P.M. Mankiewich, M.L. O'Malley: J. Appl. Phys. *70*, 2238, (1991)
D.40 D. Grischkowsky, N. Katzenellenbogen: Femtosecond pulses of terahertz radiation: Physics and applications in *Picosecond Electronics and Optoelectronics*, Technical Digest Series (Optical Society of America, Washington, DC 1991) pp. 9–14
D.41 M. van Exter, D.R. Grischkowsky: IEEE Trans. MTT-*38*, 1684–1691 (1990)
D.42 N. Katzenellenbogen, D. Grischkowsky: Appl. Phys. Lett. *58*, 222–224 (1991)
D.43 X.-C. Zhang, J.T. Darrow, B.B. Hu, D.H. Auston, M.T. Schmidt, P. Tham, E.S. Yang: Appl. Phys. Lett. *56*, 2228–2230 (1990)
D.44 S.L. Chuang, S. Schmidt-Rink, B.I. Greene, P.N. Saeta, F.J. Levi: Phys. Rev. Lett. *68*, 102 (1992)
D.45 K. Meyer, M. Pessot, G. Mourou, R. Grondin, S. Chamoun: Appl. Phys. Lett. *53*, 2254 (1988)
D.46 A.E. Iverson, G.M. Wysin, D.L. Smith, A. Redondo: Appl. Phys. Lett. *62*, 2148–2150 (1988)
D.47 X.-C. Zhang, B.B Hu, X.H. Xin, D.H. Auston: Appl. Phys. Lett. *57*, 753 (1990)
D.48 X.-C. Zhang, J.T. Darrow, B.B. Hu, D.H. Auston, M.T. Schmidt, P. Tham, E.S. Yang: Appl. Phys. Lett. *56*, 2228 (1990)
D.49 B.B. Hu, X.-C. Zhang, D.H. Auston: Phys. Rev. Lett. *67*, 2709 (1991)
D.50 X.-C. Zhang, D.H. Auston: J. Appl. Phys. *71*, 326 (1992)
D.51 Li Xu, A. Hasegawa, D.H. Auston: Phys. Rev. A *45*, 3184 (1992)
D.52 B.I. Greene, J.F. Federici, D.R. Dykaar, A.F.J. Levi, L. Pfeiffer: Opt. Lett. *16*, 4850 (1991)
D.53 D. Grischkowsky, S. Keiding, M. van Exter, Ch. Fattinger: J. Opt. Soc. Am. B 7, 200–2015 (1990)
D.54 C. Fattinger, D. Grischkowsky: Appl. Phys. Lett. *53*, 1480 (1988)
D.55 C. Fattinger, D. Grischkowsky: Phys. Rev. Lett. *62*, 2961 (1989)
D.56 D. Grischkowsky, S. Keiding: Appl. Phys. Lett. *57*, 1055 (1990)
D.57 M. van Exeter, C. Fattinger, D. Grischkowsky: Opt. Lett. *14*, 1128 (1989)
D.58 J.M. Chwalek, C. Uher, J.F. Whitaker, G.A. Mourou: Appl. Phys. Lett. *58*, 980 (1991)
D.59 G.C. Cho, W. Kutt, H. Kurtz: Phys. Rev. Lett. *65*, 764 (1990)
D.60 H.G. Roskos, M.C. Nuss, J. Shah, K. Leo, D.A.B. Miller: Phys. Rev. Lett. *68*, 2216 (1992)

Addendum E: Ultrafast Coherent Spectroscopy

Wolfgang Zinth and Wolfgang Kaiser

The recent technological developments since the writing of the original Chapter 6 in the generation of light pulses with a duration in the range of 10^{-14} s allows systematic investigations in a time domain approaching the oscillation time of visible light. A substantial number of recent publications is briefly summarized in four sections: On the time-scale of femtoseconds the standard "incoherent" optical probing techniques start to monitor the coherent response of the medium. A clear distinction between coherent and incoherent spectroscopy is no longer possible. From this point of view it is not surprising that oscillations on the signal curve, often called quantum beats, are ,observed in a variety of ultrafast experiments (Sect. E.1). Valuable information is obtained with typical coherent experimental techniques like three-pulse echo or hole-burning experiments using the shortest available light pulses (Sect. E.2). Based on the large bandwidth of the femtosecond light pulses a new excitation scheme is possible: The impulsive stimulated Raman scattering leads to the population of low-frequency vibrational modes (Sect. E.3). Several research groups have presented interesting improvements and applications of conventional coherent techniques towards unprecedented time resolution and ultrahigh sensitivity (Sect. E.4).

E.1 Oscillations in Femtosecond Experiments—Wave-Packet Motion on Electronically Excited Potential Surfaces

The short duration of femtosecond pulses is related, via a Fourier transformation, with a broad spectral width. The broad spectrum of short pulses opens interesting new possibilities for the study of excited electronic levels of large organic molecules: During the excitation process with femtosecond pulses a number of (low-frequency) vibronic levels in the S_1 vibrational manifold is excited simultaneously. This fact leads to a coherent superpositon of vibrational excitations which can be visualized as a wave packet. After excitation, this wave packet evolves in time. Its motion is determined by the action of the different resonance frequencies and dephasing times of the vibrational modes, and by the shape of the potential-energy surface. Excitation, motion and detection of these wave packets have been described in a number of publications. While *Heller* [E.1] discussed the basic principles, *Pollard* et al. applied the ideas to the interpretation of femtosecond experiments [E.2–6]. A detailed theoretical description of various forms of time- and frequency-resolved absorption experiments on complex molecular systems was presented by *Stock* and *Domcke* [E.7–11]. The investigations were subsequently extended to the analysis of time-resolved ionization spectroscopy [E.12, 13].

The first experimental indications of fast oscillations in time-resolved absorption experiments were presented by *Tang* et al. [E.14, 15]. They found oscillations in time-resolved equal-pulse absorption studies in solutions of the dye molecules malachite green and nile blue. More recently oscillations were observed in standard pump-probe and fluorescence experiments. Results were reported for semiconductors, for semiconductor quantum-well structures and even for the photosynthetic reaction center at low temperatures [E.16–18]. Of special interest are experiments where terahertz quantum beats were found in the fluorescence of dye solutions [E.19, 20].

In the case of reactive molecules the motion of the initially prepared wave packet may reflect the dynamics of the light-induced reaction. It was shown by *Zewail* et al. [E.21–25] that femtosecond spectroscopy allows to follow in real time the reaction i. e. the dissociation of small molecules in the gas phase. *Gerber* et al. recently reported first experimental studies of time-resolved molecular multiphoton ionization [E.26]. Wave-packet motion was also suggested to occur in the initial absorption transients of the protein bacteriorhodopsin. Here a fast photo-isomerization of the retinal molecule depopulates the excited electronic state within 500 fs [E.27]. Prior to this internal conversion process one finds rapid absorption transients which point to a very fast decay of an initially prepared wave packet [E.27, 28].

E.2 Femtosecond Hole-Burning and Echo Experiments

In condensed-matter physics the fastest electronic and molecular processes are of the order of one femtosecond. With recent experimental techniques these extremely short time constants are now accessible. In general, broad and structureless absorption bands hide the dynamics. The first successful hole-burning experiments on dye molecules in room-temperature liquids were presented by *Shank* et al. in 1986 [E.30, 31]. In these experiments a 10 fs probe pulse monitored the absorption change induced by a 60 fs excitation pulse. The data revealed modulations in the spectral transmittance at early delay times. These "findings" and the implication on the dephasing time T_2 were analyzed in detail [E.31]. Similar experimental results were reported for excitons of CdSe and GaAs where pronounced oscillations of the "induced transmission" were found at negative delay times [E.32, 33]. These oscillations allow to determine the dephasing time of the medium.

The first observations of photon echos for large organic molecules in liquids and for GaAs were reported by *Shank* et al. [E.34, 35]. For dye molecules in solutions it was shown that the echo decays in a first step due to the level multiplicity. Afterwards pronounced quantum beats appear. The underlying dephasing times are in order of 20 to 80 fs. From these data it was realized that the initial dephasing process reveals non-Markovian behaviour due to the stochastic coupling of the molecules to the heat bath [E.36, 37]. Similar experimental techniques provided very interesting results on carrier-carrier scattering times in GaAs. The concentration dependence gave time constants of several femtoseconds for electrons of $N_0 = 5 \cdot 10^{18}$ cm^{-3}.

E.3 Impulsive Stimulated Raman Scattering

The large bandwidth of femtosecond pulses not only leads to vibrational wave packets on the S_1-potential energy surface but also excites low-frequency vibrations in the electronic ground state. This impulsive excitation of vibrational modes can be understood in terms of stimulated Raman scattering: Both exciting fields, the Stokes and the laser field are contained in the bandwidth of the femtosecond pulse [E.38]. With pulses of a duration of $tp \simeq 100$ fs it is possible to drive coherently vibrational modes up to $v/c = 200$ cm^{-1}. Impulsive stimulated scattering generates a coherent vibrational excitation which can be probed by a delayed femtosecond light pulse. The femtosecond technique—introduced by *Nelson* in 1987—was demonstrated in a variety of experimental studies on molecular liquids, solids and even on reactive systems [E.38–48]. Impulsive stimulated excitation of vibrational modes occurs in a stimulated Raman set-up with two exciting pulses of the same frequency. It may also happen when a single, intense light pulse travels through a medium. Under the latter conditions [E.38, 49–53] a femtosecond probing pulse may be effected in several ways: the polarization of the probe pulse is changed [E.52, 53] and the spectral position of the probe pulse oscillates with the period of the molecular vibration. The later effect can be observed with suitable spectral detection. It should be noted that one pulse impulsive stimulated scattering occurs always when a short light pulse passes a Raman active medium. As a consequence special care has to be taken in femtosecond excite-and-probe absorption experiments to minimize the contributions from ground-state vibrational motion [E.51].

Impulsive stimulated scattering couples nonselectively to any Raman active low-frequency vibrational mode. The excitation of one specific low-lying vibrational mode is possible with a special sequence of exciting light pulses. *Weiner* et al. [E.54, 55] have presented the experimental realization of this idea: They applied multiple femtosecond light pulses with intervals chosen in such a way that the excitation by the different pulses added constructively. By this technique they obtained strong excitation of one specific vibrational mode.

E.4 New Results Via Conventional Coherent Techniques

In the early 80's it was suggested to use intense and incoherent light in order to measure fast coherent processes in condensed matter [E.56]. The technique is based on the idea that incoherent light can be understood as a series of ultrashort light pulses which have durations determined by the coherence time. The application of broad-band nanosecond pulses of dye lasers with small coherence times made the time domain of 10^{-13} s accessible with relatively simple experimental systems. Recently vibrational dephasing [E.57] and electronic dephasing in small semiconductor particles were studied on the time scale of 10^{-13} s [E.58, 59].

Conventional coherent Raman techniques also approached the time domain of several 10 fs. With the use of femtosecond dye lasers fast dephasing processes could be studied with a time resolution of better than 50 fs [E.60]. Simultaneously, the precision of the determination of dephasing times was strongly improved [E.60–64]. In a study of femtosecond CARS on simple liquids *Laubereau* et al. demonstrated that dephasing times may be determined with a precision of 1 % and that weak contributions due to inhomogeneous broadening may be detected [E.61]. This extremely high precision of time-resolved experiments makes them superior to spontaneous spectroscopy for the study of dynamical processes in liquids. It has been demonstrated that the rotational motion of small molecules can be investigated via time-resolved CARS [E.61, 65–68]. Different experimental groups, e.g., *Goebel* et al. have applied time-resolved coherent spectroscopy to the study of fast processes in semiconductor systems. Quantum beats of light hole and heavy-hole excitons in quantum wells were observed [E.69, 70]. Unexpectedly long dephasing times up to several hundred picoseconds were found for localized excitons in $CdS_x\ Se_{1-x}$ mixed crystals [E.71].

References

E.1 E.J. Heller: Accounts Chem. Res. *14*, 368 (1981) and references there in

E.2 S.Y. Lee, W.T. Pollard, R.A. Mathies: Chem. Phys. Lett. *160*, 531 (1989)

E.3 W.T. Pollard, S.Y. Lee, R.A. Mathies: J. Chem. Phys. *92*, 4012 (1990)

E.4 W.T. Pollard, H.L. Fragnito, J.-Y. Bigot, C.V. Shank, R.A. Mathies: Chem. Phys. Lett. *168*, 239 (1990)

E.5 S.Y. Lee, W.T. Pollard, R.A. Mathies: J. Chem. Phys. *90*, 6146 (1989)

E.6 W.T. Pollard, R.A. Mathies: In *Ultrafast Phenomena VII*, ed. by C.B. Harris, E.P. Ippen, G.A. Mourou, A.H. Zewail, Springer Ser. Chem. Phys., Vol. 53 (Springer, Berlin, Heidelberg 1990), p. 154

E.7 G. Stock, W. Domcke: Chem. Phys. *124*, 227 (1988)

E.8 G. Stock, R. Schneider, W. Domcke: J. Chem. Phys. *90*, 7184 (1989)

E.9 G. Stock, W. Domcke: J. Opt. Soc. Am. B 7, 1970 (1990)

E.10 G. Stock, W. Domcke: J. Chem. Phys. *93*, 5496 (1990)
E.11 G. Stock, W. Domcke: Phys. Rev. *A 45*, 3052 (1992)
E.12 M. Seel, W. Domcke: Chem. Phys. *151*, 59 (1991)
E.13 M. Seel, W. Domcke: J. Chem. Phys. *95*, 7806 (1991)
E.14 M.J. Rosker, F.W. Wise, C.L. Tang: Phys. Rev. Lett. *57*, 321 (1986)
E.15 F.W. Wise, M.J. Rosker, C.L. Tang: J. Chem. Phys. *86*, 2827 (1987)
E.16 J.M. Smith, C. Lakshminarayan, J.L. Knee: J. Chem. Phys. *93*, 4475 (1990)
E.17 K. Leo, J. Shah, E.O. Göbel, T.C. Damen: Phys. Rev. Lett. *66*, 201 (1991)
E.18 M.H. Vos, J.-C. Lambry, St.J. Robles, D.C. Jouvan, J. Breton, J.-L. Martin: Proc. Nat'l. Acad. Sci. (USA) *88*, 8885 (1991)
E.19 A. Mokhtari, A. Chebira, J. Chesnoy: J. Opt. Soc. Am. B 7, 1551 (1990)
E.20 A.E.A. Mokhtari, J. Chesnoy: IEEE J. QE-*25*, 2528 (1989)
E.21 M.J. Rosker, M. Dantus, A.H. Zewail: J. Chem. Phys. *89*, 6113 (1988)
E.22 M. Dantus, M.J. Rosker, A.H. Zewail: J. Chem. Phys. *89*, 6128 (1988)
E.23 M.J. Rosker, T.S. Rose, A.H. Zewail: Chem. Phys. Lett. *146*, 175 (1988)
E.24 R.M. Bowman, M. Dantus, A.H. Zewail: Chem. Phys. Lett. *161*, 297 (1989)
E.25 J.J. Gerdy, M. Dantus, R.M. Bowman, A.H. Zewail: Chem. Phys. Lett. *171*, 1 (1990)
E.26 T. Baumert, M. Grosser, R. Thalweiser, G. Gerber: Phys. Rev. Lett. *67*, 3753 (1991)
E.27 J. Dobler, W. Zinth, W. Kaiser, D. Oesterhelt: Chem. Phys. Lett. *144*, 215 (1988)
E.28 R.A. Mathies, C.H. BritoCruz, W.T. Pollard, C.V. Shank: Science *240*, 777 (1988)
E.29 W. Zinth, J. Dobler, K. Dressler, W. Kaiser: In *Ultrafast Phenomena VI*, ed. by T. Yajima, K. Yoshihara, C.B. Harris, S. Shionoya, Springer Ser. Chem. Phys., Vol. 48 (Springer, Berlin, Heidelberg 1988) p. 581
E.30 C.H. BritoCruz, R.L. Fork, W.H. Knox, C.V. Shank: Chem. Phys. Lett. *132*, 341 (1986)
E.31 C.H. BritoCruz, J.P. Gordon, P.C. Becker, R.L. Fork, C.V. Shank: IEEE J. QE-*24*, 261 (1988)
E.32 B. Fluegel, N. Peyghambarian, G. Olbright, M. Lindberg, S.W. Koch, M. Joffre, D. Hulin, A. Migus, A. Antonetti: Phys. Rev. Lett. *59*, 2588 (1987)
E.33 M. Joffre, D. Hulin, A. Migus, A. Antonetti, C. Benoit à la Guillaume, N. Peyghambarian, M. Lindberg, S.W. Koch: Opt. Lett. *13*, 276 (1988)
E.34 C.V. Shank, P.C. Becker, H.L. Fragnito, R.L. Fork: In *Ultrafast Phenomena VI*, ed. by T. Yajima, K. Yoshihara, C.B. Harris, S. Shionoya, Springer Ser. Chem. Phys., Vol. 48, (Springer, Berlin, Heidelberg 1988) p. 344
E.35 P.C. Becker, H.L. Fragnito, J.Y. Bigot, C.H. BritoCruz, R.L. Fork, C.V. Shank: Phys. Rev. Lett. *63*, 505 (1989)
E.36 W. Vogel, D.-G. Welsch, B. Wilhelmi: Phys. Rev. A *37*, 3825 (1988)
E.37 J.-Y. Bigot, M.T. Portella, R.W. Schoenlein, C.J. Bardeen, A. Migus, C.V. Shank: Phys. Rev. Lett. *66*, 1138 (1991)
E.38 K.A. Nelson, E.P. Ippen: Adv. Chem. Phys. *75*, 1 (1989)
E.39 Y.-X. Yan, K.A. Nelson: J. Chem. Phys. *87*, 6240 (1987)
E.40 S. Ruhman, A.G. Joly, B. Kohler, L.R. Williams, K.A. Nelson: Revue Phys. Appl. *22*, 1717 (1987)
E.41 K.A. Nelson, L.R. Williams: Phys. Rev. Lett. *58*, 745 (1987)
E.42 Y.-X. Yan, K.A. Nelson: J. Chem. Phys. *87*, 6257 (1987)
E.43 S. Ruhman, A.G. Joly, K.A. Nelson: IEEE J. QE-*24*, 460 (1988)
E.44 S. Ruhman, B. Kohler, A.G. Joly, K.A. Nelson: IEEE J. QE-*24*, 470 (1988)
E.45 S. Ruhman, B. Kohler, A.G. Joly, K.A. Nelson: Chem. Phys. Lett. *141*, 16 (1987)
E.46 S. Ruhman, K.A. Nelson: J. Chem Phys. *94*, 859 (1991)
E.47 S. Ruhman, L.R. Williams, A.G. Joly, B. Kohler, K.A. Nelson: J. Phys. Chem. *91*, 2237 (1987)
E.48 D. McMorrow, W.T. Lotshaw, G.A. Kenney-Wallace: IEEE J. QE-*24*, 443 (1988)
E.49 Y.-X. Yan, E.B. Gamble, K.A. Nelson: J. Chem. Phys. *83*, 5391 (1985)
E.50 S. Ruhman, A.G. Joly, K.A. Nelson: J. Chem. Phys. *86*, 6563 (1987)
E.51 J. Dobler, W. Zinth: private communication (1987)

E.52 A. Mokhtari, J. Chesnoy: Europhys. Lett. *5*, 523 (1988)
E.53 J. Chesnoy, A. Mokhtari: Phys. Rev. A *38*, 3566 (1988)
E.54 A.M. Weiner, D.E. Leaird, G.P. Wiederrecht, K.A. Nelson: Science *247*, 4948, 1317 (1990)
E.55 A.M. Weiner, D.E. Leaird, G.P. Wiederrecht, K.A. Nelson: J. Opt. Soc. Am. B *8*, 1264 (1991)
E.56 N. Morita, T. Yajima: Phys. Rev. A *30*, 2525 (1984)
E.57 T. Kobayashi, T. Hattori, A. Terasaki, K. Kurokawa: Revue Phys. Appl. *22*, 1773 (1987)
E.58 T. Tokizaki, Y. Ishida, T. Yajima: In *Ultrafast Phenomena VI*, ed. by T. Yajima, K. Yoshihara, C.B. Harris, S. Shionoya, Springer Ser. Chem. Phys. Vol. 48, (Springer, Berlin, Heidelberg 1988) p. 372
E.59 K. Misawa, T. Hattori, Y. Ohashi, H. Itoh, T. Kobayashi: In *Ultrafast Phenomena VI*, ed. by T. Yajima, K. Yoshihara, C.B. Harris, S. Shionoya, Springer Ser. Chem. Phys. Vol. 48 (Springer, Berlin, Heidelberg 1988) p. 384
E.60 W. Zinth, W. Holzapfel, R. Leonhardt: In *Ultrafast Phenomena VI*, ed. by T. Yajima, K. Yoshihara, C.B. Harris, S. Shionoya, Springer Ser. Chem. Phys. Vol. 48 (Springer, Berlin, Heidelberg 1988) p. 461
E.61 M. Fickenscher, A. Laubereau: J. Raman spectroscopy *21*, 857 (1990)
E.62 P. Aechtner, M. Fickenscher, A. Laubereau: Ber. Bunsenges. Phys. Chem. *94*, 399 (1990)
E.63 P. Aechtner, A. Laubereau: Chem. Phys. *419*, 419 (1991)
E.64 W. Zinth, R. Leonhardt, W. Holzapfel, W. Kaiser: IEEE J. QE-*24*, 455 (1988)
E.65 N. Kohles, A. Laubereau: Chem. Phys. Lett. *138*, 365 (1987)
E.66 J. Etchepare, G. Grillon, J. Arabat: Appl. Phys. B *49*, 425 (1989)
E.67 N. Kohles, P. Aechtner, A. Laubereau: Optics Commun. *65*, 391 (1988)
E.68 For more recent results on CARS see G. Marowsky, V.V. Smirnov (eds.): *Coherent Raman Spectroscopy: Recent Advances*. Springer Proc. Phys. *63* (Springer, Berlin, Heidelberg 1992)
E.69 K. Leo, T.C. Damen, J. Shah, E.O. Göbel, K. Köhler: Appl. Phys. Lett. *57*, 19 (1990)
E.70 E.O. Göbel, K. Leo, T.C. Damen, J. Shah, S. Schmitt-Rink, W. Schäfer, J.F. Müller, K. Köhler: Phys. Rev. Lett. *64*, 1801 (1990)
E.71 G. Noll, U. Siegner, S.G. Shevel, E.O. Göbel: Phys. Rev. Lett. *64*, 792 (1990)

Addendum F: Ultrashort Intramolecular and Intermolecular Vibrational Energy Transfer of Polyatomic Molecules in Liquids

Alois Seilmeier

Progress in the investigation of vibrational energy transfer in liquids since the writing of the original Chapter 7 was made particularly by the development of more sophisticated experimental techniques.

The potential of an infrared bleaching technique with independently tunable excitation and probing frequencies (double-resonance spectroscopy) was demonstrated by several groups. Transient infrared spectra taken at various delay times allow to investigate intra- and intermolecular energy-transfer processes in more detail. The technique was applied to small molecules to monitor the relaxation pathway.

Experimental data on $CHBr_3$ dissolved in CCl_4 revealed an intramolecular energy transfer from the excited CH-stretching mode to intermediate vibrational levels within 40 ps [F.1,2]. The depopulation of the intermediate state which was suggested to be the ν_2-mode at 540 cm^{-1} occurs with a time constant of $\sim$200 ps. On a nanosecond time-scale an increase in temperature by 12 K was observed due to the equilibration of the excess energy. Similar relaxation pathways were reported for $CHCl_3$ and CH_3I solutions, which were investigated by single-frequency infrared bleaching experiments [F.3]. Depopulation lifetimes of the CH-stretching mode of polyalkenes dissolved in CCl_4 were determined to be in the order of several tens of picoseconds [F.4].

A series of infrared double-resonance experiments were performed to study relaxation processes after excitation of OH-stretching vibrations of hydrogen bonds. In ethanol diluted in CCl_4 a vibrational depopulation due to rapid bond breaking within 5 ps was observed followed by a partial reassociation with a time constant of 20 ps [F.5]. After 100 ps a new thermal equilibrium with an elevated temperature is established. A detailed picture on energy transfer rates between CH-and OH-stretching modes and dissociation processes was developed from investigations at various excitation frequencies and from model calculations [F.6].

In HDO, spectral hole burning was observed in the broad absorption band of OH-stretching modes around 3400 cm^{-1}, which is composed of a discrete set of subcomponents [F.7]. Various subcomponents, which represent different molecular configurations were excited selectively. The depopulation time T_1 of one of the prominent components of the OH mode was found to be 8 ps. After 50 ps a transmission change due to an increased sample temperature was measured.

Data on the vibrational depopulation of carbonyl vibrations investigated by double-resonance experiments were reported in several publications. In acetyl bromide (H_3COBr) dissolved in CCl_4 the depopulation lifetime of the CO mode at 1812 cm^{-1} was determined to be 10 ps [F.8]. In Rh $(CO)_2$ in n-hexane a rapid energy equilibration (≤ 10 ps) between the symmetric and asymmetric CO-stretching mode is followed by a decay of the coupled modes within $T_1 = 64$ ps [F.9]. A similar strong coupling between the CO-stretching modes was found in $Co(CO)_3NO$. Excitation of the CO stretching vibration ($\tilde{\nu}_8 = 2048$ cm^{-1}) and the NO-stretching vibration ($\tilde{\nu}_2 = 1822$ cm^{-1}) showed, however, only weak CO-NO coupling; a T_1-time of 148 ps and 37 ps was obtained. respectively [F.9]. Transient infrared spectra at the CO frequencies were also used to detect vibrational heating during photochemical processes in hemoglobin [F.8,10].

New information on the vibrational-energy transfer in large molecules was obtained from several experiments. Infrared pulses tunable in the wavelength range from 4 to 10 μm allow the investigation of vibrational relaxation after excitation of skeletal modes and C = O modes at frequencies between 1000 and 2000 cm^{-1} [F.11]. A comparison of time-resolved data with linewidth measurements reveals that most of the vibrational lines in this spectral range are lifetime broadened, i.e. the intramolecular energy redistribution proceeds with time constants of several 100 fs. Vibrational energy redistribution and vibrational-energy transfer in Rhodamin 6G solutions were observed by a novel polarization-dependent amplification technique [F.12]. Depopulation of manifolds of vibrational levels at approximately 1000 cm^{-1} occured within a few picoseconds .

Vibrational relaxation in the S_1 state was studied with femtosecond time resolution. Subpicosecond time constants were found in dye solutions using an induced dichroism technique [F.13] or a fluorescence up-conversion technique [F.14].

High-lying S_n levels were resonantly excited by femtosecond ultraviolet pulses. The intramolecular energy redistribution was monitored by measuring the onset of fluorescent emission, which originates from low-lying vibronic levels of the electronic S_1 state [F.15].

A detailed theoretical model for intra- and intermolecular transfer of vibrational energy is still not existing. Progress has been made by a model which combines statistical elements with the collisional dynamics of the interaction [F.16]. The theory describes many aspects of the intermolecular-energy transfer of large molecules in solution fairly well.

Recently a resonance Raman technique was introduced for the time-resolved investigation of vibrational levels [F.17,18]. The technique does not require samples which are transparent in the infrared. The method was applied to investigate the excitation and relaxation of vibrational levels during the photocycle of bacteriorhodopsin.

For a more detailed discussion on vibrational and vibronic relaxation of large polyatomic molecules in liquids the reader is referred to the recent review paper [F.19].

References

F.1 H. Graener, R. Dohlus, A. Laubereau: Chem. Phys. Lett. *140*, 306 (1987)
F.2 H. Graener: Chem. Phys. Lett. *165*, 110 (1990)
F.3 H.J. Bakker, P.C.M. Planken, L. Kuipers, A. Lagendijk: J. Chem. Phys. *94*, 1730 (1991)
F.4 H. Graener, T.Q. Ye, A. Laubereau: J. Phys. Chem. *93*, 7044 (1989)
F.5 H. Graener, T.Q. Ye, A. Laubereau: J. Chem. Phys. *90*, 3413 (1989)
F.6 H. Graener, T.Q. Ye, A. Laubereau: J. Chem. Phys. *91*, 1043 (1989)
F.7 H. Graener, G. Seifert, A. Laubereau: Phys. Rev. Lett. *66*, 2092 (1991)
F.8 R.M. Hochstrasser, P.A. Anfinrud, R. Diller, C. Han, M. Jannone, T. Lian, B. Locke: In *Ultrafast Phenomena VII*, ed. by C.B. Harris, E.P. Ippen, G.A. Mourou, A.H. Zewail, Springer Ser. Chem. Phys., Vol. 53, (Springer Berlin, Heidelberg 1990) p. 429
F.9 S.A. Angel, P.A. Hansen, E.J. Heilweil, J.C. Stephenson: In *Ultrafast Phenomena VII*, ed. by C.B. Harris, E.P. Ippen, G.A. Mourou, A.H. Zewail, Springer Ser. Chem. Phys., Vol. 53 (Springer Berlin, Heidelberg 1990) p. 480
F.10 P.A. Anfinrud, C. Han, R.M. Hochstrasser: Proc. Nat'l. Acad. Sci. (USA) *86*, 8387 (1989)
F.11 H.-J. Hübner, M. Wörner, W. Kaiser, A. Seilmeier: Chem. Phys. Lett. *182*, 315 (1991)
F.12 G. Angel, R. Gagel, A. Laubereau: Chem. Phys. Lett. *156*, 169 (1989)
F.13 G. Angel, R. Gagel, A. Laubereau: Chem. Phys. *131*, 129 (1989)
F.14 A. Mokhtari, J . Chesnoy, A. Laubereau: Chem. Phys . Lett. *155*, 593 (1989)
F.15 F. Laermer, T. Elsaesser, W. Kaiser: Chem. Phys. Lett. *156*, 381 (1989)
F.16 U. Sukowski, A. Seilmeier, T. Elsaesser, S.F. Fischer: J. Chem. Phys. *93*, 4094 (1990)
F.17 T.L. Brack, G.H. Atkinson: J. Phys. Chem. *95*, 2351 (1991)
F.18 D. Blanchard, D.A. Gilmore, T.L. Brack, H. Lemaire, D. Hughes, G.H. Atkinson: Chem. Phys. *154*, 155 (1991)
F.19 T. Elsaesser, W. Kaiser: Annu. Rev. Phys. Chem *42*, 83 (1991)

Addendum G: Ultrafast Chemical Reactions in the Liquid State

P.F. Barbara* and Kenneth B. Eisenthal

With 1 Figure

Research on ultrafast chemical reactions in liquids has expanded greatly in the last five years since the writing of the original Chapter 8, in scientific scope, in the number of workers in the field, and in the fundamental impact of this work. Recent general progress in the area of experimental and theoretical research on chemical reactions in liquids have been the subject of a few special issues and reviews [G.1–5]. The specific fields of ultrafast studies on electron transfer [G.6,7], solvation dynamics [G.6–8] and proton transfer [G.9–10] have also been reviewed.

Solvation Dynamics

There has been intense activity in the last few years in the measurement of solvation dynamics by the transient Stokes-shift technique using subpicosecond and longer timescale fluorescence spectroscopy on polar fluorescent probes [G.6–8]. The solvation dynamics of most common liquids, including water, are now known for these time scales. The combination of theoretical research [G.11–14], especially molecular dynamics calculations [G.15–18], *and* experimental measurements [G.19–23] have led to new insight into the molecular scale motion of liquids. The collective relaxation effect of the solvent polarization that is predicted by continuum theory is roughly supported by experiment, since a significant portion of the solvation dynamics of several liquids occur on the order of t_L, the longitudinal relaxation time, which is the simple continuum-theory prediction for the solvation time [G.6–8]. But, deviations from continuum theory, especially at early times, indicate that a molecular-scale theoretical treatment is necessary for an accurate description [G.15]

Electron Transfer

A related area of active research in ultrafast spectroscopy has been the study of the dynamic solvent effect on electron-transfer reactions with small energy

* Department of Chemistry, University of Minnesota, Minneapolis, MN 55455-0431, USA

barriers [G.6,7]. For the intramolecular excited-state electron transfer of 9,9′-bianthryl it has been observed that the electron rate constant k_{et} is very close to the inverse solvation time t_s in a variety of solvents [G.24–26]. This is in qualitative agreement with simple theoretical models for the dynamic solvent effect which predict that (i) k_{et} should be roughly proportional to t_s^{-1} for activated electron transfer, and (ii) k_{et} should be approximately equal to t_s^{-1} for barrierless electron transfers. The results on 9,9′-bianthryl also agree with more sophisticated theoretical simulations [G.25]. Evidence for the dynamic solvent effect has also been observed in ultrafast experiments on the excited-state electron transfer of bis-(4,4′-dimethylaminophenyl)-sulfone DMAPS [G.27], 4-(9-anthryl)-N,N-dimethylanitine ADMA [G.28,29], and other systems [G.6,7,30,31].

A very recent focus of ultrafast studies on intramolecular electron transfer is how intramolecular vibrational promotion of electron transfer can effect electron-transfer rates, and, in particular, how the dynamic solvent effect is altered when vibrational promotion is significant. For example, the dependence of k_{et} on solvent and temperature variations for DMAPS [G.27] are consistent with a theory that combines vibrational promotion and the dynamic solvent effect [G.32]. An even more dramatic example is the inverted-regime $S_1 \rightarrow S_o$ electron transfer of the betaine compounds [G.33]. k_{et} for the betaines varies almost linearly with t_s for quickly relaxing solvents, but a much milder dependence on t_s is seen for slowly relaxing solvents where $k_{et} \gg t_s^{-1}$ (see fig. G.1). The experimental k_{et} is only weakly temperature dependent. The experimental results are in reasonable agreement with a theoretical model that includes a solvent coordinate and two types of intramolecular vibrational modes, a low-

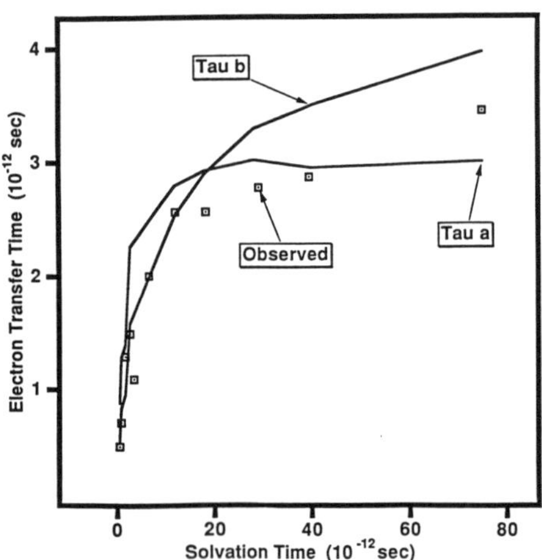

Figure G.1. The experimentally observed electron transfer time (squares) and the electron-transfer times (τ_a and τ_b) predicted by a theoretical model [G.33]. For fast solvents the predicted and observed times are a strong function of solvent dynamics. For slower solvents intramolecular degrees of freedom control the electron transfer rate.

frequency promoting mode and a high-frequency accepting mode. All three elements are apparently necessary to adequately describe ultrafast inverted-regimes electron transfers like the betaines, since theories [G.32,34] with less degrees of freedom are in disagreement with experiment by many orders-of-magnitude [G.33]. The results on the betaines strongly indicate that the dynamic solvent effect on electron transfer is particularly muted by vibrational promotion of electron transfer for inverted-regime reactions. Another example of an inverted-regime electron transfer that exhibits $k_{et} \gg t_s^{-1}$ due to vibrational promotion is found in the excited-state electron transfer of ADMA [G.28,35,36]. Analogous results have been observed in the intermolecular electron transfer of nile blue in aniline solvents [G.37].

Solvated Electrons

There has been a great deal of activity concerning the aqueous or hydrated electron, a subject of great interest in the physical and biological sciences. Besides being the simplest anion, it has a rich variety of chemical and physical properties that continue to stimulate extensive research since its spectral identification in 1962 [G.38–40]. In the last few years there have been significant advances in our understanding of the dynamics of electron solvation due to the application of femtosecond laser techniques and theoretical work using computer-simulation methods [G.41–56]. The key issue in the water solvation dynamics is whether the solvation dynamics can be described by a two-state model. The two-state model means that there are essentially two-electron species that are important in the dynamics. One is the presolvated localized electron, often called the wet electron, and the final, the equilibrium solvated electron to which the wet electron makes a transition. There are thus two absorbing species, the wet and solvated electrons rather than a series of wet electrons having their own absorption, which evolve into the hydrated electron. There are, of course, different wet electrons, which, for example, are responsible for the inhomogeneous spectral width, but which would not be important in the kinetics if the rate of spectral relaxation between the different species is rapid compared to the time scale of the wet to solvated transition. The discovery of an isosbestic point at about 820 nm strongly supported the two-state model in agreement with other studies [G.43]. The application of the two-state model permitted the determination of the wet-electron spectrum for the first time, and rate constants for the formation of the wet electron (110 fs [G.42,50,54], 300 fs [G.43a]), and its transition to the solvated electron (240 fs [G.42,50,54], 540 fs [G.43a]). The experimental results and success of the two-state model are consistent with theoretical simulations that showed that nonadiabatic transitions are the key process in electron solvation in water [G.51–53]. A proposed physical basis for the two-state model is that the wet electron is the lowest excited state of the solvated electron [G.43,47–49,51–53]. The solvation dynamics can therefore be viewed as an internal conversion process following electron localization.

Geminate Recombination of Electrons with Parent Ions and Products

Using femtosecond methods the geminate recombination dynamics of the solvated electron with what is most probably the reaction products of the photogenerated parent ion, $H_2O^+ + H_2O \rightarrow H_3O^+ + OH$, has been observed. In addition to the recombination dynamics in neat water and deuterated water, where isotope effects were observed in the solvation as well as the recombination dynamics, studies in liquid alkanes were carried out [G.56–61]. Following the advances made with picosecond lasers the further improved time resolution made possible with femtosecond lasers permitted the subpicosecond recombination dynamics to be studied with neat alkanes at room temperature. In various alkanes (octane, isooctane, hexane, and isohexane), it was found that the experimental kinetic data are in a good agreement with the Onsager model of electron diffusion in a Coulomb field [G.58,59,61–63]. One of the interesting findings was that although the thermalization distances obtained for n-octane and isooctane (2,2,4 trimethyl pentane) on three photon excitation with a 312-nm pump, were similar, 48 and 54 Å, the observed kinetics were very different [G.57,58]. Even though their densities and viscosities are very similar, the recombination in the isooctane was much more rapid, being completed in less than a picosecond whereas in n-octane it took tens of picoseconds. The difference in the dynamics was attributed to the difference in the molecular structure of the hydrocarbons. In the linear octane there is more disorder due to the entangling chains than is the case for the more spherical isooctane molecules, and thus results in a slower recombination.

Proton Transfer

Recent ultrafast experiments on proton transfer [G.9,10] have given detailed information on the kinetics and the microscopic mechanisms of those reactions. Intramolecular proton transfer in the excited state was studied in 3-hydroxyflavone [G.64], HBT [G.65–67], tinuvin [G.68–70], azaindole dimers [G.71,72] and related compounds [G.73]. In HBT [G.66] its deuterated analogue DBT [G.67], and in tinuvin [G.69], time constants of excited-state proton transfer of 100 to 200 fs were directly measured by femtosecond spectroscopy. In tinuvin, also the back-transfer of the proton in the electronic ground state occurs on a subpicosecond time scale (time constant $\tau \simeq 700$ fs), creating the (original) enol species in a highly excited vibrational ground state [G.70]. The measured time constants suggest proton transfer via intramolecular low-frequency large-amplitude modes. Novel experiments on solvent/solute complexes of proton transfer molecules have been reported in the gas-phase [G.74] and in a matrix-isolation [G.75]. A molecule with a triple fluorscence due to proton transfer and the TICT phenomenon has been reported [G.73]. Photo-induced proton injection to the solvent and geminate recombination has extensively been investigated [G.76].

A new proton-transfer system, the 1-(acylamino)-anthraquinones, has been discovered [G.77]. This reaction has the unique feature that the excited-state proton-transfer equilibrium constant can be varied systematically over a large range by changing the solvent and/or the R group on the nitrogen. Ultrafast results on this system, coupled with static spectroscopy, indicate that the proton-transfer process occurs by quantum-mechanical nuclear tunneling. Furthermore, the fact that the proton-transfer rate constant k_{PT} far exceeds t_s^{-1} for polar solvents suggests that intramolecular modes are responsible for promotion of the proton transfer reaction.

Isomerization

Ultrafast studies on isomerization reactions in recent years have been focused on the multidimensional nature of the reaction coordinate particularly in polar solvents [G.78,81]. For the barrierless isomerization of cis-stilbene, femtosecond studies indicate that the reaction is only weakly dependent on solvent friction [G.82–84], but the shape of the excited state potential energy surface is solvent dependent. Medium effects on isomerization have been investigated in clusters of cis-stilbene with solvents such as rare gases, alcohols, and hydrocarbons [G.85].

Photodissociation

Another interesting reaction type that is under active investigation is the photo-dissociation of metal carbanyl compounds, such as $Cr(CO)_6$ [G.86–92]. In these studies, the identity of the reactive intermediates and the role of the solvent have been elucidated. Transient vibrational spectroscopies, both, ir [G.89,91,92] and Raman [G.87] have particularly been revealing.

Progress has also been made on the I_2 dissociation problem, particularly with regard to understanding of vibrational relaxation in liquids [G.93].

Liquid Interfaces

In recent years the application of time-resolved second-harmonic spectroscopy has made it possible to investigate ultrafast molecular processes at liquid interfaces. The key feature of the molecular second-harmonic method that makes this possible is that it is an interface sensitive technique [G.94]. With other and more conventional spectroscopic techniques, the signal arising from the interface molecules would be overwhelmed by the much larger number of solute and solvent molecules in the bulk media. However, the bulk molecules in a centrosymmetric media do not generate a second-harmonic signal in the dipole approximation and thus the second-harmonic signal arising from the far fewer molecules in the asymmetric interfacial region can be detected.

Indeed, it is the asymmetry in forces at the interface that is responsible for its unique chemical and physical properties. The orientational structure and composition of the interface differ from the bulk media. For example, at a liquid interface there can be a preferred molecular orientation whereas in the bulk it is isotropic on the average. Similarly, the dielectric properties such as polarity, and transport properties such as viscosity, can be very different in the interface and bulk regions. One of the manifestations of these differences is in dynamic processes ranging from molecular motions to chemical reactions. In the following section we restrict the discussion to ultrafast processes at liquid interfaces, consistent with the theme of this chapter. It should be noted, however, that there has been considerable exciting research on ultrafast processes at solid surfaces using a variety of methods [G.94–103], including nonlinear ones, as well as important studies of processes that are not in the ultrafast time domain.

The approach generally used is to photoexcite molecules in the interface using a short duration pump pulse. A second pulse then selectively probes the interface by measuring the second harmonic generated at a given time after the photoexcitation step. Using the pump second-harmonic probe technique, the dynamics of intermolecular electronic-energy transfer [G.104], photoisomerization [G.105], and molecular rotation [G.106], have been investigated at the air/water interface. In the energy-transfer experiment, the enhanced decay of excited rhodamine 6 G due to the presence of cyanine dye acceptor molecules was observed. The transfer from excited donor to acceptor molecules was believed to be interfacial and not to involve the bulk acceptor molecules since the average distance from an interfacial donor molecule to a bulk acceptor molecule was large (~ 130 Å). By assuming a random distribution in density and orientation at the interface (at best a first approximation) it was found that the decay kinetics could be fit to both a two-dimensional and three-dimensional dynamics, with the fitting parameters being more resonable for the two dimensional case. The dynamics of cis-trans photoisomerization of 3,3′-diethyloxacarbocyanine iodide (DODCI), a laser dye, at the air/water interface (220 ± 40 ps) was found, using a pump second-harmonic probe method, to be significantly faster than in the bulk water (520 ± 60 ps), the latter kinetics being obtained from fluorescence-decay measurements [G.105–107]. These results suggest that the friction for motion along a reactive coordinate could be lower at the interface. Studies of molecular rotation of rhodamine 6 G at low air/water interfacial densities, were obtained and appear to be slower than in the bulk solution. In addition, effects attributed to a difference in the orientational distribution of excited-versus ground-state molecules, arising from different charge distributions are described.

References

G.1 Rate Processes in Dissipative Systemes: 50 Years after the Kramers. Special Issue, Ber. Bunsenges. Phys. Chem. **95**, 225 (1991)

G.2 Liquids, Special Issue, Chem. Phys. **149**, 1 (1990)

G.3 Friction in Liquid State Reactions. Special Issue, Chem. Phys. **152**, 1 (1991)

G.4 P. Hanggi, P. Talkner, M. Borkovec: Rev. Mod. Phys. **62**, 251 (1990); B. Bagchi, G.R. Fleming: J. Phys. Chem. **94**, 9 (1990)

G.5 B.J. Berne, M. Borkovec, J.E. Straub: J. Phys. Chem. **92**, 3711 (1988)

G.6 P.F. Barbara, W. Jarzeba: Adv. Photochem. **15**, 1 (1990)

G.7 M. Maroncelli, J. MacInnis, G.R. Fleming: Science **243**, 1674 (1989)

G.8 J.D. Simon: Acc. Chem. Res. **21**, 128 (1988)

G.9 P.F. Barbara, P.K. Walsh, L.E. Brus: J. Phys. Chem. **93**, 29 (1989)

G.10 Excited-state proton-transfer reactions are extensively discussed in the special issue: Spectroscopy and Dynamics of Elementary Proton Transfer in Polyatomic Systems. Chem. Phys. **136**, 153 (1989)

G.11 G. Vander Zwan, J.T. Hynes: J. Chem. Phys. **76**, 2993 (1982)

G.12 D.F. Calef, P.G. Wolnes: J. Phys. Cehm. **87**, 3387 (1983)

G.13 I. Rips, J. Klafter, J. Jortner: J. Chem. Phys. **88**, 3246 (1988)

G.14 B. Bagchi: Ann. Rev. Chem. **40**, 115 (1989)

G.15 M. Maroncelli, G.R. Fleming: J. Chem. Phys. **89**, 5044 (1988)
S. Rosenthal, X. Xie, M. Du, G.R. Fleming: J. Chem. Phys. *95*, 4715 (1991)

G.16 O.A. Karim, A.D.J. Haymet, M.J. Banet, J.D. Simon: J. Phys. Chem. **92**, 3391 (1988)

G.17 M. Maroncelli: J. Chem. Phys. **94**, 2084 (1991)

G.18 T. Fonseca, B.M. Ladanyi: J. Phys. Chem. **95**, 2116 (1991)

G.19 M.A. Kahlow, T.J. Kang, P.F. Barbara: J. Chem. Phys. **88**, 2372 (1988)

G.20 M.A. Kahlow, W. Jarzeba, T.J. Kang, P.F. Barbara: J. Chem. Phys. **90**, 151 (1989)

G.21 W. Jarzeba, G.C. Walker, A.E. Johnson, M.A. Kahlow, P.F. Barbara: J. Phys. Chem. **92**, 7039 (1988)

G.22 M. Maroncelli, G.R. Fleming: J. Chem. Phys. **86**, 6221 (1987)

G.23 S.G. Su, J.D. Simon: J. Phys. Chem. **91**, 2693 (1987)

G.24 T.J. Kang, M.A. Kahlow, D. Giser, S. Swallen, V. Nagarajan, W. Jarzeba, P.F. Barbara: J. Phys. Chem. **92**, 6800 (1989). M.A. Kahlow, T.J. Kang, P.F. Barbara: J. Phys. Chem. **91**, 6452 (1987)

G.25 T.J. Kang, W. Jarzeba, P.F. Barbara, T. Fonseca: Chem. Phys. **149**, 1 (1990)

G.26 N. Mataga, H. Yao, T. Okada, W. Rettig: J. Phys. Chem. **93**, 3383 (1989)

G.27 J.D. Simon, S.G. Su: Chem. Phys. **152**, 143 (1991) and reference therein.

G.28 K. Tominaga, G.C. Walkrer, T.J. Kang, P.F. Barbara: J. Phys. Chem. *95*, 10485 (1991)

G.29 T. Okada, S. Nishikawa, K. Kanaj, N. Mataga: In *Ultrafast Phenomena VII*, ed. by C.B. Harris, E.P. Ippen, G.A. Mourou, A.H. Zewail, Springer Ser. Chem. Phys., Vol. 53 (Springer, Berlin, Heidelberg, 1990) p.397

G.30 N. Mataga, S. Nishikawa, T. Asahi, T. Okada: J. Phys. Chem. **94**, 1443 (1990)
H. Leuck, M.W. Windsor, W. Rettig: J. Phys. Chem. **94**, 4550 (1990)
G.31 W. Rettig: Ber. Bunsenges. Phys. Chem. **95**, 259 (1990)
G.32 H. Sumi, R.A. Marcus: J. Chem. Phys. **84**, 4894 (1986)
W. Nadler, R.A. Marcus: J. Chem. Phys. **86**, 3906 (1987)
G.33 E. Akesson, G.C. Walker, P.F. Barbara, J. Chem. Phys. *95*, 4188 (1991)
E. Akesson, A.E. Johnson, N.E. Levinger, G.C. Walker, T.P. DuBruil, P.F. Barbara: J. Chem. Phys. *96*, 7859 (1992)
G.34 J. Jortner, M. Bixon: J. Chem. Phys. **88**, 167 (1988)
G.35 G.C. Walker, W. Jarzeba, T.J. Kang, A.E. Johnson, P.F. Barbara: J. Opt. Soc. Am. B 7, 1521 (1990)
G.36 D. Huppert, P.M. Rentzepis: J. Phys. Chem. **92**, 5466 (1988)
G.37 T. Kabyashi, Y. Takagi, H. Kandori, K. Kemnitz, K. Yoshihara: Chem. Phys. Lett. **180**, 146 (1991)
G.38 E.J. Hart, J.W. Boag: J. Am. Chem. Soc. **84**, 4090 (1962)
G.39 *Radiation Chemistry Principles and Applications*, ed. by Farhataziz and Michel A.J. Rodgers (VCH, Weinheim 1987)
G.40 *Kinetics of Inhomogeneous Processes*, ed. by G.R. Freeman, (Wiley, New York 1987)
G.41 J.M. Wiesenfeld, E.P. Ippen: Chem. Phys. Lett. **73**, 47 (1980)
G.42 A. Migus, Y. Gaudel, J.L. Matrin, A. Antonetti: Phys. Rev. Lett. **58**, 1559 (1987)
G.43 F.H. Long, H. Lu, K.B. Eisenthal: Phys. Rev. Lett. **64**, 1469 (1990)
G.43a F.H. Long, H. Lu, Xuelong Shi, K.B. Eisenthal: Chem. Phys. Lett. **185**, 47 (1991)
G.44 J. Schnitker, K. Motakabbir, P.J. Rossky, R. Friesner: Phys. Rev. Lett. **60**, 456 (1988)
G.45 A. Wallquist, G. Martyna, B.J. Berne: J. Phys. Chem. **92**, 4277 (1988)
G.46 M. Maroncelli, G.R. Fleming: J. Chem. Phys. **89**, 5044 (1988)
G.47 P.J. Rossky, J. Schnitker: J. Phys. Chem. **92**, 4277 (1989)
G.48 R.N. Barnett, U. Landman, A. Nitzan: J. Chem. Phys. **90**, 4413 (1989)
G.49 M.C. Messmer, J.D. Simon: J. Phys. Chem. **94**, 1220 (1990)
G.50 Y. Gaudel, S. Pommeret, A. Migus, N. Yamada, A. Antonetti: J. Am Chem. Soc. **112**, 2925 (1990)
G.51 F. Webster, J. Schmitker, P.J. Rossky, R. Freisner: Phys. Rev. Lett. **66**, 3172 (1991)
G.52 B. Space, D.F. Coker: J. Chem. Phys. **94**, 1976 (1991)
G.53 E. Neria, A. Nitzan, R.N. Barnett, U. Landman: Phys. Rev. Lett. **67**, 1011 (1991)
G.54 Y. Gaudel, S. Pommeret, A. Migus, A. Antonetti: J. Phys. Chem. **95**, 533, (1991)
G.55 H. Lu, F.H. Long, R.M. Bowman, K.B. Eisenthal: J. Phys. Chem. **93**, 27, (1989)
G.56 F.H. Long, H. Lu, K.B. Eisenthal: Chem. Phys. Lett. **160**, 464 (1989)
G.57 R.M. Bowman, H. Lu, K.B. Eisenthal: J. Chem. Phys. **89** (1988)
G.58 H. Lu, F.H. Long, K.B. Eisenthal: J. Opt. Soc. Am. **7**, 1511 (1990)
G.59 C.L. Braun, T.W. Scott: J. Phys. Chem. **91**, 4436 (1987)
G.60 A.M. Brearley, D.B. McDonald: Chem. Phys. Lett. **155**, 83 (1989)
G.61 Y. Kirata, N. Mataga: J. Phys. Chem. **95**, 1640 (1991)
G.62 L. Onsage: Phys. Rev. **54**, 554 (1938)
G.63 K.M. Hong, J. Noolandi: J. Chem. Phys. **68**, 5163 (1978)
G.64 G.A. Brucker, D.F. Kelley: J. Phys. Chem. **91**, 2856 (1987)
G.65 T. Elsaesser, W. Kaiser: Chem. Phys. Lett. **128**, 231 (1986)
G.66 F. Laermer, T. Elsaesser, W. Kaiser: Chem. Phys. Lett. **148**, 119 (1988)
G.67 W. Frey, F. Laermer, T. Elsaesser: J. Phys. Chem. **95**, 10391 (1991)
G.68 Y.R. Kim, J.T. Yardley, R.M. Hochstrasser: Chem. Phys. **136**, 311 (1989)
G.69 M. Wiechmann, H. Port, W. Frey, F. Laermer, T. Elsaesser: J. Phys. Chem. **95**, 1918 (1991)
G.70 W. Frey, T. Elsaesser: Chem. Phys. Lett. *189*, 565 (1992)
G.71 R.S. Moog, S.C. Bovino, J.D. Simon: J. Phys. Chem. **92**, 581 (1988)
G.72 R.S. Moog, M. Maroncelli: J. Phys. Chem. *95*, 10359 (1991)

M.J. Pereira, P.E. Share, M.J. Sarisky, S.T. Repinec, R.M. Hochstrasser, J. Luminesc, *48–9*, 2513 (1991)
M. Negrerie, F. Gai, S.M. Bellefeuille, J.W. Petrich: J. Phys. Chem. *95*, 8663 (1991)
G.73 J. Heldt, D. Garmin, M. Kasha: Chem. Phys. **136**, 321 (1989)
G.74 J. Steadman, J.A. Syage: J. Phys. Chem. *95*, 10326 (1991)
G.75 G.A. Brucker, D.F. Kelley: Chem. Phys. **136**, 213 (1989)
G.76 E. Pines, D. Huppert, N. Agmon: J. Chem. Phys. **88**, 5620 (1988)
G.77 T.P. Smith, K.Z. Zaklika, K. Thakur, P.F. Barbara: J. Am. Chem. Soc. **113**, 4035 (1991)
G.78 For a recent review, see: D.H. Waldeck: Chem. Rev. **91**, 415 (1991)
G.79 J. Schroeder: Ber. Bunsenges. Phys. Chem. **95**, 233 (1991), J. Schroeder, J. Troe: Annu. Rev. Phys. Chem. **38**, 163 (1987)
G.80 R.M. Bowman, K.B. Eisenthal, D. Miller: J. Chem. Phys. **89**, 762 (1988)
G.81 R.M. Bowman, K.B. Eisenthal: Chem. Phys. Lett. **155**, 99 (1989)
G.82 S. Abrash, S. Repinec, R.M. Hochstrasser: J. Chem. Phys. **93**, 1041 (1990), R.J. Sension, S.T. Repinec, R.M. Hochstrasser: J. Chem. Phys. **93**, 9185 (1990)
G.83 J.M. Jean, D.C. Todd, A.J. Ruggiero, G.R. Fleming: In *Ultrafast Phenomena VII*, ed. by C.B. Harris, E.P. Ippen, G.A. Mourou, A.H. Zewail, Springer Ser. Chem. Phys., Vol. 53 (Springer, Berlin, Heidelberg 1990) p. 441
G.84 J.M. Hicks, M.T. Vandersall, E.V. Sitzmann, K.B. Eisenthal: Chem. Phys. Lett. 413 (1987)
G.85 J. Petek, K. Yoshihara, Y. Fujiwara, J.G. Frey: In *Ultrafast Phenomena VII*, ed. by C.B. Harris, E.P. Ippen, G.A. Mourou, A.H. Zewail, Springer Ser. Chem. Phys., Vol. 53 (Springer, Berlin, Heidelberg 1990) p. 444, H. Petek, Y. Fujiwara, D. Kim, K. Yoshihara: J. Am. Chem. Soc. **110**, 6269 (1988)
G.86 J.D. Simon, X. Xie: J. Phys. Chem. **93**, 921 (1989)
G.87 A.G. Joly, K.A. Nelson: J. Phys. Chem. **93**, 2876 (1989)
G.88 M. Lee, C.B. Harris: J. Am. Chem. Soc. **111**, 8963 (1989)
G.89 L. Wang, X. Zhu, K.G. Spears: J. Phys. Chem. **93**, 2 (1989)
G.90 S.C. Yu, X. Xu, R.L. Lingle, J.B. Hopkins: J. Am. Chem. Soc. **112**, 3668 (1990)
G.91 P.A. Anfinrud, C. Han, T. Lian, R.M. Hochstrasser: In *Ultrafast Phenomena VII* ed. by C.B. Harris, E.P. Ippen, G.A. Mourou, A.H. Zewail, Springer Ser. Chem. Phys., Vol. 53 (Springer, Berlin, Heidelberg 1990) p.489
G.92 E.J. Heilweil, R.R. Cavanagh, J.C. Stephenson: J. Chem. Phys. **89**, 230 (1988)
G.93 M.E. Paige, C.B. Harris: Chem. Phys. **149**, 37 (1990)
G.94 Y.R. Shen: Annu. Revs. of Physical Chemistry *40*, 327 (1989)
G.95 C.V. Shank, R. Yen, C. Hirliman: Phys. Rev. Lett. **51**, 900 (1983)
G.96 V. Shannon, D.A. Koos, J.M. Robinson, G.L. Richmond: Chem. Phys. Lett. **142**, 323 (1987)
G.97 J. Miragliotta, R.S. Polizotti, R. Rabinowitz, S.D. Cameron, R.B. Hall: Chem. Phys. **143**, 123 (1990)
G.98 P. Guyot-Sionnest, P. Dumas, Y.J. Chabal, G.S. Higashi: Phys. Rev. Lett. **64**, 2156 (1990)
G.99 A.L. Harris, L. Rothberg, L. Dhar, N.J. Levions, L.H. Dubois: J. Chem. Phys. **94**, 2438 (1991)
G.100 F. Budde, T.F. Heinz, M.M.T. Log, J.A. Misewich, B.D. Smith: In *Ultrafast Phenomena VII*, ed. by C.B. Harris, E.P. Ippen, G.A. Mourou, A.H. Zewail, Springer Ser. Chem Phys., Vol. 53 (Springer, Berlin, Heidelberg 1990) p. 377
G.101 J.D. Beckerle, M.P. Casassa, R.R. Cavanaugh, E.J. Heilweil, J.C. Stephenson: Phys. Rev. Lett. **64**, 2090 (1990)
G.102 H. Schulz, J. Bialkowski, K. Sokolowski-Tinten, D.v.d. Linde: In *Ultrafast Phenomena VII*, ed. by C.B. Harris, E.P. Ippen, G.A. Mourou, A.H. Zewail, Springer Ser. Chem Phys., Vol. 53 (Springer, Berlin, Heidelberg 1990) p. 365
G.103 D.J. Campell, D.A. Higgins, R.M. Coru: J. Phys. Chem. **94**, 3681 (1990)
G.104 E.V. Sitzmann, K.B. Eisenthal: J. Chem. Phys. **90**, 2831 (1989)
G.105 E.V. Sitzmann, K.B. Eisenthal: J. Phys. Chem. **92**, 4579 (1988)
G.106 A. Castro, E.V. Sitzmann, D. Zhang, K.B. Eisenthal: J. Phys. Chem. **95**, 6752 (1991)
G.107 D. Magde, M.W. Windsor: Chem. Phys. Lett. **27**, 31 (1974)

Addendum H: Biological Processes Studied by Ultrafast Laser Techniques, an Update of Chapter 9

Carey K. Johnson and Robin M. Hochstrasser

Hemoglobin

A significant advance, for which only the first results were available three years ago (Sect. 9.1.4), was the application of transient infrared spectroscopy to the hemoglobin dynamics. This new work has contributed to better understanding the photophysics, the dissociation of CO, the heme pocket where CO resides after photodissociation, the transfer of energy (heat) from heme to solvent and the rigidity and geometry of the heme-CO structure inside the protein.

Transient Infrared Spectroscopy

It became possible recently [H.1,2] to carry out transient ir spectroscopy of optically pumped hemoglobin with ca. 200-fs time resolution. This advance enabled studies of HbCO photodissociation by means of direct observation of the bound and free CO. Although transient optical methods are sensitive, the electronic spectrum of hemes cannot respond to changes in the location of a relatively inert diatomic molecule within the protein. Thus, experiments that can observe the ligand directly and monitor any changes in its structure or environment are expected to extend significantly our understanding of hemoprotein function. Such measurements on free CO enabled us to address some new issues for hemoglobin under ambient aqueous conditions. For example, additional information was obtained on the rate of generation of CO and on the pathways of photodissociation, the efficiency of subnanosecond geminate recombination of CO to iron could be brought into relation with results from nanosecond and microsecond studies, properties of the energy and spatial distribution of CO molecules within the protein and the associated dynamics could be inferred from the spectral shapes and positions and the departure of the CO from the pocket into the solution could be considered separately from the rebinding step. Unliganded CO from the photodissociation of MbCO was observed previously in low-temperature glasses [H.3,4]; however, the experimental techniques used lacked the time resolution needed to study the ligand dynamics at biologically relevant temperatures.

The sharp spike in the ir absorption of Hb and HbCO having about a pulse width of 200 fs was thought to indicate a rapid molecular cooling [H.2] but was recently characterized as ir-optical two-photon absorption [H.5]. This signal is stronger than that of free CO and arises from the electronic response.

Photodissociation

The photodissociation of CO from HbCO at 300 K in D_2O showed that the bleaching of the FeCO absorption and the appearance of a new ir absorption corresponding to free CO are both observed at 100 fs after optical excitation [H.6]. From considerations of the potential surface for photodissociation it was concluded that CO leaves the field of Fe after it has moved ca. 1Å. The available kinetic energy permits an exit velocity of 1–2Å per 100 fs.

Geminate Recombination

It was found [H.2] that no measureable recombination of CO occurred on the picosecond or femtosecond timescale. Given that CO does recombine in model compounds we concluded that a rapid structural relaxation, blocking the fast recombination of CO, must occur along with the photo extrusion of CO. It was also discovered that 4% of the dissociated HbCO reformed in 1 ns. This corresponds approximately to expectations based on nanosecond experiments [H.7] suggesting that the protein motions at 300 K permit a description of geminate rebinding as a simple Arrhenius process with a rate coefficient of 50 ns.

Protein Dynamics

The amount of free CO in the protein diminishes by less than $15 \pm 10\%$ during the first nanosecond after photolysis [H.2]. Again, this is consistent with nanosecond measurements. The spectrum of CO appears quite similar at 300 fs, 10 ps and 100 ps after photolysis. The analysis suggests that CO resides only a few Å from the heme during this time, since at the earliest time it certainly cannot be further away than this. These results introduce the notion of "docking sites" for CO in the heme pocket region.

Heme Cooling

A full study [H.5] of the changes in ir transmission following optical pumping of Hb and HbCO [H.2] has shown that the changes are caused by the heating of water resulting from energy being transported from the heme to the solvent surrounding the protein. The water ir spectrum has a known temperature dependence and hence the change in energy content can be accurately gauged. It was found that the water temperature rises about 0.5° C in about 11 ps following excitation of Hb. This is close to expectations from classical diffusion theory and implies that the heme is cooling more rapidly than ca. 8 ps, in reasonable agreement with calculations [H.8] and other recent work [H.9].

Heme Rigidity

Recent femtosecond studies using polarized ir pulses has generated novel information on the motion of the heme in HbCO [H.10]. The anisotropy of carbon monoxide hemoglobin at ambient temperatures following photolysis of 10% of the sample with 350 fs (fwhm) visible pulses was followed by monitoring the bound CO absorption bleach with an infrared laser from 250 fs to 2 ns. The anisotropy was modeled assuming the porphyrin undergoes small damped harmonic oscillations about in-plane axes along and perpendicular to the plane containing the CO bond and an in-plane dielectric axis of the porphyrin, and expressed in terms of the protein rotational diffusion constant, the angle between the CO absorption dipole and the heme plane, the moment of inertia of the prophyrin ring, and the diffusion coefficient of the porphyrin within the protein. The anisotropy is constant at -0.174 ± 0.01 from 250 fs to 1 ns and it appears that the heme plane is rigidly held within the protein as expected from x-ray diffraction studies of crystals. These results indicate that the magnitudc of the angle between the CO and the heme plane is $72° \pm 3°$. In addition, the geometry determined previously with picosecond time-resolved infrared spectroscopy is not affected by ultrafast heme motion.

Photosynthesis

During the past three years, intense scrutiny has been focused on the primary ultrafast events in photosynthesis in the context of the recently determined crystallographic structures of *R. sphaeroides* and *R. viridis*. The key results have been the rates and the temperature dependence of electron transfer processes. A lively interplay between theory and experiment has spurred important contributions.

Primary Electron Transfer

The nature of the primary electron-transfer in photosynthetic reaction centers is still a central issue in ultrafast studies of reaction centers. Studies continue to focus on bacterial reaction centers for which structures are known and preparations are readily available. The oxidation of the special pair (P) can be detected in 2 to 3 ps following the excitation in *R. sphaeroides* and *R. viridis* at room temperature, as noted in Chap. 9. Measurements of the rates of disappearance of P and bacteriopheophytin (BPh), and of the rates of appearance of P^+ and BPh^- were compared to kinetic models of the primary electron transfer. An analysis of the initial charge separation in *R. sphaeroides* and *R. viridis* at around 10 K showed a decay of the excited state P* with a 1.2-ps time constant in *R. sphaeroides*, and with a 0.7-ps time constant in *R. viridis* [H.11,12]. The temperature dependence of this rate was measured and compared to a single-mode theory of activationless electron transfer [H.11]. This theory was shown to be capable of describing the increase in electron-transfer rate with decreasing temperature rather well for *sphaeorides*, but not for *R. viridis*.

A key question has been the role of the accessary bacteriochlorophyll monomer BCh_L. In one interpretation, a model involving direct electron transfer from P to BPh_L is preferred. In support of this model, an analysis of the low-temperature results suggested that any contribution from a two-step process involving BPh_L^+ is negligible [H.11,12]. Data consistent with direct electron transfer to BPh were also recorded at room temperature with ~100-fs time resolution [H.13]. In another interpretation, a two-step model is invoked where primary electron transfer to BCh_L is followed by a much more rapid electron transfer to BPh_L. Transient absorption spectra and time-resolved absorption changes measured for *R. sphaeroides* at room temperature with 60-fs pulses were interpreted to favor a two-step electron transfer from P to BPh_L [H.14,15]. According to this model, electron transfer from P to BCh_L occurs in 3.5 ps, followed by electron transfer from BCh_L to BPh_L in 0.9 ps. The difference spectrum associated with this component was attributed to $P^+BCh_L^-$ [H.15]. A different response has been reported at room temperature for *R. sphaeroides* with 665-nm probe pulses by two different research groups, leading to different conclusions regarding the appropriate model for primary electron transfer [H.13,15]. Most recent work on *R. sphaeroides* [H.16] and *R. viridis* [H.17] suggests the dominance of the two-step sequantial mechanism at room temperature.

Another suggestion is based on the wavelength dependence of primary electron-transfer kinetics [H.18]. Time-resolved absorption changes were measured with 150-fs pulses. The time constant for primary electron transfer was found to vary from 1.3 to 4 ps, depending on probe wavelength. This observation raises the possibility that an inhomogeneity may contribute to observations of more than one time constant in the first few picoseconds of the charge separation. The faster electron-transfer rate at a low temperature was suggested to result from a low-temperature protein configuration that favors faster rates. A similar variation of the electron-transfer rate with probe wavelength was observed for the 100 to 300-ps time-scale electron transfer from BPh_L^- to the quinone acceptor Q at room temperature [H.18]. This variation was also attributed to the reaction-center inhomogeneity. This inhomogeneity might be related to a loss of the acceptor Q in a fraction of the reaction centers [H.19].

The mechanism of primary-electron transfer was also explored by applying an external electric field [H.20]. The effect of an electric field on the rate of formation of BPh_L^- at 77 K was analyzed and found to be inconsistent with semiclassical or single-mode models for superexchange, as well as with sequential electron transfer involving BCh_L^-.

Another approach to revealing the nature of the primary-electron and the energy-transfer events utilizes hole-burning spectroscopy [H.21]. The observation of zero-phonon holes in the primary donor bands of *R. sphaeroides* and *R. viridis* resulted in determinations of the lifetime of P* which are in agreement with time-domain measurements [H.22,23].

The question of the role of the accessary BCh_L has elicited a number of theoretical treatments as well. These treatments have become more sophisticated and more closely linked to the details of protein structure. (Representative references are listed in the reference section.) A closely related issue concerns the temperature dependence of the primary electron transfer. Like the electron transfer rate from BPh_L to quinone (Chap. 9), this rate is enhanced at lower temperatures [H.11,18]. A recent model, noting that evidence for the superexchange model was obtained at 10 K, while evidence for two-step sequential electron transfer was found at room temperature, presumes parallel sequential and superexchange pathways [H.24]. The superexchange mechanism, according to this model, dominates low-temperature electron transfer, while at room temperature both mechanisms can contribute.

Very recent detailed investigations at low temperatures revealed more complicated absorption transients than reported above. The shortest time constant of 0.9 ps was tracked to even lower temperatures and found to shorten to 0.3 ps. The corresponding spectral signature did not change with temperature [H.25]. With very short excitation pulses of 45 fs duration oscillations were found in the transient absorption data suggesting the motion of a low-frequency wave packet [H.26,27].

Although attention has focused primarily on reaction centers of *R. sphaeroides* and *R. viridis*, whose structures have been determined, other photosynthetic bacteria have also been studied. These have included *Rhodobacter capsulatus*, for which kinetics similar to those in *R. sphaeroides* and *R. viridis* were observed [H.28,29]. With reaction centers of the green bacterium *Chloroflexus aurantiacus*, which contain 3 bacteriopheophytin and 3 bacteriochlorophyll molecules, the primary electron-transfer rate is ca. 7 ps [H.30]. A non-exponential decay was reported on the picosecond time scale for samples at 10 K [H.31]. Heliobacteria reaction centers represent another system of interest because of their apparent resemblance of photosystem I in green plants [H.32]. Electron transfer from the primary acceptor to the secondary acceptor occurs in 800 ps at room temperature and 300 ps at 15 K in *Heliobacterium chlorum* [H.33].

A new dimension of research on primary-electron transfer has been added with the availability of site directed mutant strains. For example, mutation of leucine M214 to histidine in *R. sphaeroides* results in the incorporation of a BChl molecule in place of BPh_L. The result is a reduced primary-electron-transfer rate (6.5 ps) and enhanced charge recombination [H.34]. The results place a lower limit on the energy of the BCh_L monomer. The effect of replacing tyrosine M210 (which interacts with BCh_L) with a different amino acid in *R. sphaeroides* was also to slow down the electron-transfer rate [H.35]. Mutation of the corresponding residue in *R. capsulatus* has a similar effect, while replacing phenylalanine L181 with a tyrosine actually enhances the rate of electron transfer [H.16]. In *R. capsulatus*, the mutation of histidine M200 to leucine resulted in a heterodimeric special pair consisting of a bacteriochlorophyll and a bacteriopheoplytin. The result is a reduced photochemical quantum yield and a reduced primary-electron-transfer rate [H.36,37].

Energy Transfer in Antenna Systems

Ultrafast electronic excitation transfer has been studied in a number of antenna systems by both time-resolved fluorescence and pump-probe methods. Time-resolved linear dichroism measurements have been exploited to generate information on both the kinetics of energy transfer and the relative orientations of donor and acceptor [H.38,39]. This method has been applied to the bacteriochlorophyll *a* protein antenna complex of the green sulfur bacterium *Prosthecochloris aestuarii* [H.40–42]. This trimeric complex contains subunits with seven BChl *a* molecules. Incoherent excitation hopping was observed on a time scale of 2–4 ps. Hole-burned linewidth measurements demonstrated that exciton components relax on a much faster time scale, ca. 100 fs [H.43,44]. Other studies have focused on antenna systems associated with photosystems I or II [H.39,41,42,45–48]. Linear dichroism and transient absorption signals were also measured in studies of excitation transfer in the allophycocyanin antenna system in the cyanobacterium *synechococcus* 6301 [H.49].

Another antenna complex of interest has been the chlorosome of *Chloroflexus aurantiacus.* Ultrafast energy transfer has been detected from the chlorosome (consisting of an aggregate of BChl *c* oligomers) to a nearby base-plate pigment in about 16 ps [H.50–55].

Carotenoids have also been found to play a role in ultrafast energy-transfer processes in photosynthesis. (For a recent review, see [H.56]). Subpicosecond singlet energy transfer was measured from carotenoid to chlorophyll by pump-probe experiments on the light-harvesting complex of *R. sphaeroides* and in thylakoids from *P. tricornutum* and *Nannochloropsis* sp. [H.57,58].

Rhodopsin and Bacteriorhodopsin

Bacteriorhodopsin

The model of the primary photochemistry of bacteriorhodopsin outlined in Chap. 9 is upheld by more recent experiments. According to this model, the retinal chromophore isomerizes on the excited-state potential surface on the subpicosecond time scale. Transient absorption spectra were measured with 60-fs pump pulses and a 6-fs broad-band probe pulse [H.59,60]. The evolution of transient hole shapes and transient absorption over the course of 1 ps are consistent with torsional motion away from the initially prepared state within 100 fs, and formation of a ground-state intermediate (J) in 500 fs [H.61]. A broad transient hole confirms the large homogenous broadening associated with the displacement along the isomerization coordinate of the excited-state potential surface.

Subpicosecond transient Raman experiments have also supported the existing model of the primary photochemistry. Raman spectra of bacteriorhodopsin obtained with 800–900 fs pulses confirmed that isomerization occurs on a shorter time scale than these pulse widths [H.62]. The fingerprint region, however, suggested a conformation of the polyene chain different from that observed after about 10 ps. This result is in agreement with pump-probe time-resolved Raman spectra with ~5-ps time resolution obtained for the J and K intermediates of bacteriorhodopsin [H.63,64]. Time-resolved anti-Stokes Raman scattering was also measured with 6-ps time resolution [H.65]. With sufficiently intense pump pulses, anti-Stokes Raman intensities were observed that were attributed to highly excited vibrational levels of ground-state bacteriorhodopsin (BR570) engendered by pump-pulse excitation. Vibrational relaxation to thermal equilibrium was observed with a rate of ca. $(7 \text{ ps})^{-1}$.

Time-resolved polarization studies of internal motion in bacteriorhodopsin have been reported over time scales from ~50-ps to microseconds [H.66].

Very recent reports of picosecond hausieut infrared spectroscopy that show, for the first time, the protein response to isomenization are cited in the Additional Reading section.

Rhodopsin

The model of the primary photochemistry and excited-surface dynamics of rhodopsin described in Chap. 9, like that for bacteriorhodopsin, has been confirmed in femtosecond pump-probe transient absorption experiments [H.67,H68]. Decay times of 200 fs and 3 ps were observed and attributed to torsional motion on the excited-state potential surfaced, followed by formation of the all-*trans* ground-state intermediate bathorhodopsin.

References

H.1 P.A. Anfinrud, C. Han, P.A. Hansen, J.N. More, R.M. Hochstrasser: In *Ultrafast Phenomena VI*, ed. by T. Yajima, K. Yoshihara, C.B. Harris, S. Shionoya, Springer Ser. Chem. Phys., Vol.48 (Springer, Berlin, Heidelberg 1988) p.442

H.2 P.A. Anfinrud, C. Han, R.M. Hochstrasser: Proc. Nat'l Acad. Sci. (USA) **86**, 8387 (1989)

H.3 F. Siebert, W. Maentek, W. Kreutz: Can. J. Spectrosc. **26**, 119 (1981)

H.4 J.O. Alben, D. Beece, S.F. Bowne, W. Doster, L. Eisenstein, H. Frauenfelder, D. Good, J.D. McDonald, M.C. Marden, P.P. Moh, L. Reinisch, A.H. Reynolds, E. Shyamsunder, K.T. Yue: Proc. Nat'l Acad. Sci. (USA) **79**, 3744-3748 (1982)

H.5 R. Diller, M. Iannone, B. Cowen, S. Maiti, J. Owrutsky, M. Li, M. Sarisky, Y.R. Kim, R.B. Locke, T. Lian, R.M. Hochstrasser: In *Transient Vibrational Spectroscopy*, ed. by H. Takahashi, Springer Proc. Phys., Vol.68 (Springer, Berlin, Heidelberg 1992)

H.6 R.M.Hochstrasser, P.A. Anfinrud, R. Diller, C. Han, M. Iannone, T. Lian, B. Locke: In *Ultrafast Phenomena VII*, ed. by C.B. Harris, E.P. Ippen, G.A. Mourou, A.H. Zewail, Springer Ser. Chem. Phys., Vol.53 (Springer, Berlin, Heidelberg 1990) p.429

H.7 L.P. Murray, J. Hofrichter, E.R. Henry, W.A. Eaton: Biophys. Chem. **29**, 63-76 (1988)

H.8 E.R. Henry, W.A. Eaton, R.M. Hochstrasser: Proc. Nat'l Acad. Sci. (USA) **83**, 8982-8986 (1986)

H.9 R. Lingle, X. Xu, H. Zhu, S.-C. Yu, J.B. Hopkins: J. Am. Chem. Soc. **113**, 3992-3994 (1991)

H.10 T. Lian, B. Locke, R.M. Hochstrasser: Biophys. J. **59**, 289a (1991)

H.11 G.R.Fleming, J.L. Martin, J. Breton: Nature **333**, 190-192 (1988)

H.12 J. Breton, J.L. Martin, G.R. Fleming, J.C. Lambry: Biochem. **27**, 8276-8284 (1988)

H.13 C. Kirmaier, D. Holten: Biochemistry **30**, 609-613 (1991)

H.14 W. Holzapfel, U. Finkele, W. Kaiser, D. Oesterhelt, H. Scheer, H.U. Stilz, W. Zinth: Chem. Phys. Lett. **160**, 1-7 (1989)

H.15 W. Holzapfel, U. Finkele, W. Kaiser, D. Oesterhelt, H. Scheer, H.U. Stilz, W. Zinth: Proc. Nat'l Acad. Sci. (USA) **87**, 5168-5172 (1990)

H.16 C.K. Chan, L.X.Q. Chen, T.J. DiMagno, D.K. Hanson. S.L. Nance, M. Schiffer, J.R. Morris, G.R. Fleming: Chem. Phys. Lett. **176**, 366-372 (1991)

H.17 K. Dressler, E. Umlauf, S. Schmidt, P. Hamm, W. Zinth, S. Buchanan, H. Michel: Chem. Phys. Lett. **183**, 270-276 (1991)

H.18 C. Kirmaier, D. Holten: Proc. Nat'l Acad. Sci. (USA) **87**, 3552-3556 (1990)

H.19 M.E. Michel-Beyerle, R. Zengerle: Chem. Phys. Lett. **176**, 85-90 (1991)

H.20 D.J. Lockhart, R.F. Goldstein, S.G. Boxer: J. Chem. Phys. **89**, 1408-1415 (1988)

H.21 W.E. Moerner (ed.): *Persistent Spectral Hole-Burning: Science and Applications*, Topics Curr. Phys., Vol.44 (Springer, Berlin, Heidelberg 1988)

H.22 S.G.Johnson, D. Tang, R. Jankowiak, J.M. Hayes, G.J. Small, D.M. Tiede: J. Phys. Chem. **93**, 5953-5957 (1989)

H.24 M. Bixon, J. Jortner, M.E. Michel-Beyerle: Biochim. Biophys. Acta **1056**, 301-315 (1991)
H.25 C. Lauterwasser, U. Finkele, H. Scheer, W. Zinth: Chem. Phys. Lett. **183**, 471-477 (1991)
H.26 M.H. Voss, J.-C. Lambry, S.H. Robles, D.C. Youvan, J. Breton, J.-L. Martin: Proc. Nat'l Acad. Sci. (USA) **88**, 8885 (1991)
H.27 M.H. Voss, J.-C. Lambry, S.H. Robles, D.C. Youvan, J. Breton, J.-L. Martin: Proc. Nat'l Acad. Sci. (USA) **89**, 613 (1992)
H.28 C.Kirmaier, D. Holten: FEBS Lett. **239**, 211-218 (1988)
H.29 C. Kirmaier, D. Holten: Isr. J. Chem. **28**, 79-85 (1988)
H.30 M. Becker, V. Nagarajan, D. Middendorf, W.W. Parson, J.E. Martin, R.E. Blankenship: Biochim. Biophys. Acta **1057**, 299-312 (1991)
H.31 J.L. Martin, J.C. Lambry, M. Ashokkumar, M.E. Michel-Beyerle, R. Feick, J. Breton: In *Ultrafast Phenomena VII*, ed. by C.B. Harris, E.P. Ippen, G.A. Mourou, A.H. Zewail, Springer Ser. Chem. Phys., Vol.53 (Springer, Berlin, Heidelberg 1990) p.524
H.32 J.T. Trost, R.E. Blankenship: Biochem. **28**, 9898-9904 (1989)
H.33 P.J.M. Van Kan, T.J. Aartsma, J. Amesz: Photosyn. Res. **22**, 61-68 (1989)
H.34 C. Kirmaier, D. Gaul, R. DeBey, D. Holten, C.C. Schenck: Science **251**, 922-927 (1991)
H.35 U. Finkele, C. Lauterwasser, W. Zinth, K.A. Gray, D. Oesterhelt: Biochem. **29**, 8517-8521 (1990)
H.36 C. Kirmaier, D. Holten, E.J. Bylina, D.C. Youvan: Proc. Nat'l Acad. Sci. (USA) **85**, 7562-7566 (1988)
H.37 L.M. McDowell, C. Kirmaier, D. Holten: Biochim. Biophys. Acta **1020**, 239-246 (1990)
H.38 W.S. Struve: J. Opt. Soc. Am. B **6**, 1586-1594 (1990)
H.39 P.A. Lyle, W.S. Struve: Photochem. Photobiol. **53**, 359-365 (1991)
H.40 P.A. Lyle, W.S. Struve: J. Phys. Chem. **94**, 7338-7339 (1990)
H.41 T.C. Causgrove, S. Yang, W.S. Struve: J. Phys. Chem. **92**, 6121-6124 (1988)
H.42 T.C. Causgrove, S. Yang, W.S. Struve: J. Phys. Chem. **92**, 6790-6795 (1988)
H.43 S.G. Johnson, G.J. Small: Chem. Phys. Lett. **155**, 371-375 (1989)
H.44 S.G. Johnson, G.J. Small: J. Phys. Chem. **95**, 471-479 (1991)
H.45 T.P. Causgrove, S. Yang, W.S. Struve: J. Phys. Chem. **93**, 6844-6850 (1989)
H.46 D.D. Eads, E.W. Castner, Jr., R.S. Alberte, L. Mets, G.R. Fleming, J. Phys. Chem. **93**, 8271-8275 (1989)
H.47 T.G. Owens, S.P. Wepp, R.S. Alberte, L. Mets, G.R. Fleming: Biophys. J. **53**, 733-745 (1988)
H.48 T.G. Owens, S.P. Wepp, L. Mets, R.S. Alberte, G.R. Fleming: Biophys. J. **56**, 95-106 (1989)
H.49 W.F. Beck, M. Debreczeny, X. Yan, K. Sauer: In *Ultrafast Phenommena VII*, ed. by C.B. Harris, E.P. Ippen, G.A. Mourou, A.H. Zewail, Springer Ser. Chem. Phys., Vol.53 (Springer, Berlin, Heidelberg 1990) p.535
H.50 M. Mimuro, T. Nozawa, N. Tamai, K. Shimada, I. Yamazaki, S. Lin, R.S. Knox, B.P. Wittmershaus, D.C. Brune, R.E. Blankenship: J. Phys. Chem. **93**, 7503-7509 (1989)
H.51 T.P. Causgrove, D.C. Brune, R.E. Blankenship, J.M. Olson: Photosynth. Res. **25**, 1-10 (1990)
H.52 T.P. Causgrove, D.C. Brune, J. Wang, B.P. Wittmershaus, R.E. Blankenship: Photosynth. Res. **26**, 39-48 (1990)
H.53 A.R.Holzwarth, E. Bittersmmann, W. Reuter, W. Wehrmeyer: Biophys. J. **57**, 133-145 (1990)
H.54 A.R. Holzwarth, M.G. Mueller, K. Griebenow: J. Photochem. Photobiol. B **5**, 457-465 (1990)

H.55 M. Miller, R.P. Cox, T. Gillbro: Biochim. Biophys. Acta **1057**, 187-194 (1991)
H.56 H.A. Frank, C.A. Violette, J.K. Trautmann, A.P. Shreve, T.G. Owens, A.C. Albrecht: Pure Appl. Chem. **63**, 109-114 (1991)
H.57 J.K. Trautmann, A.P. Shreve, T.G. Owens, A.C. Albrecht: Chem. Phys. Lett. **166**, 369-374 (1990)
H.58 J.K. Trautmann, A.P. Shreve, C.A. Violette, H.A. Frank, T.G. Owens, A.C. Albrecht: Proc. Nat'l Acad. Sci. (USA) **87**, 215-219 (1990)
H.59 R.A. Mathies, C.H. Brito Cruz, W.T. Pollard, C.V. Shank: Science **240**, 777-779 (1988)
H.60 W.T. Pollard, C.H. Brito Cruz, C.V. Shank, R.A. Mathies: J. Chem. Phys. **99**, 199-208 (1989)
H.61 W.T. Pollard, S.Y. Lee, R.A. Mathies: J. Chem. Phys. **92**, 4012-4029 (1990)
H.62 R. van den Berg, D.J. Jang, H.C. Bitting, M.A. El-Sayed: Biophys. J. **58**, 135-141 (1990)
H.63 G.H. Atkinson, T.L. Brack, D. Blanchard, G. Rumbles: Chem. Phys. **131**, 1-15 (1989)
H.64 T.L. Brack, G.H. Atkinson: J. Mol. Struct. **214**, 289-303 (1989)
H.65 T.L. Brack, G.H. Atkinson: J. Phys. Chem. **95**, 2351-2356 (1991)
H.66 C. Wan, J. Qian, C.K. Johnson: J. Phys. Chem. **94**, 8417-8423 (1990)
H.67 M. Yan, D. Manor, G. Weng, H. Chao, L. Rothberg, T. Jedju, R.R. Alfano, R.H. Callender: Proc. Nat'l Acad. Sci. (USA) **88**, 9809-9812 (1991)
H.68 R.W. Schoenlein, L.A. Peteanu, R.R. Mathies, C.V. Shank: Science **254**, 412-415 (1991)

Additional Reading ...

on Antenna Systems and Energy Transfer

Beck W.F., K. Sauer: J. Phys. Chem. **96**, 4658-4666 (1992)
Griebenov K., M.G. Mueller, A.R. Holzwarth: Biochim. Biophys. Acta **1059**, 226-232 (1991)
Kwa S.L.S., H. Van Amerongen, S. Lin, J.P. Dekker, R. Van Grondelle, W.S. Struve: Biochim. Biophys. Acta **1102**, 202-212 (1992)
Lin S., H. Van Amerongen, W.S. Struve: Biochim. Biophys. Acta **1060**, 13-24 (1991)
Lyle P.A., W.S. Struve: J. Phys. Chem. **95**, 4152-4158 (1991)
Roelofs T.A., M. Gilbert, V.A. Shuvalov, A.R. Holzwarth: Biochim. Biophys. Acta **1060**, 237-244 (1991)
Shreve A.P., J.K. Trautman, H.A. Frank, T.G. Owens, A.C. Albrecht: Biochim. Biophys. Acta **1058**, 280-288 (1991)
Sundstroem V., R. Van Grondelle: J. Photochem. Photobiol. B **15**, 141-150 (1992)

on Photosynthetic Antenna Systems and Energy Transfer

Abdurakhmanov I.A., R. Danielius, A.P. Mineev, A.P. Razzhivin: Biofizika **35**, 435-439 (1990)
Bergstroem H., W.H.J. Westerhuis, V. Sundstroem, R. van Grondelle, R.A. Niederman, T. Gillbro: FEBS Lett. **233**, 12-16 (1988)
Bittersmann E., A.R. Holzwarth, H. Senger: J. Photochem. Photobiol. B **1**, 247-260 (1987)
Bittersmann E., A.R. Holzwarth, G. Agel, W. Nultsch: Photochem. Photobiol. **47**, 101-105 (1988)

Freiberg A., V.I. Godik, T. Pullerits, K.E. Timpmann: Chem. Phys. **128**, 227-235 (1988)
Freiberg A., V.I. Godik, T. Pullerits, K. Timpmann: Biochim. Biophys. Acta **973**, 93-104 (1989)
Godik A.I., K. Timpmann, A.M. Freiberg: Dokl. Akad. Nauk SSSR **298**, 1469-1473 (1988)
Hashimoto H., Y. Koyama: Biochim. Biophys. Acta **1017**, 181-186 (1990)
Kuki M., H. Hashimoto, Y. Koyama: Chem. Phys. Lett. **165**, 417-422 (1990)
Mimuro M., I. Yamazaki, S. Itoh, N. Tamai, K. Satoh: Biochim. Biophys. Acta **933**, 478-486 (1988)
Mullineaux G.W., E. Bittersmann, J.F. Allen, A.R. Holzwarth: Biochim. Biophys. Acta **1015**, 231-241 (1990)
Mullineaux G.W., A.R. Holzwarth: FEBS Lett. **260**, 245-248 (1990)
Niederman R.A., T. Gillbro: FEBS Lett. **233**, 12-16 (1988)
Nultsch W., E. Bittersmann, A.R. Holzwarth, G. Agel: J. Photochem. Photobiol. B **5**, 481-494 (1990)
Pashchenko V.Z.: Izv. Akad. Nauk SSSR, Ser. Biol. **3**, 343-352 (1990)
Pullerits T., A. Freiberg: Chem. Phys. **149**, 409-418 (1991)
Reddy N.R.S., G.J. Small: J. Chem. Phys. **94**, 7545-7546 (1991)
Roelofs T.A., A.R. Holzwarth: Biophys. J. **57**, 1141-1153 (1990)
Sandstrom A., T. Gillbro, V. Sundstrom, J. Wendler, A.R. Holzwarth: Biochim. Biophys. Acta **933**, 54-64 (1988)
Schatz G.H., H. Brock, A.R. Holzwarth: Proc. Nat'l Acad. Sci. (USA) **84**, 8414-8418 (1987)
Visscher K.J., H. Bergstroem, V. Sundstroem, C.N. Hunter, R. Van Grondelle: Photosynth. Res. 22, 211-217 (1989)
Wang J., D.C. Brune, R.E. Blankenship: Biochim. Biophys. Acta **1015**, 457-463 (1990)

on Reaction Centers

Bixon M., J. Jortner: Chem. Phys. Lett. **159**, 17 (1989)
Bixon M., J. Jortner, M.E. Michel-Beyerle, A. Ogrodnik: Biochim. Biophys. Acta **977**, 273-286 (1989)
Chekalin S.V., Ya.A. Matveets, A.Ya. Shkuropatov, V.A. Shuvalov, A.P. Yartsev: FEBS Lett. **216**, 245-248 (1987)
Chekalin S.V., Yu.A. Matveets, A.P. Yartsev: Rev. Phys. Appl. **22**, 1761-1771 (1987)
Chekalin S.V., Yu.A. Matveets, A.P. Yartsev: Izv. Akad. Nauk SSSR, Ser. Fiz. **53**, 1462-1466 (1989)
Du M., S.J. Rosenthal, X. Xie, T.J. DiMagno, M. Schmidt, D.K. Hanson, M. Schiffer, J.R. Norris, G.R. Fleming: Proc. Nat'l Acad. Sci. (USA) **89**, 8517-8521 (1992)
Friesner R.A., Y. Won: Biochim. Biophys. Acta **977**, 99-122 (1989)
Hamm P., K.A. Gray, D. Oesterhelt, R. Feick, H. Scheer, W. Zinth: Biochim. Biophys. (1993) in press
Hastings G., J.R. Durrant, J. Barber, G. Porter, D.R. Klug: Biochemistry **31**, 7638-7647 (1992)
Hochstrasser R.M., R. Diller, S. Maiti, T. Lian, B. Locke, C. Moser, P.L. Dutton, B.R. Cowen, G.C. Walker: In *Ultrafast Phenomena VIII*, ed. by J.-L. Martin, A. Migus, G. Mourou, A.H. Zewail, Springer Ser. Chem. Phys., Vol.55 (Springer, Berlin, Heidelberg 1993)
Hu Y., S. Mukamel: Chem. Phys. Lett. **160**, 410 (1989)
Jankowiak R., D. Tang, G. Small, M. Seibert: J. Phys. Chem. **93**, 1649-1654 (1989)
Jean J.M., C.K. Chan, G.R. Fleming: Isr. J. Chem. **28**, 169-175 (1988)
Klitzing E.v., H. Kuhn: J. Phys. Chem. **94**, 1699-1702 (1990)

Lockhart D.J., C. Kirmaier, D. Holten, S.G. Boxer: J. Phys. Chem. **94**, 6987-6995 (1990)
Logunov S.L., P.P. Noks, N.I. Zakharova, B.N. Korvatovskii, V.Z. Pashchenko, A.A. Kononenko: Dokl. Akad. Nauk SSSR **307**, 238-242 (1989)
Logunov S.L., P.P. Noks, N.I. Zakharova, B.N. Korvatovskii, V.Z. Pashchenko, A.A. Kononenko: J. Photochem. Photobiol. B **5**, 41-47 (1990)
Matveets Yu.A., S.V. Chekalin, A.Yu. Shkuropatov, V.A. Shuvalor, A.P. Yartsev: Dokl. Akad. Nauk SSSR **294**, 1480-1485 (1987)
Matveets Yu.A., S.V. Chekalin, A.P. Yartsev: Dokl. Akad. Nauk SSSR **292**, 724-728 (1987)
Marveets Yu.A., M.E. Michel-Beyerle, M. Plato, J. Deisenhofer, H. Michel, M. Bixon, J. Jortner: Biochim. Biophys. Acta **932**, 52-70 (1988)
Mueller M.G., K. Griebenow, A.R. Holzwarth: Biochim. Biophys. Acta **1098**, 1-12 (1991)
Pashchenko V.Z., B.N. Korvatovskii, S.L. Logunov, A.A. Kononenko, P.P. Noks, N.I. Zakharova, N.P. Grishanova, A.B. Rubin: FEBS Lett. **214**, 28-34 (1987)
Schatz G.H., H. Brock, A.R. Holzwarth: Biophys. J. **54**, 397-405 (1988)
Scherer P.O.J., S.F. Fischer: Chem. Phys. **131**, 115-127 (1989)
Tang D., R. Jankowiak, J.K. Gillie, G.J. Small, D.M. Tiede: J. Phys. Chem. **92**, 4012-4015 (1988)
Tang D., R. Jankowiak, G.J. Small, D.M. Tiede: Chem. Phys. **131**, 99-113 (1989)
Tang D., R. Jankowiak, M. Seibert, C.F. Yocum, G.J. Small: J. Phys. Chem. **94**, 6519-6522 (1990)
Warshel A., S. Creighton, W. Parson: J. Phys. Chem. **92**, 2696-2701 (1988)
Won Y., R.A. Friesner: Biochim. Biophys. Acta **935**, 9-18 (1988)
Xie X., J.D. Simon: Biochim. Biophys. Acta **1057**, 131-139 (1991)

on Mutant Reaction Centers

Bylina E.J., C. Kirmaier, L. McDowell, D. Molteu, D.C. Youvan: Nature **336**, 182-184 (1988)
Chan Chi-Kiu, T.J. DiMagno, Lin X.O. Chan, J.R. Norrin, G.R. Fleming: Proc. Nat'l Acad. Sci. (USA) **88**, 11202 (1991)
Finkele U., C. Lauterwasser, A. Struck, H. Scheer, W. Zinth: Proc. Nat'l Acad. Sci. (USA) **89**, 9514-9518 (1992)
Kirmaier C., E.J. Bylina, D.C. Youvan, D. Holten: Chem. Phys. Lett. **159**, 251-257 (1989)
McDowell L.M., C. Kirmaier, D. Holten: Biochim. Biophys. Acta **1020**, 239-246 (1990)
McDowell L.M., D. Gaul, C. Kirmaier, D. Holten, C.C. Schenck: Biochemistry **30**, 8315-8322 (1991)
Wang X., J. Cao, P. Maroti, H.U. Stilz, U. Finkele, C. Lauterwasser, W. Zinth, D. Oesterhelt, Govindjee, C.A. Wraight: Biochim. Biophys. Acta **1100**, 1-8 (1992)

on Bacteriorhodopsin

Blanchard D., D.A. Gilmore, T.L. Brack, H. Lemaire, D. Hughes, G.H. Atkinson: Chem. Phys. **154**, 155-170 (1991)
Dexheimer S.L., Q. Wang, L.A. Peteanu, W.T. Pollard, R.A. Mathies, C.V. Shank: Chem. Phys. Lett. **188**, 61-66 (1992)
Diller R., M. Iannone, R. Bogomolni, R.M. Hochstrasser: Biophys. J. **60**, 286-289 (1991)

Diller R., M. Iannone, B.R. Cowen, S. Maiti, R.A. Bogomolni, R.M. Hochstrasser: Biochemistry **31**, 5567-5572 (1992)
Doig S.J., P.J. Reid, R.A. Mathies: J. Phys. Chem. **95**, 6372-6379 (1991)
Groma C.I., F. Raski, G. Szabo, C. Varo: Biophys. J. **54**, 77-80 (1988)
Noguchi T., S. Kolaczkowski, W. Gaertner, G.H. Atkinson: J. Phys. Chem. **94**, 4920-4926 (1990)
Ohtani H., M. Ishikawa, H. Itoh, Y. Takiguchi, T. Urakami, Y. Tsuchiya: Chem. Phys. Lett. **168**, 493-498 (1990)
Trissl H.W., W. Gaertner, W. Leibl: Chem. Phys. Lett. **158**, 515-518 (1989)
Zinth W.: Naturwissenschaften **75**, 173-177 (1988)

on Rhodopsin

Kandori H., S. Matuoka, H. Nagai, Y. Shichida, T. Yoshizawa: Photochem. Photobiol. **48**, 93-97 (1988)
Kandori H., S. Matuoka, T. Shichida, T. Yoshizawa, M. Ito, K. Tsukida, V. Balogh-Nair, K. Nakai: Biochem. **28**, 6460-6467 (1989)
Ohtani H., T. Kobayashi, M. Tsuda, T.G. Ebrey: Biophys. J. **53**, 17-24 (1988)

For further reading see the Proceedings of the 8th Int'l Conference (Antibes Juan-les Pins, France) entitled *Ultrafast Phenomena VIII*, ed. by J.-L. Martin, A. Migus, G.A. Mourou, A.H. Zewail, Springer Ser. Chem. Phys., Vol.55 (Springer, Berlin, Heidelberg 1993)

Subject Index

Topics in Applied Physics Founded by Helmut K. V. Lotsch

1 **Dye Lasers** 3rd Ed. Editor: F. P. Schäfer

2 **Laser Spectroscopy** of Atoms and Molecules. Editor: H. Walther

3 **Numerical and Asymptotic Techniques in Electromagnetics** Editor: R. Mittra

4 **Interactions on Metal Surfaces** Editor: R. Gomer

5 **Mössbauer Spectroscopy** Editor: U. Gonser

6 **Picture Processing and Digital Filtering** 2nd Ed. Editor: T. S. Huang

7 **Integrated Optics** 2nd Ed. Editor: T. Tamir

8 **Light Scattering in Solids I** 2nd Ed. Editor: M. Cardona

9 **Laser Speckle** and Related Phenomena 2nd Ed. Editor: J. C. Dainty

10 **Transient Electromagnetic Fields** Editor: L. B. Felsen

11 **Digital Picture Analysis** Editor: A. Rosenfeld

12 **Turbulence** 2nd Ed. Editor: P. Bradshaw

13 **High-Resolution Laser Spectroscopy** Editor: K. Shimoda

14 **Laser Monitoring of the Atmosphere** Editor: E. D. Hinkley

15 **Radiationless Processes** in Molecules and Condensed Phases. Editor: F. K. Fong

16 **Nonlinear Infrared Generation** Editor: Y.-R. Shen

17 **Electroluminescence** Editor: J. I. Pankove

18 **Ultrashort Light Pulses** Picosecond Techniques and Applications 2nd Ed. Editor: S. L. Shapiro

19 **Optical and Infrared Detectors** 2nd Ed. Editor: R. J. Keyes

20 **Holographic Recording Materials** Editor: H. M. Smith

21 **Solid Electrolytes** Editor: S. Geller

22 **X-Ray Optics.** Applications to Solids Editor: H.-J. Queisser

23 **Optical Data Processing.** Applications Editor: D. Casasent

24 **Acoustic Surface Waves** Editor: A. A. Oliner

25 **Laser Beam Propagation in the Atmosphere** Editor: J. W. Strohbehn

26 **Photoemission in Solids I.** General Principles Editors: M. Cardona and L. Ley

27 **Photoemission in Solids II.** Case Studies Editors: L. Ley and M. Cardona

28 **Hydrogen in Metals I.** Basic Properties Editors: G. Alefeld and J. Völkl

29 **Hydrogen in Metals II** Application-Oriented Properties Editors: G. Alefeld and J. Völkl

30 **Excimer Lasers** 2nd Ed. Editor: Ch. K. Rhodes

31 **Solar Energy Conversion.** Solid State Physics Aspects. Editor: B. O. Seraphin

32 **Image Reconstruction from Projections** Implementation and Applications Editor: G. T. Herman

33 **Electrets** 2nd Ed. Editor: G. M. Sessler

34 **Nonlinear Methods of Spectral Analysis** 2nd Ed. Editor: S. Haykin

35 **Uranium Enrichment** Editor: S. Villani

36 **Amorphous Semiconductors** 2nd Ed. Editor: M. H. Brodsky

37 **Thermally Stimulated Relaxation in Solids** Editor: P. Bräunlich

38 **Charge-Coupled Devices** Editor: D. F. Barbe

39 **Semiconductor Devices** for Optical Communication. 2nd Ed. Editor: H. Kressel

40 **Display Devices** Editor: J. I. Pankove

41 **The Computer in Optical Research** Methods and Applications. Editor: B. R. Frieden

42 **Two-Dimensional Digital Signal Processing I** Linear Filters. Editor: T. S. Huang

43 **Two-Dimensional Digital Signal Processing II** Transforms and Median Filters. Editor: T. S. Huang

44 **Turbulent Reacting Flows** Editors: P. A. Libby and F. A. Williams

45 **Hydrodynamic Instabilities and the Transition to Turbulence** 2nd Ed. Editors: H. L. Swinney and J. P. Gollub

46 **Glassy Metals I** Editors: H.-J. Güntherodt and H. Beck

47 **Sputtering by Particle Bombardment I** Editor: R. Behrisch

48 **Optical Information Processing** Fundamentals. Editor: S. H. Lee

49 **Laser Spectroscopy of Solids** 2nd Ed. Editors: W. M. Yen and P. M. Selzer

50 **Light Scattering in Solids II.** Basic Concepts and Instrumentation Editors: M. Cardona and G. Güntherodt

51 **Light Scattering in Solids III.** Recent Results Editors: M. Cardona and G. Güntherodt

52 **Sputtering by Particle Bombardment II** Sputtering of Alloys and Compounds, Electron and Neutron Sputtering, Surface Topography Editor: R. Behrisch

53 **Glassy Metals II.** Atomic Structure and Dynamics, Electronic Structure, Magnetic Properties Editors: H. Beck and H.-J. Güntherodt

54 **Light Scattering in Solids IV.** Electronic Scattering, Spin Effects, SERS, and Morphic Effects Editors: M. Cardona and G. Güntherodt

55 **The Physics of Hydrogenated Amorphous Silicon I** Structure, Preparation, and Devices Editors: J. D. Joannopoulos and G. Lucovsky

56 **The Physics of Hydrogenated Amorphous Silicon II** Electronic and Vibrational Properties Editors: J. D. Joannopoulos and G. Lucovsky